# CONVERSION TABLES

## U.S. Customary and Metric Equivalents

In the following tables, the equivalents are rounded off.

| U.S. to Metric | Metric to U.S. |
| --- | --- |
| *Length Equivalents* | |
| 1 in. = 2.54 cm (exact) | 1 cm = .394 in. |
| 1 ft = 0.305 m | 1 m = 3.28 ft |
| 1 yd = 0.914 m | 1 m = 1.09 yd |
| 1 mi = 1.61 km | 1 km = 0.62 mi |
| *Area Equivalents* | |
| 1 in.$^2$ = 6.45 cm$^2$ | 1 cm$^2$ = 0.155 in.$^2$ |
| 1 ft$^2$ = 0.093 m$^2$ | 1 m$^2$ = 10.764 ft$^2$ |
| 1 yd$^2$ = 0.836 m$^2$ | 1 m$^2$ = 1.196 yd$^2$ |
| 1 acre = 0.405 ha | 1 ha = 2.47 acres |
| *Volume Equivalents* | |
| 1 in.$^3$ = 16.387 cm$^3$ | 1 cm$^3$ = .394 in.$^3$ |
| 1 ft$^3$ = 0..28 m$^3$ | 1 m$^3$ = 35.315 ft$^3$ |
| 1 qt = 0.946 L | 1 L = 1.06 qt |
| 1 gal = 3.785 L | 1 L = 0.264 gal |
| *Mass Equivalents* | |
| 1 oz = 28.35g | 1 g = 0.035 oz |
| 1 lb = 0.454 kg | 1 kg = 2.205 lb |

## Celsius and Fahrenheit Equivalents

Conversion formulas: $F = \frac{9}{5}C + 32$ $\qquad C = \frac{5}{9}(F - 32)$

| Celsius | Fahrenheit | |
| --- | --- | --- |
| 100° | 212° | ← water boils at sea level |
| 95° | 203° | |
| 90° | 194° | |
| 85° | 185° | |
| 80° | 176° | |
| 75° | 167° | |
| 70° | 158° | |
| 65° | 149° | |
| 60° | 140° | |
| 55° | 131° | |
| 50° | 122° | |
| 45° | 113° | |
| 40° | 104° | |
| 35° | 95° | |
| 30° (comfort range) | 86° (comfort range) | |
| 25° (comfort range) | 77° (comfort range) | |
| 20° (comfort range) | 68° (comfort range) | |
| 15° | 59° | |
| 10° | 50° | |
| 5° | 41° | |
| 0° | 32° | ← water freezes at sea level |

# PREALGEBRA

## A WORKTEXT

THIRD EDITION

**D. Franklin Wright**
*Cerritos College*

**Hawkes Publishing**
A Division of Quant Systems, Inc.

**Book Team**

Editor *Marcel Prevuznak*
Editorial Assistants *Shaun Gibson, Ann Lucius, Mindy Jacobs, Julie Bennett, Greg Hill, Amanda Matlock*
Answer Key Editors *Shaun Gibson, Mindy Jacobs*
Layout *Barri Srihari, D. Sridhar, R. Sri Vishnu, P. S. Rajkumar, U. Nagesh*
Illustrator *Margaret Linton*

**Hawkes Publishing**
A Division of Quant Systems, Inc.



Library of Congress Catalog Number:



Printed in the United States of America by RR Donnelley & Sons Company

ISBN:
Text: 0-918091-51-9
Instructor's Annotated Edition: 0-918091-52-7
Student Solution Manual: 0-918091-53-5

To Michael and Stacey, who continue to
make me a proud father.

# CONTENTS

# Chapter 2 Integers 93

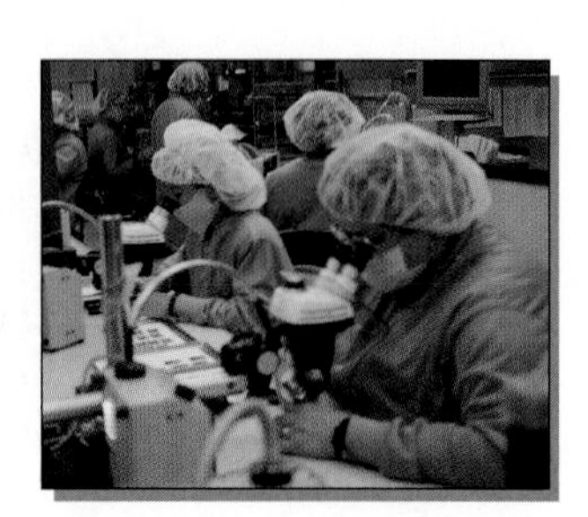

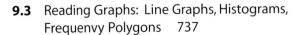

# PREFACE

## Purpose and Style

The purpose of *Prealgebra: A Worktext*, Third Edition, is to provide students with both an introduction to algebra and a learning tool for reviewing basic arithmetic skills. Students will develop reasoning and problem-solving skills and achieve satisfaction in learning, which will encourage them to continue their education in mathematics. Selected topics from algebra are integrated with topics from arithmetic in such a way that students can strengthen their arithmetic skills and become comfortable with the use of variables, negative numbers, algebraic expressions, and techniques for solving equations.

The style of writing gives carefully worded, thorough explanations that are direct, easy to understand, and mathematically accurate. The use of subheadings and shaded boxes helps students understand and reference important topics. An increased amount of white space makes it possible for students to work out each problem within the text. The new, full-color design as well as the addition of high quality images and graphics create an inviting atmosphere for the reader. Overall, these added features make the Third Edition easier to read and more user friendly for the students.

Students are expected to participate actively by working margin exercises and completion examples as they progress through each section. The answers to the margin exercises are in the back of the text and the answers to the completion examples are at the end of each section, providing the students with immediate reinforcement and the instructor with immediate feedback as to how well the students are understanding the material. Students can calculate and write answers on the text pages themselves.

Solving equations is an important topic developed throughout the text. Featured in increasingly difficult levels (a spiral approach), solving equations is discussed in Sections 1.4, 2.7, 3.7, 4.6, 5.5, 7.2, and 7.3. Topics from geometry are integrated within the discussions and problems; for example, perimeter and area in Section 1.1, area and volume in Section 4.3, similar triangles in Section 4.6, circles in Section 5.5, and right triangles and the Pythagorean Theorem in Section 5.6. Chapter 10 contains detailed coverage of measurement—in both the U.S. customary system and the metric system—and a more formal approach of geometry topics presented in earlier chapters.

The flexible style of the text allows for teaching based on the standard lecture method and for programs based on cooperative learning (small group studies) or independent study in a mathematics laboratory setting.

The NCTM curriculum standards have been taken into consideration in the development of the topics throughout the text.

## Important Topics and Ideas

A central consideration throughout this text is the development of an understanding of the interrelationship between arithmetic and algebra. With this in mind, special efforts have been made to help students gain skill and understanding of mathematics in the following ways:

- Techniques for understanding equations and solving equations are developed in a spiral approach of increasing difficulty throughout the text.
- Emphasis has been placed on the development of reading and writing skills as they relate to mathematics.

- A special effort has been made to make the exercises motivating and interesting. Applications are varied and practical and contain many facts of interest.
- The use of calculators is encouraged from the beginning. However, students are cautioned that "calculators do not think." A section on scientific notation helps students understand this type of display on their calculators.
- Estimating is an integral part of many discussions.
- Geometric concepts such as finding perimeter and area and recognizing geometric figures are integrated throughout.

## Organization in This Edition

- More topics from algebra are discussed and integrated throughout the text.
- Use of both scientific and graphing calculators is now encouraged in the third edition.
- Solving equations is introduced in Chapter 1 and developed in a spiral approach of increasing difficulty throughout the text.
- Exponents and order of operations are introduced in Chapter 1 to help in evaluating polynomials.
- Discussion of polynomials begins in Chapter 1 as an early introduction to algebraic expressions and use of order of operations.
- Combining like terms is introduced in Section 2.6 so these ideas can be used in solving equations.
- The section on factorial notation has been removed from Chapter 3.
- Work with fractions is split into two chapters to emphasize the use of prime factorizations in multiplication and division with fractions.
- A new section on factoring polynomials has been included in Chapter 8, replacing the previous section, division of polynomials, which has been moved to the Appendix.
- Chapter 9 (statistics) includes a section on creating graphs.
- Sections devoted solely to geometry are included in Chapter 10, and the topics are integrated with measurement.
- Chapter 10 combines measurement in both the U.S. customary system and the metric system with a formal introduction to geometry.

## New Design

- The appearance of the Third Edition has been enhanced by a new, full-color design.
- High quality images and drawings have been added throughout the text to illustrate mathematical concepts and create a relaxed atmosphere inviting to the reader.

- An increased amount of white space in the Third Edition gives students enough room to work out each problem in the Worktext.
- More real-world situations are applied to mathematical concepts, helping students relate math to their everyday lives.
- The Third Edition provides step-by-step instructions as well as screen-shot images of graphing calculators to evaluate selected problems.

## Special Features

In each chapter:

- *Mathematics at Work!* presents a brief discussion related to an upcoming concept from the chapter ahead and an example of mathematics used in daily life situations.
- *What to Expect in This Chapter* opens each chapter and offers an overview of the topics that will be covered.
- *Learning Objectives* are listed at the beginning of each section.
- *Cumulative Review* exercises appear in each chapter beyond Chapter 1 to provide continuous, cumulative review.

In the exercise sets:

- *Writing and Thinking About Mathematics* exercises encourage students to express their ideas, interpretations, and understanding in writing.
- *Collaborative Learning Exercises* are designed to be done in interactive groups.
- *The Recycle Bin* is a skills refresher that provides maintenance exercises on important topics from previous chapters.

## Pedagogical Features

In each section, the presentation and development are based on the following format for learning and teaching:

1. Each section begins with a list of learning objectives for that section.
2. Subsection topics are introduced by boldface subheadings for easy reference.
3. Thorough and mathematically accurate discussions feature several detailed examples completely worked out with step-by-step explanations.
4. Completion examples help students to reinforce thinking patterns developed in previous examples. (answers at the end of the section)
5. Students are prompted to complete margin exercises as they read the lessons, which provide them with immediate reinforcement of the basic ideas presented. (answers in the back of the text)
6. Graded exercises offer variety in style and level of difficulty including drill, multiple choice, matching, applications, written responses, estimating, and group discussion exercises.
7. Real-life applications appear in most sections, with several sections devoted solely to applications.
8. A full-color design provides visual support to the text's pedagogy.

Each chapter contains:

1. *What to Expect in This Chapter*
2. Chapter-opening *Mathematics at Work!*
3. An *Index of Key Ideas and Terms* with page references
4. A *Chapter Test*
5. A *Cumulative Review* of topics discussed in that chapter (except for Chapter 1) and previous chapters

## Content

There is sufficient material for a three- or four-semester-hour course. The topics in Chapters 1-7 form the core material for a three-hour course. The topics in Chapters 8, 9, and 10 provide for additional flexibility in the course depending on students' background and the goals of the course.

**Chapter 1, Whole Numbers and Exponents**, reviews the fundamental operations of addition, subtraction, multiplication, and division with whole numbers. Estimation is used to develop better understanding of whole number concepts, and word problems involving consumer items, checking accounts, and geometry help to reinforce the need for these ideas and skills. In keeping with the idea of an earlier and more integrated use of algebraic topics, variables, solving equations with whole numbers, exponents and order of operations, and an introduction to polynomials complete the discussions in Chapter 1.

**Chapter 2, Integers**, develops the ideas of number lines and absolute value and teaches the rules for operating with integers. This early introduction of integers allows positive and negative numbers to be used throughout the text in evaluations, in solving equations, and in various applications. New sections in this chapter include combining like terms, evaluating polynomials, and solving equations with integers.

**Chapter 3, Prime Numbers and Fractions**, discusses the concepts of divisibility and factors and their relationship to finding prime factorizations, which are, in turn, used to develop the skills needed for finding the least common multiple (LCM) of a set of numbers or algebraic expressions. These ideas form the basis for the work with fractions in Chapter 3 and Chapter 4. In particular, reducing, raising to higher terms, and multiplication and division with fractions, as developed in this chapter, are based on a knowledge of prime numbers and prime factorization. The last section is on solving equations, this time with fractions.

**Chapter 4, Fractions and Mixed Numbers**, shows how to add and subtract with fractions (included are fractions with variables), how to operate with mixed numbers, and how to simplify complex fractions. Area and volume are discussed, and the rules for order of operations are emphasized again. The last section discusses ratios and proportions, with some applications involving similar triangles.

**Chapter 5, Decimal Numbers and Square Roots**, covers the basic operations with decimal numbers, estimating, and the use of calculators (including scientific notation and finding square roots). To emphasize number concepts, Section 5.4 presents operating with decimal numbers, fractions, and scientific notation. In Section 5.5, Solving Equations with Decimal Numbers, the formulas for finding the circumference and area of a circle are discussed. Square roots, real numbers, and the Pythagorean Theorem are presented in Section 5.6. These topics provide an excellent opportunity for students to become more familiar with the power and convenience of their calculators.

**Chapter 6, Percent with Applications**, deals with the ideas and applications of percent, one of the most useful and most important mathematical concepts in everyday life. The idea of percent as hundredths is used to find equivalent numbers in the form of percents, decimals, and fractions and to introduce the concept of percent of profit. The applications with percent are developed using the formula $R \cdot B = A$ to indicate that there are only three basic types of percent problems. The applications in Sections 6.4 and 6.5 emphasize the use of Pòlya's four-step process for problem solving. In Section 6.6, the use of calculators is seen as necessary in using the formula for compound interest: $A = P(1 + r/n)^{nt}$. The topics of inflation and depreciation are included along with simple and compound interest.

**Chapter 7, Algebraic Topics I**, is designed to give students a running start into algebra. Section 7.1 shows how to translate English phrases into algebraic expressions and asks students to create their own word problems that might be solved using given equations. Sections 7.2 and 7.3 present methods for solving equations with more steps than those in earlier chapters. Applications involving number problems and consecutive integers in Section 7.4 illustrate translating abstract ideas into algebraic equations. Working with formulas in Section 7.5, students substitute numbers and solve for unknown variables and solve formulas for specified variables. Section 7.6 introduces the real number line and discusses methods for solving inequalities and drawing the solutions as intervals on a number line.

**Chapter 8, Algebraic Topics II**, investigates more topics from algebra. Section 8.1 discusses the properties of integer exponents and simplifying expressions using these properties. Sections 8.2 and 8.3 explain how to add, subtract, multiply, and divide with polynomials. Section 8.4 deals with factoring polynomials. The chapter closes with two sections on graphing in two dimensions, graphing ordered pairs of real numbers, and graphing linear equations.

**Chapter 9, Statistics**, includes the concepts of mean, median, mode, and range and how to read various types of graphs: bar graphs, circle graphs, pictographs, line graphs, histograms, and frequency polygons. Section 9.4 shows how to construct bar graphs and circle graphs appropriate for given databases.

**Chapter 10, Geometry and Measurement**, provides a more formal and thorough coverage of geometry. Included are the following tropics: angles and triangles, length and perimeter, area, and volume. All the discussions cover the related measures in both the U.S. customary and metric systems of measurement, explaining how to change units of measure within each system and between the systems.

## Practice and Review

There are more than 3800 exercises, carefully chosen and graded, proceeding from easy to more difficult, plus more than 350 margin exercises, and two or three completion examples in most sections. The Recycle Bin feature, which appears in many sections, generally has six to ten review exercises from previous chapters. Each chapter includes a Chapter Test, and a Cumulative Review follows every chapter after Chapter 1.

Writing and Thinking About Mathematics and Collaborative Learning Exercises are an important part of the text and provide a chance for students to improve communication skills, develop understanding of general concepts, and communicate their ideas to the instructor. Many of these questions are designed for students to investigate ideas other than those presented in the text, with responses based on their own experiences and perceptions. In many cases there is no one right answer.

Answers to the odd-numbered exercises, all Chapter Test questions, all Cumulative Review questions, and all margin exercises appear in the Answer Key at the back of the book. Answers to completion examples are given at the end of the section in which they appear.

**Some Possible Course Offerings**

| Shorter Course (Chapters 1-7) | Longer Course (Chapters 1-8) | Optional Course (Chapters 2-10) |
|---|---|---|
| Whole Numbers and Exponents | Whole Numbers and Exponents | Integers |
| Integers | Integers | Prime Numbers and Fractions |
| Prime Numbers and Fractions | Prime Numbers and Fractions | Fractions and Mixed Numbers |
| Fractions and Mixed Numbers | Fractions and Mixed Numbers | Decimal Numbers and Square Roots |
| Decimal Numbers and Square Roots | Decimal Numbers and Square Roots | Percent with Applications |
| Percent with Applications | Percent with Applications | Algebraic Topics I |
| Algebraic Topics I | Algebraic Topics I | Algebraic Topics II |
| | Algebraic Topics II | Statistics |
| | | Geometry and Measurement |

## Supplements

*Prealgebra: A Worktext*, Third Edition, is accompanied by a comprehensive supplement support package, with each item designed and created especially for this text in order to provide maximum benefit to the students and instructors who use them.

**For the Student**

Student Solutions Manual

 Videotapes

 Adventure Learning Systems' Courseware

**For the Instructor**

Instructor's Edition

 Adventure Learning Systems' Course Management System

NetTest

*Student Solution's Manual:* Worked-out solutions to odd-numbered section exercises. All Chapter Test and Cumulative Review items are included.

*Videotapes:* A series of ten tapes, one per text chapter, provide concept review and additional examples to reinforce the text presentation.

*Adventure Learning Systems' Courseware:* Multimedia courseware allows students to become better problem-solvers by creating a mastery level of learning in the classroom. The software includes a "demonstrate," "instruct," "practice," and "certify" mode in each lesson, allowing students to learn through step-by-step interactions with the software. These automated Homework System's tutorial and assessment modes extend instructional influence beyond the classroom. Intelligence is what makes the tutorials so unique. By offering intelligent tutoring and mastery level testing to measure what has been learned, the software extends the instructor's ability to influence students to solve problems. This courseware can be ordered either seperately or bundled together with the Worktext.

*Instructor's Annotated Edition:* This is a special version of the text that provides in-place answers to all exercises, tests, and review items. It also includes extensive margin annotations to the instructor from the author.

*Adventure Learning Systems' Course Management System:* An instructor's supplement to the Adventure Learning Systems' Courseware, this database allows the instructor greater control of the problem solving students do outside of the classroom. The sytem is loaded onto the server of the instructor's local area network. The instructor is then able to assign lessons with appropriate due dates, give course credit for doing homework, and assign penalties for assignments overdue. Grades for quizzes, exams, and projects may also be entered in this database.

*NetTest:* A FREE add-on to the Course Management System, NetTest creates tests with problems that have algorithmically generated parameters. It allows instructors to give tests either electronically or on paper. It will automatically grade electronic tests as well as record the score in the grade book.

## Acknowledgements

I would like to thank Editor Marcel D. Prevuznak and Editorial Assistants Shaun Gibson, Ann Lucius, Mindy Jacobs, and Julie Bennett for their hard work and invaluable assistance in the development and production of this text.

Many thanks go to the following manuscript reviewers who offered their constructive and critical comments: Alton Amidon at Pamlico Community College; Bethany Chandler at Butler County Community College; Jennifer Smiley and Steve Schwartz at Walla Walla Community College; Karen Anglin and Stan Kubicek at Blinn College; Susan Working at Grossmont College; Thom Clark at Trident Technical College; Timothy McKenna at University of Michigan—Dearborn; Nicole Sifford and Kevin Wheeler at Three Rivers Community College; Vale Biddix, Mike Bradshaw, Melissa Davis, Michelle Powell-Reece and Alisa Williams at Caldwell Community College.

Finally, special thanks go to James Hawkes for his faith in this third edition and his willingness to commit so many resources to guarantee a top-quality product for students and teachers.

D. Franklin Wright

**About the Annotations in the Instructor's Edition**

The annotations in the margin are my ideas for on-the-spot helpful classroom suggestions. I am sure you are familiar with most of them; but many of us, myself included, have thought just after a class has been dismissed, "Oh, I wish I had mentioned _____ for interest or motivation." These annotations include extra examples, historical comments, more in-depth ideas, ideas for possible class discussions, and other comments that I thought you might find interesting or helpful. Obviously, they are for use at your own discretion.

Thank you for using my text and I look forward to receiving any suggestions, comments, or errata that you might bring to my attention for future printings and editions.

Frank

# To the Student

The goal of this text and of your instructor is for you to succeed in prealgebra. Certainly, you should make this your goal as well. What follows is a brief discussion about developing good work habits and using the features of this text to your best advantage. For you to achieve the greatest return on your investment of time and energy you should practice the following three rules of learning.

1. **Reserve a block of time for study every day.**
2. **Study what you don't know.**
3. **Don't be afraid to make mistakes.**

## How to Use This Book

The following ten-step guide will not only make using this book a more worthwhile and efficient task, but it will also help you benefit more from classroom lectures or the assistance that you receive in a math lab.

1. Before you begin a chapter, read the discussion of What to Expect in This Chapter. It will give you some insight on the direction that the chapter will take, the topics that you will be learning about, and how those topics are interrelated.
2. Try to look over the assigned section(s) before attending class or lab. In this way new ideas may not sound so foreign when you hear them mentioned again. This will also help you see where you need to ask questions about material that seems difficult to you.
3. Read examples carefully. They have been chosen and written to show you all of the problem-solving steps that you need to be familiar with. You might even try to solve example problems on your own before studying the solutions that are given.
4. Work the completion examples whenever they appear. If you cannot complete the steps in one of these special examples, go back to review the standard examples that precede it, then try again. When you are satisfied with your solution to a completion example, compare your work to the answers given at the end of the section. Your ability to work completion examples successfully means that you are learning the problem-solving methods being developed in that section.
5. Work margin exercises when asked to do so throughout the lessons. They play an important role in reinforcing the ideas of each lesson and preparing you to work the section exercises. If the margin exercises seem too difficult for you, review the discussion and examples that precede them, then try again. Check your answers against those given in the Answer Key at the back of the book.
6. Work the section exercises faithfully as they are assigned. Problem-solving practice is the single most important element in achieving success in any math class, and there is no good substitute for actually doing this work yourself. Demonstrating that you can think independently through each step of each type of problem will also give you confidence in your ability to answer questions on quizzes and exams. Check the Answer Key periodically while working section exercises to be sure that you have the right ideas and are proceeding in the right manner.

7. Use the Writing and Thinking About Mathematics questions as an opportunity to explore the way that you think about math. A big part of learning and understanding mathematics is being able to talk about mathematical ideas and communicate the thinking that you do when you approach new concepts and problems. These questions can help you analyze your own approach to mathematics and, in class or group discussions, learn from ideas expressed by your fellow students.
8. Use the Chapter Index of Key Ideas and Terms as a recap when you begin to prepare for a Chapter Test. It will reference all the major ideas that you should be familiar with from that chapter and indicate where you can turn if review is needed. You can also use the Chapter Index as a final checklist once you feel you have completed your review and are prepared for the Chapter Test.
9. Chapter Tests are provided so that you can practice for the tests that are actually given in class or lab. To simulate a test situation, block out a one-hour, uninterrupted period in a quiet place where your only focus is on accurately completing the Chapter Test. Use the Answer Key at the back of the book as a self-check only after you have completed all of the questions on the test.
10. Cumulative Reviews will help you retain the skills that you acquired in studying earlier chapters. They appear after every chapter beginning with Chapter 2. Approach them in much the same manner as you would the Chapter Tests in order to keep all of your skills sharp throughout the entire course.

## How to Prepare for an Exam

### Gaining skill and Confidence

The stress that many students feel while trying to succeed in mathematics is what you have probably heard called "math anxiety." It is a real-life phenomenon, and many students experience such a high level of anxiety during mathematics exams in particular that they simply cannot perform to the best of their abilities. It is possible to overcome this stress simply by building your confidence in your ability to do mathematics and by minimizing your fears of making mistakes.

No matter how much it may seem that in mathematics you must either be right or wrong, with no middle ground, you should realize that you can be learning just as much from the times that you make mistakes as you can from the times that your work is correct. Success will come. Don't think that making mistakes at first means that you'll never be any good at mathematics. Learning mathematics requires lots of practice. Most importantly, it requires a true confidence in yourself and in the fact that with practice and persistence the mistakes will become fewer, the successes will become greater, and you will be able to say, "I *can* do this."

## Showing What You Know

If you have attended class or lab regularly, taken good notes, read your textbook, kept up with homework exercises, and asked for help when it was needed, then you have already made significant progress in preparing for an exam and conquering any anxiety. Here are a few other suggestions to maximize your preparedness and minimize your stress.

1. Give yourself enough time to review. You will generally have several days advance notice before an exam. Set aside a block of time each day with the goal of reviewing a manageable portion of the material that the text will cover. Don't cram!
2. Work lots of problems to refresh your memory and sharpen you skills. Go back to redo selected exercises from all of your homework assignments.
3. Reread you text and your notes, and use the Chapter Index of Key Ideas and Terms and the Chapter Text to recap major ideas and do a self-evaluated test simulation.
4. Be sure that you are well-rested so that you can be alert and focused during the exam.
5. Don't study up to the last minute. Give yourself some time to wind down before the exam. This will help you to organize your thoughts and feel more calm as the test begins.
6. As you take the test, realize that its purpose is not to trick you, but to give you and your instructor an accurate idea of what you have learned. Good study habits, a positive attitude, and confidence in your own ability will be reflected in your performance on any exam.
7. Finally, you should realize that you responsibility does not end with *taking* the exam. When your instructor returns your corrected exam, you should review your instructor's comments and any mistakes that you might have made. Take the opportunity to learn from this important feedback about what you have accomplished, where you could work harder, and how you can best prepare for future exams.

# 1

# Whole Numbers and Exponents

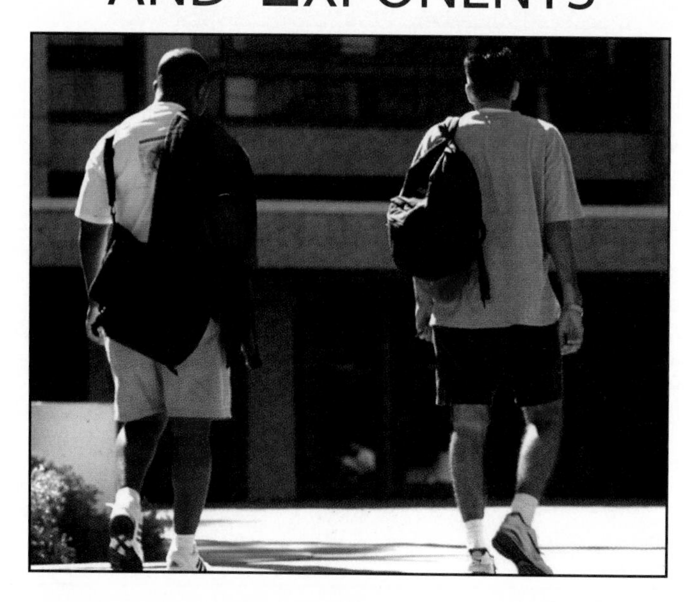

## Chapter 1 Whole Numbers and Exponents



## WHAT TO EXPECT IN CHAPTER 1

Chapter 1 provides a review of operations with whole numbers and an introduction to some fundamental algebraic concepts. Section 1.1 deals with the basic operations of addition, subtraction, multiplication, and division. The use of calculators is introduced early (Section 1.2) and throughout the text as aids in speed and accuracy. However, students are cautioned to develop an understanding of ideas and basic skills of operating with whole numbers because "Calculators do not think."

The valuable skill of estimating results is part of Section 1.2 and part of the applications in Section 1.3. Word problems appear in almost every section in the text, even though the word "applications" may not appear in the section heading. Variables are introduced early (Section 1.1) so that they may be used in stating general properties and rules.

Equations with whole number coefficients and whole number solutions are discussed in Section 1.4. In Section 1.5, the topics of exponents and rules for order of operations lead to an introductory discussion of polynomials and use of the rules for order of operations in evaluating polynomials in Section 1.6.

# Chapter 1 Whole Numbers and Exponents

## Mathematics at Work!

As a student you may be interested in information about universities and some of the related costs of attending. The following list shows tuition, room, and board expenses for 12 colleges from around the country. Costs are given as annual figures (rounded to the nearest hundred dollars and subject to change) for the year 2000. (*Note*: If you want this type of information for any college or university, write to the registrar of that school or go to Collegenet.com.)

| | | | Tuition | |
|---|---|---|---|---|
| **Institution** | **Enrollment** | **Room/Board** | **Resident** | **Nonresident** |
| Univ. of Arizona | 35,000 | $5,500 | $ 2,300 | $ 9,400 |
| Boston College | 14,700 | 8,600 | 21,700 | 21,700 |
| Univ. of Colorado | 25,100 | 5,200 | 3,100 | 15,900 |
| Florida A & M | 11,800 | 2,800 | 1,800 | 6,700 |
| Univ. of Indiana | 35,600 | 3,700 | 4,100 | 12,300 |
| Johns Hopkins | 3,700 | 7,870 | 23,700 | 23,700 |
| MIT | 4,400 | 7,200 | 26,100 | 26,100 |
| Ohio State Univ. | 48,000 | 5,100 | 4,100 | 12,100 |
| Princeton Univ. | 4,600 | 7,000 | 24,600 | 24,600 |
| Rice Univ. | 4,300 | 6,600 | 15,400 | 15,400 |
| Univ. of Tennessee | 20,300 | 4,100 | 3,100 | 9,200 |
| Univ. of Wisconsin | 29,000 | 4,000 | 3,400 | 11,600 |

For these universities and colleges:

(a) Find the average of the yearly expenses for room and board. $ 5639.17

(b) Find the average difference between the tuition for the nonresident students and resident students. $ 4608.33

(c) Determine which of these universities and colleges is the most expensive and which is the least expensive (including room and board and tuition) for nonresident students.
Least expensive: Florida A & M ($9500); Most expensive: MIT ($33,300)

# 1.1 Operations with Whole Numbers

## General Remarks to the Student

This text is written for your enjoyment! That's right, you are to have a good time studying mathematics because you are going to be successful. A positive attitude will be your greatest asset. Algebra is just arithmetic with letters, and this text will reinforce your basic understanding of arithmetic concepts with whole numbers, fractions, decimals, and percents and show how these same concepts can be applied to algebra. Then you will be ready and primed to be successful in future courses in mathematics. Hopefully, you will also see how important and useful mathematics is in your daily life. The main goals are to teach you how to think and how to solve problems. All **you** have to do to attain these goals is:

1. **Read the book.**
2. **Study every day.**
3. **Do not be afraid to make mistakes.**

Good luck! (Hint: If you follow these three procedures you won't need luck.)

## Whole Numbers Defined

The **whole numbers** are the number 0 and the natural numbers (also called the *counting numbers*). A whole number may be written with a **decimal point**, which serves as a beginning point, with the digits to the left of the decimal point. For example,

167. = 167 and 1998. = 1998

**Note:** The decimal point is mentioned here because most students already know about it, and some may ask if they can use it. The decimal point is not necessary until we deal with decimal numbers in Chapter 5, when digits are written to the right of the decimal point.

**Definition:**

The **whole numbers** are the natural (or counting) numbers and the number 0.

**Natural numbers = $N$ = { 1, 2, 3, 4, 5, 6, 7, 8, 9, 10, 11, . . . }**
**Whole numbers = $W$ = { 0, 1, 2, 3, 4, 5, 6, 7, 8, 9, 10, 11, . . . }**

Note that 0 is a whole number but not a natural number.

The three dots (called an *ellipsis*) in the definition indicate that the pattern continues without end.

## Addition with Whole Numbers

The operation of addition with whole numbers is indicated either by writing the numbers horizontally, separated by (+) signs, or by writing the numbers vertically in columns with instructions to add. For example,

$$17 + 4 + 132 \quad \text{or} \quad \begin{array}{r} 17 \\ 4 \\ \underline{+\ 132} \end{array} \quad \text{or} \quad \text{Add:} \ \begin{array}{r} 17 \\ 4 \\ \underline{+\ 132} \end{array}$$

The numbers being added are called **addends**, and the result of the addition is called the **sum**.

$$\text{Add:} \ \begin{array}{rl} 17 & \text{addend} \\ 4 & \text{addend} \\ \underline{+\ 132} & \text{addend} \\ 153 & \text{sum} \end{array}$$

Be sure to keep the digits aligned (in column form) so that you will be adding units to units, tens to tens, and so on. The speed and accuracy with which you add depend on your having memorized the basic addition facts of adding single-digit numbers.

The sum of several numbers can be found by writing the numbers vertically so that the place values are lined up in columns. Note the following:

1. If the sum of the digits in one column is more than 9,
   (a) Write the ones digits in that column, and
   (b) **carry** the other digits to the next column to the left.
2. Look for combinations of digits that total 10.

**Objectives**

① Be able to operate with whole numbers.

② Know the following properties of addition: **commutative, associative, and additive identity.**

③ Understand the following terms: **factor, product, divisor, dividend, quotient, and remainder.**

④ Know the properties of multiplication: **commutative, associative, identity, and zero factor law.**

⑤ Know that division by 0 is not possible.

⑥ Understand that when the remainder is 0, the divisor and quotient are **factors** of the dividend.

To stimulate interest and to emphasize the ease of computation with the decimal system, you might ask the students to add (and later multiply) in the Roman numeral system, i.e.:

$$\begin{array}{r} \text{XXI} \\ \underline{+\ \text{XV}} \end{array} \qquad \begin{array}{r} \text{XLV} \\ \underline{+\text{XVI}} \end{array}$$

(Note that Appendix III contains discussions and exercises related to several ancient numeration systems.)

### Example 1

To add the following numbers, we note the combinations that total 10 to find the sums quickly.

```
  2
 217        Add 7 + 9 + 4 + 6 = 7 + 9 + 10
 389                          = 26
 634 ┐10
+536 ┘      Carry the 2 from the 26.
   6
```

```
 12
 217 ┐10    Add 2 + 1 + 8 + 3 + 3 = 10 + 1 + 3 + 3
 389 ┘                            = 17
 634
+536        Carry the 1 from the 17.
  76
```

```
     12
     217    Add 1 + 2 + 3 + 6 + 5 = 10 + 2 + 5
10 → 389                          = 17
   → 634
    +536
    1776
```

*Now Work Exercises 1 & 2 in the Margin.*

Find each sum.

**1.** $81{,}675 + 43{,}248$

124,923

**2.** $245 + 463 + 905 + 344 + 438$

2395

## Properties of Addition

There are several properties of the operation of addition. To state the general form of each property, we introduce the notation of a **variable.** As we will see, variables not only allow us to state general properties, they also enable us to set up equations to help solve many types of applications.

**Definition:**

A **variable** is a symbol (generally a letter of the alphabet) that is used to represent an unknown number or any one of several numbers.

In the following statements of the properties of addition, the set of whole numbers is the **replacement set** for each variable. (The replacement set for a variable is the set of all possible values for that variable.)

**Commutative Property of Addition**

For any whole numbers $a$ and $b$, $\quad \boldsymbol{a + b = b + a}$

For example, $33 + 14 = 14 + 33$

(The **order** of the numbers in addition can be reversed.)

### Associative Property of Addition

For any whole numbers $a$ and $b$ and $c$, $\mathbf{(a+b)+c=a+(b+c)}$

For example, $(6+12)+5=6+(12+5)$

(The **grouping** of the numbers can be changed.)

### Additive Identity

The number 0 is called the **additive identity**.

For any whole number $a$, $\mathbf{a+0=a}$

For example, $8+0=8$

(The sum of a number and 0 is that same number.)

The use of variables here and in similar situations throughout Chapter 1 is designed to provide an early introduction to equations as well as to develop an understanding of the basic properties of addition.

## Example 2

Each of the properties of addition is illustrated.

**a.** $40+3=3+40$ Commutative property of addition
As a check, we see that $40+3=43$ and $3+40=43$

**b.** $2+(5+9)=(2+5)+9$ Associative property of addition
As a check, we see that $2+(14)=(7)+9=16$

**c.** $86+0=86$ Additive identity

## *Completion Example 3*

Use your knowledge of the properties of addition to find the value of the variable that will make each statement true. State the name of the property used.

| | | Value | Property |
|---|---|---|---|
| **a.** | $14+n=14$ | $n=$ ______ | ______ |
| **b.** | $3+x=5+3$ | $x=$ ______ | ______ |
| **c.** | $(1+y)+2=1+(7+2)$ | $y=$ ______ | ______ |

**NOTE:** Completion Examples are answered at the end of each section.

*Now Work Exercises 3 - 6 in the Margin.*

Find each sum and tell which property of addition is illustrated.

**3.** $(9+1)+7=9+(1+7)$
17; Associative Property of Addition

**4.** $15+8=8+15$
23; Commutative Property of Addition

Find the value of each variable that will make each statement true and name the property used.

**5.** $17+n=17$
$n=0$; Additive Identity

**6.** $36+x=12+36$
$x=12$; Commutative Property of Addition

## The Concept of Perimeter

The **perimeter** of a geometric figure is the distance around the figure. The perimeter is found by adding the lengths of the sides of the figure and is measured in linear units such as inches, feet, yards, miles, centimeters, and meters. The perimeter of the triangle (a geometric figure with three sides) shown in Figure 1.1 is found by adding the measures of the three sides. Be sure to label the answers with the correct units.

### Example 4

To find the perimeter, find the sum of the lengths of the sides:

$$\begin{array}{r} 9 \text{ in.} \\ 12 \text{ in.} \\ +\ 15 \text{ in.} \\ \hline 36 \text{ in.} \end{array} \text{ perimeter}$$

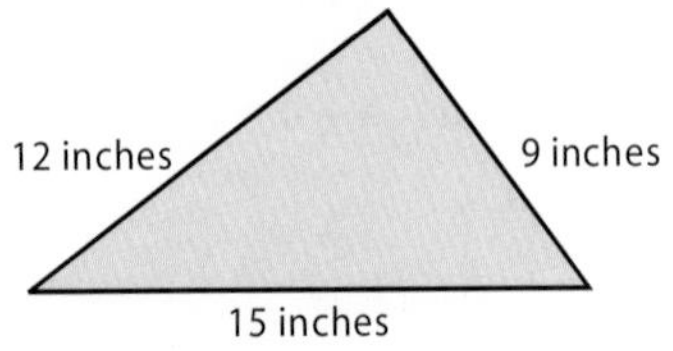

12 inches
9 inches
15 inches

**Figure 1.1**

## Subtraction with Whole Numbers

Suppose we know that the sum of two numbers is 20 and one of the numbers is 17. What is the other number? We can represent the problem in the following format:

$$\underset{\text{addend}}{17} + \underset{\text{missing addend}}{\_\_\_\_} = \underset{\text{sum}}{20}$$

This kind of "addition" problem is called **subtraction** and can be written

$$\underset{\text{sum}}{20} \underset{\text{minus}}{-} \underset{\text{addend}}{17} = \underset{\text{missing addend (or difference)}}{\_\_\_\_}$$

or,

$$\begin{array}{r} 20 \leftarrow \text{minuend} \\ -17 \leftarrow \text{subtrahend} \\ \hline 3 \leftarrow \text{difference or missing addend} \end{array}$$

In other words, subtraction is reverse addition. To subtract, we must know how to add. The missing addend is called the **difference** between the sum (now called the **minuend**) and one addend (now called the **subtrahend**). In this case, we know that the difference is 3, since

$$17 + 3 = 20$$

and we have

$$20 - 17 = 3$$

However, finding a difference such as $742 - 259$, where the numbers are larger, is more difficult, because we do not know readily what to add to 259 to get 742. To find such a difference, we use the technique for subtraction developed in Example 5 that uses place values and borrowing.

## Example 5

Find the difference.

$$\begin{array}{r} 742 \\ -\,259 \\ \hline \end{array}$$

### Solution

**Step 1:** Since 2 is smaller than 9, borrow 10 from 40 and add this 10 to 2 to get 12. This leaves 30 in the tens place; cross out 4 and write 3.

10 borrowed from 40

$$\begin{array}{r} 7\,\overset{3}{\cancel{4}}\,{}^{1}2 \\ -\,2\;5\;9 \\ \hline \end{array}$$

**Step 2:** Since 3 is smaller than 5, borrow 100 from 700, This leaves 600, so cross out 7 and write 6.

$$\begin{array}{r} \overset{6}{\cancel{7}}\,\overset{^{1}3}{\cancel{4}}\,{}^{1}2 \\ -\,2\;5\;9 \\ \hline \end{array}$$

**Step 3:** Now subtract.

$$\begin{array}{r} \overset{6}{\cancel{7}}\,\overset{^{1}3}{\cancel{4}}\,{}^{1}2 \\ -\,2\;5\;9 \\ \hline 4\;8\;3 \end{array}$$

For interest, you might discuss the technique of subtraction by adding in the subtrahend rather than borrowing in the minuend. For example,

$$\begin{array}{r} 637 \\ -\,358 \\ \hline \end{array} \qquad \begin{array}{r} 6\;{}^{1}3\;{}^{1}7 \\ -\;{}^{4}\cancel{3}\;{}^{6}\cancel{5}\;8 \\ \hline 2\;7\;9 \end{array}$$

This method is taught in many other countries, and some of your students may prefer to use it. In any case, they will probably be surprised to find that there are other methods other than what they have learned for performing the basic operations.

*Now Work Exercises 7-9 in the Margin.*

Find the following differences

**7.** $\begin{array}{r} 646 \\ -\,259 \\ \hline 387 \end{array}$

**8.** $\begin{array}{r} 7455 \\ -\,2678 \\ \hline 4777 \end{array}$

**9.** $\begin{array}{r} 7000 \\ -\,4956 \\ \hline 2044 \end{array}$

## Example 6

In pricing a new car, Jason found that he would have to pay a base price of $15,200 plus $1025 in taxes and $575 for license fees. If the bank loaned him $10,640, how much cash would Jason need to buy the car?

### Solution

The solution involves both addition and subtraction. First we add Jason's expenses and then we subtract the amount of the bank loan.

**Expenses**

| | |
|---:|---|
| $15,200 | base price |
| 1,025 | taxes |
| + 575 | license fees |
| $16,800 | total expenses |

**Cash Needed**

| | |
|---:|---|
| $ 16,800 | total expenses |
| – 10,640 | bank loan |
| $ 6,160 | cash |

Jason would need $6,160 in cash to buy the car.

## Properties of Multiplication

The process of repeated addition with the same addend can be shortened considerably by learning to multiply and **memorizing** the basic facts of multiplication with single-digit numbers. (If you know that you have trouble with basic addition and/or multiplication facts, consider going to a teachers' supply store and purchasing a set of flash cards. These cards can be surprisingly effective.)

For example, we can write the following repeated addition as multiplication:

$$175 + 175 + 175 + 175 = 4 \cdot 175 = 700$$

↑ The raised dot indicates multiplication

The result of multiplication is called a **product**, and the two numbers being multiplied are called **factors** of the product. In this example, the repeated addend ( 175 ) and the number of times it is used ( 4 ) are both **factors** of the **product** 700.

$$175 + 175 + 175 + 175 = \underset{\text{factor}}{4} \cdot \underset{\text{factor}}{175} = \underset{\text{product}}{700}$$

Several notations can be used to indicate multiplication. In this text, to avoid possible confusion between the letter $x$ (used as a variable) and the × sign, we will use the raised dot and parentheses much of the time.

### Symbols for Multiplication

| Symbol | | Example |
|---|---|---|
| $\cdot$ | raised dot | $4 \cdot 175$ |
| ( ) | numbers inside or next to parentheses | 4( 175 ) or ( 4 )175 or ( 4 )( 175 ) |
| × | cross sign | $4 \times 175$ or $\begin{array}{r} 175 \\ \underline{\times 4} \end{array}$ |

As with addition, the operation of multiplication has several properties. Multiplication is **commutative** and **associative**, and the number **1** is the **multiplicative identity**. Also, multiplication by 0 always gives a product of 0, and this fact is called the **zero factor law**.

### Commutative Property of Multiplication

For any whole number $a$ and $b$, $\boldsymbol{a \cdot b = b \cdot a}$

For example, $3 \cdot 4 = 4 \cdot 3$

(The **order** of multiplication can be reversed.)

### Associative Property of Multiplication

For any whole numbers $a$ and $b$ and $c$, $(a \cdot b) \cdot c = a \cdot (b \cdot c)$

For example, $(7 \cdot 2) \cdot 5 = 7 \cdot (2 \cdot 5)$

(The **grouping** of the numbers can be changed.)

### Multiplicative Identity

The number 1 is called the **multiplicative identity**.

For any whole number $a$, $a \cdot 1 = a$

For example, $6 \cdot 1 = 6$

(The product of any number and 1 is that same number.)

### Zero Factor Law

For any whole number $a$, $a \cdot 0 = 0$

For example, $63 \cdot 0 = 0$

(The product of a number and 0 is always 0.)

## Example 7

Each of the properties of multiplication is illustrated.

**a.** $5 \times 6 = 6 \times 5$ Commutative property of multiplication
As a check, we see that $5 \cdot 6 = 30$ and $6 \cdot 5 = 30$.

**b.** $2 \cdot (5 \cdot 9) = (2 \cdot 5) \cdot 9$ Associative property of multiplication
As a check, we see that $2 \cdot (45) = 90$ and $(10) \cdot 9 = 90$.

**c.** $8 \cdot 1 = 8$ Multiplicative identity

**d.** $196 \cdot 0 = 0$ Zero factor law

## *Completion Example 8*

Use your knowledge of the properties of multiplication to find the value of the variable that will make each statement true. State the name of the property used.

| | | Value | Property |
|---|---|---|---|
| **a.** | $24 \cdot n = 24$ | $n =$ ________ | ____________ |
| **b.** | $3 \cdot x = 5 \cdot 3$ | $x =$ ________ | ____________ |
| **c.** | $(7 \cdot y) \cdot 2 = 7 \cdot (5 \cdot 2)$ | $y =$ ________ | ____________ |
| **d.** | $82 \cdot t = 0$ | $t =$ ________ | ____________ |

To make sure that the students understand why we say "multiplication over addition," ask them to describe a "Distributive Property of Addition over Multiplication" and discuss its validity:

$a + (b \cdot c) = (a + b) \cdot (a + c)$

Tell which property of multiplication is illustrated.

**10.** $6 \cdot 17 = 17 \cdot 6$
Commutative Property of Multiplication

**11.** $(3 \cdot 8) \cdot 7 = 3 \cdot (8 \cdot 7)$
Associative Property of Multiplication

Find the value of each variable that will make each statement true and name the property of multiplication that is used.

**12.** $25 \cdot 1 = x$
$x = 25$; Multiplicative Identity

**13.** $14 \cdot 5 = 5 \cdot n$
$n = 14$; Commutative Property of Multiplication

Find each product mentally.

**14.** $7 \cdot 9000$
63, 000

**15.** $500 \cdot 300$
150, 000

### Example 9

The following products illustrate how the commutative and associative properties of multiplication are related to the method of multiplication of numbers that end with 0's.

**a.** $7 \cdot 90 = 7(9 \cdot 10) = (7 \cdot 9) \cdot 10 = 63 \cdot 10 = 630$

**b.** $6 \cdot 800 = 6(8 \cdot 100) = (6 \cdot 8) \cdot 100 = 48 \cdot 100 = 4800$

**c.** $50 \cdot 700 = (5 \cdot 10)(7 \cdot 100) = (5 \cdot 7)(10 \cdot 100) = 35 \cdot 1000 = 35{,}000$

**d.** $200 \cdot 4000 = (2 \cdot 100)(4 \cdot 1000) = (2 \cdot 4)(100 \cdot 1000) = 8 \cdot 100{,}000 = 800{,}000$

*Now Work Exercises 10-15 in the Margin.*

## Multiplication with Whole Numbers

To understand the technique for multiplying two whole numbers, we use expanded notation, the method of multiplication by powers of 10, and the following property, called the **distributive property of multiplication over addition**.

**Distributive Property of Multiplication Over Addition**

If $a$, $b$, and $c$ are whole numbers, then

$$a(b + c) = a \cdot b + a \cdot c$$

For example, to multiply $3(50 + 2)$, we can add first and then we can multiply:

$$3(50 + 2) = 3(52) = 156$$

But we can also multiply first and then add, in the following manner:

$$3(50 + 2) = 3 \cdot 50 + 3 \cdot 2 = 150 + 6 = 156$$

This step is called the distributive property. 150 and 6 are called partial products.

Or, vertically,

$$\begin{array}{rcl} 52 & = & 50 + 2 \\ \times\ 3 & = & \times\ \ \ \ \ 3 \\ \hline & & 150 + 6 = 156 \end{array}$$

partial products (150, 6); product (156)

Use the distributive property to multiply $3 \cdot 2$ and $3 \cdot 50$.

Multiplication is explained step by step in Example 10 and then shown in the standard form in Example 11.

### Example 10

The steps in multiplication are shown in finding the product $37 \cdot 27$.

**Step 1:**

$$\begin{array}{r} {}^{4}\\ 37 \\ \times\ 27 \\ \hline 9 \end{array}$$

4 carried from 49

Multiply $7 \cdot 7 = 49$.

**Step 2:**

$$\begin{array}{r} {}^{4}\\ 37 \\ \times\ 27 \\ \hline 259 \end{array}$$

Now multiply $7 \cdot 3 = 21$ and add the 4: $4 + 21 = 25$.

**Step 3:**

$$\begin{array}{r} {}^{1}\\ 37 \\ \times\ 27 \\ \hline 259 \\ 4\phantom{0} \end{array}$$

1 carried from the 14

Next multiply $2 \cdot 7 = 14$.
Write the 4 in the tens column because you are actually multiplying $20 \cdot 7 = 140$. Generally, the 0 is not written.

**Step 4:**

$$\begin{array}{r} {}^{1}\\ 37 \\ \times\ 27 \\ \hline 259 \\ 74\phantom{0} \\ \hline \end{array}$$

Now multiply $2 \cdot 3 = 6$ and add 1: $6 + 1 = 7$.

**Step 5:**

$$\begin{array}{r} 37 \\ \times\ 27 \\ \hline 259 \\ 74\phantom{0} \\ \hline 999 \end{array}$$

Add to find the final product.

### Example 11

The standard form of multiplication is used here to find the product $93 \cdot 46$.

$$\begin{array}{r} {}^{1\,1}\phantom{0}\\ 93 \\ \times\ 46 \\ \hline 558 \\ 372\phantom{0} \\ \hline 4278 \end{array}$$

558 ← $6 \cdot 3 = 18$. Write 8, carry 1. $6 \cdot 9 = 54$.
Add 1: $54 + 1 = 55$.

372 ← $4 \cdot 3 = 12$. Write 2, carry 1. $4 \cdot 9 = 36$.
Add 1: $36 + 1 = 37$.

4278 ← Product

## The Concept of Area

**Area** is the measure of the interior, or enclosed region, of a plane surface and is measured in **square units**. The concept of area is illustrated in Figure 1.2 in terms of square inches.

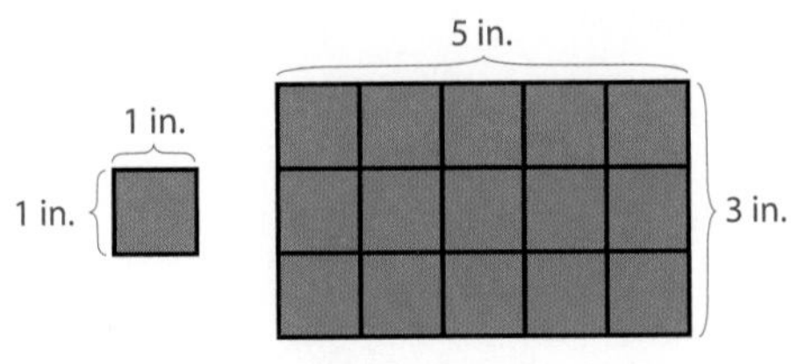


**Figure 1.2**

Some of the units of area in the metric system are square meters, square decimeters, square centimeters, and square millimeters. In the U.S. customary system, some of the units of area are square feet, square inches, and square yards.

### Example 12

The area of a rectangle ( measured in square units ) is found by multiplying its length by its width. Find the area of a rectangular plot of land with dimensions as shown here.

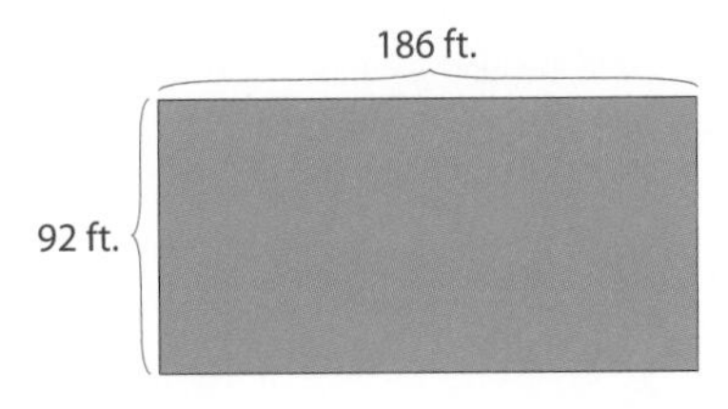


**Solution**

To find the area, we multiply $186 \cdot 92$.

$$\begin{array}{rr} & 186 \\ \times & 92 \\ \hline & 372 \\ & 1674\phantom{0} \\ \hline & 17,112 \end{array} \text{ square feet}$$

*Now Work Exercises 16-19 in the Margin.*

Find the following products.

**16.** 9( 86 )
774

**17.** ( 63 )( 75 )
4725

**18.** 256 × 31
7936

**19.** Find the area of an elementary school playground that is in the shape of a rectangle 52 yards wide and 73 yards long.
3796 square yards

## Division with Whole Numbers

We know that $8 \cdot 10 = 80$ and that 8 and 10 are **factors** of 80. They are also called **divisors** of 80. In division, we want to find how many times one number is contained in another. How many 8's are in 80? There are 10 eights in 80, and we say that 80 **divided** by 8 is 10 (or $80 \div 8 = 10$). In this case we know the results of the division because we are already familiar with the related multiplication, $8 \cdot 10$. Thus, division can be thought of as the reverse of multiplication.

| Division | | | | | | Multiplication | | |
|---|---|---|---|---|---|---|---|---|
| Dividend | | Divisor | | Quotient | | Factors | | Product |
| 24 | ÷ | 6 | = | 4 | since | $6 \cdot 4$ | = | 24 |
| 35 | ÷ | 7 | = | 5 | since | $7 \cdot 5$ | = | 35 |

As the table indicates, the number being divided is called the **dividend**; the number doing the dividing is called the **divisor**, and the result of division is called the **quotient**.

Division does not always involve factors (or exact divisors), and we need a more general idea and method for performing division. The more general method is a process known as the **division algorithm*** (or the method of **long division**) and is illustrated in Example 13. The **remainder** must be less than the divisor.
(Note: If the remainder is 0, then both the divisor and quotient are **factors** of the dividend. See Example 14.)

*An algorithm is a process or pattern of steps to be followed in solving a problem or, sometimes, only a certain part of a problem.

## Example 13

Find the quotient and remainder for $9385 \div 45$

### Solution

**Step 1:**

```
      2
45)9385
   90
    3
```

By trial, divide 40 into 90 or 4 into 9.
Write the result, 2, in the hundreds position of the quotient.
Multiply this result by the divisor and subtract.

**Step 2:**

```
      20
45)9385
   90
    38
     0
    385
```

45 will not divide into 38. So, write
0 in the tens column and multiply $0 \cdot 45 = 0$.
**Do not forget to write the 0 in the quotient.**

**Step 3:**

```
      208   ←— quotient
45)9385   ←— dividend
 /  90
divisor  38
          0
        385
        360
         25   ←— remainder
```

Choose a trial divisor of 8 for the ones digit. Since $8 \cdot 45 = 360$ and 360 is smaller than 385, 8 is the desired number.

**Check:**

Multiply the quotient by the divisor and then add the remainder. This sum should equal the dividend.

$$\begin{array}{rl} 208 & \leftarrow \text{quotient} \\ \times \quad 45 & \leftarrow \text{divisor} \\ \hline 1040 & \\ 832\phantom{0} & \\ \hline 9360 & \end{array} \qquad \begin{array}{rl} 9360 & \\ + \quad 25 & \leftarrow \text{remainder} \\ \hline 9385 & \leftarrow \text{dividend} \end{array}$$

Find each quotient and remainder.

**20.** $5\overline{)4095}$ 819 R 0

**21.** $16\overline{)8073}$ 504 R 9

**22.** $32\overline{)73,860}$ 2308 R 4

*Now Work Exercises 20-22 in the Margin.*

If the remainder is 0 in a division problem, then both the divisor and the quotient are **factors** of the dividend. We say that both factors **divide exactly** into the dividend. As a check, note that the product of the divisor and quotient will be the dividend.

## Example 14

Show that 19 and 43 are factors (or divisors) of 817.

### Solution

We divide: Either $817 \div 19$ or $817 \div 43$ will give a remainder of 0 if indeed these numbers are factors of 817.

$$\begin{array}{r} 43 \\ 19\overline{)817} \\ \underline{76}\phantom{7} \\ 57 \\ \underline{57} \\ 0 \end{array}$$

Check:

$$\begin{array}{r} 43 \\ \times\ 19 \\ \hline 387 \\ 43\phantom{7} \\ \hline 817 \end{array}$$

Thus 19 and 43 are indeed factors of 817. (**Note:** This fact could be determined simply by multiplying $19 \cdot 43$ to see if the product is 817. But in many cases, factors are "found" during a problem involving division.)

### Example 15

A plumber purchased 32 pipe fittings. What was the price of one fitting if the bill was \$512 before taxes?

#### Solution

We need to know how many times 32 is contained in 512.

$$\begin{array}{r} 16 \\ 32\overline{)512} \\ \underline{32} \\ 192 \\ \underline{192} \\ 0 \end{array}$$

The price for one fitting was \$16.

For completeness, we close this section with two rules about division involving 0. These rules will be discussed here and again in Chapter 2.

**Division with 0**

Case 1: If $a$ is any nonzero whole number, then $0 \div a = 0$

Case 2: If $a$ is any whole number, then $a \div 0$ is **undefined**.

Many students have difficulty with division involving 0, regardless of whether 0 is the dividend or the divisor. You may want to provide the students with a few more examples of each case.

The following discussion of these two cases is based on the fact that division is defined in terms of multiplication. For example,

$$14 \div 7 = 2 \quad \text{because} \quad 14 = 2 \cdot 7$$

Since fractions can be used to indicate division, we can also write:

$$\frac{14}{7} = 2 \quad \text{because} \quad 14 = 2 \cdot 7.$$

(Note: Fractions are discussed in detail in Chapters 3 and 4. )

Reasoning in the same manner, we can divide 0 by a nonzero number:

$$0 \div 6 = 0 \qquad \text{because} \qquad 0 = 6 \cdot 0$$

Again, in fraction form:

$$\frac{0}{6} = 0 \qquad \text{because} \qquad 0 = 6 \cdot 0$$

However, if $6 \div 0 = x$, then $6 = 0 \cdot x$. But this is not possible because $0 \cdot x = 0$ for any number $x$. Thus $6 \div 0$ is **undefined**. That is, there is no number that can be multiplied by 0 to give a product of 6.

In fraction form:

$$\frac{6}{0} \text{ is } \textit{undefined}.$$

For the problem $0 \div 0 = x$, we have $0 = 0 \cdot x$, which is a true statement for any value of $x$. Since there is not a unique value for $x$, we agree that $0 \div 0$ is undefined also.

In fraction form:

$$\frac{0}{0} \text{ is } \textit{undefined}.$$

That is, **division by 0 is not possible**.

---

## Completion Example Answers

**3.** **a.** $n = 0$; additive identity

**b.** $x = 5$; commutative property of addition

**c.** $y = 7$; associative property of addition

**8.** **a.** $n = 1$; multiplicative identity

**b.** $x = 5$; commutative property of multiplication

**c.** $y = 5$; associative property of multiplication

**d.** $t = 0$; zero factor law

---

Name ______________________ Section ______________ Date ____________

## Exercises 1.1

**1.** The number ______________ is called the additive identity.

**2.** Give two examples of the commutative property of addition.

**3.** Give two examples of the associative property of addition.

Name the property of addition that is illustrated in each exercise.

**4.** $20 + 0 = 0$

**5.** $8 + 74 = 74 + 8$

**6.** $13 + 42 = 42 + 13$

**7.** $(3 + 7) + 11 = 3 + (7 + 11)$

**8.** $15 + (2 + 16) = (15 + 2) + 16$

Find the following sums and differences.

**9.** $\begin{array}{r} 76 \\ 18 \\ +\ 56 \\ \hline \end{array}$

**10.** $\begin{array}{r} 164 \\ 235 \\ +\ 394 \\ \hline \end{array}$

**11.** $\begin{array}{r} 875 \\ 756 \\ 206 \\ +\ 290 \\ \hline \end{array}$

**12.** $\begin{array}{r} 21,452 \\ 32,468 \\ 105,237 \\ +\ 801,702 \\ \hline \end{array}$

**13.** $\begin{array}{r} 700 \\ -\ 104 \\ \hline \end{array}$

**14.** $\begin{array}{r} 5070 \\ -\ 4375 \\ \hline \end{array}$

**15.** $\begin{array}{r} 7,045,213 \\ -\ 2,743,521 \\ \hline \end{array}$

**ANSWERS**

1. Zero
2. Answers will vary.
3. Answers will vary.
4. Additive Identity
5. Comm. Prop. of Add.
6. Comm. Prop. of Add.
7. Assoc. Prop. of Add.
8. Assoc. Prop. of Add.
9. 150
10. 793
11. 2127
12. 960,859
13. 596
14. 695
15. 4,301,692

16. 2642 mi.

17. $18,527

18. $847

19. $1750

**16.** Mr. Swanson kept the mileage records indicated in the table shown here. How many miles did he drive during the 6 months?

| Month | Mileage |
|---|---|
| January | 456 |
| February | 398 |
| March | 340 |
| April | 459 |
| May | 504 |
| June | 485 |

**17.** During 4 years of college, June estimated her total yearly expenses for tuition, books, fees, and housing as $3540, $5200, $4357, and $5430 for each year. What were her total estimated expenses for 4 years of schooling?

**18.** If you had $980 in your checking account and you wrote a check for $358 and made a deposit of $225, what would be the balance in the account?

**19.** In pricing a four–door car, Pat found she would have to pay a base price of $10,500 plus $730 in taxes and $350 for license fees. For a two–door car of the same make, she would pay a base price of $9000 plus $540 in taxes and $290 for license fees. Including all expenses, how much cheaper was the two–door model?

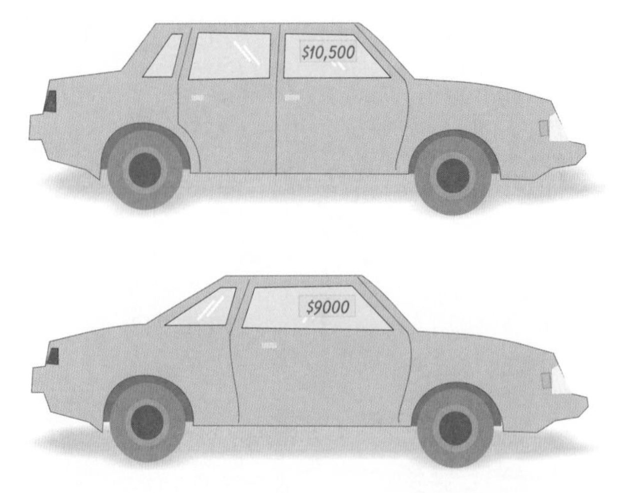

Name ______________________ Section ______________ Date ____________

**20.** The cost of repairing Sylvia's TV set would be \$350 for parts (including tax) plus \$105 for labor. To buy a new set, she would pay \$670 plus \$40 in sales tax, and the dealer would pay her \$90 for her old set. How much more would Sylvia have to pay to buy a new set than to have her old set repaired?

### Perimeter

Remember that the perimeter of a geometric figure can be found by adding the lengths of the sides of the figure. Find the perimeter of each of the following geometric figures with the given dimensions.

**21.** A square (all four sides have the same length)

**22.** A rectangle (opposite sides have the same length)

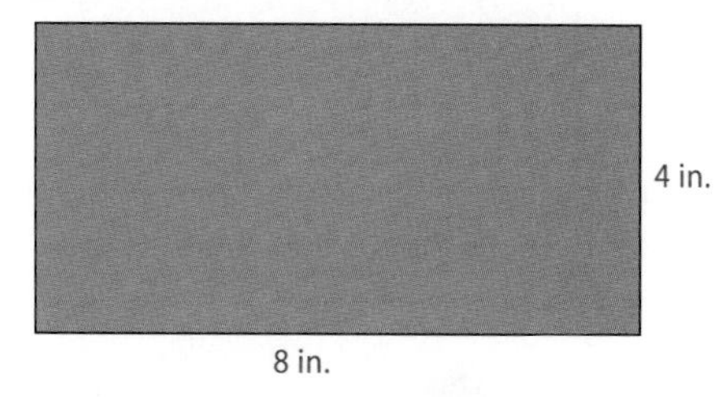


**23.** A triangle (a three sided figure)

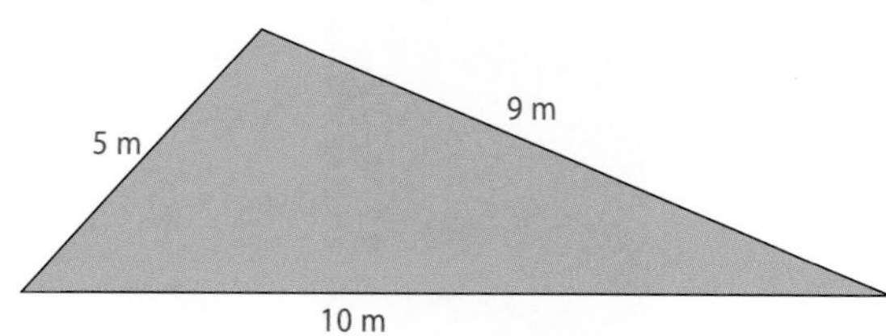


**ANSWERS**

**20.** \$165

**21.** 12 ft.

**22.** 24 in.

**23.** 24 m

24. 12 in.

25. 12 ft.

26. 32 cm

27. 250 yds.

28. 156 in.

**24.** A triangle (a three sided figure)

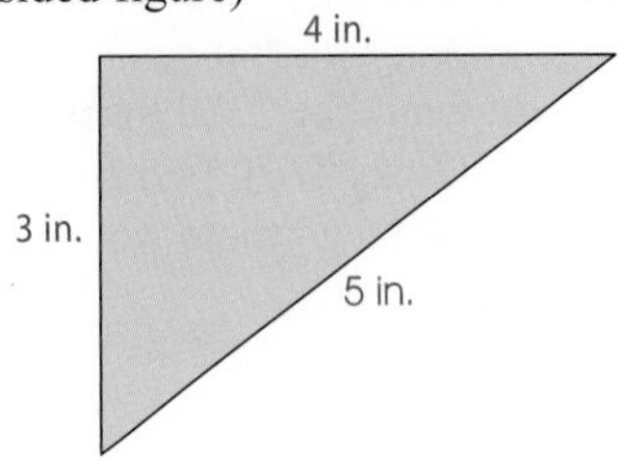


**25.** A regular hexagon (all six sides have the same length)

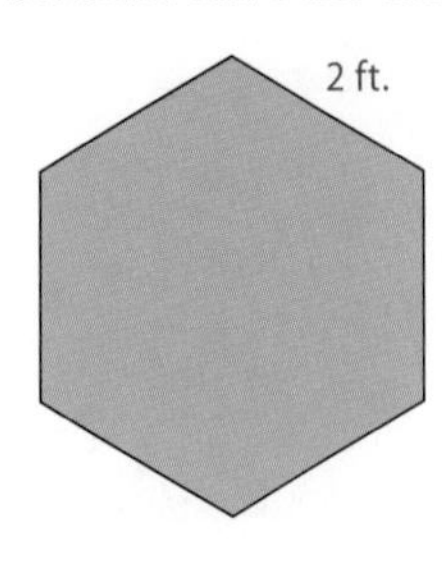


**26.** A trapezoid (two sides are parallel)

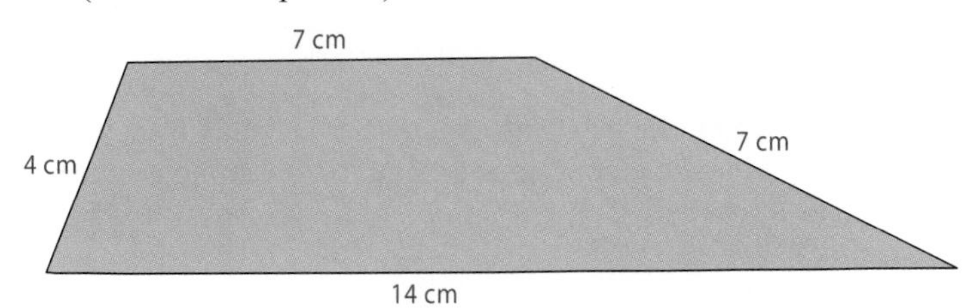


**27.** Find the perimeter of a parking lot that is in the shape of a rectangle 50 yards wide and 75 yards long.

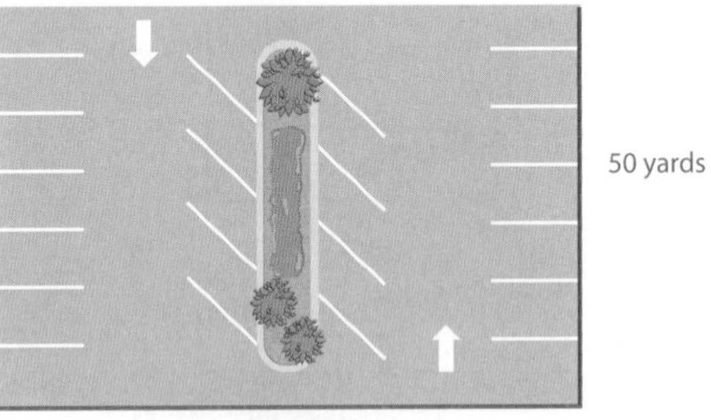


**28.** A window is in the shape of a triangle placed on top of a square. The length of each of two equal sides of the triangle is 24 inches and the third side is 36 inches long. The length of each side of the square is 36 inches long. Find the perimeter of the window.

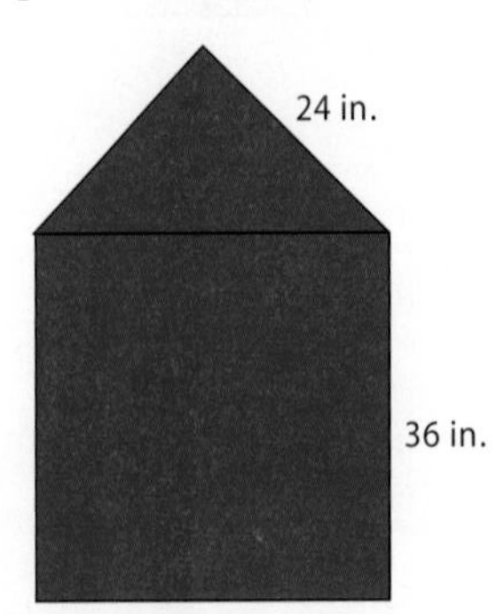

Name ______________ Section ______________ Date ______________

**29.** In your own words, describe the meaning of the word **factor**.

Answers will vary. A factor of a number is a divisor of that number.

Find the value of each variable and state the name of the property of multiplication that is illustrated.

**30.** $(3 \cdot 2) \cdot 7 = 3 \cdot (x \cdot 7)$

**31.** $(5 \cdot 10) \cdot y = 5 \cdot (10 \cdot 7)$

**32.** $25 \cdot a = 0$

**33.** $34 \cdot n = 34$

**34.** $0 \cdot 51 = n$

**35.** $8 \cdot 10 = 10 \cdot x$

Use the distributive property to find each of the following products. Show each step.

**36.** $5(2 + 6)$

$= 5 \cdot 2 + 5 \cdot 6$
$= 10 + 30$
$= 40$

**37.** $4(7 + 9)$

$= 4 \cdot 7 + 4 \cdot 9$
$= 28 + 36$
$= 64$

**38.** $2(3 + 5)$

$= 2 \cdot 3 + 2 \cdot 5$
$= 6 + 10$
$= 16$

**39.** $7(3 + 4)$

$= 7 \cdot 3 + 7 \cdot 4$
$= 21 + 28$
$= 49$

**40.** $6(4 + 8)$

$= 6 \cdot 4 + 6 \cdot 8$
$= 24 + 48$
$= 72$

**ANSWERS**

**29.** [Respond below exercise]

**30.** $x = 2$; Assoc. Prop. of Mult.

**31.** $y = 7$; Assoc. Prop. of Mult.

**32.** $a = 0$; Zero Factor Law

**33.** $n = 1$; Mult. Identity

**34.** $n = 0$; Zero Factor Law

**35.** $x = 8$; Comm. Prop. of Mult.

**36.** [Respond below exercise]

**37.** [Respond below exercise]

**38.** [Respond below exercise]

**39.** [Respond below exercise]

**40.** [Respond below exercise]

41. 140
42. 504
43. 3813
44. 3484
45. 32,640
46. 91,350
47. 6
48. 8
49. 8
50. 7
51. 1
52. 1
53. 4
54. 9
55. 0
56. 0
57. Undefined
58. Undefined
59. 0
60. Undefined
61. Undefined
62. 0

Find the following products.

**41.** $\begin{array}{r} 28 \\ \times\ \ 5 \\ \hline \end{array}$ **42.** $\begin{array}{r} 72 \\ \times\ \ 7 \\ \hline \end{array}$ **43.** $\begin{array}{r} 93 \\ \times\ 41 \\ \hline \end{array}$

**44.** $\begin{array}{r} 67 \\ \times\ 52 \\ \hline \end{array}$ **45.** $\begin{array}{r} 320 \\ \times\ 102 \\ \hline \end{array}$ **46.** $\begin{array}{r} 450 \\ \times\ 203 \\ \hline \end{array}$

Find the following quotients mentally.

**47.** $30 \div 5$ **48.** $48 \div 6$ **49.** $24 \div 3$

**50.** $63 \div 9$ **51.** $6 \div 6$ **52.** $7 \div 7$

**53.** $28 \div 7$ **54.** $72 \div 8$ **55.** $0 \div 2$

**56.** $0 \div 3$ **57.** $9 \div 0$ **58.** $11 \div 0$

**59.** $7\overline{)0}$ **60.** $0\overline{)18}$ **61.** $0\overline{)23}$

**62.** $13\overline{)0}$

Name ______________________ Section ____________ Date __________

Divide by using the division algorithm.

**63.** $16\overline{)129}$ **64.** $20\overline{)305}$ **65.** $18\overline{)207}$

**66.** $68\overline{)207}$ **67.** $102\overline{)918}$ **68.** $213\overline{)4560}$

**69.** Show that 35 and 45 are both factors of 1575 by using long division.

$$\begin{array}{r} 45 \\ 35\overline{)1575} \\ \underline{140} \\ 175 \\ \underline{175} \\ \mathbf{0} \end{array} \quad \text{or} \quad \begin{array}{r} 35 \\ 45\overline{)1575} \\ \underline{135} \\ 225 \\ \underline{225} \\ \mathbf{0} \end{array}$$

**70.** Show that 22 and 32 are both factors of 704 by using long division.

$$\begin{array}{r} 32 \\ 22\overline{)704} \\ \underline{66} \\ 44 \\ \underline{44} \\ \mathbf{0} \end{array} \quad \text{or} \quad \begin{array}{r} 22 \\ 32\overline{)704} \\ \underline{64} \\ 64 \\ \underline{64} \\ \mathbf{0} \end{array}$$

**71.** Show that 56 and 39 are both factors of 2184 by using long division.

$$\begin{array}{r} 39 \\ 56\overline{)2184} \\ \underline{168} \\ 504 \\ \underline{504} \\ \mathbf{0} \end{array} \quad \text{or} \quad \begin{array}{r} 56 \\ 39\overline{)2184} \\ \underline{195} \\ 234 \\ \underline{234} \\ \mathbf{0} \end{array}$$

**72.** Show that 125 and 157 are both factors of 19,625 by using long division.

$$\begin{array}{r} 157 \\ 125\overline{)19625} \\ \underline{125} \\ 712 \\ \underline{625} \\ 875 \\ \underline{875} \\ \mathbf{0} \end{array} \quad \text{or} \quad \begin{array}{r} 125 \\ 157\overline{)19625} \\ \underline{157} \\ 392 \\ \underline{314} \\ 785 \\ \underline{785} \\ \mathbf{0} \end{array}$$

**ANSWERS**

**63.** 8 R 1

**64.** 15 R 5

**65.** 11 R 9

**66.** 3 R 3

**67.** 9 R 0

**68.** 21 R 87

**69.** [Respond below exercise]

**70.** [Respond below exercise]

**71.** [Respond below exercise]

**72.** [Respond below exercise]

73. a. $ 21,600

b. $ 46,800

c. $ 144,000

74. a. $ 9000

b. $ 28,080

c. $ 48,600

75. 37,380 sq. ft.

76. 125 ft.

In each of the following word problems you must decide what operation (or operations) to perform. Do not be afraid of making a mistake. Problems like these help you learn to think mathematically. If an answer does not make sense to you, try another approach.

**73.** On your new job, your salary is to be \$1800 per month for the first year, and you are to get a raise each year, after the first year, of \$300 per month, and you know you are going to love your work.

**a.** What will you earn during your first year?

**b.** What total amount will you earn during the first two years on this job?

**c.** During the first five years on the job?

**74.** You rent an apartment with three bedrooms for \$750 per month, and you know that the rent will increase each year, after the first year, by \$30 per month.

**a.** What will you pay in rent for the first year?

**b.** What total amount will you pay in rent during the first 3–year period?

**c.** During the first 5–year period?

**75.** A rectangular lot for a house measures 210 feet long by 178 feet wide. Find the area of the lot in square feet. (Remember that area of a rectangle is found by multiplying length times width.)

**76.** If you know that the area of a rectangular lot is 39,000 square feet and one side measures 312 feet, what is the length of the other side of the lot? (Remember that area of a rectangle is found by multiplying length times width.)

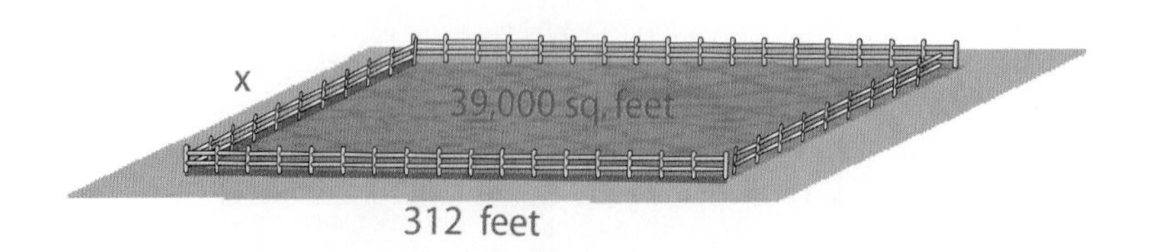

Name ______________________ Section ______________ Date __________

**ANSWERS**

77. 600 pages

78. a. 270 jars

b. 31 servings (w/ 4 oz. extra)

79. 30 sq. ft.

**77.** Three typists in one office typing pool can type a total of 30 pages of data in 2 hours. (Typing data, even on a computer, can be quite time consuming. Typing digits is slower than typing text and the digits must be in the correct columns.) How many pages can they type in a workweek of 40 hours?

**78.** A carton contains 18 jars of lemonade, and each jar contains 14 ounces of lemonade.

**a.** How many jars of lemonade are there in 15 cartons?

**b.** How many 8–ounce servings are contained in one carton?

**79.** A **triangle** is a plane figure with three sides. An **altitude** is the perpendicular distance from one tip (called a **vertex**) to the opposite side (called a **base**). The area of a triangle can be found by multiplying the length of an altitude times the length of the base and dividing by 2. Find the area of the triangle with a base of 10 feet and an altitude of 6 feet.

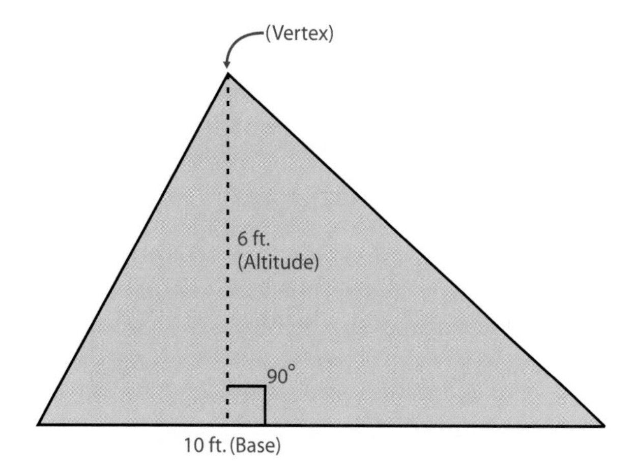

80. 30 sq. in.

81. 384 sq. in.

82. a. 100 acres

b. 50 acres

**80.** Refer to Exercise 79 for the method of finding the area of a triangle. In a **right triangle** (one in which two sides are perpendicular to each other), the two perpendicular sides can be treated as the base and altitude. Find the area of a right triangle in which the two perpendicular sides have lengths of 12 inches and 5 inches. (The longest side in a right triangle is called the **hypotenuse**.)

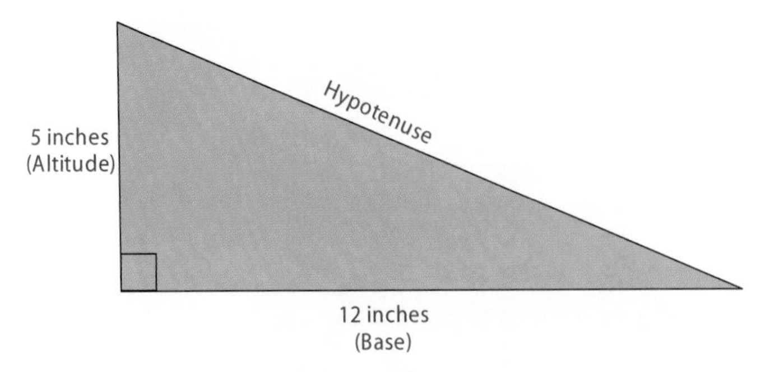


**81.** A picture is mounted in a rectangular frame (16 inches by 24 inches) and hung on a wall. How many square inches of wall space will the framed picture cover?

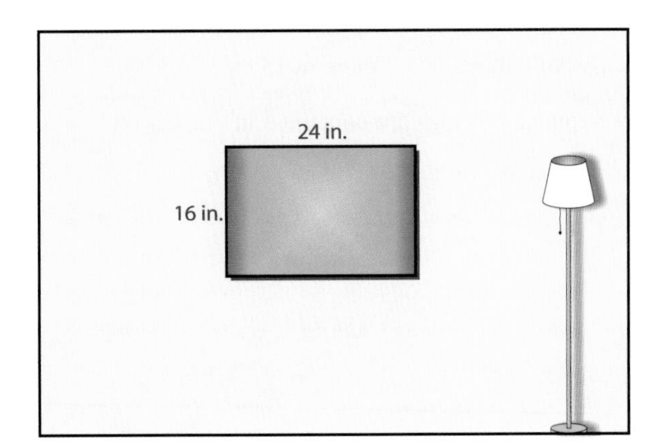


**82.** Large amounts of land area are measured in **acres**, not usually in square feet. (Metric units of land area are **ares** and **hectares**.) One acre is equal to 43,560 square feet or 4840 square yards.

**a.** Find the number of acres in a piece of land in the shape of a rectangle 4356 feet by 1000 feet.

**b.** Find the number of acres in a piece of land in the shape of a right triangle with perpendicular sides of length 500 yards by 968 yards. (See Exercises 79 and 80.)

# 1.2 Rounding Off and Estimating (Calculators Included)

## Rounding Off Whole Numbers

To **round off** a given number means to find another number close to the given number. The desired place of accuracy must be stated. For example, if you were asked to round off 762, you might say 760 or 800. Either answer could be correct, depending on whether the accuracy was to be the nearest ten or the nearest hundred. Rounded-off numbers are quite common and acceptable in many situations. For example, the census taker is interested in your approximate income: 0-$5000, 5001-$10,000, 10,001-$15,000, and so on, but not your exact income. Even the IRS allows rounding off to the nearest dollar when calculating income tax.

In Example 1, we use number lines as visual aids in understanding the rounding-off process.

**Objectives**

① Be able to operate with whole numbers to specified places of accuracy.

② Know how to use rounded-off numbers to estimate sums, differences, products, and quotients.

③ Know how to use a calculator to perform basic operations with whole numbers.

### Example 1

Round 762 (a) to the nearest ten and (b) to the nearest hundred.

**Solution**

**a.**

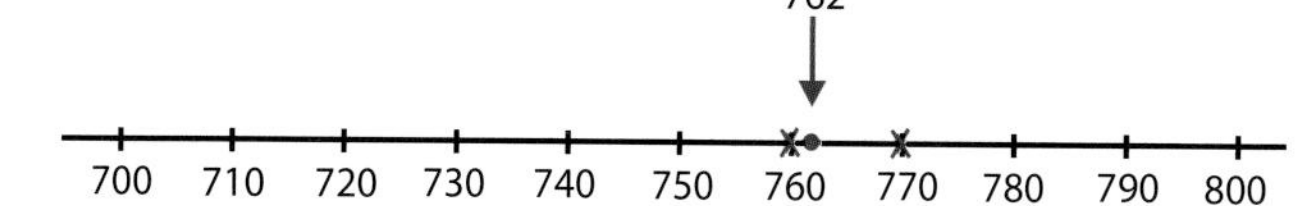


We see that 762 is closer to 760 than to 770. Thus 762 rounds off to 760 (to the **nearest ten**).

**b.**

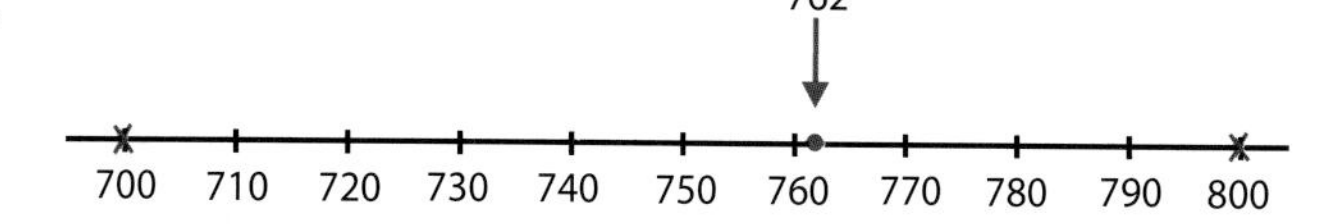


Also, 762 is closer to 800 than to 700. Thus 762 rounds off to 800 (to the **nearest hundred**).

Using number lines as aids to understanding is fine, but for practical purposes, the following rule is more useful.

### Rounding-off Rule for Whole Numbers

1. Look at the single digit just to the right of the digit that is in the place of desired accuracy.

2. **If this digit is 5 or greater,** make the digit in the desired place of accuracy one larger, and replace all digits to the right with zeros. All digits to the left remain unchanged unless a 9 is made one larger; then the next digit to the left is increased by 1.

3. **If this digit is less than 5,** leave the digit that is in the place of desired accuracy as it is, and replace all digits to the right with zeros. All digits to the left remain unchanged.

## Example 2

Round off each number as indicated:

**a.** 6849 (nearest hundred)
**b.** 3500 (nearest thousand)
**c.** 597 (nearest ten)

### Solution

**a.**

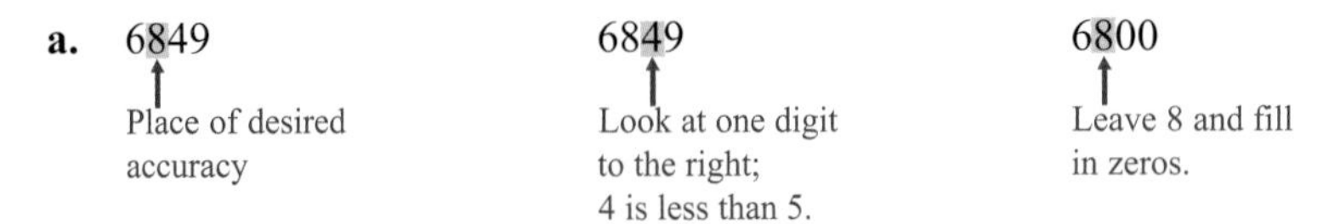


So, 6849 rounds off to 6800 (to the nearest hundred).

**b.**

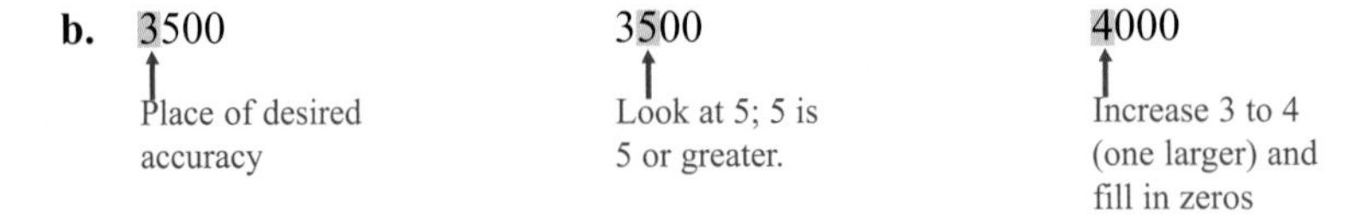


So, 3500 rounds off to 4000 (to the nearest thousand).

**c.**

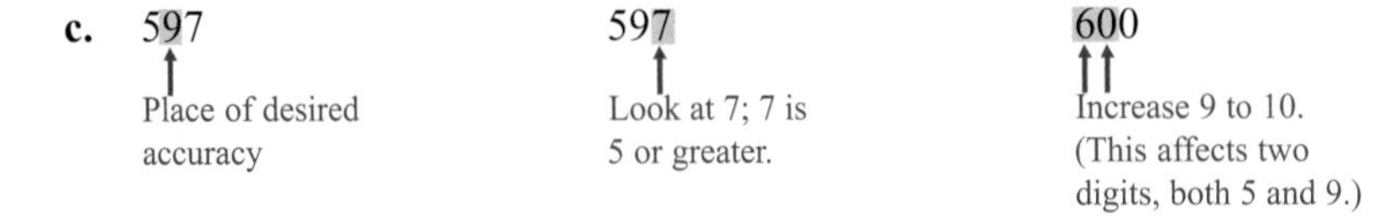


So, 597 rounds off to 600 (to the nearest ten).

## Estimating Answers

One very good use for rounded-off numbers is to **estimate** an answer (or to find an **approximate** answer) before any calculations with the given numbers are made. Thus, answers that are not reasonable (for example, because someone pushed a wrong button on a calculator or because of some other large calculation error) can be spotted. Usually, we simply repeat the calculations and find the error.

To **estimate an answer** means to use rounded-off numbers in a calculation to get some idea of what the size of the actual answer should be. This is a form of checking your work before you do it. There are some situations where an estimated answer is itself sufficient. For example, a shopper may simply estimate the total cost of purchases to be sure that he or she has enough cash to cover the cost.

**To Estimate an Answer**

1. Round off each number to the place of the **leftmost digit**.
2. Perform the indicated operation with these rounded-off numbers.

This method of estimation has the advantages of being fast and easy. The accuracy cannot be guaranteed with any method since an estimation is just that– an estimation. The student must develop some judgement skills.

## Special Note About Estimating Answers

There are several methods for estimating answers. Methods that give reasonable estimates for addition and subtraction may not be quite so appropriate for multiplication and division. You may need to use your judgment, particularly about the desired accuracy. Discuss different ideas with your instructor and fellow students. You and your instructor may choose to follow a completely different set of rules for estimating answers than those discussed here. Estimating techniques are "flexible," and we will make a slight adjustment to our leftmost digit rule with division. In any case, remember that an estimate is just that, an estimate.

### Example 3

Estimate the sum; then find the sum.

$$\begin{array}{r} 68 \\ 925 \\ +\ 487 \\ \hline \end{array}$$

#### Solution

Note that in this example numbers are rounded off to different places because they are of different sizes. That is, the leftmost digit is not in the same place for all numbers. Another method might be to round off each number to the tens place (or hundreds place). This other method might give a more accurate estimate. However, the method chosen here is relatively fast and easy to follow and will be used in the discussions throughout the text.

**a.** Estimate the sum by first rounding off each number to the place of the leftmost digit and then adding. In actual practice, many of these steps can be done mentally.

$$\begin{array}{rcrl} 68 & \rightarrow & 70 & \text{rounded-off value of } 68 \\ 925 & \rightarrow & 900 & \text{rounded-off value of } 925 \\ +\ 487 & \rightarrow & +\ \ 500 & \text{rounded-off value of } 487 \\ \cline{3-3} & & 1470 & \text{estimated sum} \end{array}$$

**b.** Now we find the sum, knowing that the answer should be close to 1470.

$$\begin{array}{r} {}^{1\,2} \\ 68 \\ 925 \\ +\ 487 \\ \hline 1480 \end{array}$$

This sum is very close to 1470.

## Example 4

Estimate the difference; then find the difference.

$$\begin{array}{r} 2783 \\ -\ \ 975 \\ \hline \end{array}$$

### Solution

**a.** Round off each number to the place of the leftmost digit and subtract using these rounded-off numbers.

$$\begin{array}{rcrl} 2783 & \rightarrow & 3000 & \text{rounded-off value of } 2783 \\ -\ 975 & \rightarrow & -1000 & \text{rounded-off value of } 975 \\ & & 2000 & \text{estimated difference} \end{array}$$

**b.** Now we find the difference, knowing that the difference should be close to 2000.

$$\begin{array}{r} \overset{1}{\not 2}\,{}^{1}7\,\overset{7}{\not 8}\,{}^{1}3 \\ -\ \ \ \ 9\,7\,5 \\ \hline 1\,8\,0\,8 \end{array}$$ This difference is close to 2000

*Now Work Exercises 1 - 4 in the Margin.*

Round off as indicated.

**1.** 9748 (nearest hundred)
9700

**2.** 9748 (nearest thousand)
10,000

**3.** Estimate the sum and then find the sum.

$$\begin{array}{r} 156 \\ 83 \\ +\ \ 75 \\ \hline \end{array}$$

Estimate: 360
Sum: 314

**4.** Estimate the difference and then find the difference.

$$\begin{array}{r} 2685 \\ -\ \ 847 \\ \hline \end{array}$$

Estimate: 2200
Difference: 1838

**Caution:** The use of rounded-off numbers to approximate answers demands some understanding of numbers in general and some judgement as to how "close" an estimate can be to be acceptable. In particular, when all numbers are rounded up or all numbers are rounded down, the estimate might not be "close enough" to detect a large error. Still, the process is worthwhile and can give "quick" and useful estimates in many cases.

## Example 5

Find the product $63 \cdot 49$, but first estimate the product.

### Solution

**a.** Estimate:

$$\begin{array}{rcrl} 63 & \rightarrow & 60 & \text{rounded-off value of 63} \\ \times 49 & \rightarrow & \times\ 50 & \text{rounded-off value of 49} \\ & & 3000 & \text{estimated product} \end{array}$$

**b.** Notice that, in rounding off, 63 was rounded down and 49 was rounded up. Thus, we can expect the actual product to be reasonably close to 3000.

$$\begin{array}{r} 63 \\ \times\ \ 49 \\ \hline 567 \\ 252\ \ \\ \hline 3087 \end{array}$$ The actual product is close to 3000.

### An Adjustment to the Process of Estimating Answers

Even though we have given a rule for rounding off to the leftmost digit for estimating an answer, this "rule" is flexible, and there are times when using two digits gives simpler calculations and more accurate estimates. Such a case is illustrated in Example 6, where 3 is easily seen to divide into 15. Thus, we say again, estimating answers does involve some basic understanding and intuitive judgement, and there is no one best way to estimate answers. However, this "adjustment" to the leftmost digit rule is more applicable to division than it is to addition, subtraction, or multiplication.

## Example 6

Estimate the quotient 148,062 ÷ 26; then find the quotient.

### Solution

**a.** Estimation:

148,062 ÷ 26 ⟶ 150,000 ÷ 30

In this case, we rounded off using the two leftmost digits for 148,062 because 150,000 can be divided evenly by 30.

```
     5000      ← approximate quotient
 30)150000
    150
    ---
      00
      00
      --
       00
       00
       --
        00
        00
        --
         0     ← remainder
```

**b.** The quotient should be near 5000.

```
     5694      ← actual quotient
 26)148,062
    130
    ---
     180
     156
     ---
      246
      234
      ---
       122
       104
       ---
        18     ← remainder
```

### Completion Example 7

Find the quotient and remainder.

```
      32
21)6857
    63
     55
   _____
   _____
   _____
   _____      remainder
```

Now estimate the quotient by mentally dividing rounded-off numbers:

___________

Is your estimate close to the actual quotient? _____________

What is the difference? _______________

*Now Work Exercises 5- 8 in the Margin.*

First estimate each product and then find the product.

**5.** $18 \times 24$

Estimate: 400

Product: 432

**6.** $129 \times 39$

Estimate:4000

Product: 5031

First estimate each quotient and then find the quotient and remainder.

**7.** $16\overline{)344}$

Estimate: 15

Quotient: 21 R 8

**8.** $41\overline{)24,682}$

Estimate: 500

Quotient: 602 R 0

## Using Calculators to Perform Calculations with Whole Numbers

**A Special Note on Calculators**

The use of calculators is discussed throughout this text. However, you should be aware that **a calculator cannot take the place of understanding mathematical concepts**. A calculator is a useful aid for speed and accuracy, but it is limited to doing only what you command it to do.

A calculator does not think!

Addition, subtraction, and multiplication with whole numbers can be accomplished accurately with hand-held calculators, as long as none of the numbers involved (including answers) has more digits than the screen on the calculator can display (eight to ten digits on most calculators ). Division that involves a remainder will give a decimal answer in which the whole number part of the quotient is accurate, but the decimal part to the right of the decimal point is generally rounded off. We will discuss decimal numbers in detail in Chapter 5.

Calculators are particularly helpful when there are several numbers involved in a problem and/or the numbers are very large. Their use has become quite common and is recommended in many classes. Some can even perform calculations with fractions.

The examples that follow illustrate the fundamental operations of addition, subtraction, multiplication, and division with two types of calculators: a scientific calculator and a TI-83 graphing calculator. There are some calculators for which the sequence of steps used here is not applicable. If you have such a calculator, you will need to study the manual that comes with it. Also, scientific calculators and the TI-83 have many keys that will not be used in this course. However, if you have a scientific calculator, it should have a key marked ($y^x$) (or ($x^y$)). The TI-83 (or TI-83 Plus or other graphing calculator) will have keys marked 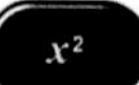 and a caret key . We will use these keys in later chapters.

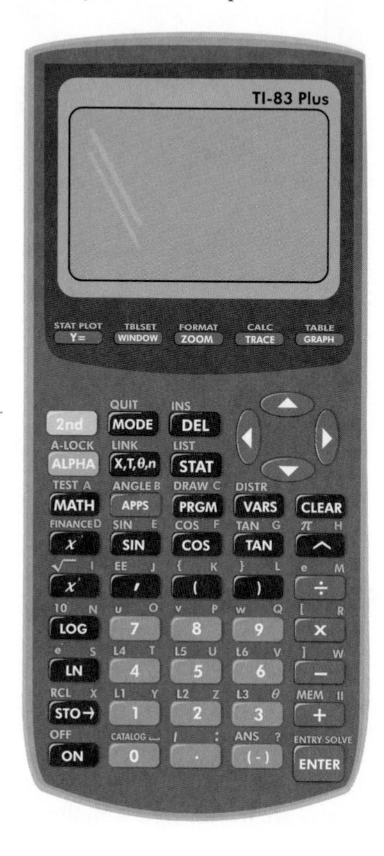


You might want to walk the students through a few calculations to be sure that each individual understands how to operate and read his or her own calculator.

**Example 8** Addition with a Calculator

$$\begin{array}{r} 899 \\ 743 \\ 625 \\ +\ 592 \\ \hline \end{array}$$

**Solution with a Scientific Calculator**

You will see only the numbers you enter and the answer. You will not see the operation symbols.

| Enter | Press | Enter | Press | Enter | Press | Enter | Press |
|---|---|---|---|---|---|---|---|
| ↓ | ↓ | ↓ | ↓ | ↓ | ↓ | ↓ | ↓ |
| 899 | (+) | 743 | (+) | 625 | (+) | 592 | (=) |

Your calculator should display the sum **2859**.

### Solution with a TI-83 Calculator

The display screen is large and you will see all the numbers you enter and all the operations and the answer.

## Example 9 Subtraction with a Calculator

Use your calculator and follow the steps indicated to find the difference:

$$\begin{array}{r} 678{,}025 \\ -\ 483{,}975 \\ \hline \end{array}$$

### Solution with a Scientific Calculator

Your calculator should display the difference **194050**.

Note that the commas are not entered with the numbers and are not shown in the displayed answer.

### Solution with a TI-83 Calculator

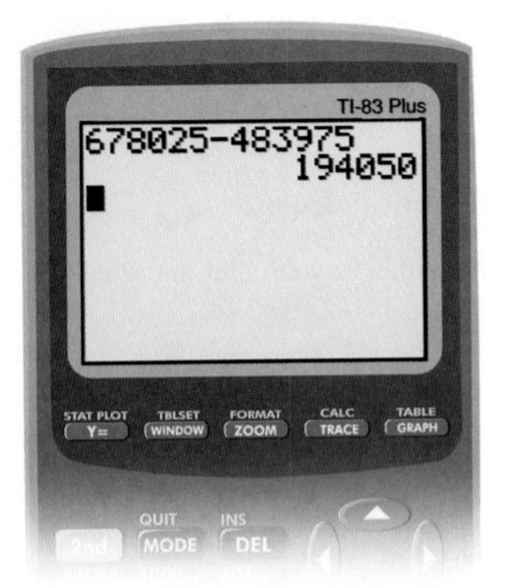

## Example 10 Multiplication with a Calculator

Use your calculator and follow the steps indicated to find the product:

$$\begin{array}{r} 572 \\ \times\ 635 \\ \hline \end{array}$$

### Solution with a Scientific Calculator

| Enter | Press | Enter | Press |
|---|---|---|---|
| ↓ | ↓ | ↓ | ↓ |
| 572 | ⊗ | 635 | ⊜ |

Your calculator should display the product **363220**.

Again, note that commas are not in the displayed answer.

### Solution with a TI-83 Calculator

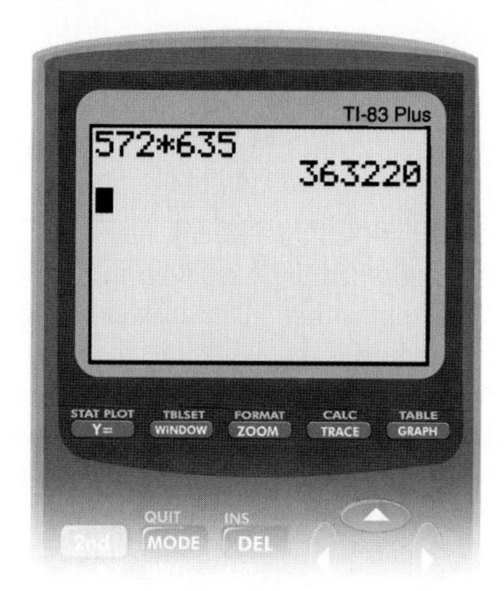


> Note that multiplication is indicated with a star (*) on the display. This is common with graphing calculators and in computer science. Probably, this symbol will become the accepted multiplication symbol and replace the raised dot and the × sign in the near future.

## Example 11 Division with a Calculator

Use your calculator and follow the steps indicated to find the quotient:

$3695 \div 48$

### Solution with a Scientific Calculator

| Enter | Press | Enter | Press |
|---|---|---|---|
| ↓ | ↓ | ↓ | ↓ |
| 3695 | ⊕÷ | 48 | ⊜ |

Your calculator should display **76.97916667**.

If the quotient is not a whole number (as is the case here), the calculator gives rounded-off decimal answers. (These numbers will be discussed in Chapter 5.) Thus, calculators do have some limitations in terms of whole numbers. The whole number part of the display (76) is the correct whole number part of the quotient, but we do not know the remainder R.

**Solution with a TI-83 Calculator**

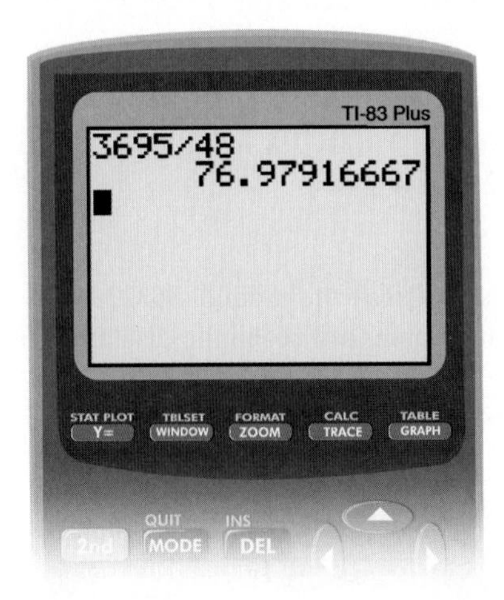


Can the remainder R be found when dividing with a calculator that does not have special keys for fractions and remainders? The answer is "Yes," and we can proceed as the following discussion indicates.

To check long division with whole numbers (see Section 1.1), we multiply the divisor and quotient and then add the remainder. This sum should be the dividend. Now consider the problem of finding the remainder R when the division gives a decimal number as in Example 11.

If we multiply the whole number in the quotient (76) times the divisor (48), we get

$$\begin{array}{rl} 48 & \text{divisor} \\ \times\ 76 & \text{whole number part of quotient} \\ \hline 288 & \\ 336\phantom{0} & \\ \hline 3648 & \text{product} \end{array}$$

This product is not equal to the dividend (3695) because we have not added on the remainder, since we do not know what it is. However, thinking in reverse, we can find the remainder by subtracting this product from the dividend. Thus, the remainder is found as follows:

$$\begin{array}{rl} 3695 & \text{dividend} \\ -\ 3648 & \text{product} \\ \hline 47 & \text{whole number remainder} \end{array}$$

## Example 12

Use a calculator to divide 1857 ÷ 17, and then find the quotient and remainder as whole numbers.

### Solution

With a calculator, 1857 ÷ 17 = 109.2352941. Multiply the whole number part of the quotient (109) by the divisor (17):

$$109 \cdot 17 = 1853$$

To find the remainder, subtract this product from the dividend:

$$1857 - 1853 = 4$$

Thus, the quotient is 109, and the remainder is 4.

*Now work Exercises 9-12 in the Margin*

## Completion Example Answer

7.
$$\begin{array}{r} 326 \\ 21\overline{)6857} \\ \underline{63}\phantom{00} \\ 55\phantom{0} \\ \underline{\mathbf{42}}\phantom{0} \\ \mathbf{137} \\ \underline{\mathbf{126}} \\ \mathbf{11} \end{array} \text{ remainder}$$

$$\begin{array}{r} \mathbf{350} \\ \mathbf{20}\overline{)\mathbf{7000}} \end{array} \text{ estimate}$$

Yes, the estimate is close. The difference is **350 – 326 = 24**.

Use a calculator to perform the indicated operations.

**9.** 579 + 346 + 942
1867

**10.** 6034(702)
4,235,868

Use a calculator to divide and find the quotient and remainder as whole numbers.

**11.** 247,512 ÷ 25
9900 R 12

**12.** 675,300 ÷ 152
4442 R 116

Name ______________ Section ______________ Date ______________

## Exercises 1.2

Round off as indicated.

To the nearest ten:

**1.** 873 **2.** 41 **3.** 92 **4.** 604

**5.** 395 **6.** 423 **7.** 968 **8.** 247

To the nearest hundred:

**9.** 4264 **10.** 5475 **11.** 575 **12.** 972

**13.** 238 **14.** 3890 **15.** 75,223 **16.** 3003

To the nearest thousand:

**17.** 7924 **18.** 5500 **19.** 6500 **20.** 6499

**21.** 14,498 **22.** 14,501 **23.** 63,365 **24.** 57,800

To the nearest ten thousand:

**25.** 78,000 **26.** 125,003 **27.** 257,000 **28.** 63,300

**29.** 119,200 **30.** 315,200 **31.** 284,900 **32.** 715,000

**ANSWERS**

1. 870
2. 40
3. 90
4. 600
5. 400
6. 420
7. 970
8. 250
9. 4300
10. 5500
11. 600
12. 1000
13. 200
14. 3900
15. 75,200
16. 3000
17. 8000
18. 6000
19. 7000
20. 6000
21. 14,000
22. 15,000
23. 63,000
24. 58,000
25. 80,000
26. 130,000
27. 260,000
28. 60,000
29. 120,000
30. 320,000
31. 280,000
32. 720,000

33. 230; 224

34. 220; 215

35. 2400; 2281

36. 17,000; 16,498

37. 8000; 7576

38. 10,000; 11,304

39. 1000; 1687

40. 40,000; 41,417

41. a. 400

b. 480

c. 300

d. 720

42. a. 30,000

b. 3000

c. 300

d. 300,000

43. a. 100

b. 6

c. 400

d. 200

44. a. 300

b. 30,000

c. 30

d. 3000

First estimate the answers using rounded-off numbers; then find the following sums and differences.

**33.** $85 + 64 + 75$

**34.** $98 + 56 + 61$

**35.** $851 + 763 + 572 + 95$

**36.** $6521 + 5742 + 3215 + 1020$

**37.** $8652 - 1076$

**38.** $63{,}504 - 52{,}200$

**39.** $10{,}435 - 8{,}748$

**40.** $74{,}503 - 33{,}086$

**41.** Round off each number and estimate the product by performing the calculations mentally.

**a.** $15 \cdot 19$

**b.** $6(82)$

**c.** $11(31)$

**d.** $(9)(77)$

**42.** Round off each number and estimate the product by performing the calculations mentally.

**a.** $57 \cdot 500$

**b.** $57(50)$

**c.** $57 \cdot 5$

**d.** $(57)(5000)$

**43.** Round off the divisor and dividend as you choose and estimate the quotient by dividing mentally.

**a.** $8\overline{)810}$

**b.** $5\overline{)32}$

**c.** $33\overline{)12{,}000}$

**d.** $18\overline{)3800}$

**44.** Round off the dividend and estimate the quotient by dividing mentally.

**a.** $3\overline{)860}$

**b.** $3\overline{)86{,}000}$

**c.** $3\overline{)86}$

**d.** $3\overline{)8600}$

Name ______________________ Section ______________ Date ____________

First estimate each product; then find the product.

**45.** $66 \times 4$

**46.** $53 \times 6$

**47.** $98 \times 12$

**48.** $39 \times 15$

**49.** $81 \times 36$

**50.** $126 \times 102$

**51.** $420 \times 204$

**52.** $508 \times 119$

First use your judgment to estimate each quotient; then find the quotient and the remainder. Understand that, depending on how the divisor and dividend are rounded off, different answers are possible.

**53.** $18\overline{)216}$

**54.** $13\overline{)260}$

**55.** $49\overline{)993}$

**56.** $50\overline{)3065}$

**57.** $716\overline{)3056}$

**58.** $630\overline{)4768}$

**59.** $414\overline{)83,276}$

**60.** $502\overline{)98,762}$

Use your calculator to perform the indicated operations and find the correct value of the variable.

**61.** $635 + 984 + 235 = n$

**62.** $98,765 + 25,436 + 205 = n$

**63.** $750,438 - 72,895 = x$

**64.** $1,095,005 - 567,890 = x$

**65.** $7045(934) = y$

**66.** $2031(745) = y$

**67.** $28,080 \div 36 = a$

**68.** $16,300 \div 25 = a$

**ANSWERS**

45. 280; 264
46. 300; 318
47. 1000; 1176
48. 800; 585
49. 3200; 2916
50. 10,000; 12,852
51. 80,000; 85,680
52. 50,000; 60,452
53. 10; 12 R 0
54. 30; 20 R 0
55. 20; 20 R 13
56. 60; 61 R 15
57. 4; 4 R 192
58. 8; 7 R 358
59. 200; 201 R 62
60. 200; 196 R 370
61. $n = 1854$
62. $n = 124,406$
63. $x = 677,543$
64. $x = 527,115$
65. $y = 6,580,030$
66. $y = 1,513,095$
67. $a = 780$
68. $a = 652$

69. $3000; $ 2548

70. $30; $ 26

71. 40,000; 36,750 sq. ft.

72. 4000 mi.; 5330 mi.

73. 237 R 13

74. 1593 R 5

75. 728 R 3

76. 1907 R 12

77. 750 R 17

78. 1316 R 6

79. 13,799 R 83

80. 10,165 R 257

For each of the following problems, first estimate the answer; then calculate the answer; then use a calculator to confirm your answer.

**69.** If your income for 1 year was $30,576, what was your monthly income?

**70.** A school bought 3045 new textbooks for a total cost of $79,170. What was the price of each book?

**71.** A rectangular lot for a house measures 210 feet long by 175 feet wide. Find the area of the lot in square feet. (The area of a rectangle is found by multiplying length times width.)

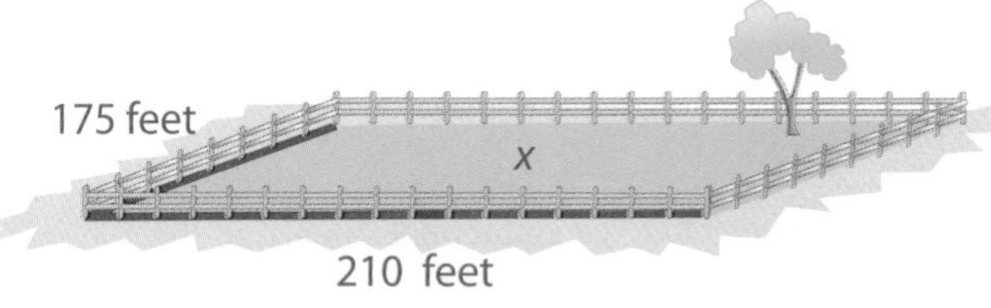


**72.** If you drive your car to work 205 days a year and the distance from your house to work is 13 miles, how many miles each year do you drive your car to and from work?

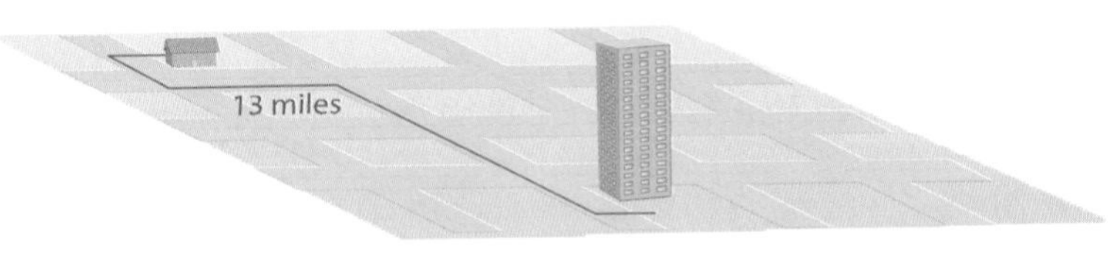


With the use of a calculator, find the quotient and remainder as whole numbers for each of the following division problems. (**Remember**: To find the whole number remainder: First multiply the divisor times the whole number part of the quotient. Then subtract this product from the dividend in the original problem.)

**73.** 3568 ÷ 15

**74.** 9563 ÷ 6

**75.** 25,483 ÷ 35

**76.** 36,245 ÷ 19

**77.** 15,767 ÷ 21

**78.** 93,442 ÷ 71

**79.** 1,835,350 ÷ 133

**80.** 2,795,632 ÷ 275

# 1.3 Applications

## Strategy for Solving Word Problems

In this section, the applications are varied and can involve combinations of the four operations of addition, subtraction, multiplication, and division. The decisions on what operations are to be used are based generally on experience and practice. **If you are not exactly sure just what operations to use, at least try something. Even by making errors in technique or judgement you are learning what does not work. If you do nothing, then you learn nothing. Do not be embarrassed by mistakes.**

The problems discussed here will come under the following headings: Number Problems, Consumer Items, Checking Accounts, and Geometry. The steps in the basic strategy listed here will help give an organized approach regardless of the type of problem.

**Objectives**

① Develop the reading skills necessary for understanding word problems.

② Be able to analyze a word problem with confidence.

③ Realize that problem solving takes time and that the learning process also involves learning from making mistakes.

**Basic Strategy for Solving Word Problems**

1. Read each problem carefully until you understand the problem and know what is being asked for.
2. Draw any type of figure or diagram that might be helpful, and decide what operations are needed.
3. Perform these operations.
4. Mentally check to see if your answer is reasonable and see if you can think of another more efficient or more interesting way to do the same problem.

Now might be a good time to explain to the students that you know that math anxiety is a real life experience, that the body reacts chemically to stressful situations, sometimes beyond the control of the person, and that some reaction is normal. Each student should be encouraged to find a way to overcome stress and anxiety related to activities such as exams and class discussions.

Number problems usually contain key words or phrases that tell what operation or operations are to be performed with given numbers. Learn to look for these key words.

**Key Words That Indicate Operations**

| Addition | Subtraction | Multiplication | Division |
|---|---|---|---|
| add | subtract (from) | multiply | divide |
| sum | difference | product | quotient |
| plus | minus | times | ratio |
| more than | less than | twice | |
| increased by | decreased by | | |

Many students think that needing to read a problem several times for understanding indicates some kind of weakness or lack of ability. Point out that this is simply not true; that reading skills in math are very important and demand a lot of practice.

## Applications Examples

### Example 1 Number Problems

Find the **difference** between the **product** of 78 and 32 and the **sum** of 135 and 776.

#### Solution

The strategy is to identify the key words (in boldface) and to perform the operations indicated by these words. **Note:** The word **and** does **not** indicate any type of operation. It is used three times in the statement of the problem, each time only as a grammatical connector (a conjunction).

Before we can find a difference, we need to find the numbers that are to be used in the subtraction: namely, the product and the sum.

**Product**

$$\begin{array}{r} 78 \\ \times 32 \\ \hline 156 \\ 234\phantom{0} \\ \hline 2496 \end{array}$$

**Sum**

$$\begin{array}{r} {}^{1}1\,{}^{1}3\,5 \\ +776 \\ \hline 911 \end{array}$$

Now we can subtract:

**Difference**

$$\begin{array}{r} {}^{1}\not{2}\,{}^{1}496 \\ -\ 911 \\ \hline 1585 \end{array} \quad \text{difference}$$

Thus, the requested difference is 1585. (As a quick mental check, calculate $80 \cdot 30 = 2400$ and $100 + 800 = 900$, and find the difference between these two estimates: $2400 - 900 = 1500$.)

### Example 2 Number Problems

Find 16 **less than** 20.

#### Solution

The phrase "less than" indicates subtraction. However, there can be some confusion as to just what is to be subtracted. In this example, we subtract 16 from 20.

**Less than**

$$\begin{array}{r} 20 \\ -\ 16 \\ \hline 4 \end{array} \quad \text{Thus 4 is 16 less than 20.}$$

To help clarify the term **less than**, note that

"7 **less than** 10" and "10 **less** 7" both indicate $10 - 7$.

## Example 3 Consumer Items

Karl bought a used car for $11,000. The salesperson added $779 for taxes and $350 for license fees. If Karl made a down payment of $2500 and financed the rest through his credit union, how much did he finance?

### Solution

Find the total cost by adding the expenses, and then subtract the down payment. (Note that the key word "added" does help here. However, only real-life experience tells us to perform both the operation of addition and the operation of subtraction.)

$$\begin{array}{rr} \$\ 11{,}000 & \\ 779 & \\ +\ \ \underline{350} & \text{total cost} \\ \$\ 12{,}129 & \end{array} \qquad \begin{array}{rl} \$\ 12{,}129 & \\ \underline{-\ 2\,500} & \\ \$\ \ \ 9\,629 & \text{total cost} \\ & \text{down payment} \\ & \text{to be financed} \end{array}$$

Karl financed $9629.

(Mental check: The car cost about $12,000, and Karl put down $2500, so an answer of about $9500 is reasonable.)

## Example 4 Checking Account

In June, Ms. Maxwell opened a checking account and deposited $5280. During the month, she made another deposit of $800 and wrote checks for $135, $450, $775, and $72. What was the balance in her account at the end of the month?

### Solution

To find the balance, we find the difference between the sum of the deposit amounts and the sum of the check amounts.

$$\begin{array}{rl} \$\ 5280 & \\ +\ \ \ 800 & \text{sum of deposits} \\ \hline \$\ 6080 & \end{array} \qquad \begin{array}{rl} \$\ \ \ 135 & \\ 450 & \\ 775 & \\ +\ \ \ \ 72 & \text{sum of checks} \\ \hline \$\ 1432 & \end{array}$$

$$\begin{array}{rl} \$\ \ {}^{5}\not{6}\ {}^{1}0\ {}^{7}\not{8}\ {}^{1}0 & \\ -\ \ \ 1\ \ 4\ \ 3\ \ 2 & \text{balance} \\ \hline \$\ \ \ 4\ \ 6\ \ 4\ \ 8 & \end{array}$$

The balance in the account was $4648 at the end of the month.

(Mental check: Using rounded-off numbers, we find that a balance of about $4500 is close.)

| MONTH OF JUNE | | | | BALANCE |
|---|---|---|---|---|
| DATE | TRANSACTION DESCRIPTION | SUBTRACTIONS | ADDITIONS | $0.00 |
| 6/4 | Deposit | | 5280 00 | 5280 00 |
| | | | | 5280 00 |
| 6/10 | Deposit | | 800 00 | 800 00 |
| | | | | 6080 00 |
| 6/16 | check- car payment | 135 00 | | 135 00 |
| | | | | 5945 00 |
| 6/17 | check- Tracy's Department Store | 450 00 | | 450 00 |
| | | | | 5495 00 |
| 6/22 | check- credit card payment | 775 00 | | 775 00 |
| | | | | 4720 00 |
| 6/30 | check- Big Fresh Supermarket | 72 00 | | 72 00 |
| | | | | 4648 00 |

One good teaching idea is to stand in front of the class and tear out (or cut out) a rectangle from a sheet of paper. This illustrates the need for subtraction to find the area of the mat, and provides the students with a good visual image of the problem.

### Example 5 Geometry

A rectangular picture is mounted in a rectangular frame with a border (called a *mat*). (A rectangle is a four-sided figure with opposite sides equal and all four angles equal, 90° each.) If the picture is 14 inches by 20 inches and the frame is 18 inches by 24 inches, what is the area of the mat?

#### Solution

In this case, a figure is very helpful. Also, we need to know that the area of a rectangle is found by multiplying length times width.

The area of the mat will be the difference between the areas of the two rectangles.

**Larger Area**

$$\begin{array}{r} 18 \\ \times\ 24 \\ \hline 72 \\ 36\phantom{0} \\ \hline 432 \end{array}$$ square inches

**Smaller Area**

$$\begin{array}{r} 14 \\ \times\ 20 \\ \hline 280 \end{array}$$ square inches

**Area of Mat**

$$\begin{array}{r} \overset{3}{\cancel{4}}\,{}^{1}3\,2 \\ -\ 2\,8\,0 \\ \hline 1\,5\,2 \end{array}$$ square inches

Name ______________ Section ______________ Date ______________

# Exercises 1.3

**ANSWERS**

## Number Problems

1. Find the **sum** of the three numbers 845, 960, and 455. Then **subtract** 690. What is the **quotient** if the **difference** is **divided by** 2?

2. The **difference** between 9000 and 1856 is **added to** 635. If this **sum** is **multiplied by** 400, what is the **product**?

3. Find the **sum** of the **product** of 23 and 47 and the **product** of 220 and 3.

4. If the **quotient** of 670 and 5 is **decreased by** 120, what is the **difference**?

5. Find 135 **less than** the **product** of 10 and 20.

6. What is 115 **less than** the **quotient** of 468 and 4?

## Consumer Items

7. To purchase a new 27-in. television set with remote control and stereo that sells for $1300 including tax, Mr. Daley paid $200 down and the remainder in five equal monthly payments. What were his monthly payments if he had been married for 3 years?

1. 785
2. 3,111,600
3. 1741
4. 14
5. 65
6. 2
7. $ 220 per month

8. $ 739

9. $ 1296

10. $ 246

11. $ 150

**8.** Ralph decided to go shopping for school clothes before college started in the fall. How much did he spend if he bought four pairs of pants for $75 a pair, five shirts for $45 each, three pairs of socks for $6 a pair, and two pairs of shoes for $98 a pair?

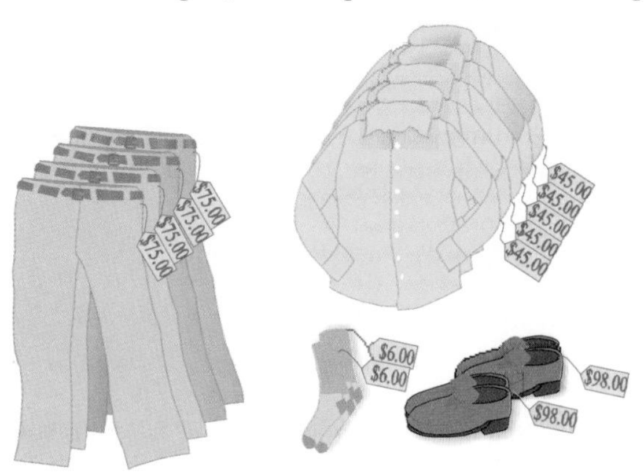


**9.** To purchase a new dining room set for $1500, Mrs. Thomas has to pay $120 in sales tax. If she made a deposit of $324, how much did she still owe? The dining room set consisted of a table and six chairs.

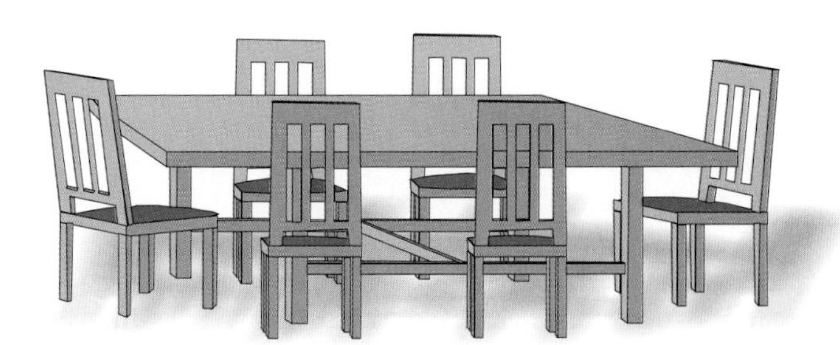

**10.** For a class in statistics, Anthony bought a new graphing calculator for $86, special graph paper for $8, computer disks for $10, a text for $105, and a workbook for $37. What did he spend for this class? The computer disks were formatted and were double sided high density to contain 1.44 MB of memory.

**11.** For her physical education class, Paula bought a pair of running shoes for $84, two pairs of socks for $5 a pair, one pair of shorts for $26, and two shirts for $15 each. What were her expenses for this class if the class was to meet at 8:00 am each morning?

Name ______________________ Section ______________ Date ____________

## ANSWERS

12. Blue car; $738

13. $ 538

14. $ 952

15. $ 7717

16. $ 2864

**12.** Lynn decided to buy a new sports car with a six–cylinder engine. She could buy a blue one for $21,500 plus $690 in taxes and $750 in fees, or she could buy a red custom model for $22,300 plus $798 in taxes and $880 in fees. If the manufacturer was giving rebates of $1200 on the red model and $900 on the blue model, which car was cheaper for her to buy? How much cheaper?

## Checking Accounts

**13.** If you opened a checking account with $975 and then wrote checks for $25, $45, $122, $8, and $237, what would be your balance?

**14.** On July 1, Mel was 24 years old and had a balance of $840 in his checking account. During the month, he made deposits of $490, $43, $320, and $49. He wrote checks for $487, $75, $82, and $146. What was the balance in his checking account at the end of the month?

**15.** During a 5–month period, Cheryl, who goes sky diving once a month, made deposits of $3520, $4560, $1595, $3650, and $2894. She wrote checks totaling $9732. If her beginning balance was $1230, what was the balance in her account at the end of the 5-month period?

| DATE | TRANSACTION DESCRIPTION | SUBTRACTIONS | ADDITIONS | BALANCE |
|---|---|---|---|---|
| | | | | $1230 00 |
| 11/4 | Deposit | | 3520 00 | |
| 11/4 | Check- Skydiving with Bill | 4874 50 | | |
| 12/15 | Extra parachute | 3210 50 | | |
| 1/3 | Deposit | | 4560 00 | |
| 1/25 | Check- skydiving | 1647 00 | | |
| 2/14 | Deposit | | 1595 00 | |
| 3/15 | Deposit | | 3650 00 | |
| 4/4 | Deposit | | 2894 00 | |
| | Ending Balance | | | |

**16.** Your friend is 5 feet tall and weighs 103 pounds. She has a checking balance of $4325 at the beginning of the week and writes checks for $55, $37,$560, $278, and $531. What is her balance at the end of the week?

17 a. 24 ft.

b. 24 sq. ft.

18 a. 86 m

b. 450 sq. m

19. 174 cm

20. 82 in.

## Geometry

**17.** A triangle has three sides: a base of 6 feet, a height of 8 feet, and a third side of 10 feet. This triangle is called a *right triangle* because one angle is 90°.

**a.** Find the perimeter of (distance around) the triangle.

**b.** Find the area of the triangle in square feet.
(To find the area, multiply base times height and then divide by 2.)

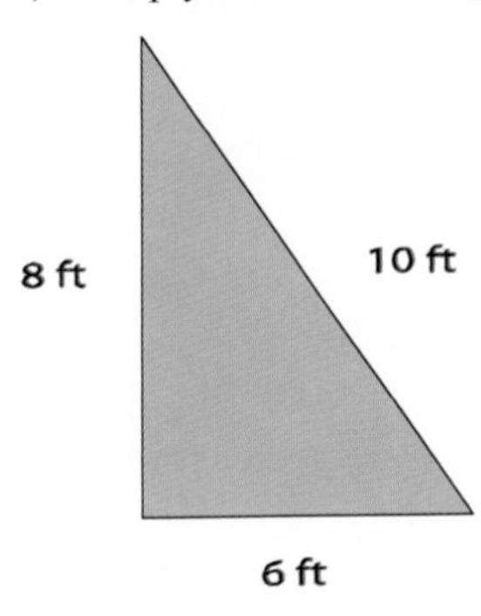


**18. a.** Find the perimeter of (distance around) a rectangle that has a width of 18 meters and a length of 25 meters.

**b.** Also find its area (multiply width times length).

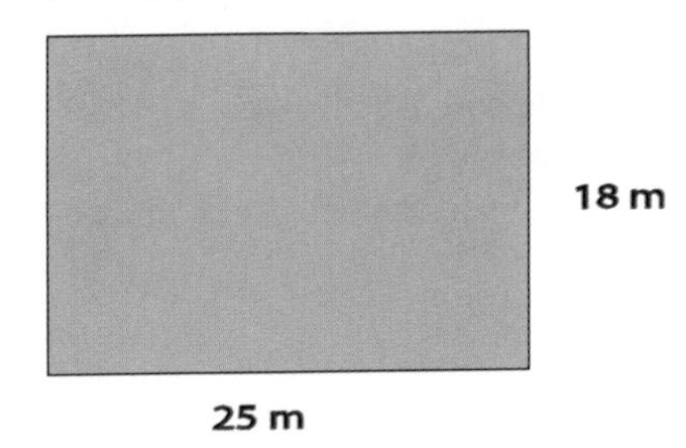


**19.** A regular hexagon is a six-sided figure with all six sides equal and all six angles equal. Find the perimeter of a regular hexagon with one side of 29 centimeters.

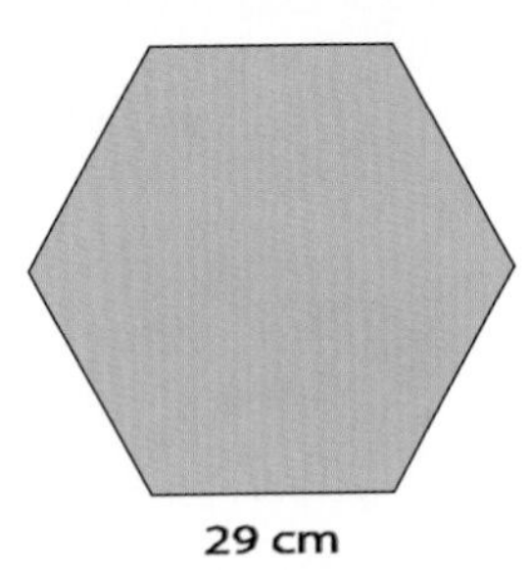


**20.** An isosceles triangle (two sides equal) is placed on top of a square to form a window, as shown in the figure. If each of the two equal sides of the triangle is 14 inches long and the square is 18 inches on each side, what is the perimeter of the window?

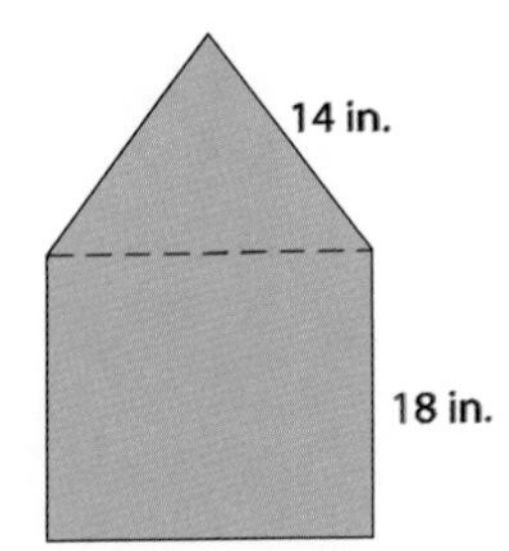

Name ______________________ Section ______________ Date ____________

**21.** A square that is 10 inches on a side is placed inside a rectangle that has a width of 20 inches and a length of 24 inches. What is the area of the region inside the rectangle that surrounds the square?
(Find the area of the shaded region in the figure.)

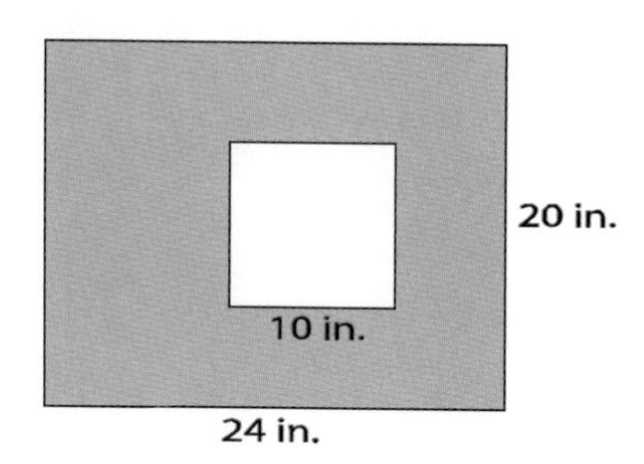


**22.** A flag is in the shape of a rectangle that is 4 feet by 6 feet with a right triangle near one corner. The right triangle is 1 foot high and 2 feet long.

**a.** If the triangle is red and the rest of the flag is blue and white, what is the area of the flag that is red?

**b.** What is the area of the flag that is blue and white?

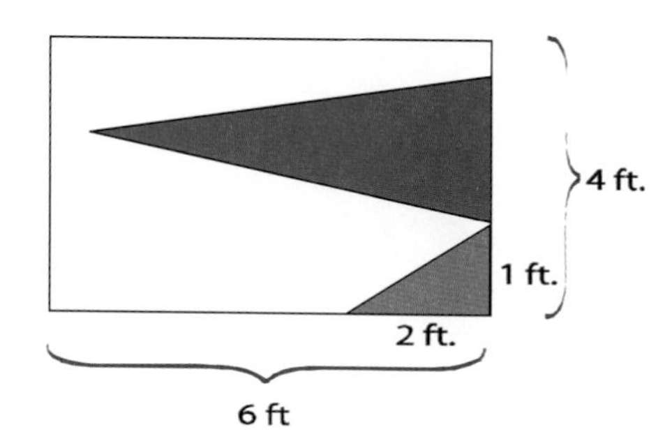


**23.** A pennant in the shape of a right triangle is 10 inches by 24 inches by 26 inches. A yellow square that is 3 inches on a side is inside the triangle and contains the team logo. The remainder of the triangle is a solid blue color. What is the area of the blue colored part of the pennant?

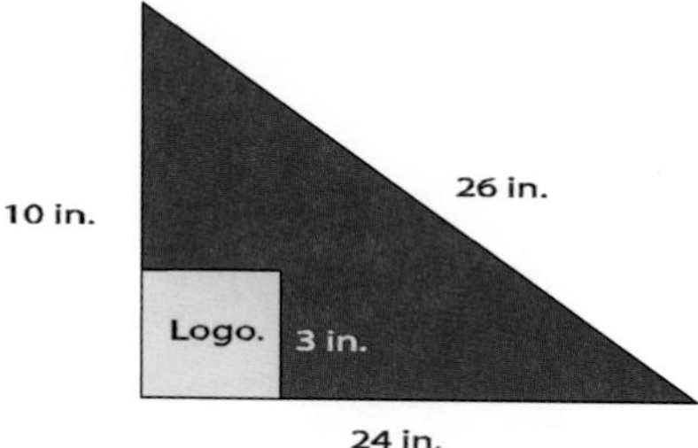


**24.** Except for the window space that is cut into the wall, one wall in a room is to be painted a light tan color. The wall is 8 feet by 15 feet and the window is 3 feet by 4 feet. There are two other walls that are 8 feet by 15 feet that are to be painted the same color. What is the total area of the three walls that is to be painted?

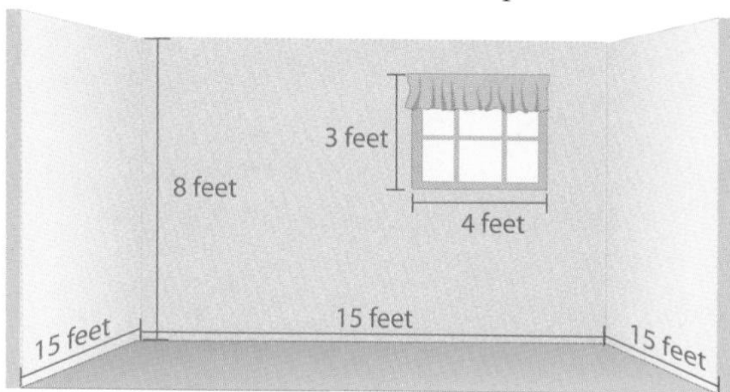


**ANSWERS**

**21.** 380 sq. in.

**22 a.** 1 sq. ft.

**b.** 23 sq. ft.

**23.** 111 sq. in.

**24.** 348 sq. ft.

25. 299 sq. ft.

26 a. 16 in.

b. 12 sq. in.

27. Octagon

stop sign

**25.** One wall of a room is 8 feet by 10 feet, and a rectangular door 3 feet by 7 feet is to be cut in the wall. The remainder of the wall is to be wallpapered. There are three other walls in the room that are 8 feet by 10 feet and these are also to be wallpapered. What is the total area that is to be wallpapered?

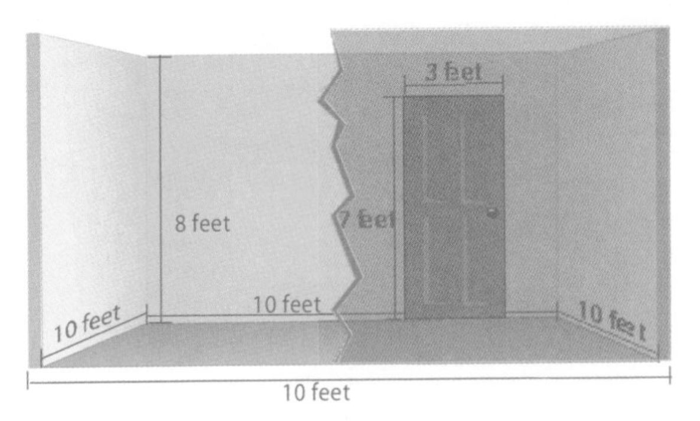


**26.** Find

**a.** the perimeter and

**b.** the area of a triangle with sides of 5 inches, 5 inches, and 6 inches and a height of 4 inches. (See Exercise 17.)

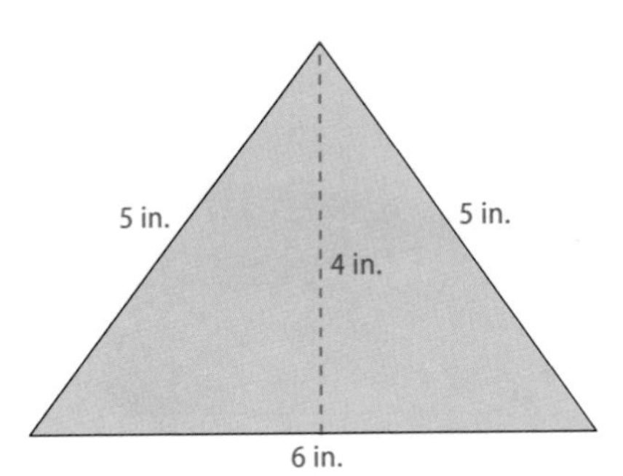


**27.** Draw a figure with eight equal sides. What do you think such a figure is called? What common sign has this shape?

Answers may vary. Students may not be familiar with names of polygons and their Latin and Greek roots.

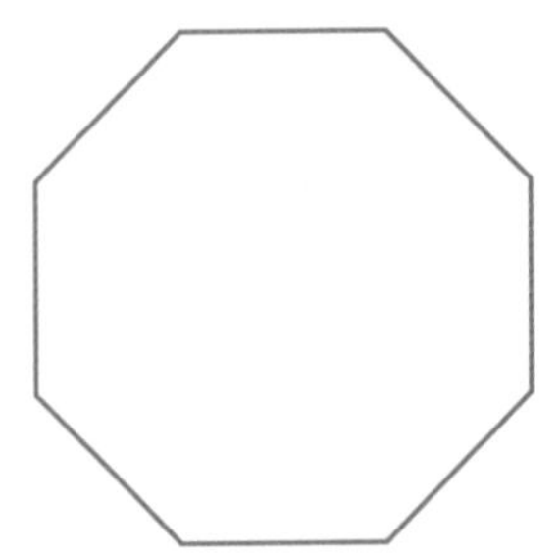

Name ______________________ Section ____________ Date __________

## WRITING AND THINKING ABOUT MATHEMATICS

**28.** In your own words (without looking at the text), write down the four steps listed as Basic Strategy for Solving Word Problems. Then compare your list with that in the text. Were you close? Do you really read the text?

1. Read each problem carefully until you understand the problem and know what is being asked for.
2. Draw any type of figure or diagram that might be helpful, and decide what operations are needed.
3. Perform these operations.
4. Mentally check to see if your answer is reasonable and see if you can think of another more efficient or more interesting way to do the same problem.

**29.** Use your calculator to find each of the following results. Explain, in your own words, what these results indicate.

**a.** $17 \div 0$ **b.** $0 \div 30$ **c.** $0 \div 21$ **d.** $23 \div 0$

Answers will vary.

**30.** Do you have a checking account? If you do, explain briefly how you balance your account at the end of each month when you receive the statement from your bank. If you do not have a checking account, then talk with a friend who does and explain how he or she balances his or her account each month.

Answers will vary.

### ANSWERS

**28.** [Respond below exercise]

**29. a.** Error

**b.** 0

**c.** 0

**d.** Error

**30.** [Respond below exercise]

# 1.4 Solving Equations with Whole Numbers ($x + b = c$, $ax = c$)

## Notation and Terminology

When a number is written next to a variable, such as $3x$ and $5y$, the number is called the **coefficient** of the variable, and the meaning is that the number and the variable are to be multiplied. Thus,

$$3x = 3 \cdot x, \qquad 5y = 5 \cdot y, \qquad \text{and} \qquad 7a = 7 \cdot a$$

If a variable has no coefficient written, then the coefficient is understood to be 1. That is,

$$x = 1x, \qquad y = 1y, \qquad \text{and} \qquad z = 1z$$

A single number, such as 10 or 36, is called a **constant.**

This type of algebraic notation and terminology is a basic part of understanding how to set up and solve equations. In fact, understanding how to set up and solve equations is one of the most important skills in mathematics, and we will be developing techniques for solving equations throughout this text.

An **equation** is a statement that two expressions are equal. For example,

$$12 = 5 + 7, \qquad 18 = 9 \cdot 2, \qquad 14 - 3 = 11, \qquad \text{and} \qquad x + 6 = 10$$

are all equations. Equations involving variables, such as

$$x + 6 = 10 \qquad \text{and} \qquad 2x = 18$$

are useful in solving word problems and applications involving unknown quantities.

If an equation has a variable, then numbers may be substituted for the variable, and any number that gives a true statement is called **a solution** to the equation.

### Objectives

① Become familiar with the terms **equation, solution, constant,** and **coefficient.**

② Learn how to solve equations by using the **Addition Principle, Subtraction Principle,** and the **Division Principle.**

### Example 1

**a.** Show that 4 **is** a solution to the equation $x + 6 = 10$.

Substituting 4 for $x$ gives

$4 + 6 = 10$ which is true.

**b.** Show that 6 **is not** a solution to the equation $x + 6 = 10$.

Substituting 6 for $x$ gives the statement

$6 + 6 = 10$ which is false.

As the discussion and examples indicate, solving equations in this section is essentially a one–operation process, and you should expect some students to "see" the solutions immediately. At this stage, they should understand that finding the solution is only one of the goals; and by getting in the habit of writing down all the steps (as shown in the text), they will have fewer problems in solving more difficult equations later.

As was stated in Section 1.1, a replacement set of a variable is the set of all possible values for that variable. In many applications the replacement set for a variable is clear from the context of the problem. For example, if a problem asks how many people attended a meeting, then the replacement set for any related variable will be the set of whole numbers. That is, there is no sense to allow a negative number (see Chapter 2) of people or a fraction (see Chapter 3) of a person at a meeting. In Example 2 the replacement set is limited to a specific set of numbers for explanation purposes.

### Example 2

Given the equation $x + 5 = 9$, determine which number from the replacement set $\{3, 4, 5, 6\}$ is the solution of the equation.

#### Solution

Each number is substituted for $x$ in the equation until a true statement is found. If no true statement is found, then the equation has no solution in this replacement set.

Substitute 3 for $x$: $3 + 5 = 9$ This statement is false. Continue to substitute values for $x$.

Substitute 4 for $x$: $4 + 5 = 9$ This statement is true. Therefore, 4 is the solution of the equation.

There is no need to substitute 5 or 6 because the solution has been found.

Generally, the replacement set of a variable is a large set, such as the set of all whole numbers, and substituting numbers one at a time for the variable would be ridiculously time consuming. In this section, and in many sections throughout the text, we will discuss methods for **finding solutions of equations** (or **solving equations**). We begin with equations such as

$x + 5 = 8,$ $x - 13 = 7,$ and $4x = 20$

or, more generally, equations of the forms

$\mathbf{x + b = c},$ $\mathbf{x - b = c},$ **and** $\mathbf{ax = c}$

where $x$ is a variable and $a$, $b$, and $c$ represent constants.

**Definitions:**

An **equation** is a statement that two expressions are equal.

A **solution** of an equation is a number that gives a true statement when substituted for the variable.

A **solution set** of an equation is the set of all solutions of the equation.

(**Note:** In this text, with a few exceptions, each equation will have only one number in its solution set. However, in future studies in mathematics, you will deal with equations that have more than one solution.)

## Basic Principles for Solving Equations

The following principles can be used to find the solution to an equation.

In the three basic principles stated here, $A$, $B$, and $C$ represent numbers (constants) or expressions that involve variables.

1. The **Addition Principle:** The equations $A = B$ and $A + C = B + C$ have the same solutions.
2. The **Subtraction Principle:** The equations $A = B$ and $A - C = B - C$ have the same solutions.
3. The **Division Principle:** The equations $A = B$ and $\frac{A}{C} = \frac{B}{C}$ (where $C \neq 0$) have the same solutions.

Essentially, these three principles say that if we perform the same operation to both sides of an equation, the resulting equation will have the same solution as the original equation. Equations with the same solution sets are said to be **equivalent.**

**Special Note:** Although we do not discuss fractions in detail until Chapter 3, we will need the concept of division in fraction form when solving equations and will assume that you are familiar with the fact that a number divided by itself is 1. For example,

$$3 \div 3 = \frac{3}{3} = 1 \qquad \text{and} \qquad 17 \div 17 = \frac{17}{17} = 1$$

In this manner, we will write expressions such as the following:

$$\frac{5x}{5} = \frac{5}{5}x = 1 \cdot x = x$$

As illustrated in the following examples, the variable may appear on the right side of an equation as well as on the left. Also, as these examples illustrate, **the objective in solving an equation is to isolate the variable on one side of the equation. That is, we want the variable on one side of the equation by itself with coefficient 1.**

## Visual Aid

A balance scales gives an excellent visual aid in solving equations. The idea is that, to keep the scales in balance, whatever is done to one side of the scales must be done to the other side. This is, of course, what is done in solving equations.

**a.**

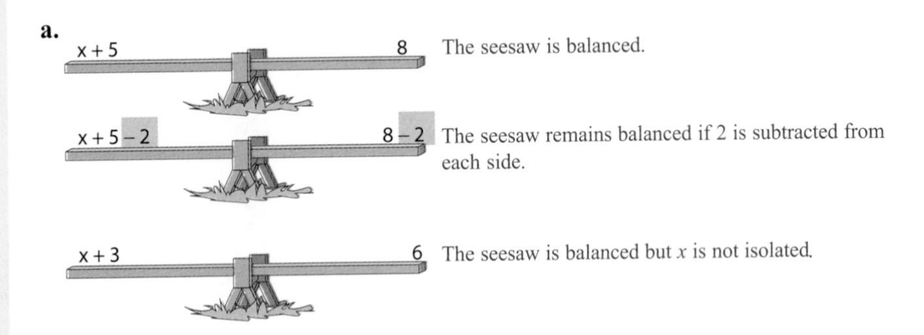


**b.**

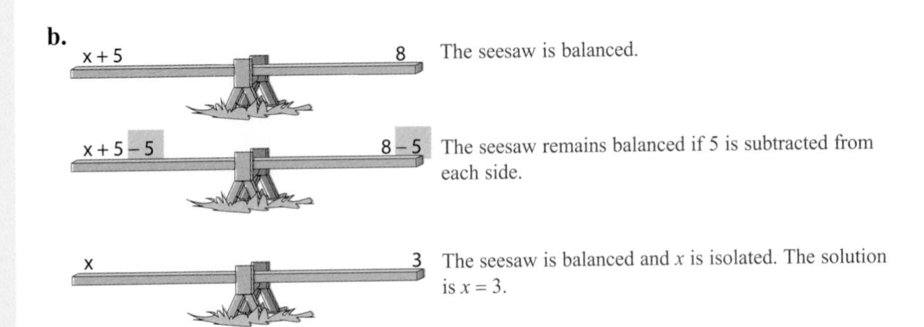

x + 5
8
The seesaw is balanced.
x + 5 – 5
8 – 5
The seesaw remains balanced if 5 is subtracted from each side.
x
3
The seesaw is balanced and x is isolated. The solution is x = 3.


### Example 3

Solve the equations **a.** $x + 5 = 14$ and **b.** $17 = x + 6$.

#### Solutions

**a.**

| | |
|---|---|
| $x + 5 = 14$ | Write the equation. |
| $x + 5 - 5 = 14 - 5$ | Using the subtraction principle, subtract 5 from both sides. |
| $x + 0 = 9$ | Simplify both sides. |
| $x = 9$ | The solution. |

**b.**

| | |
|---|---|
| $17 = x + 6$ | Write the equation. |
| $17 - 6 = x + 6 - 6$ | Using the subtraction principle, subtract 6 from both sides. |
| $11 = x + 0$ | Simplify both sides. |
| $11 = x$ | The solution. |

Note how in each step of the solution process equations are written below each other **with the equal signs aligned.** This format is generally followed in solving all types of equations throughout all of mathematics.

## Example 4

Solve the equations **a.** $y - 2 = 12$ and **b.** $25 = y - 20$

### Solutions

**a.**

| | |
|---|---|
| $y - 2 = 12$ | Write the equation. |
| $y - 2 + 2 = 12 + 2$ | Using the addition principle, add 2 to both sides. |
| $y + 0 = 14$ | Simplify both sides. Note that $-2 + 2 = 0$. |
| $y = 14$ | The solution. |

**b.**

| | |
|---|---|
| $25 = y - 20$ | Write the equation. |
| $25 + 20 = y - 20 + 20$ | Using the addition principle, add 20 to both sides. |
| $45 = y + 0$ | Simplify both sides. |
| $45 = y$ | The solution. |

## Example 5

Solve the equations **a.** $3n = 24$ **b.** $38 = 19n$ **c.** $2x + 3 + 5 = 20$

### Solutions

**a.**

| | |
|---|---|
| $3n = 24$ | Write the equation. |
| $\frac{3n}{3} = \frac{24}{3}$ | Using the division principle, divide both sides by the coefficient 3. Note that in solving equations, the fraction |
| $1 \cdot n = 8$ | Simplify by performing the division on both sides. |
| $n = 8$ | The solution. |

**b.**

| | |
|---|---|
| $38 = 19n$ | Write the equation. |
| $\frac{38}{19} = \frac{19n}{19}$ | Using the division principle, divide both sides by the coefficient 19. |
| $2 = 1 \cdot n$ | Simplify by performing the division on both sides. |
| $2 = n$ | The solution. |

**c.**

| | |
|---|---|
| $2x + 3 + 5 = 20$ | Write the equation. |
| $2x + 8 = 20$ | Simplify. |
| $2x + 8 - 8 = 20 - 8$ | Using the subtraction principle, subtract 8 from both sides. |
| $2x + 0 = 12$ | Simplify. |
| $2x = 12$ | Simplify. |
| $\frac{2x}{2} = \frac{12}{2}$ | Using the division principle, divide both sides by the coefficient 2. |
| $1 \cdot x = 6$ | Simplify. |
| $x = 6$ | The solution. |

*Now Work Exercises 1-4 in the Margin.*

Solve each of the following equations. Show each step, and keep the equal signs aligned vertically in the format used in the text.

**1.** $x + 21 = 40$

$x + 21 - 21 = 40 - 21$

$x = 19$

**2.** $9y = 36$

$\frac{9y}{9} = \frac{36}{9}$

$y = 4$

**3.** $x + 10 - 2 = 14$

$x + 8 = 14$

$x + 8 - 8 = 14 - 8$

$x = 6$

**4.** $15 + 29 - 4 = 2y$

$40 = 2y$

$\frac{40}{2} = \frac{2y}{2}$

$20 = y$

Name ______________ Section __________ Date ________

## Exercises 1.4

In each exercise, an equation and a replacement set of numbers for the variable are given. Substitute each number into the equation until you find the number that is the solution of the equation.

**1.** $x - 4 = 12$ $\{10, 12, 14, 16, 18\}$ **2.** $x - 3 = 5$ $\{4, 5, 6, 7, 8, 9\}$

**3.** $25 = y + 7$ $\{18, 19, 20, 21, 22\}$ **4.** $13 = y + 12$ $\{0, 1, 2, 3, 4, 5\}$

**5.** $7x = 105$ $\{0, 5, 10, 15, 20\}$ **6.** $72 = 8n$ $\{6, 7, 8, 9, 10\}$

**7.** $13n = 39$ $\{0, 1, 2, 3, 4, 5\}$ **8.** $14x = 56$ $\{0, 1, 2, 3, 4, 5\}$

Use either the Addition Principle, the Subtraction Principle, or the Division Principle to solve the following equations. Show each step, and keep the equal signs aligned vertically in the format used in the text.

**9.** $x + 12 = 16$ **10.** $x + 35 = 65$ **11.** $27 = x + 14$

**12.** $32 = x + 10$ **13.** $y + 9 = 9$ **14.** $y + 18 = 18$

**15.** $y - 9 = 9$ **16.** $y - 18 = 18$ **17.** $22 = n + 12$

**18.** $44 = n + 15$ **19.** $75 = n - 50$ **20.** $100 = n - 50$

**ANSWERS**

1. 16
2. 8
3. 18
4. 1
5. 15
6. 9
7. 3
8. 4
9. $x = 4$
10. $x = 30$
11. $13 = x$
12. $22 = x$
13. $y = 0$
14. $y = 0$
15. $y = 18$
16. $y = 36$
17. $10 = n$
18. $29 = n$
19. $125 = n$
20. $150 = n$

21. $x = 4$
22. $x = 8$
23. $y = 4$
24. $y = 5$
25. $4 = n$
26. $9 = n$
27. $x = 6$
28. $18 = x$
29. $47 = y$
30. $69 = y$
31. $x = 7$
32. $x = 23$
33. $x = 5$
34. $x = 5$
35. $x = 12$
36. $x = 22$
37. $3 = y$
38. $3 = y$
39. $9 = x$
40. $2 = x$
41. $n = 3$
42. $n = 14$
43. $0 = x$
44. $0 = x$
45. $0 = x$
46. $5 = x$
47. $6 = x$
48. $y = 7$
49. $n = 2$
50. $x = 0$

**21.** $8x = 32$ **22.** $9x = 72$ **23.** $13y = 52$

**24.** $15y = 75$ **25.** $84 = 21n$ **26.** $99 = 11n$

**27.** $x + 14 = 20$ **28.** $15 = x - 3$ **29.** $42 = y - 5$

**30.** $73 = y + 4$

In each of the following exercises, perform any indicated arithmetic operations before using the principles to solve the equations.

**31.** $x + 14 - 1 = 17 + 3$ **32.** $x + 20 + 3 = 6 + 40$ **33.** $7x = 25 + 10$

**34.** $6x = 14 + 16$ **35.** $3x = 17 + 19$ **36.** $2x = 72 - 28$

**37.** $21 + 30 = 17y$ **38.** $16 + 32 = 16y$ **39.** $31 + 15 - 10 = 4x$

**40.** $17 + 20 - 3 = 17x$ **41.** $3n = 43 - 40 + 6$ **42.** $2n = 26 + 3 - 1$

**43.** $23 - 23 = 5x$ **44.** $6 - 6 = 3x$ **45.** $5 + 5 - 5 - 5 = 4x$

**46.** $5 + 5 + 5 = 3x$ **47.** $6 + 6 + 6 = 3x$ **48.** $4y = 7 + 7 + 7 + 7$

**49.** $5n = 2 + 2 + 2 + 2 + 2$ **50.** $8x = 17 - 17$

# 1.5 Exponents and Order of Operations

## Terminology of Exponents

We know that repeated addition of the same number is shortened by using multiplication. For example,

$$3+3+3+3=4 \cdot 3=12$$

(4 and 3: factors; 12: product)

The result is called the **product**, and the numbers that are multiplied are called **factors** of the product.

In a similar manner, repeated multiplication by the same number can be shortened by using **exponents**. Thus, if 3 is used as a factor four times, we can write

$$3 \cdot 3 \cdot 3 \cdot 3 = 3^4 = 81$$

In an expression such as $3^4 = 81$, 3 is called the **base**, 4 is called the **exponent**, and 81 is called the **power**. (Exponents are written slightly to the right and above the base.)

$$3^4 = 81$$

(4: exponent; 3: base; 81: power)

In this case, we would say that 81 is the fourth power of 3.

### Objectives

① Understand the following terms:

**base, exponent,** and **power**

② Know how to use a calculator to find powers.

③ Know the rules for order of operations and be able to apply these rules when evaluating numerical expressions.

④ Know how to evaluate expressions with exponents, including 1 and 0.

### Example 1

| | With Repeated Multiplication | With Exponents |
|---|---|---|
| **a.** | $7 \cdot 7 = 49$ | $7^2 = 49$ |
| **b.** | $3 \cdot 3 = 9$ | $3^2 = 9$ |
| **c.** | $2 \cdot 2 \cdot 2 = 8$ | $2^3 = 8$ |
| **d.** | $10 \cdot 10 \cdot 10 \cdot 10 = 10{,}000$ | $10^4 = 10{,}000$ |

**Definition:**

A whole number $\boldsymbol{n}$ is an **exponent** if it is used to tell how many times another whole number $\boldsymbol{a}$ is used as a factor. The repeated factor $\boldsymbol{a}$ is called the **base** of the exponent. Symbolically,

$$\underbrace{a \cdot a \cdot a \cdot \ldots \cdot a \cdot a}_{n \text{ factors}} = a^n$$

**Common Error:**

**Do not** multiply the base times the exponent.

$10^2 = 10 \cdot 2$ INCORRECT
$6^3 = 6 \cdot 3$ INCORRECT

**Do** multiply the base times itself.

$10^2 = 10 \cdot 10$ CORRECT
$6^3 = 6 \cdot 6 \cdot 6$ CORRECT

In expressions with exponent 2, the base is said to be **squared**. In expressions with exponent 3, the base is said to be **cubed**.

## Example 2

**a.** $8^2 = 64$ is read "eight squared is equal to sixty-four."
**b.** $5^3 = 125$ is read "five cubed is equal to one hundred twenty-five."

Expressions with exponents other than 2 or 3 are read as the base "to the _____ power." For example,

$2^5 = 32$ is read "two to the fifth power is equal to thirty–two."

**Special Note:** There is usually some confusion about the use of the word "power." Since $2^5$ is read "two to the fifth power," it is natural to think of 5 as the power. This is not true. A power is not an exponent. A power is the product indicated by a base raised to an exponent. Thus, for the equation $2^5 = 32$, think of the phrase "two to the fifth power" in its entirety. The corresponding power is the product, 32.

Note that if there is no exponent written, then the exponent is understood to be 1. That is, any number or algebraic expression raised to the first power is equal to itself. Thus

$8 = 8^1$, $72 = 72^1$, $x = x^1$, and $6y = 6y^1$

**Definition:** For any whole number $a$, $\boldsymbol{a = a^1}$

To improve your speed in factoring (see Section 3.3) and in working with fractions and simplifying expressions, you should memorize all the squares of the whole numbers from 1 to 20. These numbers are called **perfect squares**. The following table lists these perfect squares.

Most students will know the squares of the integers from 1 to 10. Since their speed in working with fractions will improve dramatically if they are familiar with the squares of the integers from 11 to 20 and since we will study square roots in Chapter 5, students should be encouraged to memorize these squares. You might even want to give a quiz on the squares.

| Number $n$ | 1 | 2 | 3 | 4 | 5 | 6 | 7 | 8 | 9 | 10 |
|---|---|---|---|---|---|---|---|---|---|---|
| Perfect Square ($n^2$) | 1 | 4 | 9 | 16 | 25 | 36 | 49 | 64 | 81 | 100 |

| Number $n$ | 11 | 12 | 13 | 14 | 15 | 16 | 17 | 18 | 19 | 20 |
|---|---|---|---|---|---|---|---|---|---|---|
| Perfect Square ($n^2$) | 121 | 144 | 169 | 196 | 225 | 256 | 289 | 324 | 361 | 400 |

## Zero as an Exponent

The use of 0 as an exponent is a special case and, as we will see later, is needed to simplify and evaluate algebraic expressions. To understand the meaning of 0 as an exponent, study the following patterns of numbers. These patterns lead to the meaning of $2^0$, $3^0$, and $5^0$ and the general expression of the form $a^0$.

| | | |
|---|---|---|
| $2^4 = 16$ | $3^4 = 81$ | $5^4 = 625$ |
| $2^3 = 8$ | $3^3 = 27$ | $5^3 = 125$ |
| $2^2 = 4$ | $3^2 = 9$ | $5^2 = 25$ |
| $2^1 = 2$ | $3^1 = 3$ | $5^1 = 5$ |
| $2^0 = ?$ | $3^0 = ?$ | $5^0 = ?$ |

Notice that in the column of powers of 2, each number can be found by dividing the previous number by 2. Therefore, for $2^0$, we might guess that $2^0 = \frac{2}{2} = 1$. Similarly, $3^0 = \frac{3}{3} = 1$ and $5^0 = \frac{5}{5} = 1$.

Another approach to understanding the use of 0 as an exponent is to look ahead to a property of exponents used in algebra. This rule, discussed in Chapter 8, is that for exponents $m$ and $n$, $\frac{a^m}{a^n} = a^{m-n}$. Without going into a detailed discussion, applying this property with exponents $m$ and $n$ being equal gives the results just discussed with 0 as an exponent. Consider the following discussion.

We know that $\frac{2^3}{2^3} = \frac{8}{8} = 1$. Applying the property for exponents gives

$$\frac{2^3}{2^3} = 2^{3-3} = 2^0 .$$

Thus, $2^0 = 1$ makes sense, and we are lead to the following statement about 0 as an exponent.

> **The Exponent 0**
>
> If $a$ is any nonzero whole number, then
>
> $\boldsymbol{a^0 = 1}$
>
> (**Note:** The expression $0^0$ is undefined.)

*Now Work Exercises 1 - 5 in the Margin.*

With examples such as the following, you might want to use a more algebraic approach:

$$\frac{x^5}{x^2} = \frac{\cancel{x}\cdot\cancel{x}\cdot x\cdot x\cdot x}{\cancel{x}\cdot\cancel{x}} = x^3$$

$$\text{or } \frac{x^5}{x^2} = x^{5-2} = x^3 .$$

Thus, $\frac{x^4}{x^4} = x^{4-4} = x^0$

but $\frac{x^4}{x^4} = 1$.

In each of the following expressions,

**a.** Name the base
**b.** Name the exponent
**c.** Find the value of each expression.

**1.** $8^2$
**a.** 8
**b.** 2
**c.** 64

**2.** $15^0$
**a.** 15
**b.** 0
**c.** 1

Rewrite each of the following products by using exponents.

**3.** $9 \cdot 9 \cdot 9$
$9^3$

**4.** $2 \cdot 2 \cdot 2 \cdot 5 \cdot 5$
$2^3 \cdot 5^2$

**5.** Find a base and exponent form for 196 without using the exponent 1. (**Hint**: Look at the list of perfect squares.)
$14^2$

## Using a Calculator to Find Powers

### For a Scientific Calculator

A calculator with a key marked $y^x$ (or $x^y$) can be used to find powers. For example, to find $6^4$, proceed as follows:

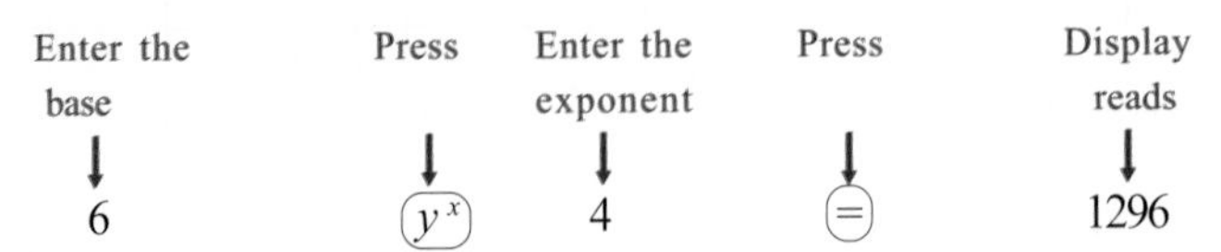


> **Note:** On some scientific calculators, the expression $y^x$ is not marked on the face of a key. You must first press another key marked INV (or 2nd) and then press the key with $y^x$ written just above it.

### For a TI-83 Graphing Calculator

The TI-83 graphing calculator has keys marked as $x^2$ and a caret key ^ . These keys can be used to find powers. The $x^2$ key will find squares quickly and the caret key ^ can be used to find other powers.

To find $6^4$ ,

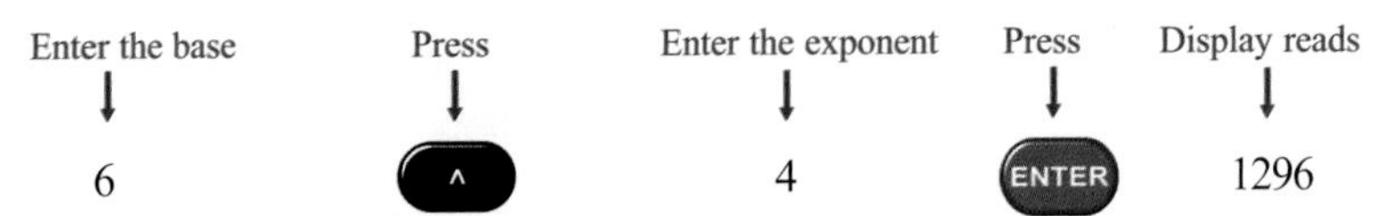


The display will actually appear as follows:

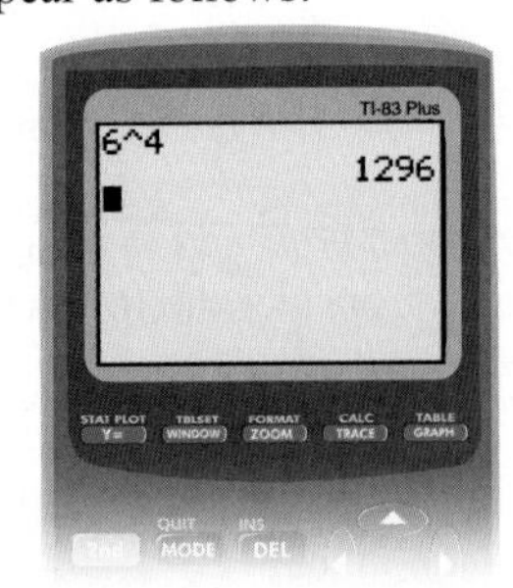


To find $27^2$,

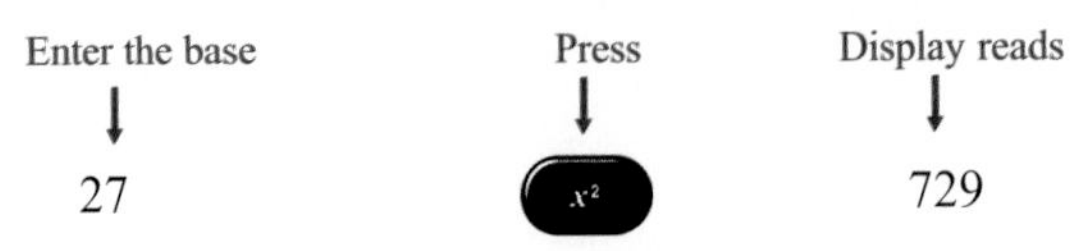


The display will actually appear as follows:

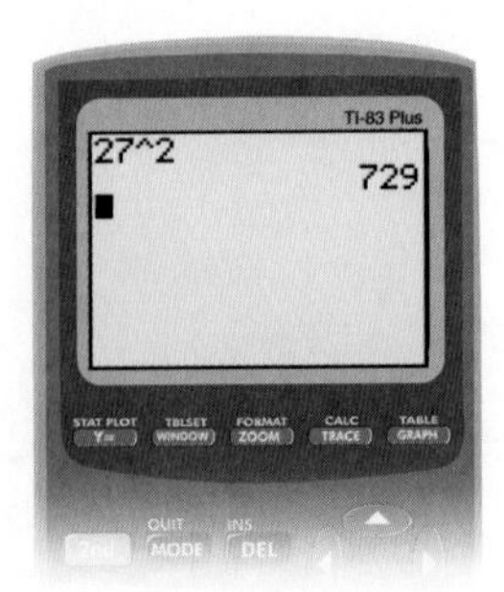

### Example 3

Use a calculator to find the value of each of the following expressions.

**a.** $9^5$ **b.** $41^4$

#### Solutions for a Scientific Calculator

**a.** **Step 1:** Enter 9. (This is the base.)
**Step 2:** Press the key $(y^x)$. (Nothing will happen. The calculator is waiting for you to enter the exponent.)
**Step 3:** Enter 5. (This is the exponent.)
**Step 4:** Press the key $(=)$.

The display should read **59049**.

**b.** **Step 1:** Enter 41.
**Step 2:** Press the key $(y^x)$. (Nothing will happen. The calculator is waiting for you to enter the exponent.)
**Step 3:** Enter 4.
**Step 4:** Press the key $(=)$.

The display should read **2825761**.

#### Solutions for a TI-83 Calculator

We show only the display screen:

*Now Work Exercises 6-9 in the Margin.*

Use a calculator to find the value of each of the following expression.

**6.** $8^5$
32,768

**7.** $56^3$
175,616

**8.** $110^4$
146,410,000

**9.** $13^7$
62,748,517

## Rules for Order of Operations

Mathematicians have agreed on a set of rules for the order of operations in simplifying (or evaluating) any numerical expression involving addition, subtraction, multiplication, division, and exponents. These rules are used in all branches of mathematics and computer science to ensure that there is only one correct answer regardless of how complicated an expression might be. For example, the expression $5 \cdot 2^3 - 14 \div 2$ might be evaluated as follows:

$$\begin{aligned} 5 \cdot 2^3 - 14 \div 2 &= 5 \cdot 8 - 14 \div 2 \\ &= 40 - 14 \div 2 \\ &= 40 - 7 \\ &= 33 \qquad \text{CORRECT} \end{aligned}$$

However, if we were simply to proceed from left to right as in the following steps, we would get an entirely different answer:

$$\begin{aligned} 5 \cdot 2^3 - 14 \div 2 &= 10^3 - 14 \div 2 \\ &= 1000 - 14 \div 2 \\ &= 986 \div 2 \\ &= 493 \qquad \text{INCORRECT} \end{aligned}$$

By the following standard set of rules for order of operations, we can conclude that **the first answer (33) is the accepted correct answer** and that **the second answer (493) is incorrect.**

## Special Note about the Difference between Evaluating Expressions and Solving Equations

Some students seem to have difficulty distinguishing between evaluating expressions and solving equations. In evaluating expressions, we are simply trying to find the numerical value of an expression. In solving equations, we are trying to find the value of some unknown number represented by a variable in an equation. These two concepts are not the same. In solving equations, we use the Addition Principle, the Subtraction Principle, and the Division Principle. In evaluating expressions, we use the following Rules for Order of Operations. These same rules are programmed into nearly every calculator.

## Rules for Order of Operations

1. First, simplify within grouping symbols, such as parentheses ( ), brackets [ ], or braces { }. Start with the innermost grouping.
2. Second, find any powers indicated by exponents.
3. Third, moving from **left to right**, perform any multiplications or divisions in the order in which they appear.
4. Fourth, moving from **left to right**, perform any additions or subtractions in the order in which they appear.

Students will find the exercises both challenging and frustrating. These frustrations generally arise from a limited ability to follow directions and pay attention to details. Common errors occur in evaluating expressions such as

$3+2(7+5)$ and

$20-15\div3+2$

where the students will add 3 + 2 first because it "looks right." You can probably help them a great deal by continually reminding them that outside parentheses, addition and subtraction are the last operations to be performed.

These rules are very explicit and should be studied carefully. Note that in Rule 3, neither multiplication nor division has priority over the other. Whichever of these operations occurs first, moving **left to right,** is done first. In Rule 4, addition and subtraction are handled in the same way. Unless they occur within grouping symbols, **addition and subtraction are the last operations to be performed**.

For example, consider the relatively simple expression $2 + 5 \cdot 6$. If we make the mistake of adding before multiplying, we get

$2+5\cdot6=7\cdot6=42$ INCORRECT

The correct procedure yields

$2+5\cdot6=2+30=32$ CORRECT

A well-known mnemonic device for remembering the Rules for Order of Operations is the following:

| Please | Excuse | My | Dear | Aunt | Sally |
|---|---|---|---|---|---|
| ↓ | ↓ | ↓ | ↓ | ↓ | ↓ |
| Parentheses | Exponents | Multiplication | Division | Addition | Subtraction |

(**Note:** Even though the mnemonic **PEMDAS** is helpful, remember that multiplication and division are performed as they appear, left to right. Also, addition and subtraction are performed as they appear, left to right.)

The following examples show how to apply these rules. In some cases, more than one step can be performed at the same time. This is possible when parts are separated by + or – signs or are within separate symbols of inclusion, as illustrated in Examples 6 – 8. **Work through each example step by step, and rewrite the examples on a separate sheet of paper.**

### Example 4

Evaluate the expression $24 \div 8 + 4 \cdot 2 - 6$.

**Solution**

$24 \div 8 + 4 \cdot 2 - 6$ Divide before multiplying in this case. Remember to move left to right.

$= 3 + 4 \cdot 2 - 6$ Multiply before adding or subtracting.

$= 3 + 8 - 6$ Add before subtracting in this case. Remember to move left to right.

$= 11 - 6$ Subtract.

$= 5$

### Example 5

Evaluate the expression $5 - 18 \div 9 \cdot 2 + 4(7)$.

**Solution**

$5 - 18 \div 9 \cdot 2 + 4(7)$ Divide.

$= 5 - 2 \cdot 2 + 4(7)$ Multiply.

$= 5 - 4 + 4(7)$ Multiply.

$= 5 - 4 + 28$ Subtract.

$= 1 + 28$ Add.

$= 29$

### Example 6

Evaluate the expression $30 \div 10 \cdot 2^3 + 3(6 - 2)$.

#### Solution

| | |
|---|---|
| $30 \div 10 \cdot 2^3 + 3\underbrace{(6-2)}$ | Operate within parentheses. |
| $= 30 \div 10 \cdot \underbrace{2^3} + 3(4)$ | Find the power. |
| $= \underbrace{30 \div 10} \cdot 8 + 3(4)$ | Divide. |
| $= \underbrace{3 \cdot 8} + \underbrace{3(4)}$ | Multiply in each part separated by +. |
| $= \underbrace{24 + 12}$ | Add. |
| $= 36$ | |

### *Completion Example 7*

Evaluate the expression $(14 - 10)[(5 + 3^2) \div 2 + 5]$.

#### Solution

| | |
|---|---|
| $(14 - 10)[(5 + 3^2) \div 2 + 5]$ | Operate within parentheses and find the power. |
| $= (\_\_\_\_\_)[(5 + \_\_\_\_\_) \div 2 + 5]$ | Operate within parentheses. |
| $= (\_\_\_\_\_)[(\_\_\_\_\_) \div 2 + 5]$ | Divide within brackets. |
| $= (\_\_\_\_\_)[\_\_\_\_\_ + 5]$ | Add within brackets. |
| $= (\_\_\_\_\_)[\_\_\_\_\_]$ | Multiply. |
| $= \_\_\_\_\_$ | |

### *Completion Example 8*

Evaluate the expression $9(2^2 - 1) - 17 + 6 \cdot 2^2$.

**Solution**

$9(2^2 - 1) - 17 + 6 \cdot 2^2$

$= 9(\_\_\_\_\_ - 1) - 17 + 6 \cdot \_\_\_\_\_$

$= 9(\_\_\_\_\_) - 17 + 6 \cdot \_\_\_\_\_$

$= \_\_\_\_\_ - 17 + \_\_\_\_\_$

$= \_\_\_\_\_ + \_\_\_\_\_$

$= \_\_\_\_\_$

*Now Work Exercises 10–14 in the Margin.*

### Completion Example Answers

**7.** $(14 - 10)[(5 + 3^2) \div 2 + 5]$

$= (\mathbf{4})[(5 + \mathbf{9}) \div 2 + 5]$

$= (\mathbf{4})[(\mathbf{14}) \div 2 + 5]$

$= (\mathbf{4})[\mathbf{7} + 5]$

$= (\mathbf{4})[\mathbf{12}]$

$= \mathbf{48}$

**8.** $9(2^2 - 1) - 17 + 6 \cdot 2^2$

$= 9(\mathbf{4} - 1) - 17 + 6 \cdot \mathbf{4}$

$= 9(\mathbf{3}) - 17 + 6 \cdot \mathbf{4}$

$= \mathbf{27} - 17 + \mathbf{24}$

$= \mathbf{10} + \mathbf{24}$

$= \mathbf{34}$

Find the value of each expression by using the rules for order of operations.

**10.** $20 + 15 \div 5$
23

**11.** $6 \cdot 4 \div 2 + 4$
16

**12.** $3 \cdot 2^3 - 10 \div 2$
19

**13.** $19 + 5(3 - 1)$
29

**14.** $2(3^2 - 1) - 4 + 3 \cdot 2^3$
36

Name ______________________ Section ____________ Date __________

**ANSWERS**

[For exercises 1-20, respond in exercise.]

## Exercises 1.5

In each of the following expressions (a) name the base, (b) name the exponent, and (c) find the value of each expression.

**1.** $2^3$
a. 2
b. 3
c. 8

**2.** $5^3$
a. 5
b. 3
c. 125

**3.** $4^2$
a. 4
b. 2
c. 16

**4.** $6^2$
a. 6
b. 2
c. 36

**5.** $9^2$
a. 9
b. 2
c. 81

**6.** $7^3$
a. 7
b. 3
c. 343

**7.** $11^2$
a. 11
b. 2
c. 121

**8.** $2^6$
a. 2
b. 6
c. 64

**9.** $3^5$
a. 3
b. 5
c. 243

**10.** $2^4$
a. 2
b. 4
c. 16

**11.** $19^0$
a. 19
b. 0
c. 1

**12.** $22^0$
a. 22
b. 0
c. 1

**13.** $1^6$
a. 1
b. 6
c. 1

**14.** $1^{57}$
a. 1
b. 57
c. 1

**15.** $24^1$
a. 24
b. 1
c. 24

**16.** $13^1$
a. 13
b. 1
c. 13

**17.** $20^3$
a. 20
b. 3
c. 8000

**18.** $15^2$
a. 15
b. 2
c. 225

**19.** $30^2$
a. 30
b. 2
c. 900

**20.** $40^2$
a. 40
b. 2
c. 1600

21. $2^4$ or $4^2$

22. $2^2$

23. $5^2$

24. $2^5$

25. $7^2$

26. $11^2$

27. $6^2$

28. $10^2$

29. $10^3$

30. $9^2$ or $3^4$

31. $2^3$

32. $5^3$

33. $6^3$

34. $7^3$

35. $3^5$

36. $10^4$ or $100^2$

37. $10^5$

38. $5^4$ or $25^2$

39. $3^3$

40. $2^6$ or $4^3$ or $8^2$

41. $6^5$

42. $7^4$

43. $11^3$

44. $13^3$

45. $2^3 \cdot 3^2$

46. $2^2 \cdot 5^3$

47. $2 \cdot 3^2 \cdot 11^2$

48. $5^3 \cdot 7^2$

49. $3^3 \cdot 7^3$

50. $2^4 \cdot 11^2 \cdot 13^2$

Find a base and exponent form for each of the following powers without using the exponent 1.

**21.** 16 **22.** 4 **23.** 25 **24.** 32

**25.** 49 **26.** 121 **27.** 36 **28.** 100

**29.** 1000 **30.** 81 **31.** 8 **32.** 125

**33.** 216 **34.** 343 **35.** 243 **36.** 10,000

**37.** 100,000 **38.** 625 **39.** 27 **40.** 64

Rewrite the following products by using exponents.

**41.** $6 \cdot 6 \cdot 6 \cdot 6 \cdot 6$ **42.** $7 \cdot 7 \cdot 7 \cdot 7$ **43.** $11 \cdot 11 \cdot 11$

**44.** $13 \cdot 13 \cdot 13$ **45.** $2 \cdot 2 \cdot 2 \cdot 3 \cdot 3$ **46.** $2 \cdot 2 \cdot 5 \cdot 5 \cdot 5$

**47.** $2 \cdot 3 \cdot 3 \cdot 11 \cdot 11$ **48.** $5 \cdot 5 \cdot 5 \cdot 7 \cdot 7$ **49.** $3 \cdot 3 \cdot 3 \cdot 7 \cdot 7 \cdot 7$

**50.** $2 \cdot 2 \cdot 2 \cdot 2 \cdot 11 \cdot 11 \cdot 13 \cdot 13$

Name ______________ Section ______________ Date ______________

Find the following perfect squares. Write as many of them as you can from memory.

**51.** $8^2$ **52.** $3^2$ **53.** $11^2$ **54.** $14^2$ **55.** $20^2$

**56.** $15^2$ **57.** $13^2$ **58.** $18^2$ **59.** $30^2$ **60.** $50^2$

Use a calculator to find each of the following powers.

[For a scientific calculator use the $y^x$ key. For a TI-83, or other graphing calculator, use the $x^2$ key or the caret key ^ .]

**61.** $52^2$ **62.** $35^2$ **63.** $25^4$ **64.** $32^4$

**65.** $125^3$ **66.** $47^5$ **67.** $5^7$ **68.** $2^{23}$

Find the value of each of the following expressions by using the rules for order of operations.

**69.** $6 + 5 \cdot 3$ **70.** $18 + 2 \cdot 5$

**71.** $20 - 4 \div 4$ **72.** $6 - 15 \div 3$

**73.** $32 - 14 + 10$ **74.** $25 - 10 + 11$

**ANSWERS**

51. 64
52. 9
53. 121
54. 196
55. 400
56. 225
57. 169
58. 324
59. 900
60. 2500
61. 2704
62. 1225
63. 390,625
64. 1,048,576
65. 1,953,125
66. 229,345,007
67. 78,125
68. 8,388,608
69. 21
70. 28
71. 19
72. 1
73. 28
74. 26

75. 2

76. 17

77. 18

78. 12

79. 0

80. 0

81. 0

82. 0

83. 46

84. 2

85. 9

86. 9

87. 74

88. 34

**75.** $18 \div 2 - 1 - 3 \cdot 2$

**76.** $6 \cdot 3 \div 2 - 5 + 13$

**77.** $2 + 3 \cdot 7 - 10 \div 2$

**78.** $14 \cdot 2 \div 7 \div 2 + 10$

**79.** $(2 + 3 \cdot 4) \div 7 - 2$

**80.** $(2 + 3) \cdot 4 \div 5 - 4$

**81.** $14(2 + 3) - 65 - 5$

**82.** $13(10 - 7) - 20 - 19$

**83.** $2 \cdot 5^2 - 8 \div 2$

**84.** $16 \div 2^4 + 9 \div 3^2$

**85.** $(2^3 + 2) \div 5 + 7^2 \div 7$

**86.** $18 - 9(3^2 - 2^3)$

**87.** $(4 + 3)^2 + (2 + 3)^2$

**88.** $(2 + 1)^2 + (4 + 1)^2$

Name ______________________ Section ______________ Date __________

**ANSWERS**

**89.** $8 \div 2 \cdot 4 - 16 \div 4 \cdot 2 + 3 \cdot 2^2$

**90.** $50 \div 2 \cdot 5 - 5^3 \div 5 + 5$

**91.** $(10+1)[(5-2)^2 + 3(4-3)]$

**92.** $(12-2)[4(6-3)+(4-3)^2]$

**93.** $100 + 2[3(4^2 - 6) + 2^3]$

**94.** $75 + 3[2(3+6)^2 - 10^2]$

**95.** $16 + 3[17 + 2^3 \div 2^2 - 4]$

**96.** $10^3 - 2[(13+3) \div 2^4 + 18 \div 3^2]$

Solve each of the following equations.

**97.** $x - 5 = 2^3 - 1 \cdot 3$

**98.** $x + 3 \cdot 2 = 4(5+1)^2$

**99.** $2x = 30 \div 2 - 11 + 2(5-1)$

**100.** $3x = 2(15 - 6 + 4) \div 13 + 1$

**89.** 20

**90.** 105

**91.** 132

**92.** 130

**93.** 176

**94.** 261

**95.** 61

**96.** 994

**97.** $x = 10$

**98.** $x = 138$

**99.** $x = 6$

**100.** $x = 1$

[Respond below
101. exercise]

[Respond below
102. exercise]

## WRITING AND THINKING ABOUT MATHEMATICS

**101.** Use your calculator to evaluate the expression $0^0$. What was the result? State, in your own words, the meaning of the result.

Error; Answers will vary.

**102. a.** What mathematics did you use today? (Yes, you did use some mathematics. I know you did.)

**b.** What mathematics did you use this week?

**c.** What mathematical thinking did you use to solve some problem (not in the text) today?

Answers will vary.

# 1.6 Evaluating Polynomials

## Definition of a Polynomial

Algebraic expressions that are numbers, powers of variables, or products of numbers and powers of variables are called **terms**. That is, an expression that involves only multiplication and/or division with variables and constants is called a **term**. Remember that a number written next to a variable indicates multiplication, and the number is called the **coefficient** of the variable. For example,

Algebraic terms in one variable:

$5x$, $25x^3$, $6y^4$, and $48n^2$

Algebraic terms in two variables:

$3xy$, $4ab$, $6xy^5$, and $10x^2y^2$

A term that consists of only a number, such as 3 or 15, is called a **constant** or a **constant term.** In this text we will discuss terms in only one variable and constants.

**Objectives**

1. Know the meanings of the following terms: **polynomial, monomial, binomial, trinomial.**
2. Know how to determine the degree of a polynomial.
3. Be able to evaluate a polynomial for a given value of the variable by following the rules for order of operations.
4. Be able to evaluate numerical expressions by using a calculator and following the rules for order of operations.

**Definition:**

A **monomial in** $\boldsymbol{x}$ is a term of the form

$kx^n$ where $\boldsymbol{k}$ and $\boldsymbol{n}$ are whole numbers.

$\boldsymbol{n}$ is called the **degree** of the monomial, and $\boldsymbol{k}$ is called the **coefficient.**

Now consider the fact that we can write constants, such as 7 and 63, in the form

$7 = 7 \cdot 1 = 7x^0$ and $63 = 63 \cdot 1 = 63x^0$

Thus, we say that a nonzero constant is a **monomial of 0 degree**. (Note that a monomial may be in a variable other than $x$.) Since we can write $0 = 0x^2 = 0x^5 = 0x^{63}$, we say that **0 is a monomial of no degree**.

In algebra, we are particularly interested in the study of expressions such as

$5x + 7$, $3x^2 - 2x + 4$, and $6a^4 + 8a^3 - 10a$

that involve sums and/or differences of terms. These algebraic expressions are called **polynomials**.

**Definition:**

A **polynomial** is a monomial or the indicated sum or difference of monomials.

The **degree of a polynomial** is the largest of the degrees of its terms.

Generally, for easy reading, a polynomial is written so that the degrees of its terms either decrease from left to right or increase from left to right. If the degrees decrease, we say that the terms are written in **descending order**. If the degrees increase, we say that the terms are written in **ascending order**. For example,

| | |
|---|---|
| $3x^4 + 5x^2 - 8x + 34$ | is a fourth–degree polynomial written in descending order. |
| $15 - 3y + 4y^2 + y^3$ | is a third–degree polynomial written in ascending order. |

For consistency, and because of the style used in operating with polynomials (see Chapter 8), the polynomials in this text will be written in descending order. Note that this is merely a preference, and a polynomial written in any order is acceptable and may be a correct answer to a problem.

**Note:** As stated earlier concerning terms, we will limit our discussions in this text to polynomials in only one variable. In future courses in mathematics you may study polynomials in more than one variable.

Some forms of polynomials are used so frequently that they have been given special names, as indicated in the following box.

**Classification of Polynomials**

| Description | Name | Example |
|---|---|---|
| Polynomial with one term | **Monomial** | $5x^3$ |
| Polynomial with two terms | **Binomial** | $7x - 23$ |
| Polynomial with three terms | **Trinomial** | $a^2 + 5a + 6$ |

### Example 1

Name the degree and type of each of the following polynomials.

**a.** $5x - 10$ **b.** $3x^4 + 5x^3 - 4$ **c.** $17y^{20}$

**Solution**

**a.** $5x - 10$ is a first-degree binomial.
**b.** $3x^4 + 5x^3 - 4$ is a fourth-degree trinomial.
**c.** $17y^{20}$ is a twentieth-degree monomial.

## Evaluating Polynomials

To evaluate a polynomial for a given value of the variable,

1. Substitute that value for the variable wherever it occurs in the polynomial and
2. Follow the rules for order of operations.

### Example 2

Evaluate the polynomial $4a^2 + 5a - 12$ for $a = 3$.

#### Solution

Substitute 3 for $a$ whenever $a$ occurs, and follow the rules for order of operations.

$$\begin{aligned} 4a^2 + 5a - 12 &= 4 \cdot 3^2 + 5 \cdot 3 - 12 \\ &= 4 \cdot 9 + 5 \cdot 3 - 12 \\ &= 36 + 15 - 12 \\ &= 39 \end{aligned}$$

### Example 3

Evaluate the polynomial $3x^3 + x^2 - 4x + 5$ for $x = 2$.

#### Solution

$$\begin{aligned} 3x^3 + x^2 - 4x + 5 &= 3 \cdot 2^3 + 2^2 - 4 \cdot 2 + 5 \\ &= 3 \cdot 8 + 4 - 4 \cdot 2 + 5 \\ &= 24 + 4 - 8 + 5 \\ &= 28 - 8 + 5 \\ &= 25 \end{aligned}$$

### Example 4

Evaluate the polynomial $x^2 + 5x - 36$ for $x = 4$.

#### Solution

$$x^2 + 5x - 36 = 4^2 + 5(4) - 36$$ Note that a raised dot or parentheses can be used to indicate multiplication.

$$\begin{aligned} &= 16 + 5(4) - 36 \\ &= 16 + 20 - 36 \\ &= 0 \end{aligned}$$

Now might be a good time to inform the students that while calculators definitely increase the speed and accuracy of many calculations, calculators are not a substitute for understanding concepts such as order of operations. If they rely solely on their calculators, particularly without using parentheses, they will incorrectly evaluate expressions such as

$(8+6)\div 2$

and $(8+6)^2$.

Also, not **all** calculators follow rules for the order of operations that are consistent with the rules used in arithmetic and algebra.

## Using a Calculator to Evaluate Numerical Expressions

### Example 5

Follow the rules for order of operations and use a calculator with parentheses keys and the $y^x$ key to evaluate each of the following expressions.

**a.** $(2^3+3^2)^2$ **b.** $(2+3\cdot 7)^3$

#### Solution with a Scientific Calculator

Remember that calculators are programmed to follow the rules for order of operations.

**a.** **Step 1:** Press the left parenthesis key ( .

**Step 2:** Enter the base 2.

**Step 3:** Press the key $y^x$.

**Step 4:** Enter the exponent 3.

**Step 5:** Press the key + .

**Step 6:** Enter the base 3.

**Step 7:** Press the key $y^x$.

**Step 8:** Enter the exponent 2.

**Step 9:** Press the right parenthesis key ) .

**Step 10:** Press the key $y^x$.

**Step 11:** Enter the exponent 2.

**Step 12:** Press the key = .

The display should read **289**.

(A quick mental check shows that the result in the parentheses is 17, and we know that $17^2 = 289$.)

#### Solution with a TI-83 graphing calculator

**b.** Enter all the symbols in the expression $(2+3\cdot 7)^3$ in the order you see them (from left to right).

The display should appear as follows:

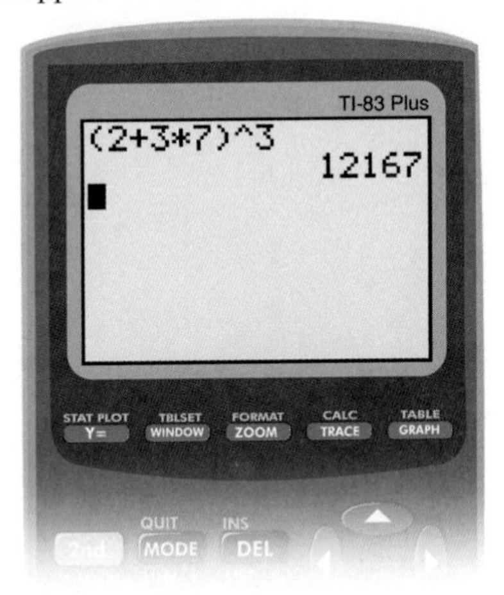

Name ______________________ Section ______________ Date __________

# Exercises 1.6

Name the degree and type of each of the following polynomials.

**1.** $x^2 + 5x + 6$

**2.** $5x^2 + 8x - 3$

**3.** $x^2 + 2x$

**4.** $3x^2 + 7x$

**5.** $13x^{12}$

**6.** $20x^5$

**7.** $y^4 + 2y^2 + 1$

**8.** $9y^3 + 8y - 10$

**9.** $a^5 + 5a^3 - 16$

**10.** $2a^4 + 5a^3 - 3a^2$

Evaluate each of the following polynomials for $x = 2$.

**11.** $x^2 + 5x + 6$

**12.** $5x^2 + 8x - 3$

**13.** $x^2 + 2x$

**14.** $3x^2 + 7x$

**15.** $13x^{12}$

**16.** $20x^5$

**ANSWERS**

**1.** 2
Trinomial

**2.** 2
Trinomial

**3.** 2
Binomial

**4.** 2
Binomial

**5.** 12
Monomial

**6.** 5
Monomial

**7.** 4
Trinomial

**8.** 3
Trinomial

**9.** 5
Trinomial

**10.** 4
Trinomial

**11.** 20

**12.** 33

**13.** 8

**14.** 26

**15.** 53,248

**16.** 640

17. 9

18. 34

19. 81

20. 18

21. 0

22. 79

23. 3

24. 15

25. 0

26. 0

27. 5

28. 9

29. 245

30. 48

Evaluate each of the following polynomials for $y = 3$.

17. $y^2 + 5y - 15$

18. $3y^2 + 6y - 11$

19. $y^3 + 3y^2 + 9y$

20. $y^2 + 7y - 12$

21. $y^3 - 9y$

22. $y^4 - 2y^2 + 6y - 2$

Evaluate each of the following polynomials for $a = 1$.

23. $2a^3 - a^2 + 6a - 4$

24. $2a^4 - a^3 + 17a - 3$

25. $3a^3 + 3a^2 - 3a - 3$

26. $5a^2 + 5a - 10$

27. $a^3 + a^2 + a + 2$

28. $a^4 + 6a^3 - 4a^2 + 7a - 1$

Evaluate each of the following polynomials for $c = 5$.

29. $c^3 + 3c^2 + 9c$

30. $c^2 + 7c - 12$

Name ______________________ Section ______________ Date ____________

**ANSWERS**

**31.** $4c^2 - 8c + 14$

**32.** $3c^2 + 15c + 150$

**33.** $4c^3 + 9c^2 - 4c + 1$

**34.** $c^3 + 8c - 20$

**35.** Evaluate the polynomial $x^2 + 5x + 4$ for each value of $x$ in the following table.

| $x$ | value of $x^2 + 5x + 4$ |
|---|---|
| **0** | 4 |
| **1** | 10 |
| **2** | 18 |
| **3** | 28 |
| **4** | 40 |
| **5** | 54 |

Use a calculator and follow the rules for order of operations to find the value of each of the following expressions.

**36.** $3(3 + 4) - 2(1 + 5)$

**37.** $(14 + 6) \div 2 + 3 \cdot 4$

**38.** $16 - 4 + 2(5 + 8)$

**39.** $14 + 6 \div 2 + 3 \cdot 4$

**40.** $(3 \cdot 4 - 2 \cdot 5)(2 \cdot 6 + 3)$

**31.** 74

**32.** 300

**33.** 706

**34.** 145

**35.** [Respond in exercise]

**36.** 9

**37.** 22

**38.** 38

**39.** 29

**40.** 30

41. 5

42. 15,525

43. 224,939

44. 500

45. 300

46. 331,969

47. 3650

48. [Respond below exercise]

49. [Respond below exercise]

50. [Respond below exercise]

**41.** $(3^2 - 3 \cdot 2)^2 - 5 + 1$

**42.** $(3^2 + 4^2)^3 - 10^2$

**43.** $(2 \cdot 8 - 3)^2(3 + 4 \cdot 2)^3$

**44.** $(5 + 3 \cdot 5)(15 \div 3 \cdot 5)$

**45.** $3 \cdot 5^2 + (3 \cdot 5)^2$

**46.** Use a calculator to help evaluate the polynomial $16x^4 + x^2 - 7x + 133$ for $x = 12$.

**47.** Use a calculator to help evaluate the polynomial $4y^3 - 4y^2 + 15y - 100$ for $y = 10$.

## Writing and Thinking About Mathematics

**48.** Of all the topics and ideas discussed in Chapter 1, which one (or more) did you find the most interesting? Explain briefly using complete sentences in paragraph form.
Answers will vary. This question is designed to provide a basis for one–on–one discussion or class discussion with your students. It may help you gain additional insight about how your students perceive this course.

**49.** Of all the topics and ideas discussed in Chapter 1, which one (or more) did you find the most difficult to learn? Explain briefly using complete sentences in paragraph form.
Answers will vary. See the annotation for no. 48 above.

**50.** Discuss how you used mathematics to solve some problem in your life today. (You did. You know you did.)
Answers will vary.

# Chapter 1 Index of Key Ideas and Terms

For Addition:

Addends: numbers being added — page 3

Sum: the result of addition — page 3

Properties: — page 4

For whole numbers $a$, $b$, and $c$

Commutative Property: $a + b = b + a$

Associative Property: $(a + b) + c = a + (b + c)$

Additive Identity: $a + 0 = a$

Perimeter: the distance around a geometric figure — page 6

For Subtraction:

Minuend, subtrahend, difference — page 6

For Multiplication — page 8

Symbols used: · raised dot

× cross sign

( ) numbers next to parentheses

Properties: — page 8

For whole numbers $a$, $b$, and $c$

Commutative Property: $a \cdot b = b \cdot a$

Associative Property: $(a \cdot b) \cdot c = a \cdot (b \cdot c)$

Multiplicative Identity: $a \cdot 1 = a$

Distributive Property: $a(b + c) = a \cdot b + a \cdot c$ — page 10

Product: the result of multiplication — page 8

Factors: Numbers being multiplied are factors of the product. — page 8

Area: a measure of the interior of a geometric figure — page 12

For Division:

Dividend, divisor, quotient — page 13

Long division, remainder — page 13

Division by 0 is not possible. — page 15

Rounding-off Rule — page 28

1. Look at the single digit just to the right of the digit that is in the place of desired accuracy.
2. If this digit is 5 or greater, make the digit in the desired place of accuracy one larger, and replace all digits to the right with zeros. All digits to the left remain unchanged unless a 9 is made one larger; then the next digit to the left is increased by 1.
3. If this digit is less than 5, leave the digit that is in the place of desired accuracy as it is, and replace all digits to the right with zeros. All digits to the left remain unchanged.

Estimating Answers — page 29

1. Round off each number to the place of the leftmost digit.
2. Perform the indicated operation with these rounded-off numbers.

Operating with Calculators page 32

The text discusses both scientific calculators and the TI-83 graphing calculator. Scientific calculators have a key marked in the form of $y^x$ and the TI-83 has keys marked  and ^ for working with exponents.

Basic Strategy for Solving Word Problems page 43

1. Read each problem carefully until you understand the problem and know what is being asked for.
2. Draw any type of figure or diagram that might be helpful, and decide what operations are needed.
3. Perform these operations.
4. Mentally check to see if your answer is reasonable, and see if you can think of another more efficient or more interesting way to do the same problem.

For Equations:

Definition, solution page 56

Basic principles for solving page 57

Addition Principle: If $A = B$, then $A + C = B + C$.

Subtraction Principle: If $A = B$, then $A - C = B - C$.

Division Principle: If $A = B$ then $\frac{A}{C} = \frac{B}{C}$ $(C \neq 0)$.

For Exponents:

Base, exponent, power page 63

The exponent 1: $a^1 = a$ page 64

The exponent 0: $a^0 = 1$ page 65

Rules for Order of Operations page 67

1. First, simplify within grouping symbols, such as parentheses ( ), brackets [ ], or braces { }. Start with the innermost grouping.
2. Second, find any powers indicated by exponents.
3. Third, moving from left to right, perform any multiplications or divisions in the order in which they appear.
4. Fourth, moving from left to right, perform any additions or subtractions in the order in which they appear.
   Mnemonic memory device PEMDAS

For Polynomials: page 79

Monomial in $x$: $kx^n$ ($k$ is the coefficient and $n$ is the degree)

Binomial: a polynomial with two terms

Trinomial: a polynomial with three terms

Polynomial: a monomial or indicated sum or difference of monomials

Name ______________ Section ______________ Date ______________

**ANSWERS**

## Chapter 1 Test

Round off as indicated.

**1.** 675 (to the nearest ten)

**2.** 13,620 (to the nearest thousand)

Find the correct value of $n$ and state the property of addition or multiplication that is illustrated.

**3.** $16 + 52 = 52 + n$

**4.** $18(20 \cdot 3) = (18 \cdot 20) \cdot n$

**5.** $74 + n = 74$

**6.** $6 + (2 + 4) = (6 + 2) + n$

First estimate each sum, and then find each sum

**7.**
$$\begin{array}{r} 9587 \\ 345 \\ +\ 2075 \\ \hline \end{array}$$

**8.**
$$\begin{array}{r} 45{,}872 \\ 589{,}651 \\ +\ 235{,}003 \\ \hline \end{array}$$

First estimate each difference, and then find each difference.

**9.**
$$\begin{array}{r} 7046 \\ -\ 4562 \\ \hline \end{array}$$

**10.**
$$\begin{array}{r} 80{,}000 \\ -\ 27{,}830 \\ \hline \end{array}$$

First estimate each product, and then find each product.

**11.**
$$\begin{array}{r} 47 \\ \times\ 25 \\ \hline \end{array}$$

**12.**
$$\begin{array}{r} 2593 \\ \times\ \ 86 \\ \hline \end{array}$$

Find the quotient and remainder.

**13.** $463\overline{)78{,}950}$

**Answers:**

1. 680
2. 14,000
3. 16; Comm. Prop. of Addition
4. 3; Assoc. Prop. of Multiplication
5. 0; Add. Identity
6. 4; Assoc. Prop. of Addition
7. Estimate: 12,300 Sum: 12,007
8. Estimate: 850,000 Sum: 870,526
9. Estimate: 2000 Difference: 2484
10. Estimate: 50,000 Difference: 52,170
11. Estimate: 1500 Product: 1175
12. Estimate: 270,000 Product: 222,998
13. 170 R 240

14. 58,000

15. a. 150 m

b. 174 m

c. 486 sq. m

16. $ 39,080

17. $x = 42$

18. $x = 35$

19. $3 = y$

20. a. 4

b. 3

c. 64

21. a. 32,768

b. 6859

c. 1

**14.** If the **quotient** of 56,000 and 7 is **subtracted from** the **product** of 22 and 3000, what is the **difference**?

**15.** A rectangular swimming pool is surrounded by a concrete border. The swimming pool is 50 meters long and 25 meters wide, and the concrete border is 3 meters wide all around the pool.

**a.** Find the perimeter of the swimming pool.

**b.** Find the perimeter of the rectangle formed by the outside of the border around the pool.

**c.** Find the area of the concrete border.

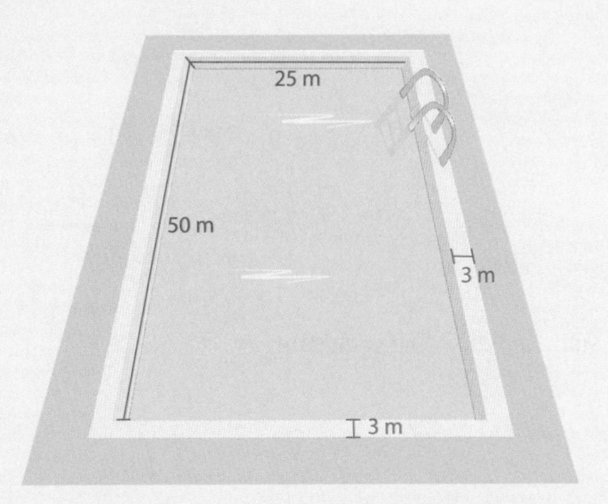


**16.** Mr. Powers is going to buy a new silver SUV for $43,000. He will pay taxes of $2500 and license fees of $1580. He plans to make a down payment of $8000 and finance the remainder at his bank. How much will he finance if he makes a salary of $85,000?

Solve the following equations.

**17.** $x - 10 = 32$

**18.** $x + 25 = 60$

**19.** $17 + 34 = 17y$

**20.** Name **a.** the base, **b.** the exponent, and **c.** find the value of the expression $4^3$.

**21.** Use a calculator to find the value of each of the following expressions.

**a.** $8^5$

**b.** $19^3$

**c.** $142^0$

Name ______________ Section ____________ Date __________

**ANSWERS**

Use the rules for order of operations to evaluate each of the following expressions.

**22.** $15 + 9 \div 3 - 10$

**23.** $18 \div 2 + 5 \cdot 2^3 - 6 \div 3$

**24.** $12 \div 3 \cdot 2 - 8 \div 2^3 + 5$

**25.** $6 + 3(7^2 - 5) - 11^2$

Name the degree and type of each of the following polynomials.

**26.** $3y^2 + 15$

**27.** $5y^4 - 2y^2 + 4$

Evaluate each of the following polynomials for $x = 2$.

**28.** $5x^2 + 6x - 8$

**29.** $2x^5 + 3x^4 - 4x^3 + 3x^2 - 5x - 1$

**30.** Tell which of the following numbers is a perfect square, and explain your reasoning.

**a.** 100 **b.** 80 **c.** 225

**31.** Explain, in your own words, why $7 \div 0$ is undefined.

Answers may vary.

**22.** 8

**23.** 47

**24.** 12

**25.** 17

**26.** Second–degree Binomial

**27.** Fourth–degree Trinomial

**28.** 24

**29.** 81

**30. a.** Yes $10^2 = 100$

**b.** No; 80 is not a perfect square

**c.** Yes $15^2 = 225$

**31.** [Respond below exercise]

# 2

# INTEGERS

## Chapter 2 Integers





## WHAT TO EXPECT IN CHAPTER 2

Chapter 2 introduces the concept of negative numbers and develops an understanding of **integers** (whole numbers and their opposites). Number lines are used as aids to provide "pictures" of integers and their relationships. One particularly important idea discussed in Section 2.1 is that of **absolute value** (or magnitude) of an integer. This idea is the foundation for the development of addition and subtraction with integers in Sections 2.2 and 2.3.

In Section 2.4, multiplication and division with integers are discussed along with a second discussion of the rules for order of operations (see Section 1.5). This time the rules are applied to expressions involving integers. Also, the fact that division by 0 is undefined is reinforced. The applications with integers in Section 2.5 relate to finding changes in values and finding the average of a set of integers. (The concept of average is developed in more detail in Chapter 9.)

The topics of combining like terms and simplifying and evaluating polynomials are covered in Section 2.6. The chapter closes with another development of the techniques for solving equations (see Section 1.4). In this case, integers are involved along with the skills related to combining like terms.

# Chapter 2 Integers

## Mathematics at Work!

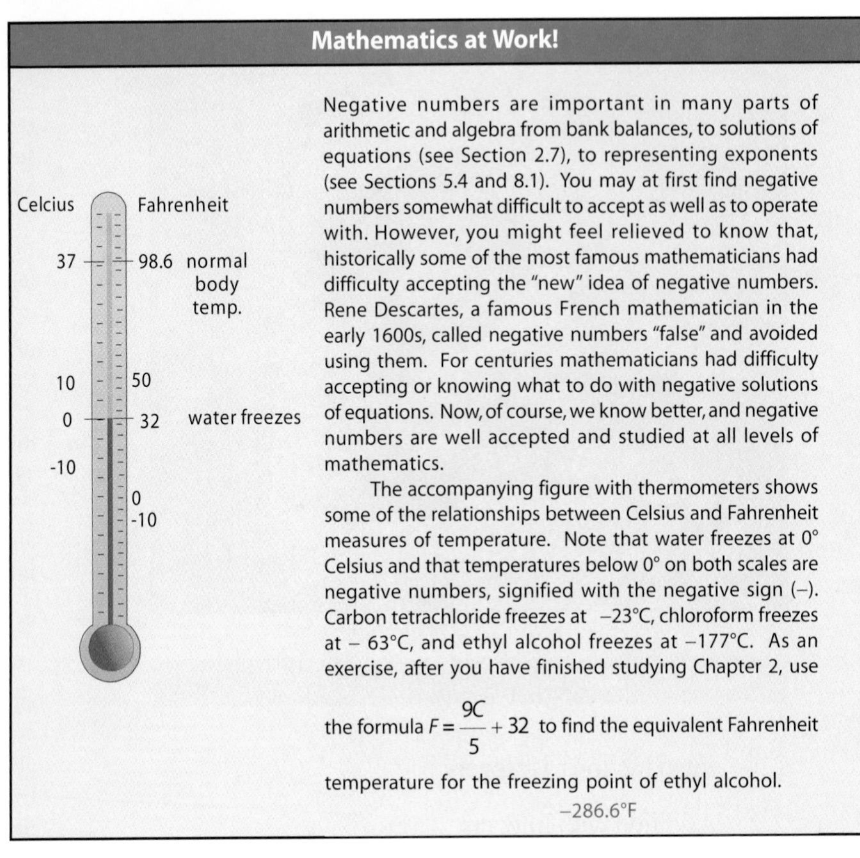


Negative numbers are important in many parts of arithmetic and algebra from bank balances, to solutions of equations (see Section 2.7), to representing exponents (see Sections 5.4 and 8.1). You may at first find negative numbers somewhat difficult to accept as well as to operate with. However, you might feel relieved to know that, historically some of the most famous mathematicians had difficulty accepting the "new" idea of negative numbers. Rene Descartes, a famous French mathematician in the early 1600s, called negative numbers "false" and avoided using them. For centuries mathematicians had difficulty accepting or knowing what to do with negative solutions of equations. Now, of course, we know better, and negative numbers are well accepted and studied at all levels of mathematics.

The accompanying figure with thermometers shows some of the relationships between Celsius and Fahrenheit measures of temperature. Note that water freezes at 0° Celsius and that temperatures below 0° on both scales are negative numbers, signified with the negative sign (–). Carbon tetrachloride freezes at –23°C, chloroform freezes at – 63°C, and ethyl alcohol freezes at –177°C. As an exercise, after you have finished studying Chapter 2, use the formula $F = \frac{9C}{5} + 32$ to find the equivalent Fahrenheit temperature for the freezing point of ethyl alcohol.

–286.6°F

## 2.1 Introduction to Integers

### Special Note from the Author about Calculators

All of the operations with integers can, of course, be done with a calculator, and there are some exercises with calculators. However, at this beginning stage, to develop a thorough understanding of operating with negative numbers, I recommend that you use a calculator as little as possible.

### Number Lines

The concepts of positive and negative numbers occur frequently in our daily lives:

| | **Negative** | **Zero** | **Positive** |
|---|---|---|---|
| *Temperatures are recorded as:* | below zero | zero | above zero |
| *The stock market will show:* | a loss | no change | a gain |
| *Altitude can be measured as:* | below sea level | sea level | above sea level |
| *Businesses will report:* | losses | no gain | profits |

In the remainder of this chapter, we will develop techniques for understanding and operating with positive and negative numbers. We begin with the graphs of numbers on **number lines**. We generally use horizontal and vertical lines for number lines. For example, choose some point on a horizontal line and label it with the number 0 (Figure 2.1)

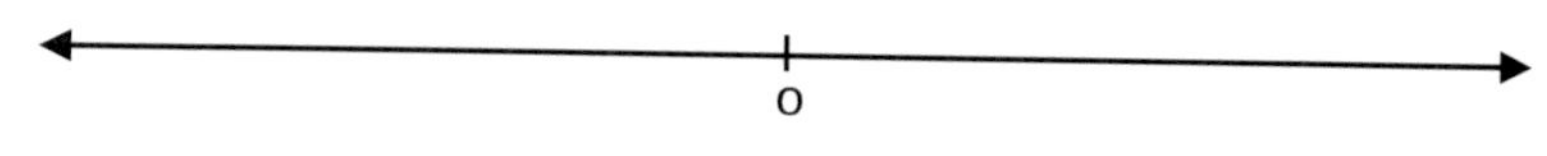


**Figure 2.1**

Now choose another point on the line to the right of 0 and label it with the number 1 (Figure 2.2).

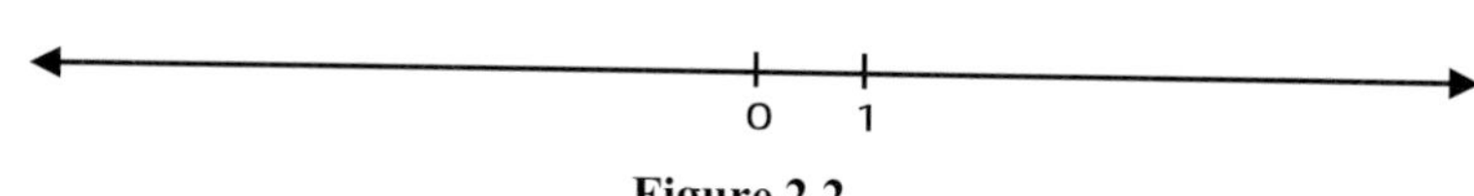


**Figure 2.2**

Points corresponding to all whole numbers are now determined and are all to the right of 0. That is, the point for 2 is the same distance from 1 as 1 is from 0, and so on (Figure 2.3).

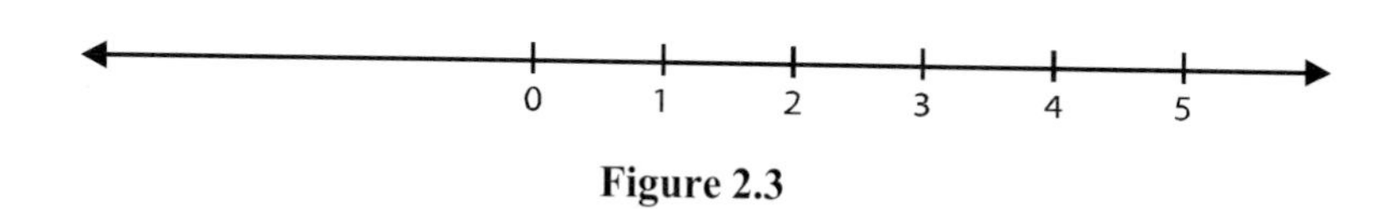


**Figure 2.3**

### Objectives

1. Know that integers are the whole numbers and their opposites.
2. Understand and be able to read inequality symbols such as < and >.
3. Know the meaning of the absolute value of an integer.
4. Be able to graph a set of integers on a number line.
5. Know that 0 is neither positive nor negative.
6. Be aware that an expression of the form $-a$ may represent a positive number or a negative number.

The graph of a number is the point on a number line that corresponds to that number, and the number is called the coordinate of the point. The terms "number" and "point" are used interchangeably. Thus, we might refer to the "point" 6. The graphs of numbers are indicated by marking the corresponding points with large dots (Figure 2.4).

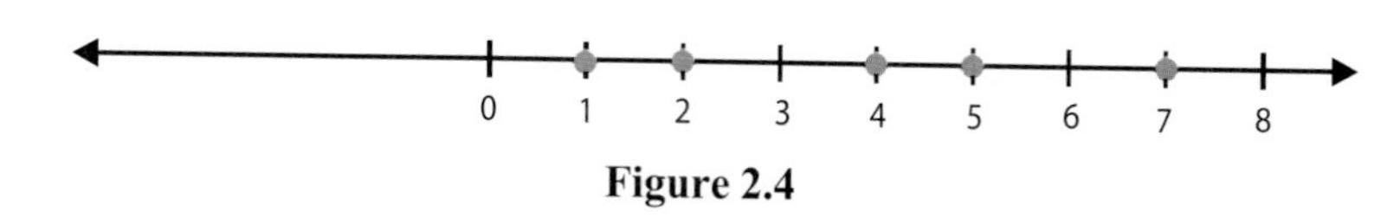


**Figure 2.4**

*The graph of the set of numbers* $S = \{1, 2, 4, 5, 7\}$ *(Note that even though other numbers are marked, only those with a large dot are considered to be "graphed.")*

The point one unit to the left of 0 is the opposite of 1. It is called negative 1 and is symbolized as $-1$. Similarly, the opposite of 2 is called negative 2 and is symbolized as $-2$, and so on (Figure 2.5).

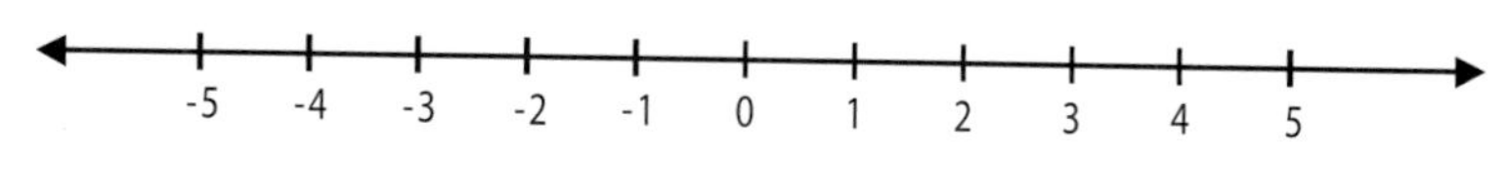


**Figure 2.5**

(**Note: Additive inverse** is the technical term for the opposite of a number. We will discuss this in more detail in Section 2.2 )

Emphasize the fact that the negative integers do not include all negative numbers. You might explain that this is the reason why we shade in dots for the graphs of sets of integers and do not shade in entire segments on the line. A few fractions can be used to illustrate the idea even though fractions are not discussed until later, in Chapters 3 and 4.

**Definition:**

The set of **integers** is the set of whole numbers and their opposites (or additive inverses).

**Integers: $\ldots, -3, -2, -1, 0, 1, 2, 3, \ldots$**

The counting numbers (all whole numbers except 0 ) are called positive integers and may be written as $+1, +2, +3$, and so on; the opposites of the counting numbers are called negative integers. 0 is neither positive nor negative (Figure 2.6).

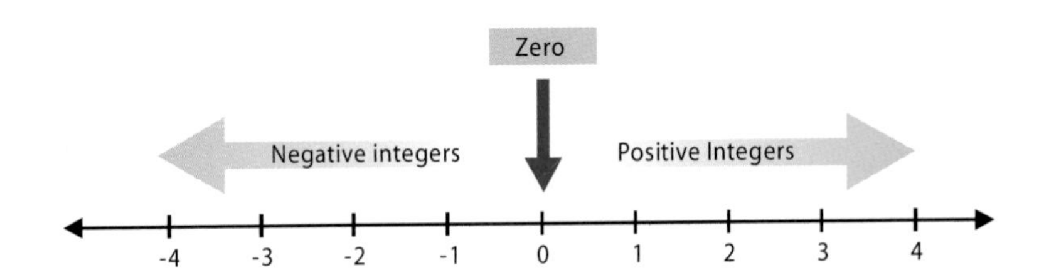


**Figure 2.6**

Note the following facts about integers:

1. The opposite of a positive integer is a negative integer.
   For example, $-(+1) = -1$ and $-(+6) = -6$
   (opposite of + 1; opposite of + 6)

2. The opposite of a negative integer is a positive integer.
   For example, $-(-2) = +2 = 2$ and $-(-8) = +8 = 8$
   (opposite of − 2; opposite of − 8)

3. The opposite of 0 is 0.
   [That is, $-0 = +0 = 0$. This shows that $-0$ should be thought of as the opposite of 0 and not as "negative 0." Remember the number 0 is neither positive nor negative, and 0 is its own additive inverse.]

**Note:** There are many types of numbers other than integers that can be graphed on number lines. We will study these types of numbers in later chapters (that is, fractions in Chapter 4 and decimal numbers and roots in Chapter 5). Therefore, you should be aware that the set of integers does not include all of the positive numbers or all of the negative numbers.

### Example 1

Find the opposite of:

**a.** $-5$ **b.** $-11$ **c.** $+14$

**Solution**

**a.** $-(-5)=5$ **b.** $-(-11)=11$ **c.** $-(+14)=-14$

### Example 2

Graph the set of integers $B=\{-3,-1,0,1,3\}$

**Solution**

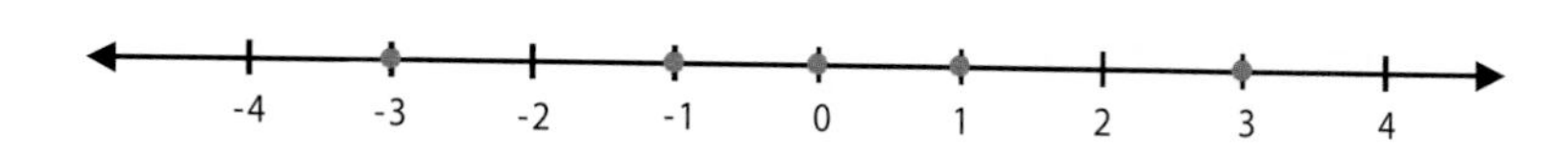


A set of numbers is infinite if it is so large that it cannot be counted. The set of whole numbers and the set of integers are both infinite. To graph an infinite set of integers, three dots are marked above the number line to indicate that the pattern shown is to continue without end. Example 3 illustrates this technique.

### Example 3

Graph the set of integers $C=\{2,4,6,8,\ldots\}$.

**Solution**

The three dots above the number line indicate that the pattern in the graph continues without end.

*Now Work Exercises 1-3 in the Margin.*

## Inequality Symbols

On a horizontal number line, **smaller numbers are always to the left of larger numbers.** Each number is smaller than any number to its right and larger than any number to its left. We use the following inequality symbols to indicate the order of numbers on the number line.

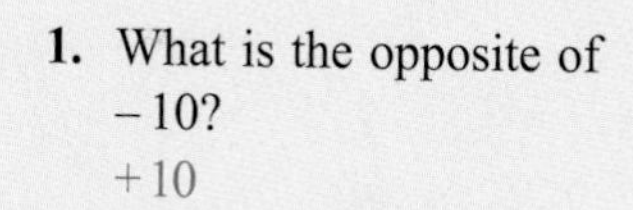

**1.** What is the opposite of $-10$?
$+10$

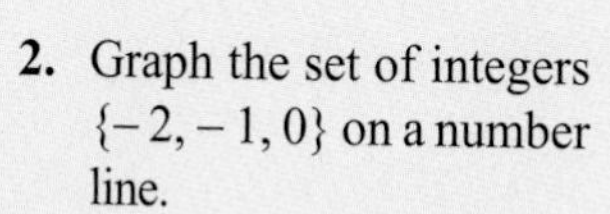

**2.** Graph the set of integers $\{-2,-1,0\}$ on a number line.

**3.** Graph the set of integers $\{\ldots,-7,-5,-3,-1\}$ on a number line.

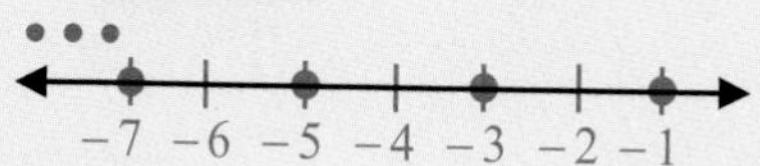

This might be a good time to begin to familiarize the students with the idea that $a < 0$ is a symbolic way of saying that $a$ is negative. Similarly, $a > 0$ implies that $a$ is positive. The relationship between the notation and the related positive and negative concepts seems to be a difficult abstraction for many students.

**Symbols for Order**

$<$ is read "is less than"
$>$ is read "is greater than"
$\leq$ is read "is less than or equal to"
$\geq$ is read "is greater than or equal to"

The following relationships can be observed on the number line in Figure 2.7.

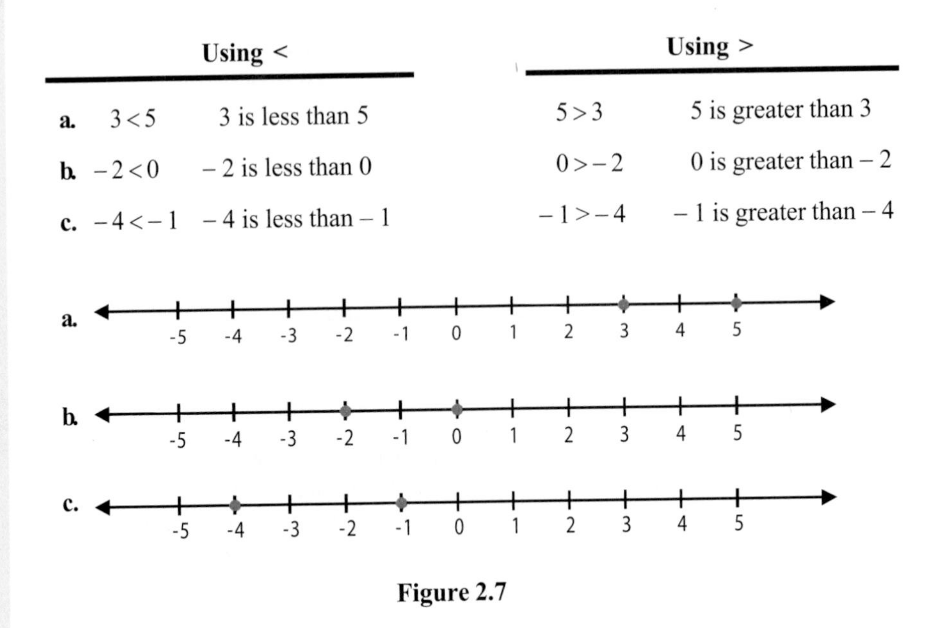

| | Using $<$ | | Using $>$ | |
|---|---|---|---|---|
| **a.** | $3 < 5$ | 3 is less than 5 | $5 > 3$ | 5 is greater than 3 |
| **b.** | $-2 < 0$ | $-2$ is less than 0 | $0 > -2$ | 0 is greater than $-2$ |
| **c.** | $-4 < -1$ | $-4$ is less than $-1$ | $-1 > -4$ | $-1$ is greater than $-4$ |



**Figure 2.7**

One useful idea implied by the previous discussion is that the symbols $<$ and $>$ can be read either from right to left or from left to right. For example, we might read

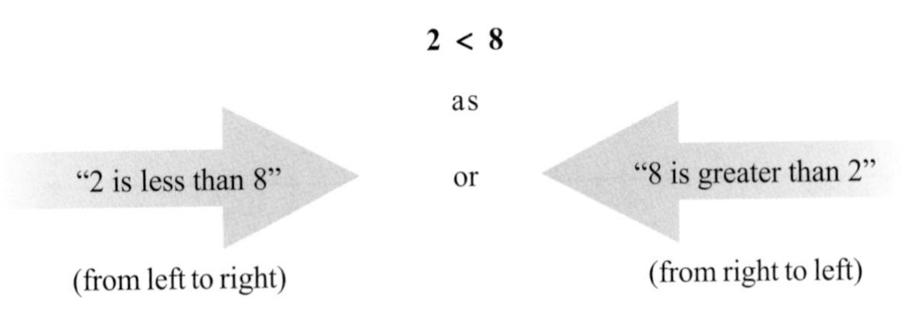


Remember that, from left to right, $\geq$ is read "greater than or equal to and $\leq$ is read "less than or equal to." Thus, the symbols $\geq$ and $\leq$ allow for both equality and inequality. That is, if $>$ **or** $=$ is true, then $\geq$ is true.

For example, $6 \geq -13$ and $6 \geq 6$ are both true statements.

$6 \geq -13$ is true $\Rightarrow$ [ $6 = -13$ is false, but $6 > -13$ is true.]

and $6 \geq 6$ is true. $\Rightarrow$ [ $6 > 6$ is false, but $6 = 6$ is true.]

### Example 4

Determine whether each of the following statements is true or false. Rewrite any false statement so that it is true. There may be more than one change that will make a false statement correct.

**a.** $4 \leq 12$ **b.** $4 \leq 4$
**c.** $4 < 0$ **d.** $-7 \geq 0$

**Solution**

**a.** $4 \leq 12$ is true, since 4 is less than 12.

**b.** $4 \leq 4$ is true, since 4 is equal to 4.

**c.** $4 < 0$ is false.
We can change the inequality to read $4 > 0$ or $0 < 4$.

**d.** $-7 \geq 0$ is false. We can write $-7 \leq 0$ or $0 \geq -7$.

*Now Work Exercises 4-7 in the Margin.*

Determine whether each of the following statements is true or false. Rewrite any false statement so that it is true. There may be more than one change that will make a false statement correct.

**4.** $6 \geq 7$

False: $6 \leq 7$

**5.** $-8 < 0$

True

**6.** $0 \geq 0$

True

**7.** $2 \leq -2$

False: $2 \geq -2$

## Absolute Value

Another concept closely related to **signed numbers** (positive and negative numbers and zero) is that of **absolute value**, symbolized with two vertical bars, $| \ |$. (**Note:** We will see later that the definition given here for the absolute value of integers is valid for any type of number on a number line.) We know, from working with number lines, that any integer and its opposite lie the same number of units from zero. For example $+5$ and $-5$ are both 5 units from 0 (Figure 2.8). The + and – signs indicate direction, and the 5 indicates distance. Thus, $|-5| = |+5| = 5$

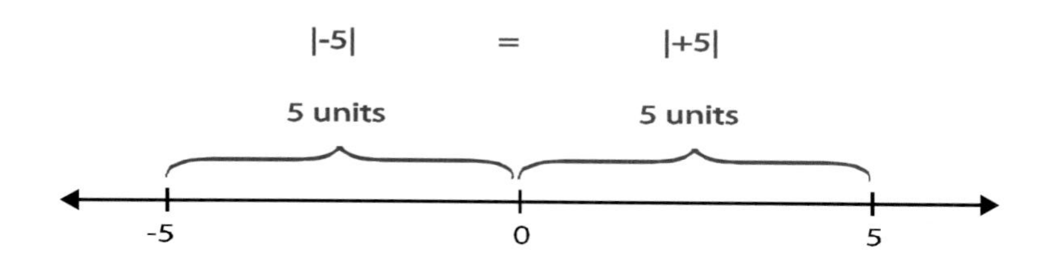


**Definition:**

The absolute value of an integer is its distance from 0. Symbolically, for any integer $a$,

$$\begin{cases} \text{If } a \text{ is a positive integer or 0,} & |a| = a. \\ \text{If } a \text{ is a negative integer,} & |a| = -a. \end{cases}$$

The absolute value of an integer is never negative.

A good class discussion can be initiated by simply writing "$-a$" on the board and asking the students if it represents a positive number, a negative number, or zero. The same idea can be reinforced by writing just "$a$" and asking the same question.

**Note:** When $a$ represents a negative number, the symbol $-a$ represents a positive number. That is, the opposite of a negative number is a positive number. For example,

If $a=-5$, then $-a=-(-5)=+5$.

Similarly,

If $x=-3$, then $-x=-(-3)=+3$.

If $y=-8$, then $-y=-(-8)=+8$.

For these examples, we have

$-a=+5$, $\quad -x=+3 \quad$ and $\quad -y=+8$

Remember to read $-x$ as "the opposite of $x$" and not "negative $x$," because $-x$ may not be a negative number. In summary,

1. If $x$ represents a positive number, then $-x$ represents a negative number.
2. If $x$ represents a negative number, then $-x$ represents a positive number.

### Example 5

$|-2|=|+2|=2$

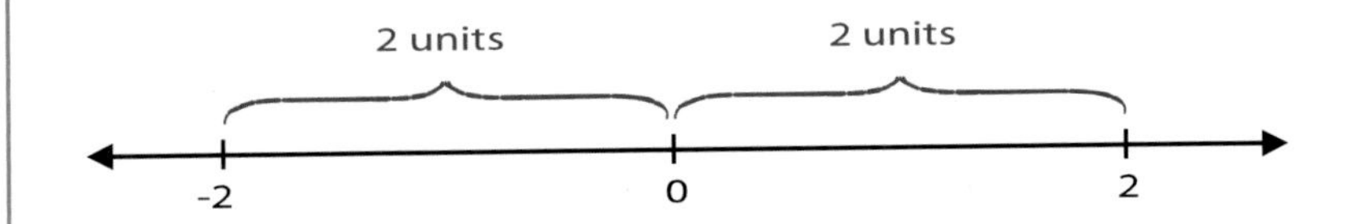


### Example 6

$|0|=-0=+0=0$

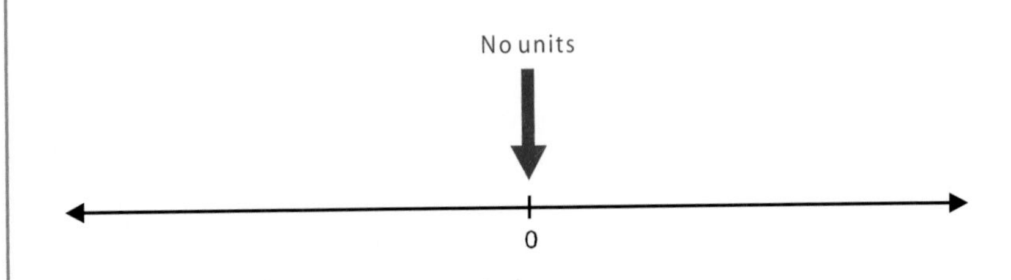


### Example 7

True or false: $|-9| \leq 9$

**Solution**

True, since $|-9|=9$ and $9 \leq 9$. (Remember that since $9=9$ the statement $9 \leq 9$ is true.)

### Example 8

If $|x| = 7$, what are the possible values for $x$?

**Solution**

Since $|-7| = 7$ and $|7| = 7$, then $x = -7$ or $x = 7$.

### Example 9

If $|a| = -2$, what are the possible values for $a$ ?

**Solution**

There are no values of $a$ for which $|a| = -2$. The absolute value of every nonzero number is positive.

## Graphing Inequalities

Inequalities, such as $x \geq 3$ and $y < 1$, where the replacement set for $x$ and $y$ is the set of integers, have an infinite number of solutions. Any integer greater than or equal to 3 is one of the solutions for $x \geq 3$ and any integer less than 1 is one of the solutions for $y < 1$. The solution sets can be represented as

$$\{3, 4, 5, 6, \ldots\} \text{ and } \{\ldots, -4, -3, -2, -1, 0\}, \text{ respectively.}$$

The graphs of the solutions to such inequalities are indicated by marking large dots over a few of the numbers in the solution sets and then putting three dots above the number line to indicate the pattern is to continue without end (see Figure 2.9).

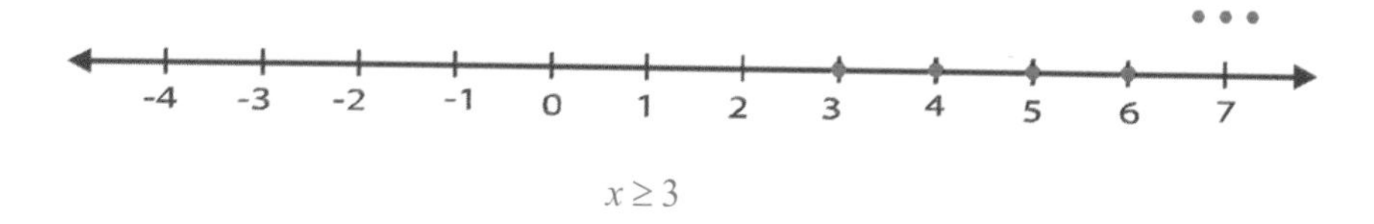


$x \geq 3$

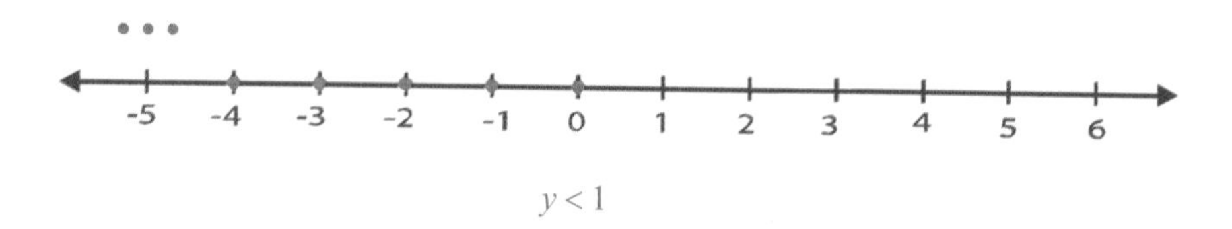


$y < 1$

**Figure 2.9**



The solution sets and the corresponding graphs for inequalities that have absolute values can be finite or infinite, depending on the nature of the inequality. Example 10 and Completion Example 11 illustrate both situations. Look these examples over carefully to fully understand the thinking involved with absolute value inequalities.

## Example 10

If $|x| \geq 4$, what are the possible integer values for $x$? Graph these integers on a number line.

### Solution

There are an infinite number of integers 4 or more units from 0, both negative and positive. These integers are $\{\ldots, -7, -6, -5, -4, 4, 5, 6, 7, \ldots\}$

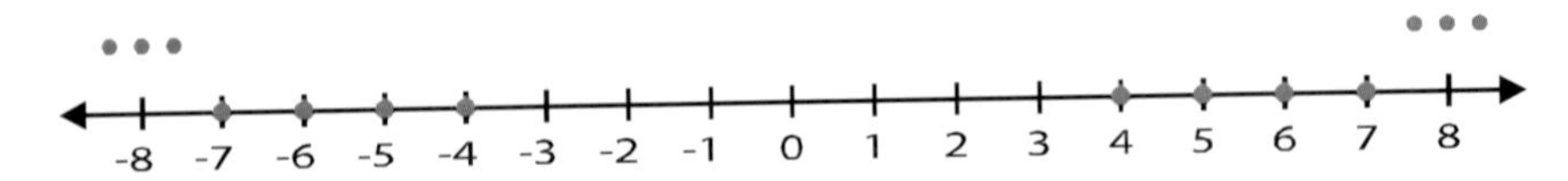


## *Completion Example 11*

If $|x| < 4$, what are the possible integer values for $x$? Graph these integers on a number line.

### Solution

The integers that are less than 4 units from 0 have absolute values less than 4. These integers are ______________________.

Graph:

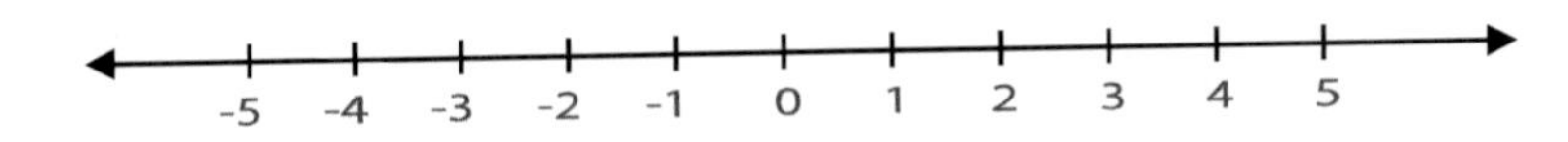


*Now Work Exercises 8-12 in the Margin.*

Determine whether each statement is true or false.

**8.** $|-3| > 3$
False

**9.** $|5| \leq |-5|$
True

In each statement, find the possible integer values for $x$ that will make the statement true.

**10.** $|x| = 2$
$2, -2$

**11.** $|x| = 0$
$0$

**12.** $|x| < 2$
$-1, 0, 1$

## Completion Example Answer

**11.** These integers are $-3, -2, -1, 0, 1, 2, 3$.

Graph:

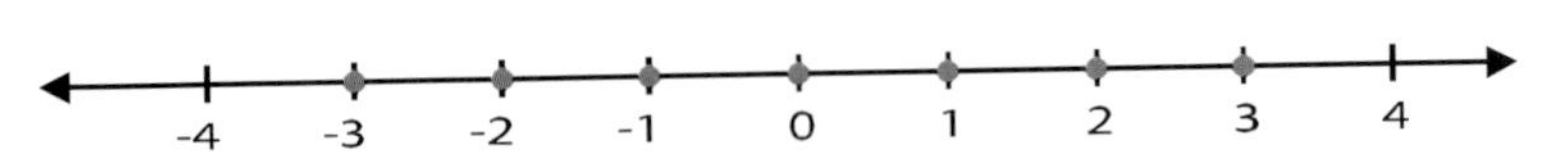

Name ____________ Section ____________ Date ____________

## Exercises 2.1

Insert the appropriate symbol in the blank that will make each statement true: <, >, or =.

1. $3 \underline{\ >\ } 0$
2. $7 \underline{\ >\ } -1$
3. $10 \underline{\ >\ } -10$
4. $-3 \underline{\ <\ } -2$
5. $-6 \underline{\ <\ } 0$
6. $-2 \underline{\ <\ } -1$
7. $-20 \underline{\ <\ } -19$
8. $-67 \underline{\ <\ } -50$
9. $|-4| \underline{\ =\ } 4$
10. $|7| \underline{\ >\ } -7$
11. $|-8| \underline{\ >\ } -8$
12. $-15 \underline{\ <\ } |-15|$

Determine whether each statement is true or false. If the statement is false, change the inequality or equality symbol so that the statement is true. (There may be more than one change that will make a false statement correct.)

13. $0 = -0$
14. $-9 < -10$
15. $-2 < 0$
16. $8 > |-8|$
17. $|2| = -2$
18. $-5 = |-5|$

### ANSWERS

1. [Respond in exercise.]
2. [Respond in exercise.]
3. [Respond in exercise.]
4. [Respond in exercise.]
5. [Respond in exercise.]
6. [Respond in exercise.]
7. [Respond in exercise.]
8. [Respond in exercise.]
9. [Respond in exercise.]
10. [Respond in exercise.]
11. [Respond in exercise.]
12. [Respond in exercise.]
13. True
14. False: $-9 > -10$
15. True
16. False: $8 = |-8|$
17. False: $|2| > -2$
18. False: $-5 < |-5|$

19. True

20. True

21. False: $|-3| = 3$

22. False: $16 = |-16|$

23. False: $-4 < |-4|$

24. True

25. [Respond in exercise.]

26. [Respond in exercise.]

27. [Respond in exercise.]

28. [Respond in exercise.]

29. [Respond in exercise.]

30. [Respond in exercise.]

31. [Respond in exercise.]

32. [Respond in exercise.]

**19.** $-3 < |-3|$ **20.** $-5 < 5$ **21.** $|-3| < 3$

**22.** $16 > |-16|$ **23.** $-4 = |-4|$ **24.** $7 = |-7|$

Graph each of the following sets of numbers on a number line.

**25.** $\{1, 3, 4, 6\}$

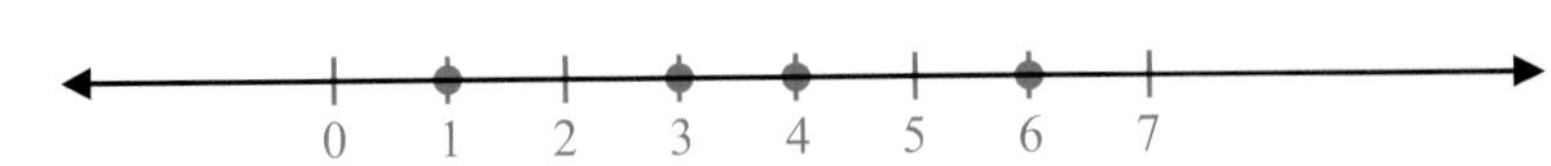


**26.** $\{-3, -2, 0, 1, 3\}$

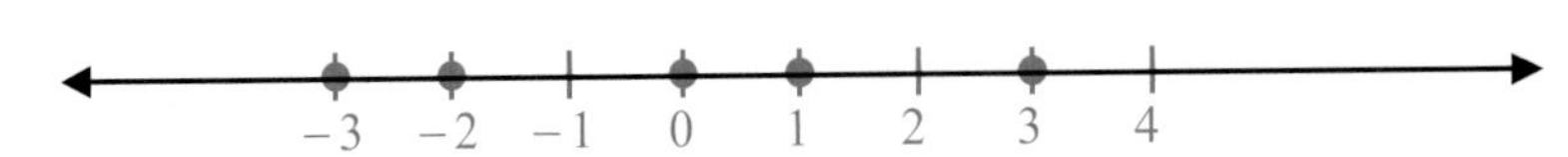


**27.** $\{-2, -1, 0, 2, 4\}$

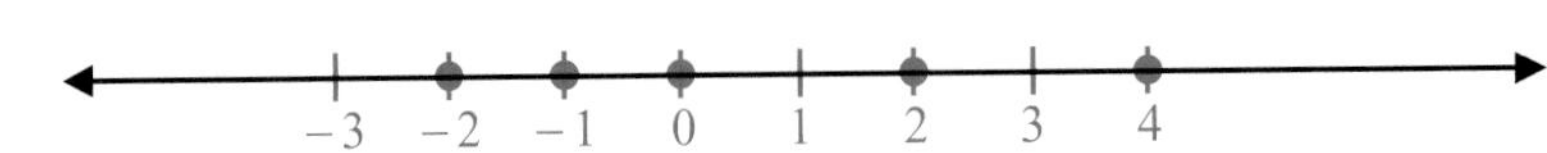


**28.** $\{-5, -4, -3, -2, 0, 1\}$

**29.** All whole numbers less than 4

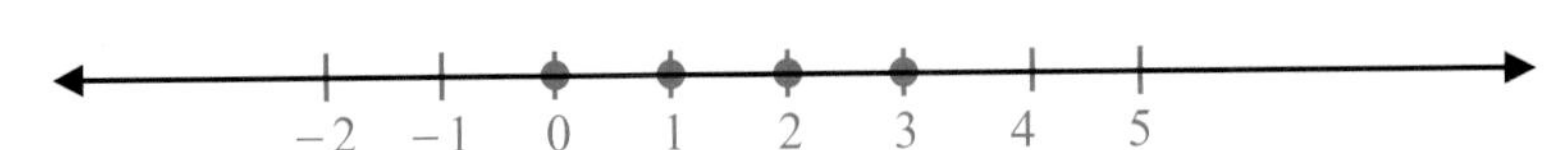


**30.** All integers less than 4

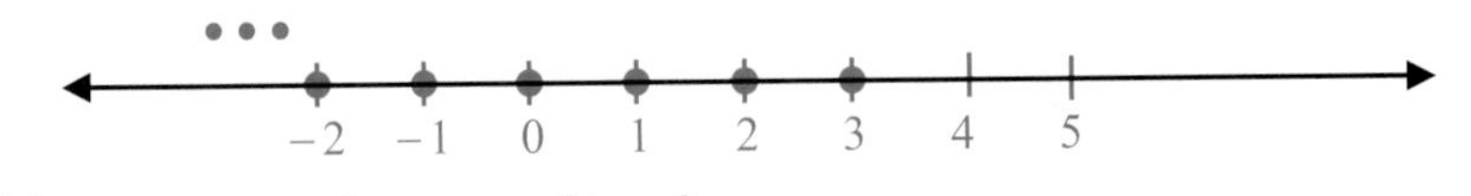


**31.** All integers greater than or equal to $-2$

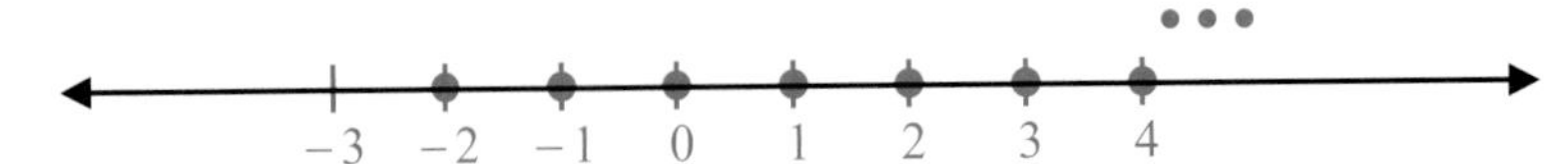


**32.** All negative integers greater than $-4$

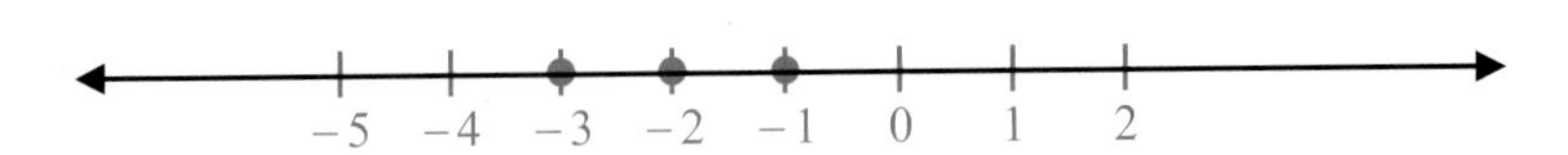

Name ______________________ Section ______________ Date __________

**33.** All whole numbers less than 0

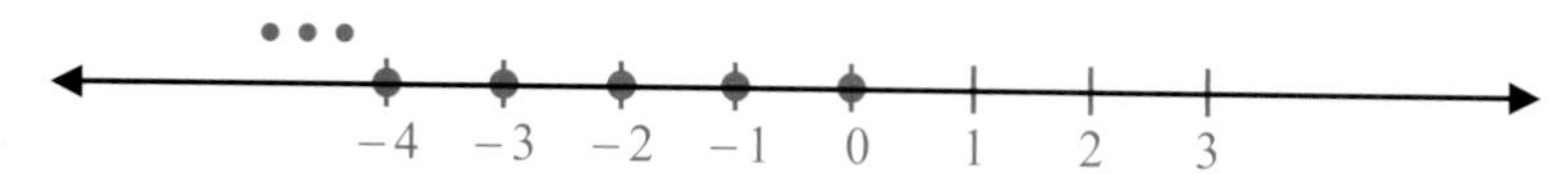


**34.** All integers less than or equal to 0

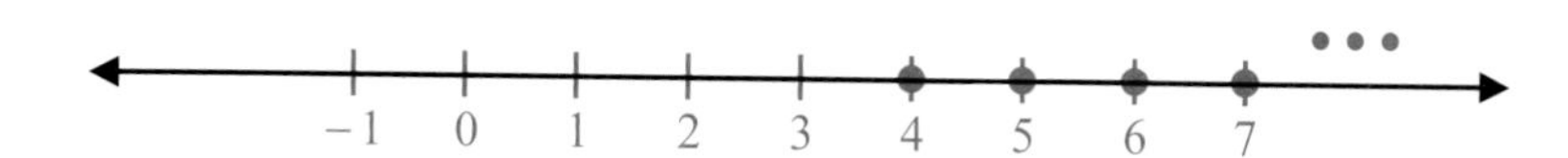


Find the integers that satisfy each of the following equations.

**35.** $|x| = 5$ **36.** $|y| = 8$ **37.** $|a| = 2$ **38.** $|a| = 0$

**39.** $|y| = -6$ **40.** $|a| = -1$ **41.** $|x| = 23$ **42.** $|x| = 105$

Graph the integers that satisfy the conditions given in each statement.

**43.** $a > 3$

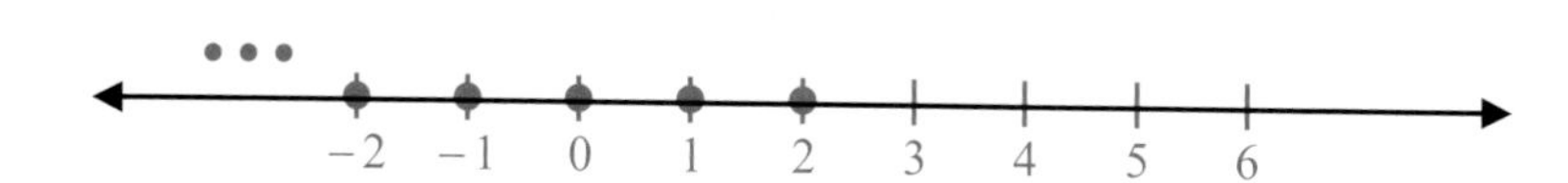


**44.** $x \leq 2$

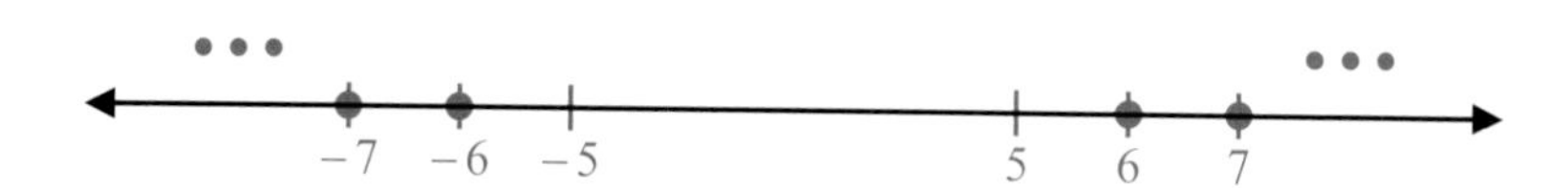


**45.** $|x| > 5$

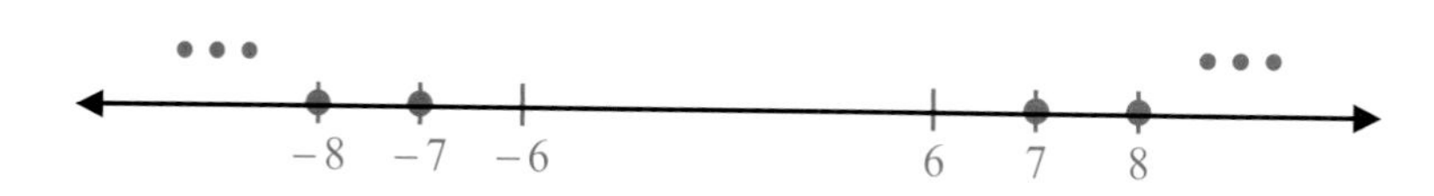


**46.** $|a| > 6$

−8 −7 −6 6 7 8

**ANSWERS**

33. [Respond in exercise.]

34. [Respond in exercise.]

35. 5, −5

36. 8, −8

37. 2, −2

38. 0

39. No Solution

40. No Solution

41. 23, −23

42. 105, −105

43. [Respond in exercise.]

44. [Respond in exercise.]

45. [Respond in exercise.]

46. [Respond in exercise.]

47. [Respond in exercise.]

48. [Respond in exercise.]

49. Sometimes

50. Sometimes

51. Sometimes

52. Sometimes

53. [Respond below exercise.]

54. [Respond below exercise.]

55. [Respond below exercise.]

**47.** $|x| \leq 2$

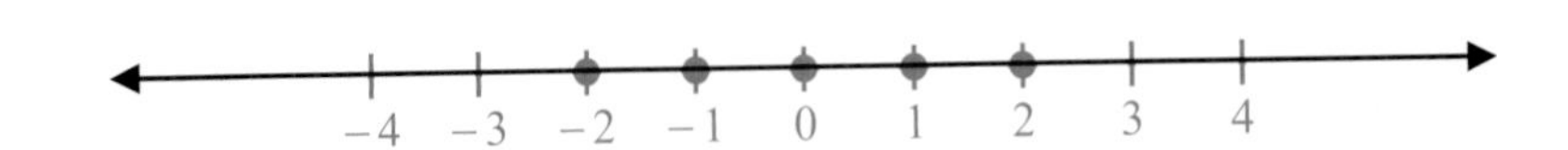


**48.** $|y| \geq 5$

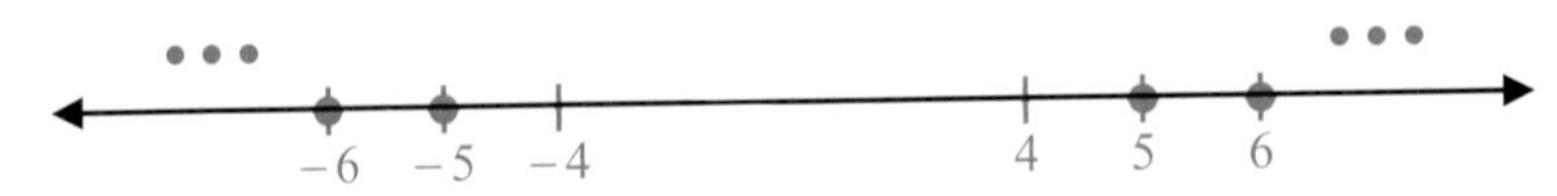


Choose the response that correctly completes each sentence. Assume that the variables represent integers. In each problem give two examples that illustrate your reasoning.

**49.** $|a|$ is (never, sometimes, always) equal to $a$.

**50.** $|x|$ is (never, sometimes, always) equal to $-x$.

**51.** $|y|$ is (never, sometimes, always) equal to a positive integer.

**52.** $|x|$ is (never, sometimes, always) greater than $x$.

## Writing and Thinking About Mathematics

**53.** Explain, in your own words, how an expression such as $-y$ might represent a positive number.

Answers will vary. If $y$ represents a negative number, then $-y$ will represent a positive number.

**54.** Explain, in your own words, the meaning of absolute value.

Answers will vary. The absolute value of an integer is its distance from zero.

$$\begin{cases} \text{If } a \text{ is a positive integer or 0,} & |a| = a. \\ \text{If } a \text{ is a negative integer,} & |a| = -a. \end{cases}$$

**55.** Explain, in your own words, why

**a.** there are no values for $x$ which $|x| < -5$.

**b.** if $|x| > -5$, then $x$ can be any integer.

Show your explanations to a friend to see if he or she can understand your reasoning.

Answers will vary.

**a.** The absolute value of an integer is never negative.

**b.** $|x| \geq 0$.

# 2.2 Addition with Integers

## An Intuitive Approach

We have discussed the operations of addition, subtraction, multiplication, and division with whole numbers. In the next three sections, we will discuss the basic rules and techniques for these same four operations with integers.

As an intuitive approach to addition with integers, consider an open field with a straight line marked with integers (much like a football field is marked every 10 yards). Imagine that a ball player stands at 0 and throws a ball to the point marked + 4 and then stands at + 4 and throws the ball 3 more units in the positive direction (to the right). Where will the ball land? (See figure 2.10.)

### Objectives

① Know how to add integers.

② Understand that the terms "opposite" and "additive inverse" have the same meaning.

③ Be able to determine whether or not a particular integer is a solution to an equation.

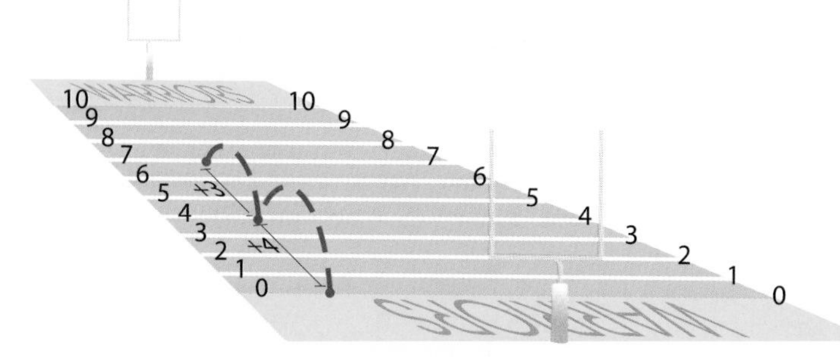


**Figure 2.10**

The ball will land at the point marked +7. We have, essentially, added the two positive integers + 4 and + 3.

$(+4)+(+3)=+7$ or $4+3=7$

Now, if the same player stands at 0 and throws the ball the same distances in the opposite direction (to the left), where will the ball land on the second throw? The ball will land at –7. We have just added two negative numbers, – 4 and – 3. (See Figure 2.11.)

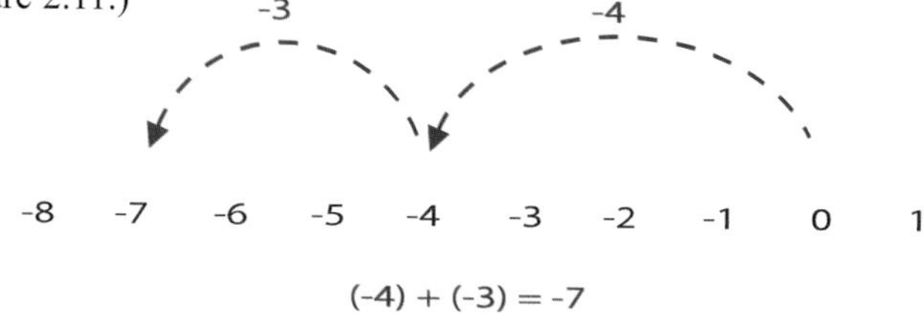


**Figure 2.11**

Sums involving both positive and negative integers are illustrated in Figures 2.12 and 2.13.

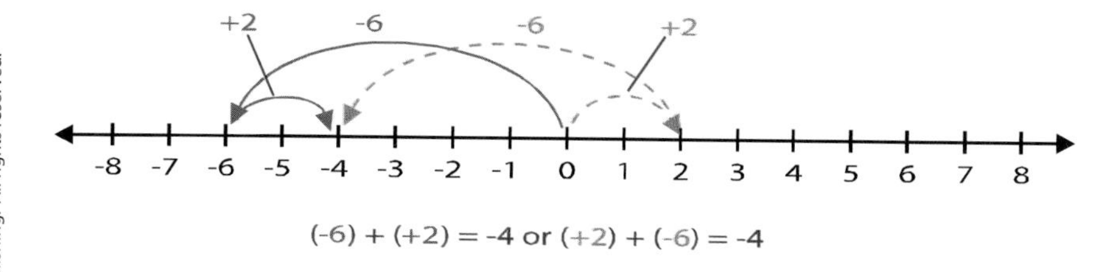


**Figure 2.12**

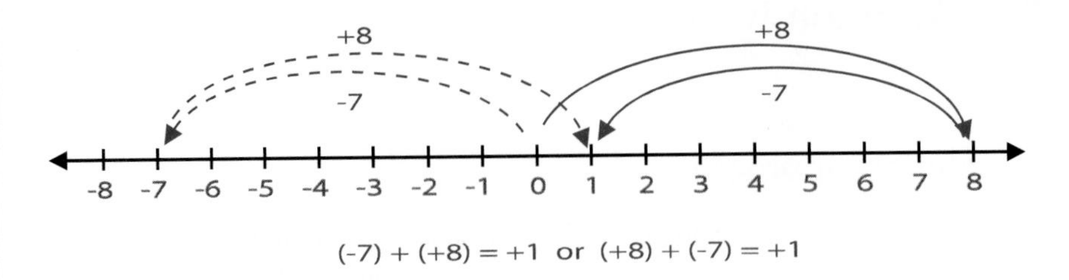


**Figure 2.13**

Although we are concerned here mainly with learning the techniques of adding integers, we note that addition with integers is **commutative** (as illustrated in Figures 2.12 and 2.13) and **associative.**

To help in understanding the rules for addition, we make the following suggestions for reading expressions with the + and – signs:

+ used as the sign of a number is read "positive."
+ used as an operation is read "plus."
– used as the sign of a number is read "negative" and "opposite."
– used as an operation is read "minus" (See Section 2.3).

In summary:

**1.** The sum of two positive integers is positive:

| $(+5)$ | $+$ | $(+4)$ | $=$ | $+9$ |
|---|---|---|---|---|
| ↑ | ↑ | ↑ | | ↑ |
| positive | plus | positive | | positive |

**2.** The sum of two negative integers is negative:

| $(-2)$ | $+$ | $(-3)$ | $=$ | $-5$ |
|---|---|---|---|---|
| ↑ | ↑ | ↑ | | ↑ |
| negative | plus | negative | | negative |

**3.** The sum of a positive integer and a negative integer may be negative or positive (or zero), depending on which number is farther from 0.

| $(+3)$ | $+$ | $(-7)$ | $=$ | $-4$ |
|---|---|---|---|---|
| ↑ | ↑ | ↑ | | ↑ |
| positive | plus | negative | | negative |

| $(+8)$ | $+$ | $(-2)$ | $=$ | $+6$ |
|---|---|---|---|---|
| ↑ | ↑ | ↑ | | ↑ |
| positive | plus | negative | | positive |

### Example 1

Find each of the following sums:

**a.** $(-15)+(-5)$

**b.** $(+10)+(-2)$

**c.** $(+4)+(+11)$

**d.** $(-12)+(+5)$

**e.** $(-90)+(+90)$

#### Solution

**a.** $(-15)+(-5)=-20$

**b.** $(+10)+(-2)=+8$

**c.** $(+4)+(+11)=+15$

**d.** $(-12)+(+5)=-7$

**e.** $(-90)+(+90)=0$

*Now Work Exercises 1-6 in the Margin*

## Rules for Addition with Integers

Saying that one number is farther from 0 than another number is the same as saying that the first number has a larger absolute value. With this basic idea, the rules for adding integers can be written out formally in terms of absolute value. If two numbers are the same distance from 0 in opposite directions, then each number is the opposite (or additive inverse) of the other and their sum is 0.

**Rules for Addition with Integers**

1. To add two integers with like signs, add their absolute values and use the common sign.

common sign ↓

$$(+6)+(+3)=+(|+6|+|+3|)=+(6+3)=+9$$

$$(-6)+(-3)=-(|-6|+|-3|)=-(6+3)=-9$$

↑ common sign

2. To add two integers with unlike signs, subtract their absolute values (the smaller from the larger) and use the sign of the integer with the larger absolute value.

because $|-15|>|+10|$ ↓

$$(-15)+(+10)=-(|-15|-|+10|)=-(15-10)=-5$$

$$(+15)+(-10)=+(|+15|-|-10|)=+(15-10)=+5$$

↑ because $|+15|>|-10|$

Find each of the following sums.

**1.** $(+3)+(+5)$
8

**2.** $(+11)+(-2)$
9

**3.** $(-4)+(-6)$
$-10$

**4.** $(-15)+(+5)$
$-10$

**5.** $(+40)+(-40)$
0

**6.** $(-20)+(-8)$
$-28$

Addition with any integer and its opposite (or additive inverse) always yields the sum of 0. The idea of "opposite" rather than "negative" is very important for understanding both addition and subtraction (see Section 2.3) with integers. The following definition and discussion clarify this idea.

**Definition**

The **opposite** of an integer is called its **additive inverse**.

The sum of any integer and its additive inverse is 0.

Symbolically, for any integer a,

$$\boldsymbol{a + (-a) = 0}$$

As an example, $(+20) + (-20) = +(|+20| - |-20|) = +(20 - 20$

**Note:** The symbol $-a$ should be read as "the opposite of $a$." Since $a$ is a variable, $-a$ might be positive, negative, or 0. For example:

If $a = 10$, then $-a = -10$ and $-a$ is **negative**.

That is, the opposite of a positive number is negative.

However,

if $a = -7$, then $-a = -(-7) = +7$ and $-a$ is **positive**.

That is, the opposite of a negative number is positive.

## Example 2

Find the additive inverse (opposite) of each integer.

**a.** 5 **b.** $-2$ **c.** $-15$

### Solution

**a.** The additive inverse of 5 is $-5$, and

$5 + (-5) = 0$

**b.** The additive inverse of $-2$ is $+2$, and

$(-2) + (+2) = 0$

**c.** The additive inverse of $-15$ is $+15$, and

$(-15) + (+15) = 0$

## Addition in a Vertical Format

Equations in algebra are almost always written horizontally, so addition (and subtraction) with integers is done much of the time in the horizontal format. However, there are situations (such as in long division) where sums (and differences) are written in a vertical format with one number directly under another, as illustrated in Example 3.

> **Note:** The positive sign ( + ) may be omitted when writing positive numbers, but the negative sign ( – ) must always be written for negative numbers. Thus, if there is no sign in front of an integer, the integer is understood to be positive.

### Example 3

Find the sums.

**a.**
$$\begin{array}{r} -20 \\ 7 \\ \underline{-30} \end{array}$$

**b.**
$$\begin{array}{r} 42 \\ -10 \\ \underline{-\ \ 3} \end{array}$$

**c.**
$$\begin{array}{r} -6 \\ -7 \\ -8 \\ \underline{-9} \end{array}$$

#### Solution

One technique for adding several integers is to mentally add the positive and negative integers separately and then add these results. (We are, in effect, using the commutative and associative properties of addition.)

**a.**
$$\begin{array}{r} -20 \\ 7 \\ \underline{-30} \\ -43 \end{array} \quad \text{or} \quad \begin{array}{r} -50 \\ \underline{7} \\ -43 \end{array}$$

**b.**
$$\begin{array}{r} 42 \\ -10 \\ \underline{-\ \ 3} \\ 29 \end{array} \quad \text{or} \quad \begin{array}{r} 42 \\ \underline{-13} \\ 29 \end{array}$$

**c.**
$$\begin{array}{r} -\ 6 \\ -\ 7 \\ -\ 8 \\ \underline{-\ 9} \\ -30 \end{array}$$

## Determining Integer Solutions to Equations

We can use our knowledge of addition with integers to determine whether a particular integer is a solution to an equation.

Find the sums:

**7.**
$$\begin{array}{r} -30 \\ 8 \\ \underline{-25} \\ -47 \end{array}$$

**8.**
$$\begin{array}{r} 52 \\ -25 \\ \underline{-\ 4} \\ 23 \end{array}$$

Determine whether $x = -5$ is a solution to each equation.

**9.** $x + 14 = 9$

Yes

**10.** $x + 5 = 0$

Yes

### Example 4

Determine whether or not the given integer is a solution to the given equation by substituting for the variable and adding.

**a.** $x + 8 = -2;\quad x = -10$

**b.** $x + (-5) = -6\,;\ x = +1$

**c.** $17 + y = 0\quad;\quad y = -17$

**Solution**

**a.**
$$x + 8 = -2$$
$$-10 + 8 \stackrel{?}{=} -2$$
$$-2 = -2$$

$-10$ **is** a solution.

**b.**
$$x + (-5) = -6$$
$$+1 + (-5) \stackrel{?}{=} -6$$
$$-4 \neq -6$$

$+1$ **is not** a solution.

**c.**
$$17 + y = 0$$
$$17 + (-17) \stackrel{?}{=} 0$$
$$0 = 0$$

$-17$ **is** a solution.

*Now Work Exercises 7-10 in the Margin.*

Name ______________________ Section ______________ Date ____________

## Exercises 2.2

Find the additive inverse (opposite) of each integer.

**1.** 15 **2.** 28 **3.** $-40$

**4.** $-32$ **5.** $-9$ **6.** 11

Find each of the indicated sums.

**7.** $8+(-10)$ **8.** $19+(-22)$ **9.** $(-3)+(-5)$

**10.** $(-4)+(-7)$ **11.** $(-1)+(-16)$ **12.** $(-6)+(-2)$

**13.** $(-9)+(+9)$ **14.** $(+43)+(-43)$ **15.** $(+73)+(-73)$

**16.** $-10+(+10)$ **17.** $-11+(-5)+(-2)$

**18.** $(-3)+(-6)+(-4)$ **19.** $-25+(-30)+(+10)$

**20.** $+36+(-12)+(-1)$ **21.** $(-35)+(+18)+(+17)$

**22.** $12+(+14)+(-16)$ **23.** $-33+(+13)+(-12)$

**24.** $-36+(-2)+(+40)$ **25.** $9+(-4)+(-5)+(+3)$

**ANSWERS**

1. $-15$
2. $-28$
3. 40
4. 32
5. 9
6. $-11$
7. $-2$
8. $-3$
9. $-8$
10. $-11$
11. $-17$
12. $-8$
13. 0
14. 0
15. 0
16. 0
17. $-18$
18. $-13$
19. $-45$
20. 23
21. 0
22. 10
23. $-32$
24. 2
25. 3

26. 5
27. – 19
28. – 13
29. – 10
30. – 155
31. – 24
32. – 34
33. 31
34. – 88
35. 76
36. 53
37. – 310
38. – 543
39. 33
40. – 57
41. 0
42. 0
43. 18
44. – 3
45. 14
46. 14
47. 35
48. 20

**26.** $12 + (-5) + (-10) + (+8)$

**27.** $-15 + (-3) + (+6) + (-7)$

**28.** $-26 + (+3) + (-15) + (+25)$

**29.** $-110 + (-25) + (125)$

**30.** $-120 + (-30) + (-5)$

Find each of the indicated sums in a vertical format.

**31.** $\begin{array}{r} -32 \\ \underline{8} \end{array}$

**32.** $\begin{array}{r} -53 \\ \underline{19} \end{array}$

**33.** $\begin{array}{r} 37 \\ \underline{-6} \end{array}$

**34.** $\begin{array}{r} -80 \\ \underline{-8} \end{array}$

**35.** $\begin{array}{r} 102 \\ -21 \\ \underline{-5} \end{array}$

**36.** $\begin{array}{r} 130 \\ -45 \\ \underline{-32} \end{array}$

**37.** $\begin{array}{r} -210 \\ -200 \\ \underline{100} \end{array}$

**38.** $\begin{array}{r} -108 \\ -105 \\ \underline{-330} \end{array}$

**39.** $\begin{array}{r} 35 \\ -2 \\ -5 \\ \underline{5} \end{array}$

**40.** $\begin{array}{r} -56 \\ -3 \\ -1 \\ \underline{3} \end{array}$

**41.** $\begin{array}{r} 28 \\ 9 \\ -9 \\ \underline{-28} \end{array}$

**42.** $\begin{array}{r} 76 \\ 12 \\ -12 \\ \underline{-76} \end{array}$

Find each of the indicated sums. Be sure to find the absolute values first.

**43.** $13 + |-5|$

**44.** $|-2| + (-5)$

**45.** $|-10| + |-4|$

**46.** $|-7| + (+7)$

**47.** $|-18| + |+17|$

**48.** $|-14| + |-6|$

Name ______________________ Section ______________ Date ___________

ANSWERS

49. No

50. No

51. Yes

52. Yes

53. Yes

54. Yes

55. Yes

56. Yes

57. No

58. No

59. Sometimes

60. Sometimes

61. Sometimes

62. Sometimes

63. Never

64. Never

65. Never

Determine whether or not the given integer is a solution of the equation by substituting the given value for the variable and adding.

**49.** $x + 5 = 7;\ x = -2$

**50.** $x + 6 = 9;\ x = -3$

**51.** $x + (-3) = -10;\ x = -7$

**52.** $-5 + x = -13;\ x = -8$

**53.** $y + (24) = 12;\ y = -12$

**54.** $y + (35) = -2;\ y = -37$

**55.** $z + (-18) = 0;\ z = 18$

**56.** $z + (27) = 0;\ z = -27$

**57.** $a + (-3) = -10;\ a = 7$

**58.** $a + (-19) = -29;\ a = 10$

Choose the response that correctly completes each statement. In each problem, give two examples that illustrate your reasoning.

**59.** If $x$ and $y$ are integers, then $x + y$ is (never, sometimes, always) equal to 0.

**60.** If $x$ and $y$ are integers, then $x + y$ is (never, sometimes, always) negative.

**61.** If $x$ and $y$ are integers, then $x + y$ is (never, sometimes, always) positive.

**62.** If $x$ is a positive integer and $y$ is a negative integer, then $x + y$ is (never, sometimes, always) equal to 0.

**63.** If $x$ and $y$ are both positive integers, then $x + y$ is (never, sometimes, always) equal to 0.

**64.** If $x$ and $y$ are both negative integers, then $x + y$ is (never, sometimes, always) equal to 0.

**65.** If $x$ is a negative integer, then $-x$ is (never, sometimes, always) negative.

66. Always ______

67. [Respond below exercise.] ______

68. [Respond below exercise.] ______

69. − 3405 ______

70. − 15,673 ______

71. − 2575 ______

72. − 1662 ______

73. − 23,094 ______

74. 10,871 ______

75. 845 ______

76. − 495 ______

**66.** If $x$ is a positive integer, then $-x$ is (never, sometimes, always) negative.

**67.** Describe in your own words the conditions under which the sum of two integers will be 0.

If they are opposites.

**68.** Explain how the sum of the absolute values of two integers might be 0. (Is this possible?)

This is possible only if both integers are zero.

Use a calculator to find the value of each expression. Use the parentheses keys on your calculator just as they are used in the expressions (from left to right). This will ensure that the operations are performed in the correct order.

**Note:** On a scientific calculator the key marked +/− is used to indicate the positive or negative sign of a number. On a TI–83 Plus graphing calculator, the key (just to the left of ENTER) marked as (−) is used to indicate a negative number. The two keys marked + and − are used for the operations of addition and subtraction, respectively.

**69.** $6790 + (-5635) + (-4560)$

**70.** $-8950 + (-3457) + (-3266)$

**71.** $-10{,}890 + (-5435) + (25{,}000) + (-11{,}250)$

**72.** $29{,}842 + (-5854) + (-12{,}450) + (-13{,}200)$

**73.** $(72{,}456 - 83{,}000) + (63{,}450 - 76{,}000)$

**74.** $(4783 + 5487 - 734) - (7125 - 8460)$

**75.** $(750 - (-320) + (-400)) - (325 - 500)$

**76.** $((-500) + (-300) - (-400)) + ((-75) + (-20))$

# 2.3 Subtraction with Integers

## Subtraction with Integers

Subtraction with whole numbers is defined in terms of addition. For example, the difference 20 – 14 is equal to 6 because 14 + 6 = 20. If we are restricted to whole numbers, then the difference 14 – 20 cannot be found because there is no whole number that we can add to 20 and get 14. Now that we are familiar with integers and addition with integers, we can reason that

$$14 - 20 = -6 \text{ because } 20 + (-6) = 14$$

That is, we can now define subtraction in terms of addition with integers so that a larger number may be subtracted from a smaller number. The definition of subtraction with integers involves the concept of additive inverses (opposites) as discussed in Section 2.2. We rewrite the definition here for convenience and easy reference.

**Objectives**

① Know how to subtract integers.

② Understand that the terms "opposite" and "additive inverse" have the same meaning.

**Definition**

The **opposite** of an integer is called its **additive inverse**. The sum of an integer and its additive inverse is 0. Symbolically, for any integer $a$,

$$\boldsymbol{a + (-a) = 0}$$

**Recall** that the symbol $-a$ should be read as "the opposite of $a$." Since $a$ is a variable, $-a$ might be positive, negative, or 0.

The term "opposite" is particularly useful in clarifying the meaning of an expression such as $-(-7)$ by reading it as "the opposite of negative seven." The idea of opposite is also fundamental to the definition of subtraction on page 6.

Subtraction with integers is defined in terms of addition, just as we defined subtraction with whole numbers in terms of addition. The distinction is that now we can subtract larger numbers from smaller numbers, and we can subtract negative and positive integers from each other. In effect, we can now study differences that are negative. For example,

| | | |
|---|---|---|
| $5 - 8 = -3$ | because | $8 + (-3) = 5$ |
| $10 - 12 = -2$ | because | $12 + (-2) = 10$ |
| $10 - (-6) = 16$ | because | $(-6) + (16) = 10$ |
| $4 - (-1) = 5$ | because | $(-1) + 5 = 4$ |
| $-5 - (-8) = 3$ | because | $(-8) + (3) = -5$ |

**Definition**

For any integers $a$ and $b$, the **difference** between $a$ and $b$ is defined as follows:

$$\boldsymbol{a - b = a + (-b)}$$

↑ minus ↑ opposite

This expression is read

"$a$ minus $b$ is equal to $a$ plus the opposite of $b$."

The students should be encouraged to read definitions carefully and to interpret the conditions given on variables. For example, in this definition, since $a$ and $b$ are "any integers," each can represent a positive or a negative integer or 0. This is probably a good time to remind the students that the symbol $-a$ (as well as $a$) can represent a positive integer, a negative integer, or 0.

The definition of subtraction is one of the most difficult for students to understand and implement. One way to develop an intuitive understanding is to emphasize the fact that the difference between any number and itself is 0. Example 1(e) illustrates this idea with a negative integer. In general, $a-(a)=0$ and $(-a)-(-a)=0$.

Since $a$ and $b$ are variables and may themselves be positive, negative, or 0, the definition automatically includes other forms as follows:

$a-(-b)=a+(+b)=a+b$ and $-a-b=-a+(-b)$

The following examples illustrate how to apply the definition of subtraction. Remember, to subtract an integer you add its opposite.

### Example 1

Find the following differences.

**a.** $(-1)-(-5)$

**b.** $(-2)-(-7)$

**c.** $(-10)-(+3)$

**d.** $(+4)-(+5)$

**e.** $(-6)-(-6)$

**Solution**

**a.** $(-1)-(-5)=(-1)+(+5)=4$

**b.** $(-2)-(-7)=(-2)+(+7)=5$

**c.** $(-10)-(+3)=(-10)+(-3)=-13$

**d.** $(+4)-(+5)=(+4)+(-5)=-1$

**e.** $(-6)-(-6)=(-6)+(+6)=0$

*Now Work Exercises 1-6 in the Margin.*

Find the following differences.

**1.** $(-4)-(-4)$
0

**2.** $(-3)-(+8)$
$-11$

**3.** $(-3)-(-8)$
5

**4.** $(+14)-(-5)$
19

**5.** $(+14)-(+5)$
9

**6.** $(-20)-(-3)$
$-17$

## Simplifying the Notation

The interrelationship between addition and subtraction with integers allows us to eliminate the use of so many parentheses. The following two interpretations of the same problem indicate that both interpretations are correct and lead to the same answer.

| $5-18=5$ | $-$ | $(+18)=$ | $-13$ |
|---|---|---|---|
| ↑ | ↑ | ↑ | ↑ |
| 5 | minus | positive 18 | negative 13 |

| $5-18=5$ | $+$ | $(-18)=$ | $-13$ |
|---|---|---|---|
| ↑ | ↑ | ↑ | ↑ |
| 5 | plus | negative 18 | negative 13 |

Thus, the expression $5-18$ can be thought of as subtraction or addition. In either case, the answer is the same. Understanding that an expression such as $5-18$ (with no parentheses) can be thought of as addition or subtraction takes some practice, but it is quite important because the notation is commonly used in all mathematics textbooks. Study the following examples carefully.

### Example 2

**a.** $5 - 18 = -13$ Think: $(+5) + (-18)$

**b.** $-5 - 18 = -23$ Think: $(-5) + (-18)$

**c.** $-16 - 3 = -19$ Think: $(-16) + (-3)$

**d.** $25 - 20 = 5$ Think: $(+25) + (-20)$

**e.** $32 - 33 = -1$ Think: $(+32) + (-33)$

In Section 2.2 we discussed the fact that addition with integers is both commutative and associative. That is, for integers a, b, and c,

$\boldsymbol{a + b = b + a}$ Commutative Property of Addition

and $\boldsymbol{a + (b + c) = (a + b) + c}$ Associative Property of Addition

However, as the following examples illustrate, subtraction is **not commutative** and **not associative**.

$13 - 5 = 8$ and $5 - 13 = -8$

So, $13 - 5 \neq 5 - 13$, and in general,

$a - b \neq b - a$ (**Note:** $\neq$ is read "is not equal to.")

Similarly,

$8 - (3 - 1) = 8 - (2) = 6$ and $(8 - 3) - 1 = 5 - 1 = 4$

So, $8 - (3 - 1) \neq (8 - 3) - 1$, and in general,

$a - (b - c) \neq (a - b) - c$

Thus, the order that we write the numbers in subtraction and the use of parentheses can be critical. Be careful when you write any expression involving subtraction.

## Subtraction in a Vertical Format

As with addition, integers can be written vertically in subtraction. One number is written underneath the other, and the sign of the integer being subtracted is changed. That is, we add the opposite of the integer that is subtracted.

### Example 3

To subtract: Add the opposite of the bottom number:

$$\begin{array}{r} -45 \\ \underline{-(+32)} \end{array} \qquad \begin{array}{rl} -45 & \\ \underline{+(-32)} & \text{sign is changed.} \\ -77 & \text{difference} \end{array}$$

### Example 4

To subtract: Add the opposite of the bottom number:

$$\begin{array}{r} -56 \\ \underline{-(+10)} \end{array} \qquad \begin{array}{rl} -56 & \\ \underline{+(-10)} & \text{sign is changed.} \\ -66 & \text{difference} \end{array}$$

Subtract the bottom number from the top number

**7.** $\begin{array}{r} -20 \\ -(+87) \\ \hline -107 \end{array}$

**8.** $\begin{array}{r} -56 \\ -(-56) \\ \hline 0 \end{array}$

**9.** $\begin{array}{r} +42 \\ -(-42) \\ \hline 84 \end{array}$

**10.** $\begin{array}{r} -32 \\ -(-69) \\ \hline 37 \end{array}$

### Example 5

To subtract: Add the opposite of the bottom number:

$$\begin{array}{r} -20 \\ -(-95) \\ \hline \end{array} \qquad \begin{array}{rl} -20 & \\ +(+95) & \text{sign is changed.} \\ \hline +75 & \text{difference} \end{array}$$

*Now Work Exercises 7–10 in the Margin*

## Determining the Solutions to Equations

Now both addition and subtraction can be used to help determine whether or not a particular integer is a solution of an equation. (This skill will prove useful in evaluating formulas and solving equations.)

### Example 6

Determine whether or not the given integer is a solution to the given equation by substituting for the variable and then evaluating.

**a.** $x-(-6)=-10\,;x=-14$

**b.** $7-y=-1;\ y=8$

**c.** $a-12=-2;\ a=-10$

**Solution**

**a.**
$$x-(-6)=-10$$
$$-14-(-6)\overset{?}{=}-10$$
$$-14+(+6)\overset{?}{=}-10$$
$$-8\neq-10$$
$-14$ **is not** a solution.

**b.**
$$7-y=-1$$
$$7-8\overset{?}{=}-1$$
$$-1=-1$$
8 **is** a solution.

**c.**
$$a-12=-2$$
$$-10-12\overset{?}{=}-2$$
$$-22\neq-2$$
$-10$ **is not** a solution.

Name ______________ Section ______________ Date ______________

## Exercises 2.3

Perform the indicated operations and write in words what to "think" in terms of addition. (See Example 3.)

1. a. $6 - 14$
   b. $-6 - 14$
2. a. $17 - 5$
   b. $-17 - 5$
3. a. $-10 - 19$
   b. $10 - 19$
4. a. $-25 - 3$
   b. $25 - 3$
5. a. $12 - 20$
   b. $-12 - 20$
6. a. $-15 - 18$
   b. $15 - 18$

Perform the indicated operations.

7. $9 - 3$
8. $10 - 7$
9. $-5 - (-10)$
10. $-8 - (-4)$
11. $-7 + (-13)$
12. $(-2) + (-15)$
13. $-14 - 20$
14. $-17 - 30$
15. $0 - (-4)$
16. $0 - (-9)$
17. $0 - 4$
18. $0 - 9$

**ANSWERS**

1. a. $-8$
   b. $-20$
2. a. $12$
   b. $-22$
3. a. $-29$
   b. $-9$
4. a. $-28$
   b. $22$
5. a. $-8$
   b. $-32$
6. a. $-33$
   b. $-3$
7. $6$
8. $3$
9. $5$
10. $-4$
11. $-20$
12. $-17$
13. $-34$
14. $-47$
15. $4$
16. $9$
17. $-4$
18. $-9$

19. 6

20. − 3

21. −15

22. − 2

23. − 22

24. 1

25. − 19

26. − 16

27. − 7

28. − 5

29. − 21

30. − 25

31. − 2

32. 0

33. 0

34. 6

35. 0

36. 9

37. − 28

38. − 80

39. − 45

**19.** $-2-3+11$

**20.** $-5-4+6$

**21.** $-6-5-4$

**22.** $-3-(-3)+(-2)$

**23.** $-7+(-7)+(-8)$

**24.** $-1+12-10$

**25.** $-8-11$

**26.** $-5-11$

**27.** $13-20$

**28.** $17-22$

**29.** $-6-7-8$

**30.** $-40+15$

**31.** $-16+15-1$

**32.** $25-13-12$

**33.** $18-11-7$

**34.** $-7-(-10)+3$

**35.** $-10-(-5)+5$

**36.** $15+(-4)+(-5)-(-3)$

**37.** $-17+(-14)-(-3)$

**38.** $-80-70-30+100$

**39.** $-18-17-9+(-1)$

Name ____________ Section ____________ Date ____________

**ANSWERS**

**40.** $-21 - 5 - 8 + (-2)$

**41.** $-4 - (-3) + (-5) + 7 - 2$

**42.** $-6 - (-8) + 10 - 3$

**43.** $26 - 13 - 17 + 5 - 20$

**44.** $35 - 22 - 8 + 7 - 30$

Subtract the bottom number from the top number.

**45.** $\begin{array}{r} 37 \\ -(45) \\ \hline \end{array}$

**46.** $\begin{array}{r} 20 \\ -(36) \\ \hline \end{array}$

**47.** $\begin{array}{r} -21 \\ -(-10) \\ \hline \end{array}$

**48.** $\begin{array}{r} -46 \\ -(-25) \\ \hline \end{array}$

**49.** $\begin{array}{r} -16 \\ -(-16) \\ \hline \end{array}$

**50.** $\begin{array}{r} -24 \\ -(-24) \\ \hline \end{array}$

**51.** $\begin{array}{r} 105 \\ -(-22) \\ \hline \end{array}$

**52.** $\begin{array}{r} 202 \\ -(-35) \\ \hline \end{array}$

Insert the symbol in the blank that will make each statement true: <, >, or =.

**53.** $-7 + (-2)$ __>__ $-4 - 6$

**54.** $-6 + (-3)$ __<__ $-2 - 2$

**55.** $0 - 8$ __<__ $0 - (-8)$

**56.** $0 - 12$ __<__ $0 - (-2)$

**57.** $9 - (-3)$ __>__ $9 - 3$

**58.** $4 - (-1)$ __>__ $4 - 1$

**59.** $-6 - 6$ __=__ $0 - 12$

**60.** $-10 - 10$ __=__ $0 - 20$

**40.** $-36$

**41.** $-1$

**42.** 9

**43.** $-19$

**44.** $-18$

**45.** $-8$

**46.** $-16$

**47.** $-11$

**48.** $-21$

**49.** 0

**50.** 0

**51.** 127

**52.** 237

**53.** [Respond in exercise]

**54.** [Respond in exercise]

**55.** [Respond in exercise]

**56.** [Respond in exercise]

**57.** [Respond in exercise]

**58.** [Respond in exercise]

**59.** [Respond in exercise]

**60.** [Respond in exercise]

61. Yes

62. Yes

63. Yes

64. Yes

65. Yes

66. Yes

67. No

68. No

69. Yes

70. Yes

71. 3°

72. 10 lb. loss; 215 lbs.

Determine whether or not the given number is a solution to the equation. Substitute the number for the variable and perform the indicated operations mentally.

**61.** $x - 7 = -9;\ x = -2$

**62.** $x - 4 = -10;\ x = -6$

**63.** $a + 5 = -10;\ a = -15$

**64.** $a - 3 = -12;\ a = -9$

**65.** $x + 13 = 13;\ x = 0$

**66.** $x + 21 = 21;\ x = 0$

**67.** $3 - y = 6;\ y = 3$

**68.** $5 - y = 10;\ y = 5$

**69.** $22 + x = -1;\ x = -23$

**70.** $30 + x = -2;\ x = -32$

**71.** Beginning with a temperature of 8° above zero, the temperature was measured hourly for 4 hours. It rose 3°, dropped 7°, dropped 2°, and rose 1°. What was the final temperature recorded?

**72.** George and his wife went on a diet plan for 5 weeks. During these 5 weeks, George lost 5 pounds, gained 2 pounds, lost 4 pounds, lost 6 pounds, and gained 3 pounds. What was his total loss or gain for these 5 weeks? If he weighed 225 pounds when he started the diet, what did he weigh at the end of the 5–week period? During the same time his wife lost 11 pounds.

Name ______________________ Section ______________ Date __________

**ANSWERS**

**73.** In a 5–day week, the NASDAQ stock market posted a gain of 145 points, a loss of 100 points, a loss of 82 points, a gain of 50 points, and a gain of 25 points. If the NASDAQ started the week at 4200 points, what was the market at the end of the week?

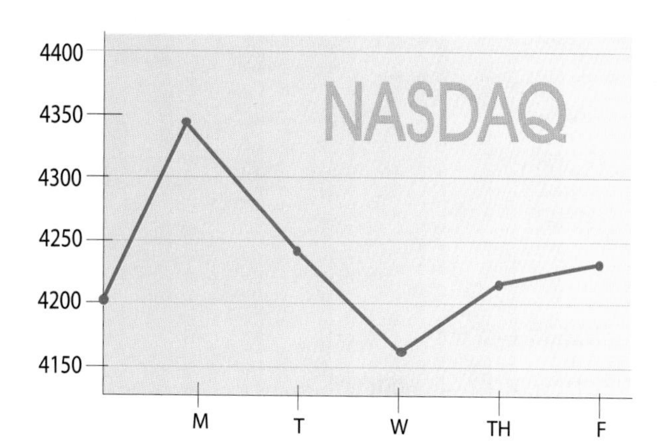


**74.** In a 5–day week, the Dow Jones stock market average showed a gain of 32 points, a gain of 10 points, a loss of 20 points, a loss of 2 points, and a loss of 25 points. What was the net change in the stock market for the week? If the Dow started the week at 10,530 points, what was the average at the end of the week?

**75.** In 10 running plays in a football game the fullback gained 2 yards, gained 12 yards, lost 5 yards, lost 3 yards, gained 22 yards, gained 3 yards, gained 7 yards, lost 2 yards, gained 4 yards, and gained 45 yards. What was his net yardage for the game? This was a good game for him, but not his best.

**73.** 4238 pts.

**74.** - 5 pts.; 10,525 pts.

**75.** 85 yds.

76.a. – 632

b. 29,942

77.a. – 12,544

b. 57,456

78.a. – 174,350

b. – 232,550

79.a. – 618,000

b. 1,618,000

80.a. 17,610

b. 672,990

81.a. – 5

b. 19

82.a. – 30

b. 10

83. – 1

84. 15

In Exercises 76 – 80, use a calculator to find the value of each expression.

**76.** **a.** 14,655 – 15,287

**b.** 14,655 – ( – 15,287 )

**77.** **a.** 22,456 – 35,000

**b.** 22,456 – (– 35,000 )

**78.** **a.** – 203,450 – 16,500 + 45,600

**b.** –203,450 – (–16,500) + (–45,600)

**79.** **a.** 500,000 – 1,043,500 – 250,000 + 175,500

**b.** 500,000 – ( – 1,043,500 ) – ( – 250,000 ) + ( – 175,500 )

**80.** **a.** 345,300 – 42,670 – 356,020 – 250,000 + 321,000

**b.** 345,300 – (–42,670) – (–356,020) – (–250,000) + (–321,000)

**81.** **a.** What should be added to – 7 to get a sum of – 12?

**b.** What should be added to –7 to get a sum of 12?

**82.** **a.** What should be subtracted from –10 to get a difference of 20?

**b.** What should be subtracted from –10 to get a difference of –20?

**83.** From the sum of – 3 and – 5, subtract the sum of – 20 and 13. What is the difference?

**84.** Find the difference between the sum of 6 and – 10 and the sum of – 15 and – 4.

Name ______________ Section ______________ Date ______________

## WRITING AND THINKING ABOUT MATHEMATICS

**85.** Explain why the expression $-x$ might represent a positive number, a negative number, or even 0.
Answers will vary. Think of $-x$ as the "opposite of $x$." If $x$ is positive, then $-x$ is negative. If x is negative, then $-x$ is positive. Also, note that 0 is its own opposite.

**86.** Under what conditions can the difference of two negative numbers be a positive number?
Answers will vary. When $x > y$, then $x - y > 0$.

### The Recycle Bin (from Section 1.5)

Find the value of each of the following expressions by using the rules for order of operations.

**1.** $22 - 2 \cdot 3^2$

**2.** $15 \div 3 + 2(10 - 7)$

**3.** $(9 - 6)^3 - 16 \div 2^3$

**4.** $8 \div 4 \cdot 2 + 12 - 15$

**5.** $1 + 6(19 - 2^2) + 17 \cdot 2 \div 34$

Use your calculator to find the value of each of the following expressions.

**6.** **a.** $6^4$ **b.** $18^3$ **c.** $25^5$

**7.** **a.** $13^0$ **b.** $1^0$ **c.** $0^0$

**8.** **a.** $(9 + 1)(5 - 2)^0$ **b.** $14 + 3(8 - 6)(12 - 9)(0)$ **c.** $62 \div 31 + 14 \div 7 - 16 \div 4$

### ANSWERS

**85.** [Respond below exercise]

**86.** [Respond below exercise]

**The Recycle Bin**

**1.** 4

**2.** 11

**3.** 25

**4.** 1

**5.** 92

**6. a.** 1296

**b.** 5832

**c.** 9765625

**7. a.** 1

**b.** 1

**c.** Undefined

**8. a.** 10

**b.** 14

**c.** 0

# 2.4 Multiplication and Division with Integers and Order of Operations

## Multiplication with Integers

Multiplication with whole numbers is a shorthand form of repeated addition. Multiplication with integers can be thought of in the same manner. For example,

$$6+6+6+6+6=5\cdot 6=30$$

and

$$(-6)+(-6)+(-6)+(-6)+(-6)=5(-6)=-30$$

and

$$(-3)+(-3)+(-3)+(-3)=4(-3)=-12$$

Repeated addition with a negative integer results in a product of a positive integer and a negative integer. Therefore, because the sum of negative integers is negative, we can reason that the **product of a positive integer and a negative integer is negative.**

### Objectives

① Know the rules for multiplication with integers.

② Know the rules for division with integers.

③ Be able to multiply and divide with integers.

④ Know that division by 0 is not defined.

⑤ Be able to apply the rules for order of operations to expressions with integers.

### Example 1

**a.** $5(-4) \quad = -20$

**b.** $3(-15) = -45$

**c.** $7(-11) = -77$

**d.** $100(-2) = -200$

**e.** $-1(9) \quad = -9$

The product of two negative integers can be explained in terms of opposites and the fact that, for any integer $a$,

$$-a=-1(a).$$

That is, the opposite of $a$ can be treated as $-1$ times $a$. Consider, for example, the product $(-2)(-5)$. We can proceed as follows:

| | |
|---|---|
| $(-2)(-5)=-1(2)(-5)$ | $-2=-1(2)$ |
| $=-1[2(-5)]$ | Associative property of multiplication |
| $=-1[-10]$ | Multiply. |
| $=-[-10]$ | $-1$ times a number is its opposite. |
| $=+10$ | Opposite of $-10$ |

Although one example does not prove a rule, this process is such that we can use it as a general procedure and come to the following correct conclusion:

**The product of two negative numbers is positive.**

### Example 2

**a.** $(-6)(-4) = +24$
**b.** $-7(-9) = +63$
**c.** $-8(-10) = +80$
**d.** $0(-15) = 0$ and $0(+15) = 0$ and $0(0) = 0$

As Example 2d. illustrates:

**The product of 0 and any integer is 0.**

The rules for multiplication with integers can be summarized as follows:

Note that the rules for multiplication with integers and the rules for division with integers are stated with $a$ and $b$ as positive integers. You might want to be sure that the students are aware of these restrictions and point out that this is convenient for making the corresponding symbolic and English statements of the rules.

**Rules for Multiplication with Integers**

If $a$ and $b$ are positive integers, then

1. The product of two positive integers is positive:
   $a \cdot b = ab$
2. The product of two negative integers is positive:
   $(-a)(-b) = ab$
3. The product of a positive integer and a negative integer is negative:
   $a(-b) = -ab$ and $(-a)(b) = -ab$
4. The product of 0 and any integer is 0:
   $a \cdot 0 = 0$ and $(-a) \cdot 0 = 0$

The **commutative** and **associative** properties for multiplication are valid for multiplication with integers, just as they are with whole numbers. Thus, to find the product of more than two integers, we can multiply any two and continue to multiply until all the integers have been multiplied.

Find each product.

1. $5(-3)$
   $-15$
2. $-8(4)$
   $-32$
3. $-6(-4)$
   $24$
4. $-19(0)$
   $0$
5. $3(-20)(-1)(-1)$
   $-60$
6. $-9(7)(-2)$
   $126$

### Example 3

**a.** $(-5)(-3)(10) = [(-5)(-3)](10)$
$= [+15](10)$
$= +150$

**b.** $(-2)(-2)(-2)(3) = +4(-2)(3)$
$= -8(3)$
$= -24$

**c.** $(-3)(5)(-6)(-1)(-1) = -15(-6)(-1)(-1)$
$= (+90)(-1)(-1)$
$= (-90)(-1)$
$= +90$

**d.** $(-4)^3 = (-4)(-4)(-4)$
$= (+16)(-4)$
$= -64$

**e.** $(-3)^2 = (-3)(-3)$
$= +9$

*Now Work Exercises 1-6 in the Margin.*

## Division with Integers

For the purposes of this discussion, we will indicate division such as $a \div b$ in the fraction form, $\frac{a}{b}$. (Operations with fractions will be discussed in detail in Chapter 4.) Thus, a division problem such as $56 \div 8$ can be indicated in the fraction form as $\frac{56}{8}$.

Since division is defined in terms of multiplication, we have

$$\frac{56}{8} = 7 \quad \text{because} \quad 56 = 8 \cdot 7$$

**Definition:**

For integers $a$, $b$, and $x$ (where $b \neq 0$),

$\boldsymbol{\frac{a}{b} = x}$ means that $\boldsymbol{a = b \cdot x}$

**Special Note:** In this section, we are emphasizing the rules for signs when multiplying and dividing with integers. With this emphasis in mind, the problems are set up in such a way that the quotients are integers. However, these rules are valid for many types of numbers, and we will use them with fractions (in Chapter 4) and decimals (in Chapter 5).

### Example 4

**a.** $\frac{+35}{+5} = +7$ because $+35 = (+5)(+7)$.

**b.** $\frac{-35}{-5} = +7$ because $-35 = (-5)(+7)$.

**c.** $\frac{+35}{-5} = -7$ because $+35 = (-5)(-7)$.

**d.** $\frac{-35}{+5} = -7$ because $-35 = (+5)(-7)$.

Find each quotient.

**7.** $\dfrac{-30}{+10}$ 
$-3$

**8.** $\dfrac{+40}{-10}$ 
$-4$

**9.** $\dfrac{32}{8}$ 
4

**10.** $\dfrac{-32}{-8}$ 
4

**11.** $\dfrac{51}{-3}$ 
$-17$

**12.** $\dfrac{12}{-4}$ 
$-3$

The rules for division with integers can be stated as follows:

**Rules for Division with Integers**

If $a$ and $b$ are positive integers, then

1. The quotient of two positive integers is positive:

$$\frac{a}{b} = +\frac{a}{b}$$

2. The quotient of two negative integers is positive:

$$\frac{-a}{-b} = +\frac{a}{b}$$

3. The quotient of a positive integer and a negative integer is negative:

$$\frac{-a}{b} = -\frac{a}{b} \quad \text{and} \quad \frac{a}{-b} = -\frac{a}{b}$$

**In summary:**
**When the signs are alike, the quotient is positive.**
**When the signs are not alike, the quotient is negative.**

*Now Work Exercises 7-12 in the Margin.*

In Section 1.1, we stated that division by 0 is undefined. This fact was explained there, and for emphasis and understanding, we explain it again here in terms of multiplication with fraction notation.

**Division by 0 Is Not Defined**

**CASE 1:** Suppose that $a \neq 0$ and $\frac{a}{0} = x$. Then, by the meaning of division, $a = 0 \cdot x$. But this is not possible because $0 \cdot x = 0$ for any value of $x$, and we stated that $a \neq 0$.

**CASE 2:** Suppose that $\frac{0}{0} = x$. Then, $0 = 0 \cdot x$ is true for all values of $x$. But we must have a unique answer for division. Therefore, we conclude that, in every case, division by 0 is not defined.

The following mnemonic device may help you remember the rules for division by 0:

$\frac{0}{K}$ is **"OK"**. 0 **can** be in the **numerator**.

$\frac{K}{0}$ is a **"KO"** (knockout). 0 **cannot** be in the **denominator**.

### Example 5

**a.** $\frac{0}{-2}=0$ because $0=-2(0)$.

**b.** $\frac{7}{0}$ is undefined. If $\frac{7}{0}=x$, then $7=0\cdot x$. But this is not possible because $0\cdot x=0$ for any value of $x$.

**c.** $\frac{-32}{0}$ is undefined. There is no number whose product with 0 is $-32$.

**d.** $\frac{0}{0}$ is undefined. Suppose you think that $\frac{0}{0}=1$ because $0=0\cdot 1$. This is certainly true. However, someone else might reason in an equally valid way that $\frac{0}{0}=92$. Since there can be no unique value, we conclude that $\frac{0}{0}$ is undefined.

## Rules for Order of Operations

The rules for order of operations were discussed in section 1.5 and are restated here for convenience and easy reference.

**Rules for Order of Operations**

1. First, simplify within grouping symbols, such as parentheses ( ), brackets [ ], or braces { }. Start with the innermost grouping.
2. Second, find any powers indicated by exponents.
3. Third, moving from **left to right**, perform any multiplications or divisions in the order in which they appear.
4. Fourth, moving from **left to right**, perform any additions or subtractions in the order in which they appear.

Now that we know how to operate with integers, we can apply these rules in evaluating expressions that involve negative numbers as well as positive numbers.

Observe carefully the use of parentheses around negative numbers in the Examples 6 and 7. Two operation signs are not allowed together. For example, do not write $+ \; -$ or $\div \; \cdot$ or $+ \; \div$ without using numbers and/or parentheses with the operations.

At some point you may want to explain to your students that "terms" separated by + or - signs can be evaluated out of the usual sequence because the addition and subtraction will come last. This will speed up the process of evaluating. For example, in Example 6, we could multiply $-3 \cdot 2$ and $4(-5)$ in the same step. I have stated this in Section 2.6.

## Example 6

Evaluate the expression $27 \div (-9) \cdot 2 - 5 + 4(-5)$.

### Solution

$$\underbrace{27 \div (-9)}_{\downarrow} \cdot 2 - 5 + 4(-5)$$ Divide.

$$= \underbrace{-3 \cdot 2}_{\downarrow} - 5 + 4(-5)$$ Multiply.

$$= -6 - 5 + \underbrace{4(-5)}_{\downarrow}$$ Multiply.

$$= \underbrace{-6 - 5}_{\downarrow} - 20$$ Add. (Remember we are adding negative numbers.)

$$= \underbrace{-11 - 20}_{\downarrow}$$ Add.

$$= -31$$

## Example 7

Evaluate the expression $9 - 11[(3 - 5^2) \div 2 + 5]$.

### Solution

$$9 - 11\left[(3 - 5^2) \div 2 + 5\right]$$ Find the power.

$$= 9 - 11\left[(3 - 25) \div 2 + 5\right]$$ Operate within parentheses.

$$= 9 - 11\left[(-22) \div 2 + 5\right]$$ Divide within brackets.

$$= 9 - 11[-11 + 5]$$ Add within brackets.

$$= 9 - 11[-6]$$ Multiply, in this case −11 times − 6.

$$= 9 + 66$$ Add.

$$= 75$$

## *Completion Example 8*

Evaluate $9(-1+2^2)-7-6 \cdot 2^2$.

**Solution**

$$
\begin{aligned}
& 9(-1+2^2)-7-6\cdot 2^2 \\
= \quad & 9(-1+\_\_\_\_)-7-6\cdot\_\_\_\_ \\
= \quad & 9(\_\_\_\_)-7-6\cdot\_\_\_\_ \\
= \quad & \_\_\_\_-7-6\cdot\_\_\_\_ \\
= \quad & \_\_\_\_-7-\_\_\_\_ \\
= \quad & \_\_\_\_-\_\_\_\_ \\
= \quad & \_\_\_\_
\end{aligned}
$$

*Now Work Exercises 13 - 15 in the Margin.*

**Completion Example Answer**

**8.**

$$
\begin{aligned}
& 9(-1+2^2)-7-6\cdot 2^2 \\
= \quad & 9(-1+\mathbf{4})-7-6\cdot\mathbf{4} \\
= \quad & 9(\mathbf{3})-7-6\cdot\mathbf{4} \\
= \quad & \mathbf{27}-7-6\cdot\mathbf{4} \\
= \quad & \mathbf{20}-7-\mathbf{24} \\
= \quad & \mathbf{20}-\mathbf{24} \\
= \quad & \mathbf{-4}
\end{aligned}
$$

Find the value of each expression by using the rules for order of operations.

**13.** $15 \div 5 + 10 \cdot 2$

23

**14.** $4 \div 2^2 + 3 \cdot 2^2$

13

**15.** $2(-1+3^2)-4-3 \cdot 2^3$

$-12$

Name ______________ Section ______________ Date ______________

## Exercises 2.4

Find the following products.

1. $4(-6)$
2. $-6(4)$
3. $-8(3)$
4. $-2(-7)$
5. $14(2)$
6. $21(3)$
7. $-2(9)$
8. $(-6)(-3)$
9. $-10(5)$
10. $9(-4)$
11. $0(-5)$
12. $0(-1)$
13. $(-2)(-1)(-7)$
14. $(-5)(3)(-4)$
15. $(-3)(7)(-5)$
16. $(-2)(-5)(-3)$
17. $(-2)(-2)(-8)$
18. $(-3)(-3)(-3)$
19. $(-5)(0)(-6)$
20. $(-9)(0)(-8)$
21. $(-1)(-1)(-1)$
22. $(-2)(-2)(-2)(-2)$
23. $(-5)(-2)(+5)(-1)$

**ANSWERS**

1. $-24$
2. $-24$
3. $-24$
4. 14
5. 28
6. 63
7. $-18$
8. 18
9. $-50$
10. $-36$
11. 0
12. 0
13. $-14$
14. 60
15. 105
16. $-30$
17. $-32$
18. $-27$
19. 0
20. 0
21. $-1$
22. 16
23. $-50$

24. 330

25. 60

26. 0

27. 0

28. a. − 1

b. 1

c. 1

d. − 1

29. a. 1

b. − 1

c. 1

d. 1

30. [Respond below exercise.]

31. − 4

32. − 3

33. − 2

34. − 4

35. − 6

36. − 10

37. − 4

38. − 7

39. − 14

40. − 3

41. − 3

42. − 4

**24.** $(-2)(-3)(-11)(-5)$

**25.** $(-10)(-2)(-3)(-1)$

**26.** $(-1)(-2)(-6)(-3)(-1)(-5)(-4)(0)$

**27.** $(-2)(-3)(-4)(-1)(-4)(-8)(-2)(0)$

**28.** **a.** $(-1)^7$

**b.** $(-1)^{10}$

**c.** $(-1)^0$

**d.** $(-1)^{31}$

**29.** **a.** $(-1)^6$

**b.** $(-1)^{11}$

**c.** $(-1)^{14}$

**d.** $(-1)^0$

**30.** Can you make a general statement relating the sign of a product and the number of negative factors?

An odd number of negative factors gives a negative product. An even number of negative factors gives a positive product. (Note: These statements assume that 0 is not a factor.)

Find the following quotients.

**31.** $\frac{-12}{3}$ **32.** $\frac{-18}{6}$ **33.** $\frac{-14}{7}$ **34.** $\frac{-28}{7}$

**35.** $\frac{-30}{5}$ **36.** $\frac{-50}{5}$ **37.** $\frac{-20}{5}$ **38.** $\frac{-35}{5}$

**39.** $\frac{39+3}{-3}$ **40.** $\frac{16+8}{-8}$ **41.** $\frac{18+9}{-9}$ **42.** $\frac{9+27}{-9}$

Name ____________ Section ____________ Date ____________

**ANSWERS**

**43.** $\frac{-36-9}{-9}$ **44.** $\frac{-13-13}{-13}$ **45.** $\frac{-28-4}{-4}$ **46.** $\frac{-13-26}{-13}$

**47.** $\frac{0}{2}$ **48.** $\frac{0}{8}$ **49.** $\frac{35}{0}$ **50.** $\frac{56}{0}$

**51.** If the product of –15 and –6 is added to the product of –32 and 5, what is the sum?

**52.** If the quotient of –56 and 8 is subtracted from the product of –12 and –3, what is the difference?

**53. a.** What number should be added to –5 to get a sum of –22?

**b.** What number should be added to +5 to get a sum of –22?

**54. a.** What number should be added to –39 to get a sum of –13?

**b.** What number should be added to +39 to get a sum of –13?

Find the value of each expression by following the rules for order of operations.

**55.** $15 \div (-5) \cdot 2 - 8$

**56.** $20 \cdot 2 \div (-10) + 4(-5)$

**57.** $3^3 \div (-9) \cdot 4 + 5(-2)$

**58.** $4^2 \div (-8)(-2) + 6(-15 + 3 \cdot 5)$

**59.** $5^2 \div (-5)(-3) - 2(4 \cdot 3)$

**60.** $-23 + 16(-3) - 10 + 2(-5)^2$

**61.** $-27 \div (-3) - 14 - 4(-2)^3$

**62.** $-36 \div (-2)^2 + 15 - 2(16 - 17)$

**43.** 5
**44.** 2
**45.** 8
**46.** 3
**47.** 0
**48.** 0
**49.** Undefined
**50.** Undefined
**51.** – 70
**52.** 43
**53. a.** – 17
**b.** – 27
**54. a.** 26
**b.** – 52
**55.** – 14
**56.** – 24
**57.** – 22
**58.** 4
**59.** – 9
**60.** – 31
**61.** 27
**62.** 8

63. 31

64. 49

65. 45

66. 157

67. 1026

68. -253

69. − 9

70. 1300

71. 4624

72. 3250

73. − 105

74. − 47,628

75. [Respond below exercise.]

76. [Respond below exercise.]

**63.** $-25 \div (-5)^2 + 35 - 3(24 - 5^2)^2$

**64.** $7 \cdot 2^3 + 3^2 - 4^2 - 4(6 - 2 \cdot 3)$

**65.** $13 - 2[(19 - 14) \div 5 + 3(-4) - 5]$

**66.** $2 - 5[(-10) \div (-5) \cdot 2 - 35]$

**67.** $6 - 20[(-15) \cdot 2 + 4(-5) - 1]$

**68.** $8 - 9[(-39) \div (-13) + 14 \cdot 2 - 7 - (-5)]$

**69.** $(7 - 10)[49 \div (-7) + 20 \cdot 3 - 4 \cdot 15 - (-10)]$

**70.** $(9 - 11)[(-10)^2 \cdot 2 + 6(-5)^2 - 10^3]$

In Exercises 71 – 74, use a calculator to find the value of each expression.

**71.** $(-72)^2 - 35(16)$

**72.** $(15)^3 - 60 + 13(-5)$

**73.** $(15 - 20)(13 - 14)^2 + 50(-2)$

**74.** $(32 - 50)^2(15 - 22)^2(17 - 20)$

## Writing and Thinking About Mathematics

**75.** Explain the conditions under which the quotient of two numbers is 0.
$x \div y = 0$ if $x = 0$ and $y \neq 0$.

**76.** Explain, in your own words, why division by 0 is not a valid arithmetic operation.
Answers will vary. The quotient must be a unique number, and the product of the quotient and 0 must be the dividend. These conditions cannot be satisfied if 0 is the divisor.

# 2.5 Applications

## Change in Value

Subtraction with integers can be used to find the change in value between two readings of measures such as temperatures, distances, and altitudes.

To calculate the change between two integer values, including direction (negative for down, positive for up), use the following rule: **First find the end value, then subtract the beginning value.**

$$(\text{ Change in Value }) = (\text{ End Value }) - (\text{ Beginning Value })$$

### Objectives

① Understand how the change between two integer values can be found by subtracting the beginning value from the end value.

② Be able to find the average of several integers.

### Example 1

On a cold day at a ski resort, the temperature dropped from a high of 25°F at 1 p.m. to a low of –10°F at 2 a.m. What was the change in temperature?

**Solution**

| –10° | – | ( 25° ) | = | –35° |
|---|---|---|---|---|
| end temperature | – | beginning temperature | = | change |

The temperature dropped 35°, so the change was –35°F.

### Example 2

A rocket was fired from a silo 1000 feet below ground level. If the rocket attained a height of 15,000 feet, what was its change in altitude?

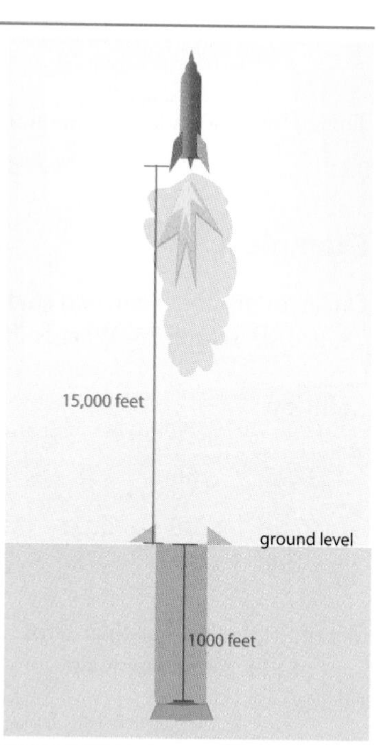


**Solution**

The end value was + 15,000 feet.
The beginning value was – 1000 feet, since the rocket was below ground level.

$$\begin{aligned} \text{Change} &= 15{,}000 - (-1000) \\ &= 15{,}000 + 1000 \\ &= 16{,}000 \text{ feet} \end{aligned}$$

The change in altitude was 16,000 feet.

## Average

A topic closely related to addition and division with integers is the **average** (also called the **arithmetic average** or **mean**) of a set of integers. Your grade in this course may be based on the average of your exam scores. Newspapers and magazines report average income, average life expectancy, averages sales, average attendance at sporting events, and so on. The average of a set of integers is one representation of the concept of the "middle" of the set. It may or may not actually be one of the numbers in the set.

**Definition:**

The **average** of a set of numbers is the value found by adding the numbers in the set, then dividing this sum by the number of numbers in the set.

In this section, we will use only sets of integers for which the average is an integer. However, in later chapters, we will remove this restriction and will find that an average can be calculated for types of numbers other than integers.

### Example 3

Find the average noonday temperature for the 5 days on which the temperature at noon was $10°$, $-2°$, $5°$, $-7°$, and $-11°$.

**Solution**

Find the sum of the temperatures and divide the sum by 5.

$$\begin{array}{r} 10 \\ -2 \\ 5 \\ -7 \\ \underline{-11} \\ -5 \end{array} \text{ sum} \qquad \frac{-5}{5} = -1° \text{ average}$$

The average noonday temperature was $-1°$ ( or $1°$ below 0 ).

### Example 4

On an art history exam, two students scored 95, five scored 86, one scored 78, and six scored 75. What is the mean score for the class on this exam?

**Solution**

$$\begin{array}{r} 95 \\ \underline{\times\ 2} \\ 190 \end{array} \quad \begin{array}{r} 86 \\ \underline{\times\ 5} \\ 430 \end{array} \quad \begin{array}{r} 78 \\ \underline{\times\ 1} \\ 78 \end{array} \quad \begin{array}{r} 75 \\ \underline{\times\ 6} \\ 450 \end{array}$$

We multiplied rather than wrote down all 14 scores. Thus, the sum of the four products represents the sum of 14 exam scores, and the sum is divided by 14.

$$\begin{array}{r} 190 \\ 430 \\ 78 \\ \underline{+\ 450} \\ 1148 \end{array} \qquad \begin{array}{r} 82 \\ 14\overline{)1148} \\ \underline{112} \\ 28 \\ \underline{28} \\ 0 \end{array} \text{ mean score}$$

The class mean is 82 points.

Name ______________________ Section ____________ Date __________

# Exercises 2.5

1. Lisa was a very nice woman with 14 grandchildren. She was born in 1888 and died on her birthday in 1960. How old was she when she died?

2. Your bank statement indicates that you are overdrawn on your checking account by \$253. How much must you deposit to bring the checking balance up to \$400? Your separate savings account balance is \$862.

3. A pilot flew a plane from an altitude of 20,000 feet to an altitude of 1500 feet. What was the change in altitude? The plane was flying over the Mojave desert at 500 miles per hour.

4. At noon the temperature at the top of a mountain was 30°F. By midnight the temperature was –6°F. What was the change in temperature from noon to midnight? Eight inches of snow fell in that same 12–hour period.

5. A test missile was fired from a submarine from 300 feet below sea level. If the missile reached a height of 16,000 feet before exploding, what was the change in altitude of the missile?

6. The members of the Math Anxiety Club sold cake at a bake sale and made a profit of \$250. If their expenses totaled \$80, what was the total amount of the sales? The chocolate cake sold out in 2 hours.

**ANSWERS**

1. 72 years old
2. \$653
3. –18,500 ft
4. –36°F
5. 16,300 ft
6. \$330

7. −105 points

8. down $8 per share

(−$8 per share)

9. 0

10. 0

11. − 6

12. − 10

13. $78 per share

14. 365 pounds

15. 85

16. 77

**7.** At the end of a 5–day week the DOW Jones stock index was at 10,495 points. If the DOW started the week at 10,600 points, what was the change in the index for that week?

**8.** On Monday, the stock of a computer company sold at a market price of $56 per share. By Friday, the stock had dropped to $48 per share. Find the change in the price of the stock. The president of the company earns a salary of $210,000 annually.

Find the average of each of the following sets of integers.

**9.** − 10, − 20, 14, 34, − 18

**10.** 56, − 64, − 38, 58, − 12

**11.** − 6, − 8, − 7, − 4, − 4, − 5, − 6, − 8

**12.** − 25, 30, − 15, − 6, − 26, − 18

**13.** Alicia bought shares of two companies on the stock market. She paid $9000 for 90 shares in one company and $6600 for 110 shares in another company. What was the average price per share for the 200 shares?

**14.** In a weight lifting program, two men bench pressed 300 pounds, three men bench pressed 350 pounds, and 5 men benched pressed 400 pounds. What was the average number of pounds that these men bench pressed?

**15.** On an English exam, two students scored 95, six students scored 90, three students scored 80, and one student scored 50. What was the mean score for the class?

**16.** In a speech class, the students graded each other on a particular assignment. (Generally students are harder on each other and themselves than the instructor is. Why, do you think, that this is the case?) On this speech, three students scored 60, three scored 70, fivescored 80, five scored 82, and four scored 85. What was the average score on this speech?

Name ____________________ Section ____________ Date ____________

**ANSWERS**

**17.** 8 degrees

**18.** –33

**19.** 92

**20.** –16

**21.a.** –33

**b.** 17

**22.** –3 lbs; 202 lbs

**17.** The temperature reading for 30 days at a ski resort were recorded as follows:

| | | | | | | | | | |
|---|---|---|---|---|---|---|---|---|---|
| 28 | 24 | 22 | 10 | – 2 | – 5 | – 2 | 12 | 10 | 15 |
| – 6 | 5 | 20 | 13 | – 2 | – 6 | – 15 | – 18 | – 10 | 8 |
| – 1 | 7 | 20 | 21 | 32 | 30 | 22 | 12 | 3 | – 7 |

What was the average temperature recorded for the 30 days?

**18.** If the product of – 12 and – 4 is added to the product of – 27 and 3, what is the sum?

**19.** If the quotient of – 72 and 9 is subtracted from the product of – 14 and – 6, what is the difference?

**20.** What number should be added to – 17 to get a sum of – 33?

**21.** **a.** What number minus – 25 gives a difference of – 8?

**b.** What number minus 25 gives a difference of – 8?

**22.** In 6 months of dieting, Bill lost 2 pounds, lost 5 pounds, lost 3 pounds, lost 10 pounds, gained 4 pounds, and lost 2 pounds. What was his average monthly change in weight? If he weighed 220 pounds before starting the diet, what was his weight at the end of the 6 months?

In Exercises 23–28, first mentally estimate the value of each expression by using rounded–off numbers. Then use a calculator to find the value of each expression. Indicate both your estimate and the calculated value. (Estimates may vary.)

23. 100; 80

24. –700; – 680

25. –16,250;– 18,152

26. –25,000; – 24,474

27. –1000; – 804

28. -50,000; – 61,417

**23.** $[\,270 + 257 + 300 - 507] \div 4$

**24.** $[(-650) + (-860) + (-530)] \div 3$

**25.** $[(-53{,}404) + (25{,}803) + (-10{,}005) + (-35{,}002)] \div 4$

**26.** $[(-62{,}496) + (-72{,}400) + (53{,}000) + (-16{,}000)] \div 4$

**27.** $[-21{,}540 - 10{,}200 + 36{,}300 - 25{,}000 + 12{,}400] \div 10$

**28.** $[\,317{,}164 - 632{,}000 - 427{,}000 + 250{,}500\,] \div 8$

# 2.6 Combining Like Terms and Evaluating Polynomials

## Combining Like Terms

A single number is called a **constant**. Any constant or variable or the indicated product of constants and powers of variables is called a **term** (see Section 1.6). Examples of terms are

$$17, \quad -2, \quad 3x, \quad 5xy, \quad -3x^2, \quad \text{and} \quad 14a^2b^3$$

A number written next to a variable (as in $9x$) or a variable written next to another variable (as in $xy$) indicates multiplication. In the term $4x^2$, the constant 4 is called the **numerical coefficient** of $x^2$ (or simply the **coefficient** of $x^2$).

**Like terms** (or **similar terms**) are terms that contain the same variables (if any) raised to the same powers. Whatever power a variable is raised to in one term, it is raised to the same power in other like terms. Constants are like terms.

**Objectives**

① Recognize like terms and terms that are not like terms.

② Know how to apply the distributive property when combining like terms.

③ Be able to evaluate a polynomial when an integer is substituted for the variable.

### Like Terms

| | |
|---|---|
| $-6$, 12, 132 | are like terms because each term is a constant. |
| $-2a$, $15a$, and $3a$ | are like terms because each term contains the same variable $a$, raised to the same power, 1. |
| $5xy^2$ and $3xy^2$ | are like terms because each term contains the same two variables, $x$ and $y$, with $x$ first-degree in both terms and $y$ second-degree in both terms. |

### Unlike Terms

| | |
|---|---|
| $7x$ and $8x^2$ | are unlike terms (**not** like terms) because the variable $x$ is not of the same power in both terms. |
| $6ab^2$ and $2a^2b$ | are **not** like terms because the variables are not of the same power in both terms. |

If no number is written next to a variable, then the coefficient is understood to be 1. For example,

$$x = 1 \cdot x, \quad a^3 = 1 \cdot a^3, \quad \text{and} \quad xy = 1 \cdot xy$$

If a negative sign ( $-$ ) is written next to a variable, then the coefficient is understood to be $-1$. For example,

$$-y = -1 \cdot y, \quad -x^2 = -1 \cdot x^2, \quad \text{and} \quad -xy = -1 \cdot xy$$

### Example 1

From the following list of terms, pick out the like terms:

$$6, \quad 2x, \quad -10, \quad -x, \quad 3x^2z, \quad -5x, \quad 4x^2z, \quad \text{and} \quad 0$$

**Solution**

**a.** 6, $-10$, and 0 are like terms. All are constants.
**b.** $2x$, $-x$, and $-5x$ are like terms.
**c.** $3x^2z$ and $4x^2z$ are like terms.

As we have discussed in Section 1.6, expressions such as

$$9+x, \quad 3y^3-4y, \quad \text{and} \quad 2x^2+8x-10$$

are not terms. These algebraic sums of terms are called **polynomials**, and our objective here is to learn to simplify polynomials that contain like terms. This process is called **combining like terms**. For example, by combining like terms, we can write

$$8x+3x=11x \quad \text{and} \quad 4n-n+2n=5n.$$

Beginning students are sometimes tempted to combine terms such as $1+18x$ as $19x$, but they will not combine the same terms written as $18x+1$. You might discuss the expression $1+18x$ from the point of view of order of operations in that the product $18x$ would be considered first, then 1 would be added last regardless of which form is written.

To understand how to combine like terms, we need the **distributive property** as it applies to integers (see Section 1.1). (**Note:** The distributive property is a property of addition and multiplication and not of any particular type of number. As we will see later, the distributive property applies to all types of numbers, including integers, fractions, and decimals.)

**Distributive Property of Multiplication over Addition**

For integers $a$, $b$, and $c$, $\quad \boldsymbol{a(b+c)=ab+ac}$.

Some examples of multiplication, using the distributive property, are:

$$\begin{aligned}
4(x+3) &= 4\cdot x+4\cdot 3 &&= 4x+12\\
9(a-2) &= 9\cdot a+9(-2) &&= 9a-18\\
-8(x+6) &= -8\cdot x+(-8)\cdot 6 &&= -8x-48\\
-8(y-1) &= -8\cdot y+(-8)\cdot(-1) &&= -8y+8
\end{aligned}$$

Now, consider the following analysis that leads to the method for combining like terms:

By using the commutative property of multiplication, we can write

$$\boldsymbol{a(b+c)=(b+c)a}$$

and $\quad \boldsymbol{(b+c)a=ba+ca}$

or $\quad \boldsymbol{ba+ca=(b+c)a}$

and $\quad 8x+3x=(8+3)x=11x$

Thus, by using the distributive property in a "reverse" sense, we have combined the like terms $8x$ and $3x$.

Similarly, we can write,

$$4n-n+2n=(4-1+2)n=5n$$

(Note that -1 is the coefficient of $-n$)

and $\quad 17y-14y=(17-14)y=3y$

### Example 2

Simplify each of the following polynomials by combining like terms whenever possible.

**a.** $8x + 10x + 1$ **b.** $5x^2 - x^2 + 6x^2 + 2x - 7x$ **c.** $2(x+5) + 3(x-1)$
**d.** $-3(x+2) - 4(x-2)$ **e.** $9y^3 + 4y^2 - 3y - 1$

**Solution**

**a.** $8x + 10x + 1 = (8 + 10)x + 1$ Use the distributive property with $8x + 10x$.
$= 18x + 1$ Note that the constant 1 is not combined with $18x$ because they are not like terms.

**b.** $5x^2 - x^2 + 6x^2 + 2x - 7x = (5 - 1 + 6)x^2 + (2 - 7)x$
$= 10x^2 - 5x$

**c.** $2(x+5) + 3(x-1) = 2x + 10 + 3x - 3$ Use the distributive property.
$= (2 + 3)x + 10 - 3$ Use the commutative property of addition.
$= 5x + 7$

**d.** $-3(x+2) - 4(x-2) = -3x - 6 - 4x + 8$ Use the distributive property.
$= (-3 - 4)x - 6 + 8$
$= -7x + 2$

**e.** $9y^3 + 4y^2 - 3y - 1$ This expression is already simplified, since it has no like terms to combine.

After some practice, you should be able to combine like terms by adding the corresponding coefficients mentally. The use of the distributive property is necessary in beginning development and serves to lay a foundation for simplifying more complicated expressions in later courses.

*Now Work Exercises 1-4 in the Margin.*

Simplify each polynomial by combining like terms whenever possible.

**1.** $7x + 9x$
$16x$

**2.** $6x - 7a - a$
$6x - 8a$

**3.** $3x^2 + x^2 - 5b + 2b$
$4x^2 - 3b$

**4.** $3x^2 - 5x^2 + x^2$
$-x^2$

## Evaluating Polynomials and Other Expressions

In Section 1.6, we evaluated polynomials for given whole number values of the variables. Now, since we know how to operate with integers, we can evaluate polynomials and other algebraic expressions for given integer values of the variables. Also, we can make the process of evaluation easier by first combining like terms, whenever possible.

**To Evaluate a Polynomial or Other Algebraic Expression:**

1. Combine like terms, if possible.
2. Substitute the values given for any variables.
3. Follow the rules for order of operations.

(**Note:** Terms separated by + and – signs may be evaluated at the same time. Then the value of the expression can be found by adding and subtracting from left to right.)

## Example 3

First simplify the polynomial $x^3 -5x^2 + 2x^2 + 3x - 4x - 15 + 3$ then evaluate it for $x = -2$.

### Solution

First, simplify the polynomial by combining like terms.

$$\begin{aligned}&x^3 - 5x^2 + 2x^2 + 3x - 4x - 15 + 3\\&= x^3 + (-5 + 2)x^2 + (3 - 4)x - 15 + 3\\&= x^3 - 3x^2 - x - 12\end{aligned}$$

Now, substitute - 2 for $x$ (**using parentheses around -2 to be sure the signs are correct**), and evaluate this simplified polynomial by following the rules for order of operations. (Note that terms separated by + and - signs may be evaluated at the same time.)

$$\begin{aligned}x^3 - 3x^2 - x - 12 &= (-2)^3 - 3(-2)^2 - (-2) - 12\\&= -8 - 3(4) - (-2) - 12\\&= -8 - 12 + 2 - 12\\&= -30\end{aligned}$$

## Example 4

Evaluate the expression $5ab + ab + 8a - a$ for $a = 2$ and $b = -3$.

### Solution

Combining like terms gives

$$\begin{aligned}5ab + ab + 8a - a &= (5 + 1)ab + (8 - 1)a\\&= 6ab + 7a.\end{aligned}$$

Substituting 2 for $a$ and $-3$ for $b$ in this simplified expression and following the rules for order of operations gives

$$6ab + 7a = 6(2)(-3) + 7(2) = -36 + 14 = -22.$$

(Note that terms separated by + and - signs may be evaluated at the same time.)

### *Completion Example 5*

First combine like terms then evaluate the polynomial $5x^2 - 2x^2 + x + 3x^2 - 2x$ for $x = -1$.

**Solution**

Combining like terms:

$$\begin{aligned} 5x^2 - 2x^2 + x + 3x^2 - 2x &= (\_\_\_\_\_)x^2 + (\_\_\_\_\_)x \\ &= \_\_\_\_\_\ x^2 - \_\_\_\_\_\ x \\ &= \_\_\_\_\_\ x^2 - x \end{aligned}$$

Substituting $x = -1$ and evaluating:

$$\begin{aligned} 6x^2 - x &= 6(\_\_\_\_\_)^2 - 1(\_\_\_\_\_) \\ &= 6(\_\_\_\_\_) + \_\_\_\_\_ \\ &= \_\_\_\_\_ + \_\_\_\_\_ \\ &= \_\_\_\_\_ \end{aligned}$$

*Now Work Exercises 5-7 in the Margin.*

## Placing Parentheses Around Negative Numbers

Consider evaluating the expressions $4x^2 + x^2$ and $7x^2 - 6x^2$ for $x = -2$. In both cases, we combine like terms first:

$$4x^2 + x^2 = 5x^2$$

and $$7x^2 - 6x^2 = -x^2$$ Note that for $-6x^2$ the coefficient is $-6$.

Now, evaluate for $x = -2$ by writing $-2$ in parentheses as follows:

$$5x^2 = 5(-2)^2 = 5(4) = 20$$ Note that for $5x^2$ the coefficient is 5.

and $$-x^2 = -1(-2)^2 = -1(4) = -4$$ Note that for $-x^2$ the coefficient is $-1$.

These examples illustrate the importance of using parentheses around negative numbers so that fewer errors will be made with signs. Equally important is the fact that in expressions, such as $-x$ and $-x^2$, the coefficient in each case is understood to be $-1$. That is,

$$-x = -1 \cdot x \quad \text{and} \quad -x^2 = -1 \cdot x^2$$

Thus, as just illustrated, we can evaluate both expressions for $x = -3$ by using parentheses:

$$-x = -1 \cdot x = -1 \cdot (-3) = +3$$

and $$-x^2 = -1 \cdot x^2 = -1 \cdot (-3)^2 = -1 \cdot 9 = -9$$

First simplify each algebraic expression, and then evaluate the expression for $x = -2$ and $y = 3$.

**5.** $3xy - x^2 + 5xy$

$8xy - x^2$; $-52$

**6.** $7x + 5x - 10x + 4x$

$6x$; $-12$

**7.** $4x - 3y + 2x + 4y$

$6x + y$; $-9$

The students will have difficulty understanding that $-x^2$ means $-1 \cdot x^2$ and that the exponent takes precedence over multiplication by $-1$. You might give them several more examples such as:

$$-2x^2,\ -3x^2, \text{ and } -5x^3$$

and have them evaluate the expression for $x = -1$ and $x = -2$. These examples will help reinforce the rules for order of operations.

Evaluate each of the following expressions.

**8. a.** $-2^4$
$-16$

**b.** $(-2)^4$
$16$

**c.** $-2^3$
$-8$

**d.** $(-2)^3$
$-8$

**9. a.** $(-7)^2$
$49$

**b.** $-7^2$
$-49$

**c.** $-7^3$
$-343$

**d.** $(-7)^3$
$-343$

**10. a.** $-1^6$
$-1$

**b.** $(-1)^6$
$1$

**c.** $(-1)^5$
$-1$

**d.** $-1^5$
$-1$

Notice that in the second case, $-3$ is squared first, then multiplied by the coefficient $-1$. The square is on $x$, not $-x$. In effect, the exponent 2 takes precedence over the coefficient $-1$. This is consistent with the rules for order of operations. However, there is usually some confusion when numbers are involved in place of variables. Study the differences in the following evaluations:

$-5^2 = -1 \cdot 5^2 = -1 \cdot 25 = -25$ and, with parentheses, $(-5)^2 = (-5)(-5) = +25$

$-6^2 = -1 \cdot 6^2 = -1 \cdot 36 = -36$ and, with parentheses, $(-6)^2 = (-6)(-6) = +36$

With parentheses around $-5$ and $-6$, the results are entirely different.

These examples are shown to emphasize the importance of using parentheses around negative numbers when evaluating expressions. Without the parentheses, an evaluation can be dramatically changed and lead to wrong answers. We state the following results for the exponent 2. (The results are the same for all positive even exponents.)

In general, except for $x = 0$,

1. **$-x^2$ is negative** $\quad (-5^2 = -25)$
2. **$(-x)^2$ is positive** $\quad ((-5)^2 = 25)$
3. **$-x^2 \neq (-x)^2$**

*Now Work Exercises 8 – 10 in the Margin.*

---

## Completion Example Answer

5. $5x^2 - 2x^2 + x + 3x^2 - 2x = (\mathbf{5 - 2 + 3})x^2 + (\mathbf{1 - 2})x$
$= \mathbf{6}x^2 - \mathbf{1}x$
$= \mathbf{6}x^2 - x$

$6x^2 - x = 6(\mathbf{-1})^2 - 1(\mathbf{-1})$
$= 6(\mathbf{+1}) + \mathbf{1}$
$= \mathbf{6 + 1}$
$= \mathbf{7}$

---

Name ____________ Section ________ Date ________

## Exercises 2.6

Use the distributive property to multiply each of the following expressions.

**1.** $6(x+3)$ **2.** $4(x-2)$ **3.** $-7(x+5)$

**4.** $-9(y-4)$ **5.** $-11(n+3)$ **6.** $2(n-6)$

Simplify each of the following expressions by combining like terms whenever possible.

**7.** $2x+3x$ **8.** $9x-4x$ **9.** $y+y$

**10.** $-x-x$ **11.** $-n-n$ **12.** $9n-9n$

**13.** $5x-7x+2x$ **14.** $-40a+20a+20a$ **15.** $3y-5y-y$

**16.** $12x+6-x-2$ **17.** $8x-14+x-10$

**18.** $6y+3-y+8$ **19.** $13x^2+2x^2+2x+3x-1$

**20.** $x^2-3x^2+5x-3+1$ **21.** $3x^2-4x+2+x^2-5$

**22.** $-4a+5a^2+4a+5a^2+6$ **23.** $3(x-1)+2(x+1)$

**ANSWERS**

1. $6x+18$
2. $4x-8$
3. $-7x-35$
4. $-9y+36$
5. $-11n-33$
6. $2n-12$
7. $5x$
8. $5x$
9. $2y$
10. $-2x$
11. $-2n$
12. $0$
13. $0$
14. $0$
15. $-3y$
16. $11x+4$
17. $9x-24$
18. $5y+11$
19. $15x^2+5x-1$
20. $-2x^2+5x-2$
21. $4x^2-4x-3$
22. $10a^2+6$
23. $5x-1$

24. $8y + 2$

25. $4x - 28$

26. $3ab + 7a$

27. $-xy + 11x - y$

28. $5x^2y - 3xy^2 + 2$

29. $2x^2y + 2$

30. $15ab^2 + 3ab$

$8a^2b + 7ab^2 + 3ab$

32. $7x^2 + 8x - 9$

33. $5x - a - 5$

34. $10xy^2 - 6xy + 2x$

35. $-2xyz - 6$

36. $-x^2 + 5x - 7; -21$

37. $4x^2 - 4x + 5; 29$

38. $3y^2 - y; 4$

39. $2y^2 - 6y - 5; 3$

40. $x^3 - 2x^2 - 2x; -12$

41. $-2x^3 + x^2 - x; 22$

**24.** $3(y + 4) + 5(y - 2)$

**25.** $-2(x - 1) + 6(x - 5)$

**26.** $5ab + 7a - 4ab + 2ab$

**27.** $2xy + 4x - 3xy + 7x - y$

**28.** $5x^2y - 3xy^2 + 2$

**29.** $5x^2y - 3x^2y + 2$

**30.** $7ab^2 + 8ab^2 - 2ab + 5ab$

**31.** $7ab^2 + 8a^2b - 2ab + 5ab$

**32.** $2(x^2 - x + 3) + 5(x^2 + 2x - 3)$

**33.** $2(x + a - 1) + 3(x - a - 1)$

**34.** $8xy^2 + 2xy^2 - 7xy + xy + 2x$

**35.** $3xyz + 2xyz - 7xyz + 14 - 20$

First simplify each polynomial, and then evaluate the polynomial for $x = -2$ and $y = -1$.

**36.** $2x^2 - 3x^2 + 5x - 8 + 1$

**37.** $5x^2 - 4x + 2 - x^2 + 3$

**38.** $y^2 + 2y^2 + 2y - 3y$

**39.** $y^2 + y^2 - 8y + 2y - 5$

**40.** $x^3 - 2x^2 + x - 3x$

**41.** $x^3 - 3x^3 + x + x^2 - 2x$

Name ______________________ Section ____________ Date __________

**ANSWERS**

**42.** $y^3 + 3y^3 + 5y - 4y^2 + 1$

**43.** $7y^3 + 4y^2 + 6 + y^2 - 12$

**44.** $2(x^2 - 3x - 5) + 3(x^2 + 5x - 4)$

**45.** $5(y^2 - 4y + 3) - 2(y^2 - 2y + 10)$

Simplify each of the following algebraic expressions, and then evaluate the expression for $a = -1$, $b = -2$, and $c = 3$.

**46.** $a^2 - a + a^2 - a$

**47.** $a^3 - 2a^3 - 3a + a - 7$

**48.** $5ab - 7a + 4ab + 2b$

**49.** $2ab + 4b - 3a + ab - b$

**50.** $2(a + b - 1) + 3(a - b + 2)$

**51.** $5(a - b + 1) + 4(a + b - 5)$

**52.** $4ac + 16ac - 4c + 2c$

**53.** $4abc + 2abc - 6ab + ab - bc + 3bc$

**54.** $-9abc + 3abc - abc + 2ab - bc + 3bc + 15$

**Answers:**

**42.** $4y^3 - 4y^2 + 5y + 1;\ -12$

**43.** $7y^3 + 5y^2 - 6;\ -8$

**44.** $5x^2 + 9x - 22;\ -20$

**45.** $3y^2 - 16y - 5;\ 14$

**46.** $2a^2 - 2a;\ 4$

**47.** $-a^3 - 2a - 7;\ -4$

**48.** $9ab - 7a + 2b;\ 21$

**49.** $3ab - 3a + 3b;\ 3$

**50.** $5a - b + 4;\ 1$

**51.** $9a - b - 15;\ -22$

**52.** $20ac - 2c;\ -66$

**53.** $6abc - 5ab + 2bc;\ 14$

**54.** $-7abc + 2ab + 2bc + 15;\ -35$

55. $a^2b^2c^2 + 4ab^2 + bc^2$; 2

56. $14a - 15b + 2c + 2$; 24

57. $12a + 8b - 3c - 64$; $-101$

58. $10a + 10b + 10c$; 0

59. 0; 0

60. $14a + 28b - 42c$; $-196$

61.a. $-16$
b. 16
c. $-64$
d. $-64$

62.a. $-125$
b. $-125$
c. $-625$
d. 625

63.a. $-216$
b. $-1$
c. $-343$
d. $-32$

64.a. $-4$
b. $-9$
c. 4
d. 9

65.a. $-2$
b. 3
c. $-10$
d. 21

**55.** $16bc^2 - 15bc^2 + 3ab^2 + ab^2 + a^2b^2c^2$

**56.** $14(a + 7) - 15(b + 6) + 2(c - 3)$

**57.** $12(a - 3) + 8(b - 2) - 3(c + 4)$

**58.** $20(a + b + c) - 10(a + b + c)$

**59.** $16(a - b + c) + 16(-a + b - c)$

**60.** $7(a + 2b - 3c) - 7(-a - 2b + 3c)$

**61.** Find the value of each of the following expressions.
**a.** $-4^2$ **b.** $(-4)^2$ **c.** $-4^3$ **d.** $(-4)^3$

**62.** Find the value of each of the following expressions.
**a.** $-5^3$ **b.** $(-5)^3$ **c.** $-5^4$ **d.** $(-5)^4$

**63.** Find the value of each of the following expressions.
**a.** $(-6)^3$ **b.** $(-1)^7$ **c.** $-7^3$ **d.** $-2^5$

**64.** Find the value of each of the following expressions for $x = 2$ and $y = -3$.
**a.** $-x^2$ **b.** $-y^2$ **c.** $(-x)^2$ **d.** $(-y)^2$

**65.** Find the value of each of the following expressions for $x = 2$ and $y = -3$.
**a.** $-x$ **b.** $-y$ **c.** $-5x$ **d.** $-7y$

# 2.7 Solving Equations with Integers ($x + b = c$, $ax = c$)

## Negative Constants, Negative Coefficients, and Negative Solutions

In Section 1.4, we discussed solving equations of the forms

$$x + b = c, \quad x - b = c, \quad \text{and} \quad ax = c$$

where $x$ is a variable and $a$, $b$, and $c$ represent constants. Each of these equations is called a **first–degree equation in $x$** because the exponent of the variable is 1 in each case. The equations in Section 1.4 were set up so that the coefficients, constants, and solutions to the equations were whole numbers. Now that we have discussed integers and operations with integers, we can discuss equations that have negative constants, negative coefficients and negative solutions.

For convenience and easy reference, the following definitions are repeated from Section 1.4.

**Objectives**

① Be able to solve equations with integer coefficients and integer solutions.

**Definitions**

An **equation** is a statement that two expressions are equal.
A **solution** to an equation is a number that gives a true statement when substituted for the variable.
A **solution set** of an equation is the set of all solutions of the equation.

(**Note:** In this text, with a few exceptions, each equation will have only one number in its solution set. However, in future studies in mathematics, you will deal with equations that have more than one

### Example 1

**a.** Show that $-5$ **is** a solution to the equation $x - 6 = -11$.

$$x - 6 = -11$$
$$-5 - 6 \stackrel{?}{=} -11$$
$$-11 = -11$$

Therefore, $-5$ **is** a solution.

**b.** Show that 6 **is not** a solution to the equation $x - 6 = -11$.

$$x - 6 = -11$$
$$6 - 6 \stackrel{?}{=} -11$$
$$0 \neq -11$$

Therefore, 6 **is not** a solution.

## Restating the Basic Principles for Solving Equations

Another adjustment that can be made, since we now have an understanding of integers, is that the Addition and Subtraction Principles discussed in Section 1.4 can be stated as one principle, the Addition Principle. Because subtraction can be treated as the algebraic sum of numbers and expressions, we know that

$$A - C = A + (-C)$$

That is, subtraction of a number can be thought of as addition of the opposite of that number.

The following two principles can be used to find the solution to first–degree equations of the forms

$$x + b = c, \qquad x - b = c, \qquad \text{and} \qquad ax = c$$

In the two basic principles stated here, $A$, $B$, and $C$ represent numbers (constants) or expressions that involve variables.

1. The **Addition Principle:** The equations $A = B$ and $A + C = B + C$ have the same solutions.

2. The **Division Principle:** The equations $A = B$ and $\frac{A}{C} = \frac{B}{C}$ (where $C \neq 0$) have the same solutions.

Essentially, these two principles state that if we perform the same operation to both sides of an equation, the resulting equation will have the same solution as the original equation. Equations with the same solution sets are said to be **equivalent**.

Note that in the principles just stated, the letters $A$, $B$, and $C$ can represent negative numbers, as well as, positive numbers.

*Remember:* **The objective in solving an equation is to isolate the variable with coefficient 1 on one side of the equation, left side or right side.**

### Example 2

Solve the equation **a.** $x + 5 = -14$ and **b.** $11 = x - 6$.

**a.**

| | |
|---|---|
| $x + 5 = -14$ | Write the equation. |
| $x + 5 - 5 = -14 - 5$ | Using the addition principle, add -5 to both sides. |
| $x + 0 = -19$ | Simplify both sides. |
| $x = -19$ | The solution. |

**b.**

| | |
|---|---|
| $11 = x - 6$ | Write the equation. |
| $11 + 6 = x - 6 + 6$ | Using the addition principle, add 6 to both sides. |
| $17 = x + 0$ | Simplify both sides. |
| $17 = x$ | The solution. |

Note that in both equations in Example 2, a constant was added to both sides of the equation. In part **a.**, a negative number was added, and in part **b.**, a positive number was added.

As in section 1.4, we will need the concept of division in fraction form when solving equations and will assume that you are familiar with the fact that a number divided by itself is 1. This fact includes negative numbers. For example,

$$(-3)\div(-3)=\frac{(-3)}{(-3)}=1 \text{ and } (-8)\div(-8)=\frac{(-8)}{(-8)}=1.$$

In this manner, we will write expressions such as the following:

$$\frac{-5x}{-5}=\frac{-5}{-5}x=1\cdot x$$

In Example 3, the Division Principle is illustrated with positive and negative constants and positive and negative coefficients.

### Example 3

Solve the equations **a.** $4n=-24$ and **b.** $42=-21n$.

#### Solution

**a.** $4n=-24$ Write the equation.

$\frac{4n}{4}=\frac{-24}{4}$ Using the division principle, divide both sides by the coefficent 4. Note that in solving equations, the fraction form of division is used.

$1\cdot n=-6$ Simplify by performing the division on both sides.

$n=-6$ The solution.

**b.** $42=-21n$ Write the equation.

$\frac{42}{-21}=\frac{-21n}{-21}$ Using the division principle, divide both sides by the coefficient -21.

$-2=1\cdot n$ Simplify by performing the division on both sides.

$-2=n$ The solution.

Solve each of the following equations.

1. $x + 6 = -20$
$x = -26$

2. $6n = -48$
$n = -8$

3. $14 = x - 5$
$x = 19$

4. $n + 14 - 10 = 7 - 2$
$n = 1$

In Example 4, we illustrate solving equations in which we combine like terms before applying the principles. (**Note:** In Chapters 3, 4, and 5, we discuss solving equations involving fractions and decimals and the use of more steps in finding the solutions. In every case, the basic principles and techniques are the same as those used here.)

### Example 4

Solve the equations **a.** $y - 10 + 11 = 7 - 8$ and **b.** $-15 = 4y - y$.

#### Solution

**a.**

| | |
|---|---|
| $y - 10 + 11 = 7 - 8$ | Write the equation. |
| $y + 1 = -1$ | Combine like terms. |
| $y + 1 - 1 = -1 - 1$ | Using the addition principle, add -1 to both sides. |
| $y + 0 = -2$ | Simplify both sides. |
| $y = -2$ | The solution |

**b.**

| | |
|---|---|
| $-15 = 4y - y$ | Write the equation. |
| $-15 = 3y$ | Combine like terms. |
| $\frac{-15}{3} = \frac{3y}{3}$ | Using the division principle, divide both sides by 3. |
| $-5 = 1 \cdot y$ | Simplify both sides. |
| $-5 = y$ | The solution |

*Now Work Exercises 1–4 in the Margin.*

Name ______________________ Section ______________ Date __________

ANSWERS

1. $-12$

2. $-8$

3. $-25$

4. $-1$

5. $-15$

6. $-9$

7. $-3$

8. $-4$

## Exercises 2.7

In each exercise, an equation and a replacement set of numbers for the variable are given. Substitute each number into the equation until you find the number that is the solution of the equation.

**1.** $x + 4 = -8$ $\{-10, -12, -14, -16, -18\}$

**2.** $x + 3 = -5$ $\{-4, -5, -6, -7, -8, -9\}$

**3.** $-32 = y - 7$ $\{-20, -21, -22, -23, -24, -25\}$

**4.** $-13 = y - 12$ $\{0, -1, -2, -3, -4, -5\}$

**5.** $7x = -105$ $\{0, -5, -10, -15, -20\}$

**6.** $-72 = 8n$ $\{-6, -7, -8, -9, -10\}$

**7.** $-13n = 39$ $\{0, -1, -2, -3, -4, -5\}$

**8.** $-14x = 56$ $\{0, -1, -2, -3, -4, -5\}$

9. $x = -14$

10. $x = -10$

11. $x = -37$

12. $x = -74$

13. $y = -18$

14. $y = -36$

15. $y = -9$

16. $y = 18$

17. $n = 0$

18. $n = 0$

19. $n = -105$

20. $n = -120$

21. $x = 4$

22. $x = 8$

23. $y = -5$

24. $y = -2$

25. $n = -21$

26. $n = -8$

27. $x = -21$

28. $x = -9$

Use either the Addition Principle or the Division Principle to solve the following equations. Show each step, and keep the equal signs aligned vertically as in the format used in the text. Combine like terms whenever necessary.

**9.** $x - 2 = -16$

**10.** $x - 35 = -45$

**11.** $-27 = x + 10$

**12.** $-42 = x + 32$

**13.** $y + 9 = -9$

**14.** $y + 18 = -18$

**15.** $y + 9 = 0$

**16.** $y - 18 = 0$

**17.** $12 = n + 12$

**18.** $15 = n + 15$

**19.** $-75 = n + 30$

**20.** $-100 = n + 20$

**21.** $-8x = -32$

**22.** $-9x = -72$

**23.** $-13y = 65$

**24.** $-15y = 30$

**25.** $-84 = 4n$

**26.** $-88 = 11n$

**27.** $x + 1 = -20$

**28.** $-12 = x - 3$

Name ______ Section ______ Date ______

**29.** $-83 = y + 13$

**30.** $3x - 2x + 6 = -20$

**31.** $4x - 3x + 3 = 16 + 30$

**32.** $7x + 3x = 30 + 10$

**33.** $6x + 2x = 14 + 26$

**34.** $3x - x = -27 - 19$

**35.** $2x + 3x = -62 - 28$

**36.** $-21 + 30 = -3y$

**37.** $-16 - 32 = -6y$

**38.** $-35 - 15 - 10 = 4x - 14x$

**39.** $-17 + 20 - 3 = 5x$

**40.** $5x + 2x = -32 + 12 + 20$

**41.** $x - 4x = -25 + 3 + 22$

**42.** $3y + 10y - y = -13 + 11 + 2$

**43.** $5y + y - 2y = 14 + 6 - 20$

**44.** $6y + y - 3y = -20 - 4$

**ANSWERS**

**29.** $y = -96$

**30.** $x = -26$

**31.** $x = 43$

**32.** $x = 4$

**33.** $x = 5$

**34.** $x = -23$

**35.** $x = -18$

**36.** $y = -3$

**37.** $y = 8$

**38.** $x = 6$

**39.** $x = 0$

**40.** $x = 0$

**41.** $x = 0$

**42.** $y = 0$

**43.** $y = 0$

**44.** $y = -6$

45. $x = 6$

46. $x = -27$

47. $n = -42$

48. $n = -4$

49. $n = -11$

50. [ Respond below exercise. ]

**45.** $4x - 10x = -35 + 1 - 2$

**46.** $6x - 5x + 3 = -21 - 3$

**47.** $10n - 3n - 6n + 1 = -41$

**48.** $-5n - 3n + 2 = 36 - 2$

**49.** $-2n - 6n - n - 10 = -11 + 100$

## Writing and Thinking about Mathematics

**50.** Explain, in your own words, why combining the Addition Principle and the Subtraction Principle as stated in Section 1.4 into one principle (the Addition Principle as stated in the section) is mathematically correct.
Answers will vary. Subtraction is defined in terms of addition.

# Chapter 2 Index of Key Ideas and Terms

Number Lines — page 94

On a horizontal number line, smaller numbers are always to the left of larger numbers.
Dots are used to graph integers on a number line.

Integers — page 96

Integers are the whole numbers and their additive inverses (or opposites).
The number 0 is its own opposite: $0 = +0 = -0$.

Inequality Symbols — page 97

$<$ read as "less than" from left to right
$>$ read as "greater than" from left to right
$\leq$ read as "less than or equal to" from left to right
$\geq$ read as "greater than or equal to" from left to right

Absolute Value: $|x|$

$$\begin{cases} \text{If } x \text{ is a positive integer or 0, } |x| = x. \\ \text{If } x \text{ is a negative integer, } |x| = -x. \end{cases}$$

Remember: $-x$ represents the opposite of $x$ and may be positive, negative, or 0.

Rules for Addition with Integers — page 109

To add two integers with like signs, add their absolute values and use the common sign.
To add two integers with unlike signs, subtract their absolute values (the smaller from the larger) and use the sign of the integer with the larger absolute value.

Subtraction with Integers — page 117

For any integers a and b, the difference between a and b is defined as follows:
$\boldsymbol{a - b = a + (-b)}$
That is, to subtract an integer, add its additive inverse (or opposite).

Multiplication with Integers — page 130

For positive integers a and b:
$\boldsymbol{a \cdot b = ab}$ and $\boldsymbol{(-a)(-b) = ab}$

The product of integers with like signs is positive.
For positive integers a and b:
$\boldsymbol{a(-b) = -ab}$ and $\boldsymbol{(-a)(b) = -ab}$

The product of integers with unlike signs is negative.
The product of 0 with any integer is 0.
$\boldsymbol{a \cdot 0 = 0}$ and $\boldsymbol{(-a) \cdot 0 = 0}$

Division with Integers page 131

For positive integers a and b:

$$\frac{a}{b} = +\frac{a}{b} \quad \text{and} \quad \frac{-a}{-b} = +\frac{a}{b}$$

The quotient of integers with like signs is positive.

For positive integers a and b:

$$\frac{-a}{b} = -\frac{a}{b} \quad \text{and} \quad \frac{a}{-b} = -\frac{a}{b}$$

The quotient of integers with unlike signs is negative.

Division by 0 Is Not Defined page 132

$\frac{a}{0}$ is undefined.

Rules for Order of Operations page 133

1. First, simplify within grouping symbols, such as parentheses ( ), brackets [ ], or braces { }. Start with the innermost grouping.
2. Second, find any powers indicated by exponents.
3. Third, moving from left to right, perform any multiplications or divisions in the order in which they appear.
4. Fourth, moving from left to right, perform any additions or subtractions in the order in which they appear.

Mnemonic memory device: **PEMDAS**

To Find the Change Between Two Integer Values: page 141

**First find the end value, and then subtract the beginning value.**

(Change in Value) = (End Value) – (Beginning Value)

Average (or Mean) of a Set of Integers page 142

The **average** of a set of numbers is the value found by adding the numbers in the set and then dividing this sum by the number of numbers in the set.

Like Terms and Unlike Terms page 147

**Like terms** (or **similar terms**) are terms that contain the same variables (if any) raised to the same powers. Constants are like terms.

Combining Like Terms page 147

Like terms are combined by applying the distributive property and adding the coefficients.

Evaluating Polynomials (or Other Algebraic Expressions) page 149

1. Combine like terms.
2. Substitute the values given for any variables.
3. Follow the rules for order of operations.

Remember that for nonzero $x$, $-x^2 \neq (-x)^2$.

Solving Equations with Integers page 157

To Solve Equations apply the following two principles:

1. The **Addition Principle**:

The equations $A = B$

and $A + C = B + C$

have the same solutions.

2. The **Division Principle**:

The equations $A = B$

and $\frac{A}{C} = \frac{B}{C}$ (where $C \neq 0$)

have the same solutions.

Name ______________________ Section ____________ Date ____________

## Chapter 2 Test

1. Graph the following set of integers on a number line: { – 2, – 1, 0, 3, 4 }

2. What are the possible integer values of $n$ if $|n| < 3$?

3. Graph the set of whole numbers less than 4.

0 1 2 3 4

Find each of the following algebraic sums.

4. $-15 + (-2) + (14)$

5. $-120 - 75 + 30$

6. $45 + 25 - 70 - 40$

7. $-35 + 20 - 17 + 30$

Find each of the following products or quotients.

8. $(-3)(-5)(-4)(7)$

9. $(-8)(-3)(-7)(-56)(0)$

10. $(5)(-2)(-4)(12)$

11. $\dfrac{-32}{4}$

12. $\dfrac{0}{1}$

13. $\dfrac{-48}{-12}$

14. $\dfrac{-100-5}{7}$

Find the average of each of the following sets of numbers.

15. 60, – 120, – 35, – 42, – 38

16. 56, 92, 84, – 60

17. A jet pilot flew her plane, with a wingspan of 70 feet, from an altitude of 32,000 feet to an altitude of 5000 feet. What was the change in altitude?

18. If the quotient of – 51 and 17 is subtracted from the product of – 21 and + 3, what is the difference?

**ANSWERS**

1. [Respond below exercise]
2. – 2, – 1, 0, 1, 2
3. [Respond below exercise]
4. – 3
5. – 165
6. – 40
7. – 2
8. – 420
9. 0
10. 480
11. – 8
12. 0
13. 4
14. – 15
15. – 35
16. 43
17. – 27,000 ft
18. – 60

19. 5 yd gain

20. 64 mph

21. 0

22. 76

23. $5x$

24. $-2x^2 + 8x + 15$

25. $-3y^3 + 3y^2 - 4y$

26. $5y^2 - 6$

27. $-51$

28. 96

29. $x = -2$

30. $y = 3$

31. $x = -28$

32. $y = 1$

**19.** In one football game the quarterback threw passes that gained 10 yards, gained 22 yards, lost 5 yards, lost 2 yards, gained 15 yards, gained 13 yards, and lost 3 yards. If 3 passes were incomplete, what was the average gain (or loss) for the 10 pass plays?

**20.** A highway patrol car clocked 25 cars at the following speeds:

| | | | | | | | | | |
|---|---|---|---|---|---|---|---|---|---|
| 65 | 60 | 55 | 54 | 63 | 62 | 68 | 70 | 72 | 68 |
| 64 | 60 | 55 | 70 | 75 | 69 | 58 | 57 | 68 | 65 |
| 65 | 60 | 65 | 50 | 82 | | | | | |

(The last one got a ticket.) Use a calculator to find the average speed (in miles per hour) of the 25 cars.

Follow the rules for order of operations, and evaluate each of the following expressions.

**21.** $(-19 + 10)[\,8^2 \div 2 \cdot 3 - 16(6)\,]$

**22.** $60 - 2[\,35 + 6(-3) - 10^2 + 3 \cdot 5^2\,]$

Combine like terms (if possible) in each of the following polynomials.

**23.** $7x + 3x - 5x$

**24.** $8x^2 + 9x - 10x^2 + 15 - x$

**25.** $y^2 - 3y^3 + 2y^2 + 5y - 9y$

**26.** $5y^2 + 3y - 2y - 6 - y$

Evaluate each of the following expressions for $x = -3$ and $y = 2$.

**27.** $14xy + 5x^2 - 6y$

**28.** $7x^2y + 8xy^2 - 11xy - 2x - 3y$

Solve each of the following equations.

**29.** $x - 28 = -30$

**30.** $-12y = -36$

**31.** $15x - 14x + 14 = -14$

**32.** $4y - 6y = 42 - 44$

Name ____________ Section ________ Date ________

ANSWERS

## Cumulative Review: Chapters 1 - 2

Round off the following whole numbers as indicated.

**1.** 6242 ( nearest hundred )

**2.** 25,345 ( nearest ten thousand )

**3.** 764 ( to the nearest ten )

**4.** 22,615( to the nearest thousand )

Find the correct value of $n$, and state the property of addition or multiplication that is illustrated.

**5.** $27 + 32 = 32 + n$

**6.** $48(20 \cdot 2) = (48 \cdot 20) \cdot n$

**7. a.** The additive identity is _____.

**b.** The multiplicative identity is _____.

**c.** Division by 0 is _____.

First estimate each sum, and then find the sum.

**8.**
$$\begin{array}{r} 3346 \\ 1255 \\ +\ 5070 \\ \hline \end{array}$$

**9.**
$$\begin{array}{r} 75,738 \\ 678,641 \\ +\ 127,500 \\ \hline \end{array}$$

First estimate each difference, and then find the difference.

**10.**
$$\begin{array}{r} 8064 \\ -\ 5349 \\ \hline \end{array}$$

**11.**
$$\begin{array}{r} 70,000 \\ -\ 32,421 \\ \hline \end{array}$$

First estimate each product, and then find the product.

**12.**
$$\begin{array}{r} 57 \\ \times\ 35 \\ \hline \end{array}$$

**13.**
$$\begin{array}{r} 2842 \\ \times\ 76 \\ \hline \end{array}$$

**1.** 6200

**2.** 30,000

**3.** 760

**4.** 23,000

**5.** $n = 27$; Commutative Prop. of Addition

**6.** $n = 2$; Associative Prop. of Multiplication

**7. a.** 0

**b.** 1

**c.** Undefined

**8.** Estimate: 9000
Sum: 9671

**9.** Estimate: 880,000
Sum: 881,879

**10.** Estimate: 3000
Difference: 2715

**11.** Estimate: 40,000
Difference: 37,579

**12.** Estimate: 2400
Product: 1995

**13.** Estimate: 240,000
Product: 215,992

14. 487 R 0

15. 106 R 438

16. [Respond in exercise.]

17. 5, – 5

18. [Respond in exercise.]

19. – 42

20. – 205

21. – 17

22. 0

23. – 180

24. – 12

25. 0

26. Undefined

27. 57

28. $788

Find the whole number quotient and remainder.

**14.** $25\overline{)12,175}$

**15.** $452\overline{)48,350}$

**16.** Graph the following set of integers on a number line: { – 3, – 2, 0, 1, 2 }

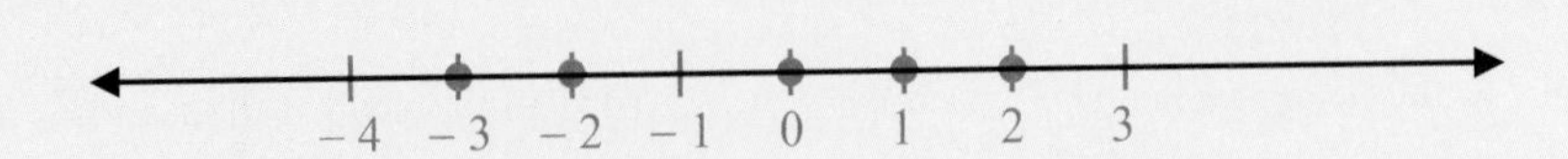


**17.** What are the possible integer values of $n$ if $|n| = 5$?

**18.** Graph the set of whole numbers less than 1.

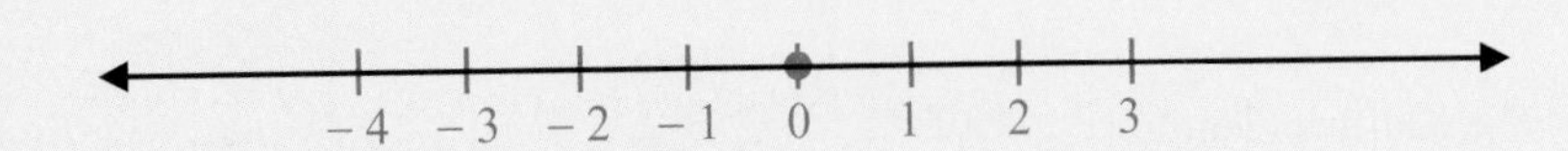


Find each of the following sums or differences.

**19.** $-16 + (-2) + (-24)$

**20.** $-220 - 65 + 80$

**21.** $-35 + 15 - (-6) + (-3)$

**22.** $|-7| + |-3| - |-10|$

Find each of the following products or quotients.

**23.** $(-2)(-6)(-3)(5)$

**24.** $\frac{-48}{4}$

**25.** $\frac{0}{-3}$

**26.** $\frac{153}{0}$

**27.** If the quotient of 105 and –7 is added to the product of –24 and –3, what is the sum?

**28.** Marlene opened a checking account with $1875. She wrote checks for $135, $284, $135, $350, and $725 and made one deposit of $542. What is her current balance?

Name ______________ Section ______________ Date ______________

**ANSWERS**

**29.** Find the perimeter of a triangle with sides of 13 inches, 20 inches, and 30 inches.

**29.** 63 in.

**30.** For a rectangular-shaped lawn that is 100 feet long and 45 feet wide, find:

**a.** the perimeter

**b.** the area

**30. a.** 290 ft.

**b.** 4500 sq. ft.

**31.** The temperature readings for 7 days in Denver were 10°, 15°, −5°, −8°, −7°, −3°, and 5°. What was the average daily temperature for these 7 days.

**31.** 1°

**32.** A jet pilot flew his plane from an altitude of 35,000 feet to an altitude of 14,000 feet. What was the change in altitude?

**32.** −21,000 ft.

Use a calculator to find the value of the variables in each of the following expressions.

**33.** $123{,}500 - 37{,}620 - 111{,}222 + 50{,}800 = n$

**33.** $n = 25{,}458$

**34.** $(-24)(-33)(-6)(-120) = x$

**34.** 570,240

**35.** $(150)^2 - 135 \cdot 16 \div 15 - 10^3 = y$

**35.** 21,356

**36.** Explain briefly why the values of $-7^2$ and $(-7)^2$ are not the same.

**36.** $-7^2 = -1 \cdot 7^2 = -49$

$(-7)^2 = 49$

**37.** In the expression $15^2$:

**a.** name the base

**b.** name the exponent

**c.** find the value of the expression

**37. a.** 15

**b.** 2

**c.** 225

38. 27

39. – 110

40. $7x^3 + 8x - 6$
trinomial
third-degree

41. $14y^2 + 2y - 1$
trinomial
second-degree

42. $37x^2 + 10$
binomial
second-degree

43. – 8

44. 22

45. – 8

46. 9

47. 6

48. – 1

49. 23

50. 754

51. [Respond below exercise.]

52. [Respond below exercise.]

Follow the rules for order of operations, and evaluate each of the following numerical expressions.

**38.** $4^3 \div 2^4 + 18 \cdot 6 \div 3 + 7 - 20$

**39.** $50 - 20(\ 3^2 \div 3 \cdot 2 + 14 \div 7\ )$

Simplify each of the following polynomials by combining like terms, and then tell what type of polynomial it is and state its degree.

**40.** $7x^3 + 8x - 15 + 9$

**41.** $9y^2 + 4y^2 + y^2 + 3y - y - 1$

**42.** $2(\ 11x^2 + 3x - 1\ ) + 3(\ 5x^2 - 2x + 4\ )$

Evaluate each of the following polynomials for $x = -2$.

**43.** $5x^2 + 6x - 16$

**44.** $-2x^2 - 5x + 20$

Solve each of the following equations.

**45.** $x - 17 = -25$

**46.** $-7x = -63$

**47.** $3x - 2x + 7 = 10 + 3$

**48.** $-52 + 48 = 6y - 2y$

Use a calculator to evaluate each of the following expressions.

**49.** $(\ 10 - 15\ )^2\ -\ 2^3\ + 80 \div 40 \cdot 3$

**50.** $17 - 3[\ (\ 13 - 11\ )^2\ +\ 42\ ] - 5^3 + 10^3$

## Writing and Thinking About Mathematics

**51.** Name one topic discussed in Chapters 1 – 2 that you found particularly interesting. Explain why you think it is important. ( Write only a short paragraph, and be sure to use complete sentences. )
Answers will vary. This question is designed to to help the instructor gain additional insight into how the students perceive the course.

**52.** Explain briefly, in your own words, how the concepts of perimeter and area are different.
Answers will vary. Perimeter is linear measure, and area is a square measure. Perimeter is a measure of a distance, and area is a measure of an interior.

# 3

# Prime Numbers and Fractions

## Chapter 3 Prime Numbers and Fractions





## WHAT TO EXPECT IN CHAPTER 3

The topics in the first half of Chapter 3 develop a foundation for the introduction of fractions in the second half of the chapter. Section 3.1 provides tests for divisibility of integers by 2, 3, 5, 6, 9, and 10. These tests can be used when factoring and finding divisors of integers without actually performing long division. Prime numbers (Section 3.2), prime factorizations (Section 3.3), and least common multiples for sets of integers and sets of algebraic terms (Section 3.4) provide the additional skills and understanding for the development of fractions and mixed numbers in Chapters 3 and 4.

**Rational numbers** (the technical name for fractions) are introduced in Section 3.5. The techniques, emphasizing prime factorizations, for multiplication and division with fractions are developed in Sections 3.5 and 3.6. Algebraic terms are included in all the work with prime factorizations and fractions. In this way, similarities between arithmetic and algebra are reinforced throughout the discussion.

The chapter ends with a section on solving equations, this time with fractions included. The text continually develops the groundwork for this most important and useful topic.

# Chapter 3 Prime Numbers and Fractions

**Mathematics at Work!**

Since the times of ancient Greece, scholars have been fascinated by numbers and their properties. Now, universities offer high-level mathematics courses and doctoral degrees in number theory.

All the divisors of a whole number less than that number are called **proper divisors** of that number. In terms of proper divisors, whole numbers fall into the following three classifications (originated by the Pythagoreans):

**abundant** → The sum of the proper divisors is more than the number (18 and 20)

**perfect** → The sum of the proper divisors is equal to the number (6 and 496)

**deficient** → The sum of the proper divisors is less than the number (15 and 21)

The first **perfect number** is 6. The divisors of 6 are 1, 2, 3, and 6. The proper divisors of 6 are 1, 2, and 3 and $1 + 2 + 3 = 6$. The third perfect number is 496.

Thus, the sum of the proper factors of 496 is

$$1 + 2 + 4 + 8 + 16 + 31 + 62 + 124 + 248 = 496$$

The next three perfect numbers are

8128, 33,550,336, and 8,589,869,056

The second perfect number is between 20 and 30. As an exercise, find this number. As another exercise, verify that 8128 is indeed a perfect number.

As of 1952, only 12 perfect numbers were known. Since then, with the advent of computers, more perfect numbers have been found, some with more than 1,000,000 digits. However, no one knows how many perfect numbers there are.

## 3.1 Tests for Divisibility

### Tests For Divisibility ( 2, 3, 5, 6, 9, 10 )

In our work with factoring (Section 3.3) and fractions (Chapters 3 and 4), we will need to be able to divide quickly and easily by small numbers. Since we will be looking for factors, we will want to know if a number is **exactly divisible** (remainder 0) by some number **before** actually dividing. There are simple tests that can be performed mentally to determine whether a number is divisible by

2, 3, 5, 6, 9, or 10

**without actually dividing.**

This chapter is designed for students to develop an understanding of factors and the related skills needed for operations with fractions. All the topics are an integral part of the development of fractions and mixed numbers in Chapters 3 and 4. Many of these ideas carry over into our work with decimals, percents, and simplification of algebraic expressions. Knowing this may help motivate some students to study more diligently than they might otherwise.

For example, can you tell (without dividing) if 585 is divisible by 2? By 3? Note that we are not trying to find the quotient, only to determine whether 2 or 3 is a factor of 585. The answer is that 585 **is not** divisible by 2 and **is** divisible by 3.

```
    292              195
 2)585            3)585
   4                3
   18               28
   18               27
    05               15
     4               15
     1 Remainder      0 Remainder
```

Thus, the number 2 **is not a factor** of 585. However, $3 \cdot 195 = 585$, and 3 and 195 **are factors** of 585.

The following rules explain how to test for divisibility by 2, 3, 5, 6, 9, and 10. There are other tests for other numbers such as 7 and 8, but the rules given here are sufficient for our purposes.

**Objectives**

① Have an understanding of the concept of **divisibility**.

② Know the rules for testing divisibility by 2, 3, 5, 6, 9, and 10

③ Be able to quickly determine whether or not an integer is divisible by 2, 3, 5, 6, 9, or 10

④ Be able to apply the concept of divisibility to products such as factorials.

**Definition:**

**Tests For Divisibility Of Integers by 2, 3, 5, 6, 9, and 10**

**For 2:** If the last digit (units digit) of an integer is 0, 2, 4, 6, or 8, then the integer is divisible by 2.

**For 3:** If the sum of the digits of an integer is divisible by 3, then the integer is divisible by 3.

**For 5:** If the last digit of an integer is 0 or 5, then the integer is divisible by 5.

**For 6:** If the integer is divisible by both 2 and 3, then it is divisible by 6.

**For 9:** If the sum of the digits of an integer is divisible by 9, then the integer is divisible by 9.

**For 10:** If the last digit of an integer is 0, then the integer is divisible by 10.

These rules of divisibility may seem somewhat trivial at this time. However, students should be encouraged to study them carefully since knowing these rules will prove to be very useful in working with prime factorizations and fractions.

**Definition:**

**Even and Odd Integers**

**Even integers** are divisible by 2.
(If an integer is divided by 2 and the remainder is 0, then the integer is even.)

**Odd integers** are not divisible by 2.
(If an integer is divided by 2 and the remainder is 1, then the integer is odd.)

The **even** integers are

**..., −10, −8, −6, −4, −2, 0, 2, 4, 6, 8, 10, ...**

The **odd** integers are

**..., −11, −9, −7, −5, −3, −1, 1, 3, 5, 7, 9, 11,...**

If the units digit of an integer is one of the even digits ( 0, 2, 4, 6, 8 ), then the integer is divisible by 2, and therefore, it is an even integer.

### Example 1

**a.** 1386 is divisible by 2 since the units digit is 6, an even digit. Thus, 1386 is an even integer.
**b.** 7701 is divisible by 3 because $7+7+0+1=15$ and 15 is divisible by 3.
**c.** 23,365 is divisible by 5 because the units digit is 5.
**d.** 9036 is divisible by 6 since it is divisible by both 2 and 3. (The sum of the digits is $9+0+3+6=18$, and 18 is divisible by 3 and so is the number. The units digit is 6 so it is an even integer. )
**e.** 9567 is divisible by 9 because $9+5+6+7=27$ and 27 is divisible by 9.
**f.** 253,430 is divisible by 10 because the units digit is 0.

*Now Work Exercises 1 - 5 in the Margin.*

In examples 2 and 3, all six tests for divisibility are used to determine which of the numbers 2, 3, 5, 6, 9, and 10 will divide into each number.

### Example 2

The number 3430 is

**a.** divisible by 2 (units digit is 0, an even digit);
**b.** not divisible by 3 ($3+4+3+0=10$ and 10 is not divisible by 3);
**c.** divisible by 5 (units digit is 0);
**d.** not divisible by 6 (To be divisible by 6 it must be divisible by both 2 and 3, but 3430 is not divisible by 3);
**e.** not divisible by 9 ($3+4+3+0=10$ and 10 is not divisible by 9);
**f.** divisible by 10 (units digit is 0).

### Example 3

The number −5718 is

**a.** divisible by 2 (units digit is 8, an even digit);
**b.** divisible by 3 ($5+7+1+8=21$ and 21 is divisible by 3);
**c.** not divisible by 5 (units digit is not 0 or 5);
**d.** divisible by 6 (divisible by both 2 and 3);
**e.** not divisible by 9 ($5+7+1+8=21$ and 21 is not divisible by 9);
**f.** not divisible by 10 (units digit is not 0).

**1.** Does 6 divide 8034? Explain why or why not.
Yes, 6 divides 8034 because the units digit is even and the sum of the digits, 15, is divisible by 3.

**2.** Does 3 divide 206? Explain why or why not.
No, 3 does not divide 206 because 3 does not divide the sum of the digits: $2+0+6=8$.

**3.** Is 4065 divisible by 3? By 9?
The sum of the digits is 15, so 3 divides the number, but 9 does not.

**4.** Is 12,375 divisible by 5? By 10?
12,375 is divisible by 5, because the units digit is 5. 12,375 is not divisible by 10, because the units digit is not 0.

**5.** Which of the numbers 2, 5, or 6 are factors of 2400? Explain.
All the numbers 2, 5, and 6 are factors of 2400. The units digit is 0, so the number is divisible by 5. Since the number is even and the sum of the digits is 6, which is divisible by 3, it is divisible by 6.

### Completion Example 4

The number –375 is

**a.** divisible by 5 because

________________________

**b.** divisible by 3 because

________________________

**c.** not divisible by 6 because

________________________

### Completion Example 5

The number 612 is

**a.** divisible by 6 because

________________________

**b.** divisible by 9 because

________________________

**c.** not divisible by 10 because

________________________

*Now Work Exercises 6 - 8 in the Margin.*

Determine which of the numbers 2, 3, 5, 6, 9, and 10 divides into each of the following numbers.

**6.** 1742
1742 is divisible by 2.

**7.** 8020
8020 is divisible by 2, 5, and 10.

**8.** 33,031
33,031 is not divisible by any of these numbers.

## Divisibility of Products

Now consider a product that is written as a product of several factors. For example,

$$3 \cdot 4 \cdot 5 \cdot 10 \cdot 12 = 7200$$

To better understand how products and factors are related, consider the problem of finding the quotient 7200 divided by 150:

$$7200 \div 150 = ?$$

By using the commutative and associative properties of multiplication (i.e. rearranging and grouping factors), we can see that

$$3 \cdot 5 \cdot 10 = 150 \quad \text{and} \quad 7200 = (3 \cdot 5 \cdot 10) \cdot (4 \cdot 12) = 150 \cdot 48$$

Thus,

$$7200 \div 150 = 48$$

In general, for any product of two or more integers, each integer is a factor of the total product. Also, the product of any combination of factors is a factor of the total product. This means that every one of the following products is a factor of 7200:

| | | |
|---|---|---|
| $3 \cdot 4 = 12$ | $3 \cdot 5 = 15$ | $3 \cdot 10 = 30$ |
| $3 \cdot 12 = 36$ | $4 \cdot 5 = 20$ | $4 \cdot 10 = 40$ |
| $4 \cdot 12 = 48$ | $5 \cdot 10 = 50$ | $5 \cdot 12 = 60$ |
| $10 \cdot 12 = 120$ | $3 \cdot 4 \cdot 5 = 60$ | $3 \cdot 4 \cdot 10 = 120$ |
| $3 \cdot 4 \cdot 12 = 144$ | $3 \cdot 5 \cdot 10 = 150$ | and so on. |

Proceeding in this manner, taking all combinations of the given factors two at a time, three at a time, and four at a time, we will find 25 combinations. Not all of these combinations lead to **different** factors. For example, the list so far shows both $5 \cdot 12 = 60$ and $3 \cdot 4 \cdot 5 = 60$. In fact, by using techniques involving prime factors (Prime numbers are discussed in Section 3.2.), we can show that there are 54 different factors of 7200, including 1 and 7200 itself.

Thus, by grouping the factors $3 \cdot 4 \cdot 5 \cdot 10 \cdot 12$ in various ways and looking at the remaining factors, we can find the quotient if 7200 is divided by any particular product of a group of the given factors. We have shown how this works in the case of $7200 \div 150 = 48$. For $7200 \div 144$, we can write

$$7200 = 3 \cdot 4 \cdot 5 \cdot 10 \cdot 12 = (3 \cdot 4 \cdot 12) \cdot (5 \cdot 10) = 144 \cdot 50$$

which shows that both 144 and 50 are factors of 7200 and $7200 \div 144 = 50$. (Also, of course, $7200 \div 50 = 144$.)

We say that any factor of a number **divides** that number.

## Example 6

Does 54 divide the product $4 \cdot 5 \cdot 9 \cdot 3 \cdot 7$? If so, how many times?

### Solution

Since $54 = 9 \cdot 6 = 9 \cdot 3 \cdot 2$, we factor 4 as $2 \cdot 2$ and rearrange the factors to find 54.

$$\begin{aligned} 4 \cdot 5 \cdot 9 \cdot 3 \cdot 7 &= 2 \cdot 2 \cdot 5 \cdot 9 \cdot 3 \cdot 7 \\ &= (9 \cdot 3 \cdot 2) \cdot (2 \cdot 5 \cdot 7) \\ &= 54 \cdot 70 \end{aligned}$$

Thus, 54 does divide the product, and it divides the product 70 times. Notice that we do not even need to know the value of the product, just the factors.

## Example 7

Does 26 divide the product $4 \cdot 5 \cdot 9 \cdot 3 \cdot 7$? If so, how many times?

### Solution

26 does **not** divide $4 \cdot 5 \cdot 9 \cdot 3 \cdot 7$ because $26 = 2 \cdot 13$ and 13 is not a factor of the product.

*Now Work Exercises 9 -11 in the Margin.*

**9.** Show that 39 divides the product $3 \cdot 5 \cdot 8 \cdot 13$ and find the quotient without actually dividing.

39 divides this product because $39 = 3 \cdot 13$ and 3 and 13 are both factors of this product. The quotient is $5 \cdot 8 = 40$.

**10.** Explain why 35 does not divide the product $2 \cdot 3 \cdot 5 \cdot 13$.

35 does not divide this product because $35 = 5 \cdot 7$ and 7 is not a factor of this product.

**11.** Show that 100 divides the product $6 \cdot 5 \cdot 4 \cdot 10 \cdot 7$ and find the quotient without actually dividing.

100 divides this product because $100 = 2 \cdot 5 \cdot 10$. Since 4 can be written as $2 \cdot 2$, the product can be written $6 \cdot 5 \cdot 2 \cdot 2 \cdot 10 \cdot 7$. So, 2, 5, and 10 are all factors of the product. The quotient is $6 \cdot 2 \cdot 7 = 84$.

## Completion Example Answers

**4.** **a.** because **the units digit is 5.**

**b.** because **the sum of the digits is 15, and 15 is divisible is 3.**

**c.** since **–375 is not even.**

**5.** **a.** because **612 is divisible by both 2 and 3.**

**b.** because **the sum of the digits is 9, and 9 is divisible by 9.**

**c.** since **the units digit is not 0.**

Name ______________________ Section ______________ Date ____________

## Exercises 3.1

For each problem, make up any four 3–digit numbers that you can think of that are divisible by all of the given numbers. (There are many possible answers.)

**1.** 2 and 3
Answers will vary.

**2.** 2 and 9
Answers will vary.

**3.** 2, 3, and 10
Answers will vary.

**4.** 5 and 6
Answers will vary.

Using the tests for divisibility, determine which of the numbers 2, 3, 5, 6, 9, and 10 (if any) will divide exactly into each of the following integers.

**5.** 82 **6.** 92 **7.** 81

**8.** 210 **9.** –441 **10.** –344

**11.** 544 **12.** 370 **13.** 575

**14.** 675 **15.** 711 **16.** –801

**17.** –640 **18.** 780 **19.** 1253

**20.** 1163 **21.** –402 **22.** –702

**23.** 7890 **24.** 6790 **25.** 777

**26.** 888 **27.** 45,000 **28.** 35,000

**ANSWERS**

1. [Respond below exercise.]
2. [Respond below exercise.]
3. [Respond below exercise.]
4. [Respond below exercise.]
5. 2
6. 2
7. 3, 9
8. 2, 3, 5, 6, 10
9. 3, 9
10. 2
11. 2
12. 2, 5, 10
13. 5
14. 3, 5, 9
15. 3, 9
16. 3, 9
17. 2, 5, 10
18. 2, 3, 5, 6, 10
19. None
20. None
21. 2, 3, 6
22. 2, 3, 6, 9
23. 2, 3, 5, 6, 10
24. 2, 5, 10
25. 3
26. 2, 3, 6
27. 2, 3, 5, 6, 9, 10
28. 2, 5, 10

29. 2
30. 3, 5, 9
31. 2, 3, 6, 9
32. 2, 5, 10
33. 2
34. 2, 3, 5, 6, 9, 10
35. 2, 3, 6
36. None
37. None
38. 2, 3, 6
39. 2
40. 2, 5, 10
41. 2, 5, 10
42. 2, 5, 10
43. None
44. 2
45. 3
46. 2, 3, 5, 6, 9, 10
47. 2, 3, 6
48. None
49. 2
50. 2, 3, 6
51. [Respond in exercise.]
52. [Respond below exercise.]

**29.** 7156 **30.** 8145 **31.** –6948

**32.** –8140 **33.** –9244 **34.** –7920

**35.** 33,324 **36.** 23,587 **37.** –13,477

**38.** –15,036 **39.** 722,348 **40.** 857,000

**41.** 635,000 **42.** 437,200 **43.** 8,737,001

**44.** 8,634,002 **45.** 4,005,303 **46.** 3,060,720

**47.** 20,506,002 **48.** 30,605,003 **49.** –15,007,004

**50.** –45,600,006

**51.** For each of the following numbers a digit is missing. Find a value for the missing digit so that the resulting number will be

**a.** divisible by 9. (There may be several correct answers or no correct answer.)

**i.** 31_1_4 **ii.** 5312_7_ **iii.** 4_7_475 **iv.** 73_8_,432

**b.** divisible by 6. (There may be several correct answers or no correct answer.)

**i.** 31_1, 4, 7_4 **ii.** 5312_4_ **iii.** 4_None_475 **iv.** 73_2, 5, 8_,432

**52.** Discuss the following statement:
An integer may be divisible by 3 and not divisible by 9. Give two examples to illustrate your point.

Answers will vary. If the integer has only one factor of 3 in its prime factorization, then it will not be divisible by 9. For example, $15 = 3 \cdot 5$ and $42 = 2 \cdot 3 \cdot 7$.

Name ______________________ Section ______________ Date __________

**ANSWERS**

53. [Respond below exercise.]

54. [Respond below exercise.]

55. [Respond below exercise.]

56. [Respond below exercise.]

57. [Respond below exercise.]

58. [Respond below exercise.]

59. [Respond below exercise.]

60. [Respond below exercise.]

61. [Respond below exercise.]

62. [Respond below exercise.]

63. [Respond below exercise.]

64. [Respond below exercise.]

65. [Respond below exercise.]

66. [Respond below exercise.]

**53.** Discuss the following statement:
An integer may be divisible by 5 and not divisible by 10. Give two examples to illustrate your point.

Answers will vary. Any integer that ends in 5 is divisible by 5 and not divisible by 10. For, example, 25 and 35.

**54.** Discuss the following statement:
If an integer is divisible by both 3 and 5, then it is divisible by 15.

Answers will vary. If an integer is divisible by both 3 and 5, then both 3 and 5 must be factors. Since they are prime, their product must also be a factor.

**55.** If an integer is divisible by both 3 and 9, will it necessarily be divisible by 27? Explain your reasoning, and give two examples to support this reasoning.

Answers will vary. Not necessarily. To be divisible by 27, an integer must have 3 as a factor three times. Two counterexamples are 9 and 18.

**56.** If an integer is divisible by both 2 and 4, will it necessarily be divisible by 8? Explain your reasoning and give two examples to support this reasoning.

Answers will vary. Not necessarily. To be divisible by 8, an integer must have 2 as a factor three times. Two counterexamples are 12 and 20.

Determine whether each of the given numbers divides (is a factor of) the given product. If it does divide the product, tell how many times. Find each product, and make a written statement concerning the divisibility of the product by the given number.

**57.** 6; $2 \cdot 3 \cdot 3 \cdot 7$
$2 \cdot 3 \cdot 3 \cdot 7 = 6 \cdot 21$. The product is 126, and 6 divides the product 21 times.

**58.** 14; $2 \cdot 3 \cdot 3 \cdot 7$
$2 \cdot 3 \cdot 3 \cdot 7 = 14 \cdot 9$. The product is 126, and 14 divides the product 9 times.

**59.** 10; $2 \cdot 3 \cdot 5 \cdot 9$
$2 \cdot 3 \cdot 5 \cdot 9 = 10 \cdot 27$. The product is 270, and 10 divides the product 27 times.

**60.** 25; $2 \cdot 3 \cdot 5 \cdot 7$
$2 \cdot 3 \cdot 5 \cdot 7 = 210$ and since this product does not have 5 as a factor twice, it is not divisible by 25.

**61.** 25; $3 \cdot 5 \cdot 7 \cdot 10$
$3 \cdot 5 \cdot 7 \cdot 10 = 25 \cdot 42$. The product is 1050, and 25 divides the product 42 times.

**62.** 35; $2 \cdot 2 \cdot 3 \cdot 3 \cdot 7$
$2 \cdot 2 \cdot 3 \cdot 3 \cdot 7 = 252$ and since this product does not have 5 as a factor , it is not divisible by 35.

**63.** 45; $2 \cdot 3 \cdot 3 \cdot 7 \cdot 15$
$2 \cdot 3 \cdot 3 \cdot 7 \cdot 15 = 45 \cdot 42$. The product is 1890, and 45 divides the product 42 times.

**64.** 36; $2 \cdot 3 \cdot 4 \cdot 5 \cdot 13$
$2 \cdot 3 \cdot 4 \cdot 5 \cdot 13 = 1560$ and since this product does not have 3 as a factor twice, it is not divisible by 36.

**65.** 40; $2 \cdot 3 \cdot 5 \cdot 7 \cdot 12$
$2 \cdot 3 \cdot 5 \cdot 7 \cdot 12 = 40 \cdot 63$. The product is 2520, and 40 divide is the product 63 times.

**66.** 45; $3 \cdot 5 \cdot 7 \cdot 10 \cdot 21$
$3 \cdot 5 \cdot 7 \cdot 10 \cdot 21 = 45 \cdot 490$. The product is 22,050, and 45 divides the product 490 times.

[Respond below exercise.]
67. a. ________

[Respond below exercise.]
b. ________

[Respond below exercise.]
68. a. ________

[Respond below exercise.]
b. ________

## WRITING AND THINKING ABOUT MATHEMATICS

**67.** The rule for divisibility by 3 is that: If the sum of the digits of a number is divisible by 3, then the number is divisible by 3. Carefully study the following analysis of this rule for the number 642.

$$\begin{aligned} 642 &= 600 + 40 + 2 \\ &= 6 \cdot 100 + 4 \cdot 10 + 2 \\ &= 6(99 + 1) + 4(9 + 1) + 2 \\ &= 6 \cdot 99 + 6 \cdot 1 + 4 \cdot 9 + 4 \cdot 1 + 2 \quad \text{The distributive property} \\ &= 6 \cdot 99 + 4 \cdot 9 + (6 + 4 + 2) \end{aligned}$$

We see that 3 is a factor of 99 and 3 is a factor of 9. Therefore, if 3 is to be a factor of the entire expression $6 \cdot 99 + 4 \cdot 9 + (6 + 4 + 2)$, 3 must be a factor of $6 + 4 + 2$ which is the sum of the digits of the number 642.

**a.** Now, follow the pattern just shown and show how the rule for divisibility for 3 applies to the number 744.

$$\begin{aligned} 744 &= 700 + 40 + 4 \\ &= 7 \cdot 100 + 4 \cdot 10 + 4 \\ &= 7(99 + 1) + 4(9 + 1) + 4 \\ &= 7 \cdot 99 + 4 \cdot 9 + (7 + 4 + 4) \end{aligned}$$

We see that 3 is a factor of both 99 and 9. Therefore, if 3 is a factor of 744, then 3 must also be a factor of $7 + 4 + 4$, which is the sum of the digits of the number 744.

**b.** Following the same pattern just discussed, show how the rule for divisibility for 3 applies to any 4–digit number in the form abcd (where a, b, c, and d are the digits).

$$\begin{aligned} abcd &= a000 + b00 + c0 + d \\ &= a \cdot 1000 + b \cdot 100 + c \cdot 10 + d \\ &= a(999 + 1) + b(99 + 1) + c(9 + 1) + d \\ &= a \cdot 999 + b \cdot 99 + c \cdot 9 + (a + b + c + d) \end{aligned}$$

We see that 3 is a factor of 999, 99, and 9. Therefore, if 3 is a factor of abcd, then 3 must also be a factor of $a + b + c + d$, which is the sum of the digits of the number abcd.

**68. a.** Follow the pattern of the discussion in Exercise 67 and show how the rule for divisibility for 9 applies to the number 7218.

$$\begin{aligned} 7218 &= 7000 + 200 + 10 + 8 \\ &= 7 \cdot 1000 + 2 \cdot 100 + 1 \cdot 10 + 8 \\ &= 7(999 + 1) + 2(99 + 1) + 1(9 + 1) + 8 \\ &= 7 \cdot 999 + 2 \cdot 99 + 1 \cdot 9 + (7 + 2 + 1 + 8) \end{aligned}$$

We see that 9 is a factor of 999, 99, and 9. Therefore, if 9 is a factor of 7218, then 9 must also be a factor of $7 + 2 + 1 + 8$, which is the sum of the digits of the number 7218.

**b.** Follow the pattern of the discussion in Exercise 67 and show how the rule for divisibility for 9 applies to any 4–digit number in the form abcd (where a, b, c, and d are the digits).

$$\begin{aligned} abcd &= a000 + b00 + c0 + d \\ &= a \cdot 1000 + b \cdot 100 + c \cdot 10 + d \\ &= a(999 + 1) + b(99 + 1) + c(9 + 1) + d \\ &= a \cdot 999 + b \cdot 99 + c \cdot 9 + (a + b + c + d) \end{aligned}$$

We see that 9 is a factor of 999, 99, and 9. Therefore, if 9 is a factor of abcd, then 9 must also be a factor of $a + b + c + d$, which is the sum of the digits of the number abcd.

Name ______________________ Section ______________ Date ____________

**ANSWERS**

69. [Respond below exercise.]

70. [Respond below exercise.]

71. [Respond below exercise.]

**69.** The text listed fourteen combinations of factors of the product $3 \cdot 4 \cdot 5 \cdot 10 \cdot 12$. There are twenty–five such combinations. Find the remaining eleven.

$3 \cdot 5 \cdot 12 = 180$ $\quad 4 \cdot 5 \cdot 10 = 200$ $\quad 3 \cdot 10 \cdot 12 = 360$ $\quad 4 \cdot 5 \cdot 12 = 240$
$4 \cdot 10 \cdot 12 = 480$ $\quad 5 \cdot 10 \cdot 12 = 600$ $\quad 3 \cdot 4 \cdot 5 \cdot 10 = 600$ $\quad 3 \cdot 4 \cdot 5 \cdot 12 = 720$
$3 \cdot 4 \cdot 10 \cdot 12 = 1440$ $\quad 3 \cdot 5 \cdot 10 \cdot 12 = 1800$ $\quad 4 \cdot 5 \cdot 10 \cdot 12 = 2400$

**70.** Discuss the following statement: Given that $n$ represents any integer, then $2n$ represents an even integer, and $2n + 1$ represents an odd integer.

Answers will vary. Regardless of whether n is even or odd, positive or negative, $2n$ has 2 as a factor and is therefore even. Since $2n + 1$ is 1 more than an even number, it must be odd.

## Collaborative Learning Exercise

**71.** In groups of three to four students, use a calculator to evaluate $10^{12}$ and $12^{10}$. Discuss what you think is the meaning of the notation on the display. (**Note:** The notation is a form of a notation called **scientific notation** and is discussed in detail in Chapter 5. Different calculators may use slightly different forms.)

Answers will vary. The question is designed to help the students become familiar with their calculators and to provide an early introduction to scientific notation.

# 3.2 Prime Numbers

## Prime Numbers and Composite Numbers

The positive integers are also called the **counting numbers** (or **natural numbers**). Every counting number, except the number 1, has at least two factors, as illustrated in the following list. Note that, in this list, every number has **at least two** factors, but 13, 19, and 7 have **exactly two** factors.

| Counting Numbers | | Factors |
|---|---|---|
| 12 | ...................................... | 1, 2, 3, 4, 6, 12 |
| 20 | ...................................... | 1, 2, 4, 5, 10, 20 |
| 13 | ...................................... | 1, 13 |
| 19 | ...................................... | 1, 19 |
| 25 | ...................................... | 1, 5, 25 |
| 7 | ...................................... | 1, 7 |
| 86 | ...................................... | 1, 2, 43, 86 |

Our work with fractions will be based on the use and understanding of counting numbers that have exactly two different factors. Such numbers (for example, 13, 19, and 7 in the list above) are called **prime numbers**. (**Note:** In later discussions involving negative integers and negative fractions, we will treat a negative integer as the product of −1 and a counting number.)

**Objectives**

① Know the definition of a **prime number**.

② Know the definition of a **composite number**.

③ Be able to list all the prime numbers less than 50.

④ Be able to determine whether a number is prime or composite.

**Definition:**

A **prime number** is a counting number greater than 1 that has only 1 and itself as factors.

or

A **prime number** is a counting number with exactly two different factors (or divisors).

As a point of historical and practical interest, Euclid proved that there are an infinite number of prime numbers. As a direct consequence, there is no largest prime number.

**Definition:**

A **composite number** is a counting number with more than two different factors (or divisors).

Thus, in the previous list, 12, 20, 25, and 36 are composite numbers.

**Note:** 1 is **neither** a prime nor a composite number. $1 = 1 \cdot 1$, and 1 is the only factor of 1. 1 does not have **exactly two different** factors, and it does not have more than two different factors.

### Example 1

Some prime numbers:

| | |
|---|---|
| 2 | 2 has exactly two different factors, 1 and 2. |
| 3 | 3 has exactly two different factors, 1 and 3. |
| 11 | 11has exactly two different factors, 1 and 11. |
| 37 | 37 has exactly two different factors, 1 and 37. |

### Example 2

Some composite numbers:

| | |
|---|---|
| 15 | 1, 3, 5, and 15 are all factors of 15. |
| 39 | 1, 3, 13, and 39 are all factors of 39. |
| 49 | 1, 7, and 49 are all factors of 49. |
| 50 | 1, 2, 5, 10, 25, and 50 are all factors of 50. |

*Now Work Exercises 1 - 6 in the Margin.*

Determine whether each of the following numbers is prime or composite.

**1.** 14
Composite

**2.** 19
Prime

**3.** 37
Prime

**4.** 2
Prime

**5.** 41
Prime

**6.** 39
Composite

## The Sieve Of Eratosthenes

There is no known formula for finding all the prime numbers. However, a famous Greek mathematician, Eratosthenes, developed a technique that helps in finding prime numbers. This technique is based on the concept of **multiples**. To find the multiples of any counting number, multiply each of the counting numbers by that number.

| **Counting Numbers:** | **1,** | **2,** | **3,** | **4,** | **5,** | **6,** | **7,** | **8, …** |
|---|---|---|---|---|---|---|---|---|
| Multiples of 2: | 2, | 4, | 6, | 8, | 10, | 12, | 14, | 16, … |
| Multiples of 5: | 5, | 10, | 15, | 20, | 25, | 30, | 35, | 40, … |
| Multiples of 7: | 7, | 14, | 21, | 28, | 35, | 42, | 49, | 56, … |

None of the multiples of a number, except possibly the number itself, can be prime, since each has that number as a factor. To sift out the prime numbers according to the **Sieve of Eratosthenes**, we proceed by eliminating multiples, as described in the following steps:

**1.** To find the prime numbers from 1 to 50, list all the counting numbers from 1 to 50 in rows of ten.

| | | | | | | | | | |
|---|---|---|---|---|---|---|---|---|---|
| 1 | 2 | 3 | 4 | 5 | 6 | 7 | 8 | 9 | 10 |
| 11 | 12 | 13 | 14 | 15 | 16 | 17 | 18 | 19 | 20 |
| 21 | 22 | 23 | 24 | 25 | 26 | 27 | 28 | 29 | 30 |
| 31 | 32 | 33 | 34 | 35 | 36 | 37 | 38 | 39 | 40 |
| 41 | 42 | 43 | 44 | 45 | 46 | 47 | 48 | 49 | 50 |

**2.** Start by crossing out 1 (since 1 is not a prime number). Next, circle 2 and cross out all the other multiples of 2; that is, cross out every second number.

| | | | | | | | | | |
|---|---|---|---|---|---|---|---|---|---|
| ~~1~~ | (2) | 3 | ~~4~~ | 5 | ~~6~~ | 7 | ~~8~~ | 9 | ~~10~~ |
| 11 | ~~12~~ | 13 | ~~14~~ | 15 | ~~16~~ | 17 | ~~18~~ | 19 | ~~20~~ |
| 21 | ~~22~~ | 23 | ~~24~~ | 25 | ~~26~~ | 27 | ~~28~~ | 29 | ~~30~~ |
| 31 | ~~32~~ | 33 | ~~34~~ | 35 | ~~36~~ | 37 | ~~38~~ | 39 | ~~40~~ |
| 41 | ~~42~~ | 43 | ~~44~~ | 45 | ~~46~~ | 47 | ~~48~~ | 49 | ~~50~~ |

**3.** The first number after 2 that is not crossed out is 3. Circle 3, and cross out all multiples of 3 that are not already crossed out; that is, after 3, every third number should be crossed out.

| ~~1~~ | (2) | (3) | ~~4~~ | 5 | ~~6~~ | 7 | ~~8~~ | ~~9~~ | ~~10~~ |
|---|---|---|---|---|---|---|---|---|---|
| 11 | ~~12~~ | 13 | ~~14~~ | ~~15~~ | ~~16~~ | 17 | ~~18~~ | 19 | ~~20~~ |
| ~~21~~ | ~~22~~ | 23 | ~~24~~ | 25 | ~~26~~ | ~~27~~ | ~~28~~ | 29 | ~~30~~ |
| 31 | ~~32~~ | ~~33~~ | ~~34~~ | 35 | ~~36~~ | 37 | ~~38~~ | ~~39~~ | ~~40~~ |
| 41 | ~~42~~ | 43 | ~~44~~ | ~~45~~ | ~~46~~ | 47 | ~~48~~ | 49 | ~~50~~ |

**4.** The next number that is not crossed out is 5. Circle 5, and cross out all multiples of 5 that are not already crossed out.

| ~~1~~ | (2) | (3) | ~~4~~ | (5) | ~~6~~ | 7 | ~~8~~ | ~~9~~ | ~~10~~ |
|---|---|---|---|---|---|---|---|---|---|
| 11 | ~~12~~ | 13 | ~~14~~ | ~~15~~ | ~~16~~ | 17 | ~~18~~ | 19 | ~~20~~ |
| ~~21~~ | ~~22~~ | 23 | ~~24~~ | ~~25~~ | ~~26~~ | ~~27~~ | ~~28~~ | 29 | ~~30~~ |
| 31 | ~~32~~ | ~~33~~ | ~~34~~ | ~~35~~ | ~~36~~ | 37 | ~~38~~ | ~~39~~ | ~~40~~ |
| 41 | ~~42~~ | 43 | ~~44~~ | ~~45~~ | ~~46~~ | 47 | ~~48~~ | 49 | ~~50~~ |

**5.** If we proceed this way, we will have the prime numbers circled and the composite numbers crossed out.

| ~~1~~ | (2) | (3) | ~~4~~ | (5) | ~~6~~ | (7) | ~~8~~ | ~~9~~ | ~~10~~ |
|---|---|---|---|---|---|---|---|---|---|
| (11) | ~~12~~ | (13) | ~~14~~ | ~~15~~ | ~~16~~ | (17) | ~~18~~ | (19) | ~~20~~ |
| ~~21~~ | ~~22~~ | (23) | ~~24~~ | ~~25~~ | ~~26~~ | ~~27~~ | ~~28~~ | (29) | ~~30~~ |
| (31) | ~~32~~ | ~~33~~ | ~~34~~ | ~~35~~ | ~~36~~ | (37) | ~~38~~ | ~~39~~ | ~~40~~ |
| (41) | ~~42~~ | (43) | ~~44~~ | ~~45~~ | ~~46~~ | (47) | ~~48~~ | ~~49~~ | ~~50~~ |

The final table shows that the prime numbers less than 50 are

**2, 3, 5, 7, 11, 13, 17, 19, 23, 29, 31, 37, 41, 43, 47**

Students should be encouraged to memorize the list of prime numbers less than 50. We will assume later that, when reducing fractions, the students readily recognize these primes.

You should note the following two facts about prime numbers:

1. 2 is the only even prime number; and
2. All other prime numbers are odd, but not all odd numbers are prime.

Rearranging the Sieve of Erathosthenes in six columns of numbers, beginning with 2, shows how the prime numbers greater than 3 appear on either side of multiples of 6. This format may make memorizing primes and/or testing for primes easier.

|  |  |  |  | Multiples of 6 ↓ |  |
|---|---|---|---|---|---|
| **2** | **3** | 4 | **5** | 6 | **7** |
| 8 | 9 | 10 | **11** | 12 | **13** |
| 14 | 15 | 16 | **17** | 18 | **19** |
| 20 | 21 | 22 | **23** | 24 | 25 |
| 26 | 27 | 28 | **29** | 30 | **31** |
| 32 | 33 | 34 | 35 | 36 | **37** |
| 38 | 39 | 40 | **41** | 42 | **43** |

## Determining Prime Numbers

An important mathematical fact is that **there is no known pattern or formula for determining prime numbers.** In fact, only recently have computers been used to show that very large numbers previously thought to be prime are actually composite. The previous discussion has helped to develop an understanding of the nature of prime numbers and showed some prime and composite numbers in list form. However, writing a list every time to determine whether a number is prime or not would be quite time consuming and unnecessary. The following procedure of dividing by prime numbers can be used to determine whether or not relatively small numbers (say, less than 1000) are prime. If a prime number smaller than the given number is found to be a factor (or divisor), then the given number is composite. (**Note**: You still should memorize the prime numbers less than 50 for convenience and ease with working with fractions.)

You might want to explain in more detail why only prime numbers are used as divisors to determine whether or not a number is prime. For example, if a number is not divisible by 2, then there is no need to try division by 4 or 6 because a number divisible by 4 or 6 would have to be even and therefore divisible by 2. Similarly, if a number is not divisible by 3, then there is no need to test 6 or 9.

**To Determine Whether a Number is Prime:**

Divide the number by progressively larger **prime numbers** (2, 3, 5, 7, 11, and so forth) until:

1. You find a remainder of 0 (meaning that the prime number is a factor and the given number is composite); or
2. You find a quotient smaller than the prime divisor (meaning that the given number has no smaller prime factors and is therefore prime itself).

**Note:** Reasoning that if a composite number were a factor, then one of its prime factors would have been found to be a factor in an earlier division, we divide only by prime numbers - that is, there is no need to divide by a composite number.

### Example 3

Is 305 a prime number?

**Solution**

Since the units digit is 5, the number 305 is divisible by 5 (by the divisibility test in Section 3.1) and is not prime. 305 is a composite number.

### Example 4

Is 247 prime?

**Solution**

Mentally, note that the tests for divisibility by the prime numbers 2, 3, and 5 fail.

$$\text{Divide by 7:} \rightarrow \begin{array}{r} 35 \\ 7\overline{)247} \\ \underline{21\phantom{0}} \\ 37 \\ \underline{35} \\ 2 \end{array}$$

Divide by 11:→ $\begin{array}{r} 22 \\ 11\overline{)247} \\ \underline{22} \\ 27 \\ \underline{22} \\ 5 \end{array}$

Divide by 13:→ $\begin{array}{r} 19 \\ 13\overline{)247} \\ \underline{13} \\ 117 \\ \underline{117} \\ 0 \end{array}$

247 is composite and not prime. In fact, $247 = 13 \cdot 19$ and 13 and 19 are factors of 247.

## Example 5

Is 109 prime?

### Solution

Mentally, note that the tests for divisibility by the prime numbers 2, 3, and 5 fail.

Divide by 7: → $\begin{array}{r} 15 \\ 7\overline{)109} \\ \underline{7} \\ 39 \\ \underline{35} \\ 4 \end{array}$

Divide by 11:→ $\begin{array}{r} 9 \\ 11\overline{)109} \\ \underline{99} \\ 10 \end{array}$

109 is a prime number.

**Note:** There is no point in dividing by larger numbers such as 13 or 17 because the quotient would only get smaller, and if any of these larger numbers were a divisor, the smaller quotient would have been found to be a divisor earlier in the procedure.

## *Completion Example 6*

Is 199 prime or composite?

### Solution

Tests for 2, 3, and 5 all fail.

Divide by 7: $7\overline{)199}$

Divide by 11: $11\overline{)199}$

Divide by 13: $13\overline{)199}$

Divide by _______: $\overline{)199}$

199 is a ____________ number.

**7.** Determine whether 221 is prime or composite. Explain each step.
Division by prime numbers shows that $221 = 13 \cdot 17$. Therefore, 221 is composite.

**8.** Determine whether 239 is prime or composite. Explain each step.
Division by prime numbers up to 17 shows no divisors. When 239 is divided by 17, the quotient is less than 17. Therefore, 239 is prime, and its only divisors are 1 and 239.

*Now Work Exercises 7 and 8 in the Margin.*

## Example 7

One interesting application of factors of counting numbers (very useful in beginning algebra) involves finding two factors whose sum is some specified number. For example, find two factors for 75 such that their product is 75 and their sum is 20.

### Solution

The factors of 75 are 1, 3, 5, 15, 25, and 75, and the pairs whose products are 75 are

$$1 \cdot 75 = 75 \qquad 3 \cdot 25 = 75 \qquad 5 \cdot 15 = 75$$

Thus, the numbers we are looking for are 5 and 15 because

$$5 \cdot 15 = 75 \qquad \text{and} \qquad 5 + 15 = 20$$

## Completion Example Answer

**6.** Divide by 7: $7\overline{)199}$ → quotient **28**; **14**, **59**, **56**, remainder **3**

Divide by 11: $11\overline{)199}$ → quotient **18**; **11**, **89**, **88**, remainder **1**

Divide by 13: $13\overline{)199}$ → quotient **15**; **13**, **69**, **65**, remainder **4**

Divide by 17: $17\overline{)199}$ → quotient **11**; **17**, **29**, **17**, remainder **12**

199 is a **prime** number.

Name ______________________ Section ______________ Date __________

## Exercises 3.2

List the first five multiples for each of the following numbers.

**1.** 3 **2.** 6 **3.** 8 **4.** 10 **5.** 11

**6.** 15 **7.** 20 **8.** 30 **9.** 41 **10.** 61

**11.** According to the arrangement related to multiples of 6, a number one more than and/or one less than a multiple of 6 is a potential prime number. Begin with the list in the text (page 191) and continue up to 100 circling the prime numbers less than 100.

| 1 | (2) | (3) | 4 | (5) | 6 | (7) | 8 | 9 | 10 |
|---|---|---|---|---|---|---|---|---|---|
| (11) | 12 | (13) | 14 | 15 | 16 | (17) | 18 | (19) | 20 |
| 21 | 22 | (23) | 24 | 25 | 26 | 27 | 28 | (29) | 30 |
| (31) | 32 | 33 | 34 | 35 | 36 | (37) | 38 | 39 | 40 |
| (41) | 42 | (43) | 44 | 45 | 46 | (47) | 48 | 49 | 50 |
| 51 | 52 | (53) | 54 | 55 | 56 | 57 | 58 | (59) | 60 |
| (61) | 62 | 63 | 64 | 65 | 66 | (67) | 68 | 69 | 70 |
| (71) | 72 | (73) | 74 | 75 | 76 | 77 | 78 | (79) | 80 |
| 81 | 82 | (83) | 84 | 85 | 86 | 87 | 88 | (89) | 90 |
| 91 | 92 | 93 | 94 | 95 | 96 | (97) | 98 | 99 | 100 |

First list all the prime numbers less than 50, and then decide whether each of the following numbers is prime or composite by dividing by prime numbers. If the number is composite, find at least two pairs of factors whose product is that number.

**12.** 23 **13.** 41 **14.** 47 **15.** 67

**16.** 55 **17.** 65 **18.** 97 **19.** 59

**20.** 205 **21.** 502 **22.** 719 **23.** 517

**24.** 943 **25.** 1073 **26.** 551 **27.** 961



**ANSWERS**

1. 3, 6, 9, 12, 15, ...
2. 6, 12, 18, 24, 30, ...
3. 8, 16, 24, 32, 40, ...
4. 10, 20, 30, 40, 50, ...
5. 11, 22, 33, 44, 55, ...
6. 15, 30, 45, 60, 75, ...
7. 20, 40, 60, 80, 100, ...
8. 30, 60, 90, 120, 150, ...
9. 41, 82, 123, 164, 205, ...
10. 61, 122, 183, 244, 305, ...
11. [Respond below exercise.]
12. Prime
13. Prime
14. Prime
15. Prime
16. Composite; 1, 55; 5, 11
17. Composite; 1, 65; 5, 13
18. Prime
19. Prime
20. Composite; 1, 205; 5, 41
21. Composite; 1, 502; 2, 251
22. Prime
23. Composite; 1, 517; 11, 47
24. Composite; 1, 943; 23, 41
25. Composite; 1, 1073; 29, 37
26. Composite; 1, 551; 19, 29
27. Composite; 1, 961; 31, 31

28. 3 and 8
29. 3 and 4
30. 2 and 8
31. 2 and 12
32. 2 and 10
33. 4 and 5
34. 4 and 9
35. 3 and 12
36. 3 and 17
37. 3 and 19
38. 5 and 5
39. 4 and 4
40. 3 and 24
41. 4 and 13
42. 5 and 11
43. 6 and 11
44. 3 and 21
45. 5 and 13
46. 3 and 25
47. 5 and 12
48. [Respond below exercise.]
49. [Respond below exercise.]
50. [Respond below exercise.]
51. [Respond below exercise.]
52. [Respond below exercise.]

In Exercises 28-47, two numbers are given. Find two factors of the first number such that their product is the first number and their sum is the second number. For example, for the numbers 45, 18, you would find two factors of 45 whose sum is 18. The factors are 3 and 15, since $3 \cdot 15 = 45$ and $3 + 15 = 18$.

**28.** 24, 11 **29.** 12, 7 **30.** 16, 10 **31.** 24, 14

**32.** 20, 12 **33.** 20, 9 **34.** 36, 13 **35.** 36, 15

**36.** 51, 20 **37.** 57, 22 **38.** 25, 10 **39.** 16, 8

**40.** 72, 27 **41.** 52, 17 **42.** 55, 16 **43.** 66, 17

**44.** 63, 24 **45.** 65, 18 **46.** 75, 28 **47.** 60, 17

**48.** Can you find two integers whose product is 60 and whose sum is –16? Explain your reasoning.

–6 and –10: $(-6)(-10) = 60$ and $(-6) + (-10) = -16$. Answers will vary as to reasoning.

**49.** Can you find two integers whose product is 75 and whose sum is –20? Explain your reasoning.

–5 and –15: $(-5)(-15) = 75$ and $(-5) + (-15) = -20$. Answers will vary as to reasoning.

**50.** Can you find two integers whose product is –20 and whose sum is –8? Explain your reasoning.

2 and –10: $2(-10) = -20$ and $2 + (-10) = -8$. Answers will vary as to reasoning.

**51.** Can you find two integers whose product is –30 and whose sum is –7? Explain your reasoning.

3 and –10: $3(-10) = -30$ and $3 + (-10) = -7$. Answers will vary as to reasoning.

**52.** Can you find two integers whose product is –50 and whose sum is +5? Explain your reasoning.

–5 and 10: $-5(10) = -50$ and $(-5) + 10 = 5$. Answers will vary as to reasoning.

Name ______________ Section __________ Date ________

**ANSWERS**

53. [Respond below exercise.]

54. [Respond below exercise.]

55. [Respond below exercise.]

## WRITING AND THINKING ABOUT MATHEMATICS

**53.** Discuss the following two statements:

**a.** All prime numbers are odd.
Not true. The number 2 is prime and even.

**b.** All positive odd numbers are prime.
Not true. Some counterexamples are 9, 15, and 21.

**54. a.** Explain why 1 is not prime and not composite.
Answers will vary. The number 1 does not have two distinct factors, or three or more factors.

**b.** Explain why 0 is not prime and not composite.
Answers will vary. Since 0 is not a counting number, it does not satisfy the definition of either prime of composite.

**55.** Numbers of the form $2^N - 1$, where N is a prime number, are sometimes prime. These prime numbers are called "Mersenne primes" (after Marin Mersenne, 1588 – 1648). Show that for N = 2, 3, 5, 7, and 13 the number $2^N - 1$ is prime, but for N = 11 the number $2^N - 1$ is composite.
(**Historical Note:** In 1978, with the use of a computer, two students at California State University at Hayward proved that the number $2^{21,701} - 1$ is prime.)

N = 2: $2^2 - 1 = 4 - 1 = 3$ 3 is prime.
N = 3: $2^3 - 1 = 8 - 1 = 7$ 7 is prime.
N = 5: $2^5 - 1 = 32 - 1 = 31$ 31 is prime.
N = 7: $2^7 - 1 = 128 - 1 = 127$ 127 is prime.
N = 13: $2^{13} - 1 = 8192 - 1 = 8191$ 8191 is prime.

N = 11: $2^{11} - 1 = 2048 - 1 = 2047$ 2047 is composite.
The factors of 2047 are: 1, 23, 89, 2047.

# 3.3 Prime Factorization

## Finding a Prime Factorization

In all our work with fractions (including multiplication, division, addition, and subtraction), we will need to find products of prime numbers called **prime factorizations**. For example, to add fractions (Section 4.1), we need to find the smallest common denominator, and we will see (Section 3.4) that this number is the **least common multiple** of the denominators.

**Objectives**

① Know the **Fundamental Theorem of Arithmetic**.

② Know the meaning of the term **prime factorization.**

③ Be able to find the prime factorization of a composite number.

The technique used in this text for finding the least common multiple of a set of integers involves finding all the prime factors of each integer. For example,

$$63 = 9 \cdot 7$$

but 9 is not prime. By factoring 9, we can write

$$63 = 9 \cdot 7 = 3 \cdot 3 \cdot 7$$

This last product ( $3 \cdot 3 \cdot 7$ ) contains all prime factors and is the **prime factorization** of 63.

Note that because multiplication is a commutative operation, the factors may be written in any order. Thus, we might write $63 = 3 \cdot 7 \cdot 3$ and still have the correct prime factorization. However, for consistency, we will generally write the factors in order, from smallest to largest.

> **Note:** For the purposes of prime factorization, a negative integer will be treated as a product of $-1$ and a positive integer. This means that only prime factorizations of positive integers need to be discussed at this time.

Regardless of the method used, you will always arrive at the same factorization for any given composite number. That is, **there is only one prime factorization for any composite number**. This fact is so important that it is called the **Fundamental Theorem of Arithmetic**.

> **The Fundamental Theorem of Arithmetic**
>
> Every composite number has exactly one prime factorization.

> **To Find the Prime Factorization of a Composite Number:**
>
> 1. Factor the composite number into any two factors.
> 2. Factor each factor that is not prime.
> 3. Continue this process until all factors are prime.
>
> The **prime factorization** is the product of all the prime factors.

The skill of finding prime factorizations is absolutely indispensable in our work with fractions. You may want to emphasize this fact several times and to be sure students understand the distinction between the terms "prime factorization" and "factors."

Many times the beginning factors needed to start the process for finding a prime factorization can be found by using the tests for divisibility by 2, 3, 5, 6, 9, and 10 discussed in Section 3.1. This was one purpose for developing these tests, and you should review them or write them down for easy reference.

### Example 1

Find the prime factorization of 90.

$$90 = 9 \cdot 10$$

Since the units digit is 0, we know that 10 is a factor.

$$= 3 \cdot 3 \cdot 2 \cdot 5$$

9 and 10 can both be factored so that each factor is a prime number. This is the prime factorization.

Or

$$90 = 3 \cdot 30$$

3 is prime, but 30 is not.

$$= 3 \cdot 10 \cdot 3$$

10 is not prime.

$$= 3 \cdot 2 \cdot 5 \cdot 3$$

All factors are prime.

Note that the final prime factorization was the same in both factor trees even though the first pair of factors was different.

Since multiplication is commutative, the order of the factors is not important. What is important is that **all the factors are prime**. Writing the factors in order, we can write

$$90 = 2 \cdot 3 \cdot 3 \cdot 5$$

or, with exponents,

$$90 = 2 \cdot 3^2 \cdot 5$$

### Example 2

Find the prime factorizations of each number : **a.** 65, **b.** 72, and **c.** 294.

**Solution**

**a.** $65 = 5 \cdot 13$

5 is a factor because the units digit is 5. Since both 5 and 13 are prime, $5 \cdot 13$ is the prime factorization.

**b.** $72 = 8 \cdot 9$

9 is a factor because the sum of the digits is 9.

$$= 2 \cdot 4 \cdot 3 \cdot 3$$

$$= 2 \cdot 2 \cdot 2 \cdot 3 \cdot 3$$

$$= 2^3 \cdot 3^2$$

using exponents

**c.** 294 = 2 · 147 — 2 is a factor because the units digits is even.

= 2 · 3 · 49 — 3 is a factor of 147 because the sum of the digits is divisible by 3.

= 2 · 3 · 7 · 7

= $2 \cdot 3 \cdot 7^2$ using exponents

Begin with the product 294 = 6 · 49, and show that you can arrive at the same prime factorization. (Note that 6 is a factor of 294 since both 2 and 3 are factors.)

### Completion Example 3

Find the prime factorization of 60.

**Solution**

60 = 6 · ____

= 2 · 3 · ____ · ____

= ________ using exponents

### Completion Example 4

Find the prime factorization of 308.

**Solution**

308 = 4 · ____

= ____ · ____ · ____ · ____

= ________ using exponents

*Now Work Exercises 1- 4 in the Margin.*

Find the prime factorization of each of the following composite numbers.

**1.** 74
2 · 37

**2.** 52
$2^2 \cdot 13$

**3.** 460
$2^2 \cdot 5 \cdot 23$

**4.** 616
$2^3 \cdot 7 \cdot 11$

## Factors of Composite Numbers

Once a prime factorization of a number is known, all the factors (or divisors) of that number can be found. For a number to be a factor of a composite number, it must be either 1, the number itself, one of the prime factors, or the product of two or more of the prime factors. (See the discussion of divisibility of products in Section 3.1 for a similar analysis.)

The only factors (or divisors) of a composite number are

**1.** 1 and the number itself,
**2.** each prime factor, and
**3.** products formed by all combinations of the prime factors (including repeated factors.)

**For your information**
From number theory: The number of factors of a counting number is the product of the numbers found by adding 1 to each of the exponents in the prime factorization of the number. For example,

$$60 = 2^2 \cdot 3^1 \cdot 5^1$$
$$(2+1)(1+1)(1+1)$$
$$= (3)(2)(2) = 12$$

So, 60 has 12 factors (as shown in Example 5).

### Example 5

Find all the factors of 60.

**Solution**

Since $60 = 2 \cdot 2 \cdot 3 \cdot 5$, the factors are

**a.** 1 and 60 (1 and the number itself),

**b.** 2, 3, and 5 (each prime factor),

**c.** $2 \cdot 2 = 4$, $2 \cdot 3 = 6$, $2 \cdot 5 = 10$, $3 \cdot 5 = 15$, $2 \cdot 2 \cdot 3 = 12$, $2 \cdot 2 \cdot 5 = 20$, and $2 \cdot 3 \cdot 5 = 30$.

The twelve factors of 60 are

1, 60, 2, 3, 5, 4, 6, 10, 12, 15, 20, and 30.

These are the only factors of 60.

### *Completion Example 6*

Find all the factors of 154.

**Solution**

The prime factorization of 154 is $2 \cdot 7 \cdot 11$.

**a.** Two factors are 1 and ________.

**b.** Three prime factors are ________, ________, and ________.

**c.** Other factors are

_____ · _____ = ______     _____ · _____ = ______

_____ · _____ = ______ .

The factors of 154 are ____________________.

Using the prime factorization of each number, find all the factors of the number.

**5.** 63

1, 3, 7, 9, 21, 63

**6.** 54

1, 2, 3, 6, 9, 18, 27, 54

**7.** 88

1, 2, 4, 8, 11, 22, 44, 88

*Now Work Exercises 5- 7 in the Margin.*

### Completion Example Answers

**3.** $60 = \mathbf{6 \cdot 10}$

$= \mathbf{2 \cdot 3 \cdot 2 \cdot 5}$

$= \mathbf{2^2 \cdot 3 \cdot 5}$ using exponents

**4.** $308 = \mathbf{4 \cdot 77}$

$= \mathbf{2 \cdot 2 \cdot 7 \cdot 11}$

$= \mathbf{2^2 \cdot 7 \cdot 11}$ using exponents

**6. a.** Two factors are 1 and **154**.

**b.** Three prime factors are **2, 7,** and **11**.

**c.** Other factors are

$\mathbf{2 \cdot 7 = 14}$    $\mathbf{2 \cdot 11 = 22}$    $\mathbf{7 \cdot 11 = 77}$.

The factors of 154 are **1, 154, 2, 7, 11, 14, 22,** and **77.**

Name ____________ Section ____________ Date ____________

## Exercises 3.3

Find the prime factorization for each of the following numbers. Use the tests for divisibility for 2, 3, 5, 6, 9, and 10, whenever they help, to find beginning factors.

**1.** 20 **2.** 44 **3.** 32 **4.** 45

**5.** 50 **6.** 36 **7.** 70 **8.** 80

**9.** 51 **10.** 162 **11.** 62 **12.** 125

**13.** 99 **14.** 94 **15.** 37 **16.** 43

**17.** 120 **18.** 225 **19.** 196 **20.** 289

**21.** 361 **22.** 400 **23.** 65 **24.** 91

**25.** 1000 **26.** 10,000 **27.** 100,000 **28.** 1,000,000



**ANSWERS**

1. $20 = 2^2 \cdot 5$
2. $44 = 2^2 \cdot 11$
3. $32 = 2^5$
4. $45 = 3^2 \cdot 5$
5. $50 = 2 \cdot 5^2$
6. $36 = 2^2 \cdot 3^2$
7. $70 = 2 \cdot 5 \cdot 7$
8. $80 = 2^4 \cdot 5$
9. $51 = 3 \cdot 17$
10. $162 = 2 \cdot 3^4$
11. $62 = 2 \cdot 31$
12. $125 = 5^3$
13. $99 = 3^2 \cdot 11$
14. $94 = 2 \cdot 47$
15. 37 is prime.
16. 43 is prime.
17. $120 = 2^3 \cdot 3 \cdot 5$
18. $225 = 3^2 \cdot 5^2$
19. $196 = 2^2 \cdot 7^2$
20. $289 = 17^2$
21. $361 = 19^2$
22. $400 = 2^4 \cdot 5^2$
23. $65 = 5 \cdot 13$
24. $91 = 7 \cdot 13$
25. $1000 = 2^3 \cdot 5^3$
26. $10{,}000 = 2^4 \cdot 5^4$
27. $100{,}000 = 2^5 \cdot 5^5$
28. $1{,}000{,}000 = 2^6 \cdot 5^6$

29. $600 = 2^3 \cdot 3 \cdot 5^2$

30. $700 = 2^2 \cdot 5^2 \cdot 7$

31. 107 is prime.

32. 211 is prime.

33. $309 = 3 \cdot 103$

34. $505 = 5 \cdot 101$

35. $165 = 3 \cdot 5 \cdot 11$

36. $231 = 3 \cdot 7 \cdot 11$

37. $675 = 3^3 \cdot 5^2$

38. $135 = 3^3 \cdot 5$

39. $216 = 2^3 \cdot 3^3$

40. $1125 = 3^2 \cdot 5^3$

41. $1692 = 2^2 \cdot 3^2 \cdot 47$

42. $2200 = 2^3 \cdot 5^2 \cdot 11$

43. $676 = 2^2 \cdot 13^2$

44. $2717 = 11 \cdot 13 \cdot 19$

45. a. $42 = 2 \cdot 3 \cdot 7$
b. [Respond below exercise.]

46. a. $24 = 2^3 \cdot 3$
b. [Respond below exercise.]

47. a. $300 = 2^2 \cdot 3 \cdot 5^2$
b. [Respond below exercise.]

48. a. $700 = 2^2 \cdot 5^2 \cdot 7$
b. [Respond below exercise.]

49. a. $66 = 2 \cdot 3 \cdot 11$
b. [Respond below exercise.]

50. a. $96 = 2^5 \cdot 3$
b. [Respond below exercise.]

51. a. $78 = 2 \cdot 3 \cdot 13$
b. [Respond below exercise.]

52. a. $130 = 2 \cdot 5 \cdot 13$
b. [Respond below exercise.]

**29.** 600 **30.** 700 **31.** 107 **32.** 211

**33.** 309 **34.** 505 **35.** 165 **36.** 231

**37.** 675 **38.** 135 **39.** 216 **40.** 1125

**41.** 1692 **42.** 2200 **43.** 676 **44.** 2717

For each of the following problems:

**a.** Find the prime factorization of each number.

**b.** Find all the factors (or divisors) of each number.

**45.** 42
1, 2, 3, 6, 7, 14, 21, 42

**46.** 24
1, 2, 3, 4, 6, 8, 12, 24

**47.** 300
1, 2, 3, 4, 5, 6, 10, 12, 15, 20, 25, 30, 50, 60, 75, 100, 150, 300

**48.** 700
1, 2, 4, 5, 7, 10, 14, 20, 25, 28, 35, 50, 70, 100, 140, 175, 350, 700

**49.** 66
1, 2, 3, 6, 11, 22, 33, 66

**50.** 96
1, 2, 3, 4, 6, 8, 12, 16, 24, 32, 48, 96

**51.** 78
1, 2, 3, 6, 13, 26, 39, 78

**52.** 130
1, 2, 5, 10, 13, 26, 65, 130

Name ______________ Section ______________ Date ______________

**53.** 150

1, 2, 3, 5, 6, 10, 15, 25, 30, 50, 75, 150

**54.** 175

1, 5, 7, 25, 35, 175

**55.** 90

1, 2, 3, 5, 6, 9, 10, 15, 18, 30, 45, 90

**56.** 275

1, 5, 11, 25, 55, 275

**57.** 1001

1, 7, 11, 13, 77, 91, 143, 1001

**58.** 715

1, 5, 11, 13, 55, 65, 143, 715

**59.** 585

1, 3, 5, 9, 13, 15, 39, 45, 65, 117, 195, 585

**60.** 2310

1, 2, 3, 5, 6, 7, 10, 11, 14, 15, 21, 22, 30, 33, 35, 42, 55, 66, 70, 77, 105 110, 154, 165, 210, 231, 330, 385 462, 770, 1155, 2310

## Collaborative Learning Exercise

In groups of three to four students, discuss the theorem in Exercise 61 and answer the related questions. Discuss your answers in class.

**61.** In higher level mathematics, number theorists have proven the following theorem: If the prime factorization of a number is written in exponential form and 1 is added to each exponent, then the number of factors of that number is the product of these sums.

For example, $60 = 2^2 \cdot 3 \cdot 5 = 2^2 \cdot 3^1 \cdot 5^1$. Adding 1 to each exponent and forming the product gives

$(2+1)(1+1)(1+1) = 3 \cdot 2 \cdot 2 = 12$

and there are twelve factors of 60.

Use this theorem to find the number of factors of each of the following numbers. Find as many of these factors as you can.

**a.** 700

$700 = 2^2 \cdot 5^2 \cdot 7$

$(2+1)(1+1)(2+1) = 3 \cdot 2 \cdot 3 = 18$

There are 18 factors of 700.

These factors are: 1, 2, 4, 5, 7, 10, 14, 20, 25, 28, 35, 50, 70, 100, 140, 175, 350, and 700.

**b.** 660

$660 = 2^2 \cdot 3^1 \cdot 5^1 \cdot 11^1$

$(2+1)(1+1)(1+1)(1+1) = 3 \cdot 2 \cdot 2 \cdot 2 = 24$. There are 24 factors of 660.

These factors are: 1, 2, 3, 4, 5, 6, 10, 11, 12, 15, 20, 22, 30, 33, 44, 55, 60, 66, 110, 132, 165, 220, 330, and 660.

**c.** 450

$450 = 2^1 \cdot 3^2 \cdot 5^2$

$(1+1)(2+1)(2+1) = 2 \cdot 3 \cdot 3 = 18$

There are 18 factors of 450.

These factors are: 1, 2, 3, 5, 6, 9, 10, 15, 18, 25, 30, 45, 50, 75, 90, 150, 225, and 450 .

**d.** 148,225

$148{,}225 = 5^2 \cdot 7^2 \cdot 11^2$

$(2+1)(2+1)(2+1) = 3 \cdot 3 \cdot 3 = 27$

There are 27 factors of 148,225.

These factors are: 1, 5, 7, 11, 25, 35, 49, 55, 77, 121, 175, 245, 275, 385, 539, 605, 847, 1225, 1925, 2695, 3025, 4235, 5929, 13, 475, 21, 175, 29, 645, 148, and 225.

**ANSWERS**

**53. a.** $150 = 2 \cdot 3 \cdot 5^2$
**b.** [Respond below exercise.]

**54. a.** $175 = 5^2 \cdot 7$
**b.** [Respond below exercise.]

**55. a.** $90 = 2 \cdot 3^2 \cdot 5$
**b.** [Respond below exercise.]

**56. a.** $275 = 5^2 \cdot 11$
**b.** [Respond below exercise.]

**57. a.** $1001 = 7 \cdot 11 \cdot 13$
**b.** [Respond below exercise.]

**58. a.** $715 = 5 \cdot 11 \cdot 13$
**b.** [Respond below exercise.]

**59. a.** $585 = 3^2 \cdot 5 \cdot 13$
**b.** [Respond below exercise.]

**60. a.** $2310 = 2 \cdot 5 \cdot 3 \cdot 7 \cdot 11$
**b.** [Respond below exercise.]

**61. a.** [Respond below exercise.]
**b.** [Respond below exercise.]
**c.** [Respond below exercise.]
**d.** [Respond below exercise.]

The Recycle Bin

1. $-1$
2. $-27$
3. $-1$
4. $-56$
5. $36$
6. $-27$
7. $-10$
8. $5$

## The Recycle Bin (from Sections 2.2 and 2.3)

Find each of the following sums or differences.

**1.** $-5+(-3)-(-7)$

**2.** $15-22-30+10$

**3.** $-16-5+20$

**4.** $-28-2-5-15-6$

**5.** $36+(-4)-(-4)$

**6.** $-10-20-(-3)$

**7.** $|-5|-|-15|$

**8.** $25+|-12|-|32|$

# 3.4 Least Common Multiple ( LCM )

## Finding the LCM of a Set of Counting ( or Natural ) Numbers

The ideas discussed in this section related to factors, prime factors, and **least common multiple** are used throughout our development of fractions and mixed numbers in Chapters 3 and 4. Study these ideas and the related techniques carefully because they will make your work with fractions much easier and more understandable.

**Objectives**

① Understand the meaning of the term **multiple**.

② Be able to use prime factorizations to find the **least common multiple** of a set of numbers and a set of algebraic terms.

③ Recognize the application of the LCM concept in a word problem.

Recall from Section 3.2 that the multiples of an integer are the products of that integer with the counting numbers. Our discussion here is based entirely on multiples of positive integers (or counting numbers). Thus, the first multiple of any counting number is the number itself, and all other multiples are larger than that number. We are interested in finding common multiples and more particularly the **least common multiple** for a set of counting numbers. For example, consider the lists of multiples of 8 and 12 shown here.

| | |
|---|---|
| **Counting Numbers:** | **1, 2, 3, 4, 5, 6, 7, 8, 9, 10, 11, ...** |
| Multiples of 8: | 8, 16, (24), 32, 40, (48), 56, 64, (72), 80, 88, ... |
| Multiples of 12: | 12, (24), 36, (48), 60, (72), 84, (96), 108, (120), 132, ... |

The common multiples of 8 and 12 are 24, 48, 72, 96, 120, ... .The **least common multiple (LCM)** is 24.

Listing all the multiples, as we just did for 8 and 12, and then choosing the least common multiple (LCM) is not very efficient. The following technique involving prime factorizations is generally much easier to use.

You might want to inform students now that this is the technique for finding the least common denominator when adding and subtracting fractions. This knowledge may serve as a motivating factor for some of them.

**To Find the LCM of a Set of Counting Numbers:**

1. Find the prime factorization of each number.
2. Find the prime factors that appear in any one of the prime factorizations.
3. Form the product of these primes using each prime the most number of times it appears in any one of the prime factorizations.

Your skill with this method depends on your ability to find prime factorizations quickly and accurately. With practice and understanding, this method will prove efficient and effective. STAY WITH IT!

(**Note**: There are other methods for finding the LCM, maybe even easier to use at first. However, just finding the LCM is not our only purpose, and the method outlined here allows for a solid understanding of using the LCM when working with fractions.)

### Example 1

Find the least common multiple (LCM) of 8, 10, and 30.

**a.** Prime factorizations:

$8 = 2 \cdot 2 \cdot 2$ three 2's

$10 = 2 \cdot 5$ one 2, one 5

$30 = 2 \cdot 3 \cdot 5$ one 2, one 3, one 5

**b.** Prime factors that are present are 2, 3, and 5.

**c.** The most of each prime factor in any one factorization:

| | |
|---|---|
| Three 2's | ( in 8 ) |
| One 3 | ( in 30 ) |
| One 5 | ( in 10 and in 30 ) |

$$\text{LCM} = 2 \cdot 2 \cdot 2 \cdot 3 \cdot 5$$
$$= 2^3 \cdot 3 \cdot 5 = 120$$

120 is the smallest number divisible by 8, 10, and 30.

## Example 2

Find the LCM of 27, 30, 35, and 42.

### Solution

**a.** Prime factorizations:

| | |
|---|---|
| $27 = 3 \cdot 3 \cdot 3$ | three 3's |
| $30 = 2 \cdot 3 \cdot 5$ | one 2, one 3, one 5 |
| $35 = 5 \cdot 7$ | one 5, one 7 |
| $42 = 2 \cdot 3 \cdot 7$ | one 2, one 3, one 7 |

**b.** Prime factors present are 2, 3, 5, and 7.

**c.** The most of each prime factor in any one factorization:

| | |
|---|---|
| One 2 | ( in 30 and in 42 ) |
| Three 3's | ( in 27 ) |
| One 5 | ( in 30 and in 35 ) |
| One 7 | ( in 35 and in 42 ) |

$$\text{LCM} = 2 \cdot 3 \cdot 3 \cdot 3 \cdot 5 \cdot 7 = 2 \cdot 3^3 \cdot 5 \cdot 7$$
$$= 1890$$

1890 is the smallest number divisible by all four of the numbers 27, 30, 35, and 42.

## *Completion Example 3*

Find the LCM for 15, 24, and 36.

### Solution

**a.** Prime factorizations:

15 = ____________
24 = ____________
36 = ____________

**b.** The only prime factors are ________, ________, and ________.

c. The most of each prime factor in any one factorization:

______ ( in 24 )
______ ( in 36 )
______ ( in 15 )

LCM = ______ = ______ (using exponents)

= ______

*Now Work Exercises 1 and 2 in the Margin.*

In Example 2, we found that 1890 is the LCM for 27, 30, 35, and 42. This means that 1890 is a multiple of each of these numbers, and each is a factor of 1890. To find out how many times each number will divide into 1890, we could divide by using long division. However, by looking at the prime factorization of 1890 (which we know) and the prime factorization of each number, we find the quotients without actually dividing. We can group the factors as follows:

$$1890 = \underbrace{(3\cdot3\cdot3)}_{27}\cdot\underbrace{(2\cdot5\cdot7)}_{70} = \underbrace{(2\cdot3\cdot5)}_{30}\cdot\underbrace{(3\cdot3\cdot7)}_{63}$$

$$= \underbrace{(5\cdot7)}_{35}\cdot\underbrace{(2\cdot3\cdot3\cdot3)}_{54} = \underbrace{(2\cdot3\cdot7)}_{42}\cdot\underbrace{(3\cdot3\cdot5)}_{45}$$

So,

27 divides into 1890 70 times,
30 divides into 1890 63 times,
35 divides into 1890 54 times, and
42 divides into 1890 45 times.

## Completion Example 4

a. Find the LCM for 18, 36, and 66, and
b. tell how many times each number divides into the LCM.

### Solution

a. 18 = ______
36 = ______ } LCM = ______
66 = ______

= ______ = 396

b. 396 = ______ = ( 2 · 3 · 3 ) · ______ = 18 · ______
396 = ______ = ( 2 · 2 · 3 · 3 ) · ______ = 36 · ______
396 = ______ = ( 2 · 3 · 11 ) · ______ = 66 · ______

*Now Work Exercises 3 and 4 in the Margin.*

Find the LCM for each set of counting numbers.

**1.** 30, 40, 70
$2^3 \cdot 3 \cdot 5 \cdot 7 = 840$

**2.** 140, 168
$2^3 \cdot 3 \cdot 5 \cdot 7 = 840$

Understanding the relationship between each number in a set and the prime factorization of the LCM of these numbers will increase the ease and speed with which students can perform addition and subtraction with fractions. In these operations, each denominator (and corresponding numerator) is multiplied by the number of times that the denominator divides into the LCM of the denominators. You may want to reinforce this idea several times.

For each set of counting numbers:
**a.** find the LCM, and
**b.** tell how many times each number divides into the LCM.

**3.** 20, 35, 70
a. LCM = 140
b. 140 = 20 · 7
= 35 · 4
= 70 · 2

**4.** 24, 35, 42
a. LCM = 840
b. 840 = 24 · 35
= 35 · 24
= 42 · 20

## Finding The LCM of a Set of Algebraic Terms

Algebraic expressions that are numbers, powers of variables, or products of numbers and powers of variables are called **terms**. Note that a number written next to a variable or two variables written next to each other indicate multiplication. For example,

$$3,\ 4ab,\ 25a^3,\ 6xy^2,\ 48x^3y^4$$

are all algebraic terms. In each case where multiplication is indicated, the number is called the **numerical coefficient** of the term. A term that consists of only a number, such as 3, is called a **constant** or a **constant term**.

Another approach to finding the LCM, particularly useful when algebraic terms are involved, is to write each prime factorization in exponential form and proceed as follows:

**Step 1:** Find the prime factorization of each term in the set and write it in exponential form, including variables.

**Step 2:** Find the largest power of each prime factor present in all of the prime factorizations.
(Remember that, if no exponent is written, the exponent is understood to be 1.)

**Step 3:** The LCM is the product of these powers.

### Example 5

Find the LCM of the terms $6a$, $a^2b$, $4a^2$, and $18b^3$.

#### Solution

Write each prime factorization in exponential form (including variables) and multiply the largest powers of each prime factor, including variables.

$$\left.\begin{aligned} 6a &= 2\cdot 3\cdot a \\ a^2b &= a^2\cdot b \\ 4a^2 &= 2^2\cdot a^2 \\ 18b^3 &= 2\cdot 3^2\cdot b^3 \end{aligned}\right\}\quad \text{LCM} = 2^2\cdot 3^2\cdot a^2\cdot b^3 = 36a^2b^3$$

### Example 6

Find the LCM of the terms $16x$, $25xy^3$, $30x^2y$.

#### Solution

Write each prime factorization in exponential form (including variables) and multiply the largest powers of each prime factor, including variables.

$$\left.\begin{aligned} 16x &= 2^4\cdot x \\ 25xy^3 &= 5^2\cdot x\cdot y^3 \\ 30x^2y &= 2\cdot 3\cdot 5\cdot x^2\cdot y \end{aligned}\right\}\quad \text{LCM} = 2^4\cdot 3\cdot 5^2\cdot x^2\cdot y^3 = 1200x^2y^3$$

## Completion Example 7

Find the LCM of the terms $8xy$, $10x^2$, and $20y$.

### Solution

$$\left.\begin{aligned} 8xy &= \_\_\_\_\_\_\_\_ \\ 10x^2 &= \_\_\_\_\_\_\_\_ \\ 20y &= \_\_\_\_\_\_\_\_ \end{aligned}\right\} \text{LCM} = \_\_\_\_\_\_\_\_$$

$$= \_\_\_\_\_\_\_\_$$

*Now Work Exercises 5 and 6 in the Margin.*

Find the LCM for each set of algebraic terms.

**5.** $14xy$, $10y^2$, $25x$
$350xy^2$

**6.** $15ab^2$, $ab^3$, $20a^3$
$60a^3b^3$

### An Application

Many events occur at regular intervals of time. Weather satellites may orbit the earth once every 10 hours or once every 12 hours. Delivery trucks arrive once a day or once a week at department stores. Traffic lights change once every 3 minutes or once every 2 minutes. The periodic frequency with which such events occur can be explained in terms of the least common multiple, as illustrated in Example 8.

## Example 8

Suppose three weather satellites – A, B, and C – are orbiting the earth at different times. Satellite A takes 24 hours, B takes 18 hours, and C takes 12 hours. If they are directly above each other now, as shown in part **a.** of the figure below, in how many hours will they again be directly above each other in the position shown in part **a.** ? How many orbits will each satellite have made in that time?

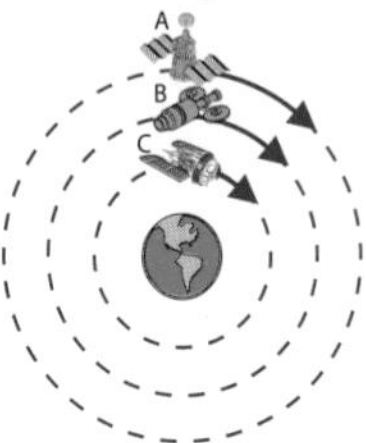


**a.** Begining positions

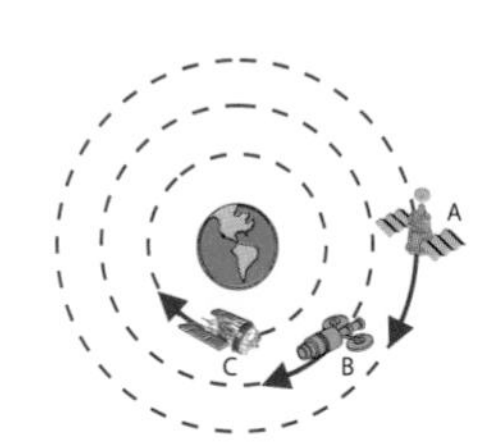


**b.** Positions after 6 hours

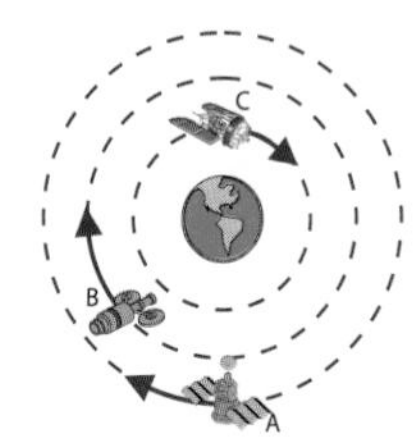


**c.** Position after 12 hours

### Solution

Study the diagrams shown on the previous page. When the three satellites are again in the position shown, each will have made some number of complete orbits. Since A takes 24 hours to make one complete orbit, the solution must be a multiple of 24. Similarly, the solution must be a multiple of 18 and a multiple of 12 to allow for complete orbits of satellites B and C.

The solution is the LCM of 24, 18, and 12.

$$\left.\begin{aligned} 24 &= 2^3 \cdot 3 \\ 18 &= 2 \cdot 3^2 \\ 12 &= 2^2 \cdot 3 \end{aligned}\right\} \text{LCM} = 2^3 \cdot 3^2 = 72$$

Thus, the satellites will align again at the position shown in 72 hours (or 3 days.) Note that:

Satellite A will have made 3 orbits: $24 \cdot \mathbf{3} = 72$
Satellite B will have made 4 orbits: $18 \cdot \mathbf{4} = 72$
Satellite C will have made 6 orbits: $12 \cdot \mathbf{6} = 72$

## Completion Example Answers

**3. a.** Prime factorizations:

$15 = \mathbf{3 \cdot 5}$
$24 = \mathbf{2 \cdot 2 \cdot 2 \cdot 3}$
$36 = \mathbf{2 \cdot 2 \cdot 3 \cdot 3}$

**b.** The only prime factors are **2, 3**, and **5**.

**c.** The most of each prime factor in any one factorization:

**Three 2's** ( in 24 )
**Two 3's** ( in 36 )
**One 5** ( in 15 )

LCM $= \mathbf{2 \cdot 2 \cdot 2 \cdot 3 \cdot 3 \cdot 5 = 2^3 \cdot 3^2 \cdot 5 = 360}$ using exponents

**4. a.**

$$\left.\begin{aligned} 18 &= \mathbf{2 \cdot 3 \cdot 3} \\ 36 &= \mathbf{2 \cdot 2 \cdot 3 \cdot 3} \\ 66 &= \mathbf{2 \cdot 3 \cdot 11} \end{aligned}\right\} \quad \text{LCM} = \mathbf{2 \cdot 2 \cdot 3 \cdot 3 \cdot 11}$$

**b.** $396 = \mathbf{2 \cdot 2 \cdot 3 \cdot 3 \cdot 11} = (2 \cdot 3 \cdot 3) \cdot (\mathbf{2 \cdot 11}) = 18 \cdot \mathbf{22}$
$396 = \mathbf{2 \cdot 2 \cdot 3 \cdot 3 \cdot 11} = (2 \cdot 2 \cdot 3 \cdot 3) \cdot (\mathbf{11}) = 36 \cdot \mathbf{11}$
$396 = \mathbf{2 \cdot 2 \cdot 3 \cdot 3 \cdot 11} = (2 \cdot 3 \cdot 11) \cdot (\mathbf{2 \cdot 3}) = 66 \cdot \mathbf{6}$

**7.**

$$\left.\begin{aligned} 8xy &= \mathbf{2^3 \cdot x \cdot y} \\ 10x^2 &= \mathbf{2 \cdot 5 \cdot x^2} \\ 20y &= \mathbf{2^2 \cdot 5 \cdot y} \end{aligned}\right\} \quad \text{LCM} = \mathbf{2^3 \cdot 5 \cdot x^2 \cdot y}$$

$$= \mathbf{40x^2y}$$

Name ______________________ Section ____________ Date __________

## Exercises 3.4

Find the LCM of each of the following sets of counting numbers.

**1.** 3, 5, 7

**2.** 2, 7, 11

**3.** 6, 10

**4.** 9, 12

**5.** 2, 3, 11

**6.** 3, 5, 13

**7.** 4, 14, 35

**8.** 10, 12, 20

**9.** 50, 75

**10.** 30, 70

**11.** 20, 90

**12.** 50, 80

**13.** 28, 98

**14.** 10, 15, 35

**15.** 6, 12, 15

**16.** 34, 51, 54

**17.** 22, 44, 121

**18.** 15, 45, 90

**19.** 14, 28, 56

**20.** 20, 50, 100

**21.** 30, 60, 120

**22.** 35, 40, 72

**23.** 144, 169, 196

**24.** 225, 256, 324

**25.** 81, 256, 361

**ANSWERS**

1. 105
2. 154
3. 30
4. 36
5. 66
6. 195
7. 140
8. 60
9. 150
10. 210
11. 180
12. 400
13. 196
14. 210
15. 60
16. 918
17. 484
18. 90
19. 56
20. 100
21. 120
22. 2520
23. 1,192,464
24. 518,400
25. 7,485,696

26. [Respond below exercise.]

27. [Respond below exercise.]

28. [Respond below exercise.]

29. [Respond below exercise.]

30. [Respond below exercise.]

31. [Respond below exercise.]

32. [Respond below exercise.]

33. [Respond below exercise.]

34. [Respond below exercise.]

35. [Respond below exercise.]

36. $200xyz$

37. $600xy^2z$

38. $100a^2b^3$

39. $112a^3bc$

40. $60x^2y$

41. $210a^2b^3$

42. $700x^2y^2z$

43. $84ab^2c^2$

In Exercises 26 – 35,

**a.** find the LCM of each set of numbers, and

**b.** state the number of times each number divides into the LCM.

**26.** 10, 15, 25
a. LCM = 150
b. $150 = 10 \cdot 15 = 15 \cdot 10 = 25 \cdot 6$

**27.** 6, 24, 30
a. LCM = 120
b. $120 = 6 \cdot 20 = 24 \cdot 5 = 30 \cdot 4$

**28.** 10, 18, 90
a. LCM = 90
b. $90 = 10 \cdot 9 = 18 \cdot 5 = 90 \cdot 1$

**29.** 12, 18, 27
a. LCM = 108
b. $108 = 12 \cdot 9 = 18 \cdot 6 = 27 \cdot 4$

**30.** 20, 28, 45
a. LCM = 1260
b. $1260 = 20 \cdot 63 = 28 \cdot 45 = 45 \cdot 28$

**31.** 99, 121, 143
a. LCM = 14,157
b. $14{,}157 = 99 \cdot 143 = 121 \cdot 117 = 143 \cdot 99$

**32.** 125, 135, 225, 250
a. LCM = 6750
b. $6750 = 125 \cdot 54 = 135 \cdot 50 = 225 \cdot 30 = 250 \cdot 27$

**33.** 40, 56, 160, 196
a. LCM = 7840
b. $7840 = 40 \cdot 196 = 56 \cdot 140 = 160 \cdot 49 = 196 \cdot 40$

**34.** 35, 49, 63, 126
a. LCM = 4410
b. $4410 = 35 \cdot 126 = 49 \cdot 90 = 63 \cdot 70 = 126 \cdot 35$

**35.** 45, 56, 98, 99
a. LCM = 194,040
b. $194{,}040 = 45 \cdot 4312 = 56 \cdot 3465 = 98 \cdot 1980 = 99 \cdot 1960$

Find the LCM of each of the following sets of algebraic terms.

**36.** $25xy, 40xyz$

**37.** $40xyz, 75xy^2$

**38.** $20a^2b, 50ab^3$

**39.** $16abc, 28a^3b$

**40.** $10x, 15x^2, 20xy$

**41.** $14a^2, 10ab^2, 15b^3$

**42.** $20xyz, 25xy^2, 35x^2z$

**43.** $12ab^2, 28abc, 21bc^2$

Name ______________________ Section ____________ Date __________

**44.** $16mn, 24m^2, 40mnp$

**45.** $30m^2n, 60mnp^2, 90np$

**46.** $13x^2y^3, 39xy, 52xy^2z$

**47.** $27xy^2z, 36xyz^2, 54x^2yz$

**48.** $45xyz, 125x^3, 150y^2$

**49.** $33ab^3, 66b^2, 121$

**50.** $15x, 25x^2, 30x^3, 40x^4$

**51.** $10y, 20y^2, 30y^3, 40y^4$

**52.** $6a^5, 18a^3, 27a, 45$

**53.** $12c^4, 95c^2, 228$

**54.** $99xy^3, 143x^3, 363$

**55.** $18abc^2, 27ax, 30ax^2y, 34a^2bc$

**56.** $25axy, 35x^5y, 55a^2by^2, 65a^3x^2$

**57.** Three security guards meet at the front gate for coffee before they walk around inspecting buildings at a manufacturing plant. The guards take 15, 20, and 30 minutes, respectively, for the inspection trip.

**a.** If they start at the same time, in how many minutes will they meet again at the front gate for coffee?

**b.** How many trips will each guard have made?

**ANSWERS**

**44.** $240m^2np$

**45.** $180m^2np^2$

**46.** $156x^2y^3z$

**47.** $108x^2y^2z^2$

**48.** $2250x^3y^2z$

**49.** $726ab^3$

**50.** $600x^4$

**51.** $120y^4$

**52.** $270a^5$

**53.** $1140c^4$

**54.** $14,157x^3y^3$

**55.** $4590a^2bc^2x^2y$

**56.** $25,025a^3bx^5y^2$

**57.a.** 60 minutes

**b.** 4, 3, and 2 trips, respectively

58. a. 90 hours
b. 6 and 5 orbits, respectively

59. a. Every 24 days
b. Every 24 days

60. a. Every 70 days
b. 13, 9, and 6 times, respectively

61. Every 180 days

62. a. 840 days
b. 105, 70, 60, and 56 trips, respectively

63. [Respond below exercise.]

64. [Respond below exercise.]

**58.** Two astronauts miss connections at their first rendezvous in space.

**a.** If one astronaut circles the earth every 15 hours and the other every 18 hours, in how many hours will they rendezvous again at the same place?

**b.** How many orbits will each astronaut have made before the second rendezvous?

**59.** Three truck drivers have dinner together whenever all three are at the routing station at the same time. The first driver's route takes 6 days, the second driver's route takes 8 days, and the third driver's route takes 12 days.

**a.** How frequently do the three drivers have dinner together?

**b.** How frequently do the first two drivers meet?

**60.** Three neighbors mow their lawns at different intervals during the summer months. The first one mows every 5 days, the second every 7 days, and the third every 10 days.

**a.** How frequently do they mow their lawns on the same day?

**b.** How many times does each neighbor mow in between the times when they all mow together?

**61.** Four ships leave the port on the same day. They take 12, 15, 18, and 30 days, respectively, to sail their routes and reload cargo. How frequently do these four ships leave this port on the same day?

**62.** Four women work for the same book company selling textbooks. They leave the home office on the same day and take 8 days, 12 days, 14 days, and 15 days, respectively, to visit schools in their own sales regions.

**a.** In how many days will they all meet again at the home office?

**b.** How many sales trips will each have made in this time?

## Writing and Thinking about Mathematics

**63.** Explain in your own words, why each number in a set divides the LCM of that set numbers.

Explanations will vary. A multiple of a number is always divisible by the number. By definition, the LCM of a set of numbers is a multiple of each of those numbers.

**64.** Explain, in your own words, why the LCM of a set of numbers is greater than or equal to each number in the set.

Explanations will vary. The multiples of any number are that number and other numbers that are larger. Therefore, the LCM of a set of numbers must be the largest number of the set or some number larger than the larger number.

Name ____________ Section ________ Date ________

ANSWERS

65. a. 300

b. $735a^3b^5c^4$

c. $756x^2y^2z^3$

## An Alternate Approach to Finding the LCM

The following alternate approach to finding the LCM is simply a different "look" at the approach discussed in the text. That is, a slightly different analysis of the prime factored forms of each number lead to the LCM.

One, rather intuitive, approach to finding the least common multiple (LCM) of a set of numbers or algebraic terms involves a step-by-step building process that relies on the concept of divisibility. In this approach, we proceed as follows:

**Step 1:** Find the prime factorization of each term in the set.
**Step 2:** Write the prime factorization of **any one** of the terms in the set.
**Step 3:** Look at a second term, and multiply what you have in step 2 by any prime factors or variables needed to ensure divisibility by this second term.
**Step 4:** Continue as in step 3 with each of the remaining terms in the set.
**Step 5:** The resulting product will be the LCM of the original set of terms.

This technique is illustrated as follows:

**Step 1:** Find the prime factorization of each term by writing the prime factorization of each coefficient and writing the powers of the variables in the form of products.

$$10x^2 = 2 \cdot 5 \cdot x \cdot x$$
$$15xy = 3 \cdot 5 \cdot x \cdot y$$
$$25xy^3 = 5 \cdot 5 \cdot x \cdot y \cdot y \cdot y$$

**Step 2:** First, write the factorization of $10x^2$, and then proceed to multiply by missing factors of $15xy$ and $25xy^3$ as follows:

$$\text{LCM} = \underline{2 \cdot 5 \cdot x \cdot x} \cdot$$

Having all the factors of $10x^2$ ensures divisibility by $10x^2$.

**Step 3:** Multiply by whatever new factors you need to ensure divisibility by $15xy$:

$$\text{LCM} = 2 \cdot \underline{5 \cdot x \cdot x \cdot 3 \cdot y} \cdot$$

Having all the factors of $15xy$ ensures divisibility by $15xy$.

**Step 4:** Multiply by whatever new factors you need to ensure divisibility by $25xy^3$.

$$\text{LCM} = 2 \cdot 5 \cdot x \cdot \underline{x \cdot 3 \cdot y \cdot 5 \cdot y} \cdot y$$

Having all the factors of $25xy^3$ ensures divisibility by $25xy^3$.

**Step 5:** The LCM is now found because it is divisible by each of the terms in the original set of terms.

$$\text{LCM} = 2 \cdot 3 \cdot 5^2 \cdot x^2 \cdot y^3 = 150x^2y^3$$

**65.** Use the method discussed above to find the LCM of each of the following sets of terms.

**a.** 15, 75, 100 **b.** $35a^3b^2, 49a^2b^5, 21abc^4$ **c.** $12xyz, 21x^2y^2z^2, 27xy^2z^3$

The Recycle Bin

1. $-15$
2. $56$
3. $-300$
4. $16$
5. $-104$
6. $6$
7. $-139$
8. $-14$

## The Recycle Bin (from Section 2.4)

Perform the indicated operations.

1. $(-5)(-3)(-1)$
2. $(-7)(+4)(-2)$
3. $10(-3)(-2)(-5)$
4. $\frac{-48}{-3}$
5. $\frac{-208}{+2}$
6. $\frac{-2(-3+15)}{-4}$

Simplify the following expressions by using the rules for order of operations.

7. $5-3[(4+2)^2-6(8-10)]$
8. $(-5)^2-5^2+7[18-4(5)]$

# 3.5 Reducing and Multiplication with Fractions

## Rational Numbers

Fractions that can be simplified so that the numerator (top number) and the denominator (bottom number) are integers, with the denominator not 0, have the technical name **rational numbers**. (**Note:** There are other types of numbers that can be written in the form of fractions in which the numerator and denominator may be numbers other than integers. These numbers are called **irrational numbers** and will be studied in Chapters 5 and 8. You may be familiar with some irrational numbers such as $\pi$, $\sqrt{2}$, and $\sqrt{3}$. See Table 4 in the appendix for an interesting discussion of $\pi$.)

In Chapters 1 and 3, we discussed the fact that **division by 0 is undefined.** This fact is indicated in the following definition of a rational number by saying that the denominator cannot be 0. We say that the denominator $b$ is **nonzero** or that $b \neq 0$ ($b$ is not equal to 0).

**Objectives**

① Know that the term **rational number** is the technical term for fraction.

② Know how to multiply fractions.

③ Be able to determine what to multiply by in order to build a fraction to higher terms.

④ Know how to reduce a fraction to lowest terms.

⑤ Be able to multiply fractions and reduce at the same time.

**Definition:**

A **rational number** is a number that can be written in the form $\frac{a}{b}$ where $a$ and $b$ are integers and $b \neq 0$.

$$\frac{a}{b} \begin{array}{l} \leftarrow \text{numerator} \\ \leftarrow \text{denominator} \end{array}$$

The following diagram illustrates the fact that whole numbers and integers are also rational numbers. However, there are rational numbers that are not integers or whole numbers.

Examples of rational numbers are 5, 0, $\frac{1}{2}, \frac{3}{4}, \frac{-9}{10}$, and $\frac{17}{-3}$. Note that in fraction form $5 = \frac{5}{1}$ and $0 = \frac{0}{1}$.

The phrase "can be written" is a key idea that might need some further explanation in class. For future reference, the students will probably be interested to know that mixed numbers (in Chapter 4) and some decimals (in Chapter 5) are also rational numbers. For example, as students will later see, the mixed number $1\frac{5}{8}$ "can be written" as $\frac{13}{8}$ and the decimal number 0.23 "can be written" as $\frac{23}{100}$.

In fact, we see that every integer can be written in fraction form with a denominator of 1. We can write:

$$0 = \frac{0}{1},\ 1 = \frac{1}{1},\ 2 = \frac{2}{1},\ 3 = \frac{3}{1},\ 4 = \frac{4}{1}, \text{ and so on}$$

$$-1 = \frac{-1}{1},\ -2 = \frac{-2}{1},\ -3 = \frac{-3}{1},\ -4 = \frac{-4}{1}, \text{ and so on}$$

That is, as illustrated in the following diagram, **every integer is also a rational number.**

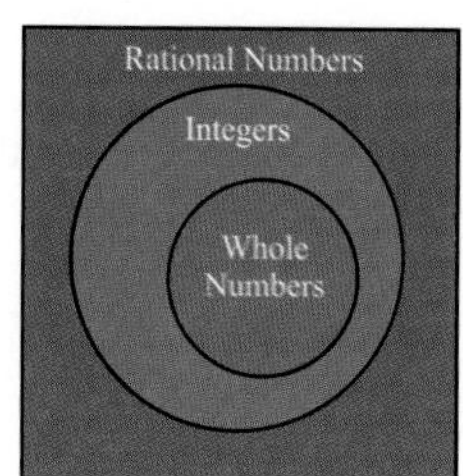

Unless otherwise stated, we will use the terms **fraction** and **rational number** to mean the same thing. Fractions can be used to indicate:

1. Equal parts of a whole or
2. Division

## Example 1

**a.** $\frac{1}{2}$ can mean 1 of 2 equal parts.

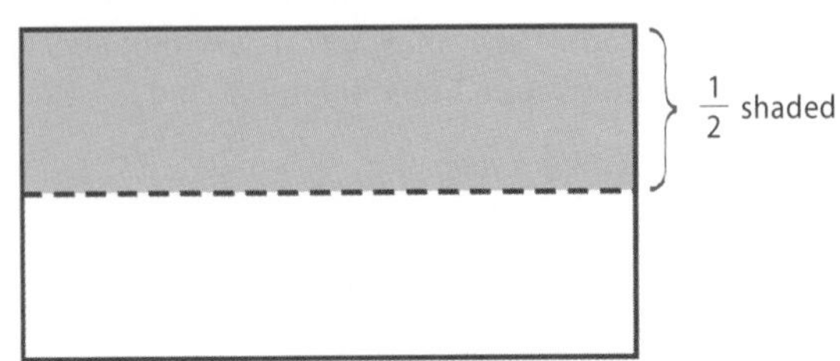


**b.** $\frac{3}{4}$ can mean 3 of 4 equal parts.

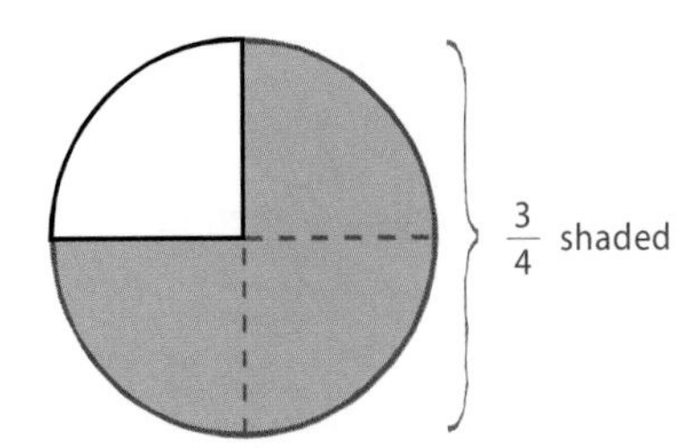


**c.** In Chapter 2, we used the fraction form of a number to indicate division. For example,

$$\frac{-35}{7} = -5 \text{ and } \frac{39}{13} = 3$$

Another closely related idea is that of **improper fractions**. These are fractions in which the numerator is greater than the denominator. For example $\frac{5}{3}, \frac{27}{15}$, and $\frac{89}{22}$ are all improper fractions. The term "improper fraction" is unfortunately somewhat misleading because there is nothing "improper" about such fractions. **In fact, in algebra and other courses in mathematics improper fractions are preferred to mixed numbers, and we will use them throughout this text.**

Some students believe that there is something "wrong" with improper fractions. You might point out that the form used for a number depends a great deal on the context of its use. For example, a carpenter would use $3\frac{1}{4}$ feet of lumber, a chemist would use 3.25 ounces of a solution, and a mathematician would substitute $\frac{13}{4}$ into an algebraic expression.

Before we begin to discuss operations with fractions, we need to clarify the use and placement of negative signs when dealing with rational numbers that are negative. Consider the following three relationships:

$$-\frac{12}{6} = -2, \qquad \frac{-12}{6} = -2, \qquad \text{and} \qquad \frac{12}{-6} = -2$$

We can conclude that a negative sign in a rational number can be placed in any one of three positions without changing the value or meaning of that number. The following statement (or rule) is true.

### Rule for the Placement of Negative Signs

If $a$ and $b$ are integers and $b \neq 0$, then

$$-\frac{a}{b} = \frac{-a}{b} = \frac{a}{-b}$$

Some students may have a question about the placement of the negative sign in a negative fraction. The fact that there are three correct positions for the placement of the negative sign can be related to division. For example,

$$-\frac{24}{3} = \frac{-24}{3} = \frac{24}{-3} = -8$$

This idea will probably need to be clarified several times throughout the course.

### Example 2

**a.** $-\frac{1}{3} = \frac{-1}{3} = \frac{1}{-3}$

**b.** $-\frac{36}{9} = \frac{-36}{9} = \frac{36}{-9} = -4$

## Multiplying Rational Numbers (or Fractions)

Now we state the rule for multiplying fractions and discuss the use of the word "of" to indicate multiplication by fractions. Remember that any integer can be written in fraction form with a denominator of 1.

### To Multiply Fractions

1. Multiply the numerators.
2. Multiply the denominators.

$$\frac{a}{b} \cdot \frac{c}{d} = \frac{a \cdot c}{b \cdot d}$$

A fraction "of" a number indicates multiplication by the fraction. Consider informing the students that this relationship is also true for decimals and percents.

Also, to relate this idea to estimating and recognizing obvious errors, if a positive number is multiplied by a positive fraction less than 1, then the product must be less than the number.

Finding the product of two fractions can be thought of as finding one fractional part **of** another fraction. For example, when we multiply

$\frac{1}{2} \cdot \frac{1}{4}$ we are finding $\frac{1}{2}$ **of** $\frac{1}{4}$

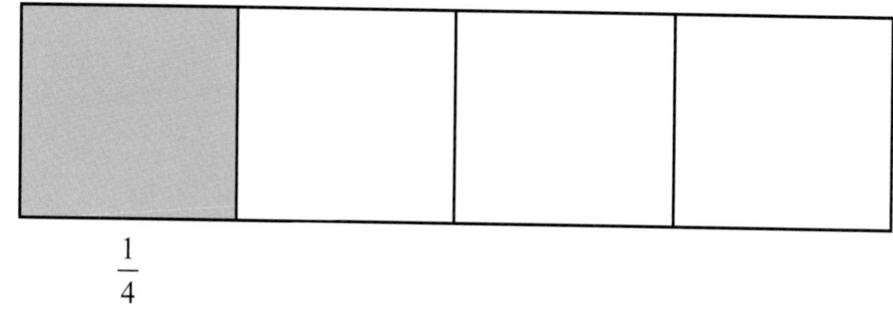

$\frac{1}{4}$

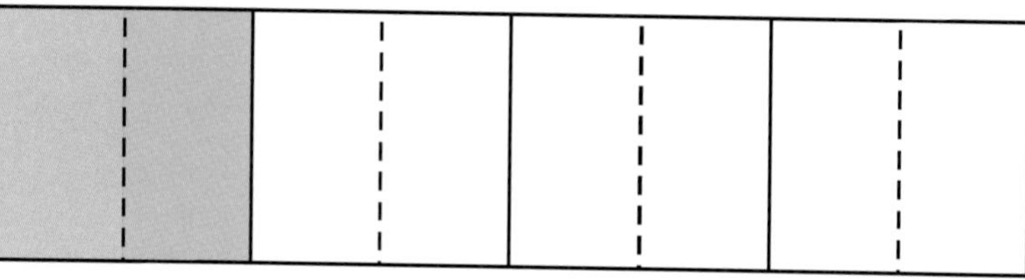

$\frac{1}{2}$ **of** $\frac{1}{4} = \frac{1}{8}$

Thus, we see from the diagram that $\frac{1}{2}$ of $\frac{1}{4}$ is $\frac{1}{8}$, and from the definition

$\frac{1}{2}\cdot\frac{1}{4}=\frac{1\cdot 1}{2\cdot 4}=\frac{1}{8}$ .

### Example 3

Find the product of $\frac{1}{3}$ and $\frac{2}{7}$.

**Solution**

$$\frac{1}{3}\cdot\frac{2}{7}=\frac{1\cdot 2}{3\cdot 7}=\frac{2}{21}$$

### Example 4

Find $\frac{2}{5}$ of $\frac{7}{3}$.

**Solution**

$$\frac{2}{5}\cdot\frac{7}{3}=\frac{14}{15}$$

### Example 5

**a.** $\frac{4}{13}\cdot 3=\frac{4}{13}\cdot\frac{3}{1}=\frac{4\cdot 3}{13\cdot 1}=\frac{12}{13}$

**b.** $\frac{9}{8}\cdot 0=\frac{9}{8}\cdot\frac{0}{1}=\frac{9\cdot 0}{8\cdot 1}=\frac{0}{8}=0$

**c.** $4\cdot\frac{5}{3}\cdot\frac{2}{7}=\frac{4}{1}\cdot\frac{5}{3}\cdot\frac{2}{7}=\frac{40}{21}$

*Now Work Exercises 1-5 in the Margin.*

Find each product.

**1.** $\frac{1}{7}\cdot\frac{3}{5}$

$\frac{3}{35}$

**2.** $\frac{0}{4}\cdot\frac{11}{16}$

$0$

**3.** $-\frac{1}{2}\left(\frac{1}{2}\right)$

$-\frac{1}{4}$

**4.** Find the product of $\frac{2}{3}$ and $\frac{8}{11}$.

$\frac{16}{33}$

**5.** Find $\frac{3}{4}$ of $\frac{5}{8}$.

$\frac{15}{32}$

Both the commutative property and the associative property of multiplication apply to rational numbers (or fractions).

**Commutative Property of Multiplication**

If $\frac{a}{b}$ and $\frac{c}{d}$ are rational numbers, then $\frac{a}{b} \cdot \frac{c}{d} = \frac{c}{d} \cdot \frac{a}{b}$

**Associative Property of Multiplication**

If $\frac{a}{b}$, $\frac{c}{d}$ and $\frac{e}{f}$ are rational numbers, then

$$\frac{a}{b} \cdot \frac{c}{d} \cdot \frac{e}{f} = \frac{a}{b} \cdot \left(\frac{c}{d} \cdot \frac{e}{f}\right) = \left(\frac{a}{b} \cdot \frac{c}{d}\right) \cdot \frac{e}{f}$$

### Example 6

$\frac{3}{5} \cdot \frac{2}{7} = \frac{2}{7} \cdot \frac{3}{5}$ Illustrates the commutative property of multiplication

$\frac{3}{5} \cdot \frac{2}{7} = \frac{6}{35}$ and $\frac{2}{7} \cdot \frac{3}{5} = \frac{6}{35}$

### Example 7

$\left(\frac{1}{2} \cdot \frac{3}{2}\right) \cdot \frac{3}{11} = \frac{1}{2} \cdot \left(\frac{3}{2} \cdot \frac{3}{11}\right)$ Illustrates the associative property of multiplication

$\left(\frac{1}{2} \cdot \frac{3}{2}\right) \cdot \frac{3}{11} = \frac{3}{4} \cdot \frac{3}{11} = \frac{9}{44}$

and

$\frac{1}{2} \cdot \left(\frac{3}{2} \cdot \frac{3}{11}\right) = \frac{1}{2} \cdot \frac{9}{22} = \frac{9}{44}$

*Now Work Exercises 6-8 in the Margin*

## Raising Fractions to Higher Terms and Reducing Fractions

We know that 1 is the multiplicative identity for integers; that is, $a \cdot 1 = a$ for any integer $a$. The number 1 is also the multiplicative identity for rational numbers, since

$$\frac{a}{b} \cdot \mathbf{1} = \frac{a}{b} \cdot \frac{\mathbf{1}}{\mathbf{1}} = \frac{a \cdot 1}{b \cdot 1} = \frac{a}{b}$$

Find each product and state the property of multiplication that is illustrated.

**6.** $\frac{3}{4} \cdot 9 = 9 \cdot \frac{3}{4}$

$\frac{27}{4}$; commutative property of multiplication

**7.** $\frac{1}{3}\left(\frac{1}{2} \cdot \frac{1}{4}\right) = \left(\frac{1}{3} \cdot \frac{1}{2}\right) \cdot \frac{1}{4}$

$\frac{1}{24}$; associative property of multiplication

**8.** $\left(-\frac{5}{6}\right) \cdot \frac{7}{8} = \frac{7}{8} \cdot \left(-\frac{5}{6}\right)$

$-\frac{35}{48}$; commutative property of multiplication

In raising to higher terms, we can multiply by any form of 1 that we choose, but we choose the form that will give us the desired denominator. With this concept in mind, you might think in terms of multiplication instead of division when raising to higher terms, particularly when variables are involved. For example, $\frac{3}{4}=\frac{?}{28a}$, think $4 \cdot 7a = 28a$ instead of $28a \div 4 = 7a$. This approach will help in developing skill in addition and subtraction with fractions.

**Multiplicative Identity**

1. For any rational number $\frac{a}{b}$, $\frac{a}{b} \cdot 1 = \frac{a}{b}$.

2. $\frac{a}{b} = \frac{a}{b} \cdot 1 = \frac{a}{b} \cdot \frac{\mathbf{k}}{\mathbf{k}} = \frac{a \cdot \mathbf{k}}{b \cdot \mathbf{k}}$ where $k \neq 0 \left(1 = \frac{k}{k}\right)$.

In effect, the value of a fraction is unchanged if both the numerator and the denominator are multiplied by the same nonzero integer $k$. This fact allows us to perform the following two operations with fractions:

1. **Raise a fraction to higher terms** (find an equal fraction with a larger denominator) and
2. **Reduce a fraction to lower terms** (find an equal fraction with a smaller denominator).

## Example 8

Raise $\frac{3}{4}$ to higher terms as indicated: $\frac{3}{4} = \frac{\mathbf{?}}{24}$.

**Solution**

We know that $4 \cdot 6 = 24$, so we use $1 = \frac{k}{k} = \frac{6}{6}$.

$$\frac{3}{4} = \frac{3}{4} \cdot \mathbf{1} = \frac{3}{4} \cdot \frac{?}{?} = \frac{3}{4} \cdot \frac{\mathbf{6}}{\mathbf{6}} = \frac{18}{24}$$

## Example 9

Raise $\frac{9}{10}$ to higher terms as indicated: $\frac{9}{10} = \frac{\mathbf{?}}{30x}$.

**Solution**

We know that $10 \cdot 3x = 30x$, so we use $k = 3x$.

$$\frac{9}{10} = \frac{9}{10} \cdot 1 = \frac{9}{10} \cdot \frac{?}{?} = \frac{9}{10} \cdot \frac{\mathbf{3x}}{\mathbf{3x}} = \frac{27x}{30x}$$

Changing a fraction to lower terms is the same as **reducing** the fraction. **A fraction is reduced to lowest terms if the numerator and the denominator have no common factor other than +1 or –1.**

**To Reduce a Fraction to Lowest Terms:**

1. Factor the numerator and denominator into prime factorizations.
2. Use the fact that $\frac{k}{k}=1$ and "divide out" all common factors.

**Note: Reduced fractions might be improper fractions.**

Encourage students to use prime factorization in reducing fractions and products as much as possible. You can point out that time is wasted by multiplying numerators and denominators and then reducing the result because the given numerators and denominators are factors of the numerator and denominator of the product. For example,

$$\frac{15}{16}\cdot\frac{3}{5}\cdot\frac{8}{12}=\frac{360}{960}$$

but 15, 3, and 8 are factors of 360, and 16, 5, and 2 are factors of 960.

## Example 10

Reduce $\frac{-16}{-28}$ to lowest terms.

### Solution

If both numerator and denominator have a factor of $-1$, then we "divide out" $-1$ and treat the fraction as positive, since $\frac{-1}{-1}=1$.

$$\frac{-16}{-28}=\frac{-1\cdot2\cdot2\cdot2\cdot2}{-1\cdot2\cdot2\cdot7}=\frac{-1}{-1}\cdot\frac{2}{2}\cdot\frac{2}{2}\cdot\frac{4}{7}$$

$$=1\cdot1\cdot1\cdot\frac{4}{7}=\frac{4}{7}$$

## Example 11

Reduce $\frac{60a^2}{72ab}$ to lowest terms.

### Solution

$$\frac{60a^2}{72ab}=\frac{6\cdot10\cdot a\cdot a}{8\cdot9\cdot a\cdot b}=\frac{2\cdot3\cdot2\cdot5\cdot a\cdot a}{2\cdot2\cdot2\cdot3\cdot3\cdot a\cdot b}$$

$$=\frac{2}{2}\cdot\frac{2}{2}\cdot\frac{3}{3}\cdot\frac{a}{a}\cdot\frac{5\cdot a}{2\cdot3\cdot b}$$

$$=1\cdot1\cdot1\cdot1\cdot\frac{5a}{6b}=\frac{5a}{6b}$$

As a shortcut, we do not generally write the number 1 when reducing the form $\frac{k}{k}$, as we did in Examples 10 and 11. We do use cancel marks to indicate dividing out common factors in numerators and denominators. **But remember that these numbers do not simply disappear. Their quotient is understood to be 1 even if the 1 is not written.**

### Example 12

Reduce $\frac{45}{36}$ to lowest terms.

#### Solution

**a.** Using prime factors, we have

$$\frac{45}{36}=\frac{\cancel{3}\cdot\cancel{3}\cdot 5}{2\cdot 2\cdot\cancel{3}\cdot\cancel{3}}=\frac{5}{4}.$$

Note that the answer is a reduced improper fraction.

**b.** Larger common factors can be divided out, but we must be sure that we have the largest common factor.

$$\frac{45}{36}=\frac{5\cdot\cancel{9}}{4\cdot\cancel{9}}=\frac{5}{4}$$

### *Completion Example 13*

Reduce $\frac{52}{65}$ to lowest terms.

#### Solution

$$\frac{52}{65}=\frac{2\cdot 2\cdot}{5\cdot}=\_\_\_\_\_\_\_\_$$

*Now Work Exercises 9-12 in the Margin.*

Reduce each fraction to lowest terms.

**9.** $\frac{45}{75}$ $\quad \frac{3}{5}$

**10.** $\frac{66x^2}{44xy}$ $\quad \frac{3x}{2y}$

**11.** $\frac{-56}{120}$ $\quad -\frac{7}{15}$

**12.** $\frac{35}{46}$ $\quad \frac{35}{46}$

## Multiplying and Reducing Fractions at the Same Time

Now we can multiply fractions and reduce all in one step by using prime factors (or other common factors). If you have any difficulty understanding how to multiply and reduce, use prime factors. By using prime factors, you can be sure that you have not missed a common factor and that your answer is reduced to lowest terms.

Examples 14, 15, and 16 illustrate how to multiply and reduce at the same time by factoring the numerators and the denominators. Note that **if all the factors in the numerator or denominator divide out, then 1 must be used as a factor**. (See Examples 15 and 16.)

### Example 14

$$-\frac{18}{35}\cdot\frac{21}{12}=-\frac{\cancel{2}\cdot\cancel{3}\cdot3\cdot\cancel{7}\cdot3}{5\cdot\cancel{7}\cdot\cancel{2}\cdot2\cdot\cancel{3}}=-\frac{9}{10}$$

### Example 15

$$\frac{17}{50}\cdot\frac{25}{34}\cdot8=-\frac{17\cdot25\cdot8}{50\cdot34\cdot1}$$

$$=\frac{\cancel{17}\cdot\cancel{25}\cdot\cancel{2}\cdot\cancel{2}\cdot2}{\cancel{2}\cdot\cancel{25}\cdot\cancel{2}\cdot\cancel{17}\cdot1}$$

In this example, 25 is a common factor that is not a prime number.

$$=\frac{2}{1}=2$$

### Example 16

$$\frac{3x}{35}\cdot\frac{15y}{9xy^3}\cdot\frac{7}{2y}=\frac{3\cdot x\cdot3\cdot5\cdot y\cdot7}{7\cdot5\cdot3\cdot3\cdot x\cdot y\cdot y\cdot y\cdot2\cdot y}$$

$$=\frac{\cancel{3}\cdot\cancel{x}\cdot\cancel{3}\cdot\cancel{5}\cdot\cancel{y}\cdot\cancel{7}}{\cancel{7}\cdot\cancel{5}\cdot\cancel{3}\cdot\cancel{3}\cdot\cancel{x}\cdot\cancel{y}\cdot y\cdot y\cdot2\cdot y}=\frac{1}{2y^3}$$

### *Completion Example 17*

Multiply and reduce to lowest terms.

$$\frac{55a^2}{26}\cdot\frac{8b}{44ab^2}\cdot\frac{91}{35}=\frac{55\cdot a\cdot a\cdot8\cdot b\cdot91}{26\cdot44\cdot a\cdot b\cdot b\cdot35}$$

= ____________

= ____________

Many students are familiar with another method of multiplying and reducing at the same time which is to divide numerators and denominators by common factors whether they are prime or not. If these factors are easily determined, then this method is probably faster. But common factors are sometimes missed with this method, whereas they are not missed with the prime factorization method. In either case, be careful and organized. The problems from Examples 14 and 15 have been worked again in Examples 18 and 19 using the division method.

You might want to discuss the fact that, while the prime factors were not used in the solution of Example 15, we did use factors. Some students may think that using prime factorizations is a waste of time because they find it easier to use the "cancel" technique. But, for others factoring will be a clearer and more organized approach. The students need to understand that the idea of factoring is an important part of algebra.

Multiply and reduce at the same time.

**13.** $\frac{15}{63} \cdot \frac{9}{5}$

$\frac{3}{7}$

**14.** $\frac{14x^2}{20} \cdot \frac{25y}{28} \cdot \frac{4}{15x}$

$\frac{xy}{6}$

**15.** $8 \cdot \frac{19}{100} \cdot \frac{24}{38} \cdot \frac{5}{18}$

$\frac{4}{15}$

You might want to introduce the idea of subtraction here. Some students may notice that an alternate solution in Example 20(b) is to subtract:

$$1 - \frac{3}{5} = \frac{2}{5}$$

## Example 18

$$-\frac{\overset{3}{\cancel{18}}}{\underset{5}{\cancel{35}}} \cdot \frac{\overset{3}{\cancel{21}}}{\underset{2}{\cancel{12}}} = -\frac{9}{10}$$

6 divides into both 18 and 12.
7 divides into both 21 and 35.

## Example 19

$$\frac{17}{50} \cdot \frac{25}{34} \cdot 8 = \frac{\overset{1}{\cancel{17}}}{\underset{2}{\cancel{50}}} \cdot \frac{\overset{1}{\cancel{25}}}{\underset{2}{\cancel{34}}} \cdot \frac{\overset{\overset{2}{\cancel{4}}}{\cancel{8}}}{1} = \frac{2}{1} = 2$$

17 divides into both 17 and 34.
25 divides into both 25 and 50.
2 divides into both 8 and 2.
2 divides into both 4 and 2.

*Now Work Exercises 13-15 in the Margin.*

## Example 20

If you had \$25 and you spend \$15 to buy computer disks, what fraction of your money did you spend for computer disks? What fraction do you still have?

### Solution

**a.** The fraction you spent is

$$\frac{15}{25} = \frac{3 \cdot \cancel{5}}{5 \cdot \cancel{5}} = \frac{3}{5}.$$

**b.** Since you still have \$10, the fraction you still have is

$$\frac{10}{25} = \frac{2 \cdot \cancel{5}}{5 \cdot \cancel{5}} = \frac{2}{5}.$$

## Completion Example Answers

**13.** $\frac{52}{65} = \frac{2 \cdot 2 \cdot \cancel{13}}{5 \cdot \cancel{13}} = \mathbf{\frac{4}{5}}$

**17.** $\frac{55a^2}{26} \cdot \frac{8b}{44ab^2} \cdot \frac{91}{35} = \frac{55 \cdot a \cdot a \cdot 8 \cdot b \cdot 91}{26 \cdot 44 \cdot a \cdot b \cdot b \cdot 35}$

$$= \frac{\cancel{5} \cdot \cancel{11} \cdot a \cdot \cancel{a} \cdot \cancel{2} \cdot \cancel{2} \cdot \cancel{2} \cdot \cancel{b} \cdot \cancel{7} \cdot \cancel{13}}{\cancel{2} \cdot \cancel{13} \cdot \cancel{2} \cdot \cancel{2} \cdot \cancel{11} \cdot \cancel{a} \cdot b \cdot \cancel{b} \cdot \cancel{5} \cdot \cancel{7}}$$

$$= \mathbf{\frac{a}{b}}$$

Name ______ Section ______ Date ______

## Exercises 3.5

1. What is the value, if any, of each of the following expressions?

   a. $\frac{0}{6}\cdot\frac{5}{7}$ b. $\frac{3}{10}\cdot\frac{0}{2}$ c. $\frac{5}{0}$ d. $\frac{16}{0}$

2. What is value, if any, of each of the following expressions?

   a. $\frac{32}{0}\cdot\frac{6}{0}$ b. $\frac{1}{5}\cdot\frac{0}{3}$ c. $\frac{0}{8}\cdot\frac{0}{9}$ d. $\frac{1}{0}\cdot\frac{0}{1}$

3. Find $\frac{1}{5}$ of $\frac{3}{5}$.
4. Find $\frac{1}{4}$ of $\frac{1}{4}$.
5. Find $\frac{2}{3}$ of $\frac{2}{3}$.
6. Find $\frac{3}{4}$ of $\frac{4}{5}$.
7. Find $\frac{6}{7}$ of $-\frac{3}{5}$.
8. Find $\frac{1}{3}$ of $-\frac{1}{4}$.
9. Find $\frac{1}{2}$ of $\frac{1}{2}$.
10. Find $\frac{1}{9}$ of $\frac{2}{3}$.

Find the following products.

11. $\frac{3}{4}\cdot\frac{3}{4}$
12. $\frac{5}{6}\cdot\frac{5}{6}$
13. $\frac{1}{9}\cdot\frac{4}{3}$
14. $\frac{4}{5}\cdot\frac{3}{7}$
15. $\frac{0}{5}\cdot\frac{2}{3}$
16. $\frac{3}{8}\cdot\frac{0}{2}$
17. $-4\cdot\frac{3}{5}$
18. $-7\cdot\frac{5}{6}$
19. $\frac{1}{3}\cdot\frac{1}{10}\cdot\frac{1}{5}$
20. $\frac{4}{7}\cdot\frac{2}{5}\cdot\frac{6}{13}$
21. $\frac{5}{8}\cdot\frac{5}{8}\cdot\frac{5}{8}$
22. $\frac{5}{11}\cdot\frac{6}{11}\cdot\frac{7}{11}$

**ANSWERS**

1. a. 0
   b. 0
   c. Undefined
   d. Undefined
2. a. Undefined
   b. 0
   c. 0
   d. Undefined
3. $\frac{3}{25}$
4. $\frac{1}{16}$
5. $\frac{4}{9}$
6. $\frac{3}{5}$
7. $-\frac{18}{35}$
8. $-\frac{1}{12}$
9. $\frac{1}{4}$
10. $\frac{2}{27}$
11. $\frac{9}{16}$
12. $\frac{25}{36}$
13. $\frac{4}{27}$
14. $\frac{12}{35}$
15. 0
16. 0
17. $-\frac{12}{5}$
18. $-\frac{35}{6}$
19. $\frac{1}{150}$
20. $\frac{48}{455}$
21. $\frac{125}{512}$
22. $\frac{210}{1331}$

23. Commutative Property of Mult.

24. Associative Property of Mult.

25. Associative Property of Mult.

26. Commutative Property of Mult.

27. $\frac{3}{4}=\frac{3}{4}\cdot\frac{3}{3}=\frac{9}{12}$

28. $\frac{2}{3}=\frac{2}{3}\cdot\frac{4}{4}=\frac{8}{12}$

29. $\frac{6}{7}=\frac{6}{7}\cdot\frac{2}{2}=\frac{12}{14}$

30. $\frac{5a}{8}=\frac{5a}{8}\cdot\frac{2}{2}=\frac{10a}{16}$

31. $\frac{3n}{-8}=\frac{3n}{-8}\cdot\frac{-5}{-5}=\frac{-15n}{40}$

32. $\frac{-3x}{16y}=\frac{-3x}{16y}\cdot\frac{5}{5}=\frac{-15x}{80y}$

33. $\frac{-5x}{13}=\frac{-5x}{13}\cdot\frac{3y}{3y}=\frac{-15xy}{39y}$

34. $\frac{-3a}{5b}=\frac{-3a}{5b}\cdot\frac{5b}{5b}=\frac{-15ab}{25b^2}$

35. $\frac{4a^2}{17b}=\frac{4a^2}{17b}\cdot\frac{2a}{2a}=\frac{8a^3}{34ab}$

36. $\frac{-3x}{1}=\frac{-3x}{1}\cdot\frac{5x}{5x}=\frac{-15x^2}{5x}$

37. $\frac{-9x^2}{10y^2}=\frac{-9x^2}{10y^2}\cdot\frac{10y}{10y}=\frac{-90x^2y}{100y^3}$

38. $\frac{-7xy^2}{-10}=\frac{-7xy^2}{-10}\cdot\frac{-10xy}{-10xy}=\frac{70x^2y^3}{100xy}$

Tell which property of multiplication is illustrated.

**23.** $\frac{a}{b}\cdot\frac{11}{14}=\frac{11}{14}\cdot\frac{a}{b}$

**24.** $\frac{3}{8}\cdot\left(\frac{7}{5}\cdot\frac{1}{2}\right)=\left(\frac{3}{8}\cdot\frac{7}{5}\right)\cdot\frac{1}{2}$

**25.** $\frac{7}{5}\cdot\left(\frac{13}{11}\cdot\frac{6}{5}\right)=\left(\frac{7}{5}\cdot\frac{13}{11}\right)\cdot\frac{6}{5}$

**26.** $7\cdot\frac{2}{3}=\frac{2}{3}\cdot 7$

Raise each fraction to higher terms as indicated. Find the values of the missing numbers.

**27.** $\frac{3}{4}=\frac{3}{4}\cdot\frac{?}{?}=\frac{?}{12}$

**28.** $\frac{2}{3}=\frac{2}{3}\cdot\frac{?}{?}=\frac{?}{12}$

**29.** $\frac{6}{7}=\frac{6}{7}\cdot\frac{?}{?}=\frac{?}{14}$

**30.** $\frac{5a}{8}=\frac{5a}{8}\cdot\frac{?}{?}=\frac{?}{16}$

**31.** $\frac{3n}{-8}=\frac{3n}{-8}\cdot\frac{?}{?}=\frac{?}{40}$

**32.** $\frac{-3x}{16y}=\frac{-3x}{16y}\cdot\frac{?}{?}=\frac{?}{80y}$

**33.** $\frac{-5x}{13}=\frac{-5x}{13}\cdot\frac{?}{?}=\frac{?}{39y}$

**34.** $\frac{-3a}{5b}=\frac{-3a}{5b}\cdot\frac{?}{?}=\frac{?}{25b^2}$

**35.** $\frac{4a^2}{17b}=\frac{4a^2}{17b}\cdot\frac{?}{?}=\frac{?}{34ab}$

**36.** $\frac{-3x}{1}=\frac{-3x}{1}\cdot\frac{?}{?}=\frac{?}{5x}$

**37.** $\frac{-9x^2}{10y^2}=\frac{-9x^2}{10y^2}\cdot\frac{?}{?}=\frac{?}{100y^3}$

**38.** $\frac{-7xy^2}{-10}=\frac{-7xy^2}{-10}\cdot\frac{?}{?}=\frac{?}{100xy}$

Name ______________ Section ______________ Date ______________

Reduce each fraction to lowest terms. Just rewrite the fraction if it is already reduced.

**39.** $\frac{24}{30}$

**40.** $\frac{14}{36}$

**41.** $\frac{22}{65}$

**42.** $\frac{60x}{75x}$

**43.** $\frac{-26y}{39y}$

**44.** $\frac{-7n}{28n}$

**45.** $\frac{-27}{56x^2}$

**46.** $\frac{-12}{35y}$

**47.** $\frac{34x}{-51x^2}$

**48.** $\frac{-30x}{45x^2}$

**49.** $\frac{51x^2}{6x}$

**50.** $\frac{6y^2}{-51y}$

**51.** $\frac{-54a^2}{-9ab}$

**52.** $\frac{-24a^2b}{-100a}$

**53.** $\frac{66xy^2}{84xy}$

**54.** $\frac{-14xyz}{-63xz}$

Multiply and reduce each product to lowest terms.

**55.** $\frac{-7}{8}\cdot\frac{4}{21}$

**56.** $\frac{-23}{36}\cdot\frac{20}{46}$

**57.** $9\cdot\frac{7}{24}$

**58.** $8\cdot\frac{5}{12}$

**ANSWERS**

**39.** $\frac{4}{5}$

**40.** $\frac{7}{18}$

**41.** $\frac{22}{65}$

**42.** $\frac{4}{5}$

**43.** $-\frac{2}{3}$

**44.** $-\frac{1}{4}$

**45.** $\frac{-27}{56x^2}$

**46.** $\frac{-12}{35y}$

**47.** $-\frac{2}{3x}$

**48.** $-\frac{2}{3x}$

**49.** $\frac{17x}{2}$

**50.** $-\frac{2y}{17}$

**51.** $\frac{6a}{b}$

**52.** $\frac{6ab}{25}$

**53.** $\frac{11y}{14}$

**54.** $\frac{2y}{9}$

**55.** $-\frac{1}{6}$

**56.** $-\frac{5}{18}$

**57.** $\frac{21}{8}$

**58.** $\frac{10}{3}$

59. $\frac{-273a}{80}$

60. $\frac{189}{52xy}$

61. $\frac{21x^2}{2}$

62. $-\frac{5m^2n}{14}$

63. $-\frac{77}{4}$

64. $\frac{189}{4}$

65. $\frac{7}{10}$; $\frac{3}{10}$

66. $\frac{4}{5}$

67. $\frac{9}{4}$ inches

**59.** $\frac{-9a}{10b} \cdot \frac{35a}{40} \cdot \frac{65b}{15a}$

**60.** $\frac{-42x}{52xy} \cdot \frac{-27}{22x} \cdot \frac{33}{9}$

**61.** $\frac{75x^2}{8y^2} \cdot \frac{-16xy}{36} \cdot 9 \cdot \frac{-7y}{25x}$

**62.** $\frac{17n}{8m} \cdot \frac{-5mn}{42n} \cdot \frac{18}{51} \cdot 4m^2$

**63.** $\frac{-3}{4} \cdot \frac{7}{2} \cdot \frac{22}{54} \cdot 18$

**64.** $\frac{-69}{15} \cdot \frac{30}{8} \cdot \frac{-14}{46} \cdot 9$

**65.** If you have \$10 and you spend \$7 for a sandwich and drink, what fraction of your money have you spent on food? What fraction do you still have?

**66.** In a class of 30 students, 6 received a grade of A. What fraction of the class did not receive an A?

**67.** A glass is 6 inches tall. If the glass is $\frac{3}{8}$ full of water, what is the height of the water in the glass?

Name ______________________ Section ____________ Date __________

**ANSWERS**

**68.** $\frac{5}{2}$ feet

**69.** $\frac{4}{5}$; 60 miles

**70.** 180 students

**68.** Suppose that a ball is dropped from a height of 20 feet and that each time the ball bounces it bounces back to $\frac{1}{2}$ the height it dropped. How high will the ball bounce on its third bounce?

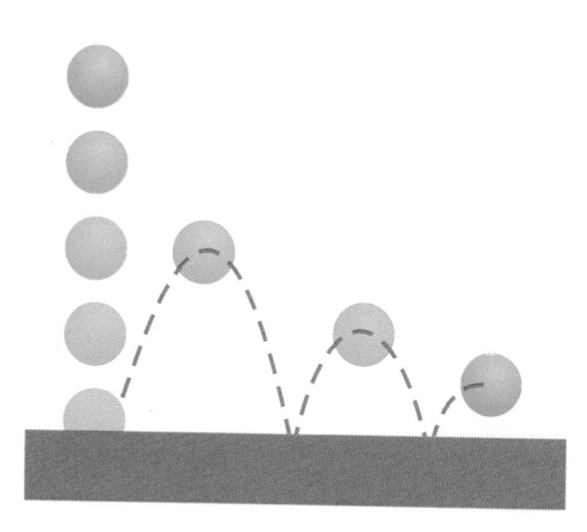

**69.** If you go on a bicycle trip of 75 miles in the mountains and $\frac{1}{5}$ of the trip is downhill, what fraction of the trip is not downhill? How many miles are not downhill?

**70.** A study showed that $\frac{3}{10}$ of the students in an elementary school were left-handed. If the school had an enrollment of 600 students, how many were left-handed?

# 3.6 Division with Fractions

## Reciprocals

If the product of two fractions is 1, then the fractions are called **reciprocals** of each other. For example,

$\frac{3}{4}$ and $\frac{4}{3}$ are reciprocals because $\frac{3}{4}\cdot\frac{4}{3}=\frac{12}{12}=1$, and

$-\frac{9}{2}$ and $-\frac{2}{9}$ are reciprocals because $\left(-\frac{9}{2}\right)\cdot\left(-\frac{2}{9}\right)=+\frac{18}{18}=1$.

**Objectives**

① Be able to recognize and find the reciprocal of a number.

② Know that division is accomplished by multiplication by the reciprocal of the divisor.

**Definition:**

The **reciprocal** of $\frac{a}{b}$ is $\frac{b}{a}$ $(a \neq 0$ and $b \neq 0)$.

The product of a nonzero number and its reciprocal is always 1.

$$\frac{a}{b}\cdot\frac{b}{a}=1$$

**Note: The number 0 has no reciprocal.** That is $\frac{0}{1}$ has no reciprocal because $\frac{1}{0}$ is undefined.

You might mention that every integer can be written in fraction form with denominator 1. Some students may need to "see" this written down to understand the nature of reciprocals of integers.

### Example 1

The reciprocal of $\frac{15}{7}$ is $\frac{7}{15}$.

$$\frac{15}{7}\cdot\frac{7}{15}=\frac{15\cdot 7}{7\cdot 15}=1$$

### Example 2

The reciprocal of $6x$ is $\frac{1}{6x}$.

$$6x\cdot\frac{1}{6x}=\frac{6x}{1}\cdot\frac{1}{6x}=1$$

*Now Work Exercises 1-3 in the Margin.*

State the reciprocal of each number of expression.

**1.** $\frac{11}{18}$

$\frac{18}{11}$

**2.** $\frac{-1}{5}$

$\frac{5}{-1}$ or $-5$

**3.** $\frac{2}{7x}$

$\frac{7x}{2}$

## Division With Fractions

To understand how to divide with fractions, we write the division in fraction form, where the numerator and denominator are themselves fractions. For example,

$$\frac{2}{3} \div \frac{5}{7} = \frac{\frac{2}{3}}{\frac{5}{7}}.$$

Now, we multiply the numerator and the denominator (both fractions) by the reciprocal of the denominator. This is the same as multiplying 1 and does not change the value of the expression. The reciprocal of $\frac{5}{7}$ is $\frac{7}{5}$, so we multiply both the numerator and the denominator by $\frac{7}{5}$.

$$\frac{2}{3} \div \frac{5}{7} = \frac{\frac{2}{3}}{\frac{5}{7}} \cdot \frac{\frac{7}{5}}{\frac{7}{5}} = \frac{\frac{2}{3} \cdot \frac{7}{5}}{\frac{5}{7} \cdot \frac{7}{5}} = \frac{\frac{2}{3} \cdot \frac{7}{5}}{1} = \frac{2}{3} \cdot \frac{7}{5}$$

Thus, a division problem has been changed into a multiplication problem:

$$\frac{2}{3} \div \frac{5}{7} = \frac{2}{3} \cdot \frac{7}{5} = \frac{14}{15}$$

**To divide by any nonzero number, multiply by its reciprocal.** In general,

$$\frac{a}{b} \div \frac{c}{d} = \frac{a}{b} \cdot \frac{d}{c}$$

where $b, c, d \neq 0$.

### Example 3

Divide: $\frac{3}{4} \div \frac{2}{3}$

#### Solution

The reciprocal of $\frac{2}{3}$ is $\frac{3}{2}$, so we multiply by $\frac{3}{2}$.

$$\frac{3}{4} \div \frac{2}{3} = \frac{3}{4} \cdot \frac{3}{2} = \frac{9}{8}$$

### Example 4

Divide: $\frac{3}{4} \div 4$

**Solution**

The reciprocal of 4 is $\frac{1}{4}$, so we multiply by $\frac{1}{4}$.

$$\frac{3}{4} \div 4 = \frac{3}{4} \cdot \frac{1}{4} = \frac{3}{16}$$

As with any multiplication problem, we reduce whenever possible by factoring numerators and denominators.

### Example 5

Divide and reduce to lowest terms: $\frac{16}{27} \div \frac{8}{9}$

**Solution**

The reciprocal of $\frac{8}{9}$ is $\frac{9}{8}$. Reduce by factoring.

$$\frac{16}{27} \div \frac{8}{9} = \frac{16}{27} \cdot \frac{9}{8} = \frac{\cancel{8} \cdot 2 \cdot \cancel{9}}{3 \cdot \cancel{9} \cdot \cancel{8}} = \frac{2}{3}$$

### Example 6

Divide and reduce to lowest terms: $\frac{9a}{10} \div \frac{9a}{10}$

**Solution**

We expect the result to be 1 because we are dividing a number by itself.

$$\frac{9a}{10} \div \frac{9a}{10} = \frac{9a}{10} \cdot \frac{10}{9a} = \frac{90a}{90a} = 1$$

### *Completion Example 7*

Divide and reduce to lowest terms: $\frac{-4x}{13} \div \frac{20}{39x}$

**Solution**

$$\frac{-4x}{13} \div \frac{20}{39x} = \frac{-4x}{13} \cdot \_\_\_\_\_\_ = \_\_\_\_\_\_\_\_ = \_\_\_\_\_\_$$

*Now Work Exercises 4-6 in the Margin.*

Divide as indicated and reduce to lowest terms.

**4.** $\frac{1}{5} \div \frac{1}{5}$

1

**5.** $\frac{x}{2} \div 2$

$\frac{x}{4}$

**6.** $\frac{-5}{13y} \div \frac{25}{52y}$

$-\frac{4}{5}$

If the product of two numbers is known and one of the numbers is also known, then the other number can be found by **dividing the product by the known number.** For example, with whole numbers, suppose the product of two numbers is 24 and one of the numbers is 8. What is the other number? Since $24 \div 8 = 3$, the other number is 3.

## Example 8

The result of multiplying two numbers is $\frac{14}{15}$. If one of the numbers is $\frac{2}{5}$, what is the other number?

**Solution**

Divide the product $\left(\text{which is } \frac{14}{15}\right)$ by the number $\frac{2}{5}$.

$$\frac{14}{15} \div \frac{2}{5} = \frac{14}{15} \cdot \frac{5}{2} = \frac{7 \cdot \cancel{2} \cdot \cancel{5}}{3 \cdot \cancel{5} \cdot \cancel{2}} = \frac{7}{3}$$

The other number is $\frac{7}{3}$.

Check by multiplying:

$$\frac{2}{5} \cdot \frac{7}{3} = \frac{14}{15}$$

### Example 9

If the product of $\frac{1}{4}$ with another number is $-\frac{11}{18}$, what is the other number?

#### Solution

Divide the product by the given number.

$$-\frac{11}{18} \div \frac{1}{4} = \frac{-11}{18} \cdot \frac{4}{1} = \frac{-11 \cdot \cancel{2} \cdot 2}{\cancel{2} \cdot 3 \cdot 3} = -\frac{22}{9}$$

The other number is $-\frac{22}{9}$.

Note that we could have anticipated that this number would be negative because the product $\left(-\frac{11}{18}\right)$ is negative and the given number $\left(\frac{1}{4}\right)$ is positive.

### Completion Example Answer

**7.** $\frac{-4x}{13} \div \frac{20}{39x} = \frac{-4x}{13} \cdot \frac{\mathbf{39x}}{\mathbf{20}} = -\frac{\cancel{\mathbf{4}} \cdot \mathbf{x} \cdot \mathbf{3} \cdot \cancel{\mathbf{13}} \cdot \mathbf{x}}{\cancel{\mathbf{13}} \cdot \cancel{\mathbf{4}} \cdot \mathbf{5}} = -\frac{\mathbf{3x^2}}{\mathbf{5}}$

Name ________________ Section ________________ Date ________________

# Exercises 3.6

1. What is the reciprocal of $\frac{13}{25}$?

2. To divide by any nonzero number, multiply by its reciprocal.

3. The quotient of $0 \div \frac{6}{7}$ is 0.

4. The quotient of $\frac{6}{7} \div 0$ is undefined.

5. Show that the phrases "ten divided by two" and "ten divided by one–half" do not have the same meaning.

   $10 \div 2 = 5$, while $10 \div \frac{1}{2} = 10 \cdot 2 = 20$

6. Show that the phrases "twelve divided by three" and "twelve divided by one–third" do not have the same meaning.

   $12 \div 3 = 4$, while $12 \div \frac{1}{3} = 12 \cdot 3 = 36$

7. Explain in your own words why division by 0 is undefined.

   Explanations will vary. Refer to the discussion on page 15 of Section 1.1.

8. Give three examples that illustrate that division is not a commutative operation.

   Answers will vary. For example, $\frac{2}{3} \div 3 = \frac{2}{3} \cdot \frac{1}{3} = \frac{2}{9}$, while $3 \div \frac{2}{3} = \frac{3}{1} \cdot \frac{3}{2} = \frac{9}{2}$. Some students may notice that the results are reciprocals of each other.

Find the following quotients. Reduce to lowest terms whenever possible.

9. $\frac{5}{8} \div \frac{3}{5}$

10. $\frac{1}{3} \div \frac{1}{5}$

11. $\frac{2}{11} \div \frac{1}{7}$

12. $\frac{2}{7} \div \frac{1}{2}$

13. $\frac{3}{14} \div \frac{3}{14}$

14. $\frac{5}{8} \div \frac{5}{8}$

15. $\frac{3}{4} \div \frac{4}{3}$

16. $\frac{9}{10} \div \frac{10}{9}$

**ANSWERS**

1. $\frac{25}{13}$
2. [Respond in exercise.]
3. [Respond in exercise.]
4. [Respond in exercise.]
5. [Respond below exercise.]
6. [Respond below exercise.]
7. [Respond below exercise.]
8. [Respond below exercise.]
9. $\frac{25}{24}$
10. $\frac{5}{3}$
11. $\frac{14}{11}$
12. $\frac{4}{7}$
13. 1
14. 1
15. $\frac{9}{16}$
16. $\frac{81}{100}$

17. $-\frac{1}{4}$

18. $-\frac{1}{10}$

19. $\frac{1}{16}$

20. $\frac{1}{20}$

21. Undefined

22. 0

23. 0

24. Undefined

25. $-\frac{8}{5}$

26. $-1$

27. $\frac{9}{20}$

28. $\frac{6}{5}$

29. $\frac{4}{3a}$

30. $\frac{4}{9}$

31. $\frac{4x^2}{3y^2}$

32. $\frac{19}{15}$

33. $98x^2$

34. $125y^2$

35. $-300a$

36. $-100b$

37. $\frac{29}{155}$

38. $\frac{22}{7a^2}$

39. $-\frac{3x}{8}$

40. $-\frac{16x}{21}$

41. a. $\frac{2}{5}$

b. $\frac{12}{25}$

42. a. Product

b. 70

17. $\frac{15}{20} \div (-3)$

18. $\frac{14}{20} \div (-7)$

19. $\frac{25}{40} \div 10$

20. $\frac{36}{80} \div 9$

21. $\frac{7}{8} \div 0$

22. $0 \div \frac{5}{6}$

23. $0 \div \frac{1}{2}$

24. $\frac{15}{64} \div 0$

25. $\frac{-16}{35} \div \frac{2}{7}$

26. $\frac{-15}{27} \div \frac{5}{9}$

27. $\frac{-15}{24} \div \frac{-25}{18}$

28. $\frac{-36}{25} \div \frac{-24}{20}$

29. $\frac{34b}{21a} \div \frac{17b}{14}$

30. $\frac{16x}{20y} \div \frac{18x}{10y}$

31. $\frac{20x}{21y} \div \frac{10y}{14x}$

32. $\frac{19a}{24b} \div \frac{5a}{8b}$

33. $14x \div \frac{1}{7x}$

34. $25y \div \frac{1}{5y}$

35. $-30a \div \frac{1}{10}$

36. $-50b \div \frac{1}{2}$

37. $\frac{29a}{50} \div \frac{31a}{10}$

38. $\frac{92}{7a} \div \frac{46a}{11}$

39. $\frac{-33x^2}{32} \div \frac{11x}{4}$

40. $\frac{-26x^2}{35} \div \frac{39x}{40}$

41. The product of $\frac{5}{6}$ with another number is $\frac{2}{5}$.
   a. Which number is the product?
   b. What is the other number?

42. The result of multiplying two numbers is 150.
   a. Is 150 a product or a quotient?
   b. If one of the numbers is $\frac{15}{7}$, what is the other one?

Name ______________________ Section ____________ Date __________

**ANSWERS**

**43.** The product of two numbers is 210.

**a.** If one of the numbers is $\frac{2}{3}$, do you expect the other number to be larger than 210 or smaller than 210?

**b.** What is the other number?

**43. a.** larger

**b.** 315

**44.** An airplane is carrying 180 passengers. This is $\frac{9}{10}$ of the capacity of the airplane.

**a.** Is the capacity of the airplane more or less than 180?

**b.** If you were to multiply 180 times $\frac{9}{10}$ would the product be more or less than 180?

**c.** What is the capacity of the airplane?

**44. a.** more

**b.** less

**c.** 200

**45.** A bus is carrying 60 passengers. This is $\frac{4}{5}$ of the capacity of the bus.

**a.** Is the capacity of the bus more or less than 60?

**b.** If you were to multiply 60 times $\frac{4}{5}$, would the product be more or less than 60?

**c.** What is the capacity of the bus?

**45. a.** more

**b.** less

**c.** 75

**46.** What is the quotient if $\frac{1}{4}$ of 36 is divided by $\frac{2}{3}$ of $\frac{5}{8}$?

**46.** $\frac{108}{5}$

**47.** What is the quotient if $\frac{7}{8}$ of 64 is divided by $\frac{3}{5}$ of $\frac{15}{16}$?

**47.** $\frac{896}{9}$

**48.** The continent of Africa covers approximately 11,707,000 square miles. This is $\frac{1}{5}$ of the land area in the world. What is the approximate total land area in the world?

**48.** 58,535,000 square miles

49. a. Yes, it would pass.

b. Pass by 3 votes.

50. a. Yes, it would pass.

b. Pass by 5 votes.

51. a. 1000

b. 2500

52. 975

**49.** The student senate has 75 members, and $\frac{7}{15}$ of these are women. A change in the senate constitution is being considered, and at the present time (before debating has begun), a survey shows that $\frac{3}{5}$ of the women and $\frac{4}{5}$ of the men are in favor of this change.

**a.** If the change requires a $\frac{2}{3}$ majority vote in favor to pass, would the constitutional change pass if the vote were taken today?

**b.** By how many votes would the change pass or fail?

**50.** The tennis club has 250 members, and they are considering putting in a new tennis court. The cost of the new court is going to involve an assessment of $200 for each member. Of the seven–tenths of the members who live quite near the club, $\frac{3}{5}$ of them are in favor of the assessment. However, $\frac{2}{3}$ of the members who live some distance away are not in favor of the assessment.

**a.** If a vote were taken today, would more than one–half of the members vote for or against the new court?

**b.** By how many votes would the question pass or fail if more than one–half of the members must vote in favor for the question to pass?

**51.** There are 4000 registered voters in Roseville, and $\frac{3}{8}$ of these voters are registered Democrats. A survey indicates that $\frac{2}{3}$ of the registered Democrats are in favor of Bond Measure A and $\frac{3}{5}$ of the other registered voters are in favor of this measure.

**a.** How many of the registered Democrats favor Measure A?

**b.** How many of the registered voters favor Measure A?

**52.** There are 3000 students in Mountain High School, and $\frac{1}{4}$ of these students are seniors. If $\frac{3}{5}$ of the seniors are in favor of the school forming a debating team and $\frac{7}{10}$ of the remaining students (not seniors) are also in favor of forming a debating team, how many students do not favor this idea?

Name ________________ Section ________ Date ________

**ANSWERS**

**53.** A manufacturing plant is currently producing 6000 steel rods per week. Because of difficulties getting materials, this number is only $\frac{3}{4}$ of the plant's potential production.

**a.** Is the potential production number more or less than 6000 rods?

**b.** If you were to multiply $\frac{3}{4}$ times 6000, would the product be more or less than 6000?

**c.** What is the plant's potential production?

**54.** A grove of orange trees was struck by an off-season frost and the result was a relatively poor harvest. This year's crop was 10,000 tons of oranges which is about $\frac{4}{5}$ of the usual crop.

**a.** Is the usual crop more or less than 10,000 tons of oranges?

**b.** If you were to multiply 10,000 times $\frac{4}{5}$, would the product be more or less than 10,000?

**c.** About how many tons of oranges are usually harvested?

**55.** Due to environmental considerations homeowners in a particularly dry area have been asked to use less water than usual. One home is currently using 630 gallons per day. This is $\frac{7}{10}$ of the usual amount of water used in this home.

**a.** Is the usual amount of water used more or less than 600 gallons?

**b.** If you were to multiply $\frac{7}{10}$ times 630, would the product be more or less than 630?

**c.** What is the usual amount of water used in this home?

**53. a.** more
**b.** less
**c.** 8000

**54. a.** more
**b.** less
**c.** 12,500

**55. a.** more
**b.** less
**c.** 900

## The Recycle Bin

1. $-2x - 6$

2. $-y^2 + 2y + 6$

3. $-11x^2 - 7x - 6$

4. 0

5. a. $8x^2 - 10x + 1$

   b. 53

6. a. $-3y^3 - 4y^2 + 20y$
   $-100$

   b. $-157$

7. a. $4y^2 + 4xy - 5x$

   b. 22

## The Recycle Bin (from Sections 2.6)

Simplify each of the following expressions by combining like terms.

1. $5x + 3x - 10x + 14 - 20$
2. $13y - y + y^2 - 2y^2 + 6 - 10y$
3. $-8x^2 - 3x^2 - 5x - 2x - 6$
4. $-6y^3 - y^3 + 7y^3 + 9y - 6y - 3y$

For each of the following problems:
**a.** Simplify each of the following expressions.
**b.** Evaluate each expression for $x = -2$ and $y = 3$

5. $3x^2 + 5x^2 - 11x + x - 10 + 11$
6. $-15y^3 + 12y^3 - 4y^2 + 18y + 2y - 100$
7. $7xy + 8x - 3xy + 4y^2 - 13x$

# 3.7 Solving Equations with Fractions ($ax + b = c$)

## Solving First-Degree Equations with Fractions

We have discussed solving first-degree equations with whole number constants and coefficients in Section 1.4 and solving first-degree equations with integer constants and integer coefficients in Section 2.7. In these sections, we dealt with equations that could be solved by applying either the Addition Principle or the Division Principle, but we did not need to apply both principles in solving any one equation. In keeping with the idea of building equation-solving skills throughout the text, we now discuss techniques for solving equations in which

**Objectives**

① Be able to solve equations with rational number (fraction) coefficients and rational number (fraction) solutions.

1. The equations have fractions (positive and negative) as constants and coefficients, and
2. the equations have fractions as solutions.

(**Note:** We will discuss equations solving again in Chapters 4, 5, and 7 using ideas related to each succeeding chapter. In this way, equation solving will become increasingly more interesting and have more applications.)

Recall that if an equation contains a variable, then a **solution** to an equation is a number that gives a true statement when substituted for the variable. And **solving the equation means to find all the solutions of the equation.** For example, if we substitute $x = -3$ in the equation

$$5x + 14 = -1$$

we get

$$5(-3) + 14 = -1,$$ which is a **true statement**.

Thus, $-3$ is the solution of the equation $5x + 14 = -1$.

**Definition:**

A **first–degree equation in $x$** (or **linear equation in $x$**) is any equation that can be written in the form

$$ax + b = c$$

where $a$, $b$, and $c$ are constants and $a \neq 0$.
(**Note:** A variable other than $x$ may be used.)

A fundamental fact of algebra, stated here without proof, is that **every first–degree equation has exactly one solution**. Therefore, if we find any one solution to a first–degree equation, then that is the only solution.

To find the solution of a first-degree equation, we apply the principles on the following page.

**Principles Used in Solving a First-Degree Equation**

In the two basic principles stated here, $A$ and $B$ represent algebraic expressions or constants. $C$ represents a constant, and $C$ is not 0 in the Multiplication Principle.

1. The **Addition Principle:** The equations $A = B$ and $A + C = B + C$ have the same solutions.

2. The **Multiplication Principle:** The equations $A = B$ and $\frac{A}{C} = \frac{B}{C}$ or $\frac{1}{C} \cdot A = \frac{1}{C} \cdot B$ (where $C \neq 0$) have the same solutions.

Essentially, these two principles say that if we perform the same operation to both sides of an equation, the resulting equation will have the same solution as the original equation.

These principles were stated in Section 2.7 as the **Addition Principle** and the **Division Principle**. The **Multiplication Principle** is the same as the **Division Principle**. As the Multiplication Principle shows, division by a nonzero number $C$ is the same as multiplication by its reciprocal, $\frac{1}{C}$. For example, to solve

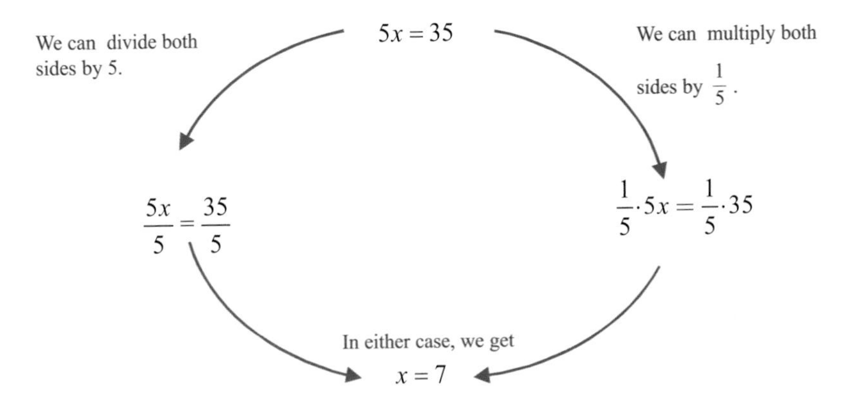


The process is further illustrated in the following examples. Study them carefully.

At this point you may want to review the concept that subtracting any number is the same as adding its opposite. In particular, as in Example 1, subtracting 17 is the same as adding -17.

## Example 1

Solve the equation $x + 17 = -12$.

**Solution**

| | |
|---|---|
| $x + 17 = -12$ | Write the equation. |
| $x + 17 - \mathbf{17} = -12 - \mathbf{17}$ | Using the Addition Principle, add −17 to both sides. |
| $x + 0 = -29$ | Simplify both sides. |
| $x = -29$ | The solution |

### Example 2

Solve the equation $8y = 248$.

#### Solution

We can divide both sides by 8, as we did in Section 2.7. However, here we show the same results by multiplying both sides by $\frac{1}{8}$.

| | |
|---|---|
| $8y = 248$ | Write the equation. |
| $\frac{1}{\cancel{8}} \cdot \overset{1}{\cancel{8}} y = \frac{1}{\cancel{8}} \cdot \overset{31}{\cancel{248}}$ | Multiply both sides by $\frac{1}{8}$ |
| $1 \cdot y = 31$ | Simplify both sides. |
| $y = 31$ | The solution |

More difficult problems may involve several steps. Keep in mind the following general process as you proceed from one step to another. That is, you must keep in mind what you are trying to accomplish.

**To Understand How to Solve Equations:**

1. Apply the distributive property to remove parentheses whenever necessary.
2. If a constant is added to a variable, add its opposite to both sides of the equation.
3. If a variable has a constant coefficient other than 1, divide both sides by that coefficient (that is, in effect, multiply both sides by the reciprocal of that coefficient).
4. Generally, use the Addition Principle first so that terms with variables are on one side and constant terms are on the other side. Then combine terms and use the Multiplication Principle.
5. Remember that the object is to isolate the variable on one side of the equation with a coefficient of 1.

### Example 3

Solve the equation $5x - 9 = 61$.

#### Solution

| | |
|---|---|
| $5x - 9 = 61$ | Write the equation. |
| $5x - 9 + 9 = 61 + 9$ | Add 9 to both sides. |
| $5x + 0 = 70$ | Simplify. |
| $5x = 70$ | Simplify. |
| $\frac{\cancel{5}x}{\cancel{5}} = \frac{70}{5}$ | Divide both sides by 5. |
| $x = 14$ | Simplify. |

## Example 4

Solve the equation $3(x - 4) + x = -20$.

### Solution

| | |
|---|---|
| $3(x-4)+x = -20$ | Write the equation. |
| $3x - 12 + x = -20$ | Apply the distributive property. |
| $4x - 12 = -20$ | Combine like terms. |
| $4x - 12 + 12 = -20 + 12$ | Add 12 to both sides. |
| $4x = -8$ | Simplify. |
| $\frac{\cancel{4}x}{\cancel{4}} = \frac{-8}{4}$ | Divide both sides by 4. |
| $x = -2$ | Simplify. |

## Checking Solutions to Equations

Checking can be done by substituting the solution found into the original equation to see if the resulting statement is true. If it is not true, then go over your procedures to look for any possible errors.

### Checking Example 3

$$5x - 9 \stackrel{?}{=} 61$$
$$5(14) - 9 \stackrel{?}{=} 61$$
$$70 - 9 \stackrel{?}{=} 61$$
$$61 = 61$$

### Checking Example 4

$$3(x-4) + x \stackrel{?}{=} -20$$
$$3(-2-4) + (-2) \stackrel{?}{=} -20$$
$$3(-6) - 2 \stackrel{?}{=} -20$$
$$-18 - 2 \stackrel{?}{=} -20$$
$$-20 = -20$$

*Now Work Exercises 1 and 2 in the Margin .*

Solve each of the following equations.

**1.** $4x + 22 = 18$

$x = -1$

**2.** $2(x - 5) + 3x = 20$

$x = 6$

## Equations With Fractions

**Special Note on Fractional Coefficients:**

An expression such as $\frac{1}{5}x$ can be thought of as a product:

$$\frac{1}{5}x = \frac{1}{5} \cdot \frac{x}{1} = \frac{1 \cdot x}{5 \cdot 1} = \frac{x}{5}$$

Thus, $\frac{1}{5}x$ and $\frac{x}{5}$ have the same meaning and $\frac{1}{9}x$ and $\frac{x}{9}$ have the same meaning.

Similarly, $\frac{2}{3}x = \frac{2x}{3}$ and $\frac{7}{8}x = \frac{7x}{8}$.

Fractional coefficients and the idea that $\frac{1}{5}x$ and $\frac{x}{5}$ have the same meaning are difficult concepts for most students. You may want to give several more examples such as

$$\frac{x}{6} = \frac{1}{6} \cdot \frac{x}{1} = \frac{1}{6}x$$
$$\frac{2x}{5} = \frac{2}{5} \cdot \frac{x}{1} = \frac{2}{5}x$$
$$-\frac{n}{3} = -\frac{1}{3} \cdot \frac{n}{1} = -\frac{1}{3}n$$

Sometimes an equation is simplified so that the variable has a fractional coefficient and all constants are on the other side of the equation. In such cases, the equation can be solved in one step by multiplying both sides of the equation by the reciprocal of the coefficient. This will give 1 as the coefficient of the variable.

### Example 5

Solve the equations **a.** $\frac{2}{3}x = 14$ and **b.** $-\frac{7n}{10} = 35$ by multiplying both sides of the equation by the reciprocal of the coefficient.

**Solution**

**a.**

$$\frac{2}{3}x = 14$$
$$\frac{3}{2} \cdot \frac{2}{3}x = \frac{3}{2} \cdot 14 \quad \text{Multiply each side by } \frac{3}{2}.$$
$$1 \cdot x = 21$$
$$x = 21$$

**b.**

$$-\frac{7n}{10} = 35$$
$$\left(-\frac{10}{7}\right)\left(-\frac{7n}{10}\right) = \left(-\frac{10}{7}\right) \cdot 35 \quad \text{Multiply each side by } \frac{-10}{7}.$$
$$\left(-\frac{10}{7}\right)\left(-\frac{7}{10}\right) \cdot n = \left(-\frac{10}{7}\right) \cdot \frac{35}{1}$$
$$1 \cdot n = -50$$
$$n = -50$$

More generally, equations with fractions can be solved by first multiplying each term in the equation by the LCM (least common multiple) of all the denominators. **The object is to find an equivalent equation with integer constants and coefficients that will be easier to solve.** That is, we generally find working with integers easier than working with fractions.

### Example 6

Solve the equation $\frac{5}{8}x - \frac{1}{4}x = \frac{3}{10}$.

**Solution**

For the denominators 8, 4, and 10 the LCM is 40. Multiply both sides of the equation by 40, apply the distributive property, and reduce to get all integer coefficients and constants because 8, 4, and 10 will all divide into 40.

$$\frac{5}{8}x - \frac{1}{4}x = \frac{3}{10}$$

$$40\left(\frac{5}{8}x - \frac{1}{4}x\right) = 40\left(\frac{3}{10}\right)$$ Multiply both sides by 40.

$$\mathbf{40}\left(\frac{5}{8}x\right) - \mathbf{40}\left(\frac{1}{4}x\right) = \mathbf{40}\left(\frac{3}{10}\right)$$ Apply the distributive property.

$$25x - 10x = 12$$

$$15x = 12$$

$$\frac{\cancel{15}x}{\cancel{15}} = \frac{12}{15}$$

$$x = \frac{4}{5}$$

### Example 7

Solve the equation $\frac{x}{2} + \frac{3x}{4} - \frac{1}{6} = 0$.

**Solution**

$$\frac{x}{2} + \frac{3x}{4} - \frac{1}{6} = 0$$

$$12\left(\frac{x}{2} + \frac{3x}{4} - \frac{1}{6}\right) = 12(0)$$ Multiply both sides by 12, the LCM of 2, 4, and 6.

$$\mathbf{12}\left(\frac{x}{2}\right) + \mathbf{12}\left(\frac{3x}{4}\right) - \mathbf{12}\left(\frac{1}{6}\right) = \mathbf{12}(0)$$ Apply the distributive property.

$$6x + 9x - 2 = 0$$

$$15x - 2 = 0$$

$$15x - 2 + 2 = 0 + 2$$

$$15x = 2$$

$$\frac{\cancel{15}x}{\cancel{15}} = \frac{2}{15}$$

$$x = \frac{2}{15}$$

### *Completion Example 8*

Explain each step in the solution process shown here.

| Equation | Explanation |
|---|---|
| $3(n-5)-n = 1$ | Write the equation. |
| $3n-15-n = 1$ | ________________ |
| $2n-15 = 1$ | ________________ |
| $2n-15+15 = 1+15$ | |
| ________________ | |
| $2n = 16$ | ________________ |
| $n = 8$ | ________________ |

*Now Work Exercises 3 and 4 in the Margin.*

Solve each of the following equation.

**3.** $\frac{3}{4}x - \frac{1}{3}x = \frac{3}{8}$

$x = \frac{9}{10}$

**4.** $\frac{x}{5} + \frac{2x}{3} - \frac{1}{2} = 0$

$x = \frac{15}{26}$

## Completion Example Answer

**8.**

| Equation | Explanation |
|---|---|
| $3(n-5)-n = 1$ | Write the equation. |
| $3n-15-n = 1$ | **Apply the distributive property.** |
| $2n-15 = 1$ | **Combine like terms.** |
| $2n-15+15 = 1+15$ | **Add 15 to both sides.** |
| $2n = 16$ | **Simplify.** |
| $n = 8$ | **Divide both sides by 2 and simplify.** |

Name ______________________ Section ____________ Date __________

## Exercises 3.7

Give an explanation (or a reason) for each step in the solution process.

**1.**

| | |
|---|---|
| $3x - x = -10$ | Write the equation. |
| $2x = -10$ | Simplify. |
| $\frac{2x}{2} = \frac{-10}{2}$ | Divide both sides by 2. |
| $x = -5$ | Simplify. |

**2.**

| | |
|---|---|
| $4x - 12 = -10$ | Write the equation. |
| $4x - 12 + 12 = -10 + 12$ | Add 12 to both sides. |
| $4x = 2$ | Simplify. |
| $\frac{4x}{4} = \frac{2}{4}$ | Divide both sides by 4. |
| $x = \frac{1}{2}$ | Simplify. |

**3.**

| | |
|---|---|
| $\frac{1}{2}x + \frac{1}{5} = 3$ | Write the equation. |
| $10\left(\frac{1}{2}x\right) + 10\left(\frac{1}{5}\right) = 10(3)$ | Multiply each term by 10. |
| $5x + 2 = 30$ | Simplify. |
| $5x + 2 - 2 = 30 - 2$ | Add –2 to both sides. |
| $5x = 28$ | Simplify. |
| $\frac{5x}{5} = \frac{28}{5}$ | Divide both sides by 5. |
| $x = \frac{28}{5}$ | Simplify. |

**4.**

| | |
|---|---|
| $\frac{y}{3} - \frac{2}{3} = 7$ | Write the equation. |
| $3\left(\frac{y}{3}\right) - 3\left(\frac{2}{3}\right) = 3(7)$ | Multiply each term by 3. |
| $y - 2 = 21$ | Simplify. |
| $y - 2 + 2 = 21 + 2$ | Add 2 to both sides. |
| $y = 23$ | Simplify. |

**ANSWERS**

1. [Respond below exercise.]
2. [Respond below exercise.]
3. [Respond below exercise.]
4. [Respond below exercise.]

5. $x = 12$

6. $y = 50$

7. $y = 55$

8. $x = -32$

9. $x = 5$

10. $x = 7$

11. $n = 7$

12. $n = 4$

13. $x = -11$

14. $x = -6$

15. $x = -6$

16. $y = -5$

17. $y = -4$

18. $x = 3$

19. $n = -\frac{14}{5}$

20. $n = -\frac{5}{2}$

21. $y = -\frac{1}{2}$

22. $x = -\frac{16}{3}$

23. $x = \frac{1}{3}$

24. $y = -4$

25. $x = -\frac{25}{2}$

26. $x = 4$

Solve each of the following equations:

**5.** $x + 13 = 25$

**6.** $y - 20 = 30$

**7.** $y - 35 = 20$

**8.** $x + 10 = -22$

**9.** $2x + 3 = 13$

**10.** $3x - 1 = 20$

**11.** $5n - 2 = 33$

**12.** $7n - 5 = 23$

**13.** $5x - 2x = -33$

**14.** $3x - x = -12$

**15.** $6x - 2x = -24$

**16.** $3y - 2y = -5$

**17.** $5y - 2 - 4y = -6$

**18.** $7x + 14 - 10x = 5$

**19.** $5(n - 2) = -24$

**20.** $2(n + 1) = -3$

**21.** $4(y - 1) = -6$

**22.** $3(x + 2) = -10$

**23.** $16x + 23x - 5 = 8$

**24.** $71y - 62y + 3 = -33$

**25.** $2(x + 9) + 10 = 3$

**26.** $4(x - 3) - 4 = 0$

Name ______________ Section ______________ Date ______________

**27.** $x - 5 - 4x = -20$

**28.** $x - 6 - 6x = 24$

**29.** $5x + 2(6 - x) = 10$

**30.** $3y + 2(y + 1) = 4$

**31.** $\frac{2}{3}y - 5 = 21$

**32.** $\frac{3}{4}x + 2 = 17$

**33.** $\frac{1}{5}x + 10 = -32$

**34.** $\frac{1}{2}y - 3 = -23$

**35.** $\frac{1}{2}x - \frac{2}{3} = 11$

**36.** $\frac{2}{3}y - \frac{1}{5} = -2$

**37.** $\frac{y}{4} + \frac{2}{3} = -\frac{1}{6}$

**38.** $\frac{x}{8} + \frac{1}{6} = -\frac{1}{10}$

**39.** $\frac{5}{8}y - \frac{1}{4} = \frac{1}{3}$

**40.** $\frac{3}{5}y - 4 = \frac{1}{5}$

**41.** $\frac{x}{7} + \frac{1}{3} = \frac{1}{21}$

**42.** $\frac{n}{3} - 6 = \frac{2}{3}$

**43.** $\frac{n}{5} - \frac{1}{5} = \frac{1}{5}$

**44.** $\frac{x}{15} - \frac{2}{15} = -\frac{2}{15}$

**45.** $\frac{3}{4} = \frac{1}{5}x - \frac{5}{8}$

**46.** $\frac{1}{2} = \frac{1}{3}x + \frac{4}{15}$

**47.** $\frac{7}{8} = \frac{3}{4}x - \frac{5}{8}$

**48.** $\frac{1}{10} = \frac{4}{5}x + \frac{3}{10}$

**ANSWERS**

**27.** $x = 5$

**28.** $x = -6$

**29.** $x = -\frac{2}{3}$

**30.** $y = \frac{2}{5}$

**31.** $y = 39$

**32.** $x = 20$

**33.** $x = -210$

**34.** $y = -40$

**35.** $x = \frac{70}{3}$

**36.** $y = -\frac{27}{10}$

**37.** $y = -\frac{10}{3}$

**38.** $x = -\frac{32}{15}$

**39.** $y = \frac{14}{15}$

**40.** $y = 7$

**41.** $x = -2$

**42.** $n = 20$

**43.** $n = 2$

**44.** $x = 0$

**45.** $x = \frac{55}{8}$

**46.** $x = \frac{7}{10}$

**47.** $x = 2$

**48.** $x = -\frac{1}{4}$

49. $n=-\frac{3}{2}$

50. $x=\frac{11}{7}$

51. $x=-\frac{1}{10}$

52. $x=\frac{4}{5}$

53. $n=-\frac{8}{9}$

54. $y=\frac{21}{5}$

55. $y=-10$

56. $x=20$

57. $x=64$

58. $y=-40$

59. $y=-45$

60. $x=-\frac{5}{6}$

61. $x=-\frac{3}{4}$

62. $n=-\frac{3}{2}$

63. $n=-\frac{5}{6}$

64. $x=-\frac{4}{7}$

65. $x=81$

**49.** $-\frac{5}{6}=\frac{2}{3}n+\frac{1}{6}$

**50.** $-\frac{2}{7}=\frac{1}{5}x-\frac{3}{5}$

**51.** $\frac{3x}{5}+\frac{2x}{5}=-\frac{1}{10}$

**52.** $\frac{5}{8}x-\frac{3}{4}x=-\frac{1}{10}$

**53.** $\frac{5n}{6}-\frac{n}{12}=-\frac{2}{3}$

**54.** $\frac{y}{7}+\frac{y}{28}=\frac{3}{4}$

**55.** $\frac{5}{6}y-\frac{7}{8}y=\frac{5}{12}$

Sometimes an equation is simplified so that the variable has a fractional coefficient and all constants are on the other side of the equation. In such cases, the equation can be solved in one step by multiplying both sides by the reciprocal of the coefficient. This will give 1 as the coefficient of the variable. For example,

$$\begin{aligned}\frac{2}{3}x&=14\\ \frac{3}{2}\cdot\frac{2}{3}x&=\frac{3}{2}\cdot 14\\ 1\cdot x&=21\\ x&=21\end{aligned}$$

Use this method to solve each of the following equations.

**56.** $\frac{3}{4}x=15$

**57.** $\frac{5}{8}x=40$

**58.** $\frac{7}{10}y=-28$

**59.** $\frac{2}{3}y=-30$

**60.** $\frac{4}{5}x=-\frac{2}{3}$

**61.** $\frac{5}{6}x=-\frac{5}{8}$

**62.** $-\frac{1}{2}n=\frac{3}{4}$

**63.** $-\frac{2}{5}n=\frac{1}{3}$

**64.** $\frac{7}{8}x=-\frac{1}{2}$

**65.** $-\frac{2}{9}x=-18$

## Chapter 3 Index of Key Ideas and Terms

Tests for Divisibility by 2, 3, 5, 6, 9, and 10 page 177

For 2: If the last digit ( units digit ) of an integer is 0, 2, 4, 6, or 8, then the integer is divisible by 2.

For 3: If the sum of the digits of an integer is divisible by 3, then the integer is divisible by 3.

For 5: If the last digit of an integer is 0 or 5, then the integer is divisible by 5.

For 6: If an integer is divisible by both 2 and 3, then it is divisible by 6.

For 9: If the sum of the digits of an integer is divisible by 9, then the integer is divisible by 9.

For 10: If the last digit of an integer is 0, then the integer is divisible by 10.

Even integers are divisible by 2. page 177

If an integer is divided by 2 and the remainder is 0, then the integer is even.

Odd integers are not divisible by 2. page 177

If an integer is divided by 2 and the remainder is 1, then the integer is odd.

Prime Numbers page 189

A prime number is a counting number greater than 1 that has only 1 and itself as factors.

A prime number is a counting number with exactly two different factors.

Composite Numbers page 189

A composite number is a counting number with more than two different factors (or divisors).

Sieve of Eratosthenes page 190

The prime numbers less than 50 are
2, 3, 5, 7, 11, 13, 17, 19, 23, 29, 31, 37, 41, 43, 47.

To Determine whether a number is prime: page 192

Divide the number by progressively larger prime numbers (2, 3, 5, 7, 11, and so forth) until:

1. You find a remainder of 0 (meaning that the prime number is a factor and the given number is composite) or
2. you find a quotient smaller than the prime divisor (meaning that the given number has no smaller prime factors and is therefore prime itself).

Finding the factors of a composite number: page 201

The only factors (or divisors) of a composite number are

1. 1 and the number, itself,
2. each prime factor, and
3. products formed by all combinations of the prime factors (including repeated factors).

Least Common Multiple (LCM): page 207

1. Find the prime factorization of each number.
2. Find the prime factors that appear in any one of the prime factorizations.
3. Form the product of these primes using each prime the most number of times it appears in any one of the prime factorizations.

How to find the LCM of a set of algebraic terms page 210

Rational Numbers: page 219

A rational number is a number that can be written in the form $\frac{a}{b}$, where $a$ and $b$ are integers and $b \neq 0$.

In this text, the terms rational numbers and fractions have the same meaning.

Improper Fractions page 220

Improper fractions are fractions in which the numerator is larger than the denominator. There is nothing wrong with improper fractions.

Multiplication with Rational Numbers page 221

Properties: commutative, associative, identity page 223

Raising to higher terms page 223

Reducing to lowest terms page 225

Multiplying and reducing at the same time page 226

Reciprocal page 235

The reciprocal of $\frac{a}{b}$ is $\frac{b}{a}$ ($a \neq 0$ and $b \neq 0$).

The product of a nonzero number and its reciprocal is always 1.

The number 0 has no reciprocal.

Division with Rational Numbers page 236

Solving First-Degree Equations ($ax + b = c$) page 247

1. The Addition Principle: The equations $A = B$ and $A + C = B + C$ have the same solutions.

2. The Multiplication Principle: The equations $A = B$

   and $\frac{A}{C} = \frac{B}{C}$ or $\frac{1}{C} \cdot A = \frac{1}{C} \cdot B$ (where $C \neq 0$)
   have the same solutions.

Name ______________ Section ______________ Date ______________

# Chapter 3 Test

Using the tests for divisibility, tell which of the numbers 2, 3, 5, 6, 9, and 10 (if any) will divide the following numbers.

**1.** 612 **2.** 190 **3.** 1169

**4.** List the multiples of 11 that are less than 100.

**5.** Find all the even prime numbers that are less than 10,000.

**6. a.** True or False: 25 divides the product $5 \cdot 4 \cdot 3 \cdot 2 \cdot 1$. Explain your answer in terms of factors.

False. The number 5 is not a factor twice of $5 \cdot 4 \cdot 3 \cdot 2 \cdot 1$.

**b.** True or False: 27 divides the product $(2 \cdot 5 \cdot 6 \cdot 7 \cdot 9)$. Explain your answer in terms of factors.

True. $2 \cdot 5 \cdot 6 \cdot 7 \cdot 9 = 2 \cdot 5 \cdot 2 \cdot 3 \cdot 7 \cdot 9 = 27 \cdot 140$.

**7.** Identify all the prime numbers in the following list:

1, 2, 4, 6, 9, 11, 13, 15, 37, 40, 42

Find the prime factorization of each number.

**8.** 80 **9.** 180 **10.** 225

**11.** Find all the factors of 90.

Find the LCM of each of the following sets of numbers or algebraic terms.

**12.** 15, 24, 35 **13.** $30xy, 40x^3, 45y^2$ **14.** $8a^4, 14a^2, 21a$

**ANSWERS**

1. 2, 3, 6, 9
2. 2, 5, 10
3. None
4. 11, 22, 33, 44, 55, 66, 77, 88, 99
5. 2
6. a. [Respond below exercise.]
   b. [Respond below exercise.]
7. 2, 11, 13, 37
8. $2^4 \cdot 5$
9. $2^2 \cdot 3^2 \cdot 5$
10. $3^2 \cdot 5^2$
11. 1, 2, 3, 5, 6, 9, 10, 15, 18, 30, 45, 90
12. 840
13. $360x^3y^2$
14. $168a^4$

15. Two long–distance runners practice on the same track and start at the same point. One takes 54 seconds and the other takes 63 seconds to go around the track once.

a. If each continues at the same pace, in how many seconds will they meet again at the starting point?

b. How many laps will each have run at that time?

c. Briefly discuss the idea of whether or not they will meet at some other point on the track as they are running.
The faster runner must pass the slower runner at some point before a full lap is gained.

16. A rational number can be written as a fraction in which the numerator and denominator are integers, and the denominator is not zero.

17. The value of $\frac{0}{-16}$ is 0, and value of $\frac{-16}{0}$ is undefined.

18. Give two examples using rational numbers that illustrate the commutative property of multiplication.

Reduce to lowest terms.

19. $\frac{108}{90}$

20. $\frac{48x}{-216x^2}$

21. $\frac{75a^3b}{60ab^2}$

Perform the indicated operations. Reduce all fractions to lowest terms.

22. $\frac{21}{26}\cdot\frac{13}{15}\cdot\frac{5}{7}$

23. $\frac{-25xy^2}{35xy}\cdot\frac{21x^2}{-30y^2}$

24. $45x\div\frac{1}{9x}$

25. $\frac{-36a^2}{50}\div\frac{63ab}{30}$

Solve each of the following equations.

26. $5x+4=-5$

27. $3(x-1)+2x=12$

28. $-\frac{7}{10}=\frac{y}{5}+\frac{1}{2}$

29. $-\frac{7}{4}n+\frac{2}{5}=\frac{1}{6}$

30. A $3\frac{1}{2}$ inch floppy disk contains 942 kilobytes of information. This amount is $\frac{2}{3}$ of the capacity of the disk. What is the capacity of the disk?

**Answers**

15. a. 378 seconds

b. 7 laps and 6 laps, respectively

c. [Respond below exercise.]

16. [Respond in exercise.]

17. [Respond in exercise.]

18. Any two examples of the form $\frac{a}{b}\cdot\frac{c}{d}=\frac{c}{d}\cdot\frac{a}{b}$

19. $\frac{6}{5}$

20. $-\frac{2}{9x}$

21. $\frac{5a^2}{4b}$

22. $\frac{1}{2}$

23. $\frac{x^2}{2y}$

24. $405x^2$

25. $-\frac{12a}{35b}$

26. $x=-\frac{9}{5}$

27. $x=3$

28. $y=-6$

29. $n=\frac{2}{15}$

30. 1413 kilobytes

Name ______________ Section ______________ Date ______________

ANSWERS

## Cumulative Review : Chapters 1 - 3

**1.** Round off 62,505 to the nearest thousand.

**2.** For the expression $7^2$ :

**a.** Name the base,

**b.** name the exponent, and

**c.** find the value of $7^2$.

Find the value of $n$ and name the property illustrated.

**3.** $36 + n = 36$

**4.** $17 + n = 14 + 17$

Perform the indicated operations.

**5.**
$$\begin{array}{r} 82 \\ 586 \\ \underline{+\ 893} \end{array}$$

**6.**
$$\begin{array}{r} 8004 \\ \underline{-\ 3459} \end{array}$$

**7.**
$$\begin{array}{r} 35 \\ \underline{\times\ 75} \end{array}$$

**8.** Estimate the quotient $3210 \div 16$; then find the quotient and the remainder.

**9.** Evaluate the expression $6 + (12 \cdot 6 + 3^2) \div 9(-5)$.

**10.** Without actually dividing, determine which of the numbers 2, 3, 5, 6, 9, and 10 (if any) will divide 27,360.

**11.** Without actually dividing, find the quotient $(7 \cdot 6 \cdot 5 \cdot 4 \cdot 3 \cdot 2 \cdot 1) \div 60$.

**12.** Use a calculator to find the value of

**a .** $7^6$ and

**b .** $9^5$.

**Answers:**

1. 63,000
2. a. 7
   b. 2
   c. 49
3. $n = 0$; Additive Identity
4. $n = 14$; Comm. Property of Add.
5. 1561
6. 4545
7. 2625
8. 150 estimate; 200 R 10
9. −39
10. 2, 3, 5, 6, 9, 10
11. 84
12. a. 117,649
    b . 59,049

13. $-58$

14. a. 5, 7, 11, 13

b. $5^3 = 125$; $7^3 = 343$;

$11^3 = 1331$; $13^3 = 2197$

15. $3 \cdot 5^3$

16. $2^2 \cdot 3 \cdot 13$

17. $2^3 \cdot 5^2$

18. 1, 2, 3, 4, 5, 6, 10, 12

15, 20, 30, 60

19. 900

20. $2100x^2y^3z^4$

21. $-18$

22. a. $2x^2 - 2x + 3$

b. 7

23. a. $-2y^3 + 6y - 2$

b. $-6$

24. $\frac{36}{5}$

25. $\frac{8b}{35a}$

26. $-\frac{5}{x}$

27. $-1$

**13.** Evaluate the expression $a^3 - 7a^2 + 6a - 10$ if $a = -2$.

**14. a.** List the prime numbers between 4 and 15.

**b.** Find the cube of each of these prime numbers.

Find the prime factorization of each of the following composite numbers.

**15.** 375 **16.** 156 **17.** 200

**18.** Find all the factors of 60.

Find the LCM for each of the following sets of numbers or terms.

**19.** 12, 45, 50 **20.** $35x^2$, $15xy^3$, $20xyz$, $25xz^4$

**21.** Find the average of $-72, -106, 59, 79$, and $-50$.

For Exercises 22 and 23,

**a.** Simplify each of the expressions by combining like terms.

**b.** Evaluate each simplified expression for $x = 2$ and $y = -1$.

**22.** $3x^2 - x^2 + 5x - 7x + 3$ **23.** $y^3 - 3y^3 + 8y - 2y - 6 + 4$

Perform the indicated operations. Reduce all fractions to lowest terms.

**24.** $\frac{11}{3} \cdot \frac{27}{22} \cdot \frac{8}{5}$ **25.** $\frac{5a}{8} \cdot \frac{32}{35a^2} \cdot \frac{6b}{15}$

**26.** $\frac{-34x^2}{51} \div \frac{2x^3}{15}$ **27.** $\frac{13y}{39} \div \frac{-13y}{39}$

Name ______________ Section ______________ Date ______________

**ANSWERS**

Solve each of the following equations.

**28.** $2(x-14)=-12$

**29.** $-3(y+16)=-17$

**30.** $20x+3=-15$

**31.** $6n-2(n+5)=46$

**32.** $\frac{2}{5}x+\frac{1}{10}=\frac{1}{2}$

**33.** $\frac{y}{4}+\frac{2y}{8}=-\frac{1}{16}$

**34.** $\frac{1}{2}x+\frac{1}{6}x=\frac{2}{3}$

**35.** $\frac{5}{7}x+\frac{1}{14}x=-\frac{3}{28}$

**36.** You have checked at the bookstore and found that you will need to buy two textbooks for a price of $47 each, one textbook for $55, three notebooks for $5 each, and other supplies for $32. If these prices include taxes, how much should you withdraw from your savings account to pay for these items and still have $50 cash left over?

**37.** Three airline pilots fly into Chicago on a regular basis. One flies in every 9 days, another flies in every 12 days, and the third flies in every 15 days. How frequently are all three pilots in Chicago at the same time?

**38.** A painting is in a rectangular frame that is 26 inches wide and 40 inches high.

**a.** How many square inches of wall space are covered when the painting is hanging?

**b.** What is the perimeter of the painting?

**28.** $x=8$

**29.** $y=-\frac{31}{3}$

**30.** $x=-\frac{9}{10}$

**31.** $n=14$

**32.** $x=1$

**33.** $y=-\frac{1}{8}$

**34.** $x=1$

**35.** $x=-\frac{3}{22}$

**36.** $246

**37.** every 180 days

**38. a.** 1040 sq. inches

**b.** 132 inches

[Respond below exercise]
39. ______

## WRITING AND THINKING ABOUT MATHEMATICS

**39.** Name one topic discussed in Chapters 1–3 that you found particularly interesting, and explain why you think it is important. (Write a short paragraph, and be sure to express your thoughts in complete sentences.)

Answers will vary. This question is designed to provide a basis for one-on-one discussion or class discussion with your students. It may help you gain additional insight as to how your students perceive this course.

# 4

# Fractions and Mixed Numbers

## WHAT TO EXPECT IN CHAPTER 4

Chapter 4 continues our work with rational numbers (fractions) and includes mixed numbers, another form of rational numbers. Each section contains word problems to reinforce the concepts as they are developed. Since mixed numbers have fraction parts, the prime factorization techniques discussed in Chapter 3 are still valuable tools. Expressions with variables are included to show some of the similarities between arithmetic and algebra and to develop confidence in working with algebraic expressions.

Section 4.1 discusses addition and subtraction with fractions, a continuation of our work with fractions in Chapter 3. Mixed numbers and operations with mixed numbers are presented in Sections 4.2 through 4.4. Positive and negative mixed numbers are included. In these sections we also review the concept of area and introduce the concept of volume and related geometric figures.

Complex fractions (fractions that contain fractions in their numerators and denominators) and a review of order of operations (this time with mixed numbers) are presented in Section 4.5. With the development of ratios and proportions in Section 4.6, we again involve the concept of solving equations as well as introduce techniques for solving new types of word problems. Also, the interesting geometric concept of similar triangles is discussed. Similar triangles provide a visual approach to the usefulness of the skills related to setting up and solving proportions.

# Chapter 4 Fractions and Mixed Numbers

**Mathematics at Work!**

VCRs can record at three different speeds: standard speed (coded SP on the screen), which records 2 hours of programming on one tape; long-play speed (coded LP on the screen), which records 4 hours of programming on one tape; and extended-play speed (coded EP on the screen), which records 6 hours of programming on one tape. The advantage of the faster speed (shorter recording time) is that it provides a much better resolution of the picture. That is, the slower the speed, the more "grainy" the picture.

Suppose that you have recorded your favorite "sit-com" for $\frac{1}{2}$ hour at standard speed and a mystery movie for 2 hours at long-play speed. A sports program is coming up that will be on the air for 2 hours. Will you be able to record the entire program on the remainder of this tape at extended-play speed ? No, you will be unable to record the entire program.

## 4.1 Addition and Subtraction with Fractions

### Addition with Fractions

To add two (or more) fractions with the same denominator, we can think of the common denominator as the "name" of each fraction. The sum has this common name. Just as 3 oranges plus 2 oranges give a total of 5 oranges, 3 eighths plus 2 eighths give a total of 5 eighths. Figure 4.1 illustrates how the sum of the two fractions $\frac{3}{8}$ and $\frac{2}{8}$ might be diagrammed.

A more formal explanation of addition with fractions involves the distributive property:

$$\frac{a}{b}+\frac{c}{b}=a\cdot\frac{1}{b}+c\cdot\frac{1}{b}$$
$$=\frac{1}{b}(a+c)$$
$$=\frac{a+c}{b}$$

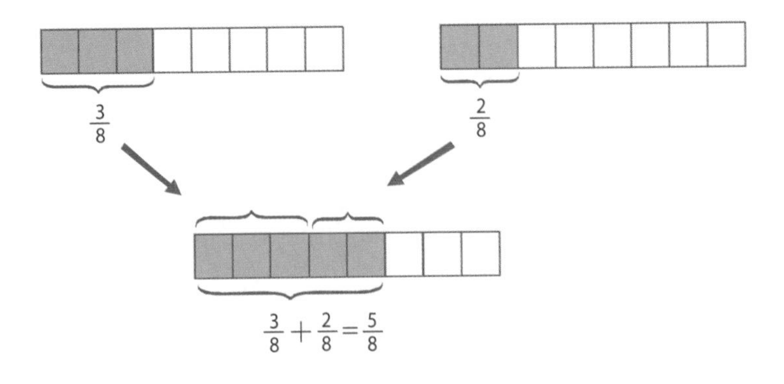


**Figure 4.1**

**To Add Two (or More) Fractions with the Same Denominator:**

1. Add the numerators. $\quad \frac{a}{b}+\frac{c}{b}=\frac{a+c}{b}$
2. Keep the common denominator.
3. Reduce, if possible.

### Example 1

$$\frac{1}{5}+\frac{2}{5}=\frac{1+2}{5}=\frac{3}{5}$$

### Example 2

$$\frac{1}{11}+\frac{2}{11}+\frac{3}{11}=\frac{1+2+3}{11}=\frac{6}{11}$$

### Objectives

① Be able to add and subtract fractions with the same denominator.

② Recall how to find the LCM.

③ Know how to add and subtract fractions with different denominators.

Of course, fractions to be added will not always have the same denominator. In these cases, the smallest common denominator should be found and used to simplify any further calculations. **The least common denominator (LCD) is the least common multiple (LCM) of the denominators. Review Section 3.4 to reinforce your knowledge of the LCM.**

**To Add Fractions with Different Denominators:**

1. Find the least common denominator (LCD).
2. Change each fraction into an equal fraction with that denominator.
3. Add the new fractions.
4. Reduce, if possible.

One way to emphasize when a common denominator is needed and to help the students remember this idea is to have them repeat out loud in class: "We need a common denominator when adding or subtracting." "We do not need a common denominator when multiplying of dividing."

Another method of distinguishing among these operations is to give examples using the same fraction twice, as in $\frac{1}{3}\cdot\frac{1}{3}=\frac{1}{9}$ and $\frac{1}{3}+\frac{1}{3}=\frac{2}{3}$.

### Example 3

Find the sum: $\frac{3}{8}+\frac{13}{12}$

**Solution**

**a.** Find the LCD. Remember that the least common denominator (LCD) is the least common multiple (LCM) of the denominators.

$$\left.\begin{aligned}8&=2\cdot2\cdot2\\12&=2\cdot2\cdot3\end{aligned}\right\}\text{LCD}=2\cdot2\cdot2\cdot3=24$$

**Note:** You might not need to use prime factorizations to find the LCD. If the denominators are numbers that are familiar to you, then you might be able to find the LCD simply by inspection.

**b.** Find fractions equal to $\frac{3}{8}$ and $\frac{13}{12}$ with denominator 24.

$\frac{3}{8}=\frac{3}{8}\cdot\frac{3}{3}=\frac{9}{24}$ Multiply by $\frac{3}{3}$ because $8\cdot3=24$

$\frac{13}{12}=\frac{13}{12}\cdot\frac{2}{2}=\frac{26}{24}$ Multiply by $\frac{2}{2}$ because $12\cdot2=24$

**c.** Add

$$\frac{3}{8}+\frac{13}{12}=\frac{9}{24}+\frac{26}{24}=\frac{9+26}{24}=\frac{35}{24}$$

**d.** The fraction $\frac{35}{24}$ is in lowest terms because 35 and 24 have only +1 and −1 as common factors.

## Example 4

Find the sum: $\frac{7}{45}+\frac{7}{36}$

### Solution

**a.** Find the LCD.

$$\left.\begin{array}{l}45=3\cdot3\cdot5\\36=2\cdot2\cdot3\cdot3\end{array}\right\}\text{LCD}=2\cdot2\cdot3\cdot3\cdot5=180$$
$$=(3\cdot3\cdot5)(2\cdot2)=45\cdot4$$
$$=(2\cdot2\cdot3\cdot3)(5)=36\cdot5$$

**b.** Steps b, c, and d from Example 3 can be written together in one process.

You might remind them that in reducing $\frac{63}{180}$ the prime factorization of 180 is already known from finding the LCD in Step (a).

$$\frac{7}{45}+\frac{7}{36}=\frac{7}{45}\cdot\frac{4}{4}+\frac{7}{36}\cdot\frac{5}{5}$$
$$=\frac{28}{180}+\frac{35}{180}=\frac{63}{180}$$
$$=\frac{\cancel{3}\cdot\cancel{3}\cdot7}{2\cdot2\cdot\cancel{3}\cdot\cancel{3}\cdot5}=\frac{7}{20}$$

Note that, in adding fractions, we also may choose to write them vertically. The process is the same.

$$\frac{7}{45}=\frac{7}{45}\cdot\frac{4}{4}=\frac{28}{180}$$
$$+\frac{7}{36}=\frac{7}{36}\cdot\frac{5}{5}=\frac{35}{180}$$
$$\frac{63}{180}=\frac{\cancel{3}\cdot\cancel{3}\cdot7}{2\cdot2\cdot\cancel{3}\cdot\cancel{3}\cdot5}=\frac{7}{20}$$

## Example 5

Find the sum: $6+\frac{9}{10}+\frac{3}{1000}$

### Solution

**a.** LCD = 1000. All the denominators are powers of 10, and 1000 is the largest. Write 6 as $\frac{6}{1}$.

**b.** $6+\frac{9}{10}+\frac{3}{1000}=\frac{6}{1}\cdot\frac{1000}{1000}+\frac{9}{10}\cdot\frac{100}{100}+\frac{3}{1000}$

$=\frac{6000}{1000}+\frac{900}{1000}+\frac{3}{1000}$

$=\frac{6903}{1000}$

Note: The only prime factors of 1000 are 2's and 5's. And since neither 2 nor 5 is a factor of 6903, the answer will not reduce.

*Now Work Exercises 1-4 in the Margin*

Add and reduce to lowest terms.

1. $\frac{1}{6}+\frac{3}{10}$

$\frac{7}{15}$

2. $\frac{1}{4}+\frac{3}{8}+\frac{7}{10}$

$\frac{53}{40}$

3. $5+\frac{3}{10}+\frac{7}{100}$

$\frac{537}{100}$

4. $\frac{1}{5}+\frac{1}{10}+\frac{3}{4}$

$\frac{21}{20}$

**Common Error:**

The following common error must be avoided:

Find the sum of $\frac{3}{2}+\frac{1}{6}$.

**Wrong Solution:**

$\frac{\not{3}^{1}}{2}+\frac{1}{\not{6}_{2}}=\frac{1}{2}+\frac{1}{2}=1$ WRONG

You **cannot** "divide out" across the + sign.

**Correct Solution:**

Use LCD = 6.

$\frac{3}{2}+\frac{1}{6}=\frac{3}{2}\cdot\frac{3}{3}+\frac{1}{6}=\frac{9}{6}+\frac{1}{6}=\frac{10}{6}$ CORRECT

Now reduce,

$\frac{10}{6}=\frac{5\cdot\not{2}}{3\cdot\not{2}}=\frac{5}{3}$ 2 is a factor in both the numerator and the denominator.

Both the commutative and associative properties of addition apply to fractions.

**Commutative Property of Addition**

If $\frac{a}{b}$ and $\frac{c}{d}$ are fractions, then $\frac{a}{b}+\frac{c}{d}=\frac{c}{d}+\frac{a}{b}$.

**Associative Property of Addition**

If $\frac{a}{b}$, $\frac{c}{d}$ and $\frac{e}{f}$ are fractions, then

$$\frac{a}{b}+\frac{c}{d}+\frac{e}{f}=\frac{a}{b}+\left(\frac{c}{d}+\frac{e}{f}\right)=\left(\frac{a}{b}+\frac{c}{d}\right)+\frac{e}{f}.$$

## Subtraction with Fractions

Finding the difference between two fractions with a common denominator is similar to finding the sum. The numerators are simply subtracted instead of added. Just as with addition, the common denominator "names" each fraction. Figure 4.2 shows how the difference between $\frac{4}{5}$ and $\frac{1}{5}$ might be diagrammed.

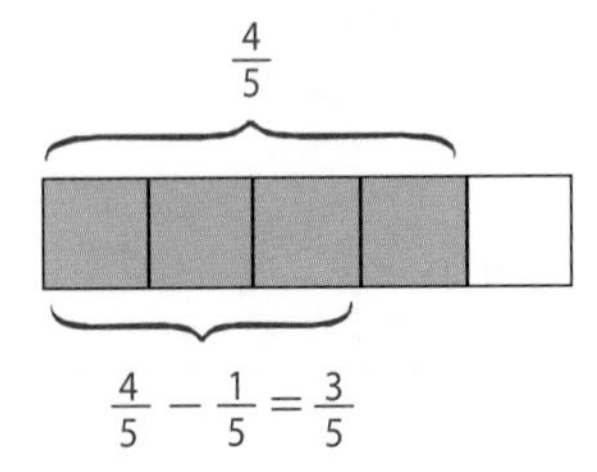


**Figure 4.2**

> **To Subtract Two Fractions with the Same Denominator:**
>
> **1.** Subtract the numerators.
>
> **2.** Keep the common denominator $\frac{a}{b}-\frac{c}{b}=\frac{a-c}{b}$.
>
> **3.** Reduce if possible.

### Example 6

Find the difference: $\frac{9}{10}-\frac{7}{10}$

**Solution**

$$\frac{9}{10}-\frac{7}{10}=\frac{9-7}{10}=\frac{2}{10}=\frac{\cancel{2}\cdot 1}{\cancel{2}\cdot 5}=\frac{1}{5}$$

The difference is reduced just as any fraction is reduced.

### Example 7

Find the difference: $\frac{6}{11}-\frac{19}{11}$

**Solution**

$$\frac{6}{11}-\frac{19}{11}=\frac{6-19}{11}=\frac{-13}{11}=-\frac{13}{11}$$

The difference is negative because a larger number is subtracted from a smaller number.

**To Subtract Fractions with Different Denominators:**

1. Find the least common denominator (LCD).
2. Change each fraction to an equal fraction with that denominator.
3. Subtract the new fractions.
4. Reduce, if possible.

### Example 8

Find the difference: $\frac{24}{55}-\frac{4}{33}$

**Solution**

**a.** Find the LCD.

$$\left.\begin{matrix}55 = 5\cdot 11\\ 33 = 3\cdot 11\end{matrix}\right\}\text{LCD} = 3\cdot 5\cdot 11 = 165$$

**b.** Find equal fractions with denominator 165, subtract, and reduce.

$$\frac{24}{55}-\frac{4}{33} = \frac{24}{55}\cdot\frac{3}{3}-\frac{4}{33}\cdot\frac{5}{5} = \frac{72-20}{165}$$

$$= \frac{52}{165} = \frac{2\cdot 2\cdot 13}{3\cdot 5\cdot 11} = \frac{52}{165}$$

The answer $\frac{52}{165}$ does not reduce because there are no common prime factors in the numerator and denominator.

Again, it will help many students to be reminded that in reducing $\frac{52}{165}$, the prime factorization of 165 is already known from finding the LCD.

### Example 9

Subtract: $1-\frac{13}{15}$

**Solution**

**a.** Since one can be written $\frac{15}{15}$ the LCD is 15.

**b.** $1-\frac{13}{15} = \frac{15}{15}-\frac{13}{15} = \frac{15-13}{15} = \frac{2}{15}$

When two or more negative numbers are added, the expression is often called an **algebraic sum**. This is a way to help avoid confusion with subtraction. Remember, (See Chapter 2) that subtraction can always be described in terms of addition.

### Completion Example 10

Find the algebraic difference as indicated: $-\frac{7}{12}-\frac{9}{20}$

**Solution**

**a.** Find the LCD.

$12 =$ ________
$20 =$ ________ } LCD = ______ = ______

**b.** $-\frac{7}{12}-\frac{9}{20}=-\frac{7}{12}\cdot\text{\_\_\_\_\_}-\frac{9}{20}\cdot\text{\_\_\_\_\_}$

$= \text{\_\_\_\_\_} - \text{\_\_\_\_\_}$

$= \text{\_\_\_\_\_} = \text{\_\_\_\_\_}$

$= \text{\_\_\_\_\_} = \text{\_\_\_\_\_}$

*Now Work Exercises 5- 8 in the Margin.*

Subtract and reduce to lowest terms.

**5.** $\frac{5}{6}-\frac{1}{6}$

$\frac{2}{3}$

**6.** $\frac{7}{10}-\frac{2}{15}$

$\frac{17}{30}$

**7.** $\frac{5}{12}-\frac{15}{16}$

$-\frac{25}{48}$

**8.** $1-\frac{11}{13}$

$\frac{2}{13}$

## Fractions Containing Variables

Now we will consider adding and subtracting fractions that contain variables in the numerators and/or the denominators. There are two basic rules that we must remember at all times:

1. Addition or subtraction can be accomplished only if the fractions have a common denominator.
2. A fraction can be reduced only if the numerator and denominator have a common factor.

### Example 11

Find the sum: $\frac{1}{a}+\frac{2}{a}+\frac{3}{a}$

**Solution**

The fractions all have the same denominator, so we simply add the numerators and keep the common denominator.

$$\frac{1}{a}+\frac{2}{a}+\frac{3}{a}=\frac{1+2+3}{a}=\frac{6}{a}$$

### Example 12

Find the sum: $\frac{3}{x}+\frac{7}{8}$

**Solution**

The two fractions have different denominators. The LCD is the product of the two denominators: $8x$.

$$\frac{3}{x}+\frac{7}{8}=\frac{3}{x}\cdot\frac{8}{8}+\frac{7}{8}\cdot\frac{x}{x}=\frac{24}{8x}+\frac{7x}{8x}=\frac{24+7x}{8x}$$

**Caution:** Do not try to reduce the answer in Example 12.

In the fraction $\frac{24+7x}{8x}$, neither 8 nor $x$ is a **factor** of the numerator.

We will see in Chapter 8, and in later courses in mathematics, how to factor and reduce algebraic fractions when there is a common factor in the numerator and denominator. For example, by using the distributive property we can find **factors** and reduce each of the following fractions:

**a.** $\frac{24+8x}{8x} = \frac{8(3+x)}{8x} = \frac{\cancel{8}(3+x)}{\cancel{8}x} = \frac{3+x}{x}$

Notice that, while 8 is a **common factor** here, $x$ is not.

**b.** $\frac{3y-15}{6y} = \frac{3(y-5)}{3 \cdot 2y} = \frac{\cancel{3}(y-5)}{\cancel{3} \cdot 2y} = \frac{y-5}{2y}$

3 is a **common factor** and $y$ is not a common factor.

The "common error" discussion on page 273 can help in discussing the algebraic expression in Example 12 and the fact that it cannot be simplified.

## Example 13

Find the difference: $\frac{y}{6} - \frac{5}{18}$

### Solution

The LCD is 18.

$$\frac{y}{6} - \frac{5}{18} = \frac{y}{6} \cdot \frac{3}{3} - \frac{5}{18} = \frac{3y-5}{18}$$

Notice that this fraction cannot be reduced because there are no common factors in the numerator and denominator.

## Example 14

Simplify the following expression: $\frac{a}{6} - \frac{1}{3} - \frac{1}{12}$

### Solution

$$\frac{a}{6} - \frac{1}{3} - \frac{1}{12} = \frac{a}{6} \cdot \frac{2}{2} - \frac{1}{3} \cdot \frac{4}{4} - \frac{1}{12} = \frac{2a}{12} - \frac{4}{12} - \frac{1}{12} = \frac{2a-4-1}{12} = \frac{2a-5}{12}$$

Notice that this fraction cannot be reduced because there are no common factors in the numerator and denominator.

### Completion Example 15

Subtract: $a-\frac{2}{7}$

**Solution**

The LCD is _____ .

$a-\frac{2}{7}=\frac{a}{1}\cdot\text{_____}-\frac{2}{7}=\text{_____}$

*Now Work Exercises 9-12 in the Margin.*

Find the indicated algebraic sums or differences.

**9.** $\frac{5}{a}+\frac{6}{a}+\frac{3}{a}$

$\frac{14}{a}$

**10.** $\frac{2}{x}+\frac{3}{8}$

$\frac{16+3x}{8x}$

**11.** $y-\frac{3}{4}-\frac{1}{2}$

$\frac{4y-5}{4}$

**12.** $\frac{n}{2}+\frac{5}{6}+\frac{1}{3}$

$\frac{3n+7}{6}$

### Example 16

If Keith's total income for the year was \$36,000 and he spent $\frac{1}{4}$ of his income on rent and $\frac{1}{12}$ of his income on his car, what total amount did he spend on these two items?

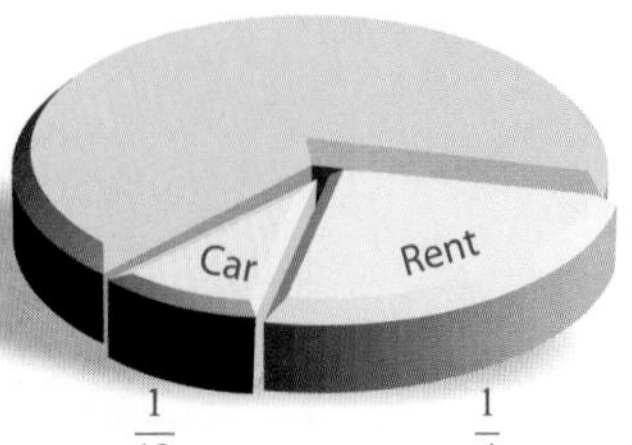


**Solution**

We can add the two fractions, and then multiply the sum by \$36,000. (Or, we can multiply each fraction by \$36,000, and then add the results. We will get the same answer either way.) The LCD is 12.

$$\frac{1}{4}+\frac{1}{12}=\frac{1}{4}\cdot\frac{3}{3}+\frac{1}{12}=\frac{3}{12}+\frac{1}{12}=\frac{4}{12}=\frac{\cancel{4}\cdot 1}{\cancel{4}\cdot 3}=\frac{1}{3}$$

Now, multiply $\frac{1}{3}$ times \$36,000.

$$\frac{1}{\cancel{3}_1}\cdot\overset{12,000}{\cancel{36,000}}=12,000$$

Keith spent a total of \$12,000 on rent and his car.

### Completion Example Answers

**10. a.** Find the LCD.

$$\left.\begin{matrix}12=\mathbf{2\cdot2\cdot3}\\20=\mathbf{2\cdot2\cdot5}\end{matrix}\right\}\text{LCD}=\mathbf{2\cdot2\cdot3\cdot5=60}$$

**b.** $$-\frac{7}{12}-\frac{9}{20}=-\frac{7}{12}\cdot\frac{\mathbf{5}}{\mathbf{5}}-\frac{9}{20}\cdot\frac{\mathbf{3}}{\mathbf{3}}$$
$$=-\frac{\mathbf{35}}{\mathbf{60}}-\frac{\mathbf{27}}{\mathbf{60}}$$
$$=\frac{\mathbf{-35-27}}{\mathbf{60}}=\frac{\mathbf{-62}}{\mathbf{60}}$$
$$=-\frac{\mathbf{2\cdot31}}{\mathbf{2\cdot30}}=-\frac{\mathbf{31}}{\mathbf{30}}$$

**15.** The LCD is **7**.

$$a-\frac{2}{7}=\frac{a}{1}\cdot\frac{\mathbf{7}}{\mathbf{7}}-\frac{2}{7}=\frac{\mathbf{7a-2}}{\mathbf{7}}$$

Name ______________________ Section ______________ Date ____________

# Exercises 4.1

Find the indicated products and sums. Reduce if possible.

**1. a.** $\frac{1}{3}\cdot\frac{1}{3}$ **b.** $\frac{1}{3}+\frac{1}{3}$ **2. a.** $\frac{1}{5}\cdot\frac{1}{5}$ **b.** $\frac{1}{5}+\frac{1}{5}$

**3. a.** $\frac{2}{7}\cdot\frac{2}{7}$ **b.** $\frac{2}{7}+\frac{2}{7}$ **4. a.** $\frac{3}{10}\cdot\frac{3}{10}$ **b.** $\frac{3}{10}+\frac{3}{10}$

**5. a.** $\frac{5}{4}\cdot\frac{5}{4}$ **b.** $\frac{5}{4}+\frac{5}{4}$ **6. a.** $\frac{7}{100}\cdot\frac{7}{100}$ **b.** $\frac{7}{100}+\frac{7}{100}$

**7. a.** $\frac{8}{9}\cdot\frac{8}{9}$ **b.** $\frac{8}{9}+\frac{8}{9}$ **8. a.** $\frac{9}{12}\cdot\frac{9}{12}$ **b.** $\frac{9}{12}+\frac{9}{12}$

**9. a.** $\frac{6}{10}\cdot\frac{6}{10}$ **b.** $\frac{6}{10}+\frac{6}{10}$ **10. a.** $\frac{5}{8}\cdot\frac{5}{8}$ **b.** $\frac{5}{8}+\frac{5}{8}$

Find the indicated sums and differences, and reduce all answers to lowest terms. (Note: In some cases you may want to reduce fractions before adding or subtracting. In this way, you may be working with a smaller common denominator.)

**11.** $\frac{3}{14}+\frac{3}{14}$ **12.** $\frac{1}{10}+\frac{3}{10}$ **13.** $\frac{4}{6}+\frac{4}{6}$

**ANSWERS**

**1. a.** $\frac{1}{9}$
**b.** $\frac{2}{3}$
**2. a.** $\frac{1}{25}$
**b.** $\frac{2}{5}$
**3. a.** $\frac{4}{49}$
**b.** $\frac{4}{7}$
**4. a.** $\frac{9}{100}$
**b.** $\frac{3}{5}$
**5. a.** $\frac{25}{16}$
**b.** $\frac{5}{2}$
**6. a.** $\frac{49}{10000}$
**b.** $\frac{7}{50}$
**7. a.** $\frac{64}{81}$
**b.** $\frac{16}{9}$
**8. a.** $\frac{9}{16}$
**b.** $\frac{3}{2}$
**9. a.** $\frac{9}{25}$
**b.** $\frac{6}{5}$
**10. a.** $\frac{25}{64}$
**b.** $\frac{5}{4}$
**11.** $\frac{3}{7}$
**12.** $\frac{2}{5}$
**13.** $\frac{4}{3}$

14. $\frac{3}{5}$

15. $\frac{8}{25}$

16. $\frac{1}{8}$

17. $-\frac{1}{5}$

18. $-\frac{1}{2}$

19. $\frac{1}{16}$

20. $\frac{7}{10}$

21. $\frac{17}{21}$

22. $\frac{9}{13}$

23. $\frac{1}{3}$

24. $\frac{5}{12}$

25. $-\frac{15}{14}$

26. $-\frac{7}{15}$

27. 0

28. $-\frac{4}{45}$

29. $-\frac{2}{25}$

30. $-\frac{1}{100}$

31. $\frac{7}{12}$

32. $\frac{1}{20}$

33. $-\frac{34}{945}$

34. $\frac{61}{45}$

**14.** $\frac{7}{15}+\frac{2}{15}$

**15.** $\frac{14}{25}-\frac{6}{25}$

**16.** $\frac{7}{16}-\frac{5}{16}$

**17.** $\frac{1}{15}-\frac{4}{15}$

**18.** $\frac{1}{12}-\frac{7}{12}$

**19.** $\frac{3}{8}-\frac{5}{16}$

**20.** $\frac{2}{5}+\frac{3}{10}$

**21.** $\frac{2}{7}+\frac{4}{21}+\frac{1}{3}$

**22.** $\frac{2}{39}+\frac{1}{3}+\frac{4}{13}$

**23.** $\frac{5}{6}-\frac{1}{2}$

**24.** $\frac{2}{3}-\frac{1}{4}$

**25.** $-\frac{5}{14}-\frac{5}{7}$

**26.** $-\frac{1}{3}-\frac{2}{15}$

**27.** $\frac{20}{35}-\frac{24}{42}$

**28.** $\frac{14}{45}-\frac{12}{30}$

**29.** $\frac{7}{10}-\frac{78}{100}$

**30.** $\frac{29}{100}-\frac{3}{10}$

**31.** $\frac{2}{3}+\frac{3}{4}-\frac{5}{6}$

**32.** $\frac{1}{5}+\frac{1}{10}-\frac{1}{4}$

**33.** $\frac{1}{63}-\frac{2}{27}+\frac{1}{45}$

**34.** $\frac{72}{105}-\frac{2}{45}+\frac{15}{21}$

Name ______________________ Section ______________ Date __________

**35.** $\frac{5}{6}-\frac{50}{60}+\frac{1}{3}$

**36.** $\frac{15}{45}-\frac{9}{27}+\frac{1}{5}$

**37.** $-\frac{1}{2}-\frac{3}{4}-\frac{1}{100}$

**38.** $-\frac{1}{4}-\frac{1}{8}-\frac{7}{100}$

**39.** $\frac{1}{10}-\left(-\frac{3}{10}\right)-\frac{4}{15}$

**40.** $\frac{1}{8}-\left(-\frac{5}{12}\right)-\frac{3}{4}$

In Exercises 41– 44, change 1 into the form $\frac{k}{k}$ so that the fractions will have a common denominator.

**41.** $1-\frac{13}{16}$

**42.** $1-\frac{3}{7}$

**43.** $1-\frac{5}{8}$

**44.** $1-\frac{2}{3}$

In Exercises 45–48, add the fractions in the vertical format.

**45.** $\frac{3}{5}$, $\frac{7}{15}$, $+\frac{5}{6}$

**46.** $\frac{4}{27}$, $\frac{5}{18}$, $+\frac{1}{9}$

**47.** $\frac{7}{24}$, $\frac{7}{16}$, $+\frac{7}{12}$

**48.** $\frac{5}{9}$, $\frac{2}{3}$, $+\frac{4}{15}$

Find the indicated algebraic sums or differences.

**49.** $\frac{x}{3}+\frac{1}{5}$

**50.** $\frac{x}{2}+\frac{3}{5}$

**51.** $\frac{y}{6}-\frac{1}{3}$

**52.** $\frac{y}{2}-\frac{2}{7}$

**53.** $\frac{1}{x}+\frac{4}{3}$

**54.** $\frac{3}{8}+\frac{5}{x}$

**ANSWERS**

35. $\frac{1}{3}$

36. $\frac{1}{5}$

37. $-\frac{63}{50}$

38. $-\frac{89}{200}$

39. $\frac{2}{15}$

40. $-\frac{5}{24}$

41. $\frac{3}{16}$

42. $\frac{4}{7}$

43. $\frac{3}{8}$

44. $\frac{1}{3}$

45. $\frac{19}{10}$

46. $\frac{29}{54}$

47. $\frac{21}{16}$

48. $\frac{67}{45}$

49. $\frac{5x+3}{15}$

50. $\frac{5x+6}{10}$

51. $\frac{y-2}{6}$

52. $\frac{7y-4}{14}$

53. $\frac{3+4x}{3x}$

54. $\frac{3x+40}{8x}$

Find the indicated algebraic sums or differences.

55. $a-\frac{7}{8}$

56. $a-\frac{5}{16}$

57. $\frac{a}{10}+\frac{3}{100}$

58. $\frac{9}{10}+\frac{a}{100}$

59. $\frac{x}{5}+\frac{1}{4}+\frac{2}{5}$

60. $\frac{x}{5}+\frac{7}{2}+\frac{5}{3}$

61. $\frac{3}{5}+\frac{2}{x}+\frac{5}{x}$

62. $\frac{6}{7}+\frac{1}{x}+\frac{7}{x}$

63. $\frac{2}{3}-\frac{2}{n}-\frac{1}{n}$

64. $\frac{n}{6}-\frac{1}{8}-\frac{2}{3}$

65. $\frac{n}{5}-\frac{1}{5}-\frac{3}{5}$

66. $\frac{x}{15}-\frac{1}{3}-\frac{2}{15}$

67. $\frac{7}{x}+\frac{3}{5}+\frac{1}{5}$

68. $\frac{1}{a}+\frac{5}{8}+\frac{1}{12}$

55. $\frac{8a-7}{8}$

56. $\frac{16a-5}{16}$

57. $\frac{10a+3}{100}$

58. $\frac{90+a}{100}$

59. $\frac{4x+13}{20}$

60. $\frac{6x+155}{30}$

61. $\frac{3x+35}{5x}$

62. $\frac{6x+56}{7x}$

63. $\frac{2n-9}{3n}$

64. $\frac{4n-19}{24}$

65. $\frac{n-4}{5}$

66. $\frac{x-7}{15}$

67. $\frac{35+4x}{5x}$

68. $\frac{24+17a}{24a}$

Name ____________ Section ____________ Date ____________

**ANSWERS**

**69.** Find the product of $-\frac{9}{10}$ and $\frac{2}{3}$. Divide the product by $-\frac{5}{3}$. What is the quotient?

**69.** $\frac{9}{25}$

**70.** Find the quotient of $-\frac{3}{4}$ and $\frac{15}{16}$. Add $-\frac{3}{10}$ to the quotient. What is the sum?

**70.** $-\frac{11}{10}$

**71.** Sam's income is \$2400 a month and he plans to budget $\frac{1}{3}$ of his income for rent and $\frac{1}{10}$ of his income for food.

**a.** What fraction of his income does he plan to spend on these two items?

**b.** What amount of money does he plan to spend each month on these two items?

**71. a.** $\frac{13}{30}$

**b.** \$1040

**72.** A watch has a rectangular-shaped display screen that is $\frac{3}{4}$ inch by $\frac{1}{2}$ inch. The display screen has a border of silver that is $\frac{1}{10}$ inch thick. What are the dimensions (length and width) of the face of the watch (including the silver border)?

**72.** $\frac{19}{20}$ in. by $\frac{7}{10}$ in.

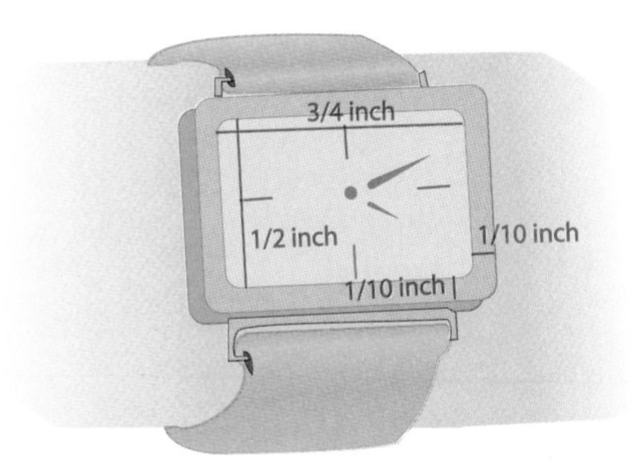

73. Four postal letters weigh $\frac{1}{2}$ ounce, $\frac{1}{5}$ ounce, $\frac{3}{10}$ ounce, and $\frac{9}{10}$ ounce. What is the total weight of the letters?

$\frac{1}{2}$ ounce

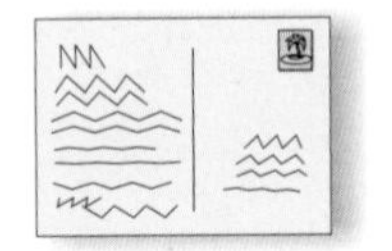

$\frac{1}{5}$ ounce

$\frac{3}{10}$ ounce

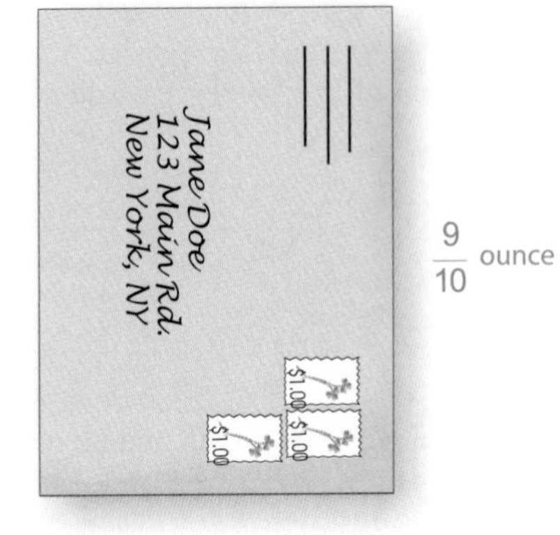


$\frac{9}{10}$ ounce

## Writing and Thinking about Mathematics

74. Pick one problem in this section that gave you some difficulty. Explain briefly why you had difficulty and why you think that you can better solve problems of this type in the future.

Responses will vary. This question is designed to provide a basis for one-on-one discussion or class discussion with your students.

### The Recycle Bin (from Section 3.5)

1. $\frac{2}{3} \cdot \frac{15}{14} \cdot \frac{21}{25}$

2. $\frac{5}{8} \cdot \frac{24}{35} \cdot \frac{14}{42}$

3. $\frac{5x^2}{4} \cdot \frac{12}{15x} \cdot \frac{11}{22}$

4. $\frac{6}{16n} \cdot \frac{8n^3}{18} \cdot \frac{9}{4n}$

5. $\frac{28a}{12a^2} \cdot \frac{13}{39a} \cdot \frac{45a}{21}$

73. $\frac{19}{10}$ oz.

74. [Respond below exercise]

The Recycle Bin

1. $\frac{3}{5}$

2. $\frac{1}{7}$

3. $\frac{x}{2}$

4. $\frac{3n}{8}$

5. $\frac{5}{3a}$

# 4.2 Introduction to Mixed Numbers

A **mixed number** is the sum of a whole number and a fraction. By convention, we usually write the whole number and the fraction side by side without the plus sign. For example,

$6+\frac{3}{5}=6\frac{3}{5}$ Read "six and three-fifths"

$11+\frac{2}{7}=11\frac{2}{7}$ Read "eleven and two-sevenths"

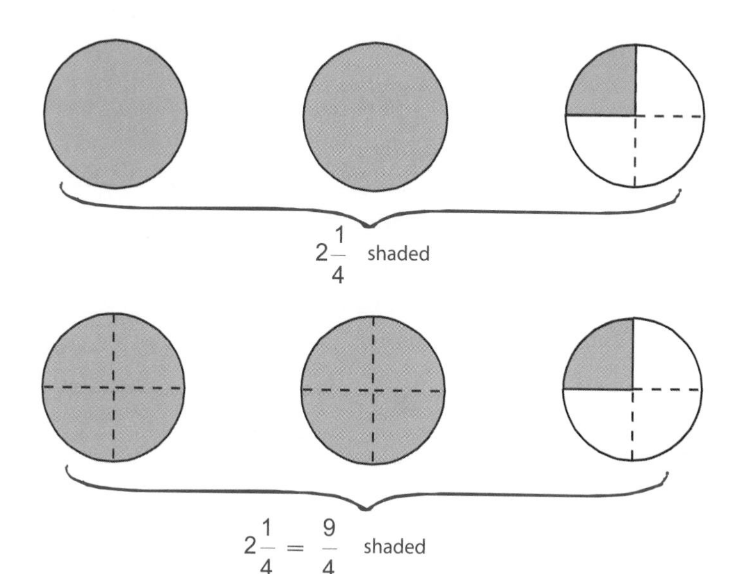


**Objectives**

1. Understand that a mixed number is the sum of a whole number and a fraction.
2. Be able to change a mixed number into an improper fraction.
3. Be able to change an improper fraction into a mixed number.
4. Know the difference between reducing an improper fraction and changing it into a mixed number.

## Changing Mixed Numbers to Fraction Form

Generally, people are familiar with mixed numbers and use them frequently. For example, "A carpenter sawed a board into two pieces, each $6\frac{3}{4}$ feet long" or "I talked with my brother for $15\frac{1}{2}$ minutes today." However, while mixed numbers are commonly used in many situations, they are not convenient for the operations of multiplication and division. These operations are more easily accomplished by first changing each mixed number to the form of an improper fraction.

Most students will know from our earlier work that multiplication is indicated when a number is written next to a variable, as in $5x$ or $-3y$. However, the fact that a mixed number indicates addition of the whole number part and the fraction part may be a source of confusion. To help them understand this idea you might use whole numbers and expanded notation. For example, 123 does not mean $1 \cdot 2 \cdot 3$, but it does mean $100+20+3$.

**Note:** To change a mixed number to an improper fraction, add the whole number and the fraction. Remember, the whole number can be written with 1 as the denominator.

### Example 1

$$6\frac{3}{5} = 6 + \frac{3}{5} = \frac{6}{1} \cdot \frac{5}{5} + \frac{3}{5} = \frac{30}{5} + \frac{3}{5} = \frac{30+3}{5} = \frac{33}{5}$$

### Example 2

$$11\frac{2}{7} = 11 + \frac{2}{7} = \frac{11}{1} \cdot \frac{7}{7} + \frac{2}{7} = \frac{77}{7} + \frac{2}{7} = \frac{77+2}{7} = \frac{79}{7}$$

There is a pattern to changing mixed numbers into improper fractions that leads to a familiar shortcut. Since the denominator of the whole number is always 1, the LCD is always the denominator of the fraction part. Therefore, in Example 1, the LCD was 5, and we multiplied the whole number 6 by 5. Similarly, in Example 2, we multiplied the whole number 11 by 7. After each multiplication, we added the numerator of the fraction part of the mixed number and used the common denominator. This process is summarized as the following shortcut:

**Shortcut for Changing Mixed Numbers to Fraction Form:**

1. Multiply the whole number by the denominator of the fraction part.
2. Add the numerator of the fraction part to this product.
3. Write this sum over the denominator of the fraction.

### Example 3

Use the shortcut method to change $6\frac{3}{5}$ to an improper fraction.

**Solution**

**Step 1:** Multiply the whole number by the denominator: $5 \cdot 6 = 30$

**Step 2:** Add the numerator: $30 + 3 = 33$

**Step 3:** Write this sum over the denominator:

$6\frac{3}{5} = \frac{33}{5}$ This is the same answer as in Example 1.

### Example 4

Change $8\frac{9}{10}$ to an improper fraction.

**Solution**

Multiply $8 \cdot 10 = 80$ and add 9:

$$80 + 9 = 89$$

Write 89 over 10 as follows:

$$8\frac{9}{10} = \frac{89}{10}$$

$$\overset{80+9}{\underset{8\cdot 10}{8 \longleftarrow}}\frac{9}{10} = \frac{8\cdot 10+9}{10} = \frac{89}{10}$$

### *Completion Example 5*

Change $11\frac{2}{3}$ to an improper fraction.

**Solution**

Multiply $11 \cdot 3 =$ _____ and add 2: _____ + _____ = _____

Write this sum, _____, over the denominator _____. Therefore,

$11\frac{2}{3} =$ _____ .

*Now Work Exercises 1-4 in the Margin.*

Change each mixed number to an improper fraction.

1. $5+\frac{1}{4}$

   $\frac{21}{4}$

2. $10+\frac{2}{3}$

   $\frac{32}{3}$

3. $7\frac{2}{9}$

   $\frac{65}{9}$

4. $3\frac{5}{32}$

   $\frac{101}{32}$

## Changing Improper Fractions to Mixed Numbers

To reverse the process (that is, to change an improper fraction to a mixed number), we use the fact that a fraction can indicate division.

**To Change an Improper Fraction to a Mixed Number:**

1. Divide the numerator by the denominator to find the whole number part of the mixed number.
2. Write the remainder over the denominator as the fraction part of the mixed number.

To emphasize the relationship between common factors and reduced fractions, you might want to point out that improper fractions can be reduced and still be in the form of improper fractions.

Change each improper fraction to a mixed number.

**5.** $\frac{20}{11}$

$1\frac{9}{11}$

**6.** $\frac{18}{5}$

$3\frac{3}{5}$

**7.** $\frac{35}{2}$

$17\frac{1}{2}$

**8.** $\frac{149}{6}$

$24\frac{5}{6}$

## Example 6

Change $\frac{67}{5}$ to a mixed number.

### Solution

Divide 67 by 5:

$$\begin{array}{r|l} 13 & \text{whole number part} \\ 5\overline{)67} & \\ \underline{5} & \\ 17 & \\ \underline{15} & \\ 2 & \text{remainder} \end{array}$$

$$\frac{67}{5} = 13 + \frac{2}{5} = 13\frac{2}{5}$$

## Example 7

Change $\frac{85}{2}$ to a mixed number.

### Solution

Divide 85 by 2:

$$\begin{array}{r|l} 42 & \text{whole number part} \\ 2\overline{)85} & \\ \underline{8} & \\ 05 & \\ \underline{4} & \\ 1 & \text{remainder} \end{array}$$

$$\frac{85}{2} = 42 + \frac{1}{2} = 42\frac{1}{2}$$

*Now Work Exercises 5 - 8 in the Margin.*

Changing an improper fraction to a mixed number is not the same as reducing it. **Reducing** involves finding common factors in the numerator and denominator. **Changing to a mixed number** involves division of the numerator by the denominator. Common factors are not involved. In any case, the fraction part of a mixed number should be in reduced form. To ensure this, we can follow either of the following two procedures:

1. Reduce the improper fraction, first, then change this fraction to a mixed number.
2. Change the improper fraction to a mixed number, first, then reduce the fraction part.

Each procedure is illustrated in Example 8.

## Example 8

Change $\frac{34}{12}$ to a mixed number with the fraction part reduced.

### Solution

**a.** Reduce first:

$$\frac{34}{12} = \frac{\cancel{2} \cdot 17}{\cancel{2} \cdot 6} = \frac{17}{6}$$

Then change to a mixed number:

$$\begin{array}{r} 2 \\ 6\overline{)17} \\ \underline{12} \\ 5 \end{array} \qquad \frac{17}{6} = 2\frac{5}{6}$$

**b.** Change to a mixed number:

$$\begin{array}{r} 2 \\ 12\overline{)34} \\ \underline{24} \\ 10 \end{array} \qquad \frac{34}{12} = 2\frac{10}{12}$$

Then reduce:

$$2\frac{10}{12} = 2\frac{\cancel{2} \cdot 5}{\cancel{2} \cdot 6} = 2\frac{5}{6}$$

Both procedures give the same result: $2\frac{5}{6}$

## Completion Example Answers

**5.** Multiply $11 \cdot 3 = \mathbf{33}$ and add 2: $\mathbf{33} + \mathbf{2} = \mathbf{35}$

Write this sum, **35**, over the denominator **3**. Therefore,

$11\frac{2}{3} = \frac{\mathbf{35}}{\mathbf{3}}$.

Name ______________________ Section ____________ Date __________

## Exercises 4.2

Write each fraction in the form of an improper fraction reduced to lowest terms.

**1.** $\frac{28}{10}$ **2.** $\frac{42}{35}$ **3.** $\frac{45}{30}$

**4.** $\frac{26}{12}$ **5.** $\frac{75}{60}$ **6.** $\frac{105}{14}$

**7.** $\frac{96}{80}$ **8.** $\frac{112}{63}$ **9.** $\frac{51}{34}$

**10.** $\frac{85}{30}$

Change each mixed number to fraction form, and reduce if possible.

**11.** $4\frac{3}{4}$ **12.** $3\frac{5}{8}$ **13.** $1\frac{2}{15}$

**14.** $5\frac{3}{5}$ **15.** $2\frac{1}{4}$ **16.** $15\frac{1}{3}$

**17.** $10\frac{2}{3}$ **18.** $12\frac{1}{2}$ **19.** $7\frac{1}{2}$



**ANSWERS**

1. $\frac{14}{5}$
2. $\frac{6}{5}$
3. $\frac{3}{2}$
4. $\frac{13}{6}$
5. $\frac{5}{4}$
6. $\frac{15}{2}$
7. $\frac{6}{5}$
8. $\frac{16}{9}$
9. $\frac{3}{2}$
10. $\frac{17}{6}$
11. $\frac{19}{4}$
12. $\frac{29}{8}$
13. $\frac{17}{15}$
14. $\frac{28}{5}$
15. $\frac{9}{4}$
16. $\frac{46}{3}$
17. $\frac{32}{3}$
18. $\frac{25}{2}$
19. $\frac{15}{2}$

20. $\frac{156}{5}$

21. $\frac{28}{3}$

22. $\frac{37}{5}$

23. $\frac{31}{10}$

24. $\frac{53}{10}$

25. $\frac{123}{20}$

26. $\frac{39}{2}$

27. $\frac{4257}{1000}$

28. $\frac{3931}{1000}$

29. $\frac{263}{100}$

30. $\frac{707}{100}$

31. $2\frac{1}{3}$

32. $1\frac{1}{2}$

33. $1\frac{1}{4}$

34. $7\frac{1}{2}$

35. $1\frac{1}{5}$

36. $1\frac{1}{4}$

37. $1\frac{1}{2}$

38. $1\frac{12}{17}$

39. $1\frac{1}{2}$

40. 5

41. 5

42. $1\frac{1}{4}$

**20.** $31\frac{1}{5}$ **21.** $9\frac{1}{3}$ **22.** $7\frac{2}{5}$

**23.** $3\frac{1}{10}$ **24.** $5\frac{3}{10}$ **25.** $6\frac{3}{20}$

**26.** $19\frac{1}{2}$ **27.** $4\frac{257}{1000}$ **28.** $3\frac{931}{1000}$

**29.** $2\frac{63}{100}$ **30.** $7\frac{7}{100}$

Change each fraction to mixed number form with the fraction part reduced to lowest terms.

**31.** $\frac{28}{12}$ **32.** $\frac{45}{30}$ **33.** $\frac{75}{60}$

**34.** $\frac{105}{14}$ **35.** $\frac{96}{80}$ **36.** $\frac{80}{64}$

**37.** $\frac{48}{32}$ **38.** $\frac{87}{51}$ **39.** $\frac{51}{34}$

**40.** $\frac{70}{14}$ **41.** $\frac{65}{13}$ **42.** $\frac{125}{100}$

Name ______________________ Section ______________ Date ____________

**ANSWERS**

**43.** $\$1\frac{1}{4}$

**44.** 2 in.

**45.** $2\frac{13}{24}$ ft.

In each of the following problems, write the answer in the form of a mixed number.

**43.** In 5 days, the price of stock in Microsoft rose $\frac{1}{4}$, fell $\frac{5}{8}$, rose $\frac{3}{4}$, rose $\frac{1}{2}$, and rose $\frac{3}{8}$ of a dollar. What was the net gain in the price of this stock over these 5 days?

**44.** Three pieces of lumber are stacked, one on top of the other. If two pieces are each $\frac{3}{4}$ inch thick and the third piece is $\frac{1}{2}$ inch thick, how many inches high is the stack?

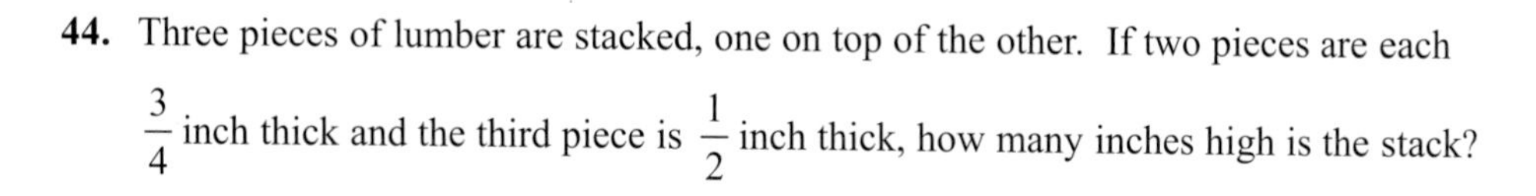

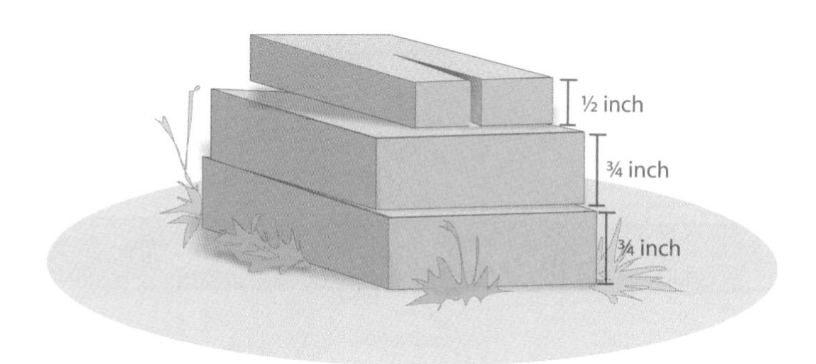


**45.** A tree in the Grand Teton National Park in Wyoming grew $\frac{1}{2}$ foot, $\frac{2}{3}$ foot, $\frac{3}{4}$ foot, and $\frac{5}{8}$ foot in 4 consecutive years. How many feet did the tree grow during these 4 years?

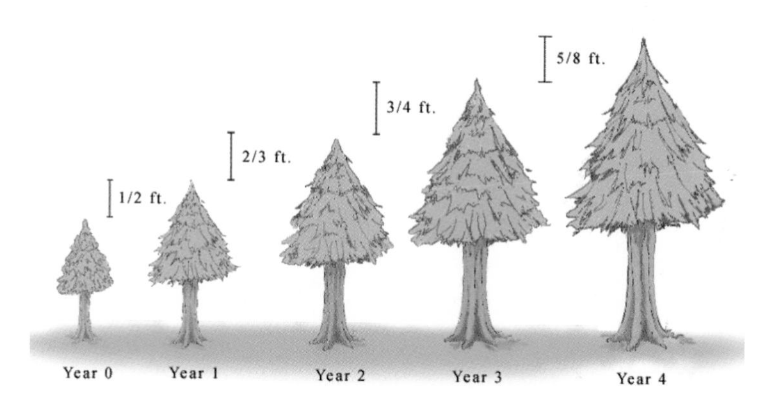

| | |
|---|---|
| **46.** | $\frac{1799}{97}$ |
| **47.** | $\frac{751}{45}$ |
| **48.** | $\frac{755}{37}$ |
| **49.** | $\frac{1338}{41}$ |
| **50.** | $\frac{918}{47}$ |
| **51.** | $\frac{2547}{35}$ |
| **52.** | $\frac{8957}{100}$ |
| **53.** | $\frac{103,334}{501}$ |
| **54.** | $\frac{248,345}{412}$ |

**The Recycle Bin**

| | |
|---|---|
| **1.** | $\frac{9}{4}$ |
| **2.** | $\frac{9}{10}$ |
| **3.** | $x^2$ |
| **4.** | $\frac{9a^2}{7}$ |
| **5.** | $\frac{3x}{10}$ |
| **6.** | $\frac{8}{35a}$ |

Use your calculator to help in changing the following mixed numbers to fraction form.

**46.** $18\frac{53}{97}$ **47.** $16\frac{31}{45}$ **48.** $20\frac{15}{37}$

**49.** $32\frac{26}{41}$ **50.** $19\frac{25}{47}$ **51.** $72\frac{27}{35}$

**52.** $89\frac{171}{300}$ **53.** $206\frac{128}{501}$ **54.** $602\frac{321}{412}$

## The Recycle Bin (from Section 3.6)

Find the indicated quotients, and reduce all answers to lowest terms

**1.** $\frac{12}{13} \div \frac{16}{39}$ **2.** $\frac{15}{42} \div \frac{25}{63}$ **3.** $\frac{32x}{20} \div \frac{24}{15x}$

**4.** $\frac{40a}{30} \div \frac{28}{27a}$ **5.** $\frac{15}{34x} \div \frac{25}{17x^2}$ **6.** $\frac{18}{65a^2} \div \frac{63}{52a}$

# 4.3 Multiplication and Division with Mixed Numbers

## Multiplication with Mixed Numbers

The fact that a mixed number is the sum of a whole number and a fraction does not help in understanding multiplication (or division) with mixed numbers. The simplest way to multiply mixed numbers is to change each mixed number to an improper fraction and then multiply. Thus, multiplication with mixed numbers is the same as multiplication with fractions. We use prime factorizations and reduce as we multiply.

**Objectives**

① Learn how to multiply with mixed numbers.

② Learn how to divide with mixed numbers.

③ Know how to find the area of a rectangle and the area of a triangle.

④ Know how to find the volume of a three-dimensional rectangular solid.

**To Multiply Mixed Numbers:**

1. Change each number to fraction form.
2. Multiply by factoring numerators and denominators, and then reduce.
3. Change the answer to a mixed number or leave it in fraction form. (The choice sometimes depends on what use is to be made of the answer.)

### Example 1

Find the product: $\left(1\frac{1}{2}\right)\left(2\frac{1}{5}\right)$

**Solution**

Change each mixed number to fraction form, then multiply the fractions.

$$\left(1\frac{1}{2}\right)\left(2\frac{1}{5}\right)=\left(\frac{3}{2}\right)\left(\frac{11}{5}\right)=\frac{3\cdot 11}{2\cdot 5}=\frac{33}{10}\text{ or }3\frac{3}{10}$$

### Example 2

Multiply and reduce to lowest terms: $4\frac{2}{3}\cdot 1\frac{1}{7}\cdot 2\frac{1}{16}$

**Solution**

$$4\frac{2}{3}\cdot 1\frac{1}{7}\cdot 2\frac{1}{16}=\frac{14}{3}\cdot\frac{8}{7}\cdot\frac{33}{16}=\frac{\cancel{2}\cdot\cancel{7}\cdot\cancel{8}\cdot\cancel{3}\cdot 11}{\cancel{3}\cdot\cancel{7}\cdot\cancel{8}\cdot\cancel{2}}=\frac{11}{1}=11$$

By truncating each mixed number (dropping the fraction part), reasonably good estimations can be found for the results of operations with mixed numbers. You might encourage your students to try this approach when they are not sure of their answers. To illustrate, in Examples 1, 2, and 3, respectively:

$\left(1\frac{1}{2}\right)\left(2\frac{1}{5}\right)$ is about $1 \cdot 2 = 2$

$4\frac{2}{3} \cdot 1\frac{1}{7} \cdot 2\frac{1}{16}$ is about $4 \cdot 1 \cdot 2 = 8$

$-\frac{5}{33} \cdot 3\frac{1}{7} \cdot 6\frac{1}{8}$ is about $-1 \cdot 3 \cdot 6 = -18$

In the third example, the fraction $-\frac{5}{33}$ is not truncated to 0. In the case of a single fraction, treat the fraction as $\pm 1$ instead of truncating. Obviously, as illustrated by Example 3, this method does not give a good estimate every time. (An alternative to truncating is rounding off. But, in rounding off, the students need a better understanding of fractions than they do in truncating.)

### Example 3

Find the product of $-\frac{5}{33}$, $3\frac{1}{7}$, and $6\frac{1}{8}$.

**Solution**

$$-\frac{5}{33} \cdot 3\frac{1}{7} \cdot 6\frac{1}{8} = \frac{-5}{33} \cdot \frac{22}{7} \cdot \frac{49}{8} = \frac{-5 \cdot \cancel{2} \cdot \cancel{11} \cdot \cancel{7} \cdot 7}{3 \cdot \cancel{11} \cdot \cancel{7} \cdot \cancel{2} \cdot 4}$$

$$= \frac{-35}{12} = -2\frac{11}{12}$$

Large mixed numbers can be multiplied in the same way as smaller mixed numbers. The products will be large numbers, and a calculator may be helpful in changing the mixed numbers to improper fractions.

**Note:** Scientific calculators are not used for multiplying fractions and mixed numbers in general because most calculators work with rounded-off decimal numbers and may not give exact answers. There are calculators that have a special notation for fractions and can perform operations with fractions.

Specifically, the TI-83 has a menu item under the key MATH. This selection is marked as ▶**Frac**.

You can add, subtract, multiply, or divide with decimal numbers and/or fractions, press the MATH key, and choose the first selection ( ▶**Frac** ) from the list. Press ENTER and ▶**Frac** will appear on the display. Press ENTER again and the answer will appear in fraction form. If you have a TI-83 or similar graphing calculator, your instructor may want you to use your calculator to help check your work.

Obviously, calculators and computers can operate quickly and accurately. However, remember that calculators and computers do not think for you. To advance in mathematics, you need to understand general concepts, and you should not rely on a calculator or computer to do your thinking for you.

## Example 4

Multiply: $32\frac{5}{6} \cdot 24\frac{1}{4}$

**Solution**

**a.** Change each mixed number into an improper fraction.

$$\begin{array}{r} 32 \\ \times\ 6 \\ \hline 192 \end{array} \qquad \begin{array}{r} 192 \\ +\ 5 \\ \hline 197 \end{array} \qquad 32\frac{5}{6} = \frac{197}{6}$$

$$\begin{array}{r} 24 \\ \times\ 4 \\ \hline 96 \end{array} \qquad \begin{array}{r} 96 \\ +\ 1 \\ \hline 97 \end{array} \qquad 24\frac{1}{4} = \frac{97}{4}$$

**b.** Now multiply the improper fractions.

$$32\frac{5}{6} \cdot 24\frac{1}{4} = \frac{197}{6} \cdot \frac{97}{4} = \frac{19{,}109}{24} = 796\frac{5}{24}$$

$$\begin{array}{r} 197 \\ \times\ 97 \\ \hline 1379 \\ 1773\phantom{0} \\ \hline 19109 \end{array} \qquad \begin{array}{r} 796\frac{5}{24} \\ 24\overline{)\,19109} \\ \underline{168}\phantom{00} \\ 230\phantom{0} \\ \underline{216}\phantom{0} \\ 149 \\ \underline{144} \\ 5 \end{array}$$

In Steps **a.** and **b.**, just shown here, the operations can be easily performed by using a calculator.

This method of multiplication with mixed numbers can involve very large numbers (as illustrated in Example 4). However, it is efficient and accurate, and you might want to convince your students of this fact by illustrating the following alternative.

By the regular method of multiplication with whole numbers, the product of two mixed numbers involves four products.

We need the sum of the following partial products:

$$\begin{array}{rl} 32\frac{5}{6} & \\ \times\ 24\frac{1}{4} & \\ \hline \frac{5}{24} & \left(\frac{1}{4} \cdot \frac{5}{6} = \frac{5}{24}\right) \\ 8 & \left(\frac{1}{4} \cdot 32 = 8\right) \\ 20 & \left(24 \cdot \frac{5}{6} = 20\right) \\ 768 & (24 \cdot 32 = 768) \\ \hline 796\frac{5}{24} & \end{array}$$

## *Completion Example 5*

Find the product and write it as a mixed number with the fraction part in lowest terms: $2\frac{1}{6} \cdot 1\frac{3}{5} \cdot 1\frac{1}{34}$

**Solution**

$$2\frac{1}{6} \cdot 1\frac{3}{5} \cdot 1\frac{1}{34} = \frac{13}{6} \cdot \underline{\qquad} \cdot \underline{\qquad}$$

$$= \frac{13 \cdot \underline{\qquad\qquad}}{2 \cdot 3 \cdot \underline{\qquad\qquad}}$$

$$= \underline{\qquad} = \underline{\qquad}$$

### Completion Example 6

Multiply: $20\frac{3}{4}\cdot 19\frac{1}{5}$

**Solution**

$$20\frac{3}{4}\cdot 19\frac{1}{5}=\frac{83}{4}\cdot \_\_\_\_$$
$$=\_\_\_\_=\_\_\_\_=\_\_\_\_$$

*Now Work Exercises 1-3 in the Margin.*

In Section 3.5, we discussed the idea that finding a fractional part of a number indicates multiplication. The key word is **of**, and we will find that this same idea is related to decimals and percents in later chapters. For example, we are familiar with expressions such as "take off 40% **of** the list price," and "three-tenths **of** our net monthly income goes to taxes." This concept is emphasized again here with mixed numbers.

> **Note:** To find a fraction **of** a number means to **multiply** the number by the fraction.

### Example 7

Find $\frac{3}{5}$ **of** 40.

**Solution**

$$\frac{3}{5}\cdot 40=\frac{3}{\cancel{5}_1}\cdot\frac{\cancel{40}^{8}}{1}=24$$

### Example 8

Find $\frac{2}{3}$ **of** $5\frac{1}{4}$.

**Solution**

$$\frac{2}{3}\cdot 5\frac{1}{4}=\frac{2}{3}\cdot\frac{21}{4}=\frac{\cancel{2}\cdot\cancel{3}\cdot 7}{\cancel{3}\cdot\cancel{2}\cdot 2}=\frac{7}{2}\text{ or }3\frac{1}{2}$$

*Now Work Exercises 4 and 5 in the Margin.*

Find each product and write the product in mixed number form with the fraction part reduced to lowest terms.

**1.** $\left(3\frac{1}{2}\right)\left(4\frac{4}{7}\right)$

16

**2.** $\left(2\frac{1}{4}\right)\left(7\frac{1}{3}\right)\left(\frac{4}{55}\right)$

$1\frac{1}{5}$

**3.** $-10\frac{2}{5}\cdot 2\frac{1}{12}$

$-21\frac{2}{3}$

**4.** Find $\frac{5}{8}$ of 72.

45

**5.** Find $\frac{1}{3}$ of $-7\frac{1}{2}$

$-2\frac{1}{2}$

## Area and Volume

**Area,** as discussed in Section 1.1, is the measure of the interior of a closed plane surface. For a rectangle, the area is the product of the length times the width. (In the form of a formula, $A = lw$.) For a triangle, the area is one-half the product of the base times its height. (In the form of a formula, $A = \frac{1}{2}bh$.) These concepts and formulas are true regardless of the type of number used for the lengths. Thus, we can use these same techniques for finding the areas of rectangles and triangles whose dimensions are mixed numbers. **Remember that area is measured in square units.**

### Example 9

Find the area of the rectangle with sides of length $5\frac{1}{2}$ feet and $3\frac{3}{4}$ feet.

**Solution**

The area is the product of the length times the width:

$5\frac{1}{2}$ ft.

$3\frac{3}{4}$ ft.

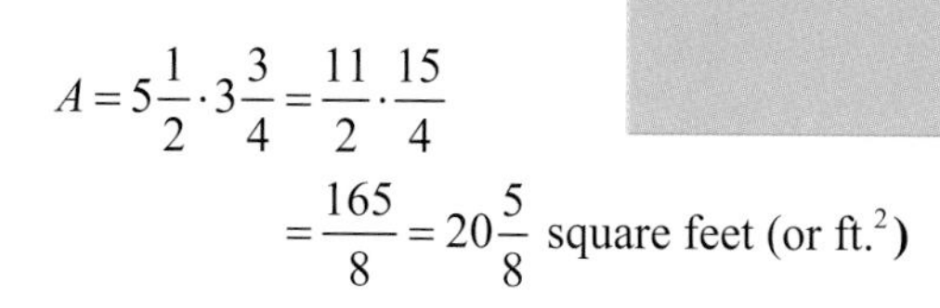

$$A = 5\frac{1}{2}\cdot 3\frac{3}{4} = \frac{11}{2}\cdot\frac{15}{4}$$

$$= \frac{165}{8} = 20\frac{5}{8} \text{ square feet (or ft.}^2\text{)}$$

Encourage the students to label each answer in the correct units. This habit will be a valuable asset for solving all types of word problems and for distinguishing among the concepts of linear measure, area measure, and volume measure.

Also, labeling answers helps in understanding the difference between abstract numbers and denominate numbers. For example, even though 3 = 3, 3 ft. ≠ 3ft.$^2$ Or, similarly, 1 + 2 = 3, but 1 ft. + 2 ft. ≠ 3ft.$^2$

**Volume** is the measure of space enclosed by a three-dimensional figure and is measured in **cubic units**. The concept of volume is illustrated in Figure 4.3 in terms of cubic inches.

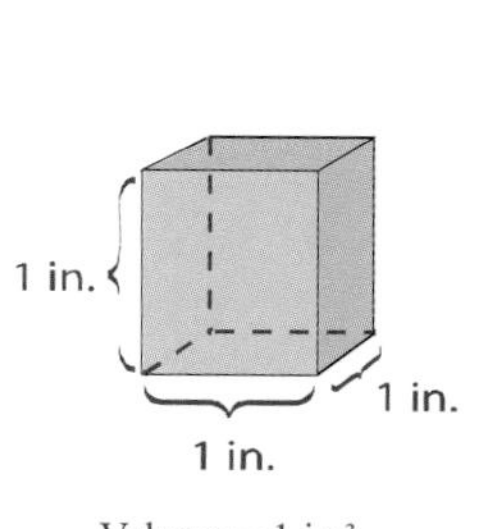


Volume = 1 in.$^3$
( or 1 cubic inch )

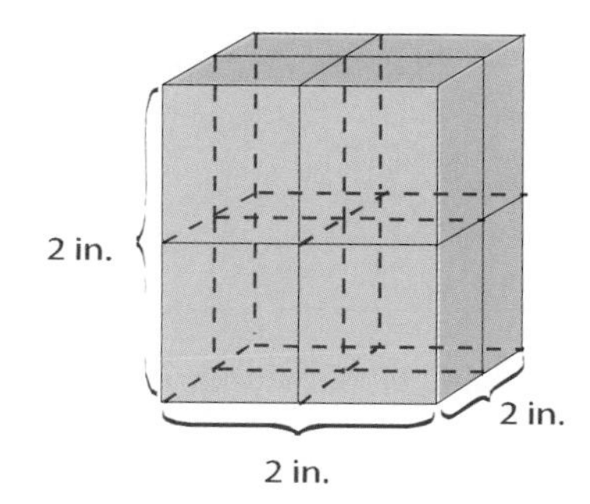


Volume = length x height x width = 2 x 2 x 2 = 8 in.$^3$
There are a total of 8 cubes that are each 1 in.$^3$ for a total of 8 in.$^3$

**Figure 4.3**

You might want to show the students how to draw their own rectangular solids in the following manner.

**STEP 1**: Draw a rectangle and a copy.

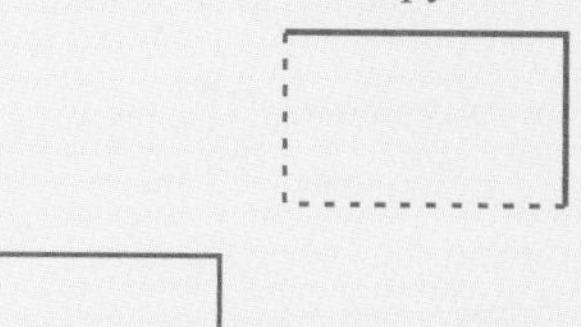


**STEP 2**: Connect the corresponding vertices.

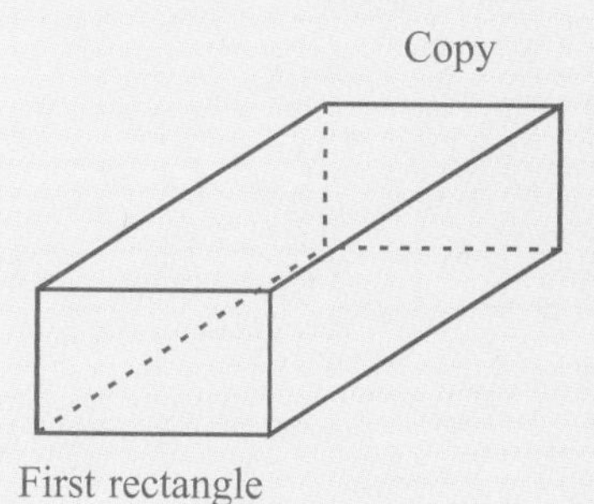

In the metric system, some of the units of volume are cubic meters ( $m^3$ ), cubic decimeters ( $dm^3$ ), cubic centimeters ( $cm^3$ ), and cubic millimeters ( $mm^3$ ). In the U.S. customary system, some of the units of volume are cubic feet ( $ft.^3$ ), cubic inches ( $in.^3$ ), and cubic yards ( $yd.^3$ ).

In this section we will discuss only the volume of a rectangular solid. The volume of such a solid is the product of its length times its width times its height. ( In the form of a formula, $V = lwh$. )

### Example 10

Find the volume of a rectangular solid with dimensions 8 inches by 4 inches by $12\frac{1}{2}$ inches.

#### Solution

$$V = lwh$$

$$V = 8 \cdot 4 \cdot 12\frac{1}{2}$$

$$= \frac{8}{1} \cdot \frac{4}{1} \cdot \frac{25}{2}$$

$$= 400 \text{ in.}^3$$

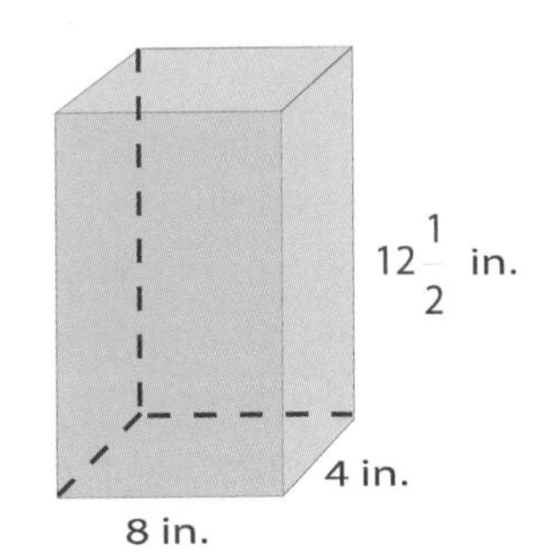


## Division with Mixed Numbers

Division with mixed numbers is the same as division with fractions, as discussed in Section 3.6. Simply change each mixed number to an improper fraction before dividing. Recall that **to divide by any nonzero number, multiply by its reciprocal.**

That is, for $c \neq 0$ and $d \neq 0$, the **reciprocal** of $\frac{c}{d}$ is $\frac{d}{c}$, and

$$\frac{a}{b} \div \frac{c}{d} = \frac{a}{b} \cdot \frac{d}{c}.$$

**To Divide with Mixed Numbers:**

1. Change each number to fraction form.
2. Multiply by the reciprocal of the divisor.
3. Reduce, if possible.

## Example 11

Divide: $6 \div 7\frac{7}{8}$

## Solution

First change the mixed number $7\frac{7}{8}$ to the improper fraction $\frac{63}{8}$ and then multiply by its reciprocal.

$$6 \div 7\frac{7}{8} = \frac{6}{1} \div \frac{63}{8} = \frac{6}{1} \cdot \frac{8}{63} = \frac{2 \cdot \cancel{3} \cdot 8}{1 \cdot \cancel{3} \cdot 3 \cdot 7} = \frac{16}{21}$$

## *Completion Example 12*

Find the quotient: $3\frac{1}{15} \div 1\frac{1}{5}$

### Solution

$$3\frac{1}{15} \div 1\frac{1}{5} = \frac{46}{15} \div \_\_\_\_ = \frac{46}{15} \cdot \_\_\_\_$$

$$= \frac{2 \cdot 23 \cdot}{3 \cdot 5 \cdot} = \_\_\_\_ = \_\_\_\_$$

*Now Work Exercises 6-8 in the Margin.*

**6.** $4\frac{2}{3} \div \frac{8}{9}$

$5\frac{1}{4}$

**7.** $-5\frac{3}{7} \div 5\frac{3}{7}$

$-1$

**8.** $24 \div 2\frac{2}{3}$

9

## Example 13

The product of two numbers is $-5\frac{1}{8}$. One of the numbers is $\frac{3}{4}$. What is the other number?

## Solution

Reasoning that $\frac{3}{4}$ is to be multiplied by some unknown number, we can write

$\frac{3}{4} \cdot ? = -5\frac{1}{8}$.

To find the missing number, **divide** $-5\frac{1}{8}$ by $\frac{3}{4}$. Do **not** multiply.

$$-5\frac{1}{8} \div \frac{3}{4} = \frac{-41}{8} \cdot \frac{4}{3} = \frac{-41 \cdot \cancel{4}}{2 \cdot \cancel{4} \cdot 3} = \frac{-41}{6} = -6\frac{5}{6}$$

The other number is $-6\frac{5}{6}$ OR, reasoning again that $\frac{3}{4}$ is to be multiplied by some unknown number, we can write an equation and solve the equation. If $x$ is the unknown number, then

$$\frac{3}{4}\cdot x = -5\frac{1}{8}$$

$$\frac{4}{3}\cdot\frac{3}{4}\cdot x = \frac{4}{3}\left(-\frac{41}{8}\right)$$

$$1\cdot x = \frac{\not{4}}{3}\left(-\frac{41}{\not{8}_2}\right) = -\frac{41}{6}$$

$$x = -\frac{41}{6} \quad (\text{ or } \quad x = -6\frac{5}{6}\ )$$

Multiply both sides by $\frac{4}{3}$, the reciprocal of the coefficient. Note that this is the same as dividing by $\frac{3}{4}$.

CHECK : $\frac{3}{4}\left(-6\frac{5}{6}\right) = \frac{3}{4}\left(\frac{-41}{6}\right) = \frac{-41}{8} = -5\frac{1}{8}$

## Completion Example Answers

**5.** $2\frac{1}{6}\cdot 1\frac{3}{5}\cdot 1\frac{1}{34} = \frac{13}{6}\cdot\mathbf{\frac{8}{5}}\cdot\mathbf{\frac{35}{34}}$

$$= \frac{13\cdot\not{2}\cdot\not{2}\cdot 2\cdot\not{5}\cdot 7}{\not{2}\cdot 3\cdot\not{5}\cdot\not{2}\cdot 17}$$

$$= \mathbf{\frac{182}{51}} = \mathbf{3\frac{29}{51}}$$

**6.** $20\frac{3}{4}\cdot 19\frac{1}{5} = \frac{83}{4}\cdot\mathbf{\frac{96}{5}}$

$$= \mathbf{\frac{7968}{20}} = \mathbf{398\frac{8}{20}} = \mathbf{398\frac{2}{5}}$$

**12.** $3\frac{1}{15}\div 1\frac{1}{5} = \frac{46}{15}\div\mathbf{\frac{6}{5}} = \frac{46}{15}\cdot\mathbf{\frac{5}{6}}$

$$= \frac{\not{2}\cdot 23\cdot\not{5}}{3\cdot\not{5}\cdot\not{2}\cdot 3} = \mathbf{\frac{23}{9}} = \mathbf{2\frac{5}{9}}$$

Name ______________________ Section ______________ Date ____________

## Exercises 4.3

Find the indicated products, and write your answers in mixed number form.

1. $\left(2\frac{4}{5}\right)\left(1\frac{1}{7}\right)$

2. $\left(2\frac{1}{3}\right)\left(3\frac{1}{2}\right)$

3. $5\frac{1}{3}\left(2\frac{1}{2}\right)$

4. $2\frac{1}{4}\left(3\frac{1}{9}\right)$

5. $\left(9\frac{1}{3}\right)3\frac{3}{4}$

6. $\left(1\frac{3}{5}\right)1\frac{1}{4}$

7. $\left(11\frac{1}{4}\right)\left(2\frac{2}{15}\right)$

8. $\left(6\frac{2}{3}\right)\left(5\frac{1}{7}\right)$

9. $4\frac{3}{8}\cdot 2\frac{4}{5}$

10. $12\frac{1}{2}\cdot 3\frac{1}{3}$

11. $9\frac{3}{5}\cdot 3\frac{3}{7}$

12. $6\frac{3}{8}\cdot 2\frac{2}{17}$

13. $5\frac{1}{4}\cdot 7\frac{2}{3}$

14. $\left(4\frac{3}{4}\right)\left(2\frac{1}{5}\right)\left(1\frac{1}{7}\right)$

15. $\left(-6\frac{3}{16}\right)\left(2\frac{1}{11}\right)\left(5\frac{3}{5}\right)$

16. $-7\frac{1}{3}\cdot 5\frac{1}{4}\cdot 6\frac{2}{7}$

17. $\left(-2\frac{5}{8}\right)\left(-3\frac{2}{5}\right)\left(-1\frac{3}{4}\right)$

18. $\left(-2\frac{1}{16}\right)\left(-4\frac{1}{3}\right)\left(-1\frac{3}{11}\right)$

**ANSWERS**

1. $3\frac{1}{5}$
2. $8\frac{1}{6}$
3. $13\frac{1}{3}$
4. 7
5. 35
6. 2
7. 24
8. $34\frac{2}{7}$
9. $12\frac{1}{4}$
10. $41\frac{2}{3}$
11. $32\frac{32}{35}$
12. $13\frac{1}{2}$
13. $40\frac{1}{4}$
14. $11\frac{33}{35}$
15. $-72\frac{9}{20}$
16. $-242$
17. $-15\frac{99}{160}$
18. $-11\frac{3}{8}$

19. $1\frac{3}{10}$

20. $2\frac{1}{10}$

21. $851\frac{1}{30}$

22. $1827\frac{1}{10}$

23. $-1291\frac{5}{42}$

24. $3016\frac{26}{75}$

25. $640\frac{1}{16}$

26. $794\frac{5}{6}$

27. 60

28. 15

29. 60

30. 200

31. $1\frac{7}{8}$

32. $\frac{9}{14}$

33. 4

34. $1\frac{6}{25}$

35. $3\frac{1}{7}$

36. $-1\frac{2}{49}$

37. $-1\frac{1}{2}$

38. $-2\frac{2}{5}$

39. $-1\frac{5}{6}$

40. $-29\frac{1}{3}$

41. $-\frac{9}{32}$

**19.** $1\frac{3}{32}\cdot 1\frac{1}{7}\cdot 1\frac{1}{25}$

**20.** $1\frac{5}{16}\cdot 1\frac{1}{3}\cdot 1\frac{1}{5}$

Use a calculator as an aid in finding the following products. Write the answers as mixed numbers.

**21.** $24\frac{1}{5}\cdot 35\frac{1}{6}$

**22.** $72\frac{3}{5}\cdot 25\frac{1}{6}$

**23.** $42\frac{5}{6}\left(-30\frac{1}{7}\right)$

**24.** $75\frac{1}{3}\cdot 40\frac{1}{25}$

**25.** $\left(-36\frac{3}{4}\right)\left(-17\frac{5}{12}\right)$

**26.** $25\frac{1}{10}\cdot 31\frac{2}{3}$

**27.** Find $\frac{2}{3}$ of 90.

**28.** Find $\frac{1}{4}$ of 60.

**29.** Find $\frac{3}{5}$ of 100.

**30.** Find $\frac{5}{6}$ of 240.

**31.** Find $\frac{1}{2}$ of $3\frac{3}{4}$.

**32.** Find $\frac{9}{10}$ of $\frac{15}{21}$.

Find the indicated quotients, and write your answers in mixed number form.

**33.** $3\frac{1}{2}\div\frac{7}{8}$

**34.** $3\frac{1}{10}\div 2\frac{1}{2}$

**35.** $6\frac{3}{5}\div 2\frac{1}{10}$

**36.** $\left(-2\frac{1}{7}\right)\div 2\frac{1}{17}$

**37.** $-6\frac{3}{4}\div 4\frac{1}{2}$

**38.** $7\frac{1}{5}\div(-3)$

**39.** $7\frac{1}{3}\div(-4)$

**40.** $7\frac{1}{3}\div\left(-\frac{1}{4}\right)$

**41.** $\left(-1\frac{1}{32}\right)\div 3\frac{2}{3}$

Name ______________________ Section ____________ Date __________

**ANSWERS**

**42.** $\left(-6\frac{3}{11}\right) \div \left(-\frac{3}{4}\right)$

**43.** $-10\frac{2}{7} \div \left(-4\frac{1}{2}\right)$

**44.** $2\frac{2}{49} \div 3\frac{3}{14}$

**45.** A telephone pole is 64 feet long and $\frac{5}{16}$ of the pole must be underground.

**a.** What fraction of the pole is above ground?

**b.** Is more of the pole above ground or underground?

**c.** How many feet of the pole must be underground?

**d.** How many feet are above ground?

(**Hint:** When dealing with fractions of a whole item, think of this item as corresponding to the whole number 1.)

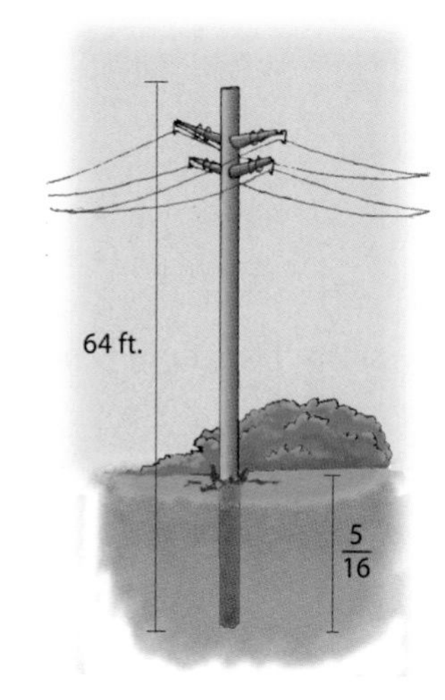


**46.** If you drive your car to work $6\frac{3}{10}$ miles one way 5 days a week, how many miles do you drive each week going to and from work?

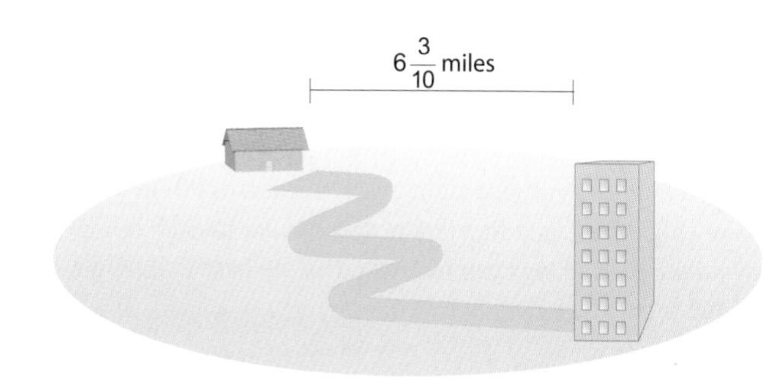


**42.** $8\frac{4}{11}$

**43.** $2\frac{2}{7}$

**44.** $\frac{40}{63}$

**45. a.** $\frac{11}{16}$

**b.** above ground

**c.** 20 ft.

**d.** 44 ft.

**46.** 63 miles

47. a. $\frac{2}{5}$

b. 180 pages

c. 15 hours

48. a. $34\frac{2}{3}$ in.

b. $75\frac{1}{9}$ in.$^2$

49. a. $20\frac{3}{4}$ m

b. $26\frac{7}{16}$ m$^2$

50. $7\frac{1}{2}$ ft.$^2$

51. 10 cm$^2$

**47.** A man can read $\frac{1}{5}$ of a book in 3 hours.

**a.** What fraction of the book can he read in 6 hours?

**b.** If the book contains 450 pages, how many pages can he read in 6 hours?

**c.** How long will he take to read the entire book?

**48.** The perimeter of a square can be found by multiplying the length of one side by 4. The area can be found by squaring the length of one side. Find

**a.** the perimeter and

**b.** the area of a square if the length of one side is $8\frac{2}{3}$ inches.

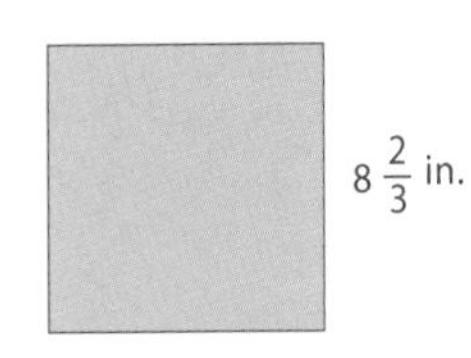


**49.** Find

**a.** the perimeter and

**b.** the area of a rectangle that has sides of length $5\frac{7}{8}$ meters and $4\frac{1}{2}$ meters.

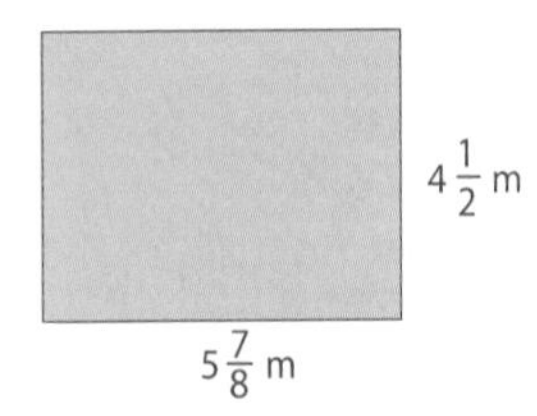


**50.** Find the area of the triangle in the figure shown here.

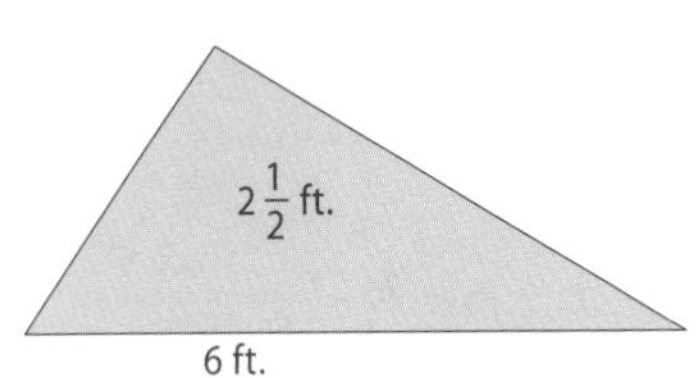


**51.** A ***right triangle*** is a triangle with one right angle (measures 90°). The two sides that form the right angle are called ***legs***, and they are perpendicular to each other. The longest side is called the ***hypotenuse***. The legs can be treated as the base and height of the triangle. Find the area of the right triangle shown here.

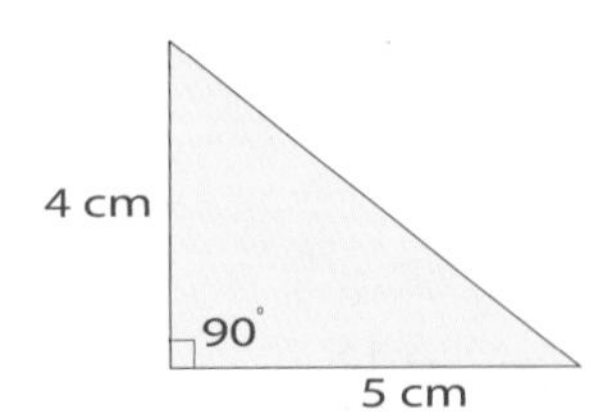

Name ______________________ Section ______________ Date ____________

**ANSWERS**

52. $388\frac{7}{10}$ in.$^3$

53. $2907\frac{44}{125}$ cm$^3$

54. a. less

b. $3000

55. a. less

b. $320

**52.** Find the volume of the rectangular solid (a box) that is $6\frac{1}{2}$ inches long, $5\frac{3}{4}$ inches wide, and $10\frac{2}{5}$ inches high.

**53.** A cereal box is $19\frac{3}{10}$ centimeters long by $5\frac{3}{5}$ centimeters wide and $26\frac{9}{10}$ centimeters high. What is the volume of the box in cubic centimeters?

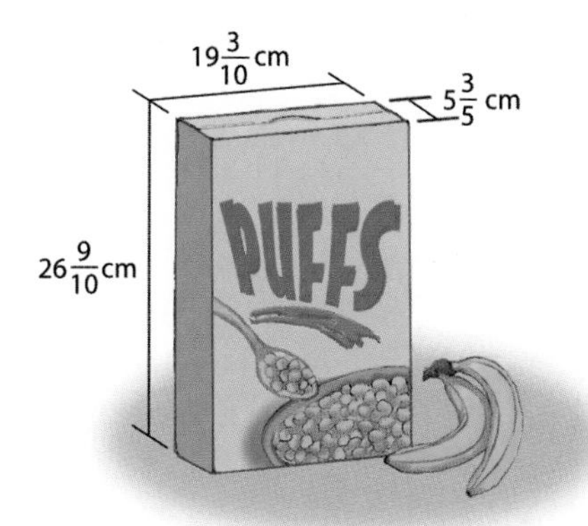


**54.** You are looking at a used car, and the salesman tells you that the price has been lowered to $\frac{4}{5}$ of its original price.

**a.** Is the current price more or less than the original price?

**b.** What was the original price if the sale price is $2400?

**55.** The sale price of a coat is $240. This is $\frac{3}{4}$ of the original price.

**a.** Is the sale price more or less than the original price?

**b.** What was the original price?

56. a. more

b. 200 passengers

57. a. more

b. 60 passengers

58. a. more

b. 208 suits

59. a. 20 gallons

b. $\$27\frac{1}{2}$ (or $27.50)

60. a. $371\frac{1}{4}$ miles

b. $\$11\frac{11}{20}$ (or $11.55)

**56.** An airplane that flies the 430-mile route between two cities is carrying 180 passengers. This is $\frac{9}{10}$ of its capacity.

**a.** Is the capacity more or less than 180?

**b.** What is the capacity of the airplane?

**57.** A bus makes several trips each day from local cities to the beach. On one such trip the bus has 40 passengers. This is $\frac{2}{3}$ of the capacity of the bus.

**a.** Is the capacity more or less than 40?

**b.** What is the capacity of the bus?

**58.** A clothing store is having a sale on men's suits. The sale rack contains 130 suits, which constitutes $\frac{5}{8}$ of the store's inventory of men's suits.

**a.** Is the inventory more or less than 130 suits?

**b.** What is the number of suits in the inventory?

**59.** You are planning a trip of and 506 miles (round trip), and you know that your car averages $25\frac{3}{10}$ miles per gallon of gas and that the gas tank on your car holds $15\frac{1}{2}$ gallons of gas.

**a.** How many gallons of gas will you use on this trip?

**b.** If the gas you buy costs $\$1\frac{3}{8}$ per gallon, how much should you plan to spend on this trip for gas?

**60.** You just drove your car 450 miles and used 20 gallons of gas, and you know that the gas tank on your car holds $16\frac{1}{2}$ gallons of gas.

**a.** What is the most number of miles you can drive on one tank of gas?

**b.** If the gas you buy costs $\$1\frac{4}{10}$ per gallon, what would you pay to fill one-half of your tank?

Name ______________ Section ______________ Date ______________

## WRITING AND THINKING ABOUT MATHEMATICS

**61.** Suppose that a fraction between 0 and 1, such as $\frac{1}{2}$ or $\frac{2}{3}$, is multiplied by some other number. Give brief discussions and several examples in answering each of the following questions.

**a.** If the other number is a positive fraction, will this product always be smaller than the other number?

**b.** If the other number is a positive whole number, will this product always be smaller than the other number?

**c.** If the other number is a negative number (integer or fraction), will this product ever be smaller than the other number?

Answers for a, b, and c will vary. The product is smaller if the other number is positive, the same if the other number is 0, and larger if the other number is negative.

### The Recycle Bin (from Section 4.1)

Find each of the indicated sums and differences.

1. $\frac{1}{4}+\frac{1}{5}+\frac{3}{10}$
2. $\frac{7}{15}-\frac{19}{20}$
3. $-\frac{4}{9}-\frac{7}{18}$
4. $\frac{3}{7}+\frac{5}{14}-\frac{5}{35}$
5. $\frac{x}{12}-\frac{1}{6}-\frac{1}{4}$
6. $\frac{3}{x}+\frac{7}{8}$

**ANSWERS**

**61. a.** [Respond below exercise]

**b.** [Respond below exercise]

**c.** [Respond below exercise]

**The Recycle Bin**

1. $\frac{3}{4}$
2. $-\frac{29}{60}$
3. $-\frac{5}{6}$
4. $\frac{9}{14}$
5. $\frac{x-5}{12}$
6. $\frac{24+7x}{8x}$

# 4.4 Addition and Subtraction with Mixed Numbers

## Addition with Mixed Numbers

Since a mixed number itself represents the sum of a whole number and a fraction part, two or more mixed numbers can be added by adding the whole numbers and the fraction parts separately.

### Objectives

① Know how to add mixed numbers.

② Be able to write the sum of mixed numbers in the form of a mixed number with the fraction part less than 1.

③ Know how to subtract mixed numbers when the fraction part of the number being subtracted is smaller than or equal to the fraction part of the other number.

④ Know how to subtract mixed numbers when the fraction part of the number being subtracted is larger than the fraction part of the other number.

⑤ Know how to add and subtract with positive and negative mixed numbers.

**To Add Mixed Numbers:**

1. Add the fraction parts.
2. Add the whole numbers.
3. Write the mixed number so that the fraction part is less than 1.

### Example 1

Find the sum: $5\frac{2}{9}+8\frac{5}{9}$

**Solution**

We can write each number as a sum and then use the commutative and associative properties of addition to treat the whole numbers and fraction parts separately.

$$\begin{aligned}5\frac{2}{9}+8\frac{5}{9} &= 5+\frac{2}{9}+8+\frac{5}{9}\\ &= (5+8)+\left(\frac{2}{9}+\frac{5}{9}\right)\\ &= 13+\frac{7}{9}=13\frac{7}{9}\end{aligned}$$

Or, vertically,

$$\begin{array}{r} 5\frac{2}{9} \\ +\ 8\frac{5}{9} \\ \hline 13\frac{7}{9} \end{array}$$

### Example 2

Add: $35\frac{1}{6}+22\frac{7}{18}$

**Solution**

In this case, the fractions do not have the same denominator. The LCD is 18.

Find each sum

1. $5\frac{1}{2}+3\frac{1}{5}$

$8\frac{7}{10}$

2. $12\frac{3}{4}+16\frac{2}{3}$

$29\frac{5}{12}$

3. $\begin{array}{r} 8 \\ +7\frac{3}{11} \\ \hline \end{array}$

$15\frac{3}{11}$

$$35\frac{1}{6}+22\frac{7}{18}=35+\frac{1}{6}+22+\frac{7}{18}$$
$$=(35+22)+\left(\frac{1}{6}\cdot\frac{3}{3}+\frac{7}{18}\right)$$
$$=57+\left(\frac{3}{18}+\frac{7}{18}\right)$$
$$=57\frac{10}{18}=57\frac{5}{9}$$

Reduce the fraction part.

### Example 3

Find the sum: $6\frac{2}{3}+13\frac{4}{5}$

**Solution**

The LCD is 15.

$$\begin{array}{rcrcr} 6\frac{2}{3} & = & 6\frac{2}{3}\cdot\frac{5}{5} & = & 6\frac{10}{15} \\ +\quad 13\frac{4}{5} & = & 13\frac{4}{5}\cdot\frac{3}{3} & = & 13\frac{12}{15} \\ \hline \end{array}$$

$$19\frac{22}{15}=19+1\frac{7}{15}=20\frac{7}{15}$$

Fraction is greater than one ↗ ; ↖ changed to a mixed number

### Example 4

Add: $5\frac{3}{4}+9\frac{3}{10}+2$

**Solution**

The LCD is 20. Since the whole number 2 has no fraction part, we must be careful to align the whole numbers if we write the numbers vertically.

$$\begin{array}{rcrcr} 5\frac{3}{4} & = & 5\frac{3}{4}\cdot\frac{5}{5} & = & 5\frac{15}{20} \\ 9\frac{3}{10} & = & 9\frac{3}{10}\cdot\frac{2}{2} & = & 9\frac{6}{20} \\ +\ 2 & = & 2 & = & 2 \\ \hline \end{array}$$

$$16\frac{21}{20}=16+1\frac{1}{20}=17\frac{1}{20}$$

Fraction is greater than one ↗ ; ↖ changed to a mixed number

*Now Work Exercises 1 -3 in the Margin.*

## Subtraction with Mixed Numbers

Subtraction with mixed numbers also involves working with the whole numbers and fraction parts, separately.

**To Subtract Mixed Numbers:**

1. Subtract the fraction parts.
2. Subtract the whole numbers.

### Example 5

Find the difference: $10\frac{3}{7}-6\frac{2}{7}$

**Solution**

$$10\frac{3}{7}-6\frac{2}{7}=(10-6)+\left(\frac{3}{7}-\frac{2}{7}\right)$$
$$=4+\frac{1}{7}=4\frac{1}{7}$$

Or, vertically,

$$\begin{array}{r} 10\frac{3}{7} \\ -\;\;6\frac{2}{7} \\ \hline 4\frac{1}{7} \end{array}$$

### Example 6

Subtract: $13\frac{4}{5}-7\frac{1}{3}$

**Solution**

The LCD for the fraction parts is 15.

$$\begin{array}{rcrcr} 13\frac{4}{5} & = & 13\frac{4}{5}\cdot\frac{3}{3} & = & 13\frac{12}{15} \\ -\;\;7\frac{1}{3} & = & 7\frac{1}{3}\cdot\frac{5}{5} & = & 7\frac{5}{15} \\ \hline & & & & 6\frac{7}{15} \end{array}$$

Sometimes the fraction part of the number being subtracted is larger than the fraction part of the first number. By "borrowing" the whole number 1 from the whole number part, we can rewrite the first number as a whole number plus an improper fraction. Then, the subtraction of the fraction parts can proceed as before.

**If the Fraction Part Being Subtracted is Larger Than the First Fraction:**

1. "Borrow" the whole number, 1, from the first whole number.
2. Add this 1 to the first fraction. (This will always result in an improper fraction, that is larger than the fraction being subtracted. )
3. Now, subtract.

## Example 7

Find the difference: $4\frac{2}{9}-1\frac{5}{9}$

**Solution**

$$\begin{array}{r} 4\frac{2}{9} \\ -\ 1\frac{5}{9} \\ \hline \end{array}$$

$\frac{5}{9}$ is larger than $\frac{2}{9}$, so "borrow" 1 from 4.

Rewrite $4\frac{2}{9}$ as $3+1+\frac{2}{9}=3+1\frac{2}{9}=3\frac{11}{9}$

$$\begin{array}{r} 4\frac{2}{9}=3+1\frac{2}{9}=3\frac{11}{9} \\ -\ 1\frac{5}{9}=1+\frac{5}{9}\ =1\frac{5}{9} \\ \hline 2\frac{6}{9}=2\frac{2}{3} \end{array}$$

## Example 8

Subtract: $10-7\frac{5}{8}$

**Solution**

The fraction part of the whole number 10 is understood to be 0. Borrow $1=\frac{8}{8}$ from 10.

$$\begin{array}{r} 10=9\frac{8}{8} \\ -7\frac{5}{8}=7\frac{5}{8} \\ \hline 2\frac{3}{8} \end{array}$$

### Example 9

Find the difference: $76\frac{5}{12} - 29\frac{13}{20}$

**Solution**

First find the LCD and then borrow 1 if necessary.

$$\left.\begin{array}{l} 12 = 2\cdot 2\cdot 3 \\ 20 = 2\cdot 2\cdot 5 \end{array}\right\} \text{LCD} = 2\cdot 2\cdot 3\cdot 5 = 60$$

$$\begin{array}{rrr} 76\frac{5}{12} = & 76\frac{25}{60} = & 75\frac{85}{60} \\ -\ 29\frac{13}{20} = & 29\frac{39}{60} = & 29\frac{39}{60} \\ \hline & & 46\frac{46}{60} = 46\frac{23}{30} \end{array}$$

borrow $1 = \frac{60}{60}$.

### *Completion Example 10*

Find the difference: $12\frac{3}{4} - 7\frac{9}{10}$

**Solution**

$$\begin{array}{rll} 12\frac{3}{4} = 12\_\_\_\_ & = 11\_\_\_\_ \\ -\ 7\frac{9}{10} = 7\_\_\_\_ & = 7\_\_\_\_ \\ \hline & \_\_\_\_ \end{array}$$

*Now Work Exercises 4-6 in the Margin.*

## Positive and Negative Mixed Numbers

The rules of addition and subtraction of positive and negative mixed numbers follow the same rules of the addition and subtraction of integers. We must be particularly careful when adding numbers with unlike signs and a negative number that has a larger absolute value than the positive number. In such a case, to find the difference between the absolute values, we must deal with the fraction parts just as in subtraction. That is, the fraction part of the mixed number being subtracted must be smaller than the fraction part of the other mixed number. This may involve "borrowing 1." The following examples illustrate two possibilities.

### Example 11

Find the algebraic sum: $-10\frac{1}{2} - 11\frac{2}{3}$

**Solution**

Both numbers have the same sign. Add the absolute values of the numbers, and use the common sign. In this example, the answer will be negative.

Find each difference.

4. $8\frac{9}{10} - 2\frac{1}{5}$

$6\frac{7}{10}$

5. $10\frac{5}{8} - 3\frac{5}{8}$

7

6. $\begin{array}{r} 21\frac{3}{4} \\ -\ 17\frac{7}{8} \\ \hline \end{array}$

$3\frac{7}{8}$

$$
\begin{array}{lll}
10\frac{1}{2} = 10\frac{1}{2}\cdot\frac{3}{3} = 10\frac{3}{6} \\
+\ 11\frac{2}{3} = 11\frac{2}{3}\cdot\frac{2}{2} = 11\frac{4}{6} \\
\hline
\qquad 21\frac{7}{6} = 22\frac{1}{6}
\end{array}
$$

Thus,

$$-10\frac{1}{2} - 11\frac{2}{3} = -22\frac{1}{6}$$

### Example 12

Find the algebraic sum: $-15\frac{1}{4} + 6\frac{2}{5}$

**Solution**

The numbers have opposite signs. Find the difference between the absolute values of the numbers, and use the sign of the number with the larger absolute value. In this example, the answer will be negative.

$$
\begin{array}{lll}
15\frac{1}{4} = 15\frac{1}{4}\cdot\frac{5}{5} = 15\frac{5}{20} = 14\frac{25}{20} \\
-\ 6\frac{2}{5} = 6\frac{2}{5}\cdot\frac{4}{4} = 6\frac{8}{20} = 6\frac{8}{20} \\
\hline
\qquad 8\frac{17}{20}
\end{array}
$$

Thus,

$$-15\frac{1}{4} + 6\frac{2}{5} = -8\frac{17}{20}$$

## Optional Approach to Adding and Subtracting Mixed Numbers

An optional approach to adding and subtracting mixed numbers is to simply do all the work with improper fractions as we did with multiplication and division in Section 4.3. By using this method, we are never concerned with the size of any of the numbers or whether they are positive or negative. You may find this approach much easier. In any case, addition and subtraction can be accomplished only if the denominators are the same. So you must still be concerned about least common denominators. For comparison purposes, Examples 13 and 14 are the same problems as in Examples 1 and 3.

### Example 13

Find the sum: $5\frac{2}{9} + 8\frac{5}{9}$

**Solution**

First, change each number into its corresponding improper fraction, then add.

$$5\frac{2}{9} + 8\frac{5}{9} = \frac{47}{9} + \frac{77}{9} = \frac{124}{9} = 13\frac{7}{9}$$

### Example 14

Find the sum: $6\frac{2}{3}+13\frac{4}{5}$

**Solution**

First, find the common denominator. Change each number into its corresponding improper fraction change each fraction to an equivalent fraction with the common denominator and add these fractions.

The LCM is 15.

$$6\frac{2}{3}+13\frac{4}{5}=\frac{20}{3}+\frac{69}{5}=\frac{20}{3}\cdot\frac{\mathbf{5}}{\mathbf{5}}+\frac{69}{5}\cdot\frac{\mathbf{3}}{\mathbf{3}}=\frac{100}{15}+\frac{207}{15}=\frac{307}{15}=20\frac{7}{15}$$

### Example 15

Subtract: $13\frac{1}{6}-17\frac{1}{4}$

**Solution**

The LCM is 12.

$$13\frac{1}{6}-17\frac{1}{4}=\frac{79}{6}-\frac{69}{4}=\frac{79}{6}\cdot\frac{\mathbf{2}}{\mathbf{2}}-\frac{69}{4}\cdot\frac{\mathbf{3}}{\mathbf{3}}=\frac{158}{12}-\frac{207}{12}=-\frac{49}{12}=-4\frac{1}{12}$$

### Completion Example Answer

**10.**

$$\begin{array}{rcrcr} 12\frac{3}{4} & = & 12\frac{\mathbf{15}}{\mathbf{20}} & = & 11\frac{\mathbf{35}}{\mathbf{20}} \\ -7\frac{9}{10} & = & 7\frac{\mathbf{18}}{\mathbf{20}} & = & 7\frac{\mathbf{18}}{\mathbf{20}} \\ \hline & & & & \mathbf{4\frac{17}{20}} \end{array}$$

Name ______________________ Section ____________ Date ____________

## Exercises 4.4

Find each sum.

**1.** $5\frac{1}{3} + 4\frac{2}{3}$

**2.** $7\frac{1}{2} + 6\frac{1}{2}$

**3.** $9 + 2\frac{3}{10}$

**4.** $18 + 1\frac{4}{5}$

**5.** $10\frac{3}{4} + \frac{1}{6}$

**6.** $3\frac{1}{4} + 5\frac{3}{8}$

**7.** $13\frac{2}{7} + 4\frac{1}{28}$

**8.** $11\frac{3}{10} + 16\frac{5}{6}$

**9.** $12\frac{3}{4} + 8\frac{5}{16}$

**10.** $13\frac{7}{10} + 2\frac{1}{6}$

**11.** $6\frac{7}{8} + 6\frac{7}{8}$

**12.** $7\frac{3}{8} + 5\frac{7}{12}$

**13.** $\frac{3}{8} + 2\frac{1}{12} + 3\frac{1}{6}$

**14.** $3\frac{5}{7} + 1\frac{1}{14} + 4\frac{3}{5}$

**15.** $13\frac{1}{20} + 8\frac{1}{15} + 9\frac{1}{2}$

**16.** $15\frac{1}{40} + 5\frac{7}{24} + 6\frac{1}{8}$

Find each difference.

**17.** $12\frac{2}{5} - 4$

**18.** $8\frac{9}{10} - 3$

**19.** $15 - 6\frac{5}{6}$

**20.** $11 - 1\frac{2}{7}$

**ANSWERS**

1. $10$
2. $14$
3. $11\frac{3}{10}$
4. $19\frac{4}{5}$
5. $10\frac{11}{12}$
6. $8\frac{5}{8}$
7. $17\frac{9}{28}$
8. $28\frac{2}{15}$
9. $21\frac{1}{16}$
10. $15\frac{13}{15}$
11. $13\frac{3}{4}$
12. $12\frac{23}{24}$
13. $5\frac{5}{8}$
14. $9\frac{27}{70}$
15. $30\frac{37}{60}$
16. $26\frac{53}{120}$
17. $8\frac{2}{5}$
18. $5\frac{9}{10}$
19. $8\frac{1}{6}$
20. $9\frac{5}{7}$

21. $2$

22. $4$

23. $3\frac{3}{10}$

24. $3\frac{1}{16}$

25. $7\frac{1}{6}$

26. $12\frac{25}{28}$

27. $2\frac{16}{21}$

28. $\frac{4}{5}$

29. $20\frac{19}{24}$

30. $15\frac{47}{48}$

31. $53\frac{41}{60}$

32. $10\frac{1}{3}$

33. $15\frac{4}{7}$

34. $\frac{1}{4}$

35. $\frac{11}{16}$

36. $1\frac{1}{8}$

37. $1\frac{2}{5}$

38. $5\frac{1}{2}$

39. $7\frac{1}{4}$

40. $9\frac{1}{10}$

41. $-12\frac{1}{4}$

42. $-5\frac{9}{20}$

43. $-5\frac{3}{8}$

44. $-11\frac{2}{5}$

45. $-6\frac{19}{24}$

46. $4\frac{1}{4}$

**21.** $8\frac{2}{3} - 6\frac{2}{3}$ **22.** $20\frac{3}{4} - 16\frac{3}{4}$ **23.** $6\frac{9}{10} - 3\frac{3}{5}$ **24.** $10\frac{5}{16} - 7\frac{1}{4}$

**25.** $21\frac{3}{4} - 14\frac{7}{12}$ **26.** $17\frac{3}{7} - 4\frac{15}{28}$ **27.** $18\frac{3}{7} - 15\frac{2}{3}$ **28.** $9\frac{3}{10} - 8\frac{1}{2}$

**29.** $26\frac{2}{3} - 5\frac{7}{8}$ **30.** $71\frac{5}{12} - 55\frac{7}{16}$ **31.** $187\frac{3}{20} - 133\frac{7}{15}$ **32.** $17-6\frac{2}{3}$

**33.** $20-4\frac{3}{7}$ **34.** $1-\frac{3}{4}$ **35.** $1-\frac{5}{16}$ **36.** $2-\frac{7}{8}$

**37.** $2-\frac{3}{5}$ **38.** $6-\frac{1}{2}$ **39.** $8-\frac{3}{4}$ **40.** $10-\frac{9}{10}$

Find the following algebraic sums and differences.

**41.** $-6\frac{1}{2}-5\frac{3}{4}$ **42.** $-2\frac{1}{4}-3\frac{1}{5}$ **43.** $-3-2\frac{3}{8}$

**44.** $-7-4\frac{2}{5}$ **45.** $-12\frac{2}{3}-\left(-5\frac{7}{8}\right)$ **46.** $-6\frac{1}{2}-\left(-10\frac{3}{4}\right)$

Name ______________________ Section ______________ Date ____________

**ANSWERS**

**47.** $-2\frac{1}{6}-\left(-15\frac{3}{5}\right)$

**48.** $-7\frac{3}{8}-\left(-4\frac{3}{16}\right)$

Find the following algebraic sums.

**49.** $-17\frac{2}{3}$
$+14\frac{2}{15}$

**50.** $-9\frac{1}{9}$
$+2\frac{5}{18}$

**51.** $-30\frac{5}{6}$
$-20\frac{2}{15}$

**52.** $-16\frac{3}{4}$
$-11\frac{5}{6}$

**53.** $-12\frac{1}{2}$
$-17\frac{2}{5}$
$-15\frac{1}{4}$

**54.** $-6\frac{1}{3}$
$-8\frac{3}{7}$
$-4\frac{1}{9}$

**55.** A board is 16 feet long. If two pieces are cut from the board, one $6\frac{3}{4}$ feet long and the other $3\frac{1}{2}$ feet long, what is the length of the remaining piece of the original board?

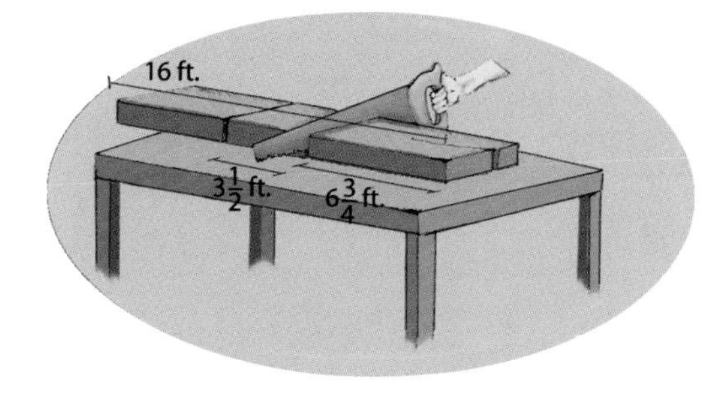


**Answers:**

47. $13\frac{13}{30}$

48. $-3\frac{3}{16}$

49. $-3\frac{8}{15}$

50. $-6\frac{5}{6}$

51. $-50\frac{29}{30}$

52. $-28\frac{7}{12}$

53. $-45\frac{3}{20}$

54. $-18\frac{55}{63}$

55. $5\frac{3}{4}$ ft.

56. $459\frac{13}{15}$ gallons

57. $10\frac{13}{24}$ inches

58. $20\frac{7}{10}$ meters

**56.** A swimming pool contains $495\frac{2}{3}$ gallons of water. If $35\frac{4}{5}$ gallons evaporate, how many gallons of water are left in the pool?

**57.** A triangle has sides of $3\frac{1}{8}$ inches, $2\frac{2}{3}$ inches, and $4\frac{3}{4}$ inches. What is the perimeter of (distance around) the triangle in inches.

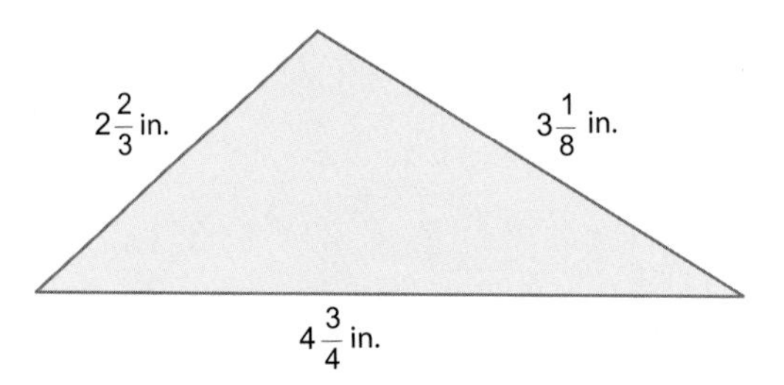


**58.** A quadrilateral has sides of $5\frac{1}{2}$ meters, $3\frac{7}{10}$ meters, $4\frac{9}{10}$ meters, and $6\frac{3}{5}$ meters. What is the perimeter of (distance around) the quadrilateral in meters?

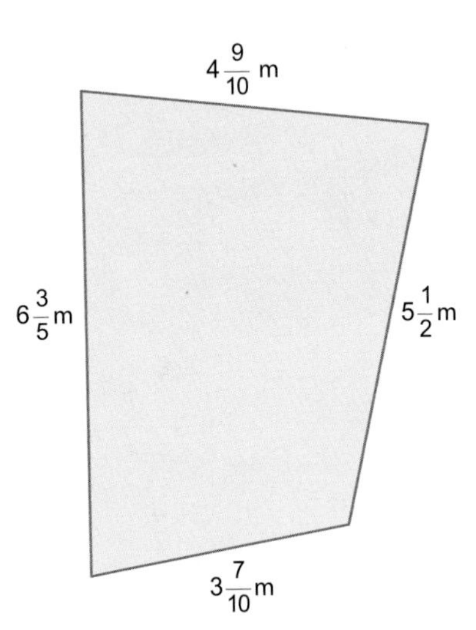

Name ______________________ Section ______________ Date __________

ANSWERS

59. A pentagon (five-sided figure) has sides of $3\frac{3}{8}$ centimeters, $5\frac{1}{2}$ centimeters, $6\frac{1}{4}$ centimeters, $9\frac{1}{10}$ centimeters, and $4\frac{7}{8}$ centimeters. What is the perimeter in centimeters of the pentagon?

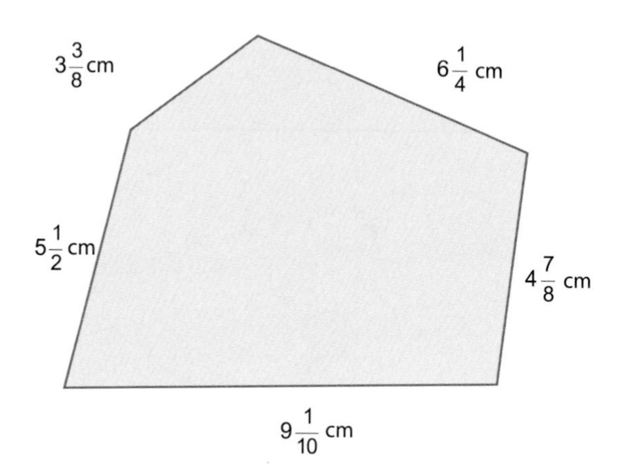


60. Among the top 25 movie gross incomes worldwide in 1999, *Star Wars: The Phantom Menace* earned \$ $922\frac{3}{5}$ million, *The Sixth Sense* earned \$ $660\frac{7}{10}$ million, *Toy Story 2* earned \$ $485\frac{7}{10}$ million, and *The Matrix* earned \$ $456\frac{2}{10}$ million. These same four movies earned domestically \$ $431\frac{1}{10}$ million, \$ $293\frac{1}{2}$ million, \$ $245\frac{4}{5}$ million, and \$ $171\frac{2}{5}$ million, respectively. What total amount did these four movies earn

**a.** in worldwide income,

**b.** in domestic income, and

**c.** in foreign income?

**59.** $29\frac{1}{10}$ cm

**60. a.** $\$2525\frac{1}{5}$ million

**b.** $\$1141\frac{4}{5}$ million

**c.** $\$1383\frac{2}{5}$ million

The Recycle Bin

1. $x=\frac{1}{2}$

2. $x=-\frac{5}{2}$ or $x=-2\frac{1}{2}$

3. $y=1$

4. $n=-5$

5. $x=\frac{3}{11}$

## The Recycle Bin (from Section 3.7)

Solve each of the following equations.

1. $\frac{2}{3}x+\frac{1}{2}=\frac{5}{6}$

2. $\frac{7}{8}x=-\frac{35}{16}$

3. $\frac{y}{4}-\frac{2}{5}=-\frac{3}{20}$

4. $\frac{n}{13}+\frac{2}{39}=-\frac{1}{3}$

5. $\frac{1}{7}x+\frac{1}{7}=\frac{2}{11}$

# 4.5 Complex Fractions and Order of Operations

## Simplifying Complex Fractions

A **complex fraction** is a fraction in which the numerator or denominator or both contain one or more fractions or mixed numbers. To simplify a complex fraction, we treat the fraction bar as a symbol of inclusion for both the numerator and denominator. The procedure is outlined as follows:

**To Simplify a Complex Fraction:**

1. Simplify the numerator so that it is a single fraction, possibly an improper fraction.
2. Simplify the denominator so that it also is a single fraction, possibly an improper fraction.
3. Divide the numerator by the denominator, and reduce if possible.

**Objectives**

① Be able to recognize and know how to simplify complex fractions.

② Be able to follow the rules for order of operations with mixed numbers.

### Example 1

Simplify the complex fraction $\dfrac{\frac{2}{3}+\frac{1}{6}}{1-\frac{1}{3}}$.

**Solution**

Simplify the numerator and denominator separately, and then divide.

$$\frac{2}{3}+\frac{1}{6}=\frac{4}{6}+\frac{1}{6}=\frac{5}{6} \quad \text{numerator}$$

$$1-\frac{1}{3}=\frac{3}{3}-\frac{1}{3}=\frac{2}{3} \quad \text{denominator}$$

So,

$$\frac{\frac{2}{3}+\frac{1}{6}}{1-\frac{1}{3}}=\frac{\frac{5}{6}}{\frac{2}{3}}=\frac{5}{6}\div\frac{2}{3}=\frac{5}{\cancel{6}_{2}}\cdot\frac{\cancel{3}^{1}}{2}=\frac{5}{4}=1\frac{1}{4}$$

**Special Note about the Fraction Bar:** The complex fraction in Example 1 could have been written in the form

$$\left(\frac{2}{3}+\frac{1}{6}\right)\div\left(1-\frac{1}{3}\right)$$

Thus, the fraction bar in a complex fraction serves the same purpose as two sets of parentheses, one surrounding the numerator and the other surrounding the denominator.

## Example 2

Simplify the complex fraction $\dfrac{3\frac{4}{5}}{\frac{3}{4}+\frac{1}{5}}$.

### Solution

$3\frac{4}{5}=\frac{19}{5}$ Change the mixed number to an improper fraction.

$\frac{3}{4}+\frac{1}{5}=\frac{15}{20}+\frac{4}{20}=\frac{19}{20}$ Add the fractions in the denominator.

So,

$$\frac{3\frac{4}{5}}{\frac{3}{4}+\frac{1}{5}}=\frac{\frac{19}{5}}{\frac{19}{20}}=\frac{19}{5}\cdot\frac{20}{19}=\frac{\overset{1}{\cancel{19}}\cdot 4\cdot\overset{1}{\cancel{5}}}{\underset{1}{\cancel{5}}\cdot\underset{1}{\cancel{19}}}=\frac{4}{1}=4$$

You might want to remind the students again that there are three correct positions for the – sign in a fraction. In Example 3, any of the following forms is acceptable:

$$-\frac{2}{15}=\frac{-2}{15}=\frac{2}{-15}$$

## Example 3

Simplify the complex fraction $\dfrac{\frac{2}{3}}{-5}$.

### Solution

In this case, division is the only operation to be performed. Note that –5 is an integer and can be written as $\frac{-5}{1}$.

$$\frac{\frac{2}{3}}{-5}=\frac{\frac{2}{3}}{\frac{-5}{1}}=\frac{2}{3}\cdot\frac{1}{-5}=-\frac{2}{15}$$

Simplify each complex fraction.

**1.** $\dfrac{\frac{3}{4}+\frac{1}{2}}{1-\frac{1}{3}}$

$\frac{15}{8}\left(\text{or } 1\frac{7}{8}\right)$

**2.** $\dfrac{2\frac{1}{3}}{\frac{1}{4}+\frac{1}{3}}$

4

**3.** $\dfrac{\frac{3}{10}}{-3}$

$-\frac{1}{10}$

*Now Work Exercises 1-3 in the Margin.*

## Order of Operations

With complex fractions, we know to simplify the numerator and denominator, separately. That is, we treat the fraction bar as a grouping symbol. More generally, expressions that involve more than one operation are evaluated by using the rules for order of operations. These rules were discussed and used in Sections 1.5 and 2.4 and are listed again here for easy reference.

**Rules for Order of Operations:**

1. First, simplify within grouping symbols, such as parentheses ( ), brackets [ ], or braces { }. Start with the innermost grouping.
2. Second, find any powers indicated by exponents.
3. Third, moving from **left to right**, perform any multiplications or divisions in the order in which they appear.
4. Fourth, moving from **left to right**, perform any additions or subtractions in the order in which they appear.

### Example 4

Use the rules for order of operations to simplify the following expression:

$$\frac{3}{5}\cdot\frac{1}{6}+\frac{1}{4}\div\left(2\frac{1}{2}\right)^2$$

**Solution**

$$\begin{aligned}\frac{3}{5}\cdot\frac{1}{6}+\frac{1}{4}\div\left(2\frac{1}{2}\right)^2 &= \frac{3}{5}\cdot\frac{1}{6}+\frac{1}{4}\div\left(\frac{5}{2}\right)^2\\ &= \frac{3}{5}\cdot\frac{1}{6}+\frac{1}{4}\div\left(\frac{25}{4}\right)\\ &= \frac{\cancel{3}}{5}\cdot\frac{1}{\underset{2}{\cancel{6}}}+\frac{1}{\cancel{4}}\cdot\frac{\cancel{4}}{25}\\ &= \frac{1}{10}+\frac{1}{25}\\ &= \frac{5}{50}+\frac{2}{50}=\frac{7}{50}\end{aligned}$$

You might want to demonstrate how + and – signs effectively separate expressions into quantities that involve the operations of multiplication and division (and exponents). This will provide students with a more efficient method for applying the rules of order of operations, since they can now perform more than one operation at each step. Remind them that the operations of addition and subtraction are the last ones to be performed. (See a similar note in Section 2.6.)

### Example 5

Simplify the expression $x+\frac{1}{3}\cdot\frac{2}{5}$ in the form of a single fraction, using the rules for order of operations.

**Solution**

First, multiply the two fractions, and then add $x$ in the form $\frac{x}{1}$.

The LCD is 15.

$$x+\frac{1}{3}\cdot\frac{2}{5}=x+\frac{2}{15}=\frac{x}{1}\cdot\frac{15}{15}+\frac{2}{15}=\frac{15x}{15}+\frac{2}{15}=\frac{15x+2}{15}$$

Another topic related to order of operations is that of **average**. As discussed in Section 2.5, the average of a set of numbers can be found by adding the numbers in the set, then dividing this sum by the number of numbers in the set. Now, we can find the average of mixed numbers as well as integers.

### Example 6

Find the average of $1\frac{1}{2}$, $\frac{5}{8}$, and $2\frac{1}{4}$.

**Solution**

Finding this average is the same as evaluating the expression.

$$\left(1\frac{1}{2}+\frac{5}{8}+2\frac{1}{4}\right)\div 3$$

Find the sum first, and then divide by 3.

$$\begin{aligned} 1\frac{1}{2} &= 1\frac{4}{8} \\ \frac{5}{8} &= \frac{5}{8} \\ +\,2\frac{1}{4} &= 2\frac{2}{8} \\ \hline 3\frac{11}{8} &= 4\frac{3}{8} \end{aligned}$$

$$4\frac{3}{8}\div 3=\frac{35}{8}\cdot\frac{1}{3}=\frac{35}{24}=1\frac{11}{24}$$

Alternatively, using improper fractions, we can write

$$\left(1\frac{1}{2}+\frac{5}{8}+2\frac{1}{4}\right)\div 3=\left(\frac{3}{2}+\frac{5}{8}+\frac{9}{4}\right)\div\frac{3}{1}=\left(\frac{3}{2}\cdot\frac{\mathbf{4}}{\mathbf{4}}+\frac{5}{8}+\frac{9}{4}\cdot\frac{\mathbf{2}}{\mathbf{2}}\right)\cdot\frac{1}{3}$$

$$=\left(\frac{12}{8}+\frac{5}{8}+\frac{18}{8}\right)\cdot\frac{1}{3}=\frac{35}{8}\cdot\frac{1}{3}=\frac{35}{24}=1\frac{11}{24}$$

*Now Work Exercises 4-6 in the Margin.*

Use the rules for order of operations to simplify each of the following expressions.

**4.** $\left(2-\frac{1}{3}\right)\div\left(1-\frac{1}{3}\right)$

$2\frac{1}{2}$

**5.** $x+\frac{1}{7}\div\frac{3}{14}$

$\frac{3x+2}{3}$

**6.** $\frac{2}{3}\div\frac{5}{12}+\frac{1}{5}\div 2^2$

$1\frac{13}{20}$

Name ______________________ Section ______________ Date __________

## Exercises 4.5

1. $\dfrac{\frac{3}{4}}{7}$

2. $\dfrac{\frac{5}{8}}{3}$

3. $\dfrac{\frac{9}{10}}{-3}$

4. $\dfrac{\frac{3}{8}}{-2}$

5. $\dfrac{\frac{4}{3x}}{\frac{8}{9x}}$

6. $\dfrac{\frac{a}{6}}{\frac{2a}{3}}$

7. $\dfrac{-2\frac{1}{3}}{-1\frac{2}{5}}$

8. $\dfrac{-7\frac{3}{4}}{-5\frac{7}{11}}$

9. $\dfrac{\frac{2}{3}+\frac{1}{5}}{4\frac{1}{2}}$

10. $\dfrac{3-\frac{1}{2}}{1-\frac{1}{3}}$

11. $\dfrac{\frac{5}{8}-\frac{1}{2}}{\frac{1}{8}-\frac{3}{16}}$

12. $\dfrac{\frac{5}{12}+\frac{1}{15}}{6}$

13. $\dfrac{3\frac{1}{5}-1\frac{1}{10}}{3\frac{1}{2}-1\frac{3}{10}}$

14. $\dfrac{6\frac{1}{10}-3\frac{1}{15}}{2\frac{1}{5}+1\frac{1}{2}}$

15. $\dfrac{1\frac{7}{12}+2\frac{1}{3}}{\frac{1}{5}-\frac{2}{15}}$

**ANSWERS**

1. $\frac{3}{28}$
2. $\frac{5}{24}$
3. $-\frac{3}{10}$
4. $-\frac{3}{16}$
5. $\frac{3}{2}\left(\text{or } 1\frac{1}{2}\right)$
6. $\frac{1}{4}$
7. $\frac{5}{3}\left(\text{or } 1\frac{2}{3}\right)$
8. $\frac{11}{8}\left(\text{or } 1\frac{3}{8}\right)$
9. $\frac{26}{135}$
10. $\frac{15}{4}\left(\text{or } 3\frac{3}{4}\right)$
11. $-2$
12. $\frac{29}{360}$
13. $\frac{21}{22}$
14. $\frac{91}{111}$
15. $\frac{235}{4}\left(\text{or } 58\frac{3}{4}\right)$

16. $-20x$

17. $24y$

18. $-\frac{15}{17}$

19. $\frac{1}{6}$

20. $\frac{625}{634}$

21. $\frac{7}{60}$

22. $-5$

23. $\frac{31}{28}\left(\text{or } 1\frac{3}{28}\right)$

24. $\frac{172}{9}\left(\text{or } 19\frac{1}{9}\right)$

25. $\frac{379}{60}\left(\text{or } 6\frac{19}{60}\right)$

26. $\frac{193}{30}\left(\text{or } 6\frac{13}{30}\right)$

27. $\frac{341}{30}\left(\text{or } 11\frac{11}{30}\right)$

28. $-\frac{27}{4}\left(\text{or } -6\frac{3}{4}\right)$

29. $-\frac{4}{21}$

30. $-\frac{15}{4}\left(\text{or } -3\frac{3}{4}\right)$

**16.** $\dfrac{-12x}{\frac{1}{2}+\frac{1}{10}}$

**17.** $\dfrac{-15y}{\frac{1}{4}-\frac{7}{8}}$

**18.** $\dfrac{\frac{1}{2}-1\frac{3}{4}}{\frac{2}{3}+\frac{3}{4}}$

**19.** $\dfrac{\frac{7}{8}-\frac{19}{20}}{\frac{1}{10}-\frac{11}{20}}$

**20.** $\dfrac{31\frac{1}{4}}{20\frac{1}{5}+11\frac{1}{2}}$

Use the rules for order of operations to simplify each of the following expressions.

**21.** $\frac{3}{5}\cdot\frac{1}{9}+\frac{1}{5}\div 2^2$

**22.** $\frac{1}{2}\div\frac{1}{4}-\frac{2}{3}\cdot 18+5$

**23.** $\frac{2}{3}\div\frac{7}{12}-\frac{2}{7}+\left(\frac{1}{2}\right)^2$

**24.** $3\frac{1}{2}\cdot 5\frac{1}{3}+\frac{5}{12}\div\frac{15}{16}$

**25.** $\frac{5}{8}\div\frac{1}{10}+\left(\frac{1}{3}\right)^2\cdot\frac{3}{5}$

**26.** $\frac{5}{9}-\frac{1}{3}\cdot\frac{2}{3}+6\frac{1}{10}$

**27.** $2\frac{1}{2}\cdot 3\frac{1}{5}\div\frac{3}{4}+\frac{7}{10}$

**28.** $\frac{4}{9}+\frac{1}{6}\div\frac{1}{2}\cdot\frac{11}{12}-7\frac{1}{2}$

**29.** $-5\frac{1}{7}\div(2+1)^3$

**30.** $\left(\frac{1}{3}-2\right)\div\left(1-\frac{1}{3}\right)^2$

Name ______________ Section __________ Date ________

**31.** $x-\frac{1}{5}-\frac{2}{3}$

**32.** $x+\frac{3}{4}+2\frac{1}{2}$

**33.** $y+\frac{1}{7}+\frac{1}{6}$

**34.** $y-\frac{4}{5}-3\frac{1}{3}$

**35.** $\frac{2a}{5}+\frac{3}{5}\cdot\frac{3}{7}$

**36.** $\frac{a}{3}\cdot\frac{1}{2}-\frac{2}{3}\div 1\frac{1}{3}$

**37.** $\frac{1}{x}\cdot\frac{3}{7}-\frac{1}{7}\div\frac{1}{2}$

**38.** $\frac{2}{x}\cdot\frac{3}{4}-\frac{5}{6}\cdot 15$

**39.** Find the average of $2\frac{1}{2}$, $3\frac{2}{5}$, $5\frac{3}{4}$, and $6\frac{1}{10}$.

**40.** Find the average of $-7\frac{3}{5}$, $-4\frac{3}{5}$, $-8\frac{1}{3}$, and $-3\frac{1}{18}$.

**41.** If $\frac{4}{5}$ of 80 is divided by $\frac{3}{4}$ of 90, what is the quotient?

**42.** If $\frac{2}{3}$ of $2\frac{7}{10}$ is added to $\frac{1}{2}$ of $5\frac{1}{4}$, what is the sum?

**ANSWERS**

**31.** $\frac{15x-13}{15}$

**32.** $\frac{4x+13}{4}$

**33.** $\frac{42y+13}{42}$

**34.** $\frac{15y-62}{15}$

**35.** $\frac{14a+9}{35}$

**36.** $\frac{a-3}{6}$

**37.** $\frac{3-2x}{7x}$

**38.** $\frac{3-25x}{2x}$

**39.** $4\frac{7}{16}$

**40.** $-5\frac{323}{360}$

**41.** $\frac{128}{135}$

**42.** $4\frac{17}{40}$

43. $-\frac{481}{1600}$

44. $\frac{97}{144}$

45. a. $\dfrac{\frac{4}{5}+\frac{2}{15}}{2\frac{1}{4}-\frac{7}{8}}$

b. $\left(\frac{4}{5}+\frac{2}{15}\right)\div\left(2\frac{1}{4}-\frac{7}{8}\right)$

46. a. $\dfrac{-4\frac{1}{2}-1\frac{3}{4}}{-5\frac{1}{10}+6\frac{1}{2}}$

b. $\left(-4\frac{1}{2}-1\frac{3}{4}\right)\div\left(-5\frac{1}{10}+6\frac{1}{2}\right)$

47. [Respond below exercise]

48. [Respond below exercise]

**43.** If the square of $\frac{5}{8}$ is subtracted from the square of $\frac{3}{10}$, what is the difference?

**44.** If the square of $\frac{1}{3}$ is added to the square of $-\frac{3}{4}$, what is the sum?

**45.** The sum of $\frac{4}{5}$ and $\frac{2}{15}$ is to be divided by the difference between $2\frac{1}{4}$ and $\frac{7}{8}$.

**a.** Write this expression in the form of a complex fraction.

**b.** Write an equivalent expression in a form using two pairs of parentheses and a division sign. Do not evaluate either expression.

**46.** The sum of $-4\frac{1}{2}$ and $-1\frac{3}{4}$ is to be divided by the sum of $-5\frac{1}{10}$ and $6\frac{1}{2}$.

**a.** Write this expression in the form of a complex fraction.

**b.** Write an equivalent expression in a form using two pairs of parentheses and a division sign. Do not evaluate either expression.

## Writing and Thinking about Mathematics

**47.** Consider any number between 0 and 1. If you square this number, will the result be larger or smaller than the original number? Is this always the case? Explain your answer.

The square of any number between 0 and 1 is always smaller than the original number. Explanations will vary.

**48.** Consider any number between $-1$ and 0. If you square this number, will the result be larger or smaller than the original number? Is this always the case? Explain your answer.

The square of any negative number will be positive number and, therefore, larger than the original number (negative) number. Explanations will vary.

# 4.6 Solving Equations: Ratios and Proportions

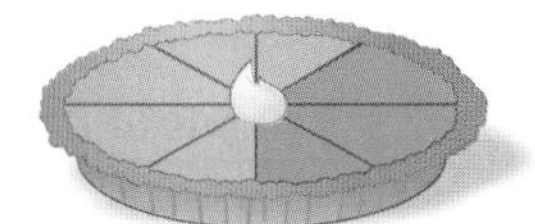

## Objectives

① Understand the meaning of a ratio.

② Understand that a proportion is an equation.

③ Be familiar with the terms **means** and **extremes**.

④ Know how to solve for the unknown quantity in a problem.

⑤ Be able to set up a proportion with an unknown term in one of the appropriate patterns that will lead to a solution of the problem.

## Understanding Ratios

We know two meanings for fractions:

1. To indicate a part of a whole:

   $\frac{3}{8}$ means $\frac{\text{3 pieces of pumpkin pie}}{\text{8 pieces in the whole pie}}$

2. To indicate division:

   $\frac{17}{5}$ means $17 \div 5$ or $5\overline{)17}$ = 3 or $3\frac{2}{5}$

   $$\begin{array}{r} 3 \\ 5\overline{)17} \\ \underline{15} \\ 2 \end{array} \text{ Remainder}$$

A third use for fractions is to compare two quantities. Such a comparison is called a **ratio**. For example, the **ratio** $\frac{5}{6}$ might mean $\frac{\text{5 feet}}{\text{6 feet}}$ or $\frac{\text{5 dimes}}{\text{6 dimes}}$.

As another example, suppose that for a pancake mix the ratio of mix to water is 2 to 1. Then, we can write

$$\frac{2 \text{ cups } \mathbf{mix}}{1 \text{ cup water}}, \text{ or } \frac{2 \text{ gallons } \mathbf{mix}}{1 \text{ gallon water}}, \text{ or } \frac{2 \text{ teaspoons } \mathbf{mix}}{1 \text{ teaspoon water}}.$$

Although the amounts would be different, we would have a pancake batter with the same consistency in every case.

> **Definition:**
>
> A **ratio** is a comparison of two quantities by division. The ratio of $a$ to $b$ can be written as
>
> $$\frac{a}{b} \quad \text{or} \quad a:b \quad \text{or} \quad a \text{ to } b$$

You might mention that "per" means to divide. For example, 50 miles per hour can be thought of as $\frac{\text{50 miles}}{\text{1 hour}}$ or 3 cents per ounce can be thought of as $\frac{\text{3 cents}}{\text{1 ounce}}$. The idea of an abstract number will probably be new to most students. However, an awareness of this concept would help them understand the need to label answers to word problems. It would also help them in solving word problems. They should also understand that labeling is absolutely necessary in setting up correct proportions for solving problems with proportions.

Remember the following characteristics of ratios:

1. Ratios can be reduced just as fractions can be reduced.
2. Whenever the units of the numbers in a ratio are the same, the ratio has no units. We say that the ratio is an **abstract number**.
3. When the numbers in a ratio have different units, then the numbers must be labeled to clarify what is being compared. Such a ratio is called a **rate.** For example, the ratio of 55 miles : 1 hour $\left(\text{or } \frac{\text{55 miles}}{\text{1 hour}}\right)$ is a rate of 55 miles per hour (or 55 mph).

## Example 1

During baseball season, major league players' batting averages are published in the newspapers. Suppose a player has a batting average of .320. (We will study decimal numbers in detail in Chapter 5.) What does this indicate?

### Solution

A batting average is a ratio (or rate) of hits to times at bat. Thus, a batting average of .320 means

$$.320 = \frac{320 \text{ hits}}{1000 \text{ times at bat}}$$

Reducing gives

$$.320 = \frac{320}{1000} = \frac{\cancel{40} \cdot 8}{\cancel{40} \cdot 25} = \frac{8 \text{ hits}}{25 \text{ times at bat}}$$

This means that we can expect this player to hit safely at a rate of 8 hits for every 25 times he comes to bat.

Whenever possible, a ratio should be an abstract number. That is, we would like a ratio to represent a ratio of two quantities with the **same units of measure**. For example, to find the ratio of 4 nickels to 2 quarters, we can, and would prefer, to write the ratio with the same units (such as pennies or nickels). Thus, the ratio is

$\frac{4 \text{ nickels}}{2 \text{ quarters}} = \frac{4 \text{ nickels}}{10 \text{ nickels}} = \frac{2}{5}$ and the ratio of 4 nickels to 2 quarters is 2 to 5.

## Example 2

What is the reduced ratio of 250 centimeters (cm) to 2 meters (m)? Centimeters and meters are units of measure in the metric system. (See Chapter 10 for more details.)

### Solution

Here we change the units so that they are the same. There are 100 centimeters in 1 meter. Therefore, 2 m = 200 cm, and the ratio is

$$\frac{250 \text{ cm}}{2 \text{ m}} = \frac{250 \text{ cm}}{200 \text{ cm}} = \frac{\cancel{50} \cdot 5}{\cancel{50} \cdot 4} = \frac{5}{4}$$

The reduced ratio can also be written as 5 : 4 or 5 to 4.

*Now Work Exercises 1-3 in the Margin.*

Write each comparison as a ratio reduced to lowest terms.

1. 3 quarters to 1 dollar.

   $\frac{3}{4}$

2. 36 inches to 5 feet.

   $\frac{3}{5}$

3. Inventory shows 5000 washers and 400 bolts. What is the ratio of washers to bolts?

   $\frac{25 \text{ washers}}{2 \text{ bolts}}$

## Understanding Proportions

The following three equations show two fractions (or ratios) to be equal:

$$\frac{6}{8}=\frac{15}{20} \quad \text{and} \quad \frac{84}{63}=\frac{204}{153} \quad \text{and} \quad \frac{3\frac{1}{2}}{7}=\frac{3\frac{1}{4}}{6\frac{1}{2}}$$

Each of these equations is called a **proportion**.

> **Definition:**
>
> A **proportion** is a statement that two ratios are equal. In symbols,
>
> $\frac{a}{b}=\frac{c}{d}$ is a proportion.

A proportion has four **terms:** The first and fourth terms are called the **extremes**. The second and third terms are called the **means**.

To better understand the terms **means** and **extremes**, we can write a proportion in the following form:

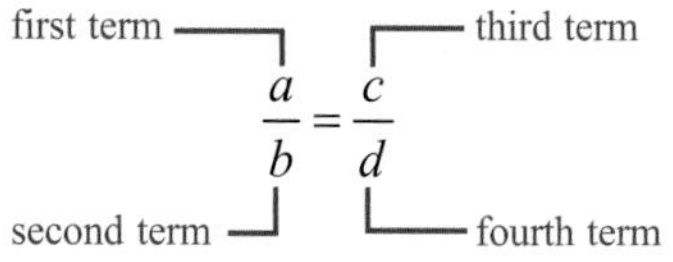

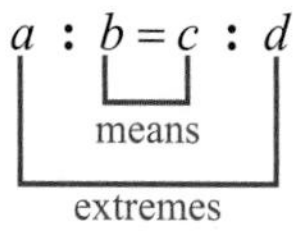

first term — $a$; second term — $b$; third term — $c$; fourth term — $d$

$$\frac{a}{b}=\frac{c}{d}$$

$$a : b = c : d$$

means: $b$, $c$

extremes: $a$, $d$

Now consider the following analysis of a proportion:

| | |
|---|---|
| $\frac{a}{b}=\frac{c}{d}$ | Write the proportion. |
| $bd\left(\frac{a}{b}\right)=\left(\frac{c}{d}\right)bd$ | Multiply both sides by $bd$. |
| $a \cdot d = b \cdot c$ | Simplify. |

The last equation states that (assuming the original proportion is true) the product of the means is equal to the product of the extremes. This fact provides a useful technique for solving problems involving proportions, and we state it formally.

**In a true proportion, the product of the extremes is equal to the product of the means.**

In symbols, $\frac{a}{b}=\frac{c}{d}$ if and only if $a \cdot d = b \cdot c$. (where $b \neq 0$ and $d \neq 0$ )

### Example 3

Determine whether each of the following proportions is true or false.

**a.** $\frac{6}{8}=\frac{15}{20}$ **b.** $\frac{84}{63}=\frac{204}{153}$ **c.** $\frac{2\frac{1}{3}}{7}=\frac{3\frac{1}{4}}{9\frac{3}{4}}$

#### Solution

**a.** For the proportion $\frac{6}{8}=\frac{15}{20}$,

the product of the extremes is $6 \times 20 = 120$, and

the product of the means is $8 \times 15 = 120$.

Since the product of the extremes is equal to the product of the means, the proportion $\frac{6}{8}=\frac{15}{20}$ is true.

**b.** For the proportion $\frac{84}{63}=\frac{204}{153}$, the calculations shown here could easily be performed with a calculator.

The product of the extremes:

$$\begin{array}{r} 153 \\ \times \quad 84 \\ \hline 612 \\ 1224\phantom{0} \\ \hline 12{,}852 \end{array}$$

The product of the means:

$$\begin{array}{r} 204 \\ \times \quad 63 \\ \hline 612 \\ 1224\phantom{0} \\ \hline 12{,}852 \end{array}$$

Since the product of the extremes is equal to the product of the means, the proportion is true.

**c.** For the proportion $\frac{2\frac{1}{3}}{7} = \frac{3\frac{1}{4}}{9\frac{3}{4}}$, change the means and extremes to improper fraction form and multiply:

The product of the means is $2\frac{1}{3} \cdot 9\frac{3}{4} = \frac{7}{3} \cdot \frac{39}{4} = \frac{7}{\cancel{3}_1} \cdot \frac{\cancel{39}^{13}}{4} = \frac{91}{4}$, and

the product of the extremes is $7 \cdot 3\frac{1}{4} = \frac{7}{1} \cdot \frac{13}{4} = \frac{91}{4}$.

Since the product of the extremes is equal to the product of the means, the proportion is true.

*Now Work Exercises 4-6 in the Margin.*

Determine whether each proportion is true or false.

**4.** $\frac{12}{16} = \frac{21}{28}$

True

**5.** $\frac{4\frac{1}{2}}{2} = \frac{3\frac{1}{2}}{4}$

False

**6.** $\frac{6}{7} = \frac{18}{21}$

True

## Applications

Proportions can be used in solving certain types of word problems. In these applications there is some unknown quantity that can be represented as one term in a proportion. The problem is solved by finding the value of this unknown term that will make the proportion a true statement.

### Solving for the Unknown Term in a Proportion

1. **If the unknown term is in a denominator**, solve by setting the product of the means equal to the product of the extremes and then solving the resulting equation.
2. **If the unknown term is in a numerator**, solve the proportion directly by multiplying both sides by the reciprocal of the denominator as in Section 3.7.

### Example 4

Find the unknown term in the following proportion: $\frac{6}{124} = \frac{3}{A}$ (In this case, the unknown term is in the denominator.)

#### Solution

$\frac{6}{124} = \frac{3}{A}$ Write the proportion.

$6 \times A = 3(124)$ Write the product of the extremes equal to the product of the means.

$\frac{\cancel{6}A}{\cancel{6}} = \frac{372}{6}$ Solve the equation.

$A = 62$

### Completion Example 5

Find the value of $x$ if $\frac{x}{100} = \frac{4}{5}$. (In this case the unknown term is in the numerator.)

**Solution**

Since the unknown term is in the numerator, multiply both sides by 100 and solve the proportion in one step.

$\frac{x}{100} = \frac{4}{5}$ Write the proportion.

$\_\_\_\_ \cdot \frac{x}{100} = \frac{4}{5} \cdot \_\_\_\_$ Multiply both sides by ________.

$x = \_\_\_\_$ Simplify.

Find the value of the unknown term in each proportion.

**7.** $\frac{x}{32} = \frac{5}{8}$

$x = 20$

**8.** $\frac{11}{100} = \frac{66}{y}$

$y = 600$

*Now Work Exercises 7 and 8 in the Margin.*

In using proportions to solve word problems, we must set up the proportion properly so that the ratios that are represented compare the units in the same order. Either pattern A or pattern B, described in the following problem, must be followed.

**Problem to Solve by Setting Up a Proportion:**

Suppose that a car will travel 500 miles on 20 gallons of gas. How far would you expect to travel on 30 gallons of gas?

**Pattern A:** Each ratio has different units, but they are in the same order. For example,

$$\frac{500 \textbf{ miles}}{20 \textit{ gallons}} = \frac{x \textbf{ miles}}{30 \textit{ gallons}} \quad \text{or} \quad \frac{20 \textit{ gallons}}{500 \textbf{ miles}} = \frac{30 \textit{ gallons}}{\text{x } \textbf{miles}}$$

**Pattern B:** Each ratio has the same units, the numerators correspond, and the denominators correspond. For example,

$$\frac{500 \textbf{ miles}}{x \textbf{ miles}} = \frac{20 \textit{ gallons}}{30 \textit{ gallons}} \quad \text{or} \quad \frac{\text{x } \textbf{miles}}{500 \textbf{ miles}} = \frac{30 \textit{ gallons}}{20 \textit{ gallons}}$$

With every one of the four equations illustrated in Pattern A and Pattern B, setting the product of the extremes equal to the product of the means will give the same equation to be solved, namely,

$20x = 500 \times 30.$

So, you may use the form that occurs to you first. The solution will be the same regardless of the form chosen.

### Example 6

An architect draws the plans for a building using a scale of $\frac{3}{4}$ inch to represent 10 feet. How many feet would 6 inches represent?

**Solution**

**Step 1:** Let $y$ represent the number of unknown feet.

**Step 2:** Set up a proportion and label the numerators and denominators to be sure that the pattern is correct.

**Step 3:** One such proportion (following Pattern B) is

$$\frac{\frac{3}{4}\ \textbf{inch}}{6\ \textbf{inches}} = \frac{10\ \textit{feet}}{y\ \textit{feet}}$$

Any of the proportions following Pattern A or Pattern B will give the same solution.

**Step 4:** Solve the proportion:

$$\frac{\frac{3}{4}\ \text{inch}}{6\ \text{inches}} = \frac{10\ \text{feet}}{y\ \text{feet}}$$

$$\frac{3}{4}y = 6 \cdot 10$$

$$\frac{\cancel{4}}{\cancel{3}} \cdot \frac{\cancel{3}}{\cancel{4}}y = \frac{4}{3} \cdot 60$$ Multiply both sides by $\frac{4}{3}$, the reciprocal of $\frac{3}{4}$.

$$y = 80$$

Therefore, on the architect's drawing, 6 inches represents 80 feet.

## Proportions in Geometry (Similar Triangles)

We will discuss angles and triangles in detail in Chapter 10. Included are facts such as the following:

1. An angle can be measured in degrees. (A protractor, as shown in Figure 4.4, can be used to find the measure of an angle.)
2. Every triangle has six parts: three sides and three angles.
3. The sum of the measures of the angles of every triangle is 180°.
4. Each endpoint of the sides of a triangle is called a **vertex** of the triangle. (Capital letters are used to label the vertices, and these letters can be used to name the angles and the sides.)

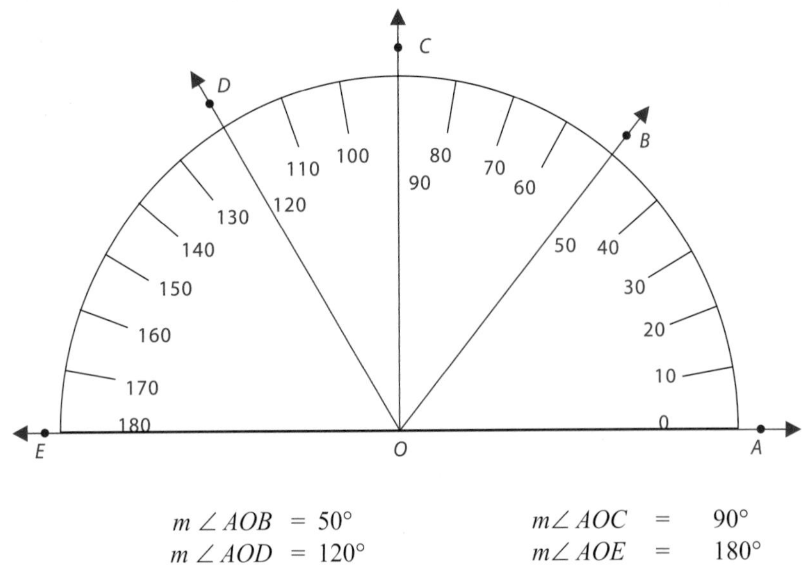


$m \angle AOB = 50°$ $\qquad$ $m\angle AOC = 90°$

$m \angle AOD = 120°$ $\qquad$ $m\angle AOE = 180°$

**Figure 4. 4**

To measure an angle with a protractor, lay the bottom edge of the protractor along one side of the angle with the vertex at the marked centerpoint on the protractor. Then, read the measure from the protractor where the other side crosses the arch part of the protractor.

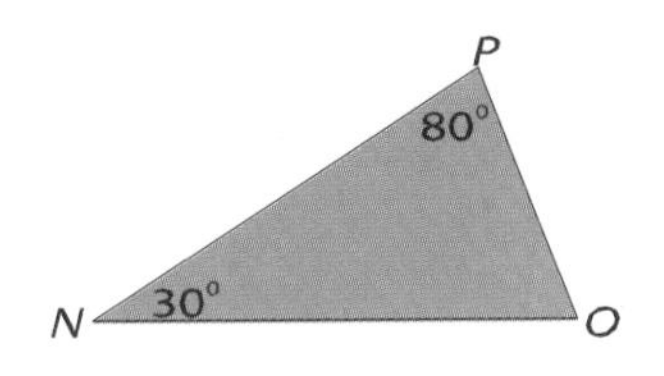


**Figure 4. 5**

To illustrate the notation and information about triangles, consider triangle ***NOP*** (symbolized $\Delta$***NOP***) in Figure 4.5. In this triangle,

$m \angle N = 30°$ $\qquad$ and $\qquad$ $m \angle P = 80°$

The sum of the measures of the three angles must be 180°. Therefore, we can set up and solve an equation for the unknown $m \angle O$ as follows:

$$\begin{aligned} m \angle N + m \angle P + m \angle O &= 180° \\ 30° + 80° + m \angle O &= 180° \\ 110° + m \angle O &= 180° \\ m \angle O &= 180° - 110° \\ m \angle O &= 70° \end{aligned}$$

In this manner, if we know the measures of two angles of a triangle, we can always find the measure of the third angle.

Two triangles are said to be **similar triangles** if they have the same "shape." They may or may not have the same "size." More formally, two triangles are similar if they have the following two properties:

You might want to show the students other types of similar geometric figures such as squares and rectangles. Related questions might be, "Are all squares similar?" or "Why are all squares similar, yet all rectangles are not similar?" You could also show them similar non-rectangular quadrilaterals and similar hexagons and develop a general discussion of the ideas related to the corresponding angles and proportionality.

**Two triangles are similar if:**

**1.** Their **corresponding angles are equal.** (The corresponding angles have the same measure.)

**2.** Their **corresponding sides are proportional.**

In similar triangles, **corresponding sides** are those sides opposite the equal angles in the respective triangles. (See Figure 4.6.)

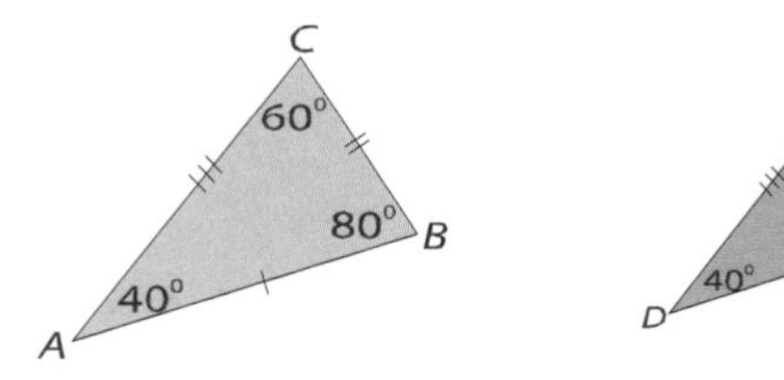


We write $\Delta ABC \sim \Delta DEF$. **(~ is read "is similar to.")** The corresponding sides are proportional, so the ratios of corresponding sides are equal and

$$\frac{AB}{DE} = \frac{BC}{EF} \text{ and } \frac{AB}{DE} = \frac{AC}{DF} \text{ and } \frac{BC}{EF} = \frac{AC}{DF}$$

**Figure 4. 6**

To say that corresponding sides are proportional means that we can set up a proportion to solve for one of the unknown sides of two similar triangles. Example 7 illustrates how this can be done.

## Example 7

Given that $\Delta ABC \sim \Delta PQR$, use the fact that corresponding sides are proportional and find the values of $x$ and $y$.

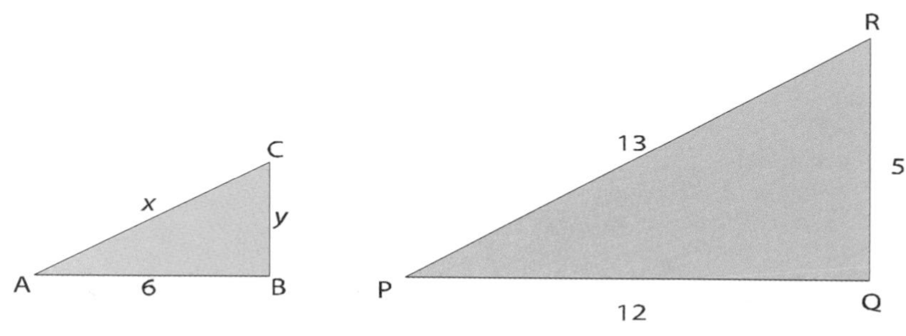


### Solution

Set up two proportions and solve for the unknown terms.

$$\frac{x}{13} = \frac{6}{12}$$

$$\frac{\cancel{13}}{1} \cdot \frac{x}{\cancel{13}} = \frac{13}{1} \cdot \frac{6}{12}$$

$$x = \frac{13}{2} \left( \text{or } x = 6\frac{1}{2} \right)$$

$$\frac{y}{5} = \frac{6}{12}$$

$$\frac{\cancel{5}}{1} \cdot \frac{y}{\cancel{5}} = \frac{5}{1} \cdot \frac{6}{12}$$

$$y = \frac{5}{2} \left( \text{or } y = 2\frac{1}{2} \right)$$

OR: Reducing first may lead to equations that are easier to solve. The answers will be the same.

$$\frac{x}{13} = \frac{6}{12}$$

$$\frac{x}{13} = \frac{1}{2}$$

$$\cancel{13} \cdot \frac{x}{\cancel{13}} = 13 \cdot \frac{1}{2}$$

$$x = \frac{13}{2}\left(\text{or } x = 6\frac{1}{2}\right)$$

$$\frac{y}{5} = \frac{6}{12}$$

$$\frac{y}{5} = \frac{1}{2}$$

$$\frac{\cancel{5}}{1} \cdot \frac{y}{\cancel{5}} = \frac{5}{1} \cdot \frac{1}{2}$$

$$y = \frac{5}{2}\left(\text{or } y = 2\frac{1}{2}\right)$$

## Example 8

A very tall magnificently decorated tree was on display in an outside patio area at the mall during the month before Christmas. Joann knew that the height of a nearby lamppost was 10 feet. At 2 p.m. on a Tuesday afternoon, she measured the shadow of the lamppost to be $3\frac{1}{2}$ feet long and the shadow of the tree to be 21 feet long. With her understanding of similar triangles, Joann was able to calculate the height of the tree. What was the height of the tree?

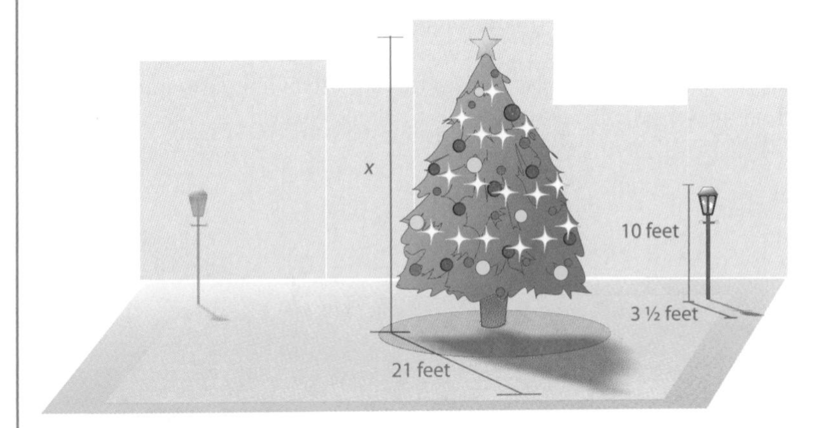

### Solution

By letting $x$ represent the height of the tree, Joann set up and solved the following proportion.

$$\frac{x \text{ ft. } \textbf{(height of tree)}}{21 \text{ ft. } \textbf{(length of tree shadow)}} = \frac{10 \text{ ft. } \textbf{(height of lamppost)}}{3\frac{1}{2} \text{ ft. } \textbf{(length of post shadow)}}$$

$$21 \cdot \frac{x}{21} = 21 \cdot \frac{10}{\frac{7}{2}}$$

$$x = \cancel{21}^{3} \times 10 \times \frac{2}{\cancel{7}}$$

$$x = 60$$

The tree was 60 feet tall.

To say that corresponding angles are equal means that the angles in the same relative positions in the two triangles are equal. Example 9 illustrates corresponding angles.

## Example 9

Find the values of $x$ and $y$ in triangles $\Delta ABC$ and $\Delta ADE$.

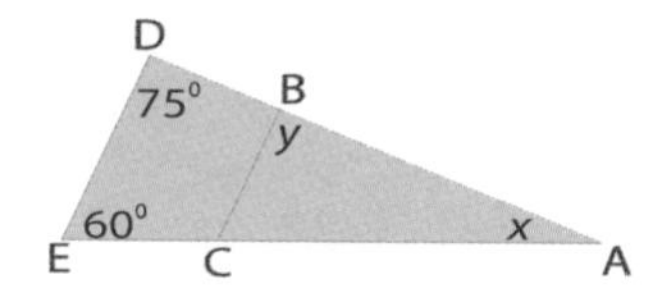


### Solution

Let $x$ be the measure of $\angle A$ and $y$ be the measure of $\angle B$ in $\Delta ABC$ as illustrated. Since $y$ is in the same relative position as 75° in $\Delta ADE$ ( $\angle B$ and $\angle D$ are corresponding angles ), we have $y = 75°$. In $\Delta ADE$, using the fact that the sum of the measures of the angles of a triangle must be 180° gives the equation $x + 75 + 60 = 180$. Solving for $x$ gives $x = 45°$.

## Completion Example Answer

**5.** In this problem the variable is in the numerator, and we can solve the proportion directly.

$$\frac{x}{100} = \frac{4}{5} \quad \text{Write the proportion.}$$

$$\mathbf{100} \cdot \frac{x}{100} = \frac{4}{5} \cdot \mathbf{100} \quad \text{Multiply both sides by 100.}$$

$$x = \mathbf{80} \quad \text{Simplify.}$$

Name ______ Section ______ Date ______

## Exercises 4.6

Write the following comparisons as ratios reduced to lowest terms. Use common units in the numerator and denominator whenever possible.

1. 1 dime to 5 nickels
2. 15 nickels to 3 quarters
3. 4 minutes to 1 hour
4. 8 hours to 1 day
5. 2 yards to 5 feet
6. 12 inches to 2 feet
7. 38 miles to 2 gallons of gas
8. 250 miles to 5 hours
9. 125 hits to 500 times at bat
10. 100 centimeters to 1 meter
11. \$300 profit to \$1000 invested
12. \$70 profit to \$1000 invested

Determine whether each proportion is true or false.

13. $\frac{5}{6}=\frac{10}{12}$
14. $\frac{-7}{21}=\frac{-4}{12}$
15. $\frac{2}{7}=\frac{5}{17}$
16. $\frac{-62}{31}=\frac{102}{-51}$
17. $\frac{8\frac{1}{2}}{2\frac{1}{3}}=\frac{4\frac{1}{4}}{1\frac{1}{6}}$
18. $\frac{6}{7}=\frac{8}{9}$

**ANSWERS**

1. $\frac{2}{5}$
2. $\frac{1}{1}$ (or 1)
3. $\frac{1}{15}$
4. $\frac{1}{3}$
5. $\frac{6}{5}$
6. $\frac{1}{2}$
7. $\frac{19\text{ mi.}}{1\text{ gal.}}$ (or 19 mpg.)
8. $\frac{50\text{ mi.}}{1\text{ hr.}}$ (or 50 mph.)
9. $\frac{1\text{ hit}}{4\text{ at-bats}}$
10. $\frac{1}{1}$ (or 1)
11. $\frac{\$3\text{ profit}}{\$10\text{ invested}}$
12. $\frac{\$7\text{ profit}}{\$100\text{ invested}}$
13. True
14. True
15. False
16. True
17. True
18. False

Solve the following proportions.

19. $\frac{3}{6}=\frac{6}{x}$

20. $\frac{3}{5}=\frac{y}{100}$

21. $\frac{1}{4}=\frac{1\frac{1}{2}}{w}$

22. $\frac{A}{20}=\frac{15}{100}$

23. $\frac{x}{6}=\frac{25}{12}$

24. $\frac{78}{13}=\frac{x}{26}$

25. $\frac{135}{B}=\frac{15}{100}$

26. $\frac{3}{10}=\frac{x}{100}$

27. $\frac{A}{1000}=\frac{18}{100}$

Solve the following problems by using proportion.

28. Investor A thinks that she should make \$9 for every \$100 she invest. How much does she expect to make on an investment of \$1500?

29. Investor B thinks that she should make \$15 for every \$150 she invest. How much does she expect to make on an investment of \$5,000?

30. Referring to Exercises 28 and 29, which investor would expect to make the most on an investment of \$10,000? How much more?

31. An architect is to draw plans for a city park. He intends to use a scale of $\frac{1}{2}$ inch to represent 25 feet. How many inches will he need to use for the length and width of a rectangular playing field that is 50 yards by 125 yards? (1 yard = 3 feet)

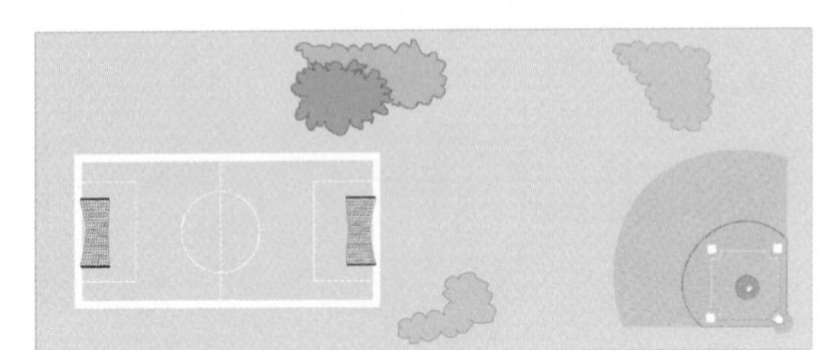

19. $x=12$

20. $y=60$

21. $w=6$

22. $A=3$

23. $x=12\frac{1}{2}$

24. $x=156$

25. $B=900$

26. $x=30$

27. $A=180$

28. \$135

29. \$500

30. Investor B; \$100 more

31. 3 in. for the width

$7\frac{1}{2}$ in. for the length

Name ______________________ Section ____________ Date __________

**ANSWERS**

32. 8 hours

33. 60 ft.

34. 259,200 revolutions

35. 7 in. by 10 in.

**32.** You know that you drove to your grandmother's house in 6 hours. How long would you estimate to drive to your cousin's house if your grandmother lives 276 miles away and your cousin lives 368 miles away?

**33.** An engineer would like to know the length of the shadow of a building that is 21 stories high at a particular time of day. However, this building is across town, and he needs the information now. So he simply goes outside and measures the length of the shadow of the building he is in, which is 14 stories tall. The shadow of his building is 40 feet long. With this information, he calculates the length of the shadow of the building across town. What is the length of that shadow?

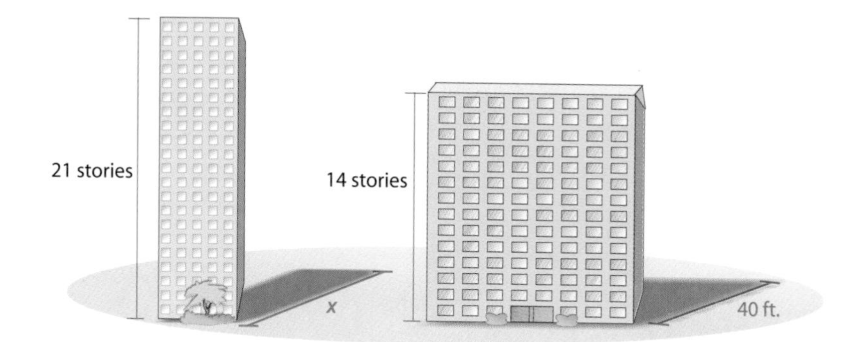


**34.** An electric fan makes 180 revolutions per minute. How many revolutions will the fan make if it runs for 24 hours?

**35.** A cartographer (mapmaker) plans to use a scale of 2 inches to represent 30 miles. What rectangular shape of paper will she need to draw on if she plans to map a region that is 75 miles wide and 120 miles long and she wants to leave a 1-inch margin around all four edges of the paper?

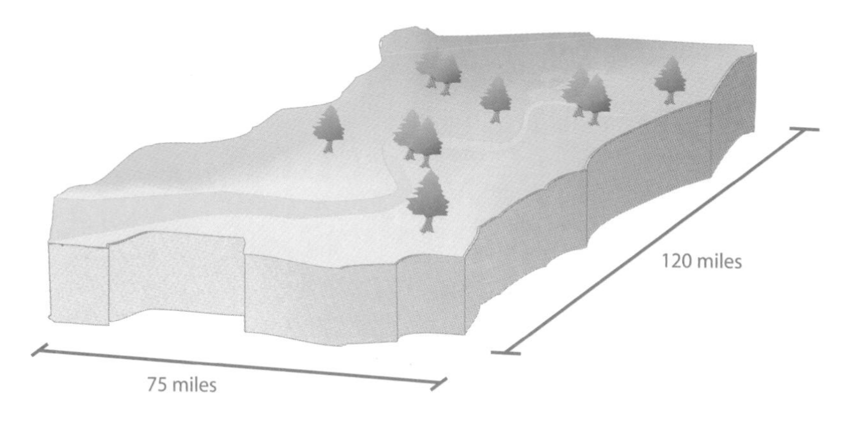

36. a. $7\frac{1}{2}$ mph.

b. 55 mph.

37. $x = 7\frac{1}{2}, y = 15$

38. $x = 3, y = 3$

39. $x = 7, y = 5$

40. $x = 4, y = 6$

41. $x = 50°, y = 60°$

42. $x = 2\frac{1}{2}, y = 2$

43. $x = 50°, y = 50°$

44. $x = 20°, y = 100°$

**36.** A test driver wants to increase the speed of the car he is driving by 3 miles per hour every 2 seconds. But he can only check his speed every 5 seconds because he is busy with other items during the test drive.

**a.** By how much should he increase his speed in 5 seconds?

**b.** If he starts checking his speed at 40 miles per hour, how fast should he be going in 10 seconds?

Exercises 37-44 each illustrate a pair of similar triangles. Find the values of $x$ and $y$ in each of these exercises.

**37.**

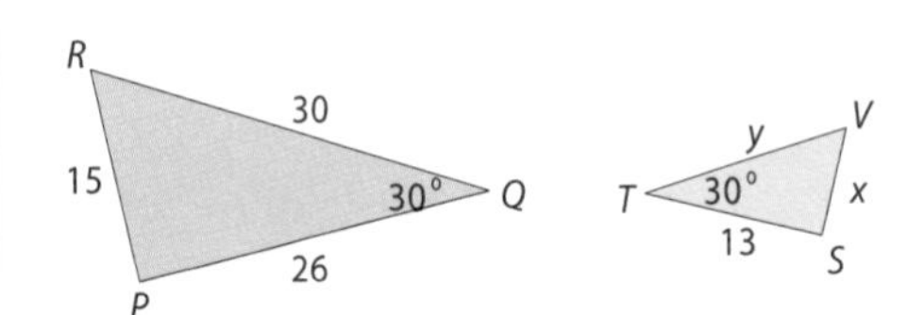


**38.**

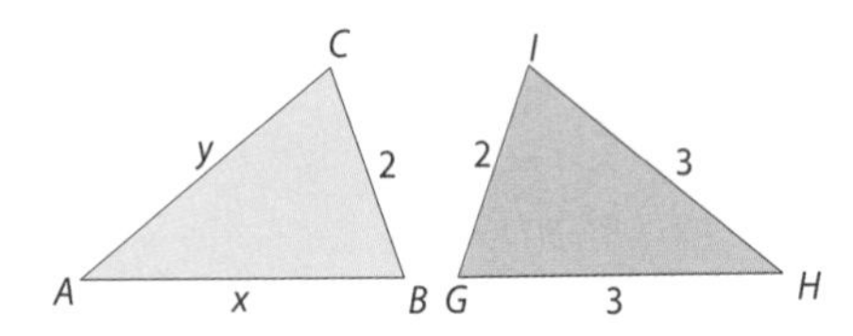


**39.**

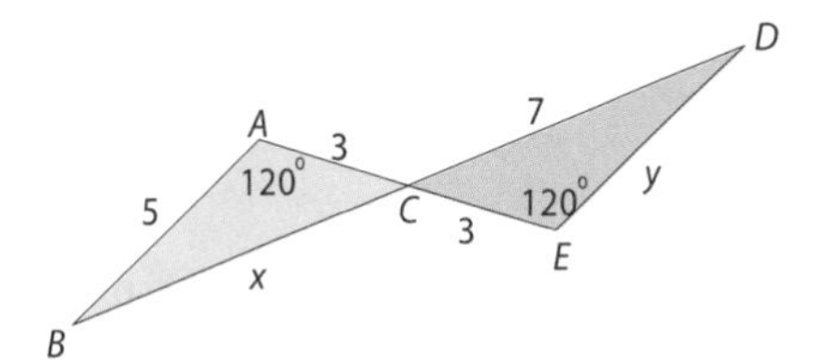


**40.**

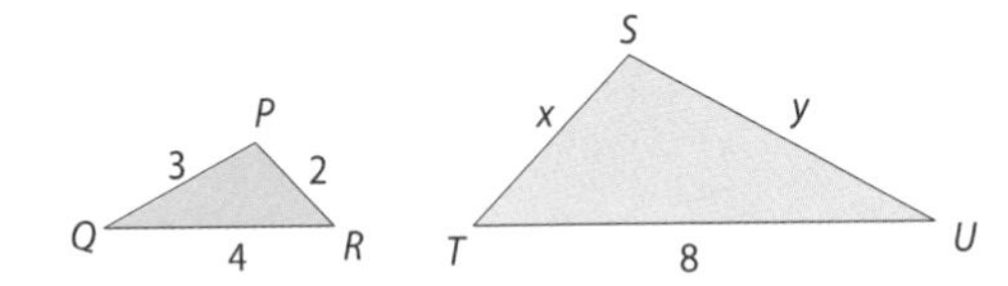


**41.**

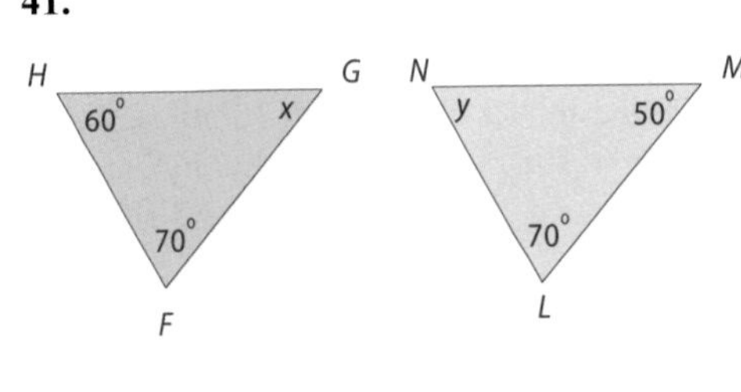


**42.**

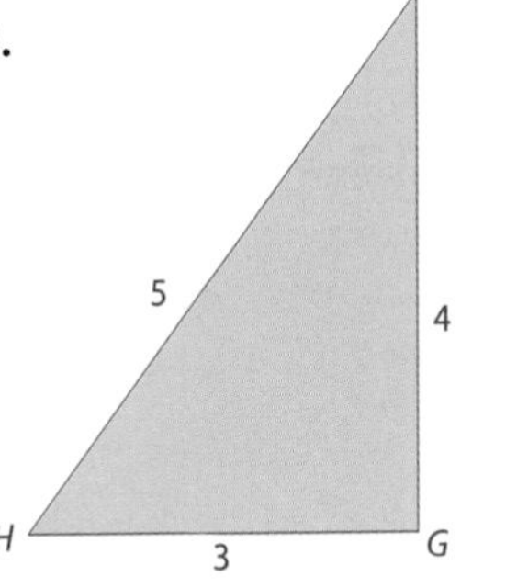


**43.**

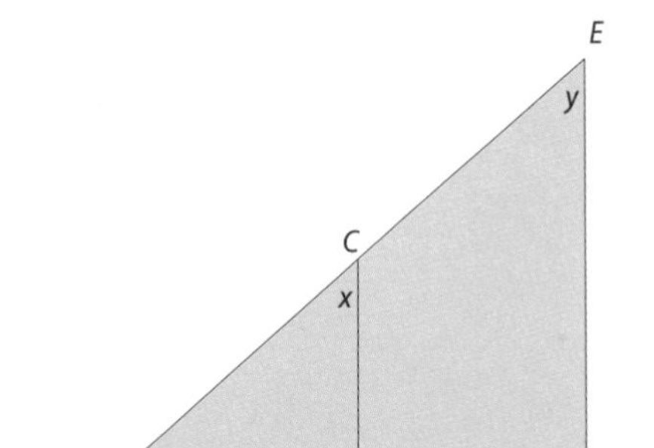


**44.**

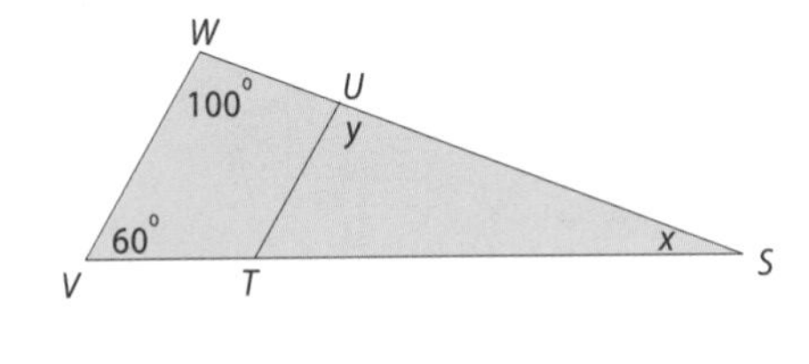

Name ______________________ Section ____________ Date __________

**ANSWERS**

45. 400 ft.

46. 320 miles

47. a. 55 lb.

b. 6 bags

c. $72

48. [Respond below exercise.]

**45.** A building casts a shadow 100 feet long and, at the same time of day, a man 6 feet tall casts a shadow $1\frac{1}{2}$ feet long. What is the height of the building?

**46.** On a road map, two cities are in a straight line 5 inches apart. You know that in fact these cities are 200 miles apart. On the same map, two other cities are in a straight line 8 inches apart. How many miles apart are these two cities?

**47.** One bag of Fertilizer & Weed Killer contains 10 pounds of fertilizer and weed treatment with a recommended coverage of 1200 square feet.

**a.** Your brother measured his lawn and asked you how many pounds of fertilizer he should buy. His lawn consists of two rectangular shapes, one 30 feet by 100 feet and the other 40 feet by 90 feet. How many pounds should he buy?

**b.** How many bags of fertilizer would you tell him to buy?

**c.** If each bag costs $12, how much would he have to pay?

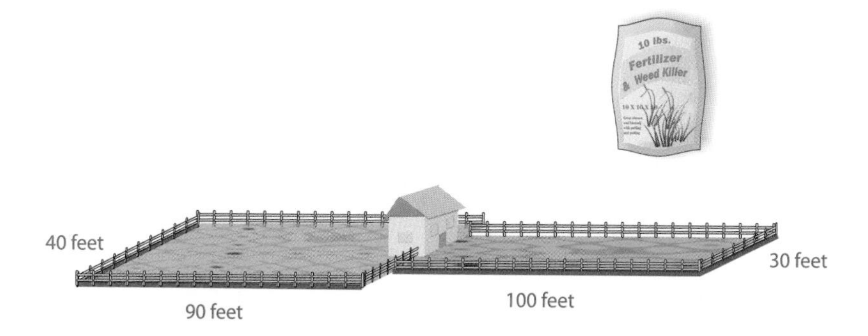


**48.** A computer manufacturer is told to expect 3 defective microchips out of every 2000 produced by a particular machine. A second, older and slower machine produces 1 defective chip in 1500 in the same amount of time. How many defective microchips should be expected from each machine in a production run of 150,000 chips?

225 defective microchips for the first computer; 100 defective microchips for the second computer.

[Respond below exercise.]
49. ______

[Respond below exercise.]
50. ______

## Writing and Thinking About Mathematics

**49. a.** Explain why the following statement is misleading:
The ratio of 4 quarters to 5 dollars is 4 : 5.

**b.** Rewrite the statement so that it is not misleading.

a. The statement is misleading because the numbers 4 and 5 are not in the same units.
b. The ratio of 4 quarters to 5 dollars is 1 : 5.

**50.** Many types of geometric figures are similar. For example, all circles are similar. For figures with line segments as sides (called polygons), corresponding sides are proportional and corresponding angles are equal (just as with triangles). Draw two figures similar to the figure shown here and explain why they are similar to this figure and to each other.

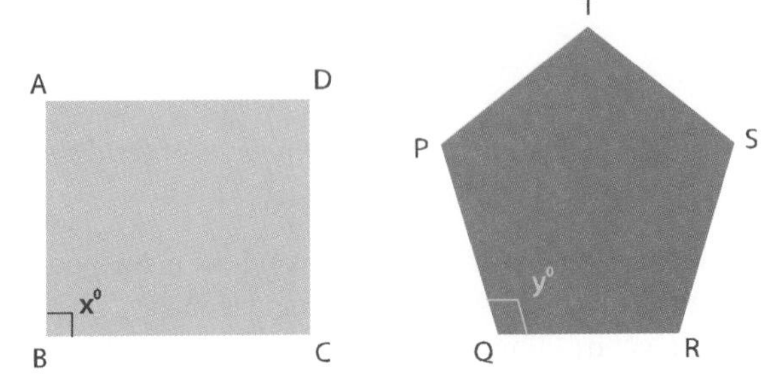


Figures drawn will vary. They are similiar to the figures shown provided their corresponding angles are equal and their corresponding sides are proportional.

## Chapter 4 Index of Key Ideas and Terms

To Change an Improper Fraction to a Mixed Number — page 287

1. Divide the numerator by the denominator to find the whole number part of the mixed number.
2. Write the remainder over the denominator as the fraction part of the mixed number.

To Multiply Mixed Numbers — page 295

1. Change each number to fraction form.
2. Multiply by factoring numerators and denominators, and then reduce.
3. Change the answer to a mixed number or leave it in fraction form.

To Divide Mixed Numbers — page 300

1. Change each number to improper fraction form.
2. Multiply by the reciprocal of the divisor.
3. Reduce if possible.

To Add Mixed Numbers — page 311

1. Add the fraction parts.
2. Add the whole numbers.
3. Write the mixed number so that the fraction part is less than 1.

To Subtract Mixed Numbers — page 313

1. Subtract the fraction parts.
2. Subtract the whole numbers.

If the Fraction Part Being Subtracted Is Larger Than the First Fraction: — page 314

1. "Borrow" the whole number 1 from the first whole number.
2. Add this 1 to the first fraction.
3. Now subtract.

For Complex Fractions

To Simplify a Complex Fraction — page 325

1. Simplify the numerator so that it is a single fraction, possibly an improper fraction.
2. Simplify the denominator so that it also is a single fraction, possibly an improper fraction.
3. Divide the numerator by the denominator, and reduce if possible.

Order of Operations page 326

Rules for Order of Operations

1. First, simplify within grouping symbols, such as parentheses ( ), brackets [ ], or braces { }. Start with the innermost grouping.
2. Second, find any powers indicated by exponents.
3. Third, moving from left to right, perform any multiplications or divisions in the order in which they appear.
4. Fourth, moving from left to right, perform any additions or subtractions in the order in which they appear.

Ratio page 333

A ratio is a comparison of two quantities by division. The ratio of $a$ to $b$ can be written as

$$\frac{a}{b} \text{ or } a : b \text{ or } a \text{ to } b$$

For Proportions

A proportion is a statement that two ratios are equal. page 335

In symbols, $\frac{a}{b} = \frac{c}{d}$ is a proportion.

A proportion has four terms. The first and fourth terms ($a$ and $d$) are called the extremes. page 335

The second and third terms ($b$ and $c$) are called the means.

In a true proportion, the product of the extremes is equal to the product of the means. In symbols,

$\frac{a}{b} = \frac{c}{d}$ if and only in $a \cdot d = b \cdot c$.

Two Triangles are Similar if: page 341

1. The corresponding angles are equal.
2. The corresponding sides are proportional.

Name ______________ Section ______________ Date ______________

## Chapter 4 Test

1. Describe, in your own words, the meaning of the term **proportion**.

   A proportion is a statement that two ratios are equal.

2. Find the reciprocal of $2\frac{1}{3}$.

Change each improper fraction to a mixed number with the fraction part reduced to lowest terms.

3. $\frac{54}{8}$

4. $\frac{342}{100}$

Change each mixed number to the form of an improper fraction.

5. $-5\frac{1}{100}$

6. $2\frac{12}{13}$

7. Find $\frac{3}{10}$ of $7\frac{2}{9}$.

8. Find the average of $3\frac{1}{3}$, $4\frac{1}{4}$ and $5\frac{1}{5}$.

Perform the indicated operations. Reduce all fractions to lowest terms.

9. $\frac{11}{20}+\frac{5}{20}$

10. $\frac{2}{9}-\frac{7}{12}$

**ANSWERS**

1. [Respond below exercise.]
2. $\frac{3}{7}$
3. $6\frac{3}{4}$
4. $3\frac{21}{50}$
5. $-\frac{501}{100}$
6. $\frac{38}{13}$
7. $2\frac{1}{6}$
8. $4\frac{47}{180}$
9. $\frac{4}{5}$
10. $-\frac{13}{36}$

11. $-\frac{11}{30}$

12. $8\frac{5}{12}$

13. $-\frac{8}{11}$

14. $-\frac{1}{6}$

15. $7\frac{11}{42}$

16. $14\frac{19}{30}$

17. $-16$

18. $11\frac{3}{5}$

19. $\frac{x-1}{9}$

20. $-\frac{1}{3}$

21. $A = 750$

22. $x = \frac{1}{96}$

**11.** $\left(\frac{1}{5}-\frac{3}{4}\right)\div 1\frac{1}{2}$

**12.** $4\frac{2}{3}-\left(-3\frac{3}{4}\right)$

**13.** $\left(-\frac{5}{6}+\frac{1}{2}\right)\div\left[\frac{3}{8}-\left(-\frac{1}{12}\right)\right]$

**14.** $\left(\frac{3}{4}-\frac{7}{8}\right)\div\left[\frac{1}{2}-\left(-\frac{1}{4}\right)\right]$

**15.**
$$\begin{array}{r} 6\frac{5}{14} \\ +\frac{19}{21} \\ \hline \end{array}$$

**16.**
$$\begin{array}{r} 23\frac{1}{10} \\ -8\frac{7}{15} \\ \hline \end{array}$$

**17.**
$$\begin{array}{r} -3\frac{5}{6} \\ -4\frac{1}{2} \\ -7\frac{2}{3} \\ \hline \end{array}$$

**18.** $2\frac{2}{7}\cdot 5\frac{3}{5}+\frac{3}{5}\div\left(-\frac{1}{2}\right)$

**19.** $\frac{x}{3}\cdot\frac{1}{3}-\frac{2}{9}\div 2$

**20.** Simplify: $\dfrac{1\frac{1}{9}+\frac{5}{18}}{-1\frac{1}{2}-2\frac{2}{3}}$

Solve the following proportions. Reduce all fractions.

**21.** $\frac{50}{A}=\frac{5}{75}$

**22.** $\dfrac{x}{\frac{3}{8}}=\dfrac{\frac{1}{4}}{9}$

Name ____________________ Section ____________ Date __________

**ANSWERS**

**23.** An artist is going to make a rectangular-shaped pencil drawing for a customer. The drawing is to be $3\frac{1}{4}$ inches wide and $4\frac{1}{2}$ inches long, and there is to be a mat around the drawing that is $1\frac{1}{2}$ inches wide.

**a.** Find the perimeter of the matted drawing.

**b.** Find the area of the matted drawing. (Sometimes an artist will charge for the size of a work as much as for the actual work itself.)

**24.** The result of multiplying two numbers is $14\frac{5}{8}$. One of the numbers is $\frac{3}{4}$.

**a.** Do you expect the other number to be smaller or larger than $14\frac{5}{8}$?

**b.** What is the other number?

**25.** Find the volume of a tissue box that measures $10\frac{9}{10}$ centimeters by $11\frac{1}{10}$ centimeters by $13\frac{5}{10}$ centimeters.

**26.** You are in charge of setting a machine that produces hairpins. The machine is to run 24 hours per day, 5 days per week.

**a.** What setting of hairpins per hour should you set the machine to if it should produce 120,000 hairpins per day?

**b.** How many hairpins will the machine produce per 5-day week?

**23. a.** $27\frac{1}{2}$ in.

**b.** $46\frac{7}{8}$ in.$^2$

**24. a.** larger

**b.** $19\frac{1}{2}$

**25.** $1633\frac{73}{200}$ cm$^3$

**26. a.** 5000 per hour

**b.** 600,000 per week

27. $\$7\frac{1}{2}$

28. $x = 3\frac{3}{4},\ y = 2\frac{1}{2}$

**27.** A manufacturing company expects to make a profit of \$3 on a product that it sells for \$8. How much profit does the company expect to make on a product that it sells for \$20 ?

**28.** The triangles shown in the figure below are similar triangles. Find the values of $x$ and $y$.

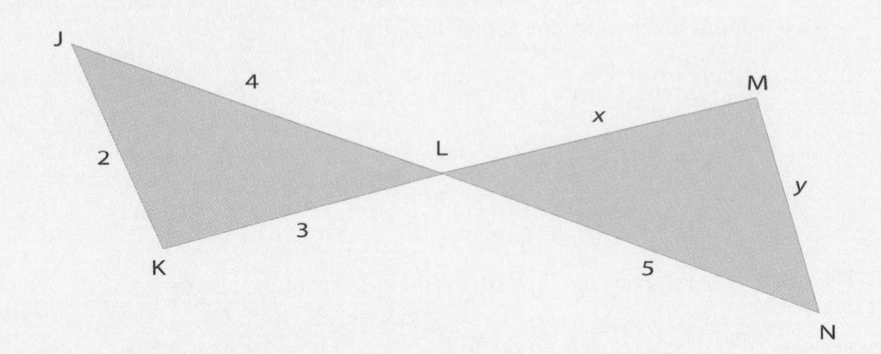

Name ______ Section ______ Date ______

ANSWERS

# Cumulative Review: Chapters 1– 4

1. Given the expression $5^3$:

   a. name the base,

   b. name the exponent, and

   c. find the value of the expression.

2. Round off 176,200 to the nearest ten thousand.

3. Match each expression with the best estimate of its value.

| | | |
|---|---|---|
| B | **a.** $175 + 92 + 96 + 125$ | A. 200 |
| A | **b.** $8465 \div 41$ | B. 500 |
| D | **c.** $32 \cdot 48$ | C. 1000 |
| C | **d.** $5484 - 4380$ | D. 1500 |

4. Multiply mentally: $70(\ 9000\ ) =$ 630,000.

5. Evaluate by using the rules for order of operations:

   $5^2 + (12 \cdot 5 \div 2 \cdot 3\ ) - 60 \cdot 4$

6. **a.** List all the prime numbers less than 25.

   **b.** List all the squares of these prime numbers.

7. Find the LCM of 45, 75, and 105.

8. The value of $0 \div 19$ is 0, whereas the value of $19 \div 0$ is undefined.

**1. a.** 5
**b.** 3
**c.** 125

**2.** 180,000

**3.** [ Respond in exercise.]

**4.** [ Respond in exercise.]

**5.** – 125

**6. a.** 2, 3, 5, 7, 11, 13, 17, 19, 23
**b.** 4, 9, 25, 49, 121, 169, 289, 361, 529

**7.** 1575

**8.** [ Respond in exercise.]

9. $44$

10. $14{,}275$

11. $588$

12. $28{,}538$

13. $204$

14. $-\frac{1}{40}$

15. $\frac{1}{63}$

16. $\frac{1}{24}$

17. $-\frac{1}{4}$

18. $68\frac{23}{60}$

19. $-6\frac{21}{40}$

20. $-\frac{121}{81}$ or $(-1\frac{40}{81})$

21. $\frac{7}{4}$ (or $1\frac{3}{4}$)

22. $\frac{123}{38}$ (or $3\frac{9}{38}$)

23. $-\frac{47}{6}$ (or $-7\frac{5}{6}$)

**9.** Find the average of 45, 36, 54, and 41.

Perform the indicated operations. Reduce all fractions to lowest terms.

**10.** $\begin{array}{r} 8376 \\ 3749 \\ +2150 \\ \hline \end{array}$

**11.** $\begin{array}{r} 1563 \\ -975 \\ \hline \end{array}$

**12.** $\begin{array}{r} 751 \\ \times 38 \\ \hline \end{array}$

**13.** $14\overline{)2856}$

**14.** $-\frac{3}{20} \div 6$

**15.** $\frac{5}{18}\left(\frac{3}{10}\right)\left(\frac{4}{21}\right)$

**16.** $-\frac{5}{6} - \left(-\frac{7}{8}\right)$

**17.** $\frac{7}{20} - \frac{3}{5}$

**18.** $\begin{array}{r} 38\frac{4}{5} \\ +29\frac{7}{12} \\ \hline \end{array}$

**19.** $\begin{array}{r} -13\frac{3}{20} \\ +6\frac{5}{8} \\ \hline \end{array}$

**20.** $3\frac{2}{3} \div \left(-2\frac{5}{11}\right)$

**21.** $-4\frac{3}{8} \div \left(-2\frac{1}{2}\right)$

Simplify each of the following expressions.

**22.** $\dfrac{1\frac{3}{10} + 2\frac{4}{5}}{2\frac{3}{5} - 1\frac{1}{3}}$

**23.** $1\frac{1}{5} - 11\frac{7}{10} + 6\frac{2}{3} \div 2\frac{1}{2}$

Name ______________ Section ______________ Date ______________

24. $\frac{x}{5} \cdot \frac{1}{3} - \frac{2}{5} \div 3$

Solve each of the following equations.

25. $x + 17 = -10$

26. $5y - 14 = 21$

27. $\frac{2}{3}x + \frac{5}{6} = \frac{1}{2}$

28. $\frac{n}{12} - \frac{1}{3} = \frac{3}{4}$

29. The discount price of a new television set is \$652. This price is $\frac{4}{5}$ of the original price.

a. Is the original price more or less than \$652?

b. What was the original price?

30. You know that the gas tank on your pickup truck holds 24 gallons, and the gas gauge reads that the tank is $\frac{3}{8}$ full.

a. How many gallons will be needed to fill the tank?

b. Do you have enough cash to fill the tank if you have \$20 with you and gas costs \$ $1\frac{3}{10}$ per gallon?

31. a. Find the perimeter of a triangle with sides of length $6\frac{5}{8}$ meters, $3\frac{3}{4}$ meters, and $5\frac{7}{10}$ meters.

b. Each meter is $3\frac{28}{100}$ feet long. What is the perimeter of triangle in feet?

32. For a rectangular-shaped swimming pool that is $35\frac{1}{2}$ feet long and $22\frac{3}{4}$ feet wide, find:

a. the perimeter and

b. the area of the pool.

**ANSWERS**

24. $\frac{x-2}{15}$

25. $x = -27$

26. $y = 7$

27. $x = -\frac{1}{2}$

28. $n = 13$

29. a. more

b. \$815

30. a. 15 gal.

b. Yes

31. a. $16\frac{3}{40}$ m

b. $52\frac{363}{500}$ ft.

32. a. $116\frac{1}{2}$ ft.

b. $807\frac{5}{8}$ ft.$^2$

33. $3\frac{1}{5}$ in.

34. a. $x = 3$

b. $y = 6\frac{3}{4}$

35. [ Respond below exercise ]

36. [ Respond below exercise ]

**33.** An architect plans to make a drawing of a house that uses the scale of 2 inches to 15 feet. If two points in the house are known to be 24 feet apart, how many inches apart on the drawing should these two points be?

**34.** The two triangles shown in the figure below are similar triangles. Find the values of $x$ and $y$.

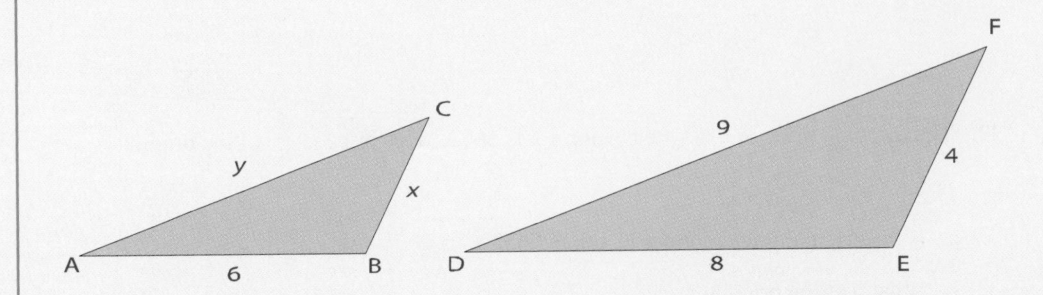


## Writing and Thinking about Mathematics

**35.** Name one topic in the first four chapters that you found particularly interesting, and explain why you think that it is important. (Write only a short paragraph, and be sure to express your thoughts in complete sentences.)

Responses will vary. This question is designed to promote class discussion and to help you gain additional insight as to how your students perceive this course.

**36.** Name one topic discussed in Chapter 4 that you found somewhat difficult. Do you now know why you had difficulty? Explain briefly.

Responses will vary. This question is designed to provide a basis for one-on-one discussions with your students.

# 5

# Decimal Numbers and Square Roots

## WHAT TO EXPECT IN CHAPTER 5

Chapter 5 develops the basic operations with decimal numbers and presents two other interesting and useful topics: scientific notation and roots. (Both are particularly relevant here because of the current emphasis on working with calculators.) Section 5.1 discusses the techniques involved in reading and writing decimal numbers and the use of the word "and" to indicate placement of the decimal point. Sections 5.2 and 5.3 deal with the operations of addition, subtraction, multiplication, and division and with estimating answers. The geometric concepts of perimeter, area, and volume show applications of decimal numbers.

As discussed in Section 5.4, decimals and fractions are closely related and can be used together in evaluating expressions by changing all the numbers in a problem to one form or the other. Scientific notation, which involves positive and negative exponents, is introduced here so that students can read results on their calculators when the calculators are in scientific mode.

The topic of solving equations with decimal numbers is introduced early in the chapter with solutions involving only one step. In Section 5.5, we complete the development of solving equations with decimals by using the techniques developed throughout the text. The formulas for the circumference and area of a circle are included as applications of equations involving decimal numbers because the infinite nonrepeating decimal number $\pi$ is used in these formulas.

The Pythagorean Theorem, discussed in Section 5.6, provides an interesting, and geometric, application of square roots. This idea helps in the mathematical transition from rational numbers to real numbers (rational and irrational numbers). In Section 5.7, we simplify expressions that contain square roots and show how to combine like radicals.

# Chapter 5 Decimal Numbers and Square Roots

**Mathematics at Work!**

As illustrated in the figure below, the shape of a baseball infield is a square 90 feet on each side. ( A square is a four-sided plane figure in which all four sides are the same length and adjoining sides meet at 90° angles. ) Do you think that the distance from home plate to second base is more than 180 feet or less than 180 feet? More than 90 feet or less than 90 feet? The distance from the pitcher's mound to home plate is $60\frac{1}{2}$ feet. Is the pitcher's mound exactly halfway between home plate and second base? Do the two diagonals of the square intersect at the pitcher's mound? What is the distance from home plate to second base ( to the nearest tenth of a foot )? No. No. 127.3 ft.

90 feet

$60\frac{1}{2}$ ft.

## 5.1 Reading, Writing, and Rounding Off Decimal Numbers

### Reading, Writing, and Rounding Off Decimal Numbers

Technically, what we write to represent numbers are symbols or notations called **numerals**. **Numbers** are abstract ideas represented by numerals. There are many notations for numbers. For example, the Romans used V to represent five and the Alexandrian Greek system used capital epsilon, E, to represent five. In the Hindu-Arabic system that we use today, the symbol 5 represents five. We will not emphasize the distinction between numbers and numerals ( or symbols ), but you should remember the fact that numbers are abstract ideas, and we use symbols to represent these ideas so that we can communicate in a meaningful manner.

The common **decimal notation** uses a **place value system** and a **decimal point**, with whole numbers written to the left of the decimal point and fractions written to the right of the decimal point. We will say that numbers represented by decimal notation are **decimal numbers**. The values of several places in this decimal system are shown in Figure 5.1.

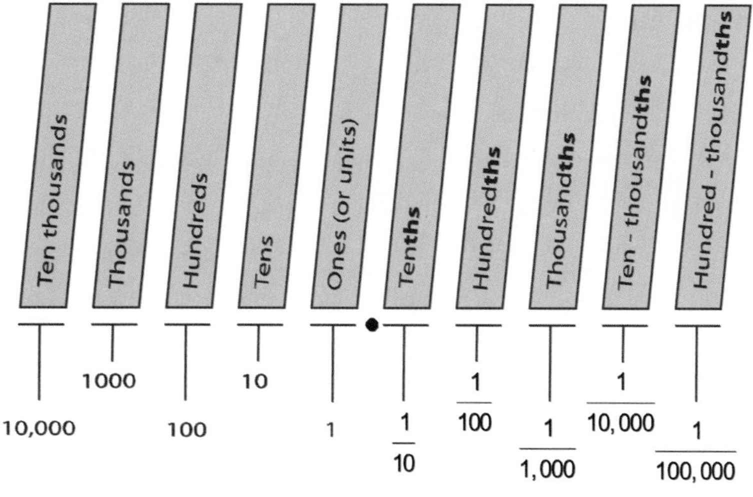


**Figure 5.1**

There are three classifications of decimal numbers:

1. **finite** ( or **terminating** ) decimals,
2. **infinite repeating** decimals, and
3. **infinite nonrepeating** decimals.

In arithmetic, we are used to dealing with finite ( or terminating ) decimals, and these are the numbers we will discuss and operate with in Sections 5.1 through 5.3. In Section 5.4, we will show how fractions are related to infinite repeating decimals and expand these ideas in Section 5.6 to include infinite nonrepeating decimals ( called **irrational numbers** ).

In reading a fraction such as $\frac{147}{1000}$, we read the numerator as a whole number ("one hundred forty seven" ) and then attach the name of the denominator ("thousand**ths**" ). Note that **ths** ( or **th** ) is used to indicate the fraction. This same procedure is followed with numbers written in decimal notation:

$\frac{147}{1000} = 0.147$ Read "one hundred forty seven thousandths."

$2\frac{36}{100} = 2.36$ Read "two and thirty six hundredths."

If there is no whole number part, as in the first example just shown, then 0 is commonly written to the left of the decimal point. Writing the 0 is not always necessary. However, writing the 0 does sometimes avoid confusion with periods at the end of sentences when decimal numbers are written in sentences. In general, decimal numbers are read ( and written ) according to the following convention:

**To Read or Write a Decimal Number:**

1. Read ( or write ) the whole number.
2. Read ( or write ) **and** in place of the decimal point.
3. Read ( or write ) the fraction part as a whole number with the name of the place of the last digit on the right.

## Objectives

① Learn to read and write decimal numbers.

② Realize the importance of the word **and** in reading and writing decimal numbers.

③ Understand that **th** indicates a fraction part of a decimal number.

④ Know how to round off decimal numbers to indicated places of accuracy.

When there is no whole number part, writing 0 for the whole number part is the standard format. This does help in aligning the decimal points for addition and subtraction and can help avoid confusion when decimal numbers are written in word problems. However, we do not read or write the word "zero" with the decimal fraction. Also, in multiplication and division with decimal numbers, the 0 can be a distraction and more confusing than helpful. You night consider treating the 0 as optional for the whole number part.

Even though students understand the use of "and" to indicate the decimal point, whole numbers are so commonly misread in daily life that students have a tendency to forget the importance and proper use of "and." You might ask them to listen (to announcers and advertisements on television) and look for misuses of "and" in numbers and report them in class.

## Example 1

Write $72\frac{8}{10}$ in decimal notation and in words.

**Solution**

72 . 8 ⟵ in decimal notation

Seventy-two and eight tenths ⟵ in words

**And** indicates the decimal point: the digit 8 is in the tenths position.

## Example 2

Write $9\frac{63}{1000}$ in decimal notation and in words.

**Solution**

One 0 must be inserted as a placeholder

9 . 063 ⟵ in decimal notation

nine and sixty-three thousandths ⟵ in words

**And** indicates the decimal point; the digit 3 is in the thousandths position.

**Special Notes:**

1. The **ths** ( or **th** ) at the end of a word indicates a fraction part ( a part to the right of the decimal point ).

   seven hundred = 700
   seven hundred**ths** = 0.07

2. The hyphen ( – ) indicates one word.

   three hundred thousand = 300,000
   three hundred-thousand**ths** = 0.00003

*Now Work Exercises 1- 4 in the Margin.*

Write each decimal number in words.

1. 30.8
   thirty and eight tenths

2. 7.06
   seven and six hundredths

3. 18.562
   eighteen and five hundred sixty-two thousandths

4. 3.0007
   three and seven ten-thousandths

## Example 3

Write fifteen hundredths in decimal notation.

**Solution**

0.15

Note that the digit 5 is in the hundredths position.

### Example 4

Write four hundred and two thousandths in decimal notation.

**Solution**

400.002 — Two 0's are inserted as placeholders. The digit 2 is in the thousandths position.

### Example 5

Write four hundred two thousandths in decimal notation.

**Solution**

0.402

Note carefully how the use of **and** in the phrase in Example 4 gives it a completely different meaning from the phrase in this example.

*Now Work Exercises 5- 8 in the Margin.*

## Rounding Off Decimal Numbers

Measuring devices such as rulers, meter sticks, speedometers, micrometers, and surveying transits, which are made by humans, give only approximate measurements. Whether the units are large ( such as miles and kilometers ) or small ( such as inches and centimeters ), there are always smaller, more accurate units ( such as eighths of an inch and millimeters ) that could be used to indicate a measurement. We are constantly dealing with approximate ( or rounded off ) numbers in our daily lives. If a recipe calls for 1.5 cups of flour and the cook puts in 1.53 cups ( or 1.47 cups ), the result will still be reasonably tasty. In fact, the measures of all ingredients will have been approximations.

There are several rules for rounding off decimal numbers. The IRS, for example, allows rounding off to the nearest dollar on income tax forms. A technique sometimes used in cases when many numbers are involved is to round to the nearest even digit at some particular place value. The rule chosen in a particular situation depends on the use of the numbers and whether there might be some sort of penalty for an error. In this text, except in the case of dollars and cents, when we will round up to the next higher cent ( a common practice in business ), we will use the following rule for rounding off:

**Rule for Rounding Off Decimal Numbers:**

1. Look at the single digit just to the right of the place of desired accuracy.
2. If this digit is 5 or greater, make the digit in the desired place of accuracy one larger and replace all digits to the right with zeros. All digits to the left remain unchanged unless a 9 is made one larger, and then the next digit to the left is increased by 1.
3. If this digit is less than 5, leave the digit in the desired place of accuracy as it is and replace all digits to the right with zeros. All digits to the left remain unchanged.
4. Trailing 0's to the right of the decimal point must be dropped so that the place of accuracy is clearly understood. If a rounded-off number has a 0 in the desired place of accuracy, then that 0 remains.

Write each decimal number in words.

5. ten and four thousandths
   10.004
6. six hundred and five tenths
   600.5
7. seven hundred and seven thousandths
   700.007
8. seven hundred seven thousandths
   0.707

## Example 6

Round off 18.649 to the nearest tenth.

### Solution

**a.** The next digit to the right is 4.

↓

18.649

↑

6 is in the tenths position.

**b.** Since 4 is less than 5, leave the 6 and replace 4 and 9 with 0's.

**c.** 18.649 rounds off to **18.6** to the nearest tenth. Note that the trailing 0's in 18.600 are dropped to indicate the position of accuracy.

## Example 7

Round off 5.83971 to the nearest thousandth.

### Solution

**a.** The next digit to the right is 7.

↓

5.83971

↑

9 is in the thousandths position.

**b.** Since 7 is greater than 5, make 9 one larger and replace 7 and 1 with 0's. ( Making 9 one larger gives 10, which affects the digit 3, too. )

**c.** 5.83971 rounds off to 5.84000, or **5.840** to the nearest thousandth, and only two trailing 0's are dropped.

## *Completion Example 8*

Round off 2.00643 to the nearest ten-thousandth.

### Solution

**a.** The digit in the ten-thousandths position is _____.

**b.** The next digit to the right is _____.

**c.** Since _____ is less than 5, leave _____ as it is and replace _____ with a 0.

**d.** 2.00643 rounds off to _____ to the nearest _____.

### Completion Example 9

Round off 9653 to the nearest hundred.

**Solution**

**a.** The decimal point is understood to be to the right of _____.

**b.** The digit in the hundreds position is _____.

**c.** The next digit to the right is _____.

**d.** Since _____ is equal to 5, change the _____ to _____ and replace _____ and _____ with 0's.

**e.** So, 9653 rounds off to _____ ( to the nearest hundred ).

**Important Note:**

The 0's must **not** be dropped in a whole number. Every 0 to the right of the desired place of accuracy to the right of the decimal point **must** be dropped.

*Now Work Exercises 9-12 in the Margin.*

### Completion Example Answers

**8. a.** The digit in the ten-thousandths position is **4**.

**b.** The next digit to the right is **3**.

**c.** Since **3** is less than 5, leave **4** as it is and replace **3** with a 0.

**d.** 2.00643 rounds off to **2.0064** to the nearest **ten-thousandth**.

**9. a.** The decimal point is understood to be to the right of **3**.

**b.** The digit in the hundreds position is **6**.

**c.** The next digit to the right is **5**.

**d.** Since **5** is equal to 5, change the **6** to **7** and replace **5** and **3** with 0's.

**e.** So 9653 rounds off to **9700** ( to the nearest hundred ).

Round off each number to the place indicated.

**9.** 8.637 ( nearest tenth )
8.6

**10.** 5.042 ( nearest hundredth )
5.04

**11.** 0.01792 ( nearest thousandth )
0.018

**12.** 239.53 ( nearest ten )
240

Name ______________ Section ______________ Date ______________

## Exercises 5.1

Write the following mixed numbers in decimal notation.

**1.** $6\frac{5}{10}$ **2.** $82\frac{3}{100}$ **3.** $19\frac{75}{1000}$

**4.** $100\frac{25}{100}$ **5.** $62\frac{547}{1000}$

Write the following decimal numbers in mixed number form. Do not reduce the fractional part.

**6.** 2.57 **7.** 13.02 **8.** 38.004

**9.** 200.6 **10.** 50.001

Write the following numbers in decimal notation.

**11.** four tenths **12.** fifteen thousandths

**13.** twenty-three hundredths **14.** five and twenty-eight hundredths

**15.** five and twenty-eight thousandths

**16.** seventy-three and three hundred forty-one thousandths

**ANSWERS**

1. 6.5
2. 82.03
3. 19.075
4. 100.25
5. 62.547
6. $2\frac{57}{100}$
7. $13\frac{2}{100}$
8. $38\frac{4}{1000}$
9. $200\frac{6}{10}$
10. $50\frac{1}{1000}$
11. 0.4
12. 0.015
13. 0.23
14. 5.28
15. 5.028
16. 73.341

17. 600.66

18. 600.066

19. 3495.342

20. 7500.0083

21. nine tenths

22. fifty-three hundredths

23. six and five hundredths

24. six and four thousandths

25. fifty and seven thousandths

26. nineteen and one hundred two thousandths

27. eight hundred and nine thousandths

28. eight hundred nine thousandths

29. five thousand and five thousandths

30. twenty-five and four thousand five hundred thirty-eight ten-thousandths

31. [ Respond in exercise. ]

**17.** six hundred and sixty-six hundredths

**18.** six hundred and sixty-six thousandths

**19.** three thousand four hundred ninety-five and three hundred forty two thousandths

**20.** seven thousand five hundred and eighty-three ten-thousandths

Write the following decimal numbers in words.

**21.** 0.9

**22.** 0.53

**23.** 6.05

**24.** 6.004

**25.** 50.007

**26.** 19.102

**27.** 800.009

**28.** 0.809

**29.** 5000.005

**30.** 25.4538

Fill in the blanks to correctly complete each statement.

**31.** Round off 34.78 to the nearest tenth.
- **a.** The digit in the tenths position is __7__.
- **b.** The next digit to the right is __8__.
- **c.** Since __8__ is greater than 5, change __7__ to __8__ and replace __8__ with 0.
- **d.** So 34.78 rounds off to __34.8__ to the nearest tenth.

Name ____________ Section ____________ Date ____________

**32.** Round off 3.00652 to the nearest ten-thousandth.

**a.** The digit in the ten-thousandths position is __5__.

**b.** The next digit to the right is __2__.

**c.** Since __2__ is less than 5, leave __5__ as it is and replace __2__ with 0.

**d.** So, 3.00652 rounds off to __3.0065__ to the nearest __ten-thoudsandths__.

Round off each of the following decimal numbers as indicated.

To the nearest tenth:

**33.** 89.016 **34.** 8.555 **35.** 18.123

**36.** 0.076 **37.** 14.332 **38.** 46.444

To the nearest hundredth:

**39.** 0.385 **40.** 0.296 **41.** 7.997

**42.** 13.1345 **43.** 0.0764 **44.** 6.0035

To the nearest thousandth:

**45.** 0.0572 **46.** 0.6338 **47.** 0.00191

**48.** 20.76963 **49.** 32.4578 **50.** 1.66666

To the nearest whole number (or nearest unit):

**51.** 479.32 **52.** 7.8 **53.** 163.5

**54.** 701.41 **55.** 300.3 **56.** 29.999

**ANSWERS**

**32.** [ Respond in exercise. ]

**33.** 89.0

**34.** 8.6

**35.** 18.1

**36.** 0.1

**37.** 14.3

**38.** 46.4

**39.** 0.39

**40.** 0.30

**41.** 8.00

**42.** 13.13

**43.** 0.08

**44.** 6.00

**45.** 0.057

**46.** 0.634

**47.** 0.002

**48.** 20.770

**49.** 32.458

**50.** 1.667

**51.** 479

**52.** 8

**53.** 164

**54.** 701

**55.** 300

**56.** 30

57. 5200

58. 6500

59. 76,500

60. 400

61. 500

62. 1600

63. 62,000

64. 75,000

65. 103,000

66. 4,501,000

67. 7,305,000

68. 573,000

69. 0.00076

70. 80,000

71. [ Respond below exercise. ]

72. [ Respond below exercise. ]

To the nearest hundred:

**57.** 5163 **58.** 6475 **59.** 76,523.2

**60.** 435.7 **61.** 453.7 **62.** 1572.36

To the nearest thousand:

**63.** 62,375 **64.** 75,445 **65.** 103,499

**66.** 4,500,766 **67.** 7,305,438 **68.** 573,333.15

**69.** 0.0007582 (nearest hundred thousandth)

**70.** 78,419 (nearest ten thousand)

In each of the following exercises, write the decimal numbers that are not whole numbers in words.

**71.** One yard is equal to 36 inches. One yard is also approximately equal to 0.914 meter. One meter is approximately equal to 1.09 yards. One meter is also approximately equal to 39.37 inches. (Thus, a meter is longer than a yard by about 3.37 inches.)

Nine hundred fourteen thousandths; one and nine hundredths; thirty-nine and thirty-seven hundredths; three and thirty-seven hundredths

**72.** One foot is equal to 12 inches. One foot is also equal to 30.48 centimeters. One square foot is approximately 0.093 square meter.

Thirty and forty-eight hundredths; ninety-three thousandths

Name ____________________ Section ____________ Date __________

**ANSWERS**

73. [ Respond below exercise. ] ________

74. [ Respond below exercise. ] ________

75. [ Respond below exercise. ] ________

76. [ Respond below exercise. ] ________

77. [ Respond below exercise. ] ________

**73.** One quart of water weighs approximately 2.0825 pounds.

Two and eight hundred twenty-five ten thousandths

**74.** The largest state in the United States is Alaska, which covers approximately 656.4 thousand square miles. The second largest state is Texas, with approximately 268.6 thousand miles. Alaska is more than 10 times the size of Wisconsin ( twenty-third in size ), with about 65.5 thousand square miles.

Six hundred fifty-six and four tenths; two hundred sixty-eight and six tenths; sixty-five and five tenths

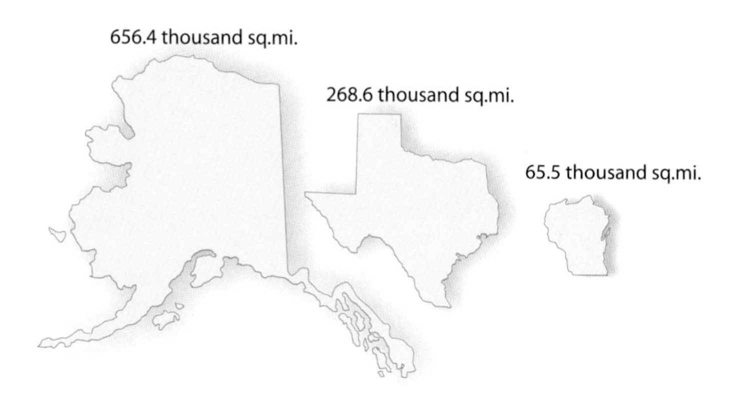


**75.** The number $\pi$ is approximately equal to 3.14159. ( See the discussion of Table 4 in the Appendix for more about $\pi$. )

Three and fourteen thousand, one hundred fifty-nine hundred thousandths

**76.** The number $e$ (used in higher-level mathematics) is approximately equal to 2.71828.

Two and seventy-one thousand, eight hundred twenty-eight hundred thousandths

**77.** An interesting fact about aging is that the longer you live, the longer you can expect to live. A white male of age 40 can expect to live 35.8 more years; of age 50, can expect to live 26.9 more years; of age 60 can expect to live 18.9 more years; of age 70 can expect to live 12.3 more years; and of age 80 can expect to live 7.2 more years. ( This same phenomenon is true of men and women of all races. )

Thirty-five and eight tenths; twenty-six and nine tenths; eighteen and nine tenths; twelve and three tenths; seven and two tenths

[ Respond below
78. exercise. ]

[ Respond below
79. exercise. ]

[ Respond below
80. exercise. ]

**The Recycle Bin**

1. a. 11,000

b. 10,891

2. a. 1900

b. 1864

3. a. 770,000

b. 784,791

4. a. 0

b. 149

5. a. 6000

b. 6488

6. a. 75,000

b. 74,784

**78.** The mean ( average ) distance from the Sun to Earth is about 92.9 million miles and from the Sun to Venus is about 67.24 million miles. One period of revolution of the Earth about the Sun takes 365.2 days, and one period of revolution of Venus about the Sun takes 224.7 days.
Ninety-two and nine tenths; sixty-seven and twenty-four hundredths; three hundred sixty-five and two tenths; two hundred twenty-four and seven tenths

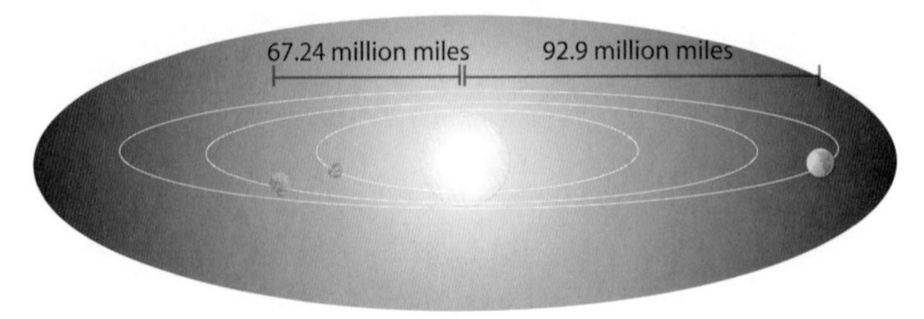


**79.** The tallest unicycle ever ridden, by Steve McPeak for 376 feet in 1980 in Las Vegas, was 101.75 feet tall.
One hundred one and seventy-five hundredths

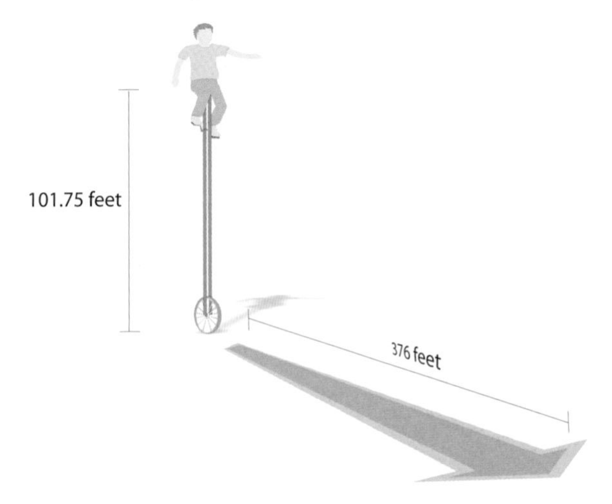


**80.** World records: 9.79 seconds for 100 meters ( by Maurice Greene, USA, 1999 ); 19.32 seconds for 200 meters ( by Michael Johnson, USA, 1996 ); 43.18 seconds for 400 meters ( by Michael Johnson, USA, 1999 ).
Nine and seventy-nine hundredths; nineteen and thirty-two hundredths; forty-three and eighteen hundredths

**The Recycle Bin ( from Section 1.2)**

First **a.** estimate each sum, and then **b.** find each sum.

**1.** 3793 + 7098

**2.** 985 + 675 + 204

**3.** 228,480 + 483,502 + 72,809

First **a.** estimate each difference, and then **b.** find each difference.

**4.** 1758 − 1609

**5.** 9355 − 2867

**6.** 80,007 − 5,223

# 5.2 Addition and Subtraction with Decimal Numbers

## Addition with Decimal Numbers

To find a sum such as 2.357 + 6.14, we can write each number in expanded notation. Then, add the whole numbers and fractions with common denominators in the following fashion:

$$\begin{aligned}2.357 + 6.14 &= 2 + \frac{3}{10} + \frac{5}{100} + \frac{7}{1000} + 6 + \frac{1}{10} + \frac{4}{100} \\ &= (2+6) + \left(\frac{3}{10} + \frac{1}{10}\right) + \left(\frac{5}{100} + \frac{4}{100}\right) + \frac{7}{1000} \\ &= 8 + \frac{4}{10} + \frac{9}{100} + \frac{7}{1000} \\ &= 8 + \frac{497}{1000} \\ &= 8.497\end{aligned}$$

**Objectives**

1. Be able to add with decimal numbers.
2. Be able to subtract with decimal numbers.
3. Know how to add and subtract with positive and negative decimal numbers.
4. Know how to estimate sums and differences with rounded-off decimal numbers.

Addition of decimal numbers can be accomplished in a much easier way by writing the decimal numbers one under the other and keeping the decimal points aligned vertically. In this way, the whole numbers will be added to whole numbers, tenths added to tenths, hundredths to hundredths, and so on ( as we did with the fractions forms in the example just given ). The decimal point in the sum is in line with the decimal points in the addends. Thus,

Decimal points are aligned vertically.

```
  2.357
+ 6.140
  8.497
```

As in the number 6.140, 0's may be written to the right of the last digit in the fraction part to help keep the digits in the correct line. This will not change the value of any number or the sum.

**To Add Decimal Numbers:**

1. Write the addends in a vertical column.
2. Keep the decimal points aligned vertically.
3. Keep digits with the same position value aligned. (Zeros may be filled in as aids.)
4. Add the numbers, keeping the decimal point in the sum aligned with the other decimal points.

## Example 1

Find the sum: 17 + 4.88 + 50.033 + 0.6

### Solution

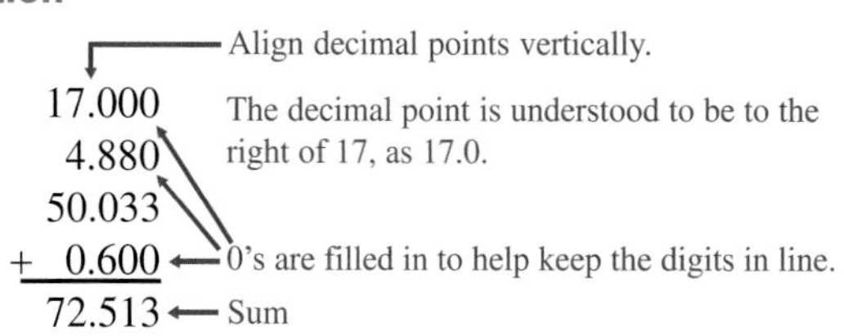
Align decimal points vertically.
17.000
4.880
50.033
+ 0.600
72.513
The decimal point is understood to be to the right of 17, as 17.0.
0's are filled in to help keep the digits in line.
Sum

**1.** Find the sum.

$$\begin{array}{r} 17 \\ 8.61 \\ 5.004 \\ +\ 29.19 \\ \hline 59.804 \end{array}$$

**2.** Combine like terms.

$45.2x + 2.08x + 3.5x$

$50.78x$

## Example 2

Simplify by combining like terms: $8.3x + 9.42x + 25.07x$.

### Solution

The method for combining like terms is to use the distributive property with decimal coefficients just as with integer coefficients.

$$\begin{aligned} 8.3x + 9.42x + 25.07x &= (8.3 + 9.42 + 25.07)x \\ &= 42.79x \end{aligned}$$

$$\begin{array}{r} 8.30 \\ 9.42 \\ +\ 25.07 \\ \hline 42.79 \end{array}$$

*Now Work Exercises 1 and 2 in the Margin.*

## Subtraction with Decimal Numbers

**To Subtract Decimal Numbers:**

1. Write the numbers in a vertical column.
2. Keep the decimal points aligned vertically.
3. Keep digits with the same position value aligned. ( Zeros may be filled in as aids. )
4. Subtract, keeping the decimal point in the difference aligned with the other decimal points.

## Example 3

Find the difference: 21.715 – 14.823

### Solution

$$\begin{array}{r} 21.715 \\ -\ 14.823 \\ \hline 6.892 \end{array} \quad \text{difference}$$

## Example 4

At the bookstore, Mrs. Gonzalez bought a text for \$35, art supplies for \$22.50, and computer supplies for \$19.25. If tax totaled \$6.14, how much change did she receive from a \$100 bill?

### Solution

**a.** Find the total of her expenses including tax.

$$\begin{array}{r} \$35.00 \\ 22.50 \\ 19.25 \\ +\ 6.14 \\ \hline \$82.89 \end{array} \quad \text{total}$$

**b.** Subtract the answer in part **a.** from $100.

$$\begin{array}{r} \$\ 100.00 \\ -\ \ 82.89 \\ \hline \$\ \ 17.11 \end{array}$$

She received $17.11 in change.

## Positive and Negative Decimal Numbers

Decimal numbers can be positive and negative, just as integers and other rational numbers can be positive and negative. Some positive and negative decimal numbers are illustrated on the number line in Figure 5.2.

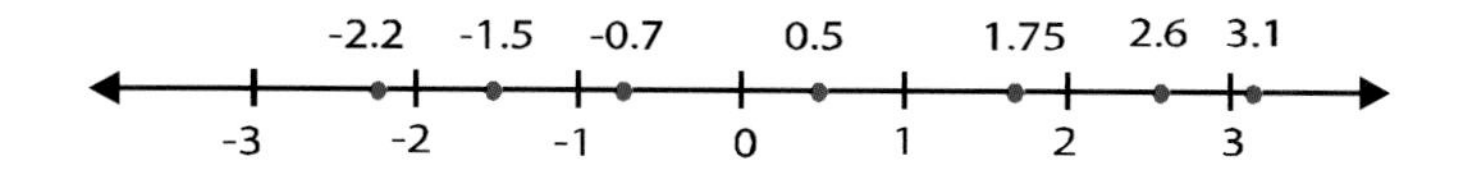


**Figure 5.2**

The rules for operating with positive and negative decimal numbers are the same as those for operating with integers ( Chapter 2 ) and fractions and mixed numbers ( Chapters 3 and 4 ). Examples 5 and 6 illustrate these ideas.

### Example 5

Find the difference: $-200.36 - (-45.87)$

**Solution**

$-200.36 - (-45.87) = -200.36 + (+45.87)$

$$\begin{array}{r} -200.36 \\ +\ 45.87 \\ \hline -154.49 \end{array}$$

We find the difference between the absolute values and use the negative sign.

### Example 6

Solve the equation: $x - 25.67 = 6.25 - 7.3$

**Solution**

| | |
|---|---|
| $x - 25.67 = 6.25 - 7.3$ | Write the equation. |
| $x - 25.67 = -1.05$ | Simplify. |
| $x - 25.67 + 25.67 = -1.05 + 25.67$ | Add 25.67 to both sides. |
| $x = 24.62$ | Simplify. |

*Now Work Exercises 3 - 6 in the Margin.*

Perform the indicated operations.

**3.** $5.63 + 16.8 + (-35.47)$
$-13.04$

**4.** $342.1 - 500$
$-157.9$

**5.** Find the difference.

$-310.5 - (-275.32)$

$-35.18$

**6.** Solve the equation.

$x + 37.6 = 4.8 - 7.25$

$x = -40.05$

## Estimating Sums and Differences

By rounding off each number to the place of the last nonzero digit on the left and adding (or subtracting) these rounded–off numbers, we can estimate (or approximate) the answer before the actual calculations are done. (See Section 1.2 for estimating answers with whole numbers.) This technique of estimating is especially helpful when working with decimal numbers, where the incorrect placement of a decimal point can change an answer dramatically.

### Example 7

First estimate the sum 84 + 3.53 + 62.71; then find the sum.

**Solution**

**a.** Estimate by adding rounded–off numbers.

| **Actual value** | **Rounded–off value** (Each number is rounded off to the place of the leftmost nonzero digit in that number.) | |
|---|---|---|
| 84.00 | 80 | |
| 3.53 | 4 | |
| 62.7 | + 60 | |
| | 144 | estimate |

**b.** Find the actual sum.

$$\begin{array}{r} 84.00 \\ 3.53 \\ +\ 62.71 \\ \hline 150.24 \end{array}$$

You might remind the students that any form of estimation involves a basic understanding of numbers and some judgment on their part. Regardless of the method used for estimating, there will be cases when the estimate is not particularly acute.

In Example 7, the estimated sum and the actual sum are reasonably close ( the difference is about 6 ). This leads us to have confidence in the answer. Although there may be some other error, at least the decimal point seems to be in the right place.

If, for example, 84 had been written as 0.84 ( instead of 84.00 ), then addition would have given a sum as follows:

decimal in the wrong place

$$\begin{array}{r} 0.84 \\ 3.53 \\ +\ 62.71 \\ \hline 67.08 \end{array}$$

Here, the difference between the estimate of 144 and the wrong sum of 67.08 is over 70 ( a relatively large amount ), and an error should be suspected.

### *Completion Example 8*

Estimate the difference 132.418 – 17.526; then find the difference.

**Solution**

**a.** Estimate:

$$\begin{array}{r} 100 \\ -\ 20 \\ \hline \end{array}$$

__________ estimate

**b.** Actual difference:

$$\begin{array}{r} 132.418 \\ -\ 17.526 \\ \hline \end{array}$$

__________ actual difference

### *Completion Example 9*

Samantha bought a pair of shoes for \$37.50, a blouse for \$15.60 ( on sale ), and a pair of slacks for \$37.75.

**a.** How much did she spend? ( Tax was included in the prices. )

**b.** What was her change from a \$100 bill?
[Mentally estimate both answers before you actually calculate them.]

**Solution**

**a.**

$$\begin{array}{r} \$37.50 \\ 15.60 \\ +\ 37.75 \\ \hline \end{array}$$

__________ expenses

**b.**

$$\begin{array}{r} \$100.00 \\ -\qquad\quad \\ \hline \end{array}$$

__________ expenses

__________ change

Did you estimate her expenses as \$100 and her change as \$0?

## Completion Example Answers

**8. a.** Estimate:

$$\begin{array}{r} 100 \\ -\ 20 \\ \hline \mathbf{80} \end{array}$$ estimate

**b.** Actual difference:

$$\begin{array}{r} 132.418 \\ -\ 17.526 \\ \hline \mathbf{114.892} \end{array}$$ actual difference

**9. a.**

$$\begin{array}{r} \$37.50 \\ 15.60 \\ +\ 37.75 \\ \hline \mathbf{\$90.85} \end{array}$$ expenses

**b.**

$$\begin{array}{r} \$100.00 \\ \mathbf{-\ 90.85} \\ \hline \mathbf{\$\ 9.15} \end{array}$$

– 90.85 expenses; \$ 9.15 change

Name ______________ Section ______________ Date ______________

## Exercises 5.2

Find each of the indicated sums. Estimate your answers, either mentally or on paper, before performing the actual calculations. Check to see that your sums are close to the estimated values.

**1.** $0.7 + 0.3 + 2.3$

**2.** $6 + 5.1 + 0.8$

**3.** $0.69 + 4.91 + 0.05$

**4.** $3.577 + 16.563 + 25.01$

**5.** $4.0085 + 0.054 + 0.7 + 0.03$

**6.** $43.655 + 9.33 + 12 + 30.1$

**7.**
$$\begin{array}{r} 47.3 \\ 42.08 \\ +\ 28.005 \\ \hline \end{array}$$

**8.**
$$\begin{array}{r} 1.007 \\ 20.332 \\ +\ 4.992 \\ \hline \end{array}$$

**9.**
$$\begin{array}{r} 107.39 \\ 64.904 \\ 335.056 \\ +\ 210.8 \\ \hline \end{array}$$

**10.**
$$\begin{array}{r} 5.0015 \\ 2.334 \\ 0.3075 \\ +\ 3.8771 \\ \hline \end{array}$$

Find each of the indicated differences. First estimate the difference mentally.

**11.** $5.3 - 3.75$

**12.** $17.82 - 8.9$

**13.** $29.6 - 13.71$

**14.** $1.0054 - 0.03$

**15.** $78.015 - 23.069$

**16.** $45.002 - 43.008$

**17.**
$$\begin{array}{r} 22.426 \\ -\ 17.538 \\ \hline \end{array}$$

**18.**
$$\begin{array}{r} 4.8 \\ -\ 0.0023 \\ \hline \end{array}$$

**19.**
$$\begin{array}{r} 31.007 \\ -\ 0.543 \\ \hline \end{array}$$

**20.**
$$\begin{array}{r} 40 \\ -\ 6.425 \\ \hline \end{array}$$

Simplify each expression by combining like terms.

**21.** $8.3x + x - 22.7x$

**22.** $9.54x - x - 12.82x$

**23.** $85.7y - 22.3y - 17.9y$

**24.** $77.5y - 34.1y - 56.3y$



**ANSWERS**

1. 3.3
2. 11.9
3. 5.65
4. 45.15
5. 4.7925
6. 95.085
7. 117.385
8. 26.331
9. 718.15
10. 11.5201
11. 1.55
12. 8.92
13. 15.89
14. 0.9754
15. 54.946
16. 1.994
17. 4.888
18. 4.7977
19. 30.464
20. 33.575
21. $-13.4x$
22. $-4.28x$
23. $45.5y$
24. $-12.9y$

25. $-0.7x$

26. $0.93x$

27. $6.3t$

28. $12.4s$

29. $10.3x - 4.1y$

30. $-12.1x + 1.9y$

31. $x = -107.3$

32. $x = 26.2$

33. $y = -1.7$

34. $y = 46.7$

35. $z = 18.39$

36. $z = -22.73$

37. $w = 3.84$

38. $w = -86.05$

39. $6950

40. $4

41. 12 in.

42. 8746.1 million bushels

**25.** $0.3x - x$

**26.** $-0.07x + x$

**27.** $13.4t - 22.7t + 15.6t$

**28.** $167.1s - 200s + 45.3s$

**29.** $2.1x + 8.2x - y - 3.1y$

**30.** $-8.4x - 3.7x + 2y - 0.1y$

Solve each of the following equations.

**31.** $x + 34.8 = 7.5 - 80$

**32.** $x - 27.9 = 18.3 - 20$

**33.** $16.5 + y = 30 - 15.2$

**34.** $20.1 + y = 16.8 + 50$

**35.** $z - 8.65 = 22.3 - 12.56$

**36.** $z - 7.64 = 15.23 - 45.6$

**37.** $w + 67.5 = 88.54 - 17.2$

**38.** $w - 47.35 = 9.6 - 143$

**39.** Martin wants to buy a new car for $18,000. He has talked to the loan officer at his credit union and knows that they will loan him $12,600. He must also pay $350 for a license fee and $1200 for taxes. What amount of cash will he need to buy the car?

**40.** Terri wants to have a haircut and a manicure. She knows that the haircut will cost $28.00 and the manicure will cost $10.50. If she plans to tip the hair stylist $7.50, how much change should she receive from a $50 bill?

**41.** An architect's scale drawing shows a rectangle that measures 2.25 inches on one side and 3.75 inches on another side. What is the perimeter ( distance around ) of the rectangle in the drawing?

**42.** In 1999, U.S. farmers produced 6344.045 millions bushels of corn and 2402.055 million bushels of wheat. What was the total combined amount of these grains produced in 1999?

Name ____________________ Section ____________ Date ____________

**ANSWERS**

43. 467.846 million

44. 0.237

45. 15.85 in.; 64.9 in.

46. $500.25

**43.** In 1998, U.S. farmers had the following amounts of livestock: 101.749 million head of cattle, 9.638 million head of dairy cows, 9.079 million sheep, 55.63 million pigs, and 291.75 million turkeys. What was the total amount of these livestock in 1998?

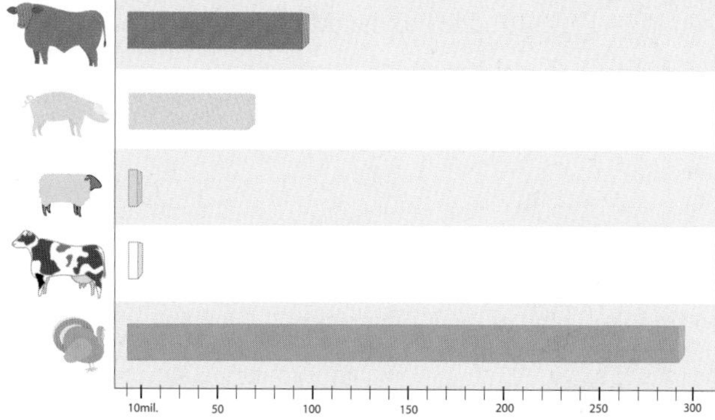


**44.** The eccentricity of a planet's orbit is the measure of how much the orbit varies from a perfectly circular pattern. Earth's orbit has an eccentricity of 0.017, and Pluto's orbit has an eccentricity of 0.254. How much greater is Pluto's eccentricity of orbit than Earth's?

**45.** Albany, New York, receives an average rainfall of 35.74 inches and 65.5 inches of snow. Charleston, South Carolina, receives an average rainfall of 51.59 inches and 0.6 inches of snow. On average, how much more rain is there in Charleston than in Albany? On average, how much more snow is there in Albany than in Charleston?

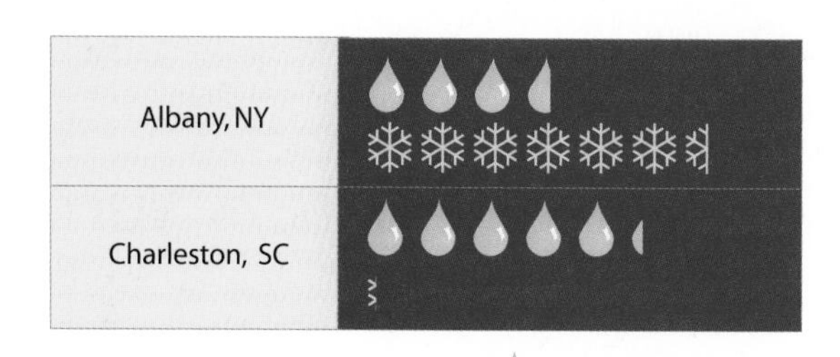


**46.** Suppose your checking account shows a balance of $382.35 at the beginning of the month. During the month, you make deposits of $580.00, $300.00, $182.50, and $45.00, and you write checks for $85.35, $210.50, $43.75, and $650. Find the end-of-the-month balance in your account.

| DATE | TRANSACTION DESCRIPTION | SUBTRACTIONS | ADDITIONS | BALANCE |
|---|---|---|---|---|
| | | | | $382 35 |
| 11/4 | Deposit | | 580 00 | 580 00<br>962 35 |
| 11/4 | check electric bill | 85 35 | | 85 35 |
| 11/6 | check Wally's Computer Shop | 210 50 | | 210 50 |
| 11/7 | Deposit | | 300 00 | 300 00 |
| 11/20 | check groceries | 43 75 | | 43 75 |
| 11/22 | Deposit | | 182 50 | 182 50 |
| 11/31 | Deposit | | 45 00 | 45 00 |
| 12/1 | check BJ's Auto Repair Shop | 650 00 | | 650 00 |

47. 74.96

48. – 21.23

49. 9.9209

50. 54.904

51. 89.9484

52. – 4.15

53. 0.660932

54. – 0.920262

55. 1406.8

56. – 118.342

57. 510.989

58. 5542.71

59. –34.35

60. – 35.415

**47.** If the sum of 33.72 and 19.63 is subtracted from the sum of 92.61 and 35.7, what is the difference?

**48.** What is the sum if – 17.43 is added to the difference between 16.3 and 20.1?

## Calculators

Use a calculator to find the value of each expression.

**49.** 6.5678 + 7.9213 – 4.5682

**50.** 67.341 – 56.329 + 43.892

**51.** 107.65 – 35.202 + 17.5004

**52.** 230 + 34.2 – 300.5 + 32.15

**53.** 0.98455 – 0.356218 + 0.0326

**54.** 0.003407 + 0.076331 – 1

**55.** 892.783 + 861.459 – 347.442

**56.** 788.383 – 806.525 – 100.2

**57.** 346.672 – ( 35.783 – 200.1 )

**58.** 5482.1 – ( 845.92 – 906.53 )

**59.** ( 28.74 – 32.569 ) + ( 45.93 – 76.451 )

**60.** ( 67.349 + 93.65 ) – ( 72.914 + 123.5 )

Name ______________________ Section ______________ Date __________

## The Recycle Bin (from Section 2.4)

Find each of the following products.

**1.** $60 \cdot 80{,}000$ **2.** $145(-27)$ **3.** $(-34)(-14)(-22)$

**4.** First estimate the following product, and then find the product.

$$\begin{array}{r} 6531 \\ \underline{\times\ 284} \end{array}$$

Find each of the following quotients.

**5.** $72 \div (-24)$ **6.** $-192 \div (-2)$ **7.** $-2076 \div 4$

**8.** Evaluate the following expression:

$-18 \div 3^2 + 42 - 3(17 - 12)^2$

### The Recycle Bin

**1.** 4,800,000

**2.** – 3915

**3.** – 10,472

**4.** 2,100,000 (estimate); 1,854,804 (product)

**5.** – 3

**6.** 96

**7.** – 519

**8.** – 35

# 5.3 Multiplication and Division with Decimal Numbers

## Multiplication with Decimal Numbers

When decimal numbers are added or subtracted, the decimal points are aligned so that digits in the tenths position are added (or subtracted) to digits in the tenths position, digits in the hundredths position are added (or subtracted) to digits in the hundredths position, and so on. In this way fractions with a common denominator are added (or subtracted). However, when fractions are multiplied, there is no need to have a common denominator. For example,

in adding $\frac{1}{2}+\frac{3}{8}$ we need the common denominator 8:

$\frac{1}{2}+\frac{3}{8}=\frac{1}{2}\cdot\frac{\mathbf{4}}{\mathbf{4}}+\frac{3}{8}=\frac{4}{8}+\frac{3}{8}=\frac{7}{8}$ but, in multiplying there is no need for a common denominator: $\frac{1}{2}\cdot\frac{3}{8}=\frac{3}{16}$

Thus, since there is no concern about common denominators when multiplying fractions, the same is true for multiplying decimals. **Therefore, there is no need to keep the decimal points in line for multiplication with decimals.**

Decimal numbers are multiplied in the same manner as whole numbers are multiplied with the added concern of the correct placement of the decimal point in the product. Two examples are shown here, in both fraction form and decimal form, to illustrate how the decimal point is to be placed in the product.

**Products in Fraction Form**

$$\frac{4}{10}\cdot\frac{6}{100}=\frac{24}{1000}$$

$$\frac{5}{1000}\cdot\frac{7}{100}=\frac{35}{100{,}000}$$

**Products in Decimal Form**

```
   .4    ← 1 place  } total of 3 places
×  .06   ← 2 places } (thousandths)
  .024

  .005   ← 3 places } total of 5 places
×  .07   ← 2 places } (hundred-thousandths)
 .00035
```

The following rule states how to multiply positive decimal numbers and place the decimal point in the product. ( **For negative decimal numbers, the rule is the same except for the determination of the sign.** )

**To Multiply Positive Decimal Numbers:**

1. Multiply the two numbers as if they were whole numbers.
2. Count the total number of places to the right of the decimal points in both numbers being multiplied.
3. Place the decimal point in the product so that the number of places to the right is the same as that found in step 2.

**Objectives**

① Be able to multiply decimal numbers and place the decimal point correctly in the product.

② Be able to divide decimal numbers and place the decimal point correctly in the quotient.

③ Be able to multiply and divide decimal numbers mentally by powers of 10.

④ Know how to estimate products and quotients with rounded-off decimal numbers.

⑤ Understand how to apply the previous skills and ideas in solving word problems.

Note that in the introductory examples, the 0 whole number part has been omitted. You might allow your students to follow the style shown here and insert the 0, if needed, in their answers for both multiplication and division with decimal numbers.

### Example 1

Multiply: 2.435 x 4.1

**Solution**

$$\left.\begin{array}{rl} 2.435 & \longleftarrow \text{3 places} \\ \times\ \ 4.1 & \longleftarrow \text{1 place} \end{array}\right\} \text{total of 4 places (ten-thousandths)}$$

$$\begin{array}{rl} \hline 2435 & \\ 97400 & \\ \hline 9.9835 & \longleftarrow \text{4 places in the product} \end{array}$$

### Example 2

Find the product: $(-0.126)(0.003)$

**Solution**

Since the signs are not alike, the product will be negative.

$$\left.\begin{array}{rl} -0.126 & \longleftarrow \text{3 places} \\ 0.003 & \longleftarrow \text{3 places} \end{array}\right\} \text{total of 6 places}$$

$$\begin{array}{rl} \hline -0.000378 & \longleftarrow \text{6 places in the product} \end{array}$$

Note that three 0's had to be inserted between the 3 and the decimal point in the product to get a total of 6 decimal places.

*Now Work Exercises 1- 4 in the Margin.*

Find each product.

**1.** $(0.9)(0.3)$

0.27

**2.** $(-2.3)(0.02)$

−0.046

**3.** $\begin{array}{r} 5.716 \\ \times\, 52.01 \\ \hline \end{array}$

297.28916

**4.** $(-0.007)(-0.536)$

0.003752

## Multiplication by Powers of 10

The following general guidelines can be used to multiply decimal numbers ( including whole numbers ) by powers of 10.

**To Multiply a Decimal Number by a Power of 10:**

1. Move the decimal point **to the right.**
2. Move it the same number of places as the number of 0's in the power of 10.

   Multiplication by **10** moves the decimal point **one** place **to the right;** Multiplication by **100** moves the decimal point **two** places **to the right;** Multiplication by **1000** moves the decimal point **three** places **to the right;** and so on.

### Example 3

The following products illustrate multiplication by powers of 10.

**a.** $10(9.35) = 93.5$ Move decimal point 1 place to the right.
**b.** $100(9.35) = 935.$ Move decimal point 2 places to the right.
**c.** $100(163) = 16{,}300$ Move decimal point 2 places to the right.
**d.** $1000(0.8723) = 872.3$ Move decimal point 3 places to the right.
**e.** $10^2(87.5) = 8750$ Exponent tells how many places to move the decimal point. Move decimal point 2 places.
**f.** $10^3(4.86591) = 4865.91$ Move decimal point 3 places to the right.

*Now Work Exercises 5- 7 in the Margin.*

## Division With Decimal Numbers

The process of division with decimal numbers is, in effect, the same as division with whole numbers with the added concern of where to place the decimal point in the quotient. This is reasonable because whole numbers are decimal numbers, and we would not expect a great change in a process as important as division. Division with whole numbers gives a quotient and possibly a remainder.

```
                 23.   ← quotient
divisor →  40)950.    ← dividend
              80
              150
              120
               30   ← remainder
```

By adding 0's onto the dividend, we can continue to divide and get a decimal quotient other than a whole number.

```
                 23.75    ← quotient is a decimal number
divisor →  40)950.00   ← 0's added on
              80
              150
              120
               300
               280
                200
                200
                  0
```

**If the divisor is a decimal number other than a whole number,** multiply both the divisor and dividend by a power of 10 so that the new divisor is a whole number. For example, we can write

$$6.2\overline{)63.86} \text{ as } \frac{63.86}{6.2}\cdot\frac{10}{10}=\frac{638.6}{62}$$

This means that

$$6.2\overline{)63.86} \text{ is } 62\overline{)638.6}$$

Similarly, we can write

$$1.23\overline{)4.6125} \text{ as } \frac{4.6125}{1.23}\cdot\frac{100}{100}=\frac{461.25}{123} \text{ or } 123\overline{)461.25}$$

Find each product by performing the operation mentally.

**5.** $10(8.36)$

83.6

**6.** $100(-0.9735)$

$-97.35$

**7.** $10^3(14.82)$

14,820

**To Divide Decimal Numbers:**

1. **1.** Move the decimal point in the divisor to the right so that the divisor is a whole number.
2. **2.** Move the decimal point in the dividend the same number of places to the right.
3. **3.** Place the decimal point in the quotient directly above the new decimal point in the dividend.
4. **4.** Divide just as with whole numbers.

**Note:**

1. **1.** In moving the decimal point, you are multiplying by a power of 10.
2. **2.** Be sure to place the decimal point in the quotient **before actually dividing.**

## Example 4

Find the quotient: 63.86 ÷ 6.2

### Solution

**a.** Write down the numbers.

$6.2\overline{)63.86}$

**b.** Move both decimal points one place to the right so that the divisor becomes a whole number. Then place the decimal point in the quotient.

$$\begin{array}{r} . \\ 6.2.\overline{)63.8.6} \end{array}$$ ⟵ decimal point in quotient

**c.** Proceed to divide as with whole numbers.

$$\begin{array}{r} 10.3 \\ 62.\overline{)638.6} \\ \underline{62\phantom{.6}} \\ 18\phantom{.6} \\ \underline{0\phantom{.6}} \\ 186 \\ \underline{186} \\ 0 \end{array}$$

## Example 5

Find the quotient: $4.6125 \div 1.23$

### Solution

**a.** Write down the numbers.

$1.23\overline{)4.6125}$

**b.** Move both decimal points two places to the right so that the divisor becomes a whole number. Then place the decimal point in the quotient.

$1.23.\overline{)4.61.25}$ ← decimal point in quotient

**c.** Proceed to divide as with whole numbers.

$$\begin{array}{r} 3.75 \\ 123.\overline{)461.25} \\ \underline{369\phantom{.25}} \\ 92\,2\phantom{5} \\ \underline{86\,1\phantom{5}} \\ 6\,15 \\ \underline{6\,15} \\ 0 \end{array}$$

*Now Work Exercises 8 - 10 in the Margin.*

If the remainder is eventually 0, then the quotient is a **terminating decimal**. We will see in Section 5.4 that if the remainder is never 0, then the quotient is an **infinite repeating decimal.** That is, the quotient will be a repeating pattern of digits. To avoid an infinite number of steps, which can never be done anyway, we generally agree to some place of accuracy for the quotient before the division is performed. If the remainder is not 0 by the time this place of accuracy is reached in the quotient, then we divide one more place and round off the quotient.

Find the indicated quotient.

**8.** $0.5\overline{)936}$ 1872

**9.** $7.2\overline{)8.208}$ 1.14

**10.** $2.13\overline{)11.928}$ 5.6

**When the Remainder is Not Zero:**

1. Decide first how many decimal places are to be in the quotient.
2. Divide until the quotient is one digit past the place of desired accuracy.
3. Using this last digit, round off the quotient to the desired place of accuracy.

## Example 6

Find the quotient $82.3 \div 2.9$ to the nearest tenth.

### Solution

Divide until the quotient is in hundredths (one place more than tenths); then round off to tenths.

hundredths

read **approximately**

```
          28.37  ≈ 28.4
2.9.)82.3.00  ←  Add 0's as needed.
     58
     243
     232
      110
       87
       230
       203
        27
```

$82.3 \div 2.9 \approx 28.4$ accurate to the nearest tenth

Find the following quotients to the nearest tenth.

11. $5.6\overline{)83.541}$ — 14.9

12. $7\overline{)34.66}$ — 5.0

## Example 7

Find the quotient $1.935 \div 3.3$ to the nearest hundredth.

### Solution

thousandths

```
       0.586
3.3.)1.9.350
     165
      285
      264
       210
       198
        12
```

$1.935 \div 3.3 \approx 0.59$ accurate to the nearest hundredth

*Now Work Exercises 11 and 12 in the Margin.*

## Division By Powers of 10

On page 390, we found that multiplication by powers of 10 can be accomplished by moving the decimal point to the right. Division by powers of 10 can be accomplished by moving the decimal point to the left.

Two general guidelines will help you to understand work with powers of 10:

1. Multiplication by a power of 10 will make a number larger, so move the decimal point to the right.
2. Division by a power of 10 will make a number smaller, so move the decimal point to the left.

**To Divide a Decimal Number by a Power of 10:**

1. Move the decimal point **to the left.**
2. Move it the same number of places as the number of 0's in the power of 10.

   Division by **10** moves the decimal point **one** place **to the left;**
   Division by **100** moves the decimal point **two** places **to the left;**
   Division by **1000** moves the decimal point **three** places **to the left;**
   and so on.

### Example 8

The following quotients illustrate division by powers of 10.

**a.** $5.23 \div 100 = \frac{5.23}{100} = 0.0523$ Move decimal point 2 places to the left.

**b.** $817 \div 10 = \frac{817}{10} = 81.7$ Move decimal point 1 place to the left.

**c.** $495.6 \div 10^3 = 0.4956$ Move decimal point 3 places to the left. The exponent tells how many places to move the decimal point.

**d.** $286.5 \div 10^2 = 2.865$ Move decimal point 2 places to the left.

*Now Work Exercises 13-15 in the Margin.*

## Estimating Products and Quotients

Estimating with multiplication and division can be used to help in placing the decimal point in the product and in the quotient to verify the reasonableness of the answer. The technique is to **round off all numbers to the place of the leftmost nonzero digit and then operate with these rounded-off numbers.**

Find each quotient by performing the operation mentally.

**13.** $73.2 \div 100$
0.732

**14.** $\frac{16}{10}$
1.6

**15.** $\frac{83.46}{1000}$
0.08346

### Example 9

First

**a.** estimate the product, and then

**b.** find the product: $(0.358)(6.2)$.

**Solution**

**a.** Estimate by multiplying rounded-off numbers.

$$\begin{array}{rl} 0.4 & \text{0.358 rounded off} \\ \times\ \ 6 & \text{6.2 rounded off} \\ \hline 2.4 & \leftarrow \text{estimate} \end{array}$$

**b.** Find the actual product.

$$\begin{array}{rl} 0.358 & \\ \times\ \ 6.2 & \\ \hline 716 & \\ 2148\phantom{0} & \\ \hline 2.2196 & \leftarrow \text{actual product} \end{array}$$

The estimated product 2.4 helps place the decimal point correctly in the product 2.2196. Thus an answer of 0.22196 or 22.196 would indicate an error in the placement of the decimal point, since the answer should be near 2.4 (or between 2 and 3).

### Example 10

First estimate the quotient $6.1 \div 0.312$; then find the quotient to the nearest tenth.

**Solution**

**a.** Using $6.1 \approx 6$ and $0.312 \approx 0.3$, estimate the quotient.

$$0.3.\overline{)6.0.} = 20. \leftarrow \text{estimate}$$

**b.** Find the quotient to the nearest tenth.

$$\begin{array}{r} 19.55 \\ 0.312.\overline{)6.100.00} \\ \underline{312\phantom{.00}} \\ 2980\phantom{.0} \\ \underline{2808\phantom{.0}} \\ 1720 \\ \underline{1560} \\ 1600 \\ \underline{1560} \\ 40 \end{array}$$

$6.1 \div 0.312 \approx 19.6$

## Applications

Some word problems may involve several operations with decimal numbers. The words do not usually say directly to add, subtract, multiply, or divide. Experience and reasoning abilities are needed to decide which operation ( if any ) to perform with the given numbers. Example 11 illustrates a problem that involves several steps and how estimating can provide a check for a reasonable answer.

Example 11 is a good illustration of the need for judgment in any estimation. In this case, if 12 is rounded off to 10, the resulting estimate will actually be a negative number. You might ask the students the implication of the negative result (paying cash would increase the cost) and why this does not make sense.

### Example 11

You can buy a car for \$8500 cash, or you can make a down payment of \$1700 and then pay \$616.67 each month for 12 months. How much can you save by paying cash?

**Solution**

**a.** Find the amount paid in monthly payments by multiplying the amount of each payment by 12. In this case, judgment dictates that we do not want to lose two full monthly payments in our estimate, so we use 12 and do not round off to 10.

| Estimate | Actual Amount | |
|---|---|---|
| \$600 | \$616.67 | |
| × 12 | × 12 | |
| 1200 | 123334 | |
| 600 | 61667 | |
| \$7200 | \$7400.04 | paid in monthly payments |

**b.** Find the total amount paid by **adding** the down payment to the answer in part **a.**

| Estimate | | Actual Amount | |
|---|---|---|---|
| \$ 2000 | down payment | \$ 1700.00 | down payment |
| + 7200 | monthly payments | + 7400.04 | monthly payments |
| \$ 9200 | total paid | \$ 9100.04 | total paid |

**c.** Find the savings by **subtracting** \$8500 from the answer in part **b.**

| Estimate | Actual Amount | |
|---|---|---|
| \$ 9200.00 | \$ 9100.04 | |
| −\$ 8500.00 | −\$ 8500.00 | |
| \$ 700.00 | \$ 600.04 | savings by paying cash |

The \$700 estimate is very close to the actual savings.

Keep reminding students that "per" means to divide. This idea will help them a great deal in deciding how to divide to find various averages. For example, miles per gallon means miles divided by gallons.

We know from Section 2.5 that the average of a set of numbers can be found by adding the numbers, and then dividing the sum by the number of addends. Another meaning of the term average is in the sense of "an average speed of 52 miles per hour" or "the average number of miles per gallon of gas." This kind of average can also be found by division. If we know the total amount of a quantity (distance, dollars, gallons of gas, etc.) and a number of units (time, items bought, miles, etc.), then we can find the average amount per unit by dividing the amount by the number of units.

Once they understand this concept, you might want to show them the technique of operating with "fractions" made up of unit labels to help them decide what operations to perform. As in Example 13,

$$\frac{\text{miles}}{\text{hour}} \times \text{hours} = \text{miles}$$

Think of "per" as indicating division. Thus, miles per hour means miles divided by hours.

## Example 12

The gas tank of a car holds 18 gallons of gasoline. Approximately how many miles per gallon does the car average if it will go 470 miles on one tank of gas?

### Solution

Miles per gallon means miles divided by gallons.
Since the question calls only for an approximate answer, rounded-off values can be used.

$18 \approx 20$ gal and $470 \approx 500$ miles

$$\begin{array}{r} 25 \\ 20\overline{)500} \\ \underline{40} \\ 100 \\ \underline{100} \\ 0 \end{array}$$ miles per gallon

The car averages about 25 miles per gallon.

If an average amount per unit is known, then a corresponding total amount can be found by multiplying. For example, if you jog an average of 5 miles per hour, then the distance you jog can be found by multiplying your average speed by the time you spend jogging.

## Example 13

If you jog at an average speed of 4.8 miles per hour, how far will you jog in 3.2 hours?

### Solution

Multiply the average speed by the number of hours.

$$\begin{array}{rl} 4.8 & \text{miles per hour} \\ \times\ 3.2 & \text{hours} \\ \hline 9\,6 & \\ 1\,4\,4\phantom{0} & \\ \hline 1\,5.3\,6 & \text{miles} \end{array}$$

You will jog 15.36 miles in 3.2 hours.

In some cases a business will advertise the total price of its items including sales tax. You will see a phrase such as "tax included." If, for example, the sales tax is figured at 0.06 times the actual price, you can find the price you are paying for the item by dividing the total price by 1.06. (The number 1.06 represents the actual price plus 0.06 times the actual price.) If you buy gas for your car, the price at the gas pump includes all types of taxes (state, federal, local, etc.). If these taxes are figured at, say 0.45 times the price of a gallon of gas, then the actual price of the gas to you can be found by dividing the price at the pump by the number 1.45.

### Example 14

Suppose that the price of a gallon of gas is stated as \$1.25 at the pump and the station owner tells you that the taxes you are paying are figured at 0.45 times the price he is actually charging for a gallon of gas. What price (to the nearest penny) is he actually charging for a gallon of gas?

### Solution

To find the actual price of a gallon of gas, before taxes, divide the total price by 1.45 .

So, the actual cost of the gas is about \$0.86 per gallon before taxes.

$$\begin{array}{r} .862 \\ 1.45.\overline{)\,1.25.000} \\ \underline{1\ 16\ 0} \\ 9\ 00 \\ \underline{8\ 70} \\ 300 \\ \underline{290} \\ 10 \end{array}$$

Name ______________ Section ______________ Date ______________

## Exercises 5.3

1. Estimate each of the following products mentally by using rounded-off values.

   **a.** ( 0.92 ) ( 0.81 )

   **b.** ( 33.6 ) ( 0.11 )

   **c.** ( 0.64 ) ( 9.71 )

   **d.** ( 0.22 ) ( 0.26 )

   **e.** ( 4.7 ) ( 1.1 )

2. Estimate each of the following products mentally by using rounded-off values.

   **a.** ( 1.62 ) ( 0.03 )

   **b.** 1.62 ( 0.003 )

   **c.** 16.2 ( 0.03 )

   **d.** 1.62 ( 3 )

3. Estimate each of the following quotients by using rounded-off values.

   **a.** $3.1\overline{)6.36}$

   **b.** $0.1\overline{)211.5}$

   **c.** $3.6\overline{)282.4}$

   **d.** $18.2\overline{)132.9}$

   **e.** $3.1\overline{)0.0636}$

4. Estimate each of the following quotients by using rounded-off values.

   **a.** $28.34 \div 0.003$

   **b.** $28.34 \div 0.03$

   **c.** $28.34 \div 0.3$

   **d.** $28.34 \div 3$

**ANSWERS**

1. a. 1.0
   b. 3.0
   c. 6.0
   d. 0.06
   e. 5.0

2. a. 0.06
   b. 0.006
   c. 0.6
   d. 6.0

3. a. 2
   b. 2000
   c. 75
   d. 5
   e. 0.02

4. a. 10,000
   b. 1000
   c. 100
   d. 10

5. 0.35
6. 0.16
7. 10.8
8. 31.5
9. 0.0004
10. 0.0009
11. −0.112
12. −1.2
13. −1
14. −0.5
15. 0.00429
16. 0.1096
17. 2.036
18. 0.01233
19. 0.002028
20. 9.335655
21. 1.632204
22. 53.067
23. 3.24
24. 0.91
25. -0.79
26. 5
27. -0.006
28. -2006
29. 0.7
30. 15.4
31. 21.3
32. 2.03
33. 5.04
34. 1.77

Find each of the indicated products.

**5.** $(0.5)(0.7)$ **6.** $(0.2)(0.8)$ **7.** $6(1.8)$

**8.** $9(3.5)$ **9.** $(0.02)(0.02)$ **10.** $(0.03)(0.03)$

**11.** $5.6(-0.02)$ **12.** $4.8(-0.25)$ **13.** $8(-0.125)$

**14.** $4(-0.125)$ **15.** $4.29(0.001)$ **16.** $5.48(0.02)$

**17.** $\begin{array}{r} 5.09 \\ \times\ 0.4 \\ \hline \end{array}$ **18.** $\begin{array}{r} 0.137 \\ \times 0.09 \\ \hline \end{array}$ **19.** $\begin{array}{r} 0.0312 \\ \times 0.065 \\ \hline \end{array}$

**20.** $\begin{array}{r} 84.105 \\ \times 0.111 \\ \hline \end{array}$ **21.** $\begin{array}{r} 16.002 \\ \times 0.102 \\ \hline \end{array}$ **22.** $\begin{array}{r} 93.1 \\ \times 0.57 \\ \hline \end{array}$

Divide.

**23.** $6.48 \div 2$ **24.** $2.73 \div 3$ **25.** $3.95 \div (-5)$

**26.** $28 \div 5.6$ **27.** $(-0.054) \div 9$ **28.** $-80.24 \div 0.04$

Find each quotient to the nearest tenth.

**29.** $9.4\overline{)6.429}$ **30.** $0.37\overline{)5.682}$ **31.** $1.64\overline{)35}$

Find each quotient to the nearest hundredth.

**32.** $2.7\overline{)5.483}$ **33.** $13\overline{)65.582}$ **34.** $3.381\overline{)6}$

Name ______________________ Section ____________ Date ____________

Find each quotient to the nearest thousandth.

**35.** $31\overline{)71}$

**36.** $1.62\overline{)0.0116}$

**37.** $0.03\overline{)6.275}$

Find each indicated product or quotient mentally by using your knowledge of multiplication and division by powers of 10.

**38.** $10(0.619)$

**39.** $100(3.76)$

**40.** $100(0.455)$

**41.** $10^3(0.95)$

**42.** $10^3(0.005)$

**43.** $10^5(7.3)$

**44.** $98.5 \div 100$

**45.** $26.483 \div 1000$

**46.** $\frac{169}{10}$

**47.** $\frac{1.78}{1000}$

**48.** $\frac{3.25}{100}$

**49.** Find

**a.** the perimeter and

**b.** the area of a square with sides 13.4 inches long.

**50. a.** If the sale price of a new television set is \$683 and sales tax is figured at 0.08 times the price, approximately what total amount is paid for the television set?

**b.** What is the exact amount paid for the television set?

**ANSWERS**

**35.** 2.290

**36.** 0.007

**37.** 209.167

**38.** 6.19

**39.** 376

**40.** 45.5

**41.** 950

**42.** 5

**43.** 730,000

**44.** 0.985

**45.** 0.026483

**46.** 16.9

**47.** 0.00178

**48.** 0.0325

**49. a.** 53.6 in.

**b.** 179.56 in.$^2$

**50. a.** \$756

**b.** \$737.64

51. $310.10

52. a. about 600 mi

b. 480.7 mi

53. a. about 20 mpg

b. 22 mpg

54. $31,500

55. $825

56. $1750

**51.** To buy a used car, you can pay $2045.50 cash or put $300 down and make 18 monthly payments of $114.20. How much would you save by paying cash?

**52.** **a.** If a car averages 25.3 miles per gallon, about how far will it go on 19 gallons of gas?

**b.** Exactly how many miles per gallon does it average?

**53.** **a.** If a car travels 330 miles on 15 gallons of gas, approximately how many miles does it travel per gallon?

**b.** Exactly how many miles per gallon does it average?

**54.** If the total price of a car was $33,075 including tax at 0.05 times the list price, you can find the list price by dividing the total price by 1.05. What was the list price? (**Note**: 1.05 represents the list price plus 0.05 times the list price.)

**55.** If the total price of a refrigerator was $874.50 including tax at 0.06 times the list price, you can find the list price by dividing the total price by 1.06. What was the list price? (**Note:** 1.06 represents the list price plus 0.06 times the list price.)

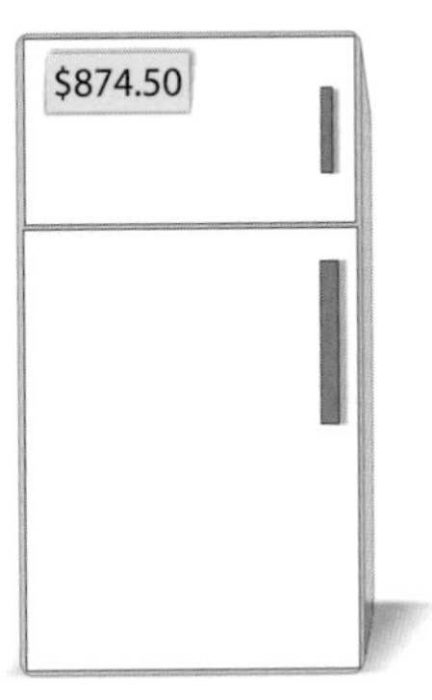


**56.** Suppose that the total interest that will be paid on a 30-year mortgage for a home loan of $150,000 is going to be $480,000. What will be the payment each month if the payments are to pay off both the loan and the interest?

Name ______________________ Section ____________ Date __________

**57.** About how long will it take an airplane to fly from Los Angeles to New York if the distance is approximately 3000 miles and the airplane averages 465 miles per hour?

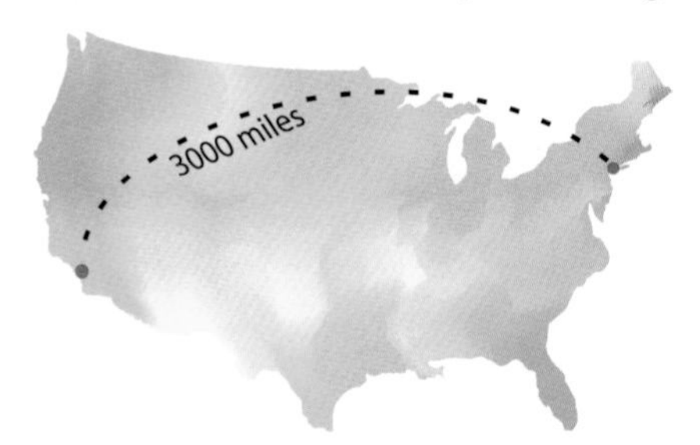


**58. a.** If a bicyclist rode 150.6 miles in 11.3 hours, about how fast did she ride in miles per hour?

**b.** What was her average speed in miles per hour (to the nearest tenth)?

**59.** Walter Payton played football for the Chicago Bears for 13 years. In those years he carried the ball 3838 times for a total of 16,726 yards. What was his average yardage per carry (to the nearest tenth)?

## Calculators

Use a calculator to find the value of each variable accurate to the nearest thousandth.

**60.** $x = 3.521(0.124)(-1.2)$

**61.** $x = 67.3(42.44)(-2.7)$

**62.** $a = -34.6(-0.02)(-3.577)$

**63.** $a = -14.5(-72.16)(-10.563)$

**64.** $t = -5.7 \div 0.214$

**65.** $t = 3.457 \div (-0.12)$

**66.** $y = \dfrac{67.152}{3.54}$

**67.** $y = \dfrac{-4580}{19.64}$

**ANSWERS**

57. about 6.45 hrs

58. a. about 15 mph

b. 13.3 mph

59. 4.4 yards per carry

60. $x = -0.524$

61. $x = -7711.772$

62. $a = -2.475$

63. $a = -11{,}052.278$

64. $t = -26.636$

65. $t = -28.808$

66. $y = 18.969$

67. $y = -233.198$

[Respond below exercise.]
68. ______

[Respond below exercise.]
69. ______

## WRITING AND THINKING ABOUT MATHEMATICS

**68.** Write a brief paragraph, with an example, discussing just what mathematics you used in your daily life this week. What was your thinking? That is, how did you decide to add, subtract, multiply, or divide?

Answers will vary.

## COLLABORATIVE LEARNING EXERCISES

**69.** Do you know how to find the gas mileage (miles per gallon) that your car is using? If you are not sure, proceed as follows and compare your mileage with other students in the class. (Your car might need some work if the mileage is not consistent or it is much worse than other similar sized cars.)

**Step 1:** Fill up your gas tank and write down the mileage indicated on the odometer.

**Step 2:** Drive the car for a few days.

**Step 3:** Fill up your gas tank again and write down the number of gallons needed to fill the tank and the new mileage indicated on the odometer.

**Step 4:** Find the number of miles that you drove by subtracting the new and old numbers indicated on the odometer.

**Step 5:** Divide the number of miles driven by the number of gallons needed to fill the tank. This number is your gas mileage (miles per gallon).

Answers will vary.

# 5.4 Decimals, Fractions, and Scientific Notation

## Decimals and Fractions

In this section, we will discuss the fact that certain decimal numbers (terminating and infinite repeating decimals ), fractions, and mixed numbers are simply different forms of the same type of number -- namely, **rational numbers**. We also will show how operations can be performed with various combinations of these numbers and review absolute value and evaluation of expressions with variables.

**Objectives**

① Know how to change decimal numbers to fraction form and/or mixed number form.

② Know how to change fractions and mixed numbers to decimal form.

③ Understand that working with decimals and fractions in the same problem may involve rounded-off numbers and approximate answers.

④ Know how to read and write very large numbers in scientific notation.

⑤ Know how to read and write very small numbers in scientific notation.

⑥ Be able to read the results on a calculator set in scientific notation mode.

From Section 5.1, we know that decimal numbers can be written in fraction form with denominators that are powers of 10. For example,

$$0.75 = \frac{75}{100} \text{ and } 0.016 = \frac{16}{1000}$$

In each case, the denominator is the value of the position of the rightmost digit. This leads to the following method for changing a decimal number to fraction form:

**Changing from Decimal Form to Fraction Form:**

A decimal number with digits to the right of the decimal point can be written in fraction form by writing a fraction with

1. A **numerator** that consists of the whole number formed by all the digits of the decimal number and
2. A **denominator** that is the power of 10 that names the rightmost digit.

In the following examples, each decimal number is changed to fraction form and then reduced by using the factoring techniques discussed in Chapter 3 for reducing fractions.

## Example 1

**a.**

$$0.75 = \frac{75}{100} = \frac{3 \cdot \cancel{5} \cdot \cancel{5}}{2 \cdot 2 \cdot \cancel{5} \cdot \cancel{5}} = \frac{3}{4}$$

↑ hundredths

**b.**

$$0.36 = \frac{36}{100} = \frac{\cancel{4} \cdot 9}{\cancel{4} \cdot 25} = \frac{9}{25}$$

↑ hundredths

**c.**

$$0.085 = \frac{85}{1000} = \frac{\cancel{5} \cdot 17}{\cancel{5} \cdot 200} = \frac{17}{200}$$

↑ thousandths

You might want to explain that a decimal number with no digits to the right of the decimal point is a whole number. And, a whole number can be written in fraction form with denominator 1. This is consistent with the method stated here.

**d.**

$$3.8 = 3\frac{8}{10} = \frac{38}{10} = \frac{\cancel{2} \cdot 19}{\cancel{2} \cdot 5} = \frac{19}{5}$$

↑ tenths

or, as a mixed number,

$$3.8 = 3\frac{8}{10} = 3\frac{4}{5}$$

**Changing from Fraction Form to Decimal Form:**

A fraction can be written in decimal form by dividing the numerator by the denominator.

1. If the remainder is 0, the decimal is said to be **terminating.**
2. If the remainder is not 0, the decimal is said to be **nonterminating.**

The following examples illustrate fractions that convert to terminating decimals.

## Example 2

Change $\frac{5}{8}$ to a decimal.

### Solution

$$\begin{array}{r} .625 \\ 8\overline{)5.000} \\ \underline{48\phantom{00}} \\ 20\phantom{0} \\ \underline{16\phantom{0}} \\ 40 \\ \underline{40} \\ 0 \end{array}$$

$$\frac{5}{8} = 0.625$$

## Example 3

Change $-\frac{4}{5}$ to a decimal.

### Solution

$-\frac{4}{5}$ also can be written $\frac{-4}{5}$ or $\frac{4}{-5}$. In any case, the result of the division will be negative. For convenience, we simply divide 4 by 5 and then write the negative sign in the quotient.

$$\begin{array}{r} .8 \\ 5\overline{)4.0} \\ \underline{4\ 0} \\ 0 \end{array} \qquad -\frac{4}{5} = -0.8$$

Thus, $-\frac{4}{5} = -0.8$ which is a terminating decimal.

Nonterminating decimals can be **repeating** or **nonrepeating.** A nonterminating repeating decimal has a repeating pattern to its digits. Every fraction with an integer numerator and a nonzero integer denominator ( called a **rational number** ) is either a terminating decimal or a repeating decimal. Nonterminating, nonrepeating decimals are called **irrational numbers** and are discussed Section 5.6.

The following examples illustrate nonterminating, repeating decimals.

## Example 4

Change $\frac{1}{3}$ to a decimal.

### Solution

$$\begin{array}{r} .333 \\ 3\overline{)1.000} \\ \underline{9\phantom{00}} \\ 10\phantom{0} \\ \underline{9\phantom{0}} \\ 10 \\ \underline{9} \\ 1 \end{array}$$

The 3 will repeat without end.

Continuing to divide will give a remainder of 1 each time.

We write

$\frac{1}{3} = 0.333...$ The ellipsis (. . . ) means "and so on" or that the digits continue without stopping.

The following discussion will help to develop an understanding of the concept of infinity as used in mathmatics and will capture the imagination and interest of many students. Consider the fact that

$$\frac{1}{9} = .11111...$$

$$\frac{2}{9} = 2 \cdot \frac{1}{9} = .22222...$$

and so until

$$\frac{9}{9} = 9 \cdot \frac{1}{9} = .99999...$$

Thus, 1.00000... = .99999...

Also, as shown by subtraction, the difference between these two numbers is 0:

$$\begin{array}{r} 1.00000... \\ -\ .99999... \\ \hline .00000... \end{array}$$

Algebraically, if $x = .99999...$, then $10x = 9.99999...$, and $10x - x = 9x = 9.00000...$, which implies that $x = 1$.

To help students understand how repeating patterns develop in the quotient when one whole number is divided by another nonzero whole number (any rational number), consider the following analysis:

The dividend has only 0's to the right of the decimal point. Thus, when a particular remainder appears for a second time (and some remainder must appear for a second time because remainders must be less than the whole number divisor), the patern of digits in the quotient is determined by the sequence of remainders.

Change each decimal to fraction form in lowest terms.

1. 0.35
   $\frac{7}{20}$

2. 0.125
   $\frac{1}{8}$

3. 2.4
   $\frac{12}{5}$

Change each fraction to decimal form accurate to the nearest hundredth.

4. $\frac{1}{7}$
   0.14

5. $-\frac{9}{16}$
   $-0.56$

## Example 5

Change $\frac{3}{7}$ to a decimal.

### Solution

$$\begin{array}{r} .428571 \\ 7\overline{)3.000000} \\ \underline{2\,8\phantom{00000}} \\ 20\phantom{0000} \\ \underline{14\phantom{0000}} \\ 60\phantom{000} \\ \underline{56\phantom{000}} \\ 40\phantom{00} \\ \underline{35\phantom{00}} \\ 50\phantom{0} \\ \underline{49\phantom{0}} \\ 10 \\ \underline{7} \\ 3 \end{array}$$

The six digits will repeat in the same pattern without end.

The remainders will repeat in sequence: 3, 2, 6, 4, 5, 1, 3, and so on. Therefore, the digits in the quotient will also repeat.

We write

$$\frac{3}{7} = 0.428571428571428571...$$

Another more convenient way to write repeating decimals is to write a **bar** over the repeating pattern of digits. Thus,

$$\frac{1}{3} = 0.\overline{3} \text{ and } \frac{3}{7} = 0.\overline{428571}$$

*Now Work Exercises 1- 5 in the Margin.*

## Operations With Both Decimals And Fractions

To evaluate expressions that contain both decimals and fractions, we can change all fractions to decimal form or change all decimals to fraction form. In cases where the decimal form of a fraction involves more than four decimal places or is a nonterminating decimal, we will agree to round off the decimal to thousandths.

## Example 6

Find the sum $10\frac{1}{2}+7.64+3\frac{4}{5}$

**a.** in decimal form and

**b.** in fraction form.

### Solution

**a.** Change each fraction to decimal form and then add. In this example, each fraction has a terminating decimal form.

$$\begin{array}{rcl} 10\frac{1}{2} & = & 10.50 \\ 7.64 & = & 7.64 \\ +\ 3\frac{4}{5} & = & 3.80 \\ \hline & & 21.94 \end{array}$$

$\frac{1}{2} = 0.50$

$\frac{4}{5} = 0.80$

**b.** Change the decimal to a fraction; then find a common denominator and add.

$$\begin{array}{rcccccc} 10\frac{1}{2} & = & 10\frac{1}{2} & = & 10\frac{1}{2} & = & 10\frac{25}{50} \\ 7.64 & = & 7\frac{64}{100} & = & 7\frac{16}{25} & = & 7\frac{32}{50} \\ +\ 3\frac{4}{5} & = & 3\frac{4}{5} & = & 3\frac{4}{5} & = & 3\frac{40}{50} \\ \hline & & & & & & 20\frac{97}{50} = 21\frac{47}{50} \end{array}$$

The answers for parts **a.** and **b.** must be equal, and they are:

$$21\frac{47}{50} = 21\frac{94}{100} = 21.94$$

## Example 7

Find the quotient $4\frac{3}{8} \div (-10)$ in decimal form.

### Solution

Since $4\frac{3}{8} = 4.375$ , we can divide mentally as

$$4\frac{3}{8} \div (-10) = \frac{4.375}{-10} = -0.4375$$

### Example 8

Evaluate the expression $x^2 + 3x - 2y$ for $x = \frac{3}{4}$ and $y = 4.2$.

**Solution**

Changing $\frac{3}{4}$ to decimal form and substituting $x = \frac{3}{4} = 0.75$ and $y = 4.2$ we have

$$\begin{aligned} x^2+3x-2y &= (0.75)^2 + 3(0.75) - 2(4.2) \\ &= 0.5625 + 2.25 - 8.4 \\ &= -5.5875 \end{aligned}$$

Changing 4.2 to fraction form and substituting $x = \frac{3}{4}$ and $y = 4.2 = \frac{42}{10} = \frac{21}{5}$

we have

$$\begin{aligned} x^2 + 3x - 2y &= \left(\frac{3}{4}\right)^2 + 3\left(\frac{3}{4}\right) - 2\left(\frac{21}{5}\right) \\ &= \frac{9}{16} + \frac{9}{4} - \frac{42}{5} \\ &= \frac{9}{16}\cdot\frac{5}{5} + \frac{9}{4}\cdot\frac{20}{20} - \frac{42}{5}\cdot\frac{16}{16} \\ &= \frac{45}{80} + \frac{180}{80} - \frac{672}{80} \\ &= -\frac{447}{80} = -5\frac{47}{80} \end{aligned}$$

Because the decimals in this example were all terminating decimals, we get exactly the same answer whether we use the fraction form or the decimal form of each number. In this case,

$$-5\frac{47}{80} = -5.5875$$

If rounded-off decimals are used, then the answers will not be exactly the same ( only close ).

*Now Work Exercises 6-10 in the Margin.*

Perform the indicated operations.

**6.** $5\frac{1}{2} + 7.46$

$12.96 = 12\frac{24}{25}$

**7.** $18.34 - 2\frac{1}{2}$

$15.84 = 15\frac{21}{25}$

**8.** $7\frac{1}{8} \div (-19)$

$-\frac{3}{8}$

**9.** Evaluate the expression $x^2 - 8x - 9$ for $x = 1.5$

$-18.75$

**10.** Evaluate the expression

$y^3 - 10y + 5$ for $y = \frac{3}{2}$

$-6.625 \left(\text{or } -\frac{53}{8} \text{ or } -6\frac{5}{8}\right)$

## Absolute Value

The definition of **absolute value** was given in Section 2.1 in terms of integers. This definition is also valid for all decimal numbers.

**Definition:**

The **absolute value** of a decimal number is its distance from 0. The absolute value of a decimal number is nonnegative. We can express the definition symbolically, for any decimal number $a$, as follows:

$$\begin{cases} \text{If } a \text{ is positive or 0, then } |a| = a. \\ \text{If } a \text{ is negative, then } |a| = -a. \end{cases}$$

### Example 9

**a.** $|-5.32| = 5.32$

**b.** $|16.7| = 16.7$

**c.** $\left|-\frac{3}{4}\right| = \frac{3}{4}$

**Special Note:** We have seen that fractions can be changed to terminating or repeating decimal forms and vice versa. These decimal forms and fractions are technically known as **rational numbers,** and the three terms **fractions, repeating decimals,** and **rational numbers** are used interchangeably. ( Terminating decimals can be considered to repeat 0's and therefore are classified as repeating decimals. )

## Scientific Notation And Calculators

Scientific calculators and graphing calculators such as the TI-83 Plushave a key labeled MODE . This key allows the calculator to display and calculate with numbers in various forms such as the common decimal form and scientific form. This scientific form, used in physics, chemistry, biology, and astronomy, where numbers are likely to be very large or very small, is called **scientific notation**. In scientific notation, decimal numbers are written as the product of a number between 1 and 10 and an integer power of 10. For example, the distance from the Earth to the Sun is about 93 million miles, and in scientific notation,

Scientific notation is introduced here, because it will invariably appear in some students' calculators either in this class or some other class, and they will ask for an explanantion or simply ignore the notation. You can tell them that a more thorough explanantion of negative exponents is developed in Chapter 8, and that they will definitely see negative exponents in algebra.

$$93,000,000 = 9.3 \times 10^7$$

between 1 and 10 ( one digit to the left of the decimal point )

In scientific notation, there is exactly one digit to the left of the decimal point.

Now, if a calculator is used to find the product 93,000,000 × 50,000, the display on the calculator will be similar to one of the following:

4.65 12 or 4.65 $^{12}$ or 4.65 E12

In each case, 12 is understood to be the exponent on 10 in scientific notation:

$$4.65 \times 10^{12} = 4,650,000,000,000$$

Thus, multiplication by $10^{12}$ moves the decimal point 12 places to the right:

$$4.65 \times 10^{12} = 4.6\,5\,0,0\,0\,0,0\,0\,0,0\,0\,0.$$

12 places right

Although we have not yet discussed negative exponents, the meaning for scientific notation is the same as division by a power of 10. (See Chapter 8 for a more detailed discussion of negative exponents.) That is, a negative exponent on 10 indicates the number of places to move the decimal point to the left. For example,

| Scientific Notation | | Decimal Notation |
| --- | --- | --- |
| $7.8 \times 10^{-5}$ | = | $0.00007.8 = 0.000078$ |
| between 1 and 10 | | 5 places left |

## Example 10

Each of the following numbers is written in both decimal notation and scientific notation.

| Decimal Notation | | Scientific Notation (One digit to the left of the decimal point) |
|---|---|---|
| **a.** 65,000 | = | $6.5 \times 10^4$ |
| **b.** 843,000 | = | $8.43 \times 10^5$ |
| **c.** 0.000843 | = | $8.43 \times 10^{-4}$ |
| **d.** 0.07 | = | $7.0 \times 10^{-2}$ |

## Example 11

With a calculator set in the mode for scientific notation, perform the following operations.

**a.** $6.3 \div 360$

**b.** $82{,}000 \times 75{,}000$

### Solution

**a.** $6.3 \div 360 = 1.75\text{ E} - 02$

This answer is the same as $1.75 \times 10^{-2}$ ( or 0.0175 ).

**b.** $82{,}000 \times 75{,}000 = 6.15\ \ 09$

This answer is the same as $6.15 \times 10^9$ ( or 6,150,000,000 ).

*Now Work Exercises 11-14 in the Margin.*

Write each number in scientific notation.

**11.** 8470

$8.47 \times 10^3$

**12.** 0.000026

$2.6 \times 10^{-5}$

Perform the following operations with a calculator set in the mode for scientific notation.

**13.** $72{,}000 \times 93{,}000$

$6.696 \times 10^9$

**14.** $3.1 \div 180$

$1.7\overline{2} \times 10^{-2}$

Name ______________ Section ______________ Date ______________

# Exercises 5.4

Change each decimal to fraction form. Do not reduce.

**1.** 0.7 **2.** 0.6 **3.** 0.5 **4.** 0.48

**5.** 0.016 **6.** 0.125 **7.** 8.35 **8.** 7.2

Change each decimal to fraction form ( or mixed number form ), and reduce it if possible.

**9.** 0.125 **10.** 1.25 **11.** 1.8 **12.** 0.375

**13.** −2.75 **14.** −3.45 **15.** 0.33 **16.** 0.029

Change each fraction to decimal form. If the decimal is nonterminating, write it using the bar notation over the repeating pattern of digits.

**17.** $\frac{2}{3}$ **18.** $\frac{5}{16}$ **19.** $\frac{-1}{7}$ **20.** $\frac{-5}{9}$

**21.** $\frac{11}{16}$ **22.** $\frac{9}{16}$ **23.** $\frac{7}{11}$ **24.** $\frac{5}{18}$

**ANSWERS**

1. $\frac{7}{10}$
2. $\frac{6}{10}$
3. $\frac{5}{10}$
4. $\frac{48}{100}$
5. $\frac{16}{1000}$
6. $\frac{125}{1000}$
7. $\frac{835}{100}$
8. $\frac{72}{10}$
9. $\frac{1}{8}$
10. $1\frac{1}{4}$
11. $1\frac{4}{5}$
12. $\frac{3}{8}$
13. $-2\frac{3}{4}$
14. $-3\frac{9}{20}$
15. $\frac{33}{100}$
16. $\frac{29}{1000}$
17. $0.\overline{6}$
18. $0.3125$
19. $-0.\overline{142857}$
20. $-0.\overline{5}$
21. $0.6875$
22. $0.5625$
23. $0.\overline{63}$
24. $0.2\overline{7}$

25. 0.208

26. 0.031

27. 0.917

28. 0.813

29. 5.714

30. 2.222

31. −2.273

32. − 0.739

33. 1.5

34. 4

35. 1.888

36. 15.59

37. 71.22

38. 10.38

39. −2.875

40. 0.97

41. − 0.625

42. − 0.375

43. 0.0567

44. 9.9225

Change each fraction to decimal form rounded off to the nearest thousandth.

**25.** $\frac{5}{24}$ **26.** $\frac{1}{32}$ **27.** $\frac{11}{12}$ **28.** $\frac{13}{16}$

**29.** $\frac{40}{7}$ **30.** $\frac{20}{9}$ **31.** $\frac{25}{-11}$ **32.** $\frac{17}{-23}$

Perform the indicated operations by writing all the numbers in decimal form. Round off to the nearest thousandth if the decimal is nonterminating.

**33.** $\frac{3}{4}+0.35+\frac{2}{5}$

**34.** $\frac{1}{4}+\frac{3}{10}+3.45$

**35.** $\frac{7}{8}+\frac{3}{5}+0.413$

**36.** $7+6\frac{37}{100}+2\frac{11}{50}$

**37.** $25.03+35+6\frac{3}{10}+4\frac{89}{100}$

**38.** $\frac{3}{100}+8\frac{3}{5}+1\frac{3}{4}$

**39.** $1-0.125-3.75$

**40.** $3-0.78-1.25$

**41.** $0.375-1$

**42.** $\frac{5}{8}-1$

**43.** $\left(\frac{3}{10}\right)^2(0.63)$

**44.** $\left(2\frac{1}{10}\right)^2(1.5)^2$

Name ______________________ Section ____________ Date __________

**45.** $\left(-1\frac{3}{8}\right)(2.1)(3.6)$

**46.** $\left(2\frac{1}{4}\right)\left(-3\frac{1}{2}\right)(4.1)$

**47.** $72.186 \div \frac{3}{5}$

**48.** $917 \div \frac{1}{4}$

**49.** $\left|-3\frac{3}{4}\right| - |21.3|$

**50.** $\left|22\frac{4}{5}\right| + |-5.8|$

**51.** $\left|-5\frac{1}{2}\right| + \left|3\frac{1}{2}\right| - |-10.7|$

**52.** $|-4.72| \div \frac{8}{9}$

Evaluate each of the following expressions for $x = 1.5 = \frac{3}{2}$ and $y = \frac{2}{3}$. ( Do not use $\frac{2}{3}$ in decimal form because you will get only an approximate answer depending on the number of decimal places you use. ) Leave the answers in fraction or mixed number form.

**53.** $x^2 + 3x - 4$

**54.** $y^2 - 5y + 6$

**55.** $2x^2 - x - 3$

**56.** $3y^2 - 5y + 2$

**57.** $xy^3 - xy$

**58.** $x^2y^2 + 2xy - 5$

**59.** $x^3 + 3x^2 + 3x + 1$

**60.** $x^2 + 6xy + 9y^2$

**ANSWERS**

**45.** $-10.395$

**46.** $-32.2875$

**47.** $120.31$

**48.** $3668$

**49.** $-17.55$

**50.** $28.6$

**51.** $-1.7$

**52.** $5.31$

**53.** $2\frac{3}{4}$

**54.** $3\frac{1}{9}$

**55.** $0$

**56.** $0$

**57.** $-\frac{5}{9}$

**58.** $-2$

**59.** $15\frac{5}{8}$

**60.** $12\frac{1}{4}$

61. $1.8 \times 10^5$

62. $7.62 \times 10^4$

63. $1.24 \times 10^{-4}$

64. $9.12 \times 10^{-7}$

65. $8.9 \times 10^2$

66. $4.321 \times 10^{-1}$

67. $5.88 \times 10^{12}$

68. a. 186,000

b. $1.116 \times 10^7$

c. $6.696 \times 10^8$

69. 40,000,000,000,000,000,000,000,000

70. 861,300,000; $8.613 \times 10^8$

71. 49,068,000,000; $4.9068 \times 10^{10}$

72. 149,553,000 $= 1.49553 \times 10^8$; 8,900,000 $8.9 \times 10^6$

Write each of the following numbers in scientific notation.

**61.** 180,000 **62.** 76,200 **63.** 0.000124

**64.** 0.000000912 **65.** 890 **66.** 0.4321

**67.** A light–year (the distance traveled by light in 1 year) is 5,880,000,000,000 miles. Write this number in scientific notation.

**68.** Light travels approximately $1.86 \times 10^5$ miles per second.
**a.** Write this number in decimal notation.
**b.** Write the equivalent number of miles per minute in scientific notation.
**c.** Write the equivalent number of miles per hour in scientific notation.

**69.** Nematode sea worms are the most numerous of all sea and land animals, with an estimated population of $4 \times 10^{25}$. Write this number in decimal notation.

**70.** The United States produces about 3.3 pounds of solid waste per capita per day. The U.S. population is about 261,000,000 people. Translate these facts into the number of pounds of solid waste produced each day in the United States. Write the number in both decimal notation and scientific notation.

**71.** The citizens of the United States use 188 gallons of water per capita per day for home use. The U.S. population is about 261,000,000 people. Translate these facts into the number of gallons of water used each day in homes in the United States. Write the number in both decimal notation and scientific notation.

**72.** The number of cars in the United States in 573 per 1000 people, and the number of cars in Australia is 500 per 1000 people. The population of the United States is about 261,000,000, and the population of Australia is about 17,800,000. Write the numbers of cars in the United states and in Australia in both decimal notation and in scientific notation.

Name ______________________ Section ______________ Date __________

**ANSWERS**

**73.** In 1992, the state of Texas consumed $9.915 \times 10^{12}$ Btu ( British thermal units ) of energy, and all the states consumed a total of $8.2128 \times 10^{13}$ Btu. Write both these numbers in decimal notation.

**74.** About how many minutes does it take for light to reach Earth from the Sun if the distance is about 93,000,000 miles and light travels 186,000 miles per second?

## Calculators

Set your calculator in scientific mode, and use it to perform the following operations. Write your answers

**a.** in the scientific notation you read on the display and

**b.** in decimal notation.

**75.** $8500 \times 90{,}000$

**76.** $2500 \times 2500$

**77.** $0.81 \div 9000$

**78.** $560 \div 0.00025$

**79.** $4000 \times 75{,}000 + 5000 \times 3000$

**80.** $700 \times 6000 + 200 \times 350{,}000$

**Answers:**

**73.** 9,915,000,000,000; 82,128,000,000,000

**74.** about 8.3 minutes

**75. a.** $7.65 \times 10^{8}$ **b.** 765,000,000

**76. a.** $6.25 \times 10^{6}$ **b.** 6,250,000

**77. a.** $9.0 \times 10^{-5}$ **b.** 0.00009

**78. a.** $2.24 \times 10^{6}$ **b.** 2,240,000

**79. a.** $3.15 \times 10^{8}$ **b.** 315,000,000

**80. a.** $7.42 \times 10^{7}$ **b.** 74,200,000

## WRITING AND THINKING ABOUT MATHEMATICS

81. [Respond below exercise.]

82. [Respond below exercise.]

83. [Respond below exercise.]

**81.** Give five examples of values for $x$ and $y$ for which

**a.** $|x+y| < |x| + |y|$ and for which

**b.** $|x+y| = |x| + |y|$

**c.** and **d.** Make a general statement that you believe to be true about the value of $x$ and $y$ in statements **a.** and **b.**

a. and b. Examples will vary.

c. $|x+y| < |x| + |y|$ when $x$ and $y$ are opposite in sign.

d. $|x+y| = |x| + |y|$ when $x$ and $y$ have the same sign or one or both are 0.

**82.** If someone made a statement that "$|x+y| > |x| + |y|$ is never possible for any value of $x$ and $y$," would you believe this statement? Explain why or why not.

Yes, because of the results of Exercise 81. Explanations will vary.

With a., b., and c. as examples, explain in your own words how you can tell quickly when one decimal number is larger ( or smaller ) than another decimal number.

**83.** **a.** The decimal number 2.765274 is larger than the decimal number 2.763895.

**b.** The decimal number 17.345678 is larger than the decimal number 17.345578.

**c.** The decimal number 0.346973 is larger than the decimal number 0.346972.

Explanations will vary. Digits are compared, place by place, until one digit is larger (or smaller.)

# 5.5 Solving Equations with Decimal Numbers

## Solving Equations with Decimals

Techniques for solving equations have been discussed in Sections 1.4, 2.7, 3.7, and 4.6. In this section, we continue our development of this most important topic by considering equations that have decimals as constants and coefficients. The two principles, the **Addition Principle** and the **Multiplication Principle**, as discussed in Sections 1.4, 2.7, and 3.7, still form the foundation for the skills needed to solve equations, regardless of the types of numbers used as constants and coefficients. **Remember that the object in solving an equation is to isolate the variable on one side of the equation.**

**Objectives**

① Learn how to solve equations with decimals.

② Know how to find the circumference and area of a circle.

### Example 1

Solve the equation $5(x + 3.7) = 27.05$.

**Solution**

| | |
|---|---|
| $5(x+3.7)=27.05$ | Write the equation. |
| $5x+18.5=27.05$ | Use the distributive property. |
| $5x+18.5-18.5=27.05-18.5$ | Add $-18.5$ to both sides. |
| $5x=8.55$ | Simplify. |
| $\frac{1}{\cancel{5}}\cdot\cancel{5}x=\frac{1}{5}\cdot 8.55$ | Multiply both sides by $\frac{1}{5}$. |
| $x=1.71$ | Simplify. |

### *Completion Example 2*

Supply the reasons for each step in the following solution of the equation: $0.6y + 18.4 = 1.5y + 22.9$.

**Solution**

| | |
|---|---|
| $0.6y+18.4=1.5y+22.9$ | ________________ |
| $0.6y+18.4-18.4=1.5y+22.9-18.4$ | ________________ |
| $0.6y=1.5y+4.5$ | ________________ |
| $0.6y-1.5y=1.5y+4.5-1.5y$ | ________________ |
| $-0.9y=4.5$ | ________________ |
| $\frac{\cancel{-0.9}y}{\cancel{-0.9}}=\frac{4.5}{-0.9}$ | ________________ |
| $y=-5$ | ________________ |

Solve the following equations.

**1.** $0.5x + 20.7 = 14.3$
$x = -12.8$

**2.** $3(1.7y + 3.6) = 12.33$
$y = 0.3$

**3.** $0.02x + 52.8 = 0.06x - 20$
$x = 1820$

### *Completion Example 3*

Solve the equation $0.1x + 3.8 = 5.72 - 0.3x$.

**Solution**

$$\begin{aligned} 0.1x + 3.8 &= 5.72 - 0.3x \\ 0.1x + 3.8 - \_\_\_\_ &= 5.72 - 0.3x - \_\_\_\_ \\ 0.1x &= \_\_\_\_ - 0.3x \\ 0.1x + \_\_\_\_ &= \_\_\_\_ - 0.3x + \_\_\_\_ \\ \_\_\_\_ x &= \_\_\_\_ \\ \_\_\_\_ x &= \_\_\_\_ \\ x &= \_\_\_\_ \end{aligned}$$

*Now Work Exercises 1-3 in the Margin.*

### Example 4

A student bought a textbook and a calculator for a total of $184.90 (including tax). If the calculator cost $15.50 more than the book, what was the cost of each item?

**Solution**

Let $x$ = the cost of the book.
Then $x + 15.50$ = the cost of the calculator.
The equation to be solved is:

$$\begin{aligned} \overbrace{x + 15.50}^{\text{cost of calculator}} + \overbrace{x}^{\text{cost of book}} &= \overbrace{184.90}^{\text{total spent}} \\ 2x + 15.50 &= 184.90 \\ 2x + 15.50 - 15.50 &= 184.90 - 15.50 \\ 2x &= 169.40 \\ x &= 84.70 \quad \text{cost of textbook} \\ x + 15.50 &= 100.20 \quad \text{cost of calculator} \end{aligned}$$

The textbook costs $84.70 and the calculator costs $100.20, with tax included in each price.

### Example 5

A car rental agency charges $14.50 per day plus $0.15 per mile. How many miles was the car driven if the charge for one day was $48.25?

**Solution**

Let $x$ = the number of miles driven.
Then $0.15x$ = the charge for the miles driven.
The related equation is:

$$14.50 + 0.15x = 48.25$$
$$14.50 + 0.15x - 14.50 = 48.25 - 14.50$$
$$0.15x = 33.75$$
$$x = 225$$

The car was driven 225 miles.

## Circumference and Area of Circles

A **formula** is a general statement ( usually an equation ) that relates two or more variables. In this section, we will discuss the formulas ( and methods of using these formulas ) for the circumference ( perimeter ) and area of a circle.

We will need the following definitions and terms related to circles.

**Important Terms and Definitions for Circles:**

**Circle:** The set of all points in a plane that are some fixed distance from a fixed point called the **center** of the circle.

**Radius:** The fixed distance from the center of a circle to any point on the circle. ( The letter $r$ is used to represent the radius of a circle. )

**Diameter:** The distance from one point on a circle to another point on the circle measured through the center. ( The letter $d$ is used to represent the diameter of a circle. )

**Circumference:** Perimeter of a circle.

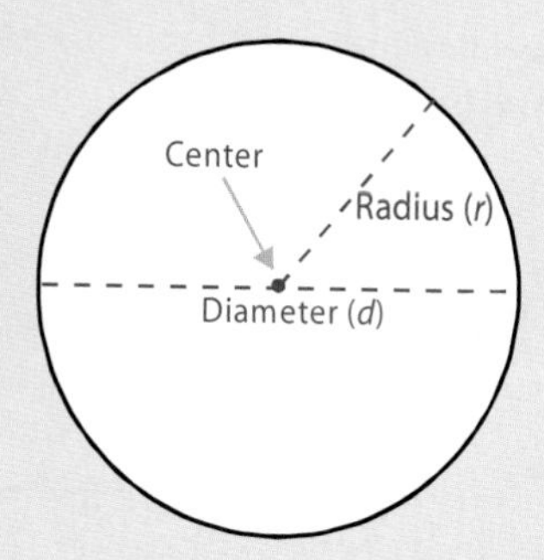


**Note:** $\pi$ is the symbol used for the constant 3.1415926535 . . . . This number is an infinite nonrepeating decimal (an **irrational number**). (See Table 4 in the Appendix for a more detailed discussion.) For our purposes, we will use $\pi = 3.14$ (accurate to hundredths). However, you should always be aware that 3.14 is only an approximation for $\pi$ and that the related answers are only approximations.

For the formulas related to circles, we need to note that a diameter is twice as long as a radius. That is, $\boldsymbol{d = 2r}$.

**Formulas for Circles:**

For circumference: $\boldsymbol{C = 2\pi r}$ and $\boldsymbol{C = \pi d}$

For area: $\boldsymbol{A = \pi r^2}$

## Example 6

Find the circumference of a circle with a diameter of 3 meters.

### Solution

Sketch the circle and label a diameter.

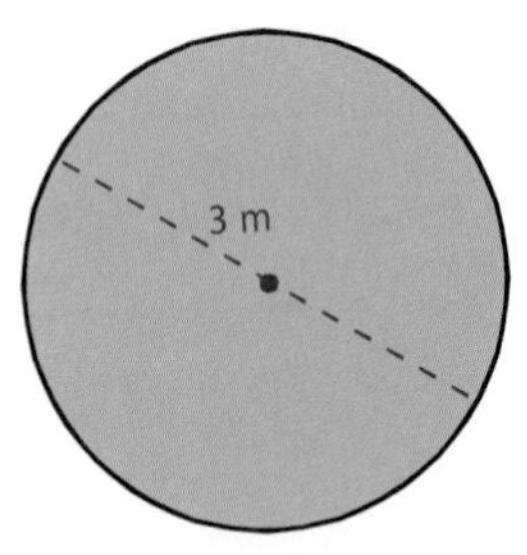


$$C = \pi d$$

$$C = 3.14(\ 3\ )$$

$$C = 9.42 \text{ m}$$

The circumference is 9.42 meters. (Remember that 9.42 m is only an approximate answer because 3.14 is only an approximation for $\pi$.)

## Example 7

Find the circumference of a circle with a radius of 6 feet.

### Solution

Sketch the circle and label a radius.

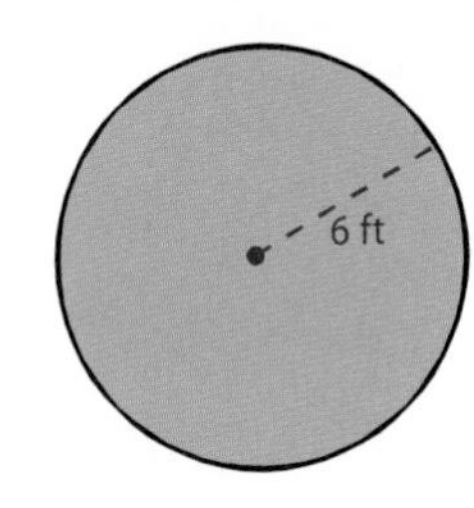


$$C = 2\pi r$$

$$C = 2(\ 3.14\ )(\ 6\ )$$

$$C = 37.68 \text{ ft.}$$

The circumference is 37.68 feet.

## Example 8

Find the area of the washer (shaded portion) with dimensions as shown.

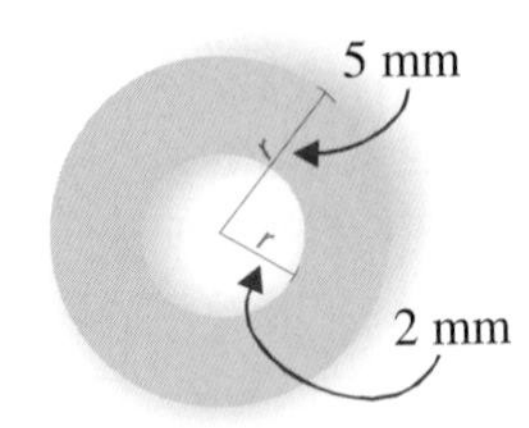


### Solution

Subtract the area of the inside (smaller) circle from the area of the outside (larger) circle. This difference will be the area of the washer (shaded portion).

**Area of Larger Circle**

$A = \pi r^2$
$A = 3.14(5^2)$
$A = 3.14(25)$
$A = 78.5 \text{ mm}^2$

**Area of Smaller Circle**

$A = \pi r^2$
$A = 3.14(2^2)$
$A = 3.14(4)$
$A = 12.56 \text{ mm}^2$

| Difference: | | |
|---|---|---|
| | $78.50 \text{ mm}^2$ | area of outer circle |
| | $- 12.56 \text{ mm}^2$ | area of inner circle |
| | $65.94 \text{ mm}^2$ | area of washer |

The area of the washer is $65.94 \text{ mm}^2$.

**Alternate Approach**

In this approach, we leave $\pi$ in the calculations and evaluate at the end.

For the larger circle: $A = \pi(5^2) = 25\pi$

For the smaller circle: $A = \pi(2^2) = 4\pi$

The difference is: $25\pi - 4\pi = 21\pi = 21(3.14) = 65.94 \text{ mm}^2$

## Example 9

Find the radius of a circle if its circumference is 31.4 inches.

### Solution

Use the formula $C = 2\pi r$, substitute 31.4 for $C$, and solve for $r$.

$$\begin{aligned} C &= 2\pi r \\ 31.4 &= 2(3.14)r \\ 31.4 &= 6.28r \\ \frac{31.4}{6.28} &= \frac{\cancel{6.28}r}{\cancel{6.28}} \\ 5 &= r \end{aligned}$$

The radius is 5 inches.

## Completion Example Answers

**2.**

| | |
|---|---|
| $0.6y + 18.4 = 1.5y + 22.9$ | Write the equation. |
| $0.6y + 18.4 - 18.4 = 1.5y + 22.9 - 18.4$ | Add – 18.4 to both sides. |
| $0.6y = 1.5y + 4.5$ | Simplify. |
| $0.6y - 1.5y = 1.5y + 4.5 - 1.5y$ | Add $-1.5y$ to both sides. |
| $-0.9y = 4.5$ | Simplify. |
| $\frac{\cancel{-0.9}y}{\cancel{-0.9}} = \frac{4.5}{-0.9}$ | Divide both sides by – 0.9. |
| $y = -5$ | Simplify. |

**3.**

$$
\begin{aligned}
0.1x + 3.8 &= 5.72 - 0.3x \\
0.1x + 3.8 - \mathbf{3.8} &= 5.72 - 0.3x - \mathbf{3.8} \\
0.1x &= \mathbf{1.92} - 0.3x \\
0.1x + \mathbf{0.3x} &= \mathbf{1.92} - 0.3x + \mathbf{0.3x} \\
\mathbf{0.4x} &= \mathbf{1.92} \\
\frac{\mathbf{0.4x}}{\mathbf{0.4}} &= \frac{\mathbf{1.92}}{\mathbf{0.4}} \\
x &= \mathbf{4.8}
\end{aligned}
$$

Name ______ Section ______ Date ______

## Exercises 5.5

Solve the following equations.

1. $0.7x + 10.2 = 31.9$

2. $0.3x + 17.8 = 57.1$

3. $0.5x - 22.7 = 14.35$

4. $0.6x - 34.6 = 15.2$

5. $8x + 6.45 = -20.03$

6. $9x + 8.57 = -124.9$

7. $0.2(y + 6.3) = 17.5$

8. $0.3(y + 5.4) = 22.77$

9. $\frac{1}{10}y - 3.99 = 12.6$

10. $\frac{1}{10}y - 4.56 = 27.8$

11. $\frac{1}{2}t + 16.3 = 21.45$

12. $\frac{1}{3}t - 0.5 = -17.8$

13. $0.01t - 10.3 = 14.73$

14. $0.02t + 16.5 = 22.8$

15. $0.1x + 0.2(x - 7) = 0.142$

16. $0.3x + 0.1(x - 15) = 0.116$

17. $-0.5(x + 8.5) = -0.3x - 10.3$

18. $-0.4(x - 12.7) = -0.2x + 11.65$

19. $0.12y + 0.25y - 5.895 = 4.3y$

20. $0.15y + 32y - 21.0005 = 10.5y$

21. $3y - 4y = -22.88 + y$

22. $8y + 7y = -35.45 + 20y$

**ANSWERS**

1. $x = 31$
2. $x = 131$
3. $x = 74.1$
4. $x = 83$
5. $x = -3.31$
6. $x = -14.83$
7. $y = 81.2$
8. $y = 70.5$
9. $y = 165.9$
10. $y = 323.6$
11. $t = 10.3$
12. $t = -51.9$
13. $t = 2503$
14. $t = 315$
15. $x = 5.14$
16. $x = 4.04$
17. $x = 30.25$
18. $x = -32.85$
19. $y = -1.5$
20. $y = 0.97$
21. $y = 11.44$
22. $y = 7.09$

23. $z = 0$

24. $z = 0$

25. $z = -4.1\overline{66}$

26. $z = -2.038$

27. $x = 0$

28. $x = 0$

29. $x = 20.3$

30. $x = -42.5$

31. $17.50 (manual); $12.95 (disk)

32. $16.62 (shirt); $37.12 (shoes)

33. −13.1

34. −5.57

35. 240 miles

**23.** $0.7(z + 14.1) = 0.3(z + 32.9)$

**24.** $0.8(z - 6.21) = 0.2(z - 24.84)$

**25.** $1.3 - 0.4z + z = -1.5$

**26.** $8.5 - 1.6z - z = 13.8$

**27.** $3.2x + 0.11 = 5(x + 0.022)$

**28.** $7.2x - 0.33 = 4(x - 0.0825)$

**29.** $2.45x + 3.56x = 142.303 - x$

**30.** $8.79x - 14.23x = 273.7 + x$

**31.** The total cost of a box of computer disks and a manual was $30.45. If the price of the box of disks was $4.55 less than the price of the manual, what was the price of each item?

**32.** A pair of tennis shoes and a shirt cost a total of $53.74. If the price of the tennis shoes was $20.50 more than the price of the shirt, what was the price of each item?

**33.** If 14.66 plus twice a number equals the sum of the number and 1.56, what is the number?

**34.** If six times a number is increased by 15.35, the result is 12.5 less than the number. What is the number?

**35.** A truck rental agency rents trucks for $45.50 per day plus $0.25 per mile. If the fee for one day's use was $105.50, how many miles was the truck driven?

Name ____________________ Section ____________ Date __________

**36.** If the rental on a luxury car is \$35 per day plus \$0.18 per mile, what would be the total paid if the car was driven for 150 miles in one day?

Find **a.** the circumference and **b.** the area of each of the following circles with the indicated dimensions. (Use $\pi = 3.14$.)

**37.**

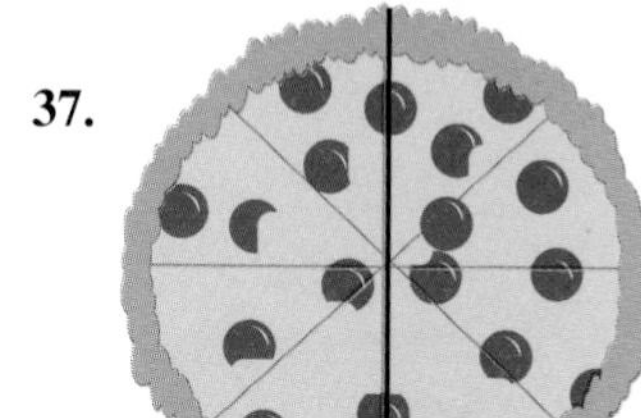


**38.**

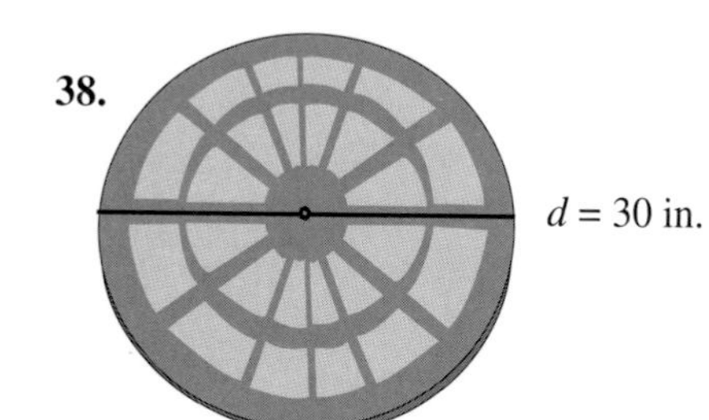


**39.**

**40.**

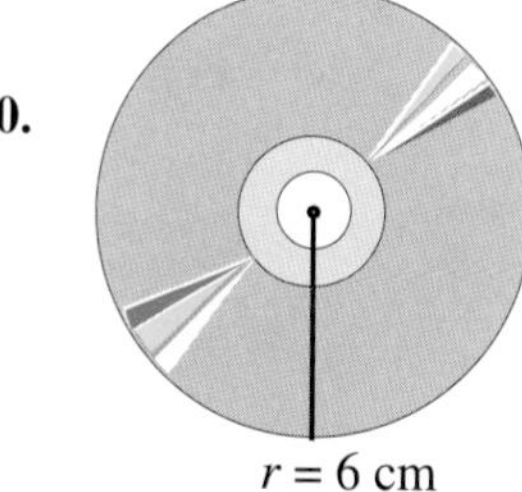


Find **a.** the circumference and **b.** the area of each of the following shapes with the indicated dimensions. (Use $\pi = 3.14$.)

**41.**

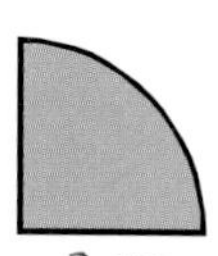


**42.**

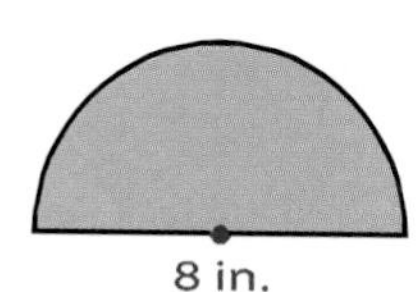


**43.**

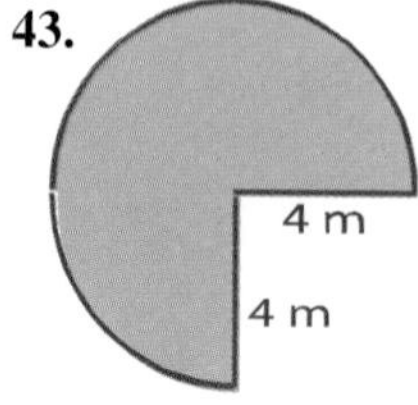


**44.**

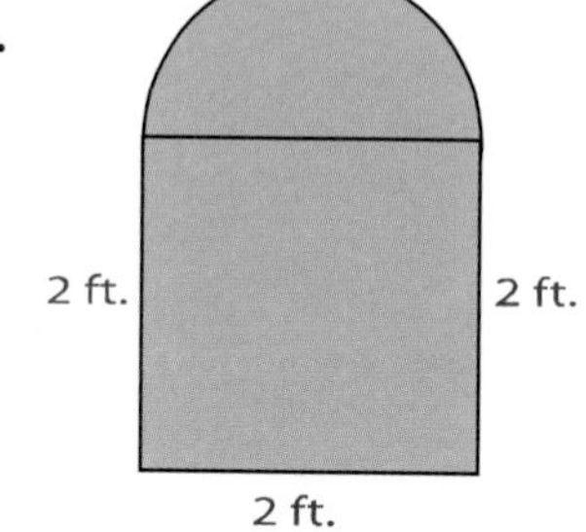


**ANSWERS**

**36.** \$62

**37. a.** 50.24 in.

**b.** 200.96 in.$^2$

**38. a.** 94.2 in.

**b.** 706.5 in.$^2$

**39. a.** 150.72 mm

**b.** 1808.64 mm$^2$

**40. a.** 37.68 cm

**b.** 113.04 cm$^2$

**41. a.** 10.71 cm

**b.** 7.065 cm$^2$

**42. a.** 20.56 in.

**b.** 25.12 in.$^2$

**43. a.** 26.84 m

**b.** 37.68 m$^2$

**44. a.** 9.14 ft.

**b.** 5.57 ft.$^2$

45. d = 16.8 in.

r = 8.4 in.

46. 14.86 m

47. 2.95 in.

48. 4.6 ft., 4.6 ft., 8.2 ft.

49. [Respond below exercise.]

**45.** Suppose that the circumference of a circle is known to be 52.752 inches. Find both the length of a diameter and the length of a radius of this circle.

**46.** If the perimeter of an equilateral triangle (all three sides equal) is decreased by 11.58 meters, the result is 33 meters. What is the length of one side of the triangle?

**47.** If the width of a rectangle is 2.45 inches and the perimeter of the rectangle is 10.8 inches, what is the length of the rectangle? (*Hint:* Use the formula $P = 2l + 2w$.)

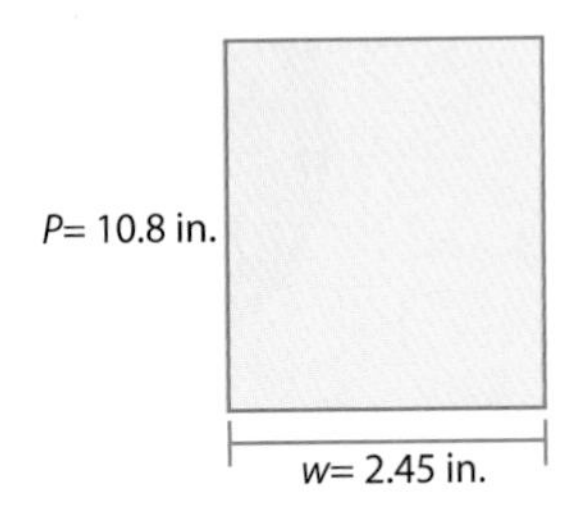


**48.** If the perimeter of a rectangle is 25.6 feet and one side is 8.2 feet long, what are the lengths of the other three sides? (*Hint*: Use the formula $P = 2l + 2w$.)

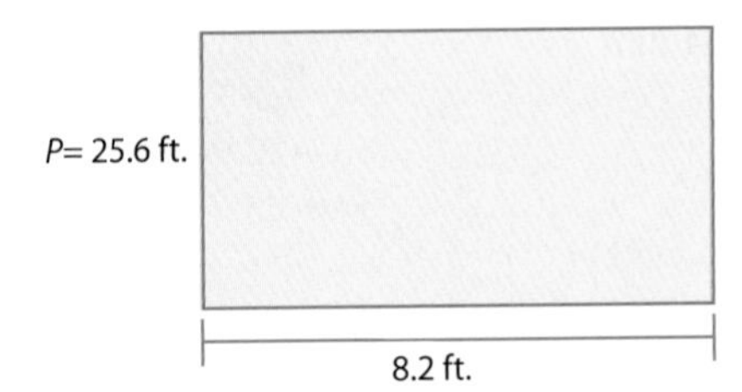


## WRITING AND THINKING ABOUT MATHEMATICS

**49.** First draw a square with sides that are each 2 inches long. Next, draw a circle inside the square that just touches each side of the square. (This circle is said to be inscribed in the square.)

**a.** What is the length of a diameter of this circle?

**b.** Do you think that the circumference is less than 8 inches or more than 8 inches? Why?

a. Diameter is 2 inches.

b. The perimeter of the square is 8 inches. Since the circle fits inside the square, its circumference must be less than 8 inches.

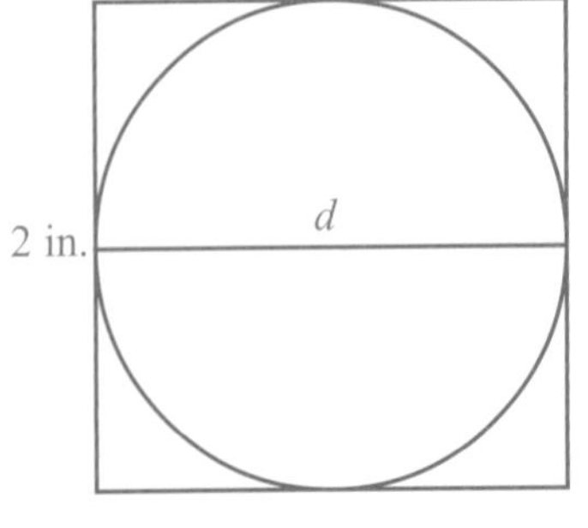

# 5.6 Square Roots and the Pythagorean Theorem

## Square Roots and Real Numbers

A number is **squared** when it is multiplied by itself. For example,

$$5^2 = 5 \cdot 5 = 25$$
$$(-20)^2 = (-20)(-20) = 400$$
$$\text{and } (3.2)^2 = (3.2)(3.2) = 10.24$$

If an integer is squared, the result is called a **perfect square.** Thus, 25 and 400 are both perfect squares and 10.24 is not a perfect square. Table 5.1 shows the perfect squares found by squaring the positive integers from 1 to 20. A more complete table is located at the back of the text.

**Table 5.1 Squares of Positive Integers from 1 to 20 (Perfect Squares)**

| | | | | |
|---|---|---|---|---|
| $1^2 = 1$ | $5^2 = 25$ | $9^2 = 81$ | $13^2 = 169$ | $17^2 = 289$ |
| $2^2 = 4$ | $6^2 = 36$ | $10^2 = 100$ | $14^2 = 196$ | $18^2 = 324$ |
| $3^2 = 9$ | $7^2 = 49$ | $11^2 = 121$ | $15^2 = 225$ | $19^2 = 361$ |
| $4^2 = 16$ | $8^2 = 64$ | $12^2 = 144$ | $16^2 = 256$ | $20^2 = 400$ |

**Objectives**

1. Memorize the squares of the whole numbers from 1 to 20.
2. Memorize the perfect square whole numbers from 1 to 400.
3. Understand the terms **square root, radical sign, radicand,** and **radical**.
4. Know how to use a calculator to find square roots.
5. Understand the terms **right triangle, hypotenuse,** and **leg**.
6. Know and be able to use the Pythagorean Theorem.

Now, we want to reverse the process of squaring. That is, given a number, we want to find a number that when squared will result in the given number. This is called **finding the square root** of the given number. For example, given 49, find the **square root** of 49. What number squared will result in 49? Since

$7^2 = 49$, the number 7 is called the **square root** of 49,

and since

$10^2 = 100$, the number 10 is called the **square root** of 100.

We write

$$\sqrt{49} = 7 \quad \text{and} \quad \sqrt{100} = 10$$

Similarly,

$$\sqrt{25} = 5 \quad \text{since} \quad 5^2 = 25$$
$$\sqrt{121} = 11 \quad \text{since} \quad 11^2 = 121$$
$$\text{and} \quad \sqrt{1} = 1 \quad \text{since} \quad 1^2 = 1$$

The symbol $\sqrt{\ }$ is called a **radical sign**.

The number under the radical sign is called the **radicand**.

The complete expression, such as $\sqrt{49}$, is called a **radical**.

Each perfect square has two square roots, one positive and one negative. For example, since $(-6)^2 = 36$ and $6^2 = 36$, both $-6$ and 6 are square roots of 36. However, to distinguish between the two square roots, the positive square root, 6, is called the **principal square root**. And, unless otherwise stated, the term **square root** is understood to mean the positive (or principal) square root. Thus,

$$\sqrt{36} = 6 \qquad \text{and} \qquad -\sqrt{36} = -6$$

As a special case, for 0,

$$\sqrt{0} = -\sqrt{0} = 0$$

Table 5.2 contains the square roots of the perfect square numbers from 1 to 400. (Note that the values shown in Table 5.2 are found by reversing the process illustrated in Table 5.1.)

**Table 5.2 Squares Roots of Perfect Squares from 1 to 400**

| | | | | |
|---|---|---|---|---|
| $\sqrt{1} = 1$ | $\sqrt{25} = 5$ | $\sqrt{81} = 9$ | $\sqrt{169} = 13$ | $\sqrt{289} = 17$ |
| $\sqrt{4} = 2$ | $\sqrt{36} = 6$ | $\sqrt{100} = 10$ | $\sqrt{196} = 14$ | $\sqrt{324} = 18$ |
| $\sqrt{9} = 3$ | $\sqrt{49} = 7$ | $\sqrt{121} = 11$ | $\sqrt{225} = 15$ | $\sqrt{361} = 19$ |
| $\sqrt{16} = 4$ | $\sqrt{64} = 8$ | $\sqrt{144} = 12$ | $\sqrt{256} = 16$ | $\sqrt{400} = 20$ |

### Example 1

Refer to Table 5.2 to find the following square roots.

**a.** $\sqrt{100}$

**b.** $\sqrt{169}$

**c.** $-\sqrt{121}$

**d.** $-\sqrt{361}$

#### Solution

**a.** $\sqrt{100} = 10$

**b.** $\sqrt{169} = 13$

**c.** $-\sqrt{121} = -11$

**d.** $-\sqrt{361} = -19$

Many square roots, such as $\sqrt{2}$, $\sqrt{3}$, and $\sqrt{24}$ (square roots of numbers other than perfect squares), are **irrational numbers**, and their decimal forms are **infinite nonrepeating decimals**. In Section 5.5, we discussed the fact that $\pi$ is also an irrational number. Table 4 in the Appendix illustrates $\pi$ accurate to over 3700 decimal places. In fact, computers have been used to calculate $\pi$ accurate to over 1 billion decimal places!

**Real Numbers** are numbers that are either rational or irrational. **That is, every rational number and every irrational number is a real number**. All the types of numbers discussed in this text are real numbers. The relationships among the various types of numbers are shown in the diagram in Figure 5.3.

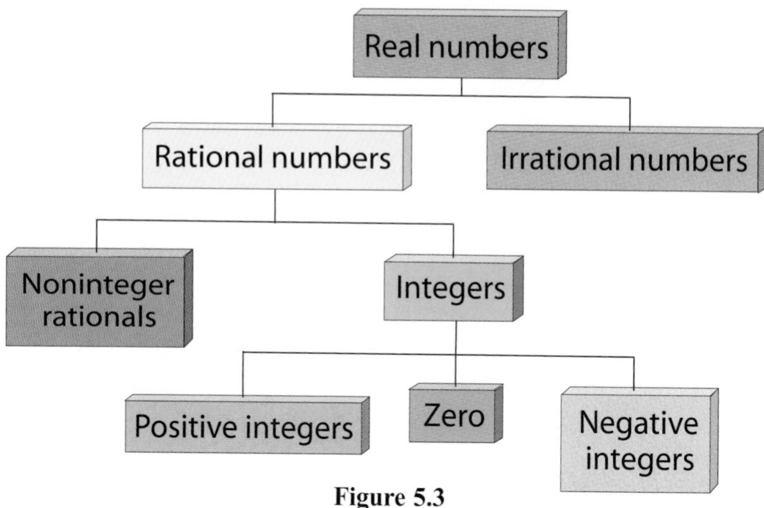


**Figure 5.3**

The definition of square root can be stated in terms of real numbers.

> **Definition:**
>
> For any real number $a$ and any nonnegative real number $b$, $a$ is a **square root** of $b$ if $a^2 = b$. If $a$ is positive, then we write $a = \sqrt{b}$. Thus, $\left(\sqrt{b}\right)^2 = b$.

## Example 2

Find the value of each of the following expressions.

**a.** $\left(\sqrt{81}\right)^2$

**b.** $\left(\sqrt{25}\right)^2$

**c.** $\left(\sqrt{3}\right)^2$

### Solution

**a.** $\left(\sqrt{81}\right)^2 = (9)^2 = 81$

**b.** $\left(\sqrt{25}\right)^2 = (5)^2 = 25$

**c.** Even though we do not know the exact decimal value of $\sqrt{3}$, its square must be 3. Thus, $\left(\sqrt{3}\right)^2 = 3$.

As a point of interest and for understanding the concept of an irrational number, $\pi$ is shown accurate to over 3700 decimal places in Table 4 in the back of the text. Some students may find the following square root algorithm interesting. For $\sqrt{2}$,

STEP 1: Find the closest square number to 2 and divide by this number:

```
      1.
   1)2.00 00 00
     1
     ---
     1 00
```

STEP 2: Double the quotient and write this number off to the left followed by a blank; bring down two 0's and divide again by this new number to the left, with the condition that the blank must be filled in by the same number used in the quotient:

```
quotient       1. 4
doubled     1)2.00 00 00
   ↓           1
               ---
  2 4          1 00
     ↑          96 ← 4 × 24 = 96
               ---
   blank          4
```

STEP 3: Continue the process until the desired accuracy is reached.

```
          1. 4  1
       1)2.00 00 00
          1
          ---
  24      1 00
           96
          ---
  28  1     400
            281 ← 281 x 1 = 281
            119
```

The values of most square roots can be only approximated with decimals. Consider the fact that 5 is between 4 and 9. Using inequality signs we can write

$$4 < 5 < 9$$

Thus, it seems reasonable (and it is true) that

$$\sqrt{4} < \sqrt{5} < \sqrt{9}$$

and we have

$$2 < \sqrt{5} < 3$$

Therefore, even though we may not know the exact value of $\sqrt{5}$, we do know that it is between 2 and 3.

Decimal approximations to $\sqrt{5}$ are shown here to illustrate more accurate estimates.

$$(2.2)^2 = 4.84 < 5$$
$$(2.23)^2 = 4.9729 < 5$$
$$(2.236)^2 = 4.999696 < 5$$
$$(2.237)^2 = 5.004169 > 5$$

Thus, we can now say that $\sqrt{5}$ is between 2.236 and 2.237. Or, symbolically,

$$2.236 < \sqrt{5} < 2.237.$$

## Example 3

Use a calculator and show by squaring 2.44 and 2.45 that $\sqrt{6}$ is between these two numbers.

### Solution

$$(2.44)^2 = 5.9536 < 6$$
$$(2.45)^2 = 6.0025 > 6$$

Thus, $\quad 2.44 < \sqrt{6} < 2.45$

## Example 4

Since $(4.898)^2 = 23.990404$ and $(4.899)^2 = 24.000201$, what can be said about $\sqrt{24}$ ?

### Solution

$\sqrt{24}$ is between 4.898 and 4.899, or

$$4.898 < \sqrt{24} < 4.899$$

Calculators can be used to approximate square roots accurate to several decimal places. To find, or approximate, a square root with a calculator, use the key labeled $\sqrt{x}$ . Example 5 illustrates the technique for both a scientific calculator and a TI–83 Plus graphing calculator. Note that the sequence of steps depends on the calculator.

## Example 5

Use a calculator to find the following square roots accurate to nine decimal places.

**a.** $\sqrt{2}$

**b.** $\sqrt{5}$

## Solution

### Using a Scientific Calculator

**a.** **Step 1:** Enter 2.

**Step 2:** Press the key $\sqrt{x}$

(Note: You may need to press a **2nd** key first and then the $\sqrt{x}$ key.)

**Step 3:** The display will show

$\sqrt{2} = 1.414213562$ accurate to nine decimal places

### Using a TI–83 Plus graphing calculator

**b.** **Step 1:** Press the 2nd key and then the $x^2$ key. This will display the symbol $\sqrt{\ }($.

**Step 2:** Enter 5 and ). This will display 5).

**Step 3:** Press ENTER and the display will show $\sqrt{(5)}$

2.236067977 accurate to nine decimal places

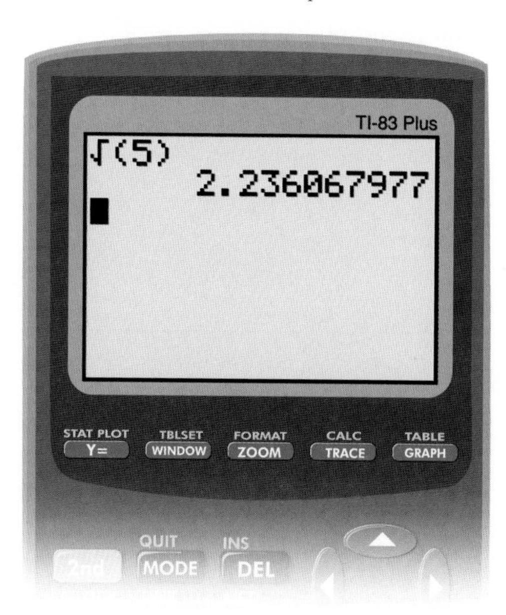


*Now Work Exercises 1-4 in the Margin.*

1. Find the value of each expression.

   **a.** $\left(\sqrt{64}\right)^2$
   64

   **b.** $\left(\sqrt{10}\right)^2$
   10

2. Use a calculator and show by squaring 2.82 and 2.83 that $\sqrt{8}$ is between these two numbers.
   $(2.82)^2 = 7.9524 < 8$ and $(2.83)^2 = 8.0089 > 8$

First estimate each square root between two integers. Then, use a calculator to find each square root accurate to nine decimal places.

3. $\sqrt{75}$
   8.660254038

4. $\sqrt{17}$
   4.123105626

## The Pythagorean Theorem

The following discussion involves right triangles and provides a variety of applications using squares and square roots.

### Terms Related to Right Triangles

**Right Triangle:** A triangle with a right angle (90°)

**Hypotenuse:** The longest side of a right triangle: the side opposite the right angle.

**Leg:** Each of the other two sides of a right triangle.

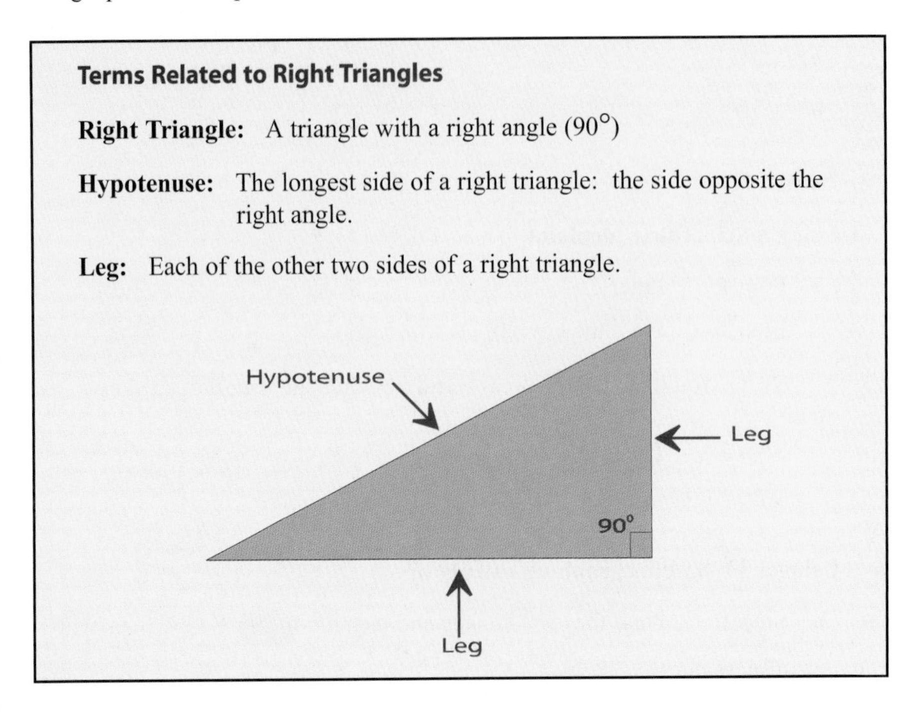


Pythagoras, (c. 585 - 501 B.C. ) a famous Greek mathematician, is given credit for discovering the following theorem (even though historians have found that the facts of the theorem were known before the time of Pythagoras).

### The Pythagorean Theorem

In a right triangle, the square of the hypotenuse is equal to the sum of the squares of the two legs:

$$c^2 = a^2 + b^2$$

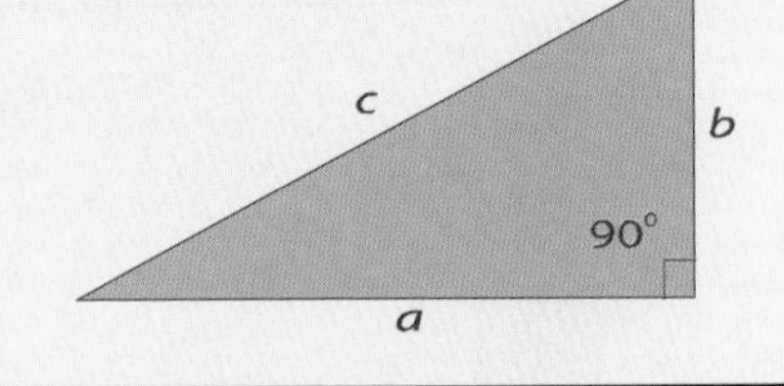


## Example 6

Show that a triangle with sides of 3 inches, 4 inches, and 5 inches must be a right triangle.

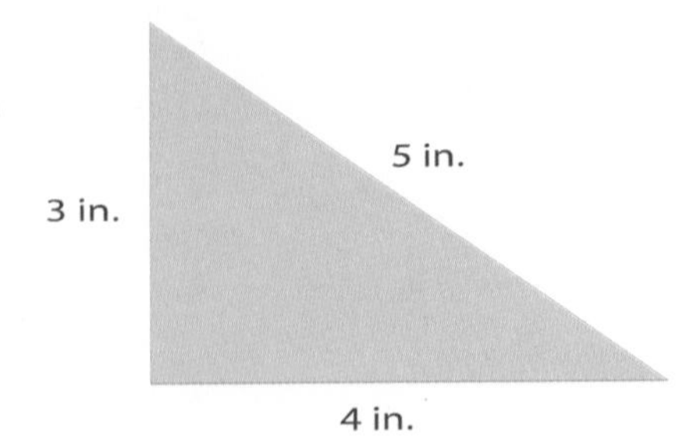

### Solution

If the triangle is a right triangle, then the longest side (5 inches) must be the hypotenuse, and the relationship

$5^2 = 3^2 + 4^2$ must be true.

Since $25 = 9 + 16$, the triangle is a right triangle. That is, the square of the hypotenuse is equal to the sum of the squares of the legs.

## Example 7

Use a calculator and the Pythagorean Theorem to find the length of the hypotenuse (accurate to two decimal places) of a right triangle with legs 1 centimeter long and 3 centimeters long.

### Solution

Let $c$ = the length of the hypotenuse.
Then, by the Pythagorean Theorem,

$$c^2 = 1^2 + 3^2$$
$$c^2 = 1 + 9$$
$$c^2 = 10$$
$$c = \sqrt{10}$$
$$c = 3.16 \text{ cm}$$

3 cm
c
1 cm

The length of the hypotenuse is about 3.16 cm long.

## Example 8

Use a calculator and the Pythagorean Theorem to find the length of a leg (accurate to two decimal places) of a right triangle with hypotenuse 20 inches long and the other leg 13 inches long.

### Solution

In this case, the length of the hypotenuse is known and the unknown is one of the legs.

Let $x$ = the length of the unknown leg.
Then, by the Pythagorean Theorem,

$$x^2 + 13^2 = 20^2$$
$$x^2 + 169 = 400$$
$$x^2 + 169 - 169 = 400 - 169$$
$$x^2 = 231$$
$$x = \sqrt{231}$$
$$x = 15.20 \quad \text{accurate to hundredths}$$

20 in.
13 in.
90°
x

So, the leg is 15.20 inches long (accurate to two decimal places).

## Example 9

A guy wire is attached to the top of a telephone pole and anchored in the ground 20 feet from the base of the pole. If the pole is 40 feet high, what is the length of the guy wire (accurate to two decimal places)?

### Solution

Let $x$ = the length of the guy wire.
Then, by the Pythagorean Theorem,

$$x^2 = 20^2 + 40^2$$
$$x^2 = 400 + 1600$$
$$x^2 = 2000$$
$$x = \sqrt{2000}$$
$$x = 44.72 \text{ ft} \quad \text{accurate to two decimal places}$$

40 ft.
x
20 feet

The length of the guy wire is about 44.72 ft. long.

*Now Work Exercises 5 -7 in the Margin.*

**5.** Find the length of the hypotenuse of a right triangle with two legs each 3 feet long.

$\sqrt{18}$ ft. $\approx$ 4.24 ft.

**6.** Find the length of the hypotenuse of a right triangle with legs of length 4 centimeters and 6 centimeters.

$\sqrt{52}$ in. $\approx$ 7.211 in.

**7.** Given the length of the hypotenuse of a right triangle is 15 inches and one leg is 5 inches, find the length of the other leg (accurate to two decimal places).

$\sqrt{200}$ in. $\approx$ 14.14 in.

Name ______________ Section ______________ Date ______________

## Exercises 5.6

State whether or not each number is a perfect square.

**1.** 121 **2.** 144 **3.** 48 **4.** 16

**5.** 400 **6.** 256 **7.** 225 **8.** 45

**9.** 40 **10.** 196

Find the value of each of the following expressions.

**11.** $\left(\sqrt{7}\right)^2$ **12.** $\left(\sqrt{8}\right)^2$ **13.** $\left(\sqrt{36}\right)^2$ **14.** $\left(\sqrt{100}\right)^2$

**15.** $\left(\sqrt{16}\right)^2$ **16.** $\left(\sqrt{49}\right)^2$ **17.** $\left(\sqrt{21}\right)^2$ **18.** $\left(\sqrt{39}\right)^2$

**19.** $\left(\sqrt{206}\right)^2$ **20.** $\left(\sqrt{352}\right)^2$

**21. a.** Do you think that $\sqrt{39}$ is more than 6 or less than 6? Explain your reasoning.

Answers will vary. $\sqrt{39} > \sqrt{36} = 6$, so $\sqrt{39} > 6$.

**b.** Do you think that $\sqrt{39}$ is more than 7 or less than 7? Explain your reasoning.

Answers will vary. $\sqrt{39} < \sqrt{49} = 7$, so $\sqrt{39} < 7$.

**22. a.** Do you think that $\sqrt{18}$ is more than 4 or less than 4? Explain your reasoning.

Answers will vary. $\sqrt{18} > \sqrt{16} = 4$, so $\sqrt{18} > 4$.

**b.** Do you think that $\sqrt{18}$ is more than 5 or less than 5? Explain your reasoning.

Answers will vary. $\sqrt{18} < \sqrt{25} = 5$, so $\sqrt{18} < 5$.

### ANSWERS

**1.** Yes, $121 = 11^2$

**2.** Yes, $144 = 12^2$

**3.** Not a perfect square

**4.** Yes, $16 = 4^2$

**5.** Yes, $400 = 20^2$

**6.** Yes, $256 = 16^2$

**7.** Yes, $225 = 15^2$

**8.** Not a perfect square

**9.** Not a perfect square

**10.** Yes, $196 = 14^2$

**11.** 7

**12.** 8

**13.** 36

**14.** 100

**15.** 16

**16.** 49

**17.** 21

**18.** 39

**19.** 206

**20.** 352

**21.** [Respond below exercise.]

**22.** [Respond below exercise.]

23. [Respond below exercise.]

24. [Respond below exercise.]

25. a. 4 and 5

b. 4.7958

26. a. 5 and 6

b. 5.2915

27. a. 3 and 4

b. 3.6056

28. a. 3 and 4

b. 3.3166

29. a. 8 and 9

b. 8.4853

**23.** Show that $(1.4142)^2 < 2$ and $(1.4143)^2 > 2$. Tell what these two facts indicate about $\sqrt{2}$.

$(1.4142)^2 = 1.99996164 < 2$ and $(1.4143)^2 = 2.0024449 > 2$. So, $1.4142 < \sqrt{2} < 1.4143$.

**24.** Show that $(2.236)^2 < 5$ and $(2.237)^2 > 5$. Tell what these two facts indicate about $\sqrt{5}$.

$(2.236)^2 = 4.999696 < 5$ and $(2.237)^2 = 5.004169 > 5$. So, $2.236 < \sqrt{5} < 2.237$.

For part **a.**, use your understanding of square roots to estimate the value of each square root. Then use your calculator to find the value of each square root accurate to four decimal places.

**25.** **a.** The nearest integers to $\sqrt{23}$ are _____ and _____ .

**b.** Find the value of $\sqrt{23}$ accurate to four decimal places.

**26.** **a.** The nearest integers to $\sqrt{28}$ are _____ and _____ .

**b.** Find the value of $\sqrt{28}$ accurate to four decimal places.

**27.** **a.** $\sqrt{13}$ is between the two integers _____ and _____ .

**b.** Find the value of $\sqrt{13}$ accurate to four decimal places.

**28.** **a.** $\sqrt{11}$ is between the two integers _____ and _____ .

**b.** Find the value of $\sqrt{11}$ accurate to four decimal places.

**29.** **a.** The nearest integers to $\sqrt{72}$ are _____ and _____ .

**b.** Accurate to four decimal places, the value of $\sqrt{72}$ is _______ .

Name ______________________ Section ______________ Date ____________

**ANSWERS**

**30. a.** The nearest integers to $\sqrt{50}$ are _____ and _____ .

**b.** Accurate to four decimal places, the value of $\sqrt{50}$ is ________ .

**31. a.** The nearest integers to $\sqrt{60}$ are _____ and _____ .

**b.** The value of $\sqrt{60}$ , accurate to four decimal places, is ________ .

**32. a.** $\sqrt{80}$ is between the integers _____ and _____ .

**b.** Accurate to four decimal places, the value of $\sqrt{80}$ is ________ .

**33. a.** $\sqrt{95}$ is between the integers _____ and _____ .

**b.** The value of $\sqrt{95}$ , accurate to four decimal places, is ________ .

Use your calculator to find the value of each of the following expressions accurate to four decimal places.

(**Note**: The sequence that you perform the operations will depend on your calculator. With a scientific calculator, you will need to find the square root first. With a TI-83 graphing calculator, you can perform the operations just as you see them in the problem.)

**34.** $1+\sqrt{2}$ **35.** $1-\sqrt{2}$ **36.** $2-\sqrt{5}$ **37.** $3+2\sqrt{3}$

**38.** $4-3\sqrt{2}$ **39.** $1+2\sqrt{6}$ **40.** $\sqrt{2}+\sqrt{3}$ **41.** $\sqrt{25}-\sqrt{36}$

**42.** $\sqrt{16}+\sqrt{4}$

**30. a.** 7 and 8

**b.** 7.0711

**31. a.** 7 and 8

**b.** 7.7460

**32. a.** 8 and 9

**b.** 8.9443

**33. a.** 9 and 10

**b.** 9.7468

**34.** 2.4142

**35.** – 0.4142

**36.** – 0.2361

**37.** 6.4641

**38.** – 0.2426

**39.** 5.8990

**40.** 3.1463

**41.** –1

**42.** 6

43. Yes,
$5^2 + 12^2 = 13^2$

44. Yes,
$20^2 + 21^2 = 29^2$

45. Yes,
$6^2 + 8^2 = 10^2$

46. No, $2^2 + 3^2 \neq 4^2$

47. $c = 2.24$

48. $c = 5.66$

49. $x = 14.14$

50. $x = 20$

Use the Pythagorean Theorem to determine whether or not each of the following triangles is a right triangle.

**43.**

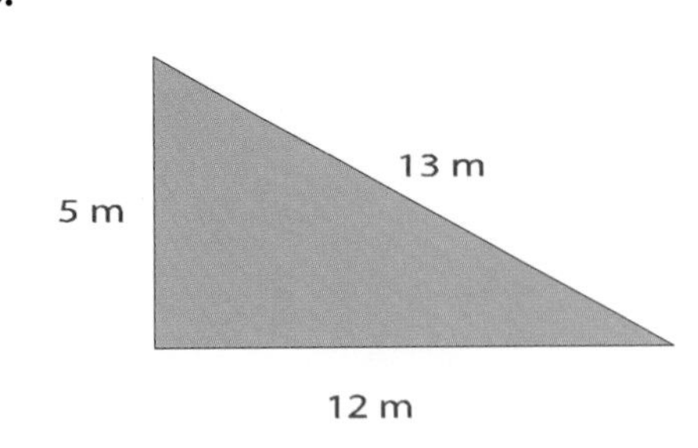


**44.**

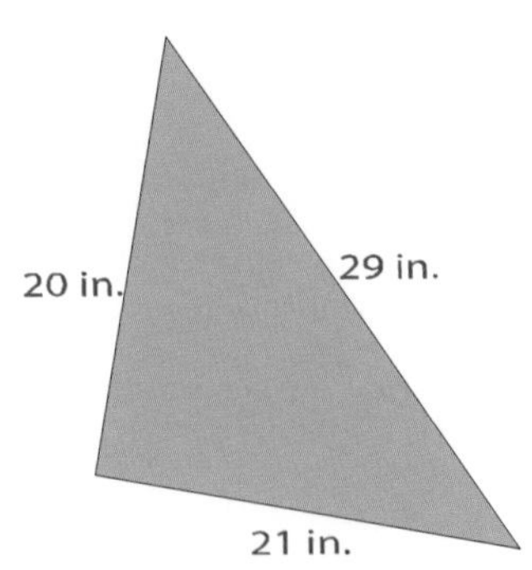


**45.**

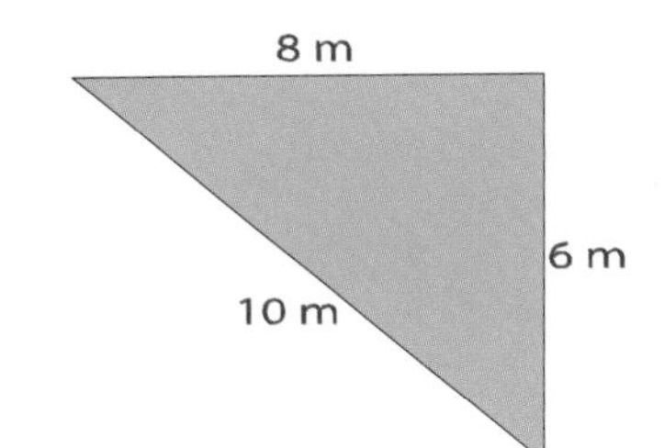


**46.**

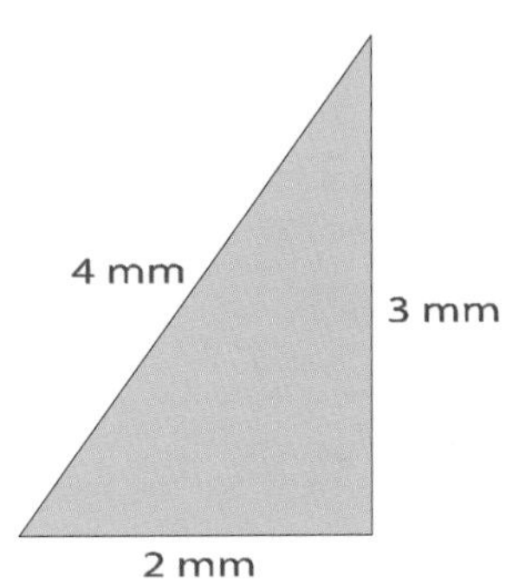


Find the length of the hypotenuse (accurate to two decimal places) of each of the following right triangles.

**47.**

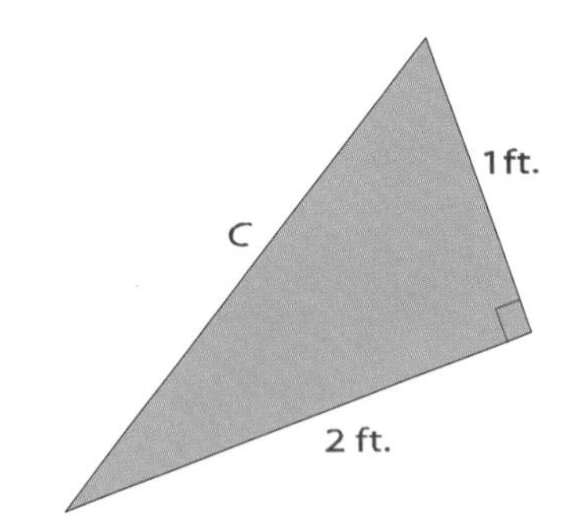


**48.**

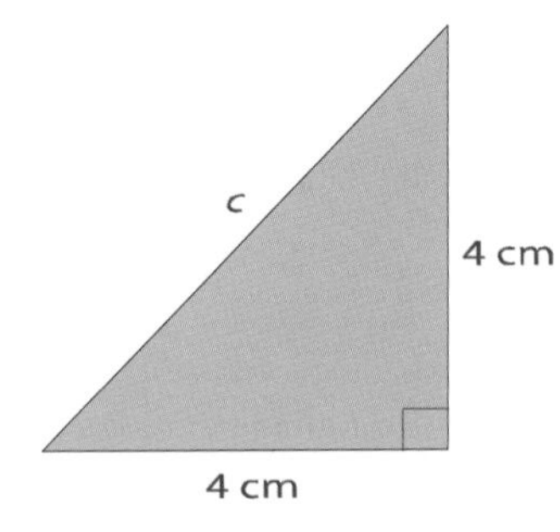


**49.**

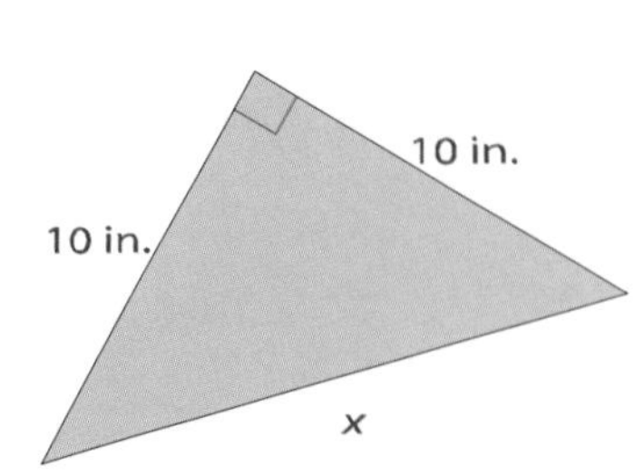


**50.**

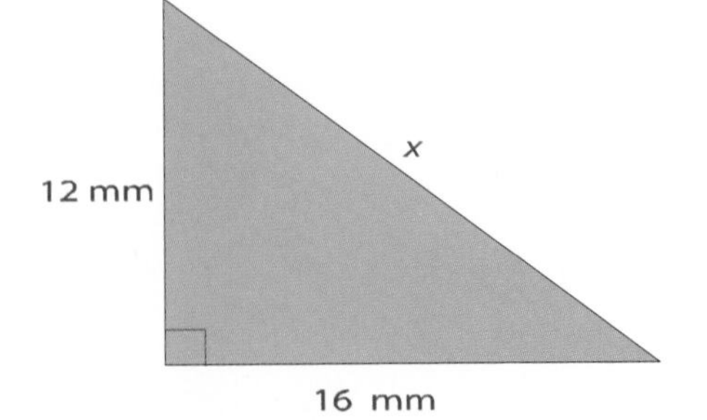

Name ______________________ Section ____________ Date ____________

**ANSWERS**

**51.**

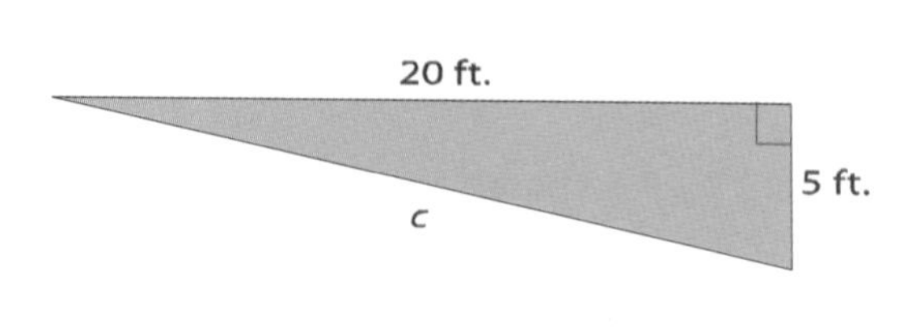


**52.**

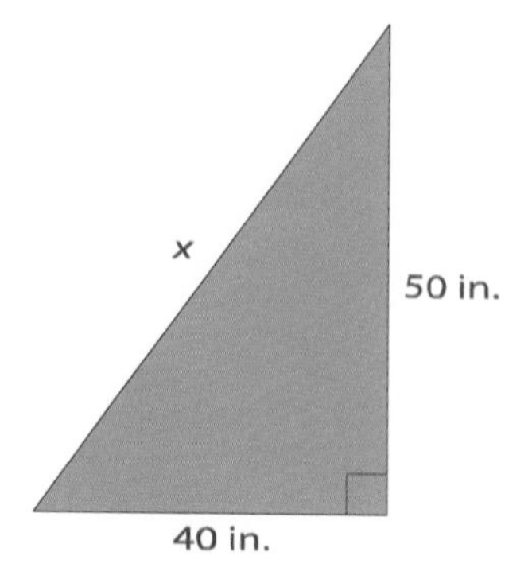


Find the length of the missing leg (accurate to two decimal places) of each of the following right triangles.

**53.**

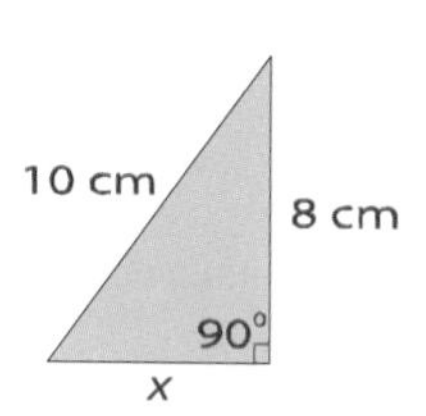


**54.**

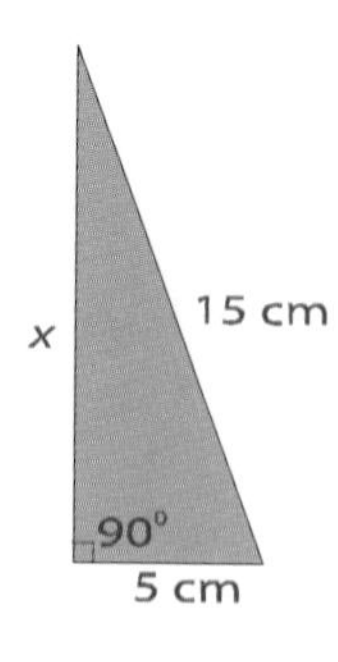


**55.**

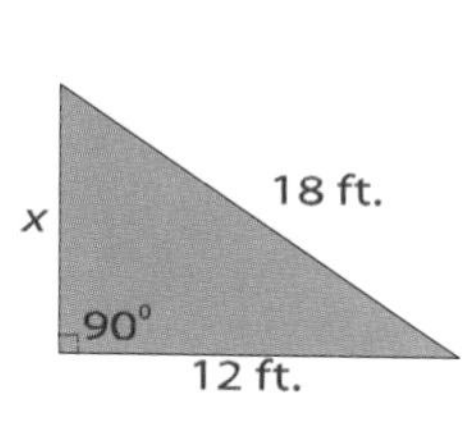


**56.**

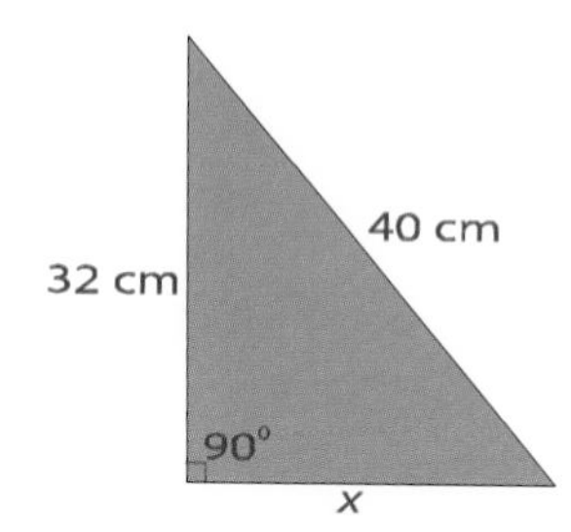


**57.** A square is said to be inscribed in a circle if each corner of the square lies on the circle. (Use $\pi = 3.14$.)

**a.** Find the circumference and area of a circle with diameter 30 feet.

**b.** Find the perimeter and area of a square inscribed in the circle.

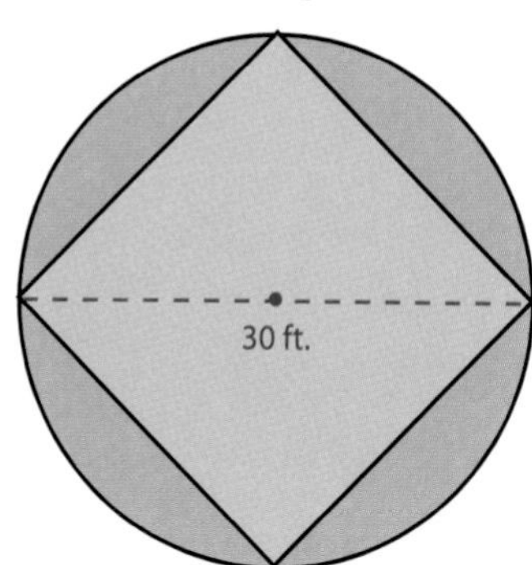

**51.** $c = 20.62$

**52.** $x = 64.03$

**53.** $x = 6.00$ cm

**54.** $x = 14.14$ cm

**55.** $x = 13.42$ ft

**56.** $x = 24$ cm

**57. a.** $C = 94.2$ ft

$A = 706.5\text{ft}^2$

**b.** $P = 84.85$ ft

$A = 450\text{ ft}^2$

58. a. 127.3 ft

b. 127.3 ft

59. a. no

b. home

c. no

60. 855.9 ft

**58.** The shape of a baseball infield is a square with sides 90 feet long.

**a.** Find the distance ( to the nearest tenth of a foot ) from home plate to second base.

**b.** Find the distance ( to the nearest tenth of a foot ) from first base to third base.

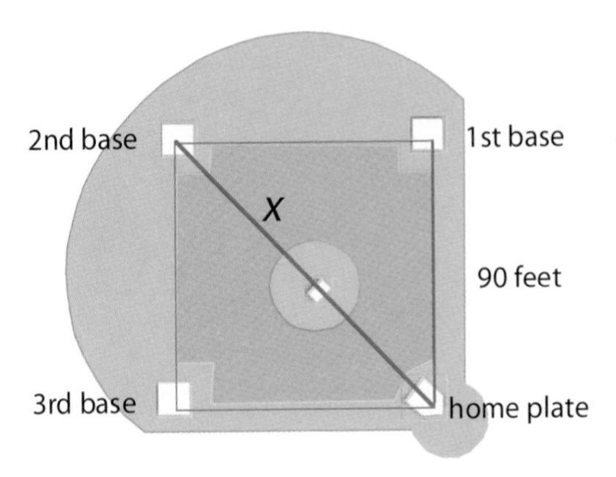


**59.** The distance from home plate to the pitcher's mound is 60.5 feet.

**a.** Is the pitcher's mound exactly half way between home plate and second base?

**b.** If not, which base is it closer to, home plate or second base?

**c.** Do the two diagonals of the square intersect at the pitcher's mound?

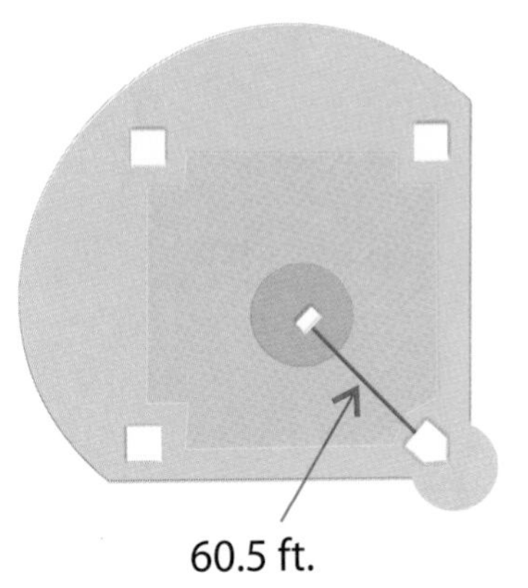


**60.** The GE Building in New York is 850 feet tall ( 70 stories). At a certain time of day, the building casts a shadow 100 feet long. Find the distance from the top of the building to the tip of the shadow (to the nearest tenth of a foot).

Name ______________________ Section ____________ Date __________

**ANSWERS**

**61.** 8.5 cm

**62.** 2.24 miles

**63.** 58.31 ft

**64.** 1, 8, 27, 64, 125, 216, 343, 512, 729, 1000

**61.** To create a square inside a square, a quilting pattern requires four triangular pieces like the one shaded in the figure shown here. If the square in the center measures 12 centimeters on a side, and the two legs of each triangle are of equal length, how long are the legs of each triangle, to the nearest tenth of a centimeter?

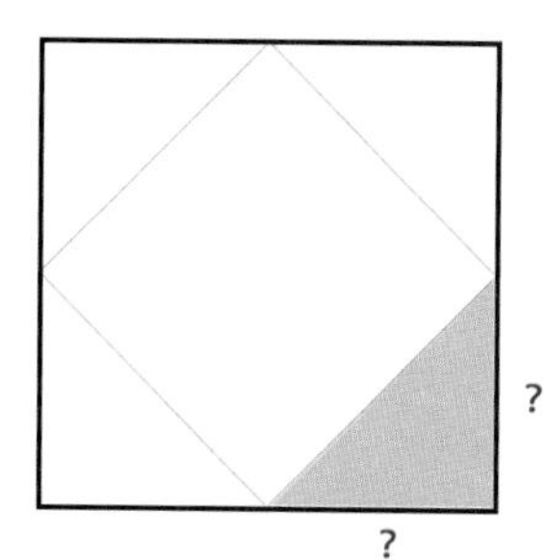


**62.** If an airplane passes directly over your head at an altitude of 1 mile, how far ( to the nearest hundredth of a mile ) is the airplane from your position after it has flown 2 miles farther at the same altitude?

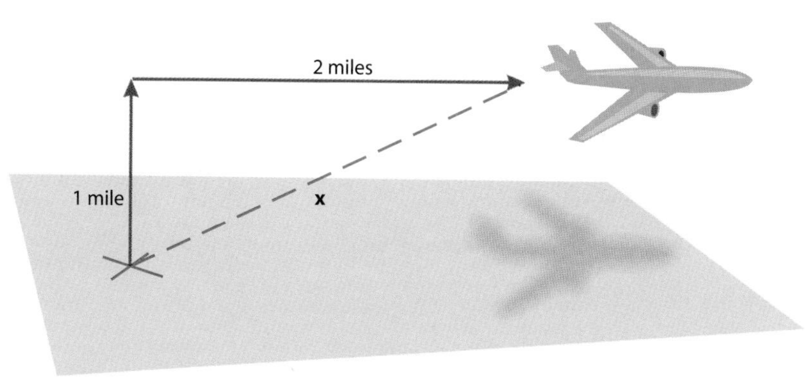


**63.** The base of a fire engine ladder is 30 feet from a building and reaches to a third floor window 50 feet above ground level, how long ( to the nearest hundredth of a foot ) is the ladder extended?

## Writing and Thinking about Mathematics

**64.** Cubes of integers are called **perfect cubes.** For example, since $2^3 = 8$, $3^3 = 27$, and $5^3 = 125$, the numbers 8, 27, and 125 are perfect cubes. Find all the perfect cubes from 1 to 1000.

65. a. 2

b. 3

c. 4

d. 7

**65.** Cube roots are symbolized by $\sqrt[3]{\phantom{x}}$. For example, since $5^3 = 125$, $\sqrt[3]{125} = 5$.

Find the following cube roots.

**a.** $\sqrt[3]{8}$

**b.** $\sqrt[3]{27}$

**c.** $\sqrt[3]{64}$

**d.** $\sqrt[3]{343}$

66. a. 2.0801

b. 4.2172

c. 4.6416

d. 12.5992

**66.** Roots in general can be defined in terms of fractional exponents. Thus, $\sqrt[3]{a} = a^{\frac{1}{3}}$. With this definition, a **scientific calculator** can be used to find (or approximate) cube roots of numbers by using the key $x^y$ (or a similar key) and the fraction $\frac{1}{3}$ as the exponent.

For example, the value of $\sqrt[3]{6}$ can be found as follows:

**Step 1:** Enter 6.

**Step 2:** Press the $x^y$ key.

**Step 3:** Press the left parenthesis parenthesis ( key.

**Step 4:** Enter the exponent in the form $1 \div 3$.

**Step 5:** Press the right parenthesis ) key.

**Step 6:** Press the = key.

The display should read 1.817120593.

Thus, $\sqrt[3]{6} = 1.8171$ accurate to four decimal places.

Use the technique just shown for finding the value of $\sqrt[3]{6}$ to find the following cube roots accurate to four decimal places.

**a.** $\sqrt[3]{9}$

**b.** $\sqrt[3]{75}$

**c.** $\sqrt[3]{100}$

**d.** $\sqrt[3]{2000}$

Name ______________________ Section ______________ Date __________

**ANSWERS**

**67. a.** 3.107232506

**b.** 4.562902635

**c.** 7.488872387

**d.** 20.000000000

**67.** A TI-83 Plus graphing calculator has specific commands for finding various roots under the MATH menu. Specifically, to find cube roots, press MATH and note that the fourth item under MATH in the menu is $\sqrt[3]{(}$. This item can be used to find cube roots.

For example, the value of $\sqrt[3]{16}$ can be found as follows:

**Step 1:** Press the MATH key.

**Step 2:** Press 4 so that $\sqrt[3]{(}$ will appear on the display.

**Step 3:** Enter 1, 6, and ). The display will now appear as $\sqrt[3]{(16)}$

**Step 4:** Press ENTER.

The display will now read:

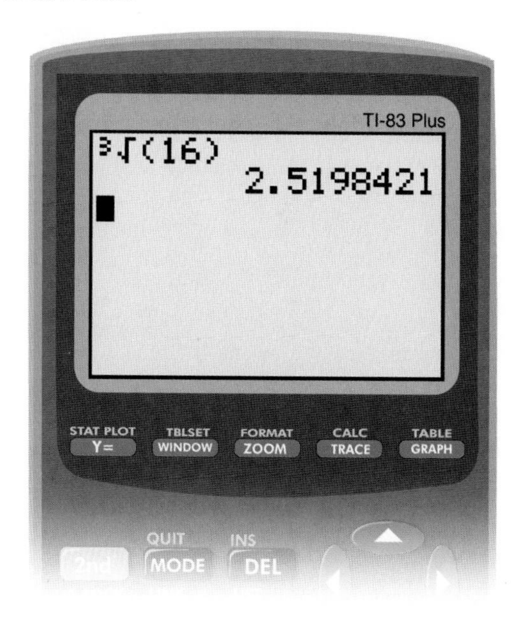


Thus, $\sqrt[3]{16} = 2.519842100$ accurate to nine decimal places.

Use the technique just shown for finding the value of $\sqrt[3]{16}$ to find the following cube roots accurate to nine decimal places.

**a.** $\sqrt[3]{30}$

**b.** $\sqrt[3]{95}$

**c.** $\sqrt[3]{420}$

**d.** $\sqrt[3]{8000}$

# 5.7 Simplifying Square Roots

## Properties of Square Roots

As we discussed in Section 5.6, the decimal values ( or approximations ) to square roots can be found by using a calculator. However, there are situations in mathematics where a simplified expression, rather than a decimal value of a square root, ( or other radical expression ) is required. If the radical expressions involve square roots, the following two properties of square roots can be used to simplify the expressions.

**Objectives**

① Know two properties of square roots.

② Be able to simplify expressions containing square roots.

**Properties of Square Roots**

For positive real numbers $a$ and $b$,

**1.** $\sqrt{ab} = \sqrt{a} \cdot \sqrt{b}$ **2.** $\sqrt{\frac{a}{b}} = \frac{\sqrt{a}}{\sqrt{b}}$

A radical expression with square roots is considered to be simplified when the radicand has no perfect square factor.

**In simplifying a radical expression,**

1. Find the largest perfect square factor of each radicand.
2. Do not leave any perfect square factor under the radical sign.

Study the following examples carefully to understand the technique for simplifying radical expressions. ( Refer to Table 5.2 for the perfect square numbers from 1 to 400. You should **memorize** these numbers and be able to recognize them quickly. )

### Example 1

Simplify each of the following radical expressions.

**a.** $\sqrt{50}$ **b.** $\sqrt{500}$ **c.** $-\sqrt{72}$ **d.** $\sqrt{\frac{9}{64}}$

**Solution**

**a.** We could write $\sqrt{50} = \sqrt{5 \cdot 10}$ , but neither 5 nor 10 is a perfect square. However, 25 is a perfect square factor of 50, and we can simplify as follows:

$$\sqrt{50} = \sqrt{25 \cdot 2} = \sqrt{25} \cdot \sqrt{2} = 5\sqrt{2}$$

**b.** Using the same technique and the fact that 100 is a perfect square factor of 500, we have

$$\sqrt{500} = \sqrt{100 \cdot 5} = \sqrt{100} \cdot \sqrt{5} = 10\sqrt{5}$$

**c.** Similarly, 36 is a perfect square factor of 72, and

$$-\sqrt{72} = -\sqrt{36 \cdot 2} = -\sqrt{36} \cdot \sqrt{2} = -6\sqrt{2}$$

Another approach to simplifying radical expressions is to find the prime factorization of the radicand and look for pairs of prime factors. These pairs of prime factors represent perfect square factors of the radicand.

$$\begin{aligned} -\sqrt{72} &= -\sqrt{2 \cdot 2 \cdot 2 \cdot 3 \cdot 3} = -\sqrt{2 \cdot 2 \cdot 3 \cdot 3} \cdot \sqrt{2} \\ &= -2 \cdot 3 \cdot \sqrt{2} = -6\sqrt{2} \end{aligned}$$

**d.** Since 9 and 64 are both perfect squares, $\sqrt{\frac{9}{64}} = \frac{\sqrt{9}}{\sqrt{64}} = \frac{3}{8}$.

## Simplifying Like Radicals

The technique of combining like radicals is similar to combining like terms in that the numbers multiplying the radicals are treated as coefficients. For example,

| **Combining Like Terms** | | **Combining Like Radicals** |
|---|---|---|
| $5x + 4x = 9x$ | and | $5\sqrt{3} + 4\sqrt{3} = 9\sqrt{3}$ |
| $13a - 6a - 2a = 5a$ | and | $13\sqrt{5} - 6\sqrt{5} - 2\sqrt{5} = 5\sqrt{5}$ |

### Example 2

Simplify each of the following radical expressions by combining like radicals (or like terms).

**a.** $5\sqrt{3} - 12\sqrt{3}$ **b.** $3\sqrt{7} + 9\sqrt{7}$ **c.** $14\sqrt{10} - 2\sqrt{10} - 8\sqrt{10}$

#### Solution

**a.** $5\sqrt{3} - 12\sqrt{3} = (5 - 12)\sqrt{3} = -7\sqrt{3}$

**b.** $3\sqrt{7} + 9\sqrt{7} = (3 + 9)\sqrt{7} = 12\sqrt{7}$

**c.** $14\sqrt{10} - 2\sqrt{10} - 8\sqrt{10} = (14 - 2 - 8)\sqrt{10} = 4\sqrt{10}$

### Example 3

Simplify the radical expression $\sqrt{45}+8\sqrt{20}$.

#### Solution

In this case, the square roots in the given expression are not like square roots. However, by first simplifying each radical expression, we find that there are like square roots present and the expression can be simplified.

$$\begin{aligned}\sqrt{45}+8\sqrt{20} &= \sqrt{9\cdot 5}+8\cdot\sqrt{4\cdot 5}\\ &= \sqrt{9}\cdot\sqrt{5}+8\cdot\sqrt{4}\cdot\sqrt{5}\\ &= 3\sqrt{5}+8\cdot 2\sqrt{5}\\ &= 3\sqrt{5}+16\sqrt{5}\\ &= 19\sqrt{5}\end{aligned}$$

*Now Work Exercises 1 - 4 in the Margin.*

#### Comment on Square Roots of Negative Numbers

Because the square of every real number is nonnegative, the square roots of negative numbers are not real numbers. Such numbers are called **imaginary numbers** or **complex numbers** and are studied in higher–level courses in mathematics. As Examples, $\sqrt{-4}$ and $\sqrt{-36}$ are **nonreal complex numbers**.

Simplify each of the following radical expressions.

**1.** $-\sqrt{810}$

$-9\sqrt{10}$

**2.** $\sqrt{\frac{49}{25}}$

$\frac{7}{5}$

**3.** $2\sqrt{3}+6\sqrt{3}$

$8\sqrt{3}$

**4.** $\sqrt{450}-\sqrt{50}$

$10\sqrt{2}$

Name ______ Section ______ Date ______

## Exercises 5.7

Simplify the radical expressions in each of the following exercises.

1. a. $\sqrt{8}$ $2\sqrt{2}$
   b. $\sqrt{12}$ $2\sqrt{3}$
   c. $\sqrt{16}$ $4$
   d. $\sqrt{20}$ $2\sqrt{5}$
   e. $\sqrt{24}$ $2\sqrt{6}$

2. a. $\sqrt{18}$ $3\sqrt{2}$
   b. $\sqrt{27}$ $3\sqrt{3}$
   c. $\sqrt{36}$ $6$
   d. $\sqrt{45}$ $3\sqrt{5}$
   e. $\sqrt{54}$ $3\sqrt{6}$

3. a. $\sqrt{50}$ $5\sqrt{2}$
   b. $\sqrt{75}$ $5\sqrt{3}$
   c. $\sqrt{100}$ $10$
   d. $\sqrt{125}$ $5\sqrt{5}$
   e. $\sqrt{150}$ $5\sqrt{6}$

4. a. $\sqrt{72}$ $6\sqrt{2}$
   b. $\sqrt{108}$ $6\sqrt{3}$
   c. $\sqrt{144}$ $12$
   d. $\sqrt{180}$ $6\sqrt{5}$
   e. $\sqrt{216}$ $6\sqrt{6}$

5. a. $\sqrt{98}$ $7\sqrt{2}$
   b. $\sqrt{147}$ $7\sqrt{3}$
   c. $\sqrt{196}$ $14$
   d. $\sqrt{245}$ $7\sqrt{5}$
   e. $\sqrt{294}$ $7\sqrt{6}$

6. a. $\sqrt{128}$ $8\sqrt{2}$
   b. $\sqrt{192}$ $8\sqrt{3}$
   c. $\sqrt{256}$ $16$
   d. $\sqrt{320}$ $8\sqrt{5}$
   e. $\sqrt{384}$ $8\sqrt{6}$

7. $-\sqrt{75}$

8. $-\sqrt{800}$

9. $-\sqrt{250}$

10. $-\sqrt{363}$

11. $\sqrt{432}$

12. $\sqrt{405}$

13. $\sqrt{1000}$

14. $\sqrt{200}$

15. $\sqrt{361}$

16. $\sqrt{\frac{36}{25}}$

17. $\sqrt{\frac{121}{144}}$

18. $\sqrt{\frac{64}{225}}$

19. $\sqrt{\frac{27}{16}}$

20. $-\sqrt{\frac{72}{25}}$

21. $-\sqrt{\frac{1}{9}}$

22. $\sqrt{16}+\sqrt{4}$

23. $\sqrt{25}+\sqrt{36}$

24. $\sqrt{9}+\sqrt{1}$



**ANSWERS**

1. [Respond beside exercise.]
2. [Respond beside exercise.]
3. [Respond beside exercise.]
4. [Respond beside exercise.]
5. [Respond beside exercise.]
6. [Respond beside exercise.]
7. $-5\sqrt{3}$
8. $-20\sqrt{2}$
9. $-5\sqrt{10}$
10. $-11\sqrt{3}$
11. $12\sqrt{3}$
12. $9\sqrt{5}$
13. $10\sqrt{10}$
14. $10\sqrt{2}$
15. $19$
16. $\frac{6}{5}$
17. $\frac{11}{12}$
18. $\frac{8}{15}$
19. $\frac{3\sqrt{3}}{4}$
20. $-\frac{6\sqrt{2}}{5}$
21. $-\frac{1}{3}$
22. $6$
23. $11$
24. $4$

25. $6\sqrt{2}$

26. $-5\sqrt{3}$

27. $7\sqrt{5}$

28. $8\sqrt{3}$

29. $-11\sqrt{6}$

30. $-5\sqrt{10}$

31. $6\sqrt{11}$

32. $8\sqrt{13}$

33. $9\sqrt{15}$

34. $7\sqrt{29}$

35. 37

36. 15

37. −1

38. 16

39. $11\sqrt{5}$

40. $49\sqrt{2}$

41. $17\sqrt{5}$

42. $20\sqrt{3}$

43. $50\sqrt{2}$

44. $13\sqrt{5}$

45. $42\sqrt{6}$

Simplify by combining like radicals.

**25.** $\sqrt{2}+5\sqrt{2}$

**26.** $2\sqrt{3}-7\sqrt{3}$

**27.** $3\sqrt{5}+4\sqrt{5}$

**28.** $6\sqrt{3}+2\sqrt{3}$

**29.** $2\sqrt{6}-13\sqrt{6}$

**30.** $12\sqrt{10}-17\sqrt{10}$

**31.** $7\sqrt{11}-\sqrt{11}$

**32.** $14\sqrt{13}-6\sqrt{13}$

**33.** $6\sqrt{15}+3\sqrt{15}$

**34.** $8\sqrt{29}-\sqrt{29}$

In each of the following expressions, simplify each radical and then combine like radicals if possible.

**35.** $3\sqrt{81}+2\sqrt{25}$

**36.** $2\sqrt{100}-\sqrt{25}$

**37.** $\sqrt{9}-\sqrt{16}$

**38.** $\sqrt{49}+\sqrt{81}$

**39.** $3\sqrt{5}+2\sqrt{80}$

**40.** $\sqrt{162}+4\sqrt{200}$

**41.** $7\sqrt{125}-6\sqrt{45}$

**42.** $3\sqrt{300}-2\sqrt{75}$

**43.** $6\sqrt{50}+5\sqrt{32}$

**44.** $\sqrt{125}+4\sqrt{20}$

**45.** $10\sqrt{150}-2\sqrt{96}$

# Chapter 5 Index of Key Ideas and Terms

Decimal notation — page 364

Reading and writing decimal numbers — page 365

1. Read (or write) the whole number.
2. Read (or write) "and" in place of the decimal point.
3. Read (or write) the fraction part as a whole number with the name of the place of the last digit on the right.

Rules for rounding off decimal numbers — page 367

1. Look at the single digit just to the right of the place of desired accuracy.
2. If this digit is 5, or greater, make the digit in the desired place of accuracy one larger and replace all digits to the right with zeros. All digits to the left remain unchanged unless a 9 is made one larger, and then the next digit to the left is increased by 1.
3. If this digit is less than 5, leave the digit in the desired place of accuracy as it is and replace all digits to the right with zeros. All digits to the left remain unchanged.
4. Trailing 0's to the right of the decimal point must be dropped so that the place of accuracy is clearly understood. If a rounded-off number has a 0 in the desired place of accuracy, then that 0 remains.

To Add (or Subtract) Decimal Numbers: — page 377

1. Write the numbers in a vertical column.
2. Keep the decimal points aligned vertically.
3. Keep digits with the same position value aligned. (Zeros may be filled in as aids.)
4. Add (or Subtract), keeping the decimal point in the difference aligned with the other decimal points.

To Multiply Positive Decimal Numbers: — page 389

1. Multiply the two numbers as if they were whole numbers.
2. Count the total number of places to the right of the decimal points in both numbers being multiplied.
3. Place the decimal point in the product so that the number of places to the right is the same as that found in Step 2.

To Multiply a Decimal Number by a Power of 10: page 390

1. Move the decimal point to the right.
2. Move it the same number of places as the number of 0's in the power of 10.

To Divide Decimal Numbers: page 392

1. Move the decimal point in the divisor to the right so that the divisor is a whole number.
2. Move the decimal point in the dividend the same number of places to the right.
3. Place the decimal point in the quotient directly above the new decimal point in the dividend.
4. Divide just as with whole numbers.

To Divide a Decimal Number by a Power of 10: page 395

1. Move the decimal point to the left.
2. Move it the same number of places as the number of 0's in the power of 10.

Changing from Decimals to Fractions: page 407

A decimal number with digits to the right of the decimal point can be written in fraction form by writing a fraction with

1. A numerator that consists of the whole number formed by all the digits of the decimal number and
2. A denominator that is the power of 10 that names the rightmost digit.

Changing from Fractions to Decimals: page 408

A fraction can be written in decimal form by dividing the numerator by the denominator.

1. If the remainder is 0, the decimal is said to be terminating.
2. If the remainder is not 0, the decimal is said to be nonterminating.

Scientific Notation page 413

A decimal number is written in scientific notation if it is written as the product of a number between 1 and 10 and the exponent form of a power of 10. (Exponents may be positive or negative.)

For Circles: page 423

Circle: the set of all points in a plane that are some fixed distance from a fixed point called the center of the circle

Radius: The fixed distance from the center of a circle to any point on the circle.

Diameter: The distance from one point on a circle to another point on the circle measured through the center.

Circumference: perimeter of a circle.

Formulas for Circles: page 424

For Circumference: $\mathbf{C = 2\pi r}$ and $\mathbf{C = \pi d}$

For Area: $\mathbf{A = \pi r^2}$

Perfect Squares page 431

Real Numbers page 433

Rational Numbers: terminating or infinite repeating decimals
Irrational Numbers: infinite nonrepeating decimals

Square Roots page 433

For any real number $a$ and any positive real number $b$, $a$ is a square root of $b$ if $a^2 = b$.

If $a$ is positive, then we write $a = \sqrt{b}$

Right Triangles page 436

Right Triangle: A triangle with a right angle (90°).

Hypotenuse: The longest side of a right triangle: the side opposite the right angle

Leg: Each of the other two sides of a right triangle

The Pythagorean Theorem page 436

In a right triangle, the square of the hypotenuse is equal to the sum of the square of the two legs: $c^2 = a^2 + b^2$

Properties of Square Roots page 449

For positive real number $a$ and $b$,

1. $\sqrt{ab} = \sqrt{a} \cdot \sqrt{b}$

2. $\sqrt{\frac{a}{b}} = \frac{\sqrt{a}}{\sqrt{b}}$

Name ______________________ Section ______________ Date ____________

## Chapter 5 Test

1. Write 32.064
   a. in words and
   b. as a mixed number with the fraction part reduced to lowest terms.

2. Write two and thirty-two thousandths in decimal notation.

Round off each number as indicated.

| | Nearest Tenth | Nearest Hundredth | Nearest Thousandth | Nearest Ten |
|---|---|---|---|---|
| 3. 216.7049 | a. 216.7 | b. 216.70 | c. 216.705 | d. 220 |
| 4. 73.01485 | a. 73.0 | b. 73.01 | c. 73.015 | d. 70 |

Perform the indicated operation.

5. $85.9 + 36.963$

6. $946.75 - 1073.24$

7. $|82.1| + |-200| - |-76.83|$

8. $13\frac{2}{5} + 10 + 27.316$

9. $(1.93)(-2.75)$

10. $(0.1)(0.02)(0.03)$

**ANSWERS**

1. a. thirty-two and sixty-four thousandths
   b. $32\frac{8}{125}$
2. 2.032
3. [Respond in exercise.]
4. [Respond in exercise.]
5. 122.863
6. −126.49
7. 205.27
8. 50.716
9. −5.3075
10. 0.00006

11. 0.08217

12. 0.2182

13. 82,170

14. 10.80667

15. 25.81

16. 16.49

17. a. $6.7 \times 10^7$

b. 0.000083

18. a. 14.12

b. −1.9875 or $\left(-1\frac{79}{80}\right)$

19. $x = 2.5$

20. $x = -10$

21. $y = -1.1$

22. a. 121

b. 5

**11.** $\frac{82.17}{10^3}$

**12.** $1\frac{1}{1000} - \frac{3}{4} - 0.0328$

**13.** $1000(82.17)$

**14.** $\begin{array}{r} 18.41 \\ \times\ 0.587 \\ \hline \end{array}$

Find each quotient to the nearest hundredth.

**15.** $0.31\overline{)8}$

**16.** $5.6\overline{)92.35}$

**17. a.** Write 67,000,000 in scientific notation.

**b.** Write $8.3 \times 10^{-5}$ in decimal notation.

**18.** Use $x = -2.1$ and $y = \frac{3}{4}$ to evaluate each of the following expressions.

**a.** $2x^2 - 3x - 1$

**b.** $y^2 - 5y + 1.2$

Solve each of the following equations.

**19.** $3x - 16.2 = -8.7$

**20.** $x - 0.1x + 5 = -4$

**21.** $5.5y + 3.2y + 6 = -3.57$

**22.** Find the value of each expression.

**a.** $\left(\sqrt{121}\right)^2$

**b.** $\left(\sqrt{5}\right)^2$

Name ______________________ Section ______________ Date ____________

**23.** Use a calculator to find the value of each expression accurate to four decimal places.

**a.** $\sqrt{300}$ **b.** $1+2\sqrt{3}$

**24.** Simplify each of the following radical expressions.

**a.** $\sqrt{40}$ **b.** $2\sqrt{20}+3\sqrt{45}$ **c.** $\sqrt{32}+3\sqrt{18}$

**25. a.** What theorem do you use to determine whether or not a triangle is a right triangle?

**b.** If a triangle has sides of 24 inches, 26 inches and 10 inches, is it a right triangle? Why or why not?

**26.** A flagpole is 30 feet high. A rope is to be stretched from the top of the pole to a point on the ground 15 feet from the base of the pole. If one extra foot is needed at each end of the rope to tie knots, how long should the rope be ( to the nearest tenth of a foot )?

**27.** The local athletic club wants to build a circular practice track that has an inner diameter of 50 yards.

**a.** What will be the inner circumference of the track?

**b.** If the track is to be 5 yards wide ( all the way around ), what will be the outer circumference of the track?

**c.** How many square yards of land will be covered by the surface of the track?

**ANSWERS**

**23. a.** 17.3205

**b.** 4.4641

**24. a.** $2\sqrt{10}$

**b.** $13\sqrt{5}$

**c.** $13\sqrt{2}$

**25. a.** The Pythagorean Theorem

**b.** Yes, $10^2 + 24^2 = 26^2$

**26.** 35.5 ft

**27. a.** 157 yd

**b.** 188.4 yd

**c.** 863.5 $yd^2$

28. a. 359 mi

b. 24 gal

29. Pay cash, because he can save $5760.

30. a. 6 and 7

b. 6.7823

**28.** Your car gets 20.5 miles per gallon of gas and the tank holds 17.5 gallons of gas.

**a.** How far can your car travel ( to the nearest mile ) on a full tank of gas?

**b.** How many gallons of gas ( to the nearest gallon ) should you plan to use on a trip of 500 miles?

**29.** Your friend plans to buy a new car for $16,000. He can either save up and pay cash or make a down payment of $1600 and 48 monthly payments of $420. What would you recommend that he do? Why?

**30.** **a.** The nearest integers to $\sqrt{46}$ are _____ and _____.

**b.** Use a calculator to find $\sqrt{46}$ accurate to four decimal places.

Name ______________ Section ______________ Date ______________

## Cumulative Review: Chapters 1 - 5

**1.** Name the property illustrated, and find the value of the variable.

**a.** $21 + x = 15 + 21$

**b.** $5(2 \cdot y) = (5 \cdot 2) \cdot 3$

**c.** $6 + (3 + 17) = (a + 3) + 17$

**2.** Subtract 605 from the sum of 843 and 467; then find the quotient if the difference is divided by 5.

**3.** Estimate the product of 845 and 3200.

Evaluate each of the following expressions.

**4.** $3^2 - 5^2$

**5.** $28 + 5(2^3 + 25) \div 11 \cdot 2$

**6.** $24 + [16 - (6^2 \cdot 1 - 20)]$

**7.** $36 \div 4 - 9 \cdot 2^2$

**8.** Find $\frac{3}{4}$ of 76.

**9.** Find the prime factorization of 380.

**10.** Find the LCM for each set of terms.

**a.** 28, 70, 80

**b.** $24a^2, 15ab^3, 20a^2b^2$

### ANSWERS

**1. a.** commutative prop. of addition; $x = 15$

**b.** associative prop. of multiplication; $y = 3$

**c.** associative prop. of addition; $a = 6$

**2.** 141

**3.** 2,400,000

**4.** −16

**5.** 58

**6.** 24

**7.** −27

**8.** 57

**9.** $2 \cdot 2 \cdot 5 \cdot 19$

**10. a.** 560

**b.** $120a^2b^3$

11. a. $\frac{3}{200}$

b. $6\frac{8}{25}$

c. $-1\frac{1}{20}$

12. $12\frac{19}{24}$

13. 6

14. $157\frac{13}{20}$

15. $73.28 \left(\text{or } 73\frac{7}{25}\right)$

16. 61.65

17. $\frac{13}{15}$

18. $6\frac{3}{32}$

19. $\frac{5x-3}{30}$

20. 280.7

21. 102.91

22. a. $8.65 \times 10^5$

b. 0.00341

**11.** Change each decimal to fraction form or mixed number form, and reduce if possible.

**a.** 0.015 **b.** 6.32 **c.** −1.05

Perform the indicated operations. Reduce all answers.

**12.** $5\frac{3}{8}+7\frac{5}{12}$

**13.** $22\frac{1}{2}\div 3\frac{3}{4}$

**14.** $\begin{array}{r} 510\frac{7}{20} \\ -\ 352\frac{7}{10} \\ \hline \end{array}$

**15.** $\begin{array}{r} 15\frac{1}{2} \\ 35.68 \\ +\ 22\frac{1}{10} \\ \hline \end{array}$

**16.** $|-52.3|+|16.1|-|-6.75|$

**17.** $\left(\frac{1}{2}\right)^2\left(\frac{16}{15}\right)+\frac{6}{5}\div 2$

**18.** $\dfrac{3\frac{1}{2}+4\frac{5}{8}}{2-\frac{2}{3}}$

**19.** $\frac{x}{3}\cdot\frac{1}{2}-\frac{2}{5}\div 4$

**20.** Find the quotient of 42.1 ÷ 0.15 to the nearest tenth.

**21.** Evaluate the expression $x^2 - 5x + 17$ for $x = -7.1$

**22.** **a.** Write the number 865,000 in scientific notation.

**b.** Write the number $3.41 \times 10^{-3}$ in decimal notation.

Name ______________ Section ______________ Date ______________

**ANSWERS**

Solve the following equations.

**23.** $6x - 10x + 7 = 4x + 15$

**24.** $3.5y = y - 7.525$

**25.** $0.3x + 0.2(x + 5) = 1 - x$

**26.** $\frac{n}{14} - \frac{1}{2} = \frac{3}{7}$

**27.** Simplify the following radical expressions.

**a.** $\sqrt{150}$

**b.** $-\sqrt{2000}$

**c.** $\sqrt{\frac{52}{81}}$

**d.** $4\sqrt{54} + 2\sqrt{96}$

**28.** Find the length of the hypotenuse of the right triangle shown here.

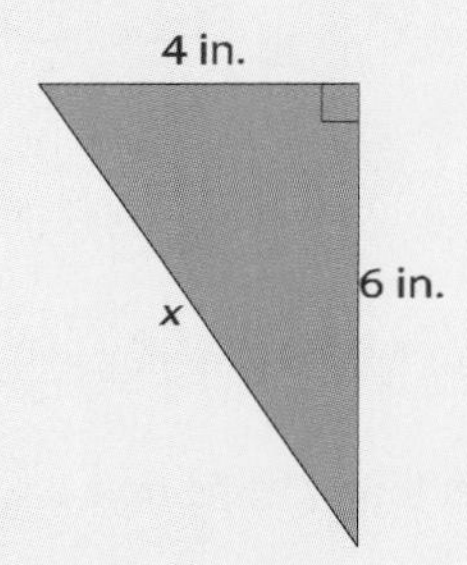


**29.** **a.** Do you think that $\sqrt{29}$ is more than or less than 5? Explain your reasoning.

**b.** Use your calculator to find the value of $\sqrt{29}$ accurate to three decimal places.

**30.** Rose was paid a salary of $600 per week and $22.50 for each hour she worked over 40 hours. How much would she make in 1 year if she averaged 42 hours of work per week except for her 3 weeks of paid vacation each year?

**31.** Find

**a.** the circumference and

**b.** the area of a circle with diameter 2.5 feet.

**23.** $x = -1$

**24.** $y = -3.01$

**25.** $x = 0$

**26.** $n = 13$

**27. a.** $5\sqrt{6}$

**b.** $-20\sqrt{5}$

**c.** $\frac{2\sqrt{13}}{9}$

**d.** $20\sqrt{6}$

**28.** $2\sqrt{13}$ in. $\approx 7.211$ in.

**29. a.** More, because $5^2 = 25 < 29$

**b.** 5.385

**30.** $33,405

**31. a.** 7.85 ft.

**b.** 4.90625 $\text{ft}^2$

32. $90\text{ cm}^3$

33. a. More than $1500

b. $2250

34. a. $x = 1.8$

b. $y = \frac{5}{3}\left(\text{or } y = 1\frac{2}{3}\right)$

35. 2 in. by 2.8 in.

36. $x = 2.4$; $y = 1.8$

37. [Respond below exercise.]

**32.** Draw a sketch of a rectangular solid with length 5 centimeters, width 3 centimeters, and height 6 centimeters. Find the volume of this solid.

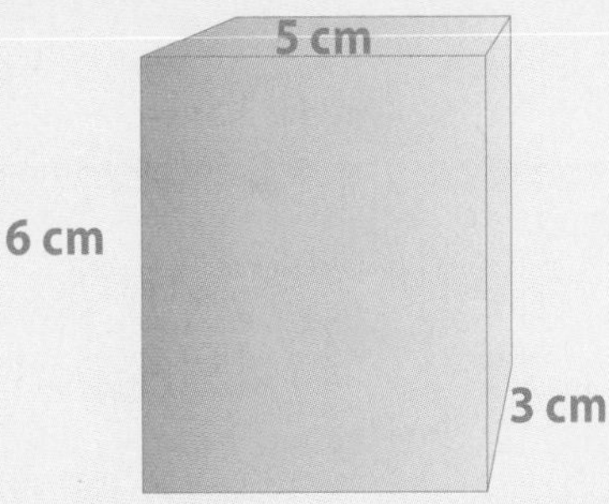


**33.** An advertised sale on computers declares $\frac{1}{3}$ off the original price. The advertised price is $1500.

**a.** Was the original price less than or more than $1500?

**b.** What was the original price?

**34.** Solve each of the following proportions.

**a.** $\frac{x}{7.2} = \frac{3.6}{14.4}$

**b.** $\frac{4\frac{1}{2}}{y} = \frac{9}{3\frac{1}{3}}$

**35.** Pat has been drawing house plans of her "dream house." The master bedroom is to be a rectangle 20 feet wide and 28 feet long. If she uses a scale of $\frac{1}{2}$ inch to represent 5 feet, what will be the dimensions of the bedroom on the drawing?

**36.** The two triangles shown here are similar triangles. Find the values of $x$ and $y$.

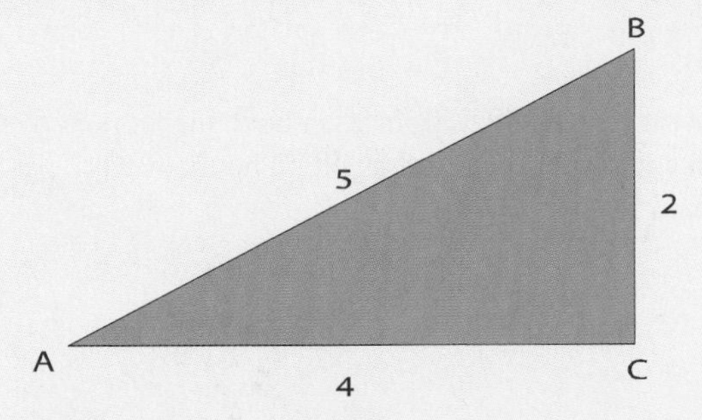


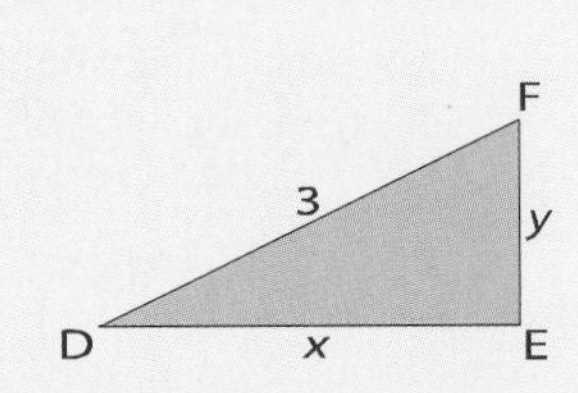


## Writing and Thinking about Mathematics

**37.** So far in this course, what one topic have you found to be the most interesting and informative? Briefly explain why.

Responses will vary. The question is designed to help the students reflect on what they are learning.

# 6

# PERCENT WITH APPLICATIONS

**Chapter 6 Percent With Applications**





## WHAT TO EXPECT IN CHAPTER 6

Chapter 6 deals with percent and applications with percent. Percent is one of the most useful and most important mathematical concepts in our daily lives. The concept of percent of profit and the interrelations among percent, decimals, and fractions are included in Section 6.1.

Section 6.2 lays the groundwork for applications with percent by using the formula $R \cdot B = A$ to indicate that there are only three basic types of percent problems, all of which can be solved by applying this one formula. Section 6.3 tests your understanding of percent through multiple-choice exercises that require mental estimations with very little actual calculation.

Sections 6.4 and 6.5 involve a variety of applications of percents. We emphasize Pòlya's four-step process. This approach to problem solving is valid for solving any real-life problem, whether it involves mathematics or not.

The chapter closes with a section on the concepts and related formulas for simple interest and compound interest. Also included are the topics of inflation and depreciation.

# Chapter 6 Percent with Applications

## Mathematics at Work!

Making money and investing money are important concerns for every individual living in a modern industrial ( and computerized ) society. Money is needed for housing, food, clothing, entertainment, and transportation. As you accumulate money through work and investments and borrow money for housing and automobiles, you must have a basic understanding of **percent** and **interest**.

Merchants and bankers assume that you, as a consumer, are responsible for understanding interest and the effects of different rates of interest. The following table illustrates the time for doubling a saving account ( or loan ) of $10,000 compounded daily at three different rates of interest.

**Value of $10,000 Compounded Daily**

| Time ( in years ) | 8% | 10% | 18% Credit Card Rate |
|---|---|---|---|
| 1 | $10,832.78 | $11,051.56 | $11,971.64 |
| 2 | $11,734.91 | $12,213.69 | $14,332.01 |
| 3 | $12,712.16 | $13,498.03 | $17,157.76 |
| 4 | $13,770.79 | $14,917.42 | $20,540.64 |
| 5 | $14,917.59 | $16,486.07 | ↑ |
| 6 | $16,159.89 | $18,219.67 | Doubled in 4 years |
| 7 | $17,505.64 | $20,135.57 | |
| 8 | $18,963.47 | ↑ | |
| 9 | $20,542.69 | Doubled in 7 years | |
| | ↑ Doubled in 9 years | | |

# 6.1 Understanding Percent

## Percent Means Hundredths

The word **percent** comes from the Latin *per centum*, meaning "per hundred." So, **percent means hundredths**, or **percent is the ratio of a number to 100**. The symbol % is called the **percent symbol** (or **percent sign**). As we shall see, this symbol can be treated as equivalent to the fraction $\frac{1}{100}$.

For example,

$$\frac{25}{100} = 25\left(\frac{1}{100}\right) = 25\% \quad \text{and} \quad \frac{7}{10} = \frac{70}{100} = 70\left(\frac{1}{100}\right) = 70\%$$

In Figure 6.1, the large square is partitioned into 100 small squares, and each small square represents 1%, or $1 \cdot \frac{1}{100}$, of the large square. Thus the shaded portion of the large square is

$$\frac{40}{100} = 40 \cdot \frac{1}{100} = 40\%$$

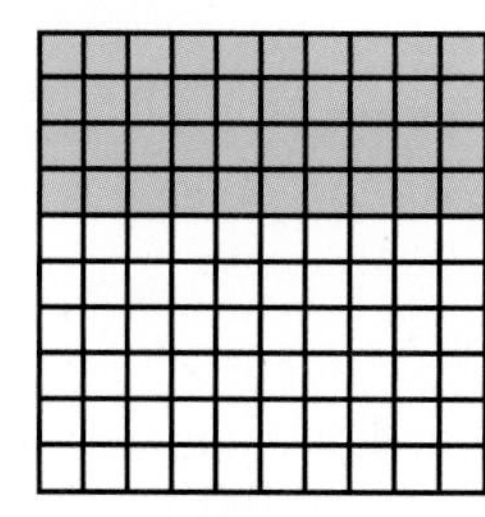

$\frac{40}{100} = 40\%$

**Figure 6.1**

If a fraction has a denominator of 100, then ( with no change in the placement of a decimal point or fraction in the numerator ) the numerator can be read as a percent by dropping the denominator, 100, and adding on the % symbol.

**Objectives**

① Understand that percent means hundredths.

② Relate percent to fractions with denominator 100.

③ Know how to compare profit with investment as a percent.

④ Be able to change percents to decimals and decimals to percents.

⑤ Be able to change fractions to percents and percents to fractions.

## Example 1

Each fraction is changed to a percent.

**a.** $\frac{30}{100} = 30 \cdot \frac{1}{100} = 30\%$ Remember that percent means hundredths.

**b.** $\frac{45}{100} = 45\%$ The % symbol indicates hundreds or $\frac{1}{100}$.

**c.** $\frac{7.3}{100} = 7.3\%$ Note that the decimal point is not moved.

**d.** $\frac{20\frac{1}{2}}{100} = 20\frac{1}{2}\%$ or $20.5\%$ Note that the fraction $\frac{1}{2}$ is part of the answer.

**e.** $\frac{250}{100} = 250\%$ If the numerator is larger than 100, then the number is larger than 1 and it is more than 100%.

Another approach in emphasizing that percent means hunderdths is to use the words in equation form as follows:

25 **hundredths** = 25 **percent**
4.3 **hundredths** = 4.3 **percent**
$37\frac{1}{2}$ **hundredths** = $37\frac{1}{2}$ **percent**

As a point of clarification, implicit in our discussion of percent is the fact that percents are the ratios of positive numbers.

## Percent Of Profit

**Percent of profit** is the ratio of money made to money invested. Generally, two investments do not involve the same amount of money, and therefore, the comparative success of each investment cannot be based on the amount of profit. In comparing investments, the investment with the greater percent of profit is considered the better investment.

The use of percent gives an effective method of comparison because each ratio has the same denominator ( 100 ).

This discussion of profit is designed to reinforce the idea that a ratio is a percent. This approach also applies to the percent problems in Section 6.4.

### Example 2

Calculate the percent of profit for both **a.** and **b.**, and tell which is the better investment.

**a.** $200 made as profit by investing $500

**b.** $270 made as profit by investing $900

**Solution**

In each case, find the ratio of dollars profit to dollars invested and reduce the ratio so that it has a denominator of 100. Do not reduce to lowest terms.

**a.** $\dfrac{\$200 \text{ profit}}{\$500 \text{ invested}} = \dfrac{\not{5} \cdot 40}{\not{5} \cdot 100} = \dfrac{40}{100} = 40\%$

**b.** $\dfrac{\$270 \text{ profit}}{\$900 \text{ invested}} = \dfrac{\not{9} \cdot 30}{\not{9} \cdot 100} = \dfrac{30}{100} = 30\%$

Investment **a.** is better than investment **b.** because 40% is larger than 30%. Obviously, $270 profit is more than $200 profit, but the money risked ( $900 ) is also greater.

Change each fraction to a percent.

1. $\dfrac{11}{100}$
   11%

2. $\dfrac{140}{100}$
   140%

3. $\dfrac{8.25}{100}$
   8.25%

4. Calculate the percent of profit if $300 is made of an investment of $6000.
   5%

### *Completion Example 3*

Which of the following is the better investment?

**a.** An investment of $1000 that makes a profit of $150

**b.** An investment of $800 that makes a profit of $128

**Solution**

**a.** $\dfrac{\$\_\_\_\_\_\_\_\_ \text{ profit}}{\$\ 1000 \text{ invested}} = \dfrac{\_\_\_\_\_\_\_\_}{\_\_\_\_\_\_ \cdot 100} = \_\_\_\_\_\_\_\_\%$

**b.** $\dfrac{\$\_\_\_\_\_\_\_\_ \text{ profit}}{\$\ 800 \text{ invested}} = \dfrac{\_\_\_\_\_\_\_\_}{\_\_\_\_\_\_ \cdot 100} = \_\_\_\_\_\_\_\_\%$

Thus, investment ( _____ ) is better because _____% is larger than _____%.

*Now Work Exercises 1-4 in the Margin.*

## Decimals and Percents

One way to change a decimal number to a percent is to first change the decimal to fraction form with denominator 100 and then change the fraction to percent form. For example:

| Decimal Form | | Fraction Form | | Percent Form |
|---|---|---|---|---|
| 0.33 | = | $\frac{33}{100} = 33 \cdot \frac{1}{100}$ | = | 33% |
| 0.74 | = | $\frac{74}{100} = 74 \cdot \frac{1}{100}$ | = | 74% |
| 0.2 | = | $\frac{2}{10} = \frac{20}{100} = 20 \cdot \frac{1}{100}$ | = | 20% |
| 0.625 | = | $\frac{62.5}{100} = 62.5 \cdot \frac{1}{100}$ | = | 62.5% |

By noting the way that the decimal point is moved in these four examples, we can make the change directly ( and more efficiently ) by using the following rule regardless of how many digits there are in the decimal number.

**To Change a Decimal to a Percent:**

**Step 1:** Move the decimal point two places to the right.

**Step 2:** Add the % symbol.

( These two steps have the effect of multiplying by 100 and then dividing by 100. Thus the number is not changed. Just the form is changed. )

If the students read the decimals and fractions out loud in class, you can emphasize to them the importance of reading both 0.33 and $\frac{33}{100}$ as "thirty-three hundredths" and not "point three three" or "33 over 100." This could provide some added insight in understanding the transition from decimals and fractions to percent for some students.

A key concept to clarify for all your students is that when a decimal is changed to a percent the number ( or the value of the number ) is not changed; rather, only its form is changed. For example, 0.254 and 25.4% are two forms of the same number.

## Example 4

Change each decimal to an equivalent percent.

**a.** 0.254

**b.** 0.005

**c.** 1.5

**d.** 0.2

### Solution

**a.** $0.254 = 25.4\%$

Decimal point moved two places to the right — % symbol added on

**b.** $0.005 = 0.5\%$ Note that this is less than 1%.

Decimal point moved two places to the right — % symbol added on

**c.** $1.5 = 150\%$ Note that this is more than 100%

Decimal point moved two places to the right — % symbol added on

**d.** $0.2 = 20\%$

Decimal point moved two places to the right — % symbol added on

To change percents to decimals, we reverse the procedure for changing decimals to percents. For example,

$$38\% = 38\left(\frac{1}{100}\right) = \frac{38}{100} = 0.38$$

As indicated in the following statement, the same result can be found by noting the placement of the decimal point.

**To Change a Percent to a Decimal:**

**Step 1:** Move the decimal point two places to the left.

**Step 2:** Delete the % symbol.

### Example 5

Change each percent to an equivalent decimal.

**a.** 38%

**b.** 16.2%

**c.** 100%

**d.** 0.25%

**Solution**

**a.** 38% = 0.38 ← % symbol deleted

↑ Understood decimal point

↑ Decimal point moved two places left

**b.** 16.2% = 0.162

**c.** 100% = 1.00 = 1

**d.** 0.25% = 0.0025 The percent is less than 1%, and the decimal is less than 0.01.

The following relationships between decimals and percents provide helpful guidelines when changing from one form to the other.

**A decimal number that is**

1. less than 0.01 is less than 1%.
2. between 0.01 and 0.10 is between 1% and 10%.
3. between 0.10 and 1.00 is between 10% and 100%.
4. more than 1.00 is more than 100%.

*Now Work Exercises 5-8 in the Margin.*

## Fractions and Percents

As we discussed earlier, if a fraction has denominator 100, the fraction can be changed to a percent by writing the numerator and adding on the % symbol. If the denominator is one of the factors of 100,

1, 2, 4, 5, 10, 20, 25, or 50 ( factors of 100 )

we can easily write it in an equivalent form with denominator 100 and then change it to a percent. For example,

$$\frac{3}{4}=\frac{3}{4}\cdot\frac{25}{25}=\frac{75}{100}=75\% \quad (\text{Note that } 4\cdot 25=100.)$$

$$\frac{2}{5}=\frac{2}{5}\cdot\frac{20}{20}=\frac{40}{100}=40\% \quad (\text{Note that } 5\cdot 20=100.)$$

$$\frac{13}{50}=\frac{13}{50}\cdot\frac{2}{2}=\frac{26}{100}=26\% \quad (\text{Note that } 50\cdot 2=100.)$$

Change each decimal to a percent.

**5.** 0.35

35%

**6.** 2.17

217%

Change each percent to a decimal.

**7.** 21.3%

0.213

**8.** 0.6%

0.006

However, most fractions do not have denominators that are factors of 100. A more general approach ( easily applied with calculators ) is to change the fraction to decimal form by dividing the numerator by the denominator, and then change the decimal to a percent.

**To Change a Fraction to a Percent:**

**Step 1:** Change the fraction to a decimal.
( Divide the numerator by the denominator. )

**Step 2:** Change the decimal to a percent.

### Example 6

Change $\frac{5}{8}$ to a percent.

**Solution**

First divide 5 by 8 to get the decimal form. ( This can be done with a calculator.)

$$\begin{array}{r} .625 \\ 8\overline{)5.000} \\ \underline{4\,8\phantom{00}} \\ 20\phantom{0} \\ \underline{16\phantom{0}} \\ 40 \\ \underline{40} \\ 0 \end{array}$$

Change 0.625 to a percent:

$$\frac{5}{8} = 0.625 = 62.5\%$$

### Example 7

Change $\frac{11}{20}$ to a percent.

**Solution**

Divide:

$$\begin{array}{r} .55 \\ 20\overline{)11.00} \\ \underline{10\,0\phantom{0}} \\ 1\,00 \\ \underline{1\,00} \\ 0 \end{array}$$

Thus,

$\frac{11}{20} = 0.55 = 55\%$

Or, we can note that 20 is a factor of 100 and write

$\frac{11}{20} = \frac{11}{20} \cdot \frac{5}{5} = \frac{55}{100} = 55\%$

### Completion Example 8

Change $3\frac{3}{5}$ to a percent.

**Solution**

Since $3\frac{3}{5}$ is larger than 1, the percent will be more than _____%.

$3\frac{3}{5} = \frac{18}{5}$

Divide:

$$\begin{array}{r} . \phantom{0} \\ 5\overline{)18.0} \\ \underline{\phantom{00}} \\ 30 \\ \underline{\phantom{0}} \\ 0 \end{array}$$

Now, $3\frac{3}{5} = \frac{18}{5}$ = _____ = _____%.

*Now Work Exercises 9-12 in the Margin.*

**Agreement for Rounding Off Decimal Quotients**

In this text,

1. Decimal quotients that are exact with four decimal places ( or less ) will be written with four decimal places ( or less ).
2. Decimal quotients that are not exact will be divided to the fourth place, and the quotient will be rounded off to the third place ( thousandths ).

Change each fraction or mixed number to a percent.

**9.** $\frac{13}{50}$

26%

**10.** $1\frac{3}{4}$

175%

**11.** $\frac{1}{10}$

10%

**12.** $\frac{3}{80}$

3.75%

## Using a Calculator

Most calculators will give answers accurate to 8 or more decimal places, so if you use a calculator to perform the long division when changing a fraction to a decimal, be sure to follow the agreement in statement 2 above.

### Example 9

Change $\frac{1}{3}$ to a percent.

#### Solution

With a calculator, $\frac{1}{3} = 0.333333333$. Rounding off gives

$\frac{1}{3} = 0.333 = 33.3\%$ This answer is not exact.

To be exact, we can divide and leave the answer with a fraction.

$$\begin{array}{r} .33\frac{1}{3} \\ 3\overline{)1.00} \\ \underline{9} \\ 10 \\ \underline{9} \\ 1 \end{array} \qquad \frac{1}{3} = 0.33\frac{1}{3} = 33\frac{1}{3}\% \qquad (\text{or } 33.\bar{3}\%)$$

$33\frac{1}{3}\%$ is exact and 33.3% is rounded off.

Both answers are acceptable. However, you should remember that 33.3% is a rounded-off answer.

### Example 10

In 1900, there were about 27,000 college graduates (received bachelor's degrees), of which about 22,000 were men. In 1990, there were about 1,050,000 college graduates, of which about 490,000 were men. Find the percentage of college graduates that were men in each of these years.

#### Solution

Divide with a calculator:

For 1900, $\frac{22{,}000}{27{,}000} = 0.814814814 \approx 81.5\%$ of college graduates were men.

For 1990, $\frac{490{,}000}{1{,}050{,}000} = 0.46666666 \approx 46.7\%$ of college graduates were men.

**To Change a Percent to a Fraction or Mixed Number:**

**Step 1:** Write the percent as fraction with denominator 100 and delete the % symbol.

**Step 2:** Reduce the fraction.

### Example 11

Change each percent to an equivalent fraction in reduced form.

**a.** 28%

**b.** $9\frac{1}{4}\%$

**c.** 735%

**Solution**

**a.** $28\% = \frac{28}{100} = \frac{\cancel{4}^{1} \cdot 7}{\cancel{4}_{1} \cdot 25} = \frac{7}{25}$

**b.** $9\frac{1}{4}\% = \frac{9\frac{1}{4}}{100} = \frac{\frac{37}{4}}{100} = \frac{37}{4} \cdot \frac{1}{100} = \frac{37}{400}$

Note that $100 = \frac{100}{1}$ and division by 100 is the same as multiplication by $\frac{1}{100}$.

**c.** $735\% = \frac{735}{100} = \frac{\cancel{5}^{1} \cdot 147}{\cancel{5}_{1} \cdot 20} = \frac{147}{20}$ or $7\frac{7}{20}$

### *Completion Example 12*

Change $37\frac{1}{2}\%$ to an equivalent fraction in reduced form.

**Solution**

$37\frac{1}{2}\% = \frac{\_\_\_\_}{100} = \frac{\_\_\_\_}{100}$

$= \_\_\_\_ \cdot \frac{1}{100} = \_\_\_\_ = \_\_\_\_$

*Now Work Exercises 13-15 in the Margin.*

Change each percent to an equivalent fraction or mixed number with all fractions reduced.

**13.** 90%

$\frac{9}{10}$

**14.** 32%

$\frac{8}{25}$

**15.** 335%

$3\frac{7}{20}$

### A Common Misunderstanding

The fractions $\frac{1}{4}$ and $\frac{1}{2}$ are often confused with the percents $\frac{1}{4}\%$ and $\frac{1}{2}\%$. The differences can be clarified by using decimals.

| Percent | Decimal | Fraction |
|---|---|---|
| $\frac{1}{4}\%$ (or 0.25%) | 0.0025 | $\frac{25}{10,000} = \frac{1}{400}$ |
| $\frac{1}{2}\%$ (or 0.5%) | 0.005 | $\frac{5}{1000} = \frac{1}{200}$ |
| 25% | 0.25 | $\frac{25}{100} = \frac{1}{4}$ |
| 50% | 0.50 | $\frac{50}{100} = \frac{1}{2}$ |

Thus,

$\frac{1}{4} = 0.25$ and $\frac{1}{4}\% = 0.0025$

$0.25 \neq 0.0025$

Similarly,

$\frac{1}{2} = 0.50$ and $\frac{1}{2}\% = 0.005$

$0.50 \neq 0.005$

You can think of $\frac{1}{4}$ as being one-fourth of a dollar ( a quarter ) and $\frac{1}{4}\%$ as being one-fourth of a penny. $\frac{1}{2}$ can be thought of as one- half of a dollar and $\frac{1}{2}\%$ as one-half of a penny.

Some percents are so common that their decimal and fraction equivalents should be memorized. Their fractional values are particularly easy to work with, and many times calculations involving these fractions can be done mentally.

You might explain to your students that familiarity with these equivalents will reinforce the importance of fractions and minimize students' reliance on calculators. For example, many calculations involving simple interest are relatively simple to perform when the percent is in fraction form.

**Common Percent - Decimal - Fraction Equivalents**

$1\% = 0.01 = \frac{1}{100}$

$25\% = 0.25 = \frac{1}{4}$

$50\% = 0.50 = \frac{1}{2}$

$75\% = 0.75 = \frac{3}{4}$

$100\% = 1.00 = 1$

$33\frac{1}{3}\% = 0.33\frac{1}{3} = \frac{1}{3}$

$66\frac{2}{3}\% = 0.66\frac{2}{3} = \frac{2}{3}$

$12\frac{1}{2}\% = 0.125 = \frac{1}{8}$

$37\frac{1}{2}\% = 0.375 = \frac{3}{8}$

$62\frac{1}{2}\% = 0.625 = \frac{5}{8}$

$87\frac{1}{2}\% = 0.875 = \frac{7}{8}$

## Completion Example Answers

**3.** **a.** $\frac{\$\mathbf{150}\text{ profit}}{\$1000\text{ invested}} = \frac{\cancel{10}\cdot\mathbf{15}}{\cancel{10}\cdot 100} = \mathbf{15\%}$

**b.** $\frac{\$\mathbf{128}\text{ profit}}{\$800\text{ invested}} = \frac{\cancel{8}\cdot\mathbf{16}}{\cancel{8}\cdot 100} = \mathbf{16\%}$

Thus, investment **(b)** is better because **16%** is larger than **15%**.

**8.** Since $3\frac{3}{5}$ is larger than 1, the percent will be more than **100%**.

$3\frac{3}{5} = \frac{18}{5}$

Divide:

$$\begin{array}{r} \mathbf{3.6} \\ 5\overline{)18.0} \\ \mathbf{15} \\ \hline 30 \\ \mathbf{30} \\ \hline 0 \end{array}$$

Now, $3\frac{3}{5} = \frac{18}{5} = \mathbf{3.6} = \mathbf{360\%}$

**12.** $37\frac{1}{2}\% = \frac{\mathbf{37\frac{1}{2}}}{100} = \frac{\frac{\mathbf{75}}{\mathbf{2}}}{100} = \frac{\mathbf{75}}{\mathbf{2}}\cdot\frac{1}{100} = \frac{\mathbf{3}\cdot\cancel{\mathbf{25}}}{\mathbf{2}\cdot\mathbf{4}\cdot\cancel{\mathbf{25}}} = \frac{\mathbf{3}}{\mathbf{8}}$

Name ____________________ Section ______________ Date ____________

## Exercises 6.1

What percent of each square is shaded?

**1.**

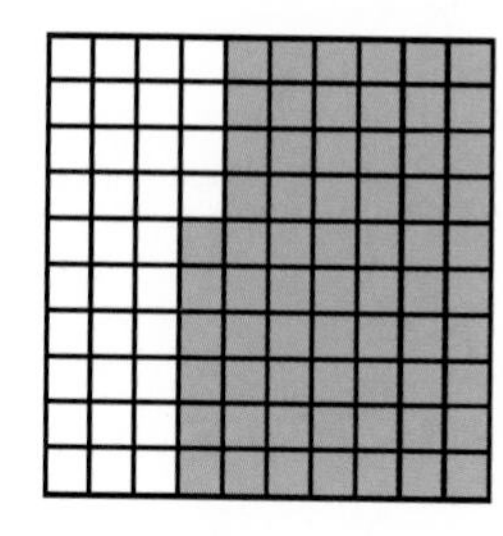

**2.**

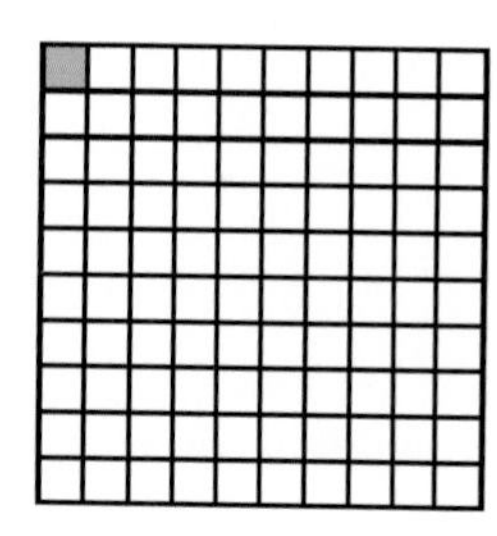

**3.**

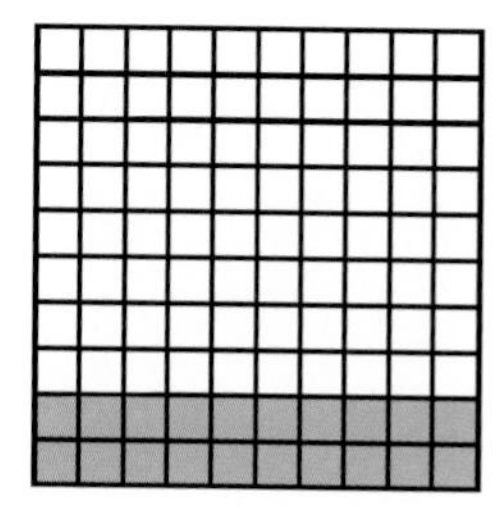

**4.**

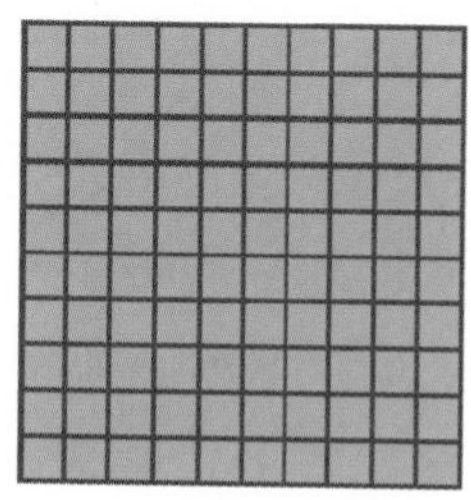

**5.**

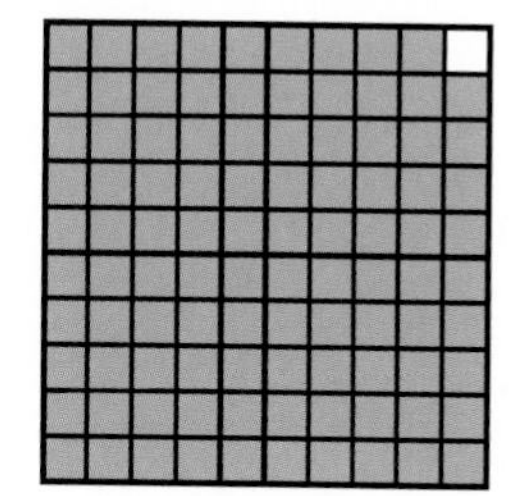

**6.**

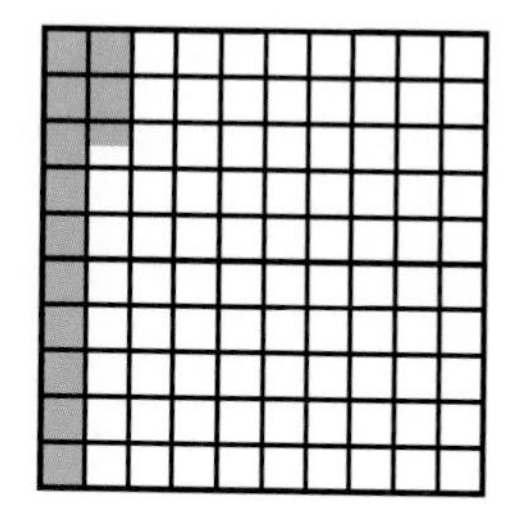

Change the following fractions to percents.

**7.** $\frac{30}{100}$ **8.** $\frac{90}{100}$ **9.** $\frac{48}{100}$ **10.** $\frac{125}{100}$

**11.** $\frac{16.3}{100}$ **12.** $\frac{0.5}{100}$ **13.** $\frac{24\frac{1}{2}}{100}$ **14.** $\frac{17\frac{3}{4}}{100}$

**ANSWERS**

1. 66%
2. 1%
3. 20%
4. 100%
5. 99%
6. 12.5%
7. 30%
8. 90%
9. 48%
10. 125%
11. 16.3%
12. 0.5%
13. $24\frac{1}{2}\%$
14. $17\frac{3}{4}\%$

15. a. $\frac{19}{100} = 19\%$

b. $\frac{17}{100} = 17\%$

c. Investment a. is better.

16. a. $\frac{10}{100} = 10\%$

b. $\frac{10}{100} = 10\%$

c. They are equal.

17. a. $\frac{5}{100} = 5\%$

b. $\frac{4}{100} = 4\%$

c. Investment a. is better.

18. a. $\frac{15}{100} = 15\%$

b. $\frac{12}{100} = 12\%$

c. Investment a. is better.

19. 3%

20. 30%

21. 300%

22. 52%

23. 5.5%

24. 0.4%

25. 175%

26. 230%

27. 108%

28. 200%

29. 36%

30. 50%

In Exercises 15-18, write the ratio of profit to investment as hundredths, and compare the percents. Tell which investment is better, **a.** or **b.**

**15. a.** A profit of $38 on a $200 investment.

**b.** A profit of $51 on a $300 investment.

**c.** Which investment is better?

**16. a.** A profit of $70 on a $700 investment.

**b.** A profit of $100 on a $1000 investment.

**c.** Which investment is better?

**17. a.** A profit of $150 on a $3000 investment.

**b.** A profit of $160 on a $4000 investment.

**c.** Which investment is better?

**18. a.** A profit of $300 on a $2000 investment.

**b.** A profit of $480 on a $4000 investment.

**c.** Which investment is better?

Change the following decimals to percents.

**19.** 0.03 **20.** 0.3 **21.** 3.0 **22.** 0.52

**23.** 0.055 **24.** 0.004 **25.** 1.75 **26.** 2.3

**27.** 1.08 **28.** 2 **29.** 0.36 **30.** 0.5

Name ______________________ Section ______________ Date __________

Change the following percents to decimals.

**31.** 2% **32.** 8% **33.** 22% **34.** 15%

**35.** 25% **36.** 50% **37.** 10.1% **38.** 12.5%

**39.** $6\frac{1}{2}\%$ **40.** $15\frac{1}{4}\%$ **41.** 80% **42.** $19\frac{3}{4}\%$

Change the following fractions and mixed numbers to percents.

**43.** $\frac{1}{20}$ **44.** $\frac{7}{10}$ **45.** $\frac{24}{25}$ **46.** $\frac{1}{9}$

**47.** $\frac{1}{7}$ **48.** $\frac{5}{6}$ **49.** $1\frac{3}{8}$ **50.** $2\frac{1}{15}$

Change the following percents to fractions or mixed numbers in reduced form.

**51.** 5% **52.** 50% **53.** $12\frac{1}{2}\%$ **54.** $16\frac{2}{3}\%$

**ANSWERS**

**31.** 0.02

**32.** 0.08

**33.** 0.22

**34.** 0.15

**35.** 0.25

**36.** 0.5

**37.** 0.101

**38.** 0.125

**39.** 0.065

**40.** 0.1525

**41.** 0.8

**42.** 0.1975

**43.** 5%

**44.** 70%

**45.** 96%

**46.** $11\frac{1}{9}\%$ or 11.1%

**47.** $14\frac{2}{7}\%$ or 14.3%

**48.** $83\frac{1}{3}\%$ or 83.3%

**49.** $137\frac{1}{2}\%$ or 137.5%

**50.** $206\frac{2}{3}\%$ or 206.7%

**51.** $\frac{1}{20}$

**52.** $\frac{1}{2}$

**53.** $\frac{1}{8}$

**54.** $\frac{1}{6}$

55. $1\frac{1}{5}$

56. $1\frac{2}{5}$

57. $\frac{1}{500}$

58. $\frac{3}{400}$

59. [Respond below exercise]

60. [Respond below exercise]

61. [Respond below exercise]

62. [Respond below exercise]

63. [Respond below exercise]

64. [Respond below exercise]

65. 0.15

66. 0.085

67. 32%

68. 0.9%

**55.** 120% **56.** 140% **57.** 0.2% **58.** 0.75%

Find the missing forms of each number and write your answers in the table below.

| | Fraction Form | Decimal Form | Percent Form |
|---|---|---|---|
| **59.** | $\frac{7}{8}$ | **b.** 0.875 | **c.** 87.5% |
| **60.** | $\frac{19}{20}$ | **b.** 0.95 | **c.** 95% |
| **61.** | **a.** $\frac{3}{50}$ | 0.06 | **c.** 6% |
| **62.** | **a.** $1\frac{4}{5}$ | 1.8 | **c.** 180% |
| **63.** | **a.** $\frac{6}{25}$ | **b.** 0.24 | 24% |
| **64.** | **a.** $\frac{2}{3}$ | **b.** $0.\overline{6}$ | $66\frac{2}{3}\%$ |

**65.** The interest rate on a loan is 15%. Change 15% to a decimal.

**66.** The sales commission for the clerk in a retail store is figured at $8\frac{1}{2}\%$. Change $8\frac{1}{2}\%$ to a decimal.

**67.** To calculate what your maximum monthly house payment should be, a banker multiplied your monthly income by 0.32. Change 0.32 to a percent.

**68.** Suppose that the state motor vehicle licensing fee is figured by multiplying the cost of your car by 0.009. Change 0.009 to a percent.

Name ______________ Section ______________ Date ______________

**ANSWERS**

**69.** $\frac{7}{20}$

**70.** 2.5%

**71.** about 12% in California; about 0.18% in Wyoming

**72.** $1\frac{2}{3}\%$ ( or $1.\overline{6}\%$ )

**73. a.** 0.000575

**b.** 20.7%

**74. a.** $\frac{3}{20}$ because 20 is a factor of 100.

**b.** $\frac{1}{10}$ because 10 is a factor of 100.

**c.** $\frac{4}{5}$ because 5 is a factor of 100.

**69.** A department store offers a 35% discount during a special sale on dresses. Change 35% to a fraction reduced to lowest terms.

**70.** The discount you earn by paying cash is found by multiplying the amount of your purchase by 0.025. Change 0.025 to a percent.

**71.** According to the 1990 census, California ranks no. 1 as the most populated state and Wyoming ranks no. 50 as the least populated state. Approximately what percent of the population of the United States lives in each of these states if there are about 250,000,000 people in the country, about 30,000,000 in California, and about 450,000 in Wyoming?

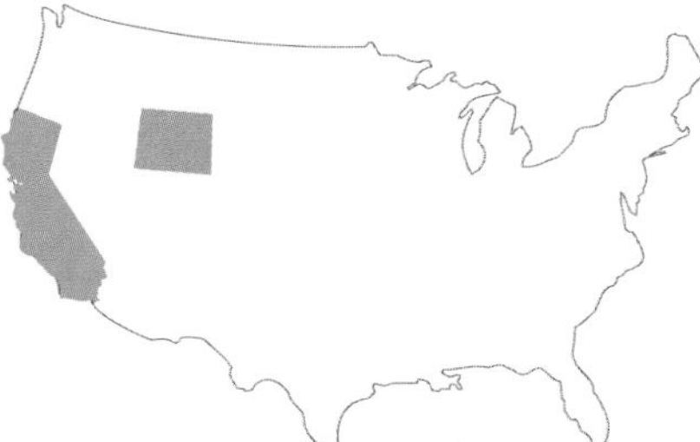

**72.** In an association of 300 local artists, 5 are on a selection committee for a special showing in a shopping mall. What percent of the association is on the committee?

**73.** A credit card company charges 0.0575% interest per day for credit.

**a.** Write this interest rate in decimal form.

**b.** What is this rate of interest as an annual rate? ( Use the business year or 360 days per year. )

**74.** Three pairs of fractions are given. In each case, tell which fraction would be easier to change to a percent mentally, and explain your reasoning.

**a.** $\frac{3}{20}$ or $\frac{1}{8}$ **b.** $\frac{1}{10}$ or $\frac{1}{9}$ **c.** $\frac{5}{12}$ or $\frac{4}{5}$

[Respond below exercise]
75. ____________

## WRITING AND THINKING ABOUT MATHEMATICS

**75.** Use your calculator to find the decimal form for each of the two fractions $\frac{3}{11}$ and $\frac{2}{3}$.

Does your calculator round off at the place of the last digit on the display? Explain how you can tell. [ **Note:** Some calculators will round off and others will "truncate" ( simply cut off ) the remaining digits. Check with your instructor to find out which calculators used in class round off and which calculators truncate. ]

Answers will depend on the student's calculator.

**The Recycle Bin**

1. $x = 2.2$
2. $y = \frac{5}{2}$ (or $y = 2.5$)
3. $w = 8$
4. $A = 0.345$
5. 10.6 gallons
6. $2.20

### The Recycle Bin ( from Section 4.6)

Solve each of the following proportions.

**1.** $\frac{x}{3.5} = \frac{4.4}{7}$

**2.** $\frac{3}{5} = \frac{1\frac{1}{2}}{y}$

**3.** $\frac{276}{6} = \frac{368}{w}$

**4.** $\frac{A}{0.75} = \frac{2.3}{5}$

**5.** If gasoline costs $1.20 per gallon, how many gallons can be bought for $12.72?

**6.** A company expects to make $1 on an item it sells for $10. What would the company expect to make on an item it sells for $22?

# 6.2 Solving Percent Problems

Objectives

① Recognize three types of percent problems.

② Solve percent problems using the formula $R \cdot B = A$.

## The Relationship Between Proportion and Percent

As stated in Section 6.1, **percent means hundredths**. Thus, percent is the **ratio** of a number to 100. We can write 25% in the ratio form $\frac{25}{100}$. And, more generally, $P\% = \frac{P}{100}$. With this ratio concept in mind, consider the following statement related to percent:

**"12% of 75 is 9."**

This statement means that the ratio of 9 to 75 is the same as 12 to 100. Thus, the statement can be written in the form of a proportion as follows:

$$\mathbf{\frac{12}{100} = \frac{9}{75}}$$

The statement **"12% of 75 is 9."** has three numbers in it. In general, solving a percent problem involves knowing two of these numbers and trying to find the third. That is, there are **three basic types of percent problems**, and we can solve each type of problem by using a proportion. For example,

"12% of 75 is what number?" corresponds to the proportion $\frac{12}{100} = \frac{A}{75}$

"12% of what number is 9?" corresponds to the proportion $\frac{12}{100} = \frac{9}{B}$

"What percent of 75 is 9?" corresponds to the proportion $\frac{P}{100} = \frac{9}{75}$

These proportions illustrate the following notation and terminology to be used throughout our discussion of percents problems:

| P% | of | B | is | A. |
|---|---|---|---|---|
| ↑ | | ↑ | | ↑ |
| **Percent** | | **Base** | | **Amount** |

leading to the proportion $\mathbf{\frac{P}{100} = \frac{A}{B}}$

Example 1 illustrates each of the three types of percent problems. In each case, using the techniques for solving proportions discussed in Section 4.6, we set up the corresponding proportion and solve for the unknown quantity.

### Example 1

**a.** What is 35% of 700?

**Solution**

In this case, the Amount is unknown and the Percent is 35 and the Base is 700.

$$\frac{35}{100} = \frac{A}{700}$$

$$35 \cdot 700 = 100 \cdot A$$ The product of the means is equal to the product of the extremes.

$$24{,}500 = 100A$$

$$\frac{24{,}500}{100} = \frac{\cancel{100}A}{\cancel{100}}$$

$$245 = A$$

Thus, 35% of 700 is **245**.

**b.** 16% of what number is 40?

**Solution**

In this case, the Base is unknown and the Percent is 16 and the Amount is 40.

$$\frac{16}{100} = \frac{40}{B}$$

$$16 \quad B = 40 \quad 100$$ The product of the means is equal to the product of the extremes.

$$\frac{\cancel{16}B}{\cancel{16}} = \frac{4000}{16}$$

$$B = 250$$

Thus, 16% of **250** is 40.

**c.** What percent of 540 is 81?

**Solution**

In this case, the Percent is unknown and the Base is 540 and the Amount is 81.

$$\frac{P}{100} = \frac{81}{540}$$

$$\frac{P}{100} = \frac{3}{20}$$ Reducing first

$$100 \cdot \frac{P}{100} = \frac{3}{20} \cdot 100$$ Multiplying both sides by 100

$$P = 15$$

Thus, **15%** of 540 is 81.

## The Basic Formula: $R \cdot B = A$

In Example 1, we have illustrated the **three basic problems related to percent** in terms of proportions and shown how to solve these proportions. You may choose to use this method of proportions to solve any or all of the problems related to percents. Another method, that relates more to algebra and setting up equations, is to change the percent into decimal form ( or fraction form ) and call this decimal form a RATE. In this method, we write

$$\frac{P}{100} = R \text{ where R is the decimal form of P\%.}$$

Now, consider the statement

16% of 50 is 8.

This statement can be translated into an equation in the following way:

| | | | | | |
|---|---|---|---|---|---|
| Sentence → | 16% | of | 50 | is | 8 |
| | ↓ | ↓ | ↓ | ↓ | ↓ |
| Equation → | 0.16 | · | 50 | = | 8 |
| | ↑ percent changed to decimal ↓ | ↑ "of" changed to times ↓ | ↕ | ↑ "is" changed to equal ↓ | ↕ |
| Basic formula → | Rate | · | Base | = | Amount |

The terms represented in the basic formula are those discussed earlier and are explained in detail in the following box.

> $R$ = **RATE** or percent ( as a decimal or fraction )
>
> $B$ = **BASE** ( number we are finding the percent of )
>
> $A$ = **AMOUNT** or percentage ( a part of the base )
>
> "of" means to multiply ( the raised dot · is used in the formula ).
>
> "is" means equal ( = ).
>
> The relationship among $R$, $B$, and $A$ is given in the formula
>
> $$\boldsymbol{R \cdot B = A}$$

## Using the Formula

Even though, as stated earlier, there are just **three basic types of problems that involve percent**, many people have a difficult time differentiating among them and determining whether to multiply or divide in a particular problem. Using the formula $\boldsymbol{R \cdot B = A}$ ( or the proportion method discussed earlier ) helps to avoid these difficulties. If the values of two of the quantities in the formula $R \cdot B = A$ are known, then these values can be substituted into the formula, and the third value can be determined by solving the resulting equation.

**Note:** The formula also can be written in the form

$$A = R \cdot B$$

As illustrated in the following box, this form is convenient for solving Type 1 percent problems.

**The Three Basic Types of Percent Problems and the Formula $R \cdot B = A$**

**Type 1:** Find the Amount, given the Percent ( rate ) and the Base.

What is 65% of 800?

$$\begin{array}{c} A = R \cdot B \\ \downarrow \quad \searrow \quad \searrow \\ A = 0.65 \cdot 800 \end{array}$$

$R$ and $B$ are known. The object is to find $A$.

**Type 2:** Find the Base, given the Percent ( rate ) and the Amount.

42% of what number is 157.5?

$$\begin{array}{c} R \cdot B = A \\ \swarrow \quad \downarrow \quad \searrow \\ 0.42 \cdot B = 157.5 \end{array}$$

$R$ and $A$ are known. The object is to find $B$.

**Type 3:** Find the Percent ( rate ), given the Base and the Amount.

What percent of 92 is 115?

$$\begin{array}{c} R \cdot B = A \\ \downarrow \quad \downarrow \quad \searrow \\ R \cdot 92 = 115 \end{array}$$

$A$ and $B$ are known. The object is to find $R$.

Many students are not sure whether to multiply or divide when solving a percent problem. This decision is made for you when you use the formula $R \cdot B = A$. Once the substitutions for the known values in the formula have been made, **the process of solving the equation for the remaining unknown value determines whether multiplication or division is needed.**

The following examples illustrate how to substitute into the formula and how to solve the resulting equations.

### Example 2

What is 65% of 800?

#### Solution

In this problem $R = 65\% = 0.65$ and $B = 800$. This is a Type 1 problem, and we want to find the value of $A$.

$$A = R \cdot B \qquad 0.65$$
$$A = 0.65 \cdot 800 \qquad \times 800$$
$$A = 520 \qquad 520.00$$

So, 65% of 800 is **520**.

Type 1 problems, as illustrated in Example 1, are the most common type of percent problems and the easiest to solve. The rate ( $R$ ) and the base ( $B$ ) are known. The unknown amount ( $A$ ) is found by multiplying the rate times the base.

**Note:** The operations in these examples can be performed with a calculator or by hand, as shown in Example 2. In either case, the equations should be written so that the = signs are aligned one above the other. Also, writing the equations and the calculated values helps you remember whether you are multiplying or dividing.

### Example 3

42% of what number is 157.5?

#### Solution

In this problem, $R = 42\% = 0.42$ and $A = 157.5$. We want to find the value of $B$. Substitution in the formula gives.

$$R \cdot B = A$$
$$0.42 \cdot B = 157.5$$
$$\frac{\cancel{0.42} \cdot B}{\cancel{0.42}} = \frac{157.5}{0.42}$$ Divide both side by 0.42, the coefficient of B. ( The division can be performed with a calculator. )
$$B = 375$$

So, 42% of **375** is 157.5.

### Example 4

What percent of 92 is 115?

#### Solution

In this problem, $B = 92$ and $A = 115$. We want to find the value of $R$. ( Do you expect $R$ to be more than 100% or less than 100%? ) Substitution in the formula gives

$$R \cdot B = A$$
$$R \cdot 92 = 115$$
$$\frac{R \cdot \cancel{92}}{\cancel{92}} = \frac{115}{92}$$ Divide both side by 92, the coefficient of $B$. ( The division can be performed with a calculator. )
$$R = 1.25$$
$$R = 125\%$$ $R$ is more than 100% because $A$ is larger than $B$.

So, **125%** of 92 is 115.

### *Completion Example 5*

Find 35% of 89.

**Solution**

In this problem, $R =$ _____% = _____ and $B =$ _____. We want to find the value of _____. Substitution in the formula gives

$A = R \cdot B$

$A =$ _____ · _____

$A =$ __________

Thus, 35% of 89 is __________.

### *Completion Example 6*

What percent of 320 is 48?

**Solution**

In this problem, $A =$ _____ and $B =$ _____. We want to find the value of _____. Substitution in the formula gives

$R \cdot B = A$

$R \cdot$ _____ = _____

$\dfrac{R \cdot \_\_\_\_\_}{\_\_\_\_\_} = \dfrac{\_\_\_\_\_}{\_\_\_\_\_}$

$R =$ _____

Thus, _____% of 320 is 48.

As we mentioned just before Example 1, deciding whether to multiply or divide in a percent problem is a common difficulty. You might reinforce the fact that this decision is "already made" once correct substitutions are made in the formula: $R \cdot B = A$ .

The multiplication or division, as shown in the examples, can be easily performed with a calculator. However, **you should always write down the equation first so that you properly identify the unknown quantity and are certain that the correct operation is being performed.**

*Now Work Exercises 1-3 in the Margin.*

Find the unknown quantity.

1. 10% of 90 is_____.
   9

2. 50% of _____ is 72.
   144

3. Find 33% of 180.
   59.4

Percents can be changed to fraction form, and in some cases, the fraction form will actually simplify the work. For example, we know that

$$75\% = \frac{3}{4}, \qquad 33\frac{1}{3}\% = \frac{1}{3}, \qquad \text{and} \qquad 12.5\% = \frac{1}{8}$$

The use of fraction equivalents in place of the decimal form of percent is illustrated in the next two examples. ( See the list in Section 6.1 for common percent-decimal equivalents. )

### Example 7

Find 75% of 56.

**Solution**

This is a Type 1 problem with $R = 75\% = \frac{3}{4}$ and $B = 56$.

$$A = R \cdot B$$

$$A = \frac{3}{\cancel{4}_1} \cdot \cancel{56}^{14} = 3 \cdot 14 = 42$$

This multiplication can be performed without a calculator since 4 is a factor of 56.

So, 75% of 56 is **42**.

### Example 8

The amount 55 is 62.5% of what number?

**Solution**

Here $R = 62.5\% = \frac{5}{8}$ and $A$ is 55. The unknown is $B$. Now,

$$R \cdot B = A$$

$$\frac{5}{8} \cdot B = 55$$

$$\frac{\cancel{8}}{\cancel{5}} \cdot \frac{\cancel{5}}{\cancel{8}} \cdot B = \frac{8}{\cancel{5}_1} \cdot \cancel{55}^{11}$$

$$B = 88$$

Multiply both sides by $\frac{8}{5}$, the reciprocal of the coefficient $\frac{5}{8}$.

So, 55 is 62.5% of **88**.

Remember the following two facts about percent and the relative sizes of the amount ( $A$ ) and the base ( $B$ ):

**1.** A percent is just another form of a fraction: $R = \frac{A}{B}$.

**2. a.** If $R < 100\%$, then $A < B$.
**b.** If $R > 100\%$, then $A > B$.

## Completion Example Answers

**5.** In this problem, $R = \mathbf{35\%} = \mathbf{0.35}$ and $B = \mathbf{89}$. We want to find the value of $\boldsymbol{A}$. Substitution in the formula gives

$$A = R \cdot B$$
$$A = 0.35 \cdot 89$$
$$A = 31.15$$

Thus, 35% of 89 is **31.15**.

**6.** In this problem, $A = \mathbf{48}$ and $B = \mathbf{320}$. We want to find the value of $\boldsymbol{R}$. Substitution in the formula gives

$$R \cdot B = A$$

$$R \cdot \mathbf{320} = \mathbf{48}$$

$$\frac{R \cdot \cancel{\mathbf{320}}}{\cancel{\mathbf{320}}} = \frac{\mathbf{48}}{\mathbf{320}}$$

$$R = \mathbf{0.15}$$

Thus, **15**% of 320 is 48.

Name ______________________ Section ____________ Date ____________

## Exercises 6.2

In Exercises 1-24, first write the related equation using $R \cdot B = A$ ( or $A = R \cdot B$ ), and then solve for the unknown quantity. ( Percents that do not come out even may be rounded off to the nearest tenth of one percent or left in the form of a fraction of a percent. )

**1.** 10% of 60 is _____.

**2.** 10% of 80 is _____.

**3.** 15% of 70 is _____.

**4.** 15% of 80 is _____.

**5.** 25% of 40 is _____.

**6.** 25% of 80 is _____.

**7.** 75% of 120 is _____.

**8.** 80% of 10 is _____.

**9.** 20% of _____ is 45.

**10.** 2% of _____ is 45.

**11.** 3% of _____ is 65.4.

**12.** 30% of _____ is 952.2.

**13.** 50% of _____ is 310.

**14.** 50% of _____ is 62.5.

**15.** 150% of _____ is 180.

**16.** 110% of _____ is 82.5.

**17.** _____% of 160 is 240.

**18.** _____% of 150 is 60.

**ANSWERS**

1. $A = 6$
2. $A = 8$
3. $A = 10.5$
4. $A = 12$
5. $A = 10$
6. $A = 20$
7. $A = 90$
8. $A = 8$
9. $B = 225$
10. $B = 2250$
11. $B = 2180$
12. $B = 3174$
13. $B = 620$
14. $B = 125$
15. $B = 120$
16. $B = 75$
17. $R = 1.5 = 150\%$
18. $R = 0.4 = 40\%$

19. $R = 2 = 200\%$

20. $R = 2 = 200\%$

21. $R = \frac{1}{3} = 33\frac{1}{3}\%$

22. $R = 0.45 = 45\%$

23. $R = 0.252 = 25.2\%$

24. $R = 0.315 = 31.5\%$

25. $A = 17.5$

26. $A = 27.52$

27. $B = 75$

28. $B = 86$

29. $R = 1.5 = 150\%$

30. $R = 1.4 = 140\%$

31. $B = 36$

32. $B = 51$

33. $A = 115$

34. $A = 58.5$

35. $A = 58.56$

36. $A = 11.34$

37. $R = 1.25 = 125\%$

38. $R = 0.2 = 20\%$

39. $B = 80$

40. $B = 2500$

**19.** _____% of 17 is 34.

**20.** _____% of 35 is 70.

**21.** _____% of 48 is 16.

**22.** _____% of 100 is 45.

**23.** _____% of 30 is 7.56.

**24.** _____% of 12 is 3.78.

In Exercises 25-40,

**a.** write the related equation, and then

**b.** solve the equation for the unknown quantity.

Note that even though the wording is changed slightly from that in Exercises 1-24, the formula $R \cdot B = A$ ( or $A = R \cdot B$ ) still applies in each problem.

**25.** _____ is 50% of 35.

**26.** _____ is 32% of 86.

**27.** 16.5 is 22% of _____.

**28.** 27.52 is 32% of _____.

**29.** 15 is _____% of 10.

**30.** 14 is _____% of 10.

**31.** 24 is $66\frac{2}{3}\%$ of _____.

**32.** 17 is $33\frac{1}{3}\%$ of _____.

**33.** _____ is 25% of 460.

**34.** _____ is 18% of 325.

**35.** Find 24% of 244.

**36.** Find 16.2% of 70.

**37.** What percent of 32 is 40?

**38.** What percent of 100 is 20?

**39.** 100 is 125% of what number?

**40.** 1250 is 50% of what number?

Name ______________ Section ________ Date ________

Use fractions to solve the following percent problems.

**41.** Find 50% of 84.

**42.** Find 75% of 224.

**43.** What is $12\frac{1}{2}\%$ of 32?

**44.** What is $37\frac{1}{2}\%$ of 160?

**45.** $33\frac{1}{3}\%$ of 75 is what number?

**46.** $62\frac{1}{2}\%$ of 1600 is what number?

**47.** What percent of 92 is 92?

**48.** What percent of 92 is 184?

**49.** 100% of what number is 61.5?

**50.** Find 25% of 64.64.

**51.** The population of Zimbabwe was estimated to be 11,200,000 in 1994, with an expected increase of 3% per year. At this rate of increase, what was the expected population of Zimbabwe for the year 1996?

**52.** In 1991, the United States produced 53.9 million metric tons of wheat and 190 million metric tons of corn. The world production was 551 million metric tons of wheat and 478.8 million metric tons of corn. What percent (to the nearest tenth of one percent) of the world's wheat and what percent of the world's corn did the United States produce in 1991?

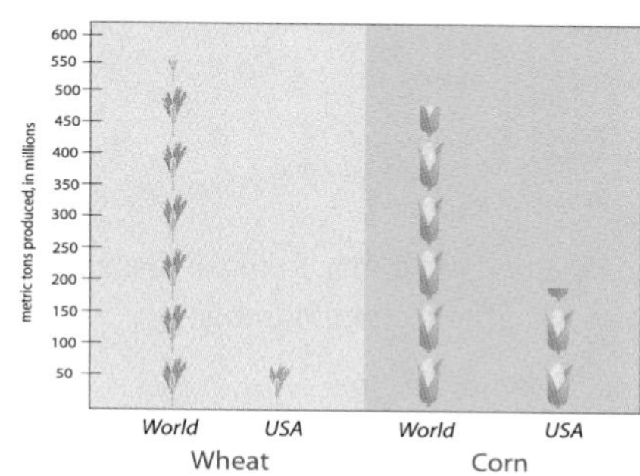


**53.** In 2000, the two largest cities in the world were Tokyo and New York, with populations of 34.8 million and 20.2 million respectively.

**a.** How many more people lived in Tokyo than in New York in 2000?

**b.** With the population of New York as the base, what percent (to the nearest tenth of one percent) was the population of Tokyo more than the population of New York?

**ANSWERS**

**41.** 42

**42.** 168

**43.** 4

**44.** 60

**45.** 25

**46.** 1000

**47.** 100%

**48.** 200%

**49.** 61.5

**50.** 16.16

**51.** 11,882,080

**52.** wheat: 9.8%; corn: 39.7%

**53. a.** 14.6 million

**b.** 72.3%

54. 784%

55. [Respond below exercise.]

56. [Respond below exercise.]

57. [Respond below exercise.]

**54.** In 1915, the Internal Revenue Service ( IRS ) had income from income and profit taxes of \$80 million and in 1993 \$627.2 million. What was the percent of increase of income for the IRS from income and profit taxes during the years of 1915 to 1993? (Use the taxes of 1915 as the base. )

## Writing and Thinking about Mathematics

**55. a.** Explain why 10% of a decimal number can be found by moving the decimal one place to the left.

Explanations will vary. The decimal equivalent of 10% is 0.10, and multiplication of a decimal number by 0.10 moves the decimal point one place to the left.

**b.** Explain why 10% of a decimal number can be found by multiplying the number by the fraction $\frac{1}{10}$.

Explanations will vary. Taking 10% of a number is the same as dividing by 10 or multiplying by $\frac{1}{10}$.

**56.** To find 20% of a decimal number, you might first find 10% by moving the decimal point one place to the left ( see Exercise 55 above ) and then multiplying this result by 2.

**a.** Use this method to find 20% of 62.

$(6.2) \cdot 2 = 12.4$

**b.** Explain how you could use a similar technique to find 30% of a decimal number.

Explanations will vary. First find 10% of the number, and then multiply by 3.

**c.** Use this method to find 30% of 62.

$(6.2) \cdot 3 = 18.6$

**57.** Write a set of rules ( or methods ) similar to those discussed in Exercises 55 and 56 for finding 1%, 2%, and 3% of a decimal number.

Answers will vary. Students may suggest finding 1% of a number by moving the decimal point two places to the left and then finding the desired multiple.

# 6.3 Estimating with Percent

## Estimating with Percent

The purpose of this section is to reinforce your understanding of percent before you begin to work with applications in Section 6.4 and the remainder of the chapter. The approach is to provide you with an opportunity to check your understanding about the relative sizes of numbers and various percents of those numbers by giving you multiple-choice problems and problems in which you must place the decimal point in the correct position. It is important for you to practice your estimating skills and to estimate answers mentally before performing actual calculations. Doing so will test your understanding. You may then check your choices by actually performing the calculations with a calculator.

**Objectives**

① Develop a clear understanding of percent.

② Develop skills for reasoning and calculating with percents.

For example, suppose that the problem is " Find 10% of 600, " and the choices are **a.** 6, **b.** 60, **c.** 600, **d.** 6000. You might arrive at the correct choice by reasoning in any of several ways:

You might reason that $10\% = \frac{1}{10}$ and $\frac{1}{10}$ of 600 is 60.

You might reason that 600 is 100% of 600, and 6000 is an even larger percentage of the base. Also, 6 is only 1% of 600. So the correct answer is 60.

Or, you might simply remember that 10% of a number can be found by moving the decimal point one place to the left.

There is no one correct method of reasoning. In fact, you may reason differently with different types of problems. Regardless of what method you use, do not be afraid of making a mistake. **What is important is that you try to learn from your mistakes.** This can help you identify and correct possible flaws in your reasoning.

### Example 1

Choose the correct answer by reasoning and calculating mentally. What is 40% of 620?

**a.** 2.48 **b.** 24.8 **c.** 248 **d.** 480

#### Solution

You might reason and approximate as follows:

40% is close but slightly less than 50%, and 50% of 620 is 310. The only choice close to but slightly less than 310 is choice **c.** 248.

Or, you might reason and calculate as follows:

10% of 620 is 62. Therefore, since 40% is 4 times 10%, 40% of 620 is 4 times 62 ( or 248 ).

In either case, the correct choice is **c.** 248.

Even though there are sample "reasonings" in Examples 1 and 2, you might ask the students to explain their reasoning in class. This will help them in working the exercises and might give some useful insights into their reasoning processes.

## Example 2

Choose the correct answer by reasoning and calculating mentally. Find 110% of 8000.

**a.** 880 **b.** 8800 **c.** 88,000 **d.** 80,000

### Solution

You might reason as follows:

Since 110% = 100% + 10%, the answer will be 100% of 8000 plus 10% of 8000,
or 8000 + 800 = 8800.

Or, you might reason more simply as follows:

110% is close to 100%, and 100% of 8000 is 8000. The only choice close to 8000 is choice **b.** 8800.

In either case, the correct choice is **b.** 8800.

Name ____________ Section ____________ Date ____________

## Exercises 6.3

In each of the following exercises, one of the choices is the correct answer to the problem. You are to work through the entire set of exercises by reasoning and calculating mentally. After you have checked your answers, you should work out the answers to any problems you missed and try to understand why your reasoning was incorrect.

**1.** Write 3% as a fraction.

**a.** $\frac{3}{1}$ **b.** $\frac{3}{10}$ **c.** $\frac{3}{100}$ **d.** $\frac{3}{1000}$

**2.** Write 7% as a fraction.

**a.** $\frac{7}{1}$ **b.** $\frac{7}{10}$ **c.** $\frac{7}{100}$ **d.** $\frac{7}{1000}$

**3.** Write $12\frac{1}{2}\%$ as a fraction, reduced.

**a.** $\frac{1}{8}$ **b.** $\frac{25}{2}$ **c.** 12.5 **d.** 125

**4.** Write $87\frac{1}{2}\%$ as a fraction, reduced.

**a.** $\frac{1}{8}$ **b.** $\frac{3}{8}$ **c.** $\frac{5}{8}$ **d.** $\frac{7}{8}$

**5.** Write $2\frac{1}{2}\%$ as a decimal.

**a.** 0.025 **b.** 0.25 **c.** 2.5 **d.** 25

**6.** Write $12\frac{1}{2}\%$ as a decimal.

**a.** 0.0125 **b.** 0.125 **c.** 1.25 **d.** 12.5

**7.** Find 10% of 530.

**a.** 5.3 **b.** 53 **c.** 10.6 **d.** 106

**ANSWERS**

**1.** c

**2.** c

**3.** a

**4.** d

**5.** a

**6.** b

**7.** b

8. c
9. b
10. b
11. b
12. b
13. d
14. c
15. a
16. c
17. b
18. d

**8.** Find 20% of 530.

**a.** 1.06 **b.** 10.6 **c.** 106 **d.** 212

**9.** Find 30% of 530.

**a.** 106 **b.** 159 **c.** 212 **d.** 265

**10.** Find 40% of 530.

**a.** 21.2 **b.** 212 **c.** 2120 **d.** 21,200

**11.** What is 100% of 8.02?

**a.** 0.802 **b.** 8.02 **c.** 80.2 **d.** 802

**12.** What is 200% of 8.02?

**a.** 1.604 **b.** 16.04 **c.** 160.4 **d.** 1604

**13.** What is 110% of 300?

**a.** 30 **b.** 33 **c.** 300 **d.** 330

**14.** What is 120% of 300?

**a.** 60 **b.** 300 **c.** 360 **d.** 630

**15.** Write 0.25% as a decimal.

**a.** 0.0025 **b.** 0.025 **c.** 0.25 **d.** 2.5

**16.** Write 75% as a decimal.

**a.** 0.0075 **b.** 0.075 **c.** 0.75 **d.** 7.5

**17.** Write $\frac{1}{3}$ as a percent.

**a.** $\frac{1}{3}\%$ **b.** $33\frac{1}{3}\%$ **c.** 66% **d.** $66\frac{2}{3}\%$

**18.** Write $\frac{2}{3}$ as a percent.

**a.** $\frac{2}{3}\%$ **b.** $33\frac{1}{3}\%$ **c.** 66% **d.** $66\frac{2}{3}\%$

Name ______________________ Section ____________ Date ________

**19.** Write $\frac{5}{8}$ as a percent.

**a.** $12\frac{1}{2}\%$ **b.** 25% **c.** $37\frac{1}{2}\%$ **d.** $62\frac{1}{2}\%$

**20.** Write $\frac{3}{8}$ as a percent.

**a.** 12.5% **b.** 25% **c.** 37.5% **d.** 62.5%

**21.** 10% of what number is 74?

**a.** 7.4 **b.** 74 **c.** 740 **d.** 7400

**22.** 10% of what number is 46?

**a.** 4.6 **b.** 46 **c.** 460 **d.** 4600

**23.** 30% of what number is 15?

**a.** 45 **b.** 50 **c.** 450 **d.** 500

**24.** 30% of what number is 21?

**a.** 63 **b.** 70 **c.** 630 **d.** 700

**25.** What percent of 836 is 83.6?

**a.** 1% **b.** 10% **c.** 100% **d.** 1000%

**26.** What percent of 1500 is 15?

**a.** 1% **b.** 10% **c.** 100% **d.** 1000%

**27.** What percent of 25 is 250?

**a.** 1% **b.** 10% **c.** 100% **d.** 1000%

**28.** What percent of 93.7 is 937?

**a.** 1% **b.** 10% **c.** 100% **d.** 1000%

**ANSWERS**

**19.** d

**20.** c

**21.** c

**22.** c

**23.** b

**24.** b

**25.** b

**26.** a

**27.** d

**28.** d

29. c

30. c

31. a

32. a

33. b

34. b

35. b

36. a

37. c

38. c

39. c

40. d

**29.** What is 150% of 50?

**a.** 25 **b.** 50 **c.** 75 **d.** 100

**30.** What is 150% of 100?

**a.** 50 **b.** 100 **c.** 150 **d.** 1500

**31.** What is 1% of 100?

**a.** 1 **b.** 10 **c.** 100 **d.** 1000

**32.** What is 1% of 200?

**a.** 2 **b.** 20 **c.** 200 **d.** 2000

**33.** What is 1% of 83.6?

**a.** 0.0836 **b.** 0.836 **c.** 8.36 **d.** 83.6

**34.** What is 1% of 94.7?

**a.** 0.0947 **b.** 0.947 **c.** 9.47 **d.** 94.7

**35.** What percent of 1000 is 250?

**a.** 2.5% **b.** 25% **c.** 250% **d.** 2500%

**36.** What percent of 1000 is 10?

**a.** 1% **b.** 10% **c.** 100% **d.** 1000%

**37.** 10% of what number is 10?

**a.** 1 **b.** 10 **c.** 100 **d.** 1000

**38.** 5% of what number is 5?

**a.** 1 **b.** 10 **c.** 100 **d.** 1000

**39.** 75% of what number is 36?

**a.** 27 **b.** 36 **c.** 48 **d.** 75

**40.** 25% of what number is 40?

**a.** 10 **b.** 20 **c.** 50 **d.** 160

Name ______________________ Section ______________ Date ____________

**ANSWERS**

**41.** Choose the best estimate of 30% of 92.17.

**a.** 9 **b.** 18 **c.** 27 **d.** 36

**42.** Choose the best estimate of 80% of 327.

**a.** 25 **b.** 33 **c.** 65 **d.** 250

**43.** Choose the best estimate of 19.3% of 516.

**a.** 10 **b.** 20 **c.** 100 **d.** 200

**44.** Choose the best estimate of 8.7% of 3500.

**a.** 9 **b.** 35 **c.** 350 **d.** 3500

**45.** Approximately what percent of 1000 is 1585?

**a.** 50% **b.** 66% **c.** 80% **d.** 150%

**46.** Approximately what percent of 2000 is 3120?

**a.** 20% **b.** 30% **c.** 150% **d.** 200%

**47.** 947 is approximately 50% of what number?

**a.** 500 **b.** 1000 **c.** 1800 **d.** 4500

**48.** 345 is approximately 50% of what number?

**a.** 34 **b.** 170 **c.** 700 **d.** 1500

**49.** 72 is approximately $33\frac{1}{3}\%$ of what number?

**a.** 21 **b.** 24 **c.** 100 **d.** 200

**50.** 53 is approximately $66\frac{2}{3}\%$ of what number?

**a.** 15 **b.** 75 **c.** 150 **d.** 200

**41.** c
**42.** d
**43.** c
**44.** c
**45.** d
**46.** c
**47.** c
**48.** c
**49.** d
**50.** b

51. 86.3

52. 172.6

53. 258.9

54. 1.26

55. 2.52

56. 3.78

57. 810.0

58. 81.0

59. 0.8456

60. 1.6912

61. 45.8

62. 54.96

63. 14.0

64. 17.5

65. 24.0

66. 48.0

67. 550.0

68. 110.0

69. 134.16

70. 178.6

In each of the following problems, the boldface digits are correct but the decimal point is missing. Without actually performing a calculation, you are to place the decimal point in the number so that the statement is true.

**51.** 10% of 863 is **8 6 3**

**52.** 20% of 863 is **1 7 2 6**

**53.** 30% of 863 is **2 5 8 9**

**54.** 10% of 12.6 is **0 1 2 6 0**

**55.** 20% of 12.6 is **0 2 5 2 0**

**56.** 30% of 12.6 is **0 3 7 8 0**

**57.** 90% of 900 is **8 1 0 0 0**

**58.** 9% of 900 is **8 1 0 0 0**

**59.** 1% of 84.56 is **0 8 4 5 6**

**60.** 2% of 84.56 is **0 1 6 9 1 2**

**61.** 100% of 45.8 is **4 5 8 0 0**

**62.** 120% of 45.8 is **5 4 9 6 0**

**63.** 200% of 7 is **1 4 0 0**

**64.** 250% of 7 is **1 7 5 0 0**

**65.** $33\frac{1}{3}\%$ of 72 is **0 2 4 0 0**

**66.** $66\frac{2}{3}\%$ of 72 is **0 4 8 0 0**

**67.** $62\frac{1}{2}\%$ of 880 is **0 5 5 0 0**

**68.** $12\frac{1}{2}\%$ of 880 is **0 1 1 0 0**

**69.** 6% of 2236 is **0 1 3 4 1 6 0**

**70.** 5% of 3572 is **0 1 7 8 6 0**

# 6.4 Applications with Percent: Discount, Sales Tax, Commission, Profit, and Tipping

## The Problem-Solving Process

George Pòlya ( 1887-1985 ), a famous professor at Stanford University, studied the process of discovery learning. Among his many accomplishments, he developed the following four-step process as an approach to problem solving:

1. Understand the problem.
2. Devise a plan.
3. Carry out the plan.
4. Look back over the results.

For a complete discussion of these ideas, see *How To Solve It* by Pòlya ( Princeton University Press, 1945, 2nd edition, 1957 ). You might enjoy the following quote by Pòlya: "The traditional mathematics professor of the popular legend is absentminded. He usually appears in public with a lost umbrella in each hand. He prefers to face a blackboard and to turn his back on the class. He writes $a$, he says $b$, he means $c$, but it should be $d$. Some of his sayings are handed down from generation to generation."

There are a variety of types of applications discussed throughout this text and subsequent courses in mathematics, and you will find these four steps helpful as guidelines for understanding and solving all of them. Applying the necessary skills to solve exercises, such as adding fractions or solving equations, is not the same as accumulating the knowledge to solve problems. **Problem solving can involve careful reading, reflection, and some original or independent thought.**

### Objectives

① Know Pòlya's Four-Step Process for Solving Problems.

② Be familiar with the terms **discount, sales tax, commission,** and **percent**.

③ Understand that percent of profit can be based on cost or on selling price.

④ Recognize and be able to solve a variety of types of percent problems.

⑤ Know how to calculate a tip for a service.

### Basic Steps for Solving Word Problems

1. Understand the problem. For example,
   - **a.** Read the problem.
   - **b.** Understand all the words.
   - **c.** If it helps, restate the problem in your own words.
   - **d.** Be sure that there is enough information.
2. Devise a plan using, for example, one or all of the following:
   - **a.** Guess, estimate, or make a list of possibilities.
   - **b.** Draw a picture or diagram.
   - **c.** Use a variable and form an equation.
3. Carry out the plan. For example,
   - **a.** Try all the possibilities you have listed.
   - **b.** Study your picture or diagram for insight into the solution.
   - **c.** Solve any equation that you may have set up.
4. Look back over the results. For example,
   - **a.** Can you see an easier way to solve the problem?
   - **b.** Does your solution actually work? Does it make sense in terms of the wording of the problems? Is it reasonable?
   - **c.** If there is an equation, check your answer in the equation.

In this section, we will be concerned with applications that involve percent and the types of percent problems that were discussed in Section 6.2. The use of a calculator is recommended, although you should keep in mind that a calculator is a tool to enhance and not replace the necessary skills and abilities related to problem solving. Also, your personal experience, knowledge, and general understanding of problem-solving situations will determine the level of difficulty of some problems. The following examples illustrate some basic strategies for solving applications with percent.

## Discount and Sales Tax

To attract customers or to sell goods that have been in stock for some time, retailers and manufacturers offer a **discount**, a reduction in the **original selling price**. The new, reduced price is called the **sale price**, and the discount is the difference between the original price and the sale price. The **rate of discount** ( or **percent of discount** ) is a percent of the original price.

Discounts are sometimes advertised as a percent "off" the selling price. For example,

"The White Sale is now 20% off the original price."
or "Computers are on sale at a discount of 15% off the list price."

**Sales tax** is a tax charged on goods sold by retailers, and it is assessed by states and cities for income to operate various services. The **rate of the sales tax** varies from state to state ( or even city to city in some cases ). In fact, some states do not have a sales tax at all.

**Terms Related to Discount and Sales Tax**

| | |
|---|---|
| **Discount:** | Reduction in original selling price |
| **Sale price:** | Original selling price minus the discount |
| **Rate of discount:** | Percent of original price to be discounted |
| **Sales tax:** | Tax based on actual selling price |
| **Rate of sales tax:** | Percent of actual selling price |

As you study the examples and work through the exercises, be sure to **label each amount of money as to what it represents ( such as original price, discount, sale price, sales tax, profit, and so forth** ). This will help in organizing the results and help in determining what operations to perform.

### Example 1

A new refrigerator that regularly sells for $1200 is on sale at a 20% discount.

**a.** What is the amount of the discount?
**b.** What is the sale price?

**Solution**

**Step 1:** Read the problem carefully. Do you understand all the words?

**Step 2:** Make a plan. The plan here is

**a.** to find the amount of the discount, then

**b.** subtract this amount from the original price to find the sale price.

**Step 3:** Carry out the plan as shown below by multiplying and then subtracting.

**Step 4:** Check to see that the answer makes sense.

( For example, "Does the sale price seem to be 20% less than the original price?" )

**a.** Find the discount: 20% of 1200 is _____.

$$0.20 \cdot 1200 = A$$
$$240 = A$$

| | |
|---|---|
| 1200 | Original price |
| × .20 | Rate of Discount |
| 240.00 | Discount |

The discount in **$240**.

**b.** Find the sale price. The problem does not say specifically how to find the sale price. We know from experience, however, the meaning of the word "discount" and that a discount is subtracted from the original price.

| | |
|---|---|
| $1200.00 | Original price |
| − 240.00 | Discount |
| 960.00 | Sale price |

The sale price is **$960.00**.

## Example 2

If sales tax is figured at $7\frac{1}{4}\%$, what is the final cost of the refrigerator in Example 1?

### Solution

**Step 1:** Read the problem carefully.

**Step 2:** Make a plan. The plan is to find the tax on the answer from Example 1 and then add the tax to the sale price to find the final cost.

**Step 3:** Carry out the plan as shown below by multiplying and then adding.

**Step 4:** Check to see that the answer is reasonable.

( For example, "Is the total of the sale price plus the tax still less than the original price?" )

**a.** Find the amount of the sales tax $\left(7\frac{1}{4}\% = 7.25\% = 0.0725\right)$

$7\frac{1}{4}\%$ of $960 is __________ .

$$0.0725 \cdot 960 = A$$
$$69.60 = A$$

| | |
|---|---|
| 960 | Sale price |
| ×0.0725 | Tax rate |
| 4800 | |
| 1920 | |
| 6720 | |
| 69.6000 | Sales tax |

The sales tax is **$69.60**.

**b.** From experience, we know to add tax to the sale price to find the final cost.

| | |
|---|---|
| $960.00 | Sale price |
| + 69.60 | Sales tax |
| $1029.60 | Final cost |

The final cost of the refrigerator is **$1029.60**.

### Example 3

Large fluffy towels were on sale at a discount of 30%. If the sale price was $8.40, what was the original price?

**Solution**

**Step 1:** Read the problem carefully. Read it two or three times until you understand the nature of the problem and all the terms. This problem involves critical thinking before any calculation.

**Step 2:** Make a plan. First realize that we are **not** trying to find the discount. We already know the sale price. We need to realize that the sale price is 70% of the original price. ( 100% – 30% = 70% )

Therefore, the plan is to set up a percent problem and solve the related equation.

( **Note**: Do **not** take 30% of $8.40. The sale price of $8.40 is **not** 30% of the original price. )

**Step 3:** Carry out the plan as shown below by setting up the equation $0.70\,B = 8.40$ and then solving the equation.

**Step 4:** Check to see that the answer makes sense.

( For example, "Is the original price more than $8.40?" )

70% of ______ is $8.40

$$0.70 \cdot B = 8.40$$
$$\frac{\cancel{0.70} \cdot B}{\cancel{0.70}} = \frac{8.40}{0.70}$$
$$B = \$12.00$$

$$\begin{array}{r} 12. \\ 0.70\overline{)8.40} \\ \underline{70} \\ 140 \\ \underline{140} \end{array}$$

The original price of the towels was **$12.00** each.

## Commissions

A **commission** is a fee paid to an agent or salesperson for a service. Commissions are usually a percent of a negotiated contract or a percent of sales. In some cases, salespeople earn a straight commission on what they sell. In other cases, as illustrated in Example 4, the salesperson earns a base salary plus a commission on sales above a certain level.

### Example 4

A saleswoman earns a salary of \$1100 a month plus a commission of 8% on whatever she sells after she has sold \$8500 in merchandise. What did she earn the month she sold \$22,500 in merchandise?

**Solution**

**Step 1:** Read the problem carefully. Do you understand all the terms?

**Step 2:** Make a plan. Find the commission and add it to the salary. But the commission is made only on the amount over \$8500.

**Step 3:** Carry out the plan as shown below by finding the base for the commission. Multiply this base by 8% and add the result to \$1100.

**Step 4:** Make sure that the answer is reasonable.

( For example, "Is the income about right for someone who makes a salary of \$1100 a month?" )

**a.** First find the base for her commission. Since the commission is based on what she sells over \$8500, we subtract \$8500 from her sales.

| | |
|---|---|
| \$22,500 | Total sales |
| − 8,500 | Amount on which she does not earn a commission |
| \$14,000 | Base for commission |

**b.** Now find the amount of the commission by finding 8% of the base.

$$A = 8\% \text{ of } \$14,000$$
$$A = 0.08 \cdot \$14,000$$
$$A = \$1120 \quad \text{Amount of commission}$$

**c.** Now add the amount of the commission to her salary to find her income for the month.

| | |
|---|---|
| \$1100.00 | Salary |
| + 1120.00 | Commission |
| \$2220.00 | Income for the month |

She earned **\$2220** for the month.

## Percent of Profit

Manufacturers and retailers are concerned with the profit on each item produced or sold. In this sense, the **profit on an item** is the difference between the selling price to the customer and the cost to the company. For example, suppose that a department store can buy a certain light fixture from a manufacturer for \$40 ( the store's cost ) and sell the fixture for \$50 ( the selling price ). The profit for the store is the difference between the selling price and the cost of the fixture. That is, in this case,

$$\text{Profit} = \$50 - \$40 = \$10$$

The **percent of profit is a ratio.** And as the following discussion indicates, this ratio can be based either on cost or on selling price.

**Terms Related to Profit**

**Profit:** The different between selling price and cost.

Profit = Selling price − Cost

**Percent of Profit:** There are two types; both are ratios with **profit in the numerator**.

1. Percent of profit **based on cost** is the ratio of profit to cost:

$$\frac{\text{Profit}}{\text{Cost}} = \%\text{ of profit based on cost} \qquad (\text{Cost is in the denominator.})$$

2. Percent of profit **based on selling price** is the ratio of profit to selling price:

$$\frac{\text{Profit}}{\text{Selling price}} = \%\text{ of profit based on selling price}$$

( Selling price is in the denominator. )

The fact that a percent of profit is a ratio can be seen by looking at the formula for percent problems: $\boldsymbol{R \cdot B = A}$. By dividing both sides of this equation by $B$, we get the following result:

$$R \cdot B = A$$

$$\frac{R \cdot \cancel{B}}{\cancel{B}} = \frac{A}{B}$$

$$\boldsymbol{R = \frac{A}{B}}$$

Thus, the rate ( or percent ) is a ratio. Remember this: $\boldsymbol{R = \frac{A}{B}}$

## Example 5

A retail store markets calculators that cost the store $45 each and are sold to customers for $60 each.

**a.** What is the profit on each calculator?

**b.** What is the percent of profit based on cost?

**c.** What is the percent of profit based on selling price?

### Solution

**a.** First find the profit.

|  |  |
|---|---|
| $60.00 | Selling price |
| − 45.00 | Cost |
| $15.00 | Profit |

The profit is **$15** per calculator.

For **b.** and **c.**, use a ratio to find each percent of profit, and change the fraction to a percent.

**b.** For profit based on cost, remember that cost is in the denominator.

$$\frac{\$15 \text{ profit}}{\$45 \text{ cost}} = \frac{1}{3} = 33\frac{1}{3}\% \qquad \text{Profit based on cost}$$

**c.** For profit based on selling price, remember that selling price is in the denominator.

$$\frac{\$15 \text{ profit}}{\$60 \text{ selling price}} = \frac{1}{4} = 25\% \qquad \text{Profit based on selling price}$$

*Now Work Exercise 1 in the Margin.*

Percent of profit **based on cost** is always higher than percent of profit **based on selling price** because the selling price is larger than the cost. The business community reports whichever percent serves it purpose better. Your responsibility as an investor or consumer is to know which percent is reported and what it means to you.

Remember to write the steps of a problem and the results in a neat, organized form even if you perform the actual calculations with a calculator.

## Tipping

**Tipping** ( or **leaving a tip** ) is the custom of leaving a percent of a bill ( usually at a restaurant ) as a payment to the server for providing good service. The amount of the tip is usually 10% to 20% of the total amount charged ( including tax ), depending on the quality of the service. The "usual" tip is 15%, but in the case of particularly bad service, no tip is required.

Since you probably do not carry your calculator with you when dining out, the calculation of a 15% tip can be an interesting mental exercise in using percents. The following Short Method uses the basic facts that 5% is half of 10% and that 10% of a decimal number can be found by moving the decimal point 1 place to the left.

### Short Method for Calculating a 15% Tip

1. For ease of calculation, round off the amount of the bill to the nearest whole dollar.
2. Find 10% of the rounded-off amount by moving the decimal point 1 place to the left.
3. Divide the answer in step 2 by 2. ( This represents 5% of the rounded-off amount or one-half of 10% ).
4. Add the two amounts found in steps 2 and 3. This sum is the amount of the tip.

**1.** Women's coats are on sale for $250 each. This is a discount of $100 from the original price.

**a.** What was the original price?

$350

**b.** What was the rate of discount?

$28\frac{4}{7}\%$ (or 28.6%)

**c.** If the coats cost the store $200 each, what was the percent of profit based on cost?

25%

**d.** What was the percent of profit based on the actual selling price?

20%

**Note:** There are ways other than that discussed here to calculate the amount of a tip. One commonly used method, especially when the sales tax is about 7% or 8%, is to simply look at the sales tax and double this amount. Another method is to round off to the nearest even dollar. This makes mental calculations easier. Or you might round off to the nearest $10. Or you might leave a tip that makes the total of your expenses a whole-dollar amount. ( This is common for businesspeople who are turning in a expense account for reimbursement. )

Consider two bills of $8.95 and $8.25. Some people might round up both bills to $10 and leave a tip of 15% as $1.00 + $0.50 = $1.50 in each case. (These people would leave a tip of $1.50 for any bill between $5 and $15.)

However, according to the Short Method in this text, we will calculate the tip as follows:

Round up $8.95 to $9.00 ( the nearest whole dollar ), and calculate 15% as

$0.90 + $0.45 = $1.35.

Round down $8.25 to $8.00 ( the nearest whole dollar ), and calculate 15% as

$0.80 + $0.40 = $1.20.

## Example 6

You take your family out for dinner and the bill comes to $36.40, including tax. If you plan to leave a 15% tip, what should be the amount of the tip?

### Solution

For ease of computation, round off $36.40 to $36.00.

Find 10% of $36.00 by moving the decimal point 1 place to the left:

10% of $36.00 = $3.60

Now find 5% of $36.00 by dividing $3.60 by 2:

$3.60 ÷ 2 = $1.80

Adding gives the amount of the tip

$$\begin{array}{r} \$3.60 \\ +\ 1.80 \\ \hline \$5.40 \end{array} \quad \text{Amount of tip}$$

**Note:** We will give $5.40 as the textbook answer. However, practically speaking ( depending on how much change you have in your pocket, how good the service was, etc. ), you might leave either $5.00 or $6.00.

Name ______________________ Section ______________ Date ____________

## Exercises 6.4

The following problems may involve several calculations, and calculators are recommended. Follow Pòlya's four-step process for problem solving.

**Step 1:** Make sure that you read the problem until you understand it. Write down the known information.

**Step 2:** Make a plan for solving the problem.

**Step 3:** Then follow through with your plan

**Step 4:** Go over your answer to see if it makes sense to you. In some cases you may think of a better plan ( or another plan ) after you have solved the problem. This type of thinking may prove useful in solving future problems.

**Note:** If calculations with dollars and cents involve three or more decimal places, round answers up to the next higher cent regardless of the digit in the thousandths place.

**1.** A realtor works on 6% commission. What is his commission on a house he sold for $125,000?

**2.** A car saleswoman earns a commission of 7% on each car she sells. How much did she earn on the sale of a car for $12,500?

**3.** A sales clerk receives a monthly salary of $500 plus a commission of 6% on all sales over $3500. What did the clerk earn the month she sold $8000 in merchandise?

**4.** A computer programmer was told that he would be given a bonus of 5% of any money his programs could save the company. How much would he have to save the company to earn a bonus of $500?

**ANSWERS**

**1.** $7500

**2.** $875

**3.** $770

**4.** $10,000

5. 1.5%

6. $8.12

7. 136

8. $9000

9. 40

10. a. 93.75%

b. 6.25%

**5.** The property taxes on a house were $1050. What was the tax rate if the house was valued at $70,000?

**6.** If sales tax is figured at 7.25%, how much tax will be added to the total purchase price of three textbooks, priced at $25.00, $35.00, and $52.00?

**7.** In one season a basketball player missed 15% of her free throws. How many free throws did she make if she attempted 160 free throws?

**8.** An auto dealer paid $8730 for a large order of special parts. This was not the original price. The amount paid reflects a 3% discount off the original price because the dealer paid cash. What was the original price of the parts?

**9.** A student missed 6 problems on a mathematics test and received a grade of 85%. If all the problems were of equal value, how many problems were on the test?

**10.** A kicker on a professional football team made 45 of 48 field goal attempts.

**a.** What percent of his attempts did he make?

**b.** What percent did he miss? Would you keep this player on your team or trade for a new kicker?
Answers will vary for second part of b.

Name ______________________ Section ____________ Date __________

**ANSWERS**

11. a. $9.38

b. $3.38

12. $97,600

13. $37,500

14. a. $700

b. $140

c. $33\frac{1}{3}$

d. 25%

15. a. $180

b. 40%

c. $28\frac{4}{7}$% (or 28.6%)

**11.** At a department store white sale, sheets were originally marked $12.50 and pillowcases were originally marked $4.50. What is the sale price of

**a.** sheets and

**b.** pillowcases if each item is discounted 25% from the original marked price?

**12.** You want to purchase a new home for $122,000. The bank will loan you 80% of the purchase price. How much will the bank loan you? This amount is called your mortgage and you will pay it off over several years with interest. For example, a 30–year loan will probably cost you a total of more than 3 times the original loan amount.

**13.** Suppose you sell your home for $100,000 and you owe the balance of the mortgage of $55,000 to the bank. You pay a real estate agent a fee of 6% of the selling price and other fees and taxes that total $1500. How much cash do you have after the sale if the home has 3 bedrooms? ( You may have to pay income taxes later. )

**14.** The Golf Pro Shop had a set of 10 golf clubs that were marked on sale for $560. This was a discount of 20% off the original selling price.

**a.** What was the original selling price?

**b.** If the clubs cost the Golf Pro Shop $420, what was its profit?

**c.** What was the shop's percent of profit based on cost?

**d.** What was the percent of profit based on the sale price?

**15.** The cost of a television set to a store owner was $450, and she sold the set for $630.

**a.** What was her profit?

**b.** What was her percent of profit based on cost?

**c.** What was her percent of profit based on selling price? The set had a 20-inch screen.

16. a. $3125

b. $3375

17. 30%; 32%; 12%

18. a. $49.40

b. 80%

19. a. 86%

b. 300

c. 42

20. $2466.02

21. a. $2550

b. $2703

**16.** A car dealer bought a used car for $2500. He marked up the price so that he would make a profit of 25% base on his cost.

**a.** What was the selling price?

**b.** If the customer paid 8% of the selling price in taxes and fees, what was the customer's total cost for the car? The car was 5 years old.

**17.** In one year Mr. Hill, who is 35 years old, earned $32,000. He spent $9600 on rent, $10,240 on food, and $3840 on taxes. What percent of his income did he spend on each of those items?

**18.** To get more subscribers, a book club offered three books with an original total selling price of $61.75 for a special price of $12.35.

**a.** What was the amount of the discount?

**b.** Based on the original selling price of these books, what was the rate of discount on these three books? Do you think this would be a reasonable thing for the book club to do? Why?
Answers will vary for the rest of part b.

**19.** The author of a book was told that she would have to cut the number of pages by 14% for the book to sell at a popular price and still make a profit.

**a.** If these cuts were made, what percent of the original number of pages was in the final version?

**b.** If the finished book contained 258 pages, how many pages were in the original form?

**c.** How many pages were cut?

**20.** A department store received a shipment of dresses together with the bill for $2856.50. Some of the dresses were not as ordered, however, and were returned at once. The value of the returned dresses was $340.15. The terms of the billing provided the store with a 2% discount if it paid cash within 2 weeks. What did the store finally pay for the dresses it kept if it paid cash within 2 weeks?

**21. a.** If the sales tax on the purchase of a computer was $153 and tax was figured at 6% of the selling price, what was the selling price?

**b.** What did the customer pay for the computer? The computer had two floppy disk drives and a 10 gigabyte hard disk drive with 128 megabytes of RAM memory.

Name ______________ Section ______________ Date ______________

**ANSWERS**

22. a. $7134.29
b. $6559.23

23. a. $875
b. $700
c. $750.75

24. a. more than $27
b. $30

25. a. $156
b. $5460

26. a. $1.80
b. $13.55

**22.** The total discount on a new car was $1499.40. This included a rebate from the manufacturer of $1000.

**a.** What was the original price of the car if the dealer discount was 7% of the original price?

**b.** What would a customer pay for the car if taxes were 5% of the final selling price and license fees were $642.60?

**23.** The discount on a new coat was $175. This was a discount of 20%.

**a.** What was the original selling price of the coat?

**b.** What was the sale price?

**c.** What would a customer pay for the coat if a 7.25% sales tax was added to the sale price?

**24.** Data cartridges for computer backup were on sale for $27.00 ( a package of two cartridges ). This price was a discount of 10% off the original price.

**a.** Was the original price more than $27 or less than $27?

**b.** What was the original price?

**25.** In the roofing business, shingles are sold by the "square," which is enough material to cover a 10 ft by 10 ft square ( or 100 square feet ). A roofing supplier has a closeout on shingles at a 40% discount.

**a.** If the original price was $260 per square, what is the sale price per square?

**b.** How much would a roofer pay for 35 squares?

In Exercises 26-30, use the rule of thumb stated in the text to calculate a 15% tip.

**26.** Walt decided to treat himself to lunch, and the bill was $11.75, including tax.

**a.** What amount did he leave as 15% tip?

**b.** What total amount, including tip, did he pay for lunch?

27. a. $10.35

b. $79.25

28. a. $2.85

b. $21.75

29. $26.45

30. $103.50

31. [ Respond below exercise. ]

**27.** A math teacher took two of her colleagues to dinner, and the bill was $65.00 plus a 6% sales tax.

**a.** What amount did she leave as a 15% tip?

**b.** What was the total amount, including tip, she paid for the dinner?

**28.** Ann paid for lunch at the local pizza parlor for three of her study partners because they were celebrating having passed a statistics exam. The bill was $18.00 plus a 5% tax.

**a.** What amount did she leave as a 15% tip?

**b.** What was the total amount, including tip, she paid for lunch?

**29.** On Monday night, Barbara and Alan decided to stay home, watch a football game, and have Chinese food delivered. If the bill was $23.00 for the food and they tipped the driver 15%. What total amount did they pay?

**30.** The coach had the basketball team over to his house after the last game of the season and ordered 6 large pizzas delivered. If the total bill was $90.00 for the pizzas and he gave the driver a 15% tip, what total amount did he pay?

## WRITING AND THINKING ABOUT MATHEMATICS

**31.** One shoe salesman worked on a straight 9% commission. His friend worked on a salary of $400 plus a 5% commission.

**a.** How much did each salesman make during the month in which each sold $7500 worth of shoes?

$675 and $775, respectively.

**b.** What percent more did the salesman who made the most make? Explain why there is more than one answer to part **b.**.

The $100 difference is 12.9% based on the second salesman's pay, or $14.8% based on the first salesman's pay.

Name ______________________ Section ____________ Date __________

ANSWERS

32. [ Respond below exercise. ]

33. [ Respond below exercise. ]

**32.** A man weighed 200 pounds. He lost 20 pounds in 3 months. Then he gained back 20 pounds 2 months later.

**a.** What percent of his weight did he lose in the first 3 months?

He loses 10% of his original weight.

**b.** What percent of his weight did he gain back? The loss and the gain are the same, but the two percents are different. Explain why

He gains back $11\frac{1}{9}\%$ of his weight, because the gain of 20 pounds is based on 180 pounds, not the original weight of 200 pounds.

## Collaborative Learning Exercise

**33.** With the class separated into teams of two to four students, each team is to analyze the following problem and decide how to answer the related questions. Then each team leader is to present the team's answers and related ideas to the class for general discussion.

Jerry works in a bookstore and gets a salary of $500 per month plus a commission of 3% on whatever he sells over $2000. Wilma works in the same store, but she has decided to work on a straight 8% commission.

**a.** At what amount of sales will Jerry and Wilma make the same amount of money?

$8800

**b.** Up to that point, who would be making more?

Jerry

**c.** After that point, who would be making more? Explain briefly. ( If you were offered a job at this bookstore, which method of payment would you choose? )

Wilma
Explanations will vary. For sales over $8800, 8% of the sales is more than $500 plus 3% of the sales over $2000.

# 6.5 Applications: Buying a Car, Buying a Home

## Buying a Car

At some time in your life, you are likely to purchase a car. This purchase will be one of the most expensive investments you will make. ( Buying a home is the greatest expense for most people. ) As with finances in general, paying cash for a car is cheaper than financing the car with a bank or savings and loan. If you are going to finance the purchase, at least be aware of the expenses involved and study all the papers so that you are an "intelligent" buyer and are aware of the total amount that you are going to be paying for the car.

**Objectives**

① Become aware of and learn how to calculate the expenses involved in buying a car.

② Become aware of and learn how to calculate the expenses involved in buying a house.

**Expenses in Buying a Car**

**Purchase price:** The selling price agreed on by the seller and the buyer

**Sales tax:** A fixed percent that varies from state to state

**License fee:** Fixed by the state, often based on the type of car and its value

### Example 1

You are going to buy a used car for \$8500. The bank will loan you 70% of the purchase price. But you must pay a 6% sales tax and a \$150 license fee. How much cash do you need to buy the car?

**Solution**

**a.** Calculate the down payment ( 30% of \$8500, since the bank will loan 70% ).

$$\begin{array}{r} \$8500 \\ \times\ \ 0.30 \\ \hline \$2550.00 \end{array} \quad \text{Down payment}$$

**b.** Calculate the sales tax ( 6% of \$8500 ).

$$\begin{array}{r} \$8500 \\ \times\ \ 0.06 \\ \hline \$510.00 \end{array} \quad \text{Sales tax}$$

**c.** Add all the cash expenses.

$$\begin{array}{rl} \$2550.00 & \text{Down payment} \\ 510.00 & \text{Sales tax} \\ +\ \ 150.00 & \text{License tax} \\ \hline \$3210.00 & \text{Total cash needed} \end{array}$$

You need \$3210 cash to buy the car.

Encourage any homeowners in your class to discuss some of their experiences in buying or owning a home. All of the details and costs involved can be an "eye opener" for prospective home buyers. Such a discussion can also provide a chance to emphasize for everyone the importance of understanding percent, particularly in the context where a difference of 1% or even $\frac{1}{2}$% can translate to hundreds or even thousands of dollars.

## Buying a Home

For first-time buyers, the process of buying a home can be an anxiety-laden experience. This purchase will probably be the most expensive investment of their lives, and the legal aspects, seemingly endless fees, and terminology used by realtors and lenders can be overwhelming. In this section, we will discuss some of the expenses that can be expected when a home is purchased and some of the related terminology. If or when the time comes for you to consider buying a home, you will be an informed and intelligent buyer.

### Expenses in Buying a Home

**Purchase price:** The selling price ( what you have agreed to pay )

**Down payment:** Cash you pay to the seller ( usually 20% to 30% of the purchase price )

**Mortgage loan ( first trust deed ):** Loan to you by a bank or a savings and loan ( difference between purchase price and down payment )

**Mortgage fee ( or points ):** Loan fee charged by the lender ( usually 1% to 3% of the mortgage loan )

**Fire insurance:** Insurance against the loss of your home by fire ( required by almost all lenders )

**Recording fees:** Local and state fees for recording you as the legal owner

**Property taxes:** Taxes that must be prepaid before the lender will give you the loan ( usually 6 months in advance )

**Legal fees:** Fees charged by a lawyer or escrow company and/or a title search company for completing all forms in a legal manner

## Example 2

You buy a home for $150,000. Your down payment is 20% of the selling price, and the mortgage fee is 2% of the new mortgage. You also have to pay $500 for fire insurance, $350 for taxes, $50 for recording fees, and $310 for legal fees. What is the amount of your mortgage? How much cash must you provide to complete the purchase?

### Solution

**a.** Find the down payment

| | | |
|---|---|---|
| | $150,000 | Selling price |
| × | 0.2 | |
| | $30,000.00 | Down payment |

**b.** Find the mortgage and mortgage fee ( 2% of mortgage ).

| | | |
|---|---|---|
| | $150,000 | Selling price |
| − | 30,000 | Down payment |
| | $120,000 | Mortgage |

| | | |
|---|---|---|
| | $120,000 | Mortgage |
| × | 0.02 | |
| | $2400.00 | Mortgage fee |

**c.** Add all cash expenses.

| | | |
|---|---|---|
| | $30,000 | Down payment |
| | 2,400 | Mortgage fee ( loan fee ) |
| | 500 | Fire insurance |
| | 350 | Taxes |
| | 50 | Recording fees |
| + | 310 | Legal fees |
| | $33,610 | Cash to complete purchase |

The mortgage will be $120,000, and you will need $33,610 in cash.

Name ______________________ Section ____________ Date __________

# Exercises 6.5

1. How much cash would you need to buy a used car for $6000 if the sales tax is calculated at 5%, the license fee is $90, and your credit union will let you borrow 80% of your expenses?

2. To buy a used car for $7500, you must pay 7% in sales taxes, a license fee of $110, and a down payment of 15% of the purchase price to the bank. How much cash will you need to buy the car?

3. Suppose that you are looking at a new red convertible for $35,000. You have talked to the salesman and know that the license fee will be $350 and sales tax will be 6% of the selling price. You then shop around at several savings and loan associations and find that the lender with the best interest rate will loan you 75% of your expenses. How much cash will you need to buy the car?

4. Your brother has been looking at a used car for $6500, and you have told him that you will buy his old car for $1500. If sales tax is figured at 6% of the selling price and the license fee is $75, how much cash ( from his own pocket ) will your brother need to purchase the car? ( He doesn't want to get a loan. )

5. A new "off road vehicle" with four-wheel drive is priced at $28,000, and your old car is worth $2500 in trade-in value. Sales tax is at 6% of the selling price, license fees are $325, and the bank will loan you 70% of your expenses before the trade-in value is deducted from what you will owe. The bank is not interested in your old car or what role it plays in your plans to purchase the new car. How much cash will you need to complete the purchase?

**ANSWERS**

1. $1278

2. $1760

3. $9362.50

4. $5465

5. $6501.50

6. $9422.50

7. a. $24,000

b. $96,000

c. $26,280

8. $24,510

9. Yes. You will need $27,300, and you have $30,000.

**6.** A used Mercedes station wagon is on sale for $8500, and the deal is just too good for your uncle to pass up. He has been wanting this particular type of station wagon for the last 2 years. You have told him that you will loan him $5000 for 1 year at 5% interest, and he will have to pay sales tax at 6.5% and a license fee of $120. How much will the station wagon cost him?

**7.** A home is sold for $120,000. The buyer has to make a down payment of 20% of the selling price and pay a loan fee of 1% of the mortgage, $200 for fire insurance, $140 for recording fees, $450 for taxes, and $530 for legal fees.

**a.** What is the amount of the down payment?

**b.** What is the amount of the mortgage?

**c.** How much cash does the buyer need to complete the purchase?

**8.** You are planning to buy a home and make an offer, accepted by the seller, of $95,000. The bank will loan you 75% of the selling price, and there is no loan fee. If the costs are $250 for fire insurance, $325 for taxes, and $185 for legal fees, how much cash will you need to make this purchase?

**9.** Your "dream" home is on special sale at a close–out price of $125,000, including some furniture and upgrades in the model home. You and your spouse have accumulated $30,000, and you go to the savings and loan to see if this is enough to make the purchase. The lender tells you that they will charge 1.5 points ( 1.5% loan fee ) and you must put down 20% of the purchase price, pay $300 for fire insurance, $225 for taxes, and $275 for legal fees. Can you afford to make the purchase?

Name ______________________ Section ______________ Date ____________

ANSWERS

**10.** Karen is planning to buy a condominium with one bedroom for $65,000, She has saved $7000 for a down payment, and the lender has told her that since she is a "first–time" buyer, she need only put down 5% of the purchase price. There will be no points, and fire insurance will be $250, legal fees will be $150, and taxes will be $450. How much cash will Karen have left in her savings if she decides to buy the condominium?

**10.** $2900

**11.** A large 5-bedroom home with an ocean view sold for $1,260,000. The buyers made a down payment of 25% of the selling price and paid a loan fee of 1% of the amount of the mortgage. They also paid $2200 for fire insurance, $1050 for recording fees, and $1800 for taxes. ( The seller paid all legal fees. )

**a.** What was the amount of the down payment?

**b.** What was the amount of the mortgage?

**c.** How much was the loan fee?

**d.** How much cash did the buyers need to complete the purchase?

**11. a.** $315,000

**b.** $945,000

**c.** $9450

**d.** $329,500

**12.** George saw a home for sale for $350,000, and since the market was in a favorable state for buyers, he made an offer of $275,000. The seller made a counteroffer of $310,000, and George accepted. He then found a lender that would charge him 1 point for a mortgage fee if he put down 20% of the selling price. He also would need to pay $345 for legal fees, $650 for taxes, $320 for recording fees, and $500 for fire insurance.

**a.** How much would he need for a down payment?

**b.** What would be the amount of the first trust deed?

**c.** What loan fee would he be paying?

**d.** What amount of cash would he need to buy the home?

**12. a.** $62,000

**b.** $248,000

**c.** $2480

**d.** $66,295

# 6.6 Simple Interest and Compound Interest

## Understanding Simple Interest ( $I = Prt$ )

The following terms and ideas are related to interest:

**Interest:** money paid for the use of money
**Principal:** money that is invested or borrowed
**Rate:** percent of interest ( stated as an annual interest rate )

Regardless of whether you are a borrower, lender, or investor, the calculations for finding interest are the same. Although interest rates can vary from year to year ( or even daily ), the concept of interest is the same throughout the world.

A **note** is a loan for a period of 1 year or less, and the interest earned ( or paid ) is called **simple interest.** A note involves only one payment at the end of the term of the note and includes both principal and interest. If the borrower wants the money for more than 1 year, terms can be arranged in which the borrower pays the interest at the end of 1 year and then renegotiates a new note ( or extension of the old note ) for some new period of time. Eventually, the borrower must pay back the original loan amount plus any interest earned.

The following formula is used to calculate simple interest:

### Formula for Calculating Simple Interest

$$I = Prt$$

Where

$I$ = **Interest** ( earned or paid )

$P$ = **Principal** ( the amount invested or borrowed )

$r$ = **Rate** of interest ( stated as an annual or yearly rate )

$t$ = **Time** ( in years or fraction of a year )

**Note:** For the purpose of calculations, we will use 360 days in one year ( 30 days a month ). This is a common practice in business and banking. Although with the advent of computers, many leading institutions now base their calculations on 365 days per year and pay or charge interest on a daily basis.

Although the rate of interest is generally given in the form of a percent, the calculations in the formula are made by changing the percent into decimal form or fraction form.

**Objectives**

① Understand the concept of simple interest.

② Know how to find simple interest by using the formula $I = Prt$

③ Understand the concept of compound interest.

④ Know how to find the compound interest by using the formula

$$A = P\left(1 + \frac{r}{n}\right)^{nt}$$

⑤ Understand the concepts of inflation and depreciation and be able to use the related formulas.

Many students use their credit cards with a view of whether they can "afford" the payments and are not aware of the cost of this convenience. To help them develop a better understanding of interest and interest rates, you might contrast the amount of interest they would pay on a credit card balance of \$2000 at 20% with the interest they would earn in a savings account of \$2000 at 3% and point out that investors that make 20% are considered very fortunate.

## Example 1

You want to borrow $2000 at 12% interest for only 90 days. How much interest would you pay?

### Solution

$$P = \$2000$$

$$r = 12\% = 0.12 \text{ (in decimal form) or } \frac{12}{100} \text{ ( in fraction form )}$$

$$t = 90 \text{ days} = \frac{90}{360} \text{ year} = \frac{1}{4} \text{ year}$$

Now, using the formula,

$$\boldsymbol{I = Prt}$$

$$I = 2000 \cdot \overset{0.03}{\cancel{0.12}} \cdot \frac{1}{\cancel{4}} = 2000 \cdot 0.03 = \$60.00$$

You would pay $60 in interest if you borrowed $2000 for 90 days at 12%.

## Example 2

Carmen loaned $500 to a friend for 6 months at an interest rate of 8%. How much will her friend pay her at the end of the 6 months?

### Solution

$$P = \$500$$

$$r = 8\%$$

$$t = 6 \text{ months} = \frac{6}{12} \text{ year} = \frac{1}{2} \text{ year}$$

The interest is found by using the formula:

$$\boldsymbol{I = Prt}$$

$$I = 500 \cdot \overset{0.04}{\cancel{0.08}} \cdot \frac{1}{\cancel{2}} = 500 \cdot 0.04 = \$20.00$$

The interest is $20 and the total amount to be paid at the end of 6 months is

$$\text{Principal } + \text{Interest} = \$500 + \$20 = \$520$$

## Example 3

What principal would you need to invest to earn \$450 in interest in 6 months if the rate of interest was 9%?

### Solution

In this problem we know the interest, rate, and time, and want to find the principal $P$.

$$I = \$450$$
$$r = 9\% = 0.09$$
$$t = 6 \text{ months} = \frac{1}{2} \text{ year}$$

Substitute into the formula $\boldsymbol{I = Prt}$ and solve for $P$.

$$450 = P \cdot 0.09 \cdot \frac{1}{2}$$

$$2 \cdot 450 = P \cdot 0.09 \cdot \frac{1}{\cancel{2}} \cdot \cancel{2}$$ Multiply both sides by 2.

$$900 = P \cdot 0.09$$

$$\frac{900}{0.09} = \frac{P \cdot \cancel{0.09}}{\cancel{0.09}}$$ Divide both sides by 0.09.

$$10,000 = P$$

You would need a principal amount of \$10,000 invested at 9% to make \$450 in 6 months.

## *Completion Example 4*

Find the rate of interest that would be paid if \$50 interest was to be earned on \$2000 in 3 months.

### Solution

The unknown quantity is ______________ .

$I = \$$ ___________ .

$t = 3$ months = ______ year.

Substituting into the formula $\boldsymbol{I = Prt}$ gives

$$50 = \_\_\_\_\_\_ \cdot r \cdot \frac{1}{4}$$

$$50 = \_\_\_\_\_\_ \cdot r$$ Simplify the right-hand side.

$$\frac{50}{\_\_\_\_} = \frac{\_\_\_\_ \cdot r}{\_\_\_\_}$$ Divide both sides by 500, the coefficient of $r$.

$$\_\_\_\_\_ = r$$

The interest rate would be _____%.

*Now Work Exercises 1-3 in the Margin.*

1. Bette borrowed \$3000 at 16% for 1 year to buy a used car. How much interest did she pay?
   \$480

2. Stacey loaned her aunt \$2000 at 10% interest for 9 months. How much interest did she earn?
   \$150

3. What interest rate would you be paying if you borrowed \$1000 for 6 months and paid \$75 in interest?
   15%

## Using the Formula $I = Prt$ Repeatedly To Calculate Compound Interest

Interest paid on interest earned is called **compound interest**. To calculate compound interest, we can calculate the simple interest for each period of time that interest is compounded, using a **new principal for each calculation**. This new principal is **the previous principal plus the earned interest**. The calculations can be performed in a step–by step manner, as indicated in the following outline:

### To Calculate Compound Interest

1. Using the formula $I = Prt$ , calculate the simple interest where $t = \frac{1}{n}$ and $n$ is the number of periods per year for compounding.

   For example,

   For compounding annually, $n = 1$ and $t = \frac{1}{1} = 1$

   For compounding semiannually, $n = 2$ and $t = \frac{1}{2}$

   For compounding quarterly, $n = 4$ and $t = \frac{1}{4}$

   For compounding monthly, $n = 12$ and $t = \frac{1}{12}$

   For compounding daily, $n = 360$ and $t = \frac{1}{360}$

2. Add this interest to the principal to create a new value for the principal.
3. Repeat steps 1 and 2 however many times that the interest is to be compounded.

**Note Again**: For the purpose of calculations, we will use 360 days in one year ( 30 days a month ). Of course, you may choose to use 365 days per year. Your answers will differ only slightly.

In Example 5 compound interest is calculated in the step-by-step manner just outlined. This process is, to say the least, somewhat laborious and time-consuming. However, it does serve to develop a basic understanding of the concept of compound interest. After this example, formulas are used for calculating compound interest, inflation, depreciation, and current value.

**Note About Rounding Off Pennies**: If calculations with dollars and cents involve three or more decimal places, round answers up to the next higher cent regardless of the digit in the thousandths place. ( This is the typical round off technique used in stores for calculating sales tax. )

**Example 5**

If an account is compounded monthly ( $n = 12$ ) at 12%, how much interest will a principal of \$5500 earn in 3 months? What will be the balance in the account?

**Solution**

Since the compounding is monthly, use $t = \frac{1}{n} = \frac{1}{12}$ in the formula $I = Prt$.

And, since the period is three months, calculate the interest three times. Calculate each time with a new principal.

**a.** First period: the principal is $P = \$5500$.

$$I = 5500 \cdot \overset{0.01}{\cancel{0.12}} \cdot \frac{1}{\cancel{12}} = 5500 \cdot 0.01 = \$55 \qquad \text{interest for first period}$$

**b.** Second period: the new principal is $P = \$5500 + \$55 = \$5555$. interest for second period

$$I = 5555 \cdot \overset{0.01}{\cancel{0.12}} \cdot \frac{1}{\cancel{12}} = 5555 \cdot 0.01 = \$55.55 \qquad \text{interest for second period}$$

**c.** Third period: the new principal is $P = \$5555 + 55.55 = \$5610.55$.

$$I = 5610.55 \cdot \overset{0.01}{\cancel{0.12}} \cdot \frac{1}{\cancel{12}} = 5610.55 \cdot 0.01 = \$56.11 \qquad \text{interest for third period}$$

\$ 55.00
55.55
+ 56.11
\$166.66 Total interest earned in 3 months

The balance in the account will be \$5500.00 + \$166.66 = \$5666.66.

4. If an account is compounded quarterly at 8%, how much interest will be earned in 9 months on an investment of \$4000? Show each quarterly calculation.

First quarter:

$$(\$4000)(0.08)\left(\frac{1}{4}\right) = \$80$$

Second quarter:

$$(\$4080)(0.08)\left(\frac{1}{4}\right) = \$81.60$$

Third quarter:

$$(\$4161.60)(0.08)\left(\frac{1}{4}\right) = \$83.24$$

Total interest = \$244.84

*Now Work Exercise 4 in the Margin.*

## Calculating Compound Interest with The Formula $A = P\left(1 + \frac{r}{n}\right)^{nt}$

The steps outlined in Example 5 illustrate how to adjust the principal for each new time period of the compounding process and how the interest increases for each new time period. The interest is greater each time period because the new adjusted principal ( the old principal plus the interest over the previous time period ) is greater. In this process, we have calculated the actual interest for each time period.

The following compound interest formula does not calculate the actual interest for each time period. In fact, this formula does not calculate the interest directly. This formula does calculate the total accumulated **amount** ( also called the **future value** of the principal ). Then, to find the total interest earned, subtract the initial principal from the accumulated amount.

Examples 6 and 8 illustrate, step-by-step, how to use a calculator in working with the formula. Example 7 illustrates how to find the total interest earned.

You might want to discuss the reasonableness of the formula with the following algebraic development using $n = 1$ for three years.

First year $A = P(1 + r)$.

For the second year, we replace P in the original formula with the new principal $P(1 + r)$ which leads to the expression

$$A = [P(1 + r)](1 + r) = P(1 + r)^2.$$

For the third year, we replace $P$ in the original formula with the new principal $P(1 + r)^2$ and arrive at

$$A = [P(1 + r)^2](1 + r) = P(1 + r)^3.$$

**Compound Interest Formula**

When interest is compounded, the total **amount** $A$ accumulated ( including principal and interest ) is given by the formula

$$A = P\left(1+\frac{r}{n}\right)^{nt}$$

where

$P$ = the principal

$r$ = the annual interest rate

$t$ = the length of time in years

$n$ = the number of compounding periods in 1 year

The evaluation of the expressions in the parentheses in Example 6 illustrates the importance of the rules for order of operations. You can point out that a correct answer is not possible, even with the use of a calculator, unless the rules are followed.

## Example 6

Max invested \$4500 at 9% interest to be compounded monthly. What will be the amount in his account in 5 years?

**Solution** $P =$ \$4500, $r = 9\% = 0.09$

$n =$ 12 times per year, $t = 5$ years

Substituting into the formula gives

$$\begin{aligned} A &= 4500\left(1+\frac{0.09}{12}\right)^{12(5)} \\ &= 4500(1+0.0075)^{60} \\ &= 4500(1.0075)^{60} \\ &= \$7045.56 \end{aligned}$$

**Steps in evaluating with a scientific calculator:**

( **Note**: Follow the rules for order of operations by working within the parentheses first. )

**Step 1:** Enter 0.09.

**Step 2:** Press ÷ .

**Step 3:** Enter 12.

**Step 4:** Press = .

**Step 5:** Press +.

**Step 6:** Enter 1.

**Step 7:** Press = . ( The display should read 1.0075.) (This is the base. )

**Step 8:** Press the key $x^y$ . ( On some calculators a Shift or 2nd key may need to be pressed first. )

**Step 9:** Enter 60. ( This is the exponent 12·5. )

**Step 10:** Press = .

The number 1.565681027 is now on the display. This is the future value of \$1. Since \$4500 was invested, multiply by 4500 to find the total amount.

**Step 11:** Press ×.

**Step 12:** Enter 4500.

**Step 13:** Press =.

The display should read 7045.564621.
The sequence of steps can be diagrammed as follows:

( Enter ) 0.09 ( press ) ÷ ( enter ) 12 ( press ) = ( press ) + ( enter ) 1 ( press ) =
( display reads ) 1.0075 ( press ) $x^y$ ( enter ) 60 ( press ) =

↑ base ↑ exponent

( press ) × ( enter ) 4500 ( press ) = 7045.564621

↑ principal ↑ amount

The amount in Max's account in 5 years will be **$7045.57**.

## Steps in Evaluating with a TI-83 Plus Calculator:

A graphing calculator automatically follows the rules for order of operations. Therefore, you can enter all the numbers and parentheses just as you see them placed in the formula. You can enter the exponent as 60 or enter it as the product of 12 times 5. The product must be in parentheses as ( 12 ∗ 5 ).

**Step 1:** Enter the numbers and parentheses so that the display appears as follows:

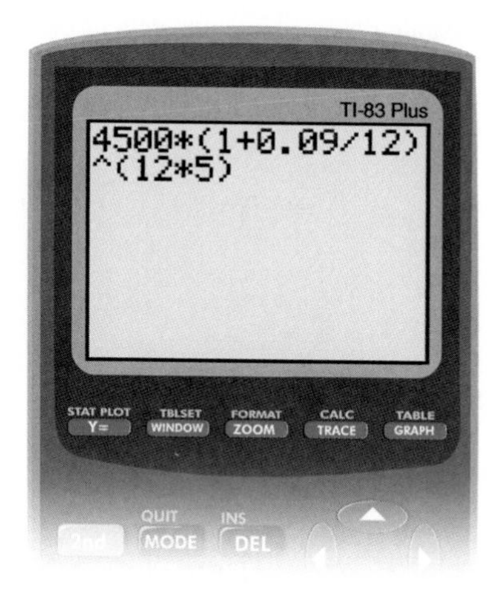


**Step 2:** Press ENTER
The display should now read:

With either type of calculator, we find that the amount in Max's account in 5 years will be $7045.56.

(Remember to round up for pennies regardless of the digit in the thousandths position.)

To find the total interest earned on an investment that has earned interest by compounding, subtract the initial principal from the accumulated amount:

$I = A - P$

### Example 7

How much interest did Max earn in the investment described in Example 6 ?

**Solution**

$$\begin{aligned} I &= A - P \\ &= 7045.57 - 4500.00 \\ &= 2545.57 \end{aligned}$$

Max earned \$2545.57 in interest.

### *Completion Example 8*

**a.** Use the compound interest formula to find the value of \$6000 invested for 4 years if it is compounded daily at 12%.

**b.** Find the amount of interest earned.

**Solution**

**a.** With a **scientific calculator**, follow the steps outlined in the first part of Example 6 with

$P = \$6000,$ $\quad r = 12\% = 0.12$ $\quad n = 360\,,$ $\quad t = 4$

( **Note**: You may also use a TI-83 graphing calculator. The answer will be the same. But, the steps will be different. )

Substituting into the formula gives

$$\begin{aligned} A &= 6000\left(1 + \_\_\_\_\right)^{360(\_\_\_\_)} \\ &= 6000\left(1 + \_\_\_\_\right)^{\_\_\_\_} \\ &= 6000\left(1. + \_\_\_\_\right)^{\_\_\_\_} \\ &= 6000(\_\_\_\_\_\_) \\ &= \_\_\_\_\_\_\_\_ \end{aligned}$$

The value ( or amount ) will be \$_________.

**b.** The interest earned will be
$I = \$\_\_\_\_\_\_\_\_\_ - \$6000.00 = \$\_\_\_\_\_\_\_\_\_$

*Now Work Exercise 5 in the margin.*

5. Use a calculator to calculate the value accumulated if \$10,000 is compounded daily at 9% for 5 years. How much interest will be earned?

   \$15,682.24; \$5682.24

## Inflation and Depreciation

**Inflation** ( also called as **cost-of-living index** ) is a measure of your relative purchasing power. For example, if inflation is at 6%, then you will need to increase your income by 6% each year to be able to afford to buy the same items ( or live in the same life-style ) that you did the year before. Many worker bargaining committees tie their salary requests to the government's cost-of-living index each year.

Inflation can be treated in the same manner as interest compounded annually ( once a year ). In Example 9 we illustrate this idea by considering the annual growth in your salary based on an inflation rate that is assumed to remain steady at 8%.

The formula for the accumulated amount $A$ due to **inflation** is the same as the formula for compound interest with $n = 1$:

$$A = P(1 + r)^t$$

### Example 9

Suppose that your income is \$1800 per month ( or \$21,600 per year ) and that you will receive a cost-of-living raise each year. If inflation is steady at 8%, in how many years will you be making \$3000 per month ( \$36,000 per year )?

#### Solution

We answer this question by forming a table of value for the amount.

$$A = 21{,}600(1 + 0.08)^t = 21{,}600(1.08)^t$$

You can check these values with your calculator. Notice that for the first year $t = 0$, since you do not receive a raise until the second year.

| Year | $t$ | Amount = $1{,}600(1.08)^t$ |
|---|---|---|
| 1 | 0 | \$21,600.00 |
| 2 | 1 | 23,328.00 |
| 3 | 2 | 25,194.24 |
| 4 | 3 | 27,209.78 |
| 5 | 4 | 29,386.57 |
| 6 | 5 | 31,737.49 |
| 7 | 6 | 34,276.49 |
| 8 | 7 | 37,018.61 |

You will make over \$36,000 during your eighth year on the job. Your relative purchasing power will be the same as it was 8 years before. Obviously, if you want to improve your standard of living, you should increase your income at a rate higher that the cost-of-living.

You might discuss the fact that the powers of positive fractions less than 1 become smaller as the size of the exponents increase. For example, $(0.6)^2 = 0.36$ and $(0.6)^3 = 0.216$. Thus, in the depreciation formula, the value $V$ becomes smaller as time increases and the formula "makes sense."

**Depreciation** is the decrease in value of an item. Depreciation is used to determine the value of property and machinery ( also called **capital goods** ) for income tax purposes, and if we know the original value and the rate of depreciation of an item, we can calculate its current market value. For example, if a car was purchased for \$15,000 and it depreciates 20% each year, then its value each year is 80% of its value the previous year.

Value after 1 year is \$15,000(0.80) = \$12,000

Value after 2 years is $\underbrace{[\$15{,}000(0.80)]}_{\text{first year}}(0.80) = \$12{,}000\,(0.80)$

$= \$9600$

or

$\$15{,}000(\ 0.80\ )^2 = 15{,}000(\ 0.64\ ) = \$9600$

The **current value** $V$ of an item due to **depreciation** can be determined from the formula

$$V = P\,(\,1 - r\,)^t$$

where

$P$ = the original value
$r$ = the annual rate of depreciation (in decimal form)
$t$ = the time in years

## Example 10

Suppose that a certain make of automobile depreciates 15% each year. Find the current market value of one of these automobiles if it is 5 years old and its original cost was \$20,000.

### Solution

$P = \$20{,}000,\quad r = 15\% = 0.15,\quad 1 - r = 0.85,\quad t = 5$ years

Substituting into the formula,

$$\begin{aligned} V &= P\,(\,1 - r\,)^t \\ &= 20{,}000\,(\,0.85\,)^5 \\ &= 20{,}000\,(\,0.443705312\,) \quad \text{Using a scientific calculator and the } x^y \text{ key.} \\ &= 8874.11 \end{aligned}$$

The current market value of the automobile is \$8874.11.

## Completion Example Answers

**4.** The unknown quantity is $\boldsymbol{r}$.

$I = \mathbf{\$50},\ P = \mathbf{\$2000},\qquad r = 3 \text{ months} = \frac{\mathbf{1}}{\mathbf{4}} \text{ year}$

Substituting into the formula $I = Prt$

$50 = \mathbf{2000} \cdot r \cdot \frac{1}{4}$

$50 = \mathbf{500} \cdot r$ Simplify the right-hand side.

$\frac{50}{\mathbf{500}} = \frac{\mathbf{500} \cdot r}{\mathbf{500}}$ Divide both sides by 500, the coefficient of $r$.

$\mathbf{0.10} = r$

The interest rate would be **10%**.

**8. a.** Follow the steps outlined in Example 6 with

$P = \$6000,\qquad r = 12\% = 0.12,\qquad n = 360,\qquad t = 4$

Substituting into the formula,

$$\begin{aligned} A &= 6000\left(1 + \frac{\mathbf{0.12}}{\mathbf{360}}\right)^{360 \cdot 4} \\ &= 6000\left(1 + \mathbf{0.000333333}\right)^{\mathbf{1440}} \\ &= 6000\left(1.\mathbf{615945149}\right) \\ &= 9695.68 \end{aligned}$$

The value (or amount) will be **$9695.68**.

**b.** The interest will be

$I = \mathbf{\$9695.68} - \$6000.00 = \mathbf{\$3695.68}$

Name ________________ Section __________ Date ________

## Exercises 6.6

( **Note**: For the answers in these exercises, we have used 360 days in a year and 30 days in a month. You may choose to use 365 days in a year in which case your answers may be slightly different. )

**1.** How much simple interest would be paid on a loan of $1000 at 15% for 3 months?

(**Note:** 3 months $= \frac{3}{12}$ year $= \frac{1}{4}$ year.)

**2.** Paul loaned his uncle $1500 at an interest rate of 10% for 9 months. How much simple interest did he earn? ( **Note:** 9 months $= \frac{9}{12}$ year $= \frac{3}{4}$ year. )

**3.** How much simple interest is earned on a loan of $5000 at 11% for a period of 6 months?

( **Note:** 6 months $= \frac{6}{12}$ year $= \frac{1}{2}$ year. )

**4.** If you borrow $750 for 30 days at 18%, how much simple interest will you pay?

( **Note:** 30 days $= \frac{30}{360}$ year $= \frac{1}{12}$ year. )

**5.** Find the simple interest paid on a savings account at $1800 for 120 days at 8%.

**6.** A savings account of $3200 is left for 90 days and draws interest at a rate of 7%.

**a.** How much interest is earned?

**b.** What is the balance in the account at the end of the 90–day period?

**7.** For how many days must you leave $1000 in a savings account at 5.5% to earn $11.00 in interest? ( *Hint:* Find the value of $t$ in the formula $I = Prt$. This value is a fraction of a year. To find the number of days, multiply this fraction by 360. )

**8.** What is the rate of interest charged if a 90-day loan of $2500 is paid off with $2562.50? ( *Hint:* The payoff is principal plus interest. So first find the interest earned. )

**ANSWERS**

1. $37.50
2. $112.50
3. $275
4. $11.25
5. $48
6. a. $56
   b. $3256
7. 72 days
8. 10%

9. 10%

10. a. $1000

b. $\frac{3}{4}$ yr (or 9 months)

11. a. $470.25

b. $29.75

c. To collect interest at 18%

12. a. $2431

b. $231

c. $69

13. [ Respond in exercise.]

**9.** What is the rate of interest charged if a 9 - month loan of $3000 is paid off with $3225.00? ( *Hint*: The payoff is principal plus interest. So first find the interest earned. )

**10.** A savings account of $25,000 is drawing interest at 8%.

**a.** How much interest will be earned in 6 months?

**b.** How long will it take for the account to earn $1500 in interest?

**11.** You buy an oven on sale with the price reduced from $500 to $450. The store has allowed you to keep the oven for 1 year without making a payment. However, they are charging you interest on what you owe at a rate of 18%.

**a.** How much will you end up paying for the oven if you pay the total amount owed in 3 months?

**b.** How much did you save by buying the oven on sale?

**c.** For what reason did the store allow you to keep the oven for 1 year before paying?

**12.** A new computer is on sale for $2200, and no payments need to be made for 6 months. The original price was $2500. The terms are that the buyer will pay the total amount owed at the end of 6 months, and the interest rate is 21%.

**a.** How much will the buyer end up paying for the computer?

**b.** How much will the buyer save by paying cash on the purchase date instead of waiting 6 months to pay what is owed?

**c.** If the buyer waits 6 months to pay, how much will he or she save from the original price by buying the computer on sale?

**13.** Determine the missing item in each row and write your answers in the table.

| Principal | Rate | Time | Interest |
|---|---|---|---|
| $ 400 | 16% | 90 days | **a.** $ 16 |
| **b.** $ 100 | 15% | 120 days | $ 5 |
| $ 560 | 12% | **c.** 30 days | $ 5.60 |
| $ 2700 | **d.** 8.5% | 40 days | $ 25.50 |

Name ______________________ Section ______________ Date ____________

**14.** Determine the missing item in each row and write your answers in the table.

| Principal | Rate | Time | Interest |
|---|---|---|---|
| \$ 1000 | $10\frac{1}{2}\%$ | 60 days | **a.** \$ 17.50 |
| \$ 800 | $13\frac{1}{2}\%$ | **b.** 60 days | \$ 18.00 |
| \$ 2000 | **c.** 11.5% | 9 months | \$ 172.50 |
| **d.** \$ 1133.34 | 7.5% | 1 year | \$ 85.00 |

In Exercise 15 - 20, show the calculations for each period of compounding in the step-by-step manner illustrated in Example 5.

**15. a.** If a bank compounds interest quarterly at 8% on a certificate of deposit, what will your \$7500 deposit be worth in 6 months?

**b.** In 1 year?

**16. a.** If an account is compounded quarterly at a rate of 6%, how much interest will be earned on \$5000 in 1 year?

**b.** What will be the total amount in the account?

**17.** You borrowed \$5000 and agreed to make equal payments of \$1000 each plus interest over the next 5 years. Interest is at a rate of 8% based only on what you owe.

**a.** How much interest will you pay?

**b.** How much interest would you have paid if you did not make the annual payments and paid only the \$5000 plus interest compounded annually at the end of 5 years?

**18.** An amount of \$9000 is deposited in a savings account and interest is compounded monthly at 10%. What will be the balance of the account in 6 months?

**19. a.** How much interest will be earned on a savings account of \$3000 in 2 years if interest is compounded annually at 7.25%?

**b.** If interest is compounded semiannually?

**ANSWERS**

**14.** [ Respond in exercise.]

**15. a.** \$7803

**b.** \$8118.25

**16. a.** \$306.82

**b.** \$5306.82

**17. a.** \$1200

**b.** \$2346.65

**18.** \$9459.48

**19. a.** \$450.77

**b.** \$459.23

20. $17,364.38

21. a. $26,620

b. $26,801.92, No

c. [ Respond in exercise.]

22. a. $1493.43

b. $93.43

23. a. $13,498.03

b. about 7 years

24. about 14 years

25. a. $24,492.29

b. $95.64

26. $540,000

**20.** If interest is compounded semiannually at 10%, what will be the value of $15,000 in $1\frac{1}{2}$ years?

In Exercise 21 – 30, use the following formulas whenever they apply.

$$A = P\left(1 + \frac{r}{n}\right)^{nt}, \quad I = A - P, \quad A = P(1+r)^t, \quad \text{and} \quad V = P(1-r)^t$$

**21. a.** What will be the value of a $20,000 savings account at the end of 3 years if interest is calculated at 10% compounded annually?

**b.** Suppose the interest is compounded semiannually. What is the value? Is the value the same?

**c.** If not, explain why not in your own words.

Explanations will vary. By compounding semiannually, interest is posted twice as frequently, so more interest is earned on interest accrued.

**22. a.** Calculate the interest earned in one year on $10,000 compounded monthly at 14%.

**b.** What is the difference between this and simple interest at 14% for 1 year?

**23. a.** What will be the value of $10,000 compounded daily at 10% for 3 years?

**b.** Use your calculator and choose values for $t$ to use in the formula until you find approximately how many years of daily compounding are needed for the value to accumulate to $20,000.

**24.** Use your calculator and choose values for $t$ to use in the formula until you find approximately how many years are needed for $3000 to triple if interest is compounded daily at 8%.

**25.** Suppose that $50,000 is invested in a certificate of deposit for 5 years and the interest rate is 8%.

**a.** What will be the interest earned if it is compounded monthly?

**b.** How much more interest would be earned if it were compounded daily?

**26.** The value of your house today is $125,000. If inflation is at 5%, what will be its value, to the nearest thousand dollars, in 30 years?

Name ______________ Section ______________ Date ______________

**ANSWERS**

**27.** The current prices of four items are given below. What will be their prices in 10 years if inflation is at 6%?

**a.** A car: $41,000 **b.** A television set: $1850

**c.** A textbook: $95.00 **d.** A cup of coffee: $1.25

**27. a.** $73,424.76

**b.** $3313.07

**c.** $170.14

**d.** $2.24

**28.** If a new pickup truck is valued at $18,000, what will be its value in 3 years if it depreciates 22% each year?

**28.** $8541.94

**29.** $1,078,673.79

**29.** Suppose that an apartment complex is purchased for $1,500,000. For property tax purposes, the land is considered to be 30% of the value of the property. For income tax purposes, the owners are allowed to depreciate the value of the buildings (capital goods ) by 5% per year. What will be the value of the apartment complex ( buildings and land ) in 10 years? ( **Note:** This will not be the market value, but it will form the basis for capital gains taxes when the property is sold. )

**30. a.** $1443.89

**b.** $1165.70

**31.** [ Respond in exercise.]

**30. a.** Calculate the interest you would pay on a 1- year loan of $7500 at 18% if the interest were compounded every 3 months and you made no monthly payments.

**b.** If you made payments of $1000 ( against principal and interest ) every 3 months and paid the balance plus interest owed at the end of the year, how much interest would you pay?

**32.** [ Respond in exercise.]

**33.** [ Respond in exercise.]

In Exercises 31-36, find the amount ( $A$ ) and the interest earned ( $I$ ) for the given information.

| Compounding Period | Principal | Annual Rate | Time | $A$ | $I = A - P$ |
|---|---|---|---|---|---|
| **31.** Quarterly | $1000 | 10% | 5 yr | **a.** $1638.62 | **b.** $638.62 |
| **32.** Monthly | $1000 | 10% | 5 yr | **a.** $1645.31 | **b.** $645.31 |
| **33.** Daily | $1000 | 10% | 5 yr | **a.** $1648.61 | **b.** $648.61 |
| **34.** Monthly | $5000 | 7.5% | 10 yr | **a.** $10,560.33 | **b.** $5560.33 |
| **35.** Daily | $25,000 | 8% | 20 yr | **a.** $123,803.81 | **b.** $98,803.81 |
| **36.** Daily | $25,000 | 12% | 20 yr | **a.** $275,469.23 | **b.** $250,469.23 |

**34.** [ Respond in exercise.]

**35.** [ Respond in exercise.]

**36.** [ Respond in exercise.]

[ Respond below exercise.]
37. ______

[ Respond below exercise.]
38. ______

You may want to tell your studenets of the "Rule of 72" for the doubling time for any principal. A good estimate for the time it takes for a principal to double compounded daily can be found by dividing the interest rate into 72.

## WRITING AND THINKING ABOUT MATHEMATICS

**37.** Use your calculator and choose the value of $t$ ( in years ) to use in the formula for compound interest until you find how many years of daily compounding at 8% are needed for an investment of \$5000 to **approximately** double in value. Write down the values you chose for $t$, why you chose those particular values, and the corresponding accumulated values of money. Explain why you agree or disagree with the idea that \$10,000 would double in a shorter time. ( **Note:** This is a type of trial-and-error exercise. There is no best way to do the problem. However, with some practice, you should develop an understanding of compound interest so that you can do this type of problem in a more efficient manner each time. )

| $t$ | $\left(1+\dfrac{0.08}{360}\right)^{360t}$ | A |
|---|---|---|
| Answers will vary. | | |
| 5 | 1.491758405 | \$7458.80 |
| 8 | 1.896346042 | \$9481.74 |
| 9 | 2.054268885 | \$10,271.35 |

The time for doubling depends on the rate of interest and the period of compounding and is independent of the principal.

## COLLABORATIVE LEARNING EXERCISE

**38.** With the class separated into teams of two to four students, each team is to analyze the following problem related to compound interest. Each team leader is to discuss the results found by the team and how the team arrived at these results. A general classroom discussion should follow with the class coming to an understanding of the concepts of **present value** and **future value**.

Suppose that you would like to set aside some money today for your child's college education. Your child is 3 years old and will be going to college ( you hope ) at the age of 18. What amount should you invest today ( called the **present value** ) at 8% compounded daily to accumulate \$40,000 ( called the **future value** ) for your child's education?

Responses will vary. Some students may use the formula $A = P\left(1 + \dfrac{r}{n}\right)^{nt}$ and solve for $P$.

$$P = \frac{A}{\left(1 + \dfrac{r}{n}\right)^{nt}} = \frac{40,000}{\left(1 + \dfrac{0.08}{360}\right)^{360 \cdot 15}} = \$12,049.38$$

# Chapter 6 Index of Key Ideas and Terms

To Change a Decimal to a Percent: page 471

Step 1: Move the decimal point two places to the right.
Step 2: Add on the % symbol.

To Change a Percent to a Decimal: page 472

Step 1: Move the decimal point two places to the left.
Step 2: Delete the % symbol.

To Change a Fraction to a Percent: page 474

Step 1: Change the fraction to a decimal.
Step 2: Change the decimal to a percent.

To Change a Percent to a Fraction or Mixed Number: page 477

Step 1: Write the percent as a fraction with denominator 100 and delete the % symbol.
Step 2: Reduce the fraction.

For the Basic Formula: $\boldsymbol{R \cdot B = A}$ page 489

R = RATE or percent ( as a decimal or fraction )
B = BASE ( number we are finding the percent of )
A = AMOUNT or percentage ( a part of the base )
"of" means to multiply.
"is" means equal.
The formula represents three types of percent problems.

Estimating with Percent page 499

Pòlya's Four-Step Process for Solving Problems page 507

1. Understand the problem.
2. Devise a plan.
3. Carry out the plan.
4. Look back over the results.

Discount and Sales Tax page 508

Commission page 510

Percent of Profit page 511

Percent of profit based on cost
Percent of profit based on selling price

Short Method for Calculating a 15% Tip page 513

Expenses in Buying a Car page 523

Purchase price: The selling price agreed on by the seller and the buyer.
Sales tax: A fixed percent that varies from state to state.
License fee: Fixed by the state, often based on the type of car and its value.

Expenses in Buying a Home page 524

Purchase price: The selling price ( what you have agreed to pay )
Down payment: Cash you pay to the seller
Mortgage loan ( or first trust deed ): Loan to you by a bank or a savings and loan
Mortgage fee ( or points ): Loan fee charged by the lender
Fire insurance: Insurance against loss of your home by fire
Recording fees: Local and state fees for recording you as the legal owner
Property taxes: Taxes that must be prepaid before the lender will give you the loan
Legal fees: Fees charged by a lawyer or escrow company and/or a title search company for completing all forms in a legal manner

Simple Interest: $I = Prt$ page 531

I = Interest ( earned or paid )
P = Principal ( the amount invested or borrowed )
r = Rate of interest ( stated as an annual or yearly rate )
t = Time ( in years or fraction of a year )

Compound Interest: $A = P\left(1 + \frac{r}{n}\right)^{nt}$ page 536

A = Amount accumulated
P = the principal
r = the annual interest rate
t = the length of time in years
n = the number of compounding periods in 1 year

Inflation: $A = P(1 + r)^t$ page 539

A = the amount ( or value after t years )
P = the principal ( or value today )
r = the annual rate of inflation ( in decimal form )
t = the length of time in years

Depreciation: $V = P(1 - r)^t$ page 540

V = the current value
P = the original value
r = the annual rate of depreciation ( in decimal form )
t = the time in years

Name ______________________ Section ______________ Date ____________

## Chapter 6 Test

In Problems 1-5, change each number to a percent.

**1.** $\frac{102}{100}$ **2.** 0.0725 **3.** 0.005 **4.** $\frac{3}{20}$ **5.** $\frac{3}{8}$

In Problems 6-8, change each percent to a decimal.

**6.** 24% **7.** 6.8% **8.** $5\frac{1}{2}\%$

In Problems 9-11, change each percent to a fraction or mixed number. Reduce to lowest terms.

**9.** 160% **10.** 9.6% **11.** $62\frac{1}{2}\%$

In Problems 12-15, solve each problem for the unknown quantity.

**12.** 10% of 92 is 9.2. **13.** 24 is 50 % of 48.

**14.** 135.45 is 35% of 387. **15.** 15.225 is 75% of 20.3.

Choose the correct answer by reasoning and calculating mentally.

**16.** Write 13% as a fraction.

**a.** $\frac{13}{1}$ **b.** $\frac{13}{10}$ **c.** $\frac{13}{100}$ **d.** $\frac{13}{1000}$

**ANSWERS**

1. 102%
2. 7.25%
3. 0.5%
4. 15%
5. 37.5%
6. 0.24
7. 0.068
8. 0.055
9. $1\frac{3}{5}$
10. $\frac{12}{125}$
11. $\frac{5}{8}$
12. [Respond in exercise]
13. [Respond in exercise]
14. [Respond in exercise]
15. [Respond in exercise]
16. c

17. d

18. b

19. $5400

20. a. more

b. 97%

c. $2500

d. $75

21. $85.30

**17.** Write $\frac{7}{8}$ as a percent.

**a.** 12.5% **b.** 37.5% **c.** 62.5% **d.** 87.5%

**18.** Find 35% of 100.

**a.** 3.5 **b.** 35 **c.** 350 **d.** 3500

**19.** A real estate company charges a commission of 6% for the sale of a home and pays the agent 60% of this fee. How much will the agent earn on the sale of a home for $150,000?

Did you read the problem at least twice? Do you understand all of the terms used?

What is your plan to solve the problem?

Carry out this plan. Show all of your work.

Does the answer seem reasonable to you?

**20.** A customer received a discount of 3% on the purchase of a new sofa because she paid cash. She paid $2425 in cash.

**a.** Was the original price more or less than $2425?

**b.** The amount $2425 is what percent of the original price?

**c.** What was the original price?

**d.** How much did she save by paying cash?

**21.** A lawyer took three of her clients out for lunch, and the bill was $70.00. If sales tax was at 6% and she left a 15% tip, how much did she pay for lunch?

Name ____________ Section ________ Date ________

**ANSWERS**

**22.** Leona bought a used car for $1500 and replaced parts and made repairs that cost her another $500. Then she sold the car for $2600.

**a.** What was her percent of profit based on cost?

**b.** What was her percent of profit based on selling price?

**23.** A note of $2000 is given for 9 months and will be paid off with a total of $2135. What is the rate of interest? ( Use 360 days in a year and 30 days in a month. )

**24.** An amount of $10,000 is deposited in a savings account, and interest is compounded monthly at 6.5%.

**a.** What will be the balance in the account in 5 years?

**b.** How much interest will have been earned? ( Use 360 days in a year and 30 days in a month. )

**25.** The purchase price of a new home is $95,000. The buyer is to make a down payment of 20% and pay a loan fee of 1 point.

**a.** What is the amount of the down payment?

**b.** What is the amount of the first trust deed?

**c.** If fire insurance is $300, legal fees are $175, recording fees are $85, and taxes are $290, what amount of cash does the buyer need to complete the purchase?

**22. a.** 30%

**b.** $23\frac{1}{13}\%$ (or 23.1%)

**23.** 9%

**24. a.** $13,828.18

**b.** $3828.18

**25. a.** $19,000

**b.** $76,000

**c.** $20,610

Name ______________________ Section ______________ Date ____________

## Cumulative Review: Chapters 1 - 6

**ANSWERS**

**1.** The number 1 is called the multiplicative identity because for any real number $a$, $a \cdot 1 = a$.

**2.** Explain why subtraction is not a commutative operation.

In general, $a - b \neq b - a$. Any counterexample will also serve as an explanation.

**3.** Round off the numbers 2.78 and –3.59 to the nearest tenth, and then round off the product of these two rounded-off numbers to the nearest tenth.

**4. a.** First estimate the quotient $945.7 \div 10.6$; then

**b.** Find the quotient accurate to the nearest hundredth.

**5.** Evaluate the polynomial $3x^2 - 5.2x + 15$ for $x = -2$.

**6.** Find 30% of the value of the expression $(3.7)^2 - (0.1)(-4.7)$.

**7.** Find the prime factorization of each number.

**a.** 65 **b.** 144 **c.** 270

**8.** List the multiples of 6 that are less than 100.

**9. a.** Does 10 divide the product $7 \cdot 6 \cdot 5 \cdot 3$?

**b.** If so, what is the quotient?

**10.** Solve the proportion $\frac{A}{16} = \frac{0.125}{8}$.

**Answers:**

1. [Respond in exercise]
2. [Respond below exercise]
3. 2.8; –3.6; –10.1
4. a. 90 (estimate)
   b. 89.22
5. 37.4
6. 4.248
7. a. $5 \cdot 13$
   b. $2^4 \cdot 3^2$
   c. $2 \cdot 3^3 \cdot 5$
8. 6, 12, 18, 24, 30, 36, 42, 48, 54, 60, 66, 72, 78, 84, 90, 96
9. a. Yes
   b. 63
10. $A = 0.25$

Evaluate ( or simplify ) each of the following expressions. Reduce all fractions.

**11.** $\frac{27}{35} \cdot \frac{2}{3} \cdot \frac{7}{18}$

**12.** $\frac{x}{3} \div \frac{2}{3} + \frac{1}{4}$

**13.** $3.6 - 7\frac{1}{2} + |-2.6| - \left|5\frac{1}{4}\right|$

Solve the following equations.

**14.** $-30 = 0.9x + 16.8$

**15.** $3(a + 5) - 8a = 4$

**16.** $\frac{1}{2}x + \frac{3}{5} = -\frac{1}{10}$

**17.** 41% of _800_ is 328.

**18.** Find **a.** the perimeter and **b.** the area of a rectangular playground that is 100 yards long and 65 yards wide.

**19.** A right triangle has legs of 18 inches and 24 inches.

**a.** How long is the hypotenuse?

**b.** What is the perimeter of the triangle?

**c.** What is the area of the triangle?

**20.** The pole for a stop sign is 12 feet long.

**a.** If $\frac{5}{16}$ of the length of the pole is to be below the ground, what fraction of the length of the pole is to be above ground?

**b.** How many feet are to be above ground?

**11.** $\frac{1}{5}$

**12.** $\frac{2x+1}{4}$

**13.** $-6.55$

**14.** $-52 = x$

**15.** $a = \frac{11}{5}$ (or $a = 2.2$)

**16.** $x = -\frac{7}{5}$

**17.** [Respond in exercise]

**18. a.** 330 yd

**b.** 6500 yd²

**19. a.** 30 in.

**b.** 72 in.

**c.** 216 in.²

**20. a.** $\frac{11}{16}$

**b.** $8\frac{1}{4}$ ft

Name ______________________ Section ______________ Date ____________

ANSWERS

**21.** Suppose that your friend weighs 100 pounds.

**a.** If she first lost 10% and then gained 10% in weight, how much would she weigh?

**b.** If she first gained 10% and then lost 10%, how much would she weigh?

**c.** Can you explain these results?

Explanations will vary. The result is the same, since (100)(0.9)(1.1) = (100)(1.1)(0.9) = 99.

**22.** What will be the interest earned in one year on a savings account of \$2300 that pays simple interest at $6\frac{1}{4}\%$ ?

**23.** If a principal of \$10,000 is invested for 720 days and interest is compounded daily at 8%, how much interest is earned? ( Use 360 days in a year and 30 days in a month. )

**24.** Determine the missing item in each row if the interest is simple interest. ( Use 360 days in a year and 30 days in a month. )

| Principal | Rate | Time | Interest |
|---|---|---|---|
| \$ 500 | 12 % | 180 days | **a.** \$30 |
| \$ 200 | 18 % | **b.** $\frac{1}{2}$ yr | \$ 18 |
| \$ 1000 | **c.** 20% | 9 months | \$ 150 |

**25.** At its annual close-out sale a clothing store had men's suits that had originally sold for \$350 marked down to \$250.

**a.** What was the rate of discount based on the original selling price?

**b.** If each suit cost the store \$200, what was the percent of profit based on cost?

**c.** What was the percent of profit based on the sale price?

**26.** On a certain part of an automobile assembly line, three out of every five employees are male.

**a.** How many female employees are there out of 450 people working this part of the line?

**b.** If this number of female employees were to be increased by 10% and there were still only 450 people on the line, then how many would be men?

**21. a.** 99 lb

**b.** 99 lb

**c.** Respond below exercise]

**22.** \$143.75

**23.** \$1734.91

**24.** [Respond below exercise]

**25. a.** $28\frac{4}{7}\%$ (or 28.6%)

**b.** 25%

**c.** 20%

**26. a.** 180

**b.** 252

27. 27 min

28. $38.92

29. a. right;

$10^2 + 24^2 = 26^2$

b. 60 cm

c. 120 $cm^2$

30. 15 ft and 18 ft

31. a. $33,750

b. $101,250

c. $36,645

32. $74.85

**27.** Marty can wash the car in $\frac{3}{4}$ of an hour, and Larry can wash the car in 1.2 hours. How many minutes faster is Marty at washing the car?

**28.** Find the average ( to the nearest cent ) of the following set of prices for table lamps: $25.50; $32.75; $56.00; $30.00; $16.50; $72.75

**29.** **a.** What type of triangle is a triangle with sides of 10 centimeters, 24 centimeters, and 26 centimeters? Why?

Find **b.** the perimeter and
**c.** the area of the triangle.

**30.** Suppose that two triangles are similar and that the sides of the first triangle are 4 feet, 10 feet, and 12 feet long. If the shortest side of the second triangle is 6 feet long, what are the lengths of the other two sides?

**31.** The purchase price of a condominium is $135,000. The buyer is to make a 25% down payment and pay a loan fee of 2% of the mortgage.

**a.** What is the amount of the down payment?

**b.** What is the amount of the first trust deed?

**c.** If legal fees are $200, fire insurance is $225, taxes are $320, and recording fees are $125, what amount of cash does the buyer need to complete the purchase?

**32.** Karl took his family out to dinner, and the bill was $62.00. If sales tax was at 5% and he left a 15% tip, what total amount did he pay for the dinner?

# 7

# ALGEBRAIC TOPICS I

Chapter 7 Algebraic Topics I





## WHAT TO EXPECT IN CHAPTER 7

Chapter 7 provides a more in-depth study of techniques for solving equations and deals with applications that can be solved by setting up and solving equations. Section 7.1 discusses the "art" of translating English phrases into algebraic expressions, and vice versa. Sections 7.2 and 7.3 develop equation-solving skills relating to equations that have variables on both sides and present more work with equations with fractions.

In Section 7.4, Pòlya's four-step process for solving problems is emphasized again. The problems in this section use the methods of translating phrases discussed in Section 7.1 and introduce the concept of consecutive integers.

In Section 7.5, we work with formulas by evaluating formulas, by using given values of the variables, and by solving formulas for specified variables in the formulas. The chapter closes with Section 7.6, which reviews real numbers and covers the topics of real number lines, intervals of real numbers, and solving linear (or first-degree) inequalities of the form $ax + b < c$.

# Chapter 7 Algebraic Topics I

**Mathematics at Work!**

Formulas are equations used to show relationships between two or more variables. Several formulas have been used already in this text. For example,

$I = Prt$ — For simple interest

$A = P\left(1+\frac{r}{n}\right)^{nt}$ — For compound interest

$C = \pi d$ and $A = \pi r^2$ — For circumference and area of a circle

Formulas illustrate an important and effective use of mathematics in many fields of study. You may be familiar with some of the following formulas, or you may study them in the future.

$C = \frac{5}{9}(F - 32)$ — For relating Celsius and Fahrenheit temperatures

$\frac{W}{L} = \frac{L}{W+L}$ — For calculating the Golden Ratio ( According to early Greek mathematicians, the "most perfect" rectangles had width and length that satisfied this formula. )

$s = \sqrt{\frac{n\sum x^2 - (\sum x)^2}{n(n-1)}}$ — For calculating a standard deviation in statistics

For class discussion, you might bring some formulas to class from other courses that you are currently studying such as economics, auto mechanics, geology, chemistry, nursing, and so on. ( See Section 7.5 for more formulas. )

# 7.1 Translating English Phrases

## Translating English Phrases into Algebraic Expressions

Translating a problem into mathematical symbolism for a solution involves translating English phrases into equivalent algebraic expressions. With this skill and understanding, we can then translate entire sentences and groups of sentences into algebraic equations that will yield solutions to the original problem. As the following examples illustrate, several English phrases may translate into the same algebraic expression. In each case, a variable is used to represent "a number." This technique of letting a variable represent an unknown number is basic to solving problems by using algebraic methods.

| English Phrase | Algebraic Expression |
|---|---|
| **1.** A number **plus** 5<br>The **sum** of a number and 5<br>5 **more than** a number<br>5 **added to** a number | $x+5$<br>Here $x$ represents "a number." |
| **2.** 6 **times** a number<br>The **product** of 6 and a number<br>6 **multiplied by** a number | $6x$<br>Here $x$ represents "a number." |
| **3.** 2 **times** the **sum** of a number and 3<br>**Twice** the **sum** of $y$ and 3<br>The **product** of 2 and the **quantity** $y$ plus 3 | $2(y+3)$<br>Here $y$ represents "a number." |
| **4.** 3 **added to 2 times** a number<br>$2y$ **increased by** 3<br>3 **more than twice** a number | $2y+3$<br>Here $y$ represents "a number." |
| **5.** The **difference** between $6n$ and 8<br>8 **less than 6 times** a number<br>6 **times** a number **minus** 8<br>$6n$ **decreased by** 8<br>8 **subtracted from** $6n$ | $6n-8$<br>Here $n$ represents "a number." |

## Objectives

① Be able to translate English phrases into algebraic expressions.

② Be able to translate algebraic expressions into English phrases.

③ Know how to make up a word problem that fits a given equation.

**Special Remark:** In example 4, the phrase "3 more than twice a number" was translated into the expression $2y + 3$. If the expression had been written $3 + 2y$, then, while the translation is not quite literal, no harm would be done mathematically because **addition is commutative** and

$$2y + 3 = 3 + 2y.$$

However, in example 5, the translation of the phrase "8 **less than** 6 times a number" must be done correctly because **subtraction is not commutative**. The part of the phrase reading "8 less than" means to subtract 8. Thus,

"8 less than 6 times a number" means $6n - 8$

and "6 times a number less than 8" means $8 - 6n$.

Therefore, be very careful when writing expressions indicating subtraction. Be sure that the subtraction is in the order indicated by the wording in the problem. The same is true when writing expressions indicating division.

**Key Words That Indicate Operations**

| Addition | Subtraction | Multiplication | Division |
|---|---|---|---|
| add | subtract (from) | multiply | divide |
| sum | difference | product | quotient |
| plus | minus | times | ratio |
| more than | less than | twice | |
| increased by | decreased by | triple, etc. | |
| | | of (with fractions and percent) | |

Again, since subtraction and division are not commutative operations, the words **"difference"** and **"quotient"** deserve a special mention because their use implies that the numbers are to be operated on **in the order given**. For example, be careful to note that we make the following agreements in mathematics:

"The **difference** between 10 and 16" means $10 - 16$ which is $-6$.

"The **difference** between 16 and 10" means $16 - 10$ which is $+6$.

So, more generally, for subtraction,

"The **difference** between $x$ and 7" means $x - 7$, **not** $7 - x$. Remember, in general $x - 7 \neq 7 - x$.

And, for division,

"The **quotient** of $y$ and 16" means $\frac{y}{16}$, not $\frac{16}{y}$. Remember, in general,

$\frac{y}{16} \neq \frac{16}{y}$.

If we did not have this agreement, then the phrases might have more than one interpretation and be considered **ambiguous**.

Ambiguity can be avoided if the approach discussed here is followed carefully. However, we know that students will forget just what approach to take and have difficulty with some translations. To help you understand how they are thinking and find the reading level of the class, consider having the students read and translate phrases and expressions out loud in class.

An **ambiguous phrase** is one whose meaning is not clear or for which there may be two or more interpretations. This is a common occurrence in ordinary everyday language, and misunderstandings occur frequently. Imagine the difficulties diplomats have in communicating ideas from one language to another trying to avoid ambiguities. Even the order of subjects, verbs, and adjectives may not be the same from one language to another. Translating grammatical phrases in any language into mathematical expressions is quite similar. To avoid ambiguous phrases in mathematics, we try to be precise in the use of terminology, to be careful with grammatical construction, and to **follow the rules for order of operations**.

## Translating Algebraic Expression into English Phrases

Consider the two different expressions

$$5(n+6) \quad \text{and} \quad 5n+6.$$

When an expression has **a quantity in parentheses**, we will indicate this with a phrase such as

"the quantity" or "the sum of" or "the difference between."

Otherwise, in translating algebraic expressions into English phrases, we agree that the operations are to be indicated in the order given. Thus,

$5(n+6)$ is translated as "five times **the sum of** a number increased by 6"
or "five times the **quantity** $n$ plus 6"

whereas,

$5n+6$ is translated as "five times a number increased by 6"
or "five times $n$ plus 6".

### Example 1

Write an English phrase that indicates the meaning of each expression. In each case the variable is translated as "a number."

**a.** $7x$

**b.** $3a+9$

**c.** $3(a-9)$

**d.** $2x+4x$

#### Solution

| Algebraic Expression | Possible English Phrase |
|---|---|
| **a.** $7x$ | The product of 7 and a number |
| **b.** $3a+9$ | 9 more than three times a number |
| **c.** $3(a-9)$ | 3 times the difference between a number and 9 |
| **d.** $2x+4x$ | The sum of twice a number and four times that number |

Write an English phrase that indicates the meaning of each expression.

1. $x + 2x$

   The sum of a number and twice the number.

2. $5x + 1$

   Five times a number increased by 1.

Change each phrase into an equivalent algebraic expression.

3. 6 less than three times a number

   $3x - 6$

4. Seven times the difference between a number and 2

   $7(x - 2)$

### Example 2

Change each phrase into an equivalent algebraic expression. Remember to use a variable to represent the unknown number.

**a.** The quotient of a number and 5

**b.** 8 less than twice a number

**c.** $\frac{2}{3}$ times a number

**d.** Three times the difference between 16 and a number

**Solution**

| Phrase | Algebraic Expression |
|---|---|
| **a.** The quotient of a number and 5 | $\frac{x}{5}$ |
| **b.** 8 less than twice a number | $2a - 8$ |
| **c.** $\frac{2}{3}$ times a number | $\frac{2}{3}n$ |
| **d.** Three times the difference between 16 and a number | $3(16 - y)$ |

*Now Work Exercises 1-4 in the Margin.*

## Translating Equations into Word Problems

Have you ever said or heard someone say something like,

"I can do the mathematics. I just have trouble solving word problems."

Maybe you are one of these people? Basically, what they mean is that they can perform simple arithmetic or algebraic manipulations. However, reasoning and translating ideas into symbols that represent the information given in a "word problem" is where they are having trouble. Well, the bad news is that the reasoning, translating, and solving are the most important parts of mathematics. The skills needed for operating with numbers in arithmetic and solving equations in algebra are just that -- skills. For sure, without these skills, you would not be able to solve word problems. However, they are not sufficient. That is, you need much more in the way of understanding abstract concepts to be able to solve word problems and to solve problems in your daily life. The good news is that with experience and practice all of this can be learned.

In Section 7.4, you will be given a chance to solve word problems that can be solved by translating English phrases into algebraic expressions and sentences into equations. The solutions to the equations are the solutions to the corresponding word problems. In this section, to help in understanding the abstract relationship between word problems and algebraic equations, we are going to reverse the process. You are going to be given an equation and asked to "make up" your own related word problem that might use this equation to find a solution. Consider the following examples.

### Example 3

For each equation, make up your own word problem that might use the equation in its solution. Remember that the variable can be translated into something like "a number," or "some number."

**a.** $5x + 10 = -10$

**b.** $3y + 25 = 2(y + 6)$

#### Solution

**a.** A mystery number is multiplied by 5 and the product is increased by 10. If the result is equal to $-10$, what is the mystery number?

**b.** If 25 is added to 3 times a real number, the result will be equal to twice the sum of the same number and 6. What is the number?

**Note:** In Example 3, the "translations" are not unique. In fact, there are many ways to make up a problem for each equation. However, you should be able to show your "word problem" to your classmates and have them agree that the related equation will give the solution to the problem.

Name ______________________ Section ____________ Date __________

## Exercises 7.1

Write an English phrase that indicates the meaning of each algebraic expression.
Answers may vary. Here are some possibilities.

**1.** $3x$
three times a number

**2.** $x + 12$
the sum of a number and 12

**3.** $x - 15$
fifteen less than a number

**4.** $-2y$
the product of −2 and a number

**5.** $6y + 4.9$
six times a number plus 4.9

**6.** $2y + 7.5$
7.5 more than two times a number

**7.** $2n + 3n$
twice a number increased by three times the number

**8.** $5n - 2n$
five times a number decreased by twice the number

**9.** $2.5n + 3.5n$
2.5 times a number plus 3.5 times the same number

**10.** $6x + x - 1$
six times a number plus the number minus one

**11.** $5x - x + 2$
the product of five and a number minus the number plus 2

**12.** $3y - 10$
ten less than 3 times a number

**13.** $1.5y + 4$
four more than the product of 1.5 and a number

**14.** $1 + 7x$
one increased by the product of seven and a number

**15.** $7 - 3x$
seven decreased by the product of 3 and a number

**16.** $2(x + 5)$
twice the sum of a number and 5

**17.** $3(x - 2)$
three times the difference of a number and 2

**18.** $5(y - 1)$
five times the difference between a number and 1

**19.** $-4(x + 3)$
the product of −4 and the sum of a number and 3

**20.** $8 + 2(n - 1)$
eight plus two times the difference between a number and 1

**21.** $\frac{n}{3} + 15$
the quotient of a number and 3 increased by 15

**22.** $\frac{x}{3} + 1$
one more than the quotient of a number and 3

**23.** $\frac{x}{5} - \frac{3}{2}$
three-halves less than a quotient of a number and 5

**24.** $\frac{y}{6} + \frac{1}{6}$
one-sixth more than a number divided by 6

**ANSWERS**

For Exercises 1 - 24, respond below exercise.

Write an algebraic expression described by each of the following English phrases.

**25. a.** 3 less than a number

**b.** a number less than three

**26. a.** a number increased by 7

**b.** 7 increased by a number

**27.** the product of a number and 6

**28.** the quotient of a number and 7

**29.** the quotient of 7 and a number

**30.** a number decreased by $\frac{3}{2}$

**31.** 4 times the sum of a number and 8

**32.** 18 more than twice a number

**33.** 6 times the difference between a number and 1

**34.** 8 divided by a number plus 3

**35.** 8 divided by the quantity $x + 3$

**36.** the sum of twice a number and three times the same number

**37.** the difference between 5 times a number and twice the number

**38.** 6 minus the sum of a number and 7

**25. a.** $n - 3$

**b.** $3 - n$

**26. a.** $n + 7$

**b.** $7 + n$

**27.** $6x$

**28.** $\frac{x}{7}$

**29.** $\frac{7}{x}$

**30.** $y - \frac{3}{2}$

**31.** $4(y + 8)$

**32.** $2x + 18$

**33.** $6(x - 1)$

**34.** $\frac{8}{x} + 3$

**35.** $\frac{8}{x+3}$

**36.** $2n + 3n$

**37.** $5n - 2n$

**38.** $6 - (x + 7)$

Name ______________ Section ______________ Date ______________

**ANSWERS**

Write an English phrase corresponding to each expression in the given pairs.
Answers may vary. Here are some possibilities.

**39. a.** $4x - 5$

**b.** $4(x - 5)$

a. four times a number decreased by five
b. four times the difference between a number and 5

**40. a.** $7x + 14$

**b.** $7(x + 14)$

a. fourteen more than 7 times a number
b. seven times the sum of a number and 14

**41. a.** $3x + 1$

**b.** $3(x + 1)$

a. the product of a number and 3 increased by 1
b. three times the sum of a number and 1

**42. a.** $9x - 1.9$

**b.** $9(x - 1.9)$

a. 1.9 less than the product of 9 and a number
b. nine times the difference between a number and 1.9

**43. a.** $2n - 7$

**b.** $2(n - 7)$

a. twice a number minus 7
b. twice the difference between a number and 7

**44. a.** $8n + \frac{1}{2}$

**b.** $8\left(n + \frac{1}{2}\right)$

a. the product of 8 and a number plus $\frac{1}{2}$
b. eight times the sum of a number and $\frac{1}{2}$

Write an algebraic expression that corresponds to each of the following descriptions.

**45.** the number of minutes in $h$ hours

**46.** the number of hours in $d$ days

**47.** the cost of $x$ gallons of gasoline if 1 gallon costs $1.25

**48.** the cost of $x$ graphing calculators if one calculator costs $60

**49.** the number of hours in $m$ minutes

**50.** the number of years in $d$ days

**51.** the number of days in $y$ years

**52.** the number of seconds in $m$ minutes

39. [Respond below exercise]
40. [Respond below exercise]
41. [Respond below exercise]
42. [Respond below exercise]
43. [Respond below exercise]
44. [Respond below exercise]
45. $60h$
46. $24d$
47. $1.25x$
48. $60x$
49. $\frac{m}{60}$
50. $\frac{d}{365}$
51. $365y$
52. $60m$

For Exercises 53 - 68, respond below exercise.

Make up your own word problem that might use the given equation in its solution. Be creative! Translate the variable into something like "a strange number," or "the age of a dog," or "an amount invested." For Exercises 53–68, the student responses will be varied and imaginative.

**53.** $2x + 3 = -4$

**54.** $3x - 2 = -5$

**55.** $1.25x = 6.25$

**56.** $5.29x = 37.03$

**57.** $n + (n + 1) = 25$

**58.** $n + (n + 2) = 135$

**59.** $7x = 3x - 22$

**60.** $8x = 2x + 18$

**61.** $2x + 3x = x$

**62.** $x = 5x - 6x$

**63.** $1.6 + 2n = n$

**64.** $2.5 + 3n = 2(n - 1)$

**65.** $3(x + 1) = 2x - 1$

**66.** $\frac{x}{4} - 3 = \frac{1}{2}$

**67.** $\frac{n}{3} + \frac{3}{4} = \frac{1}{4}$

**68.** $\frac{2}{3}x + \frac{1}{6} = \frac{1}{12}$

# 7.2 Solving Equations I

## First-Degree Equations with Variables on Both Sides

We have discussed first-degree equations and techniques for solving first-degree equations in Sections 1.4, 2.7, 3.7, 4.6, and 5.5. In each of these sections, we have progressed to slightly more advanced techniques with different types of numbers. In this section, we continue to develop the skills related to solving equations by solving equations with variables on both sides of the equations. There are no new concepts involved, just more steps.

For review and reference, we restate the definition of a first-degree equation and the fundamental principles used in solving first-degree equations.

**Objectives**

① Learn how to solve first-degree equations that have variables on both sides of the equal sign.

**Definition:**

A **first-degree equation in $x$** ( or a **linear equation in $x$** ) is any equation that can be written in the form

$$ax + b = c \quad \text{where } a, b, \text{ and } c \text{ are constants and } a \neq 0.$$

( **Note:** A variable other than $x$ may be used. )

**Principles Used in Solving a First-Degree Equation**

In the two basic principles stated here, $A$ and $B$ represent algebraic expressions or constants. In the Multiplication Principle, $C$ represents a constant, and $C$ is not 0 .

1. The **Addition Principle:** The equations $A = B$ and $A + C = B + C$ have the same solutions.

2. The **Multiplication Principle:** For $C \neq 0$, the equations $A = B$

   and $C \cdot A = C \cdot B$

   and $\frac{A}{C} = \frac{B}{C}$

   and $\frac{1}{C} \cdot A = \frac{1}{C} \cdot B$

   have the same solutions.

Essentially, these two principles say that if we perform the same operation to both sides of an equation, the resulting equation will have the same solution as the original equation. Note that the **Multiplication Principle** could also be called the **Division Principle** because dividing by a number is the same as multiplying by its reciprocal.

For interest, understanding, and motivation, you might want to show a few higher degree equations that have more than one solution. For example, $x^2 - 5x + 6 = 0$ has the two solutions $x = 2$ and $x = 3$. To validate this concept, you could have the students check both solutions. You might also tell them that second-degree equations have no more than two solutions, third-degree equations have no more than three solutions, and so on.

As we have discussed previously and the following examples illustrate, solving an equation may involve several steps. Keep in mind the following general process as you proceed from one step to another.

| **To Understand How to Solve Equations:** | **For Example:** |
| --- | --- |
| **1.** Simplify both sides of the equation. ( This includes applying the distributive property and/or combining like terms. ) | $2(x+3) = 14-24$<br>$2x+6 = -10$ |
| **2.** If a constant is added to a variable, add the opposite of the constant to both sides of the equation and simplify. | $2x+6-\mathbf{6} = -10-\mathbf{6}$<br>$2x = -16$ |
| **3.** If a variable has a constant coefficient other than 1, divide both sides by that coefficient ( that is, in effect, multiply both sides by the reciprocal of that coefficient ). | $\frac{1}{\cancel{2}}\cdot\cancel{2}x = \frac{1}{2}\cdot(-16)$<br>$x = -8$ |
| **4.** Generally, use the Addition Principle first so that terms with variables are on one side and constant terms are on the other side.Then combine like terms and use the Multiplication Principle. | |
| **5.** Remember that the object is to isolate the variable ( with coefficient 1 ) on one side of the equation. | |

### Example 1

Solve the equation $5x + 3 = 2x - 15$.

#### Solution

| | |
| --- | --- |
| $5x+3=2x-15$ | Write the equation. |
| $5x+3-3=2x-15-3$ | Add $-3$ to both sides. |
| $5x=2x-18$ | Simplify. |
| $5x-2x=2x-18-2x$ | Add $-2x$ to both sides. |
| $3x=-18$ | Simplify. |
| $\frac{\cancel{3}x}{\cancel{3}}=\frac{-18}{3}$ | Divide both sides by 3. |
| $x=-6$ | Simplify. |

### Example 2

Solve the equation $3(x-5)=-x+21$.

**Solution**

$$3(x-5)=-x+21$$ Write the equation.

$$3x-15=-x+21$$ Apply the distributive property.

$$3x-15+15=-x+21+15$$ Add 15 to both sides.

$$3x=-x+36$$ Simplify.

$$3x+x=-x+36+x$$ Add $x$ to both sides.

$$4x=36$$ Simplify.

$$\frac{\cancel{4}x}{\cancel{4}}=\frac{36}{4}$$ Divide both sides by 4.

$$x=9$$ Simplify.

### *Completion Example 3*

Supply the reasons for each step in solving the equation $3(n-4)+n=8n$.

**Solution**

$$3(n-4)+n=8n$$ Write the equation.

$$3n-12+n=8n$$ ______________________

$$4n-12=8n$$ ______________________

$$4n-12-4n=8n-4n$$ ______________________

$$-12=4n$$ ______________________

$$\frac{-12}{4}=\frac{\cancel{4}n}{\cancel{4}}$$ ______________________

$$-3=n$$ ______________________

### Completion Example Answer

**3.** $$3(n-4)+n=8n$$ Write the equation.

$$3n-12+n=8n$$ **Apply the distributive property.**

$$4n-12=8n$$ **Simplify.**

$$4n-12-4n=8n-4n$$ **Add $-4n$ to both sides.**

$$-12=4n$$ **Simplify.**

$$\frac{-12}{4}=\frac{\cancel{4}n}{\cancel{4}}$$ **Divide both sides by 4.**

$$-3=n$$ **Simplify.**

Name ______________________ Section ______________ Date ____________

## Exercises 7.2

Solve each of the following first–degree equations.

**1.** $3x = x - 12$

**2.** $5x = 2x + 33$

**3.** $6x = 2x - 24$

**4.** $3y = 2y - 5$

**5.** $4y = 3y + 22$

**6.** $7y = 6y - 1$

**7.** $5y - 2 = 4y - 6$

**8.** $7x + 14 = 10x + 5$

**9.** $5x = 4x$

**10.** $8a = 10a$

**11.** $7n = 3n$

**12.** $1.6n = 0.8n$

**13.** $6x - 5 = 3x + 2$

**14.** $2x + 6 = 5x + 10$

**15.** $5(z - 2) = 3(z - 8)$

**16.** $2(z + 1) = 3z + 3$

**17.** $4(y - 1) = 2y + 6$

**18.** $6x - 3 = 3(x + 2)$

**19.** $71t - 62t + 3 = -31t$

**20.** $16s + 23s - 5 = 14s$

**21.** $6x + 5 + 3x = 3x - 13$

**22.** $5x - 2x + 4 = 3x + x - 1$

**ANSWERS**

1. $x = -6$
2. $x = 11$
3. $x = -6$
4. $y = -5$
5. $y = 22$
6. $y = -1$
7. $y = -4$
8. $x = 3$
9. $x = 0$
10. $a = 0$
11. $n = 0$
12. $n = 0$
13. $x = \frac{7}{3}$
14. $x = -\frac{4}{3}$
15. $z = -7$
16. $z = -1$
17. $y = 5$
18. $x = 3$
19. $t = -\frac{3}{40}$
20. $s = \frac{1}{5}$
21. $x = -3$
22. $x = 5$

23. $x = 4$

24. $x = -7$

25. $y = 2$

26. $y = -2$

27. $x = \frac{9}{5}$

28. $x = \frac{1}{2}$

29. $x = \frac{2}{3}$

30. $n = -1$

31. $x = 1$

32. $x = -4$

33. $x = 6$

34. $x = -1$

35. $x = -3$

36. [Respond below exercise.]

37. [Respond below exercise.]

38. [Respond in exercise.]

**23.** $7x + x - 6 = 2(x + 9)$

**24.** $x - 5 + 4x = 4(x - 3)$

**25.** $3(-y + 6) = 3y + 2(y + 1)$

**26.** $2(-y + 1) = 5y + 2(6 - y)$

**27.** $2 - 5(x - 3) = 8$

**28.** $5x + 2(x - 1) = 3x$

**29.** $3x - 2(x - 1) = 4x$

**30.** $25 - 31 - 5n = n$

**31.** $8.2x + 13 = 1.4x + 19.8$

**32.** $6.5 + 1.2x = 0.5 - 0.3x$

**33.** $4.7 - 0.3x = 0.5x - 0.1$

**34.** $x - 0.1x + 0.9 = 0.2(x + 1)$

**35.** $0.25 + 3x + 6.5 = 0.75x$

## Writing and Thinking about Mathematics

**36.** Explain, in your own words, why every whole number is also an integer, but not every integer is a whole number.

Answers will vary. All whole numbers are integers,but not all integers are whole numbers. Integers are the whole numbers and their opposites.

**37.** Explain, in you own words, why integers are also rational numbers.

Answers will vary. Integers can be written as fractions with denominator 1.

**38.** Complete the following sentences:

**a.** Every rational number can be written in decimal form as a terminating decimal or as an infinite repeating decimal.

**b.** Every irrational number can be written in decimal form as an infinite nonrepeating decimal.

# 7.3 Solving Equations II

## First Degree Equations with Fractions

In this section, we will discuss solving equations that involve fractions and have variables on both sides and with fractions. These equations are simply a combination of the ideas discussed in Section 3.7 (equations with fractions) and those in Section 7.2 (equations with variables on both sides of the equal sign).

**Objectives**

① Learn how to solve first-degree equations that contain fractions and have variables on both sides of the equal sign.

The basic concepts, principles, and definitions related to solving first-degree equations were stated in Section 7.2, and we will not restate them here. However, you might refer to Sections 3.4 and 3.7 for methods of finding the LCM (least common multiple) and of using the LCM in solving equations with fractions.

First-degree equations with fractions can be solved by first multiplying both sides of the equation by the LCM of all the denominators and applying the distributive property. This will give an equivalent equation with **integer** constants and coefficients and, in general, integers are easier to work with than fractions. Thus, the application of the distributive property is the only difference in solving the equations here and solving those in Section 7.2.

### Example 1

Solve the equation.

$$\frac{1}{2}x+\frac{3}{4}x-1=\frac{2}{3}x$$

#### Solution

For 2, 4, and 3, the LCM is 12.

$$\frac{1}{2}x+\frac{3}{4}x-1=\frac{2}{3}x$$

$$12\left(\frac{1}{2}x+\frac{3}{4}x-1\right)=12\cdot\frac{2}{3}x \quad \text{Multiply both sides by 12.}$$

$$12\left(\frac{1}{2}x\right)+12\left(\frac{3}{4}x\right)+12(-1)=12\left(\frac{2}{3}x\right) \quad \text{Apply the distributive property.}$$

$$6x+9x-12=8x \quad \text{Coefficients and Constants are now integers.}$$

$$15x-12=8x$$

$$15x-12+12=8x+12$$

$$15x=8x+12$$

$$15x-8x=8x+12-8x$$

$$7x=12$$

$$\frac{\cancel{7}x}{\cancel{7}}=\frac{12}{7}$$

$$x=\frac{12}{7}$$

We have told the students that the method of solving by multiplying by the LCM of the denominators is used to find an equivalent equation with integer coefficients and constants. Consider showing them the method of combining like terms, even with fractional coefficients, and multiplying by the reciprocal of the coefficient of $x$ in the last step:

$$\frac{1}{2}x+\frac{3}{4}x-1=\frac{2}{3}x$$

$$\frac{1}{2}x+\frac{3}{4}x-\frac{2}{3}x=1$$

$$\frac{6}{12}x+\frac{9}{12}x-\frac{8}{12}x=1$$

$$\frac{7}{12}x=1$$

$$\frac{12}{7}\cdot\frac{7}{12}x=1\cdot\frac{12}{7}$$

$$x=\frac{12}{7}$$

## Example 2

Solve the equation.

$$\frac{1}{3}y+\frac{1}{6}=\frac{2}{5}y-\frac{7}{10}$$

### Solution

The LCM for 3, 6, 5, and 10 is 30.

$$\frac{1}{3}y+\frac{1}{6}=\frac{2}{5}y-\frac{7}{10}$$

$$30\left(\frac{1}{3}y+\frac{1}{6}\right)=30\left(\frac{2}{5}y-\frac{7}{10}\right)$$ Multiply both sides by 30.

$$30\left(\frac{1}{3}y\right)+30\left(\frac{1}{6}\right)=30\left(\frac{2}{5}y\right)-30\left(\frac{7}{10}\right)$$ Apply the distributive property.

$$10y+5=12y-21$$

$$10y+5-5=12y-21-5$$

$$10y=12y-26$$

$$10y-12y=12y-26-12y$$

$$-2y=-26$$

$$\frac{-2y}{-2}=\frac{-26}{-2}$$

$$y=13$$

Name ______________ Section ______________ Date ______________

## Exercises 7.3

Solve each of the following first-degree equations.

1. $\frac{2}{3}x - 5 = \frac{1}{3}x + 20$

2. $\frac{1}{2}x + \frac{3}{4}x = -15$

3. $\frac{3}{5}x - 4 = \frac{1}{5}x - 5$

4. $\frac{5}{8}x - \frac{1}{4} = \frac{2}{5}x + \frac{1}{3}$

5. $\frac{x}{8} + \frac{1}{6} = \frac{x}{10} + 2$

6. $\frac{y}{4} + \frac{2}{3} = \frac{y}{5} - \frac{1}{6}$

7. $\frac{2}{3}y - \frac{4}{5} = \frac{1}{15}y - \frac{4}{5}$

8. $\frac{y}{5} + \frac{3}{4} = \frac{y}{2} + \frac{3}{4}$

9. $\frac{1}{2}(x+1) = \frac{1}{3}(x-1)$

10. $\frac{3}{8}\left(y - \frac{1}{2}\right) = \frac{1}{8}\left(y + \frac{1}{2}\right)$

11. $\frac{4x}{3} + \frac{3}{4} = \frac{x}{6} + \frac{3}{2}$

12. $\frac{2x}{5} - \frac{2}{3} = \frac{x}{3} - \frac{4}{3}$

13. $\frac{3n}{14} + \frac{1}{4} = \frac{n}{7} - \frac{1}{4}$

14. $\frac{2y}{3} + \frac{y}{3} = -\frac{3}{4} + \frac{y}{2}$

**ANSWERS**

1. $x = 75$
2. $x = -12$
3. $x = -\frac{5}{2}$
4. $x = \frac{70}{27}$
5. $x = \frac{220}{3}$
6. $y = -\frac{50}{3}$
7. $y = 0$
8. $y = 0$
9. $x = -5$
10. $y = 1$
11. $x = \frac{9}{14}$
12. $x = -10$
13. $n = -7$
14. $y = -\frac{3}{2}$

15. $y = \frac{6}{5}$

16. $n = \frac{1}{6}$

17. $x = \frac{19}{27}$

18. $x = -\frac{2}{3}$

19. $y = \frac{10}{11}$

20. $y = -\frac{21}{25}$

21. $x = \frac{27}{8}$

22. $y = 2$

23. $x = -7$

24. $x = \frac{1}{4}$

25. $x = \frac{7}{60}$

**15.** $\frac{y}{3} - \frac{1}{2} = \frac{y}{6} - \frac{y}{4}$

**16.** $\frac{5n}{6} + \frac{1}{9} = \frac{3n}{2}$

**17.** $\frac{2}{5}x - \frac{1}{3} = \frac{3}{10} - \frac{1}{2}x$

**18.** $\frac{3}{4}x + \frac{1}{5}x = \frac{1}{2}x - \frac{3}{10}$

**19.** $\frac{5}{8}y - \frac{3}{4}y = \frac{1}{3}y - \frac{5}{12}$

**20.** $\frac{3}{7}y + \frac{1}{6}y + \frac{1}{2} = 0$

**21.** $\frac{x}{9} - \frac{x}{3} + \frac{3}{4} = 0$

**22.** $\frac{1}{6}y + \frac{7}{15} - \frac{2}{5}y = 0$

**23.** $\frac{x}{35} - \frac{x}{7} - \frac{4}{5} = 0$

**24.** $x + \frac{2}{3}x - 2x = \frac{x}{6} - \frac{1}{8}$

**25.** $3x + \frac{1}{2}x - \frac{2}{5}x = \frac{x}{10} + \frac{7}{20}$

# 7.4 Applications: Numbers and Consecutive Integers

## The Problem-Solving Process

Whenever you study applications, you should remember the basic four-step process developed by Pòlya that was outlined in Section 6.4.

**Pòlya's Four-Step Process for Solving Problems**

1. Understand the problem. ( Read it carefully and be sure that you understand all the terms used. )
2. Devise a plan. ( Set up an equation or a table or chart relating the information. )
3. Carry out the plan. ( Perform any indicated operations in Step 2. )
4. Look back over the results. ( Ask yourself if the answer seems reasonable and if you could solve similar problems in the future. )

**Objectives**

① Be able to read a word problem and know what is being asked for.

② Know how to set up an equation whose solution will be the solution to a word problem.

③ Develop a sense of the reasonableness of the solution to an equation as a solution to the corresponding word problem.

In this section, the word problems will be of two types:

1. Number problems involving translating phrases, such as those we discussed in Section 7.1, and
2. Problems related to the concept of consecutive integers.

## Solving Number Problems

In Section 7.1, we discussed translating English phrases into algebraic expressions. In this section, we will use those skills to read number problems and translate the sentences and phrases in the problem into a related equation. The solution of this equation will be the solution to the problem.

The process of using an equation with an unknown variable to solve a word problem is one of the most important and most useful ideas in all of algebra. Further courses in mathematics, such as statistics and calculus, use this fundamental process with more sophisticated techniques to solve progressively more complicated problems.

As you learn more and more abstract mathematical ideas, you will find that you will use these ideas and the related processes to solve a variety of everyday problems as well as problems in specialized fields of study. Generally, you may not even be aware of the fact that you are using your mathematical talents. However, these skills and ideas will be part of your thinking and problem solving techniques for the rest of your life.

Many students consider themselves poor in mathematics, because they do not quickly find solutions to word problems. They believe that the solutions and proofs given in books were developed in some easy way and "obvious manner" that they have failed to understand.

You might discuss with the class the fact that good problem solvers do not usually find solutions immediately. They may find several methods, including guessing and making lists of possible solutions to similar problems, before they develop some insight and a polished solution. That is, Step 4 in Polya's four step process is very important and should not be overlooked.

### Example 1

Seven less than four times a number is equal to twice the number increased by three. Find the number.

**Solution**

Let $n$ = the unknown number.

Translate "seven less than four times a number" to $\underline{4n - 7}$.

( Remember that "7 less than" means to subtract 7. )

Translate "twice the number increased by three" to $\underline{2n + 3}$.

Form the equation: $\underline{4n - 7 = 2n + 3}$.

Solve the equation:

$$
\begin{aligned}
4n - 7 &= 2n + 3 \\
4n - 7 + 7 &= 2n + 3 + 7 \\
4n &= 2n + 10 \\
4n - 2n &= 2n + 10 - 2n \\
2n &= 10 \\
\frac{\cancel{2}n}{\cancel{2}} &= \frac{10}{2} \\
n &= 5
\end{aligned}
$$

CHECK:

$$
\begin{aligned}
4(5) - 7 &\overset{?}{=} 2(5) + 3 \\
20 - 7 &\overset{?}{=} 10 + 3 \\
13 &= 13
\end{aligned}
$$

The number is 5.

## Completion Example 2

Fifteen plus a number is equal to $\frac{3}{2}$ times the number. What is the number?

### Solution

Let $x$ = the number.

Translate "fifteen plus a number" to ______.

Translate "$\frac{3}{2}$ times the number" to _____.

Form the equation: _____.

Solve the equation:

$$
\begin{aligned}
\_\_\_\_\_ &= \_\_\_\_\_\_ \\
15 + x - x &= \frac{3}{2}x - x \\
15 &= \frac{1}{2}x \\
\_\_\_\_\_ \cdot 15 &= \_\_\_\_\_ \cdot \frac{1}{2}x \\
\_\_\_\_\_ &= x
\end{aligned}
$$

$$\frac{3}{2}x - x = \left(\frac{3}{2} - 1\right)x = \frac{1}{2}x$$

CHECK:

"Fifteen plus the number" translates to $15 + 30 = 45$, and

"$\frac{3}{2}$ times the number" translates to $\frac{3}{2}(30) = 45$.

So ______ is the correct number.

### Example 3 ( Perimeter )

The perimeter of a triangle is 35 feet. If one side is 3 feet more than another side and the third side is 5 feet more than the sum of the other two, what are the lengths of the sides of the triangle?

**Solution**

Let

$x$ = the length of the first side.

$x + 3$ = the length of the second side.

$x + (x + 3) + 5$ = the length of the third side.

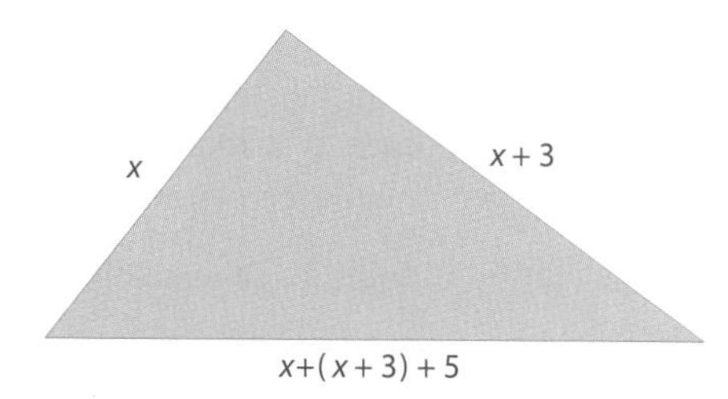


Since the perimeter is 35 feet, the sum of the three sides must be 35.

First side + Second side + Third side = Perimeter

$$x + (x+3) + x + (x+3) + 5 = 35$$

$$4x + 11 = 35$$

$$4x + 11 - 11 = 35 - 11$$

$$4x = 24$$

$$\frac{\not{4}x}{\not{4}} = \frac{24}{4}$$

$$x = 6$$

Thus, the sides are $x$ = length of first side = 6 ft

$x + 3$ = length of second side = 9 ft

$x + (x + 3) + 5$ = length of third side = 20 ft

CHECK: The perimeter is 6ft + 9ft + 20ft = 35 ft.

*Now Work Exercises 1 - 3 in the Margin.*

## Solving Consecutive Integer Problems

Remember that the set of integers consists of the whole numbers and their opposites:

$$\{\ldots, -4, -3, -2, -1, 0, 1, 2, 3, \ldots\}$$

In this section, we will be discussing only integers and these integers will have certain properties. Therefore, if you get a result that has a fraction or decimal number ( not an integer ), you will know that an error has been made and you should correct some part of your work.

The following terms and the ways of representing the integers must be understood before attempting the problems.

1. Four more than twice a number is equal to the difference between the number and 16. Find the number.

   $2n + 4 = n - 16$; $n = -20$

2. The sum of a number and 68 is equal to 35 times the number. What is the number?

   $n + 68 = 35n$; $n = 2$

3. The perimeter of a triangle is 20 inches. If two sides are of the same length and the third side is 8 inches long, what is the length of each of the other two sides?

   $2n + 8 = 20$; $n = 6$ in.

To help you determine the reading level of the class, you might have the students read the problems in the examples and exercises out loud in class. You might find that the reading habits of certain students involve relatively simple difficulties such as stopping in midsentence or at the end of a line in the text rather than at the end of a complete sentence. That is, they understand the individual words, but have failed to comprehend the problem in its entirety.

**Definitions:**

**Consecutive Integers:**

Integers are consecutive if each is 1 more than the previous integer. Three consecutive integers can be represented as

$n,\quad n+1,\quad$ and $\quad n+2$

**Consecutive Odd Integers:**

Odd integers are consecutive if each is 2 more than the previous odd integer. Three consecutive odd integers can be represented as

$n,\quad n+2,\quad$ and $\quad n+4\qquad$ Where $n$ is an **odd** integer

**Consecutive Even Integers:**

Even integers are consecutive if each is 2 more than the previous even integer. Three consecutive even integers can be represented as

$n,\quad n+2,\quad$ and $\quad n+4\qquad$ Where $n$ is an **even** integer

**Note that consecutive even and consecutive odd integers are represented in the same way:**

**$n,\quad n+2,\quad$ and $\quad n+4$**

**The value of the first integer, $n$, determines whether the remaining integers are odd or even.** For example,

| <u>*n* is odd</u> | | <u>*n* is even</u> |
|---|---|---|
| If $n = 11$ | *or* | If $n = 36$ |
| then $n+2 = 13$ | | then $n+2 = 38$ |
| and $n+4 = 15$ | | and $n+2 = 40$ |

## Example 4

Find three consecutive integers whose sum is –27.

### Solution

Let

$n$ = the first integer

$n+1$ = the second consecutive integer

$n+2$ = the third consecutive integer

The equation to be solved is

$$n+(n+1)+(n+2)=-27$$

$$n+n+1+n+2=-27$$

$$3n+3=-27$$

$$3n+3-3=-27-3$$

$$3n=-30$$

$$\frac{\cancel{3}n}{\cancel{3}}=\frac{-30}{3}$$

$n=-10$ the first integer

$n+1=-9$ the second integer

$n+2=-8$ the third integer

CHECK:

$$(-10)+(-9)+(-8)=-27$$

## Example 5

Find three consecutive odd integers such that the sum of the first and third is 23 more than the second.

### Solution

Let

$n$ = the first odd integer,

$n+2$ = the second consecutive odd integer, and

$n+4$ = the third consecutive odd integer.

The equation to be solved is

$$n+(n+4) = (n+2)+23$$

$$n+n+4 = n+2+23$$

$$2n+4 = n+25$$

$$2n+4-4 = n+25-4$$

$$2n = n+21$$

$$2n-n = n+21-n$$

$n = 21$ the first odd integer

$n+2 = 23$ the second consecutive odd integer

$n+4 = 25$ the third consecutive odd integer

CHECK:

$$\underbrace{\text{first}}_{\downarrow\ 21} + \underbrace{\text{third}}_{\downarrow\ 25} \stackrel{?}{=} \underbrace{\text{second}}_{\downarrow\ 23} + 23$$

$$21 + 25 \stackrel{?}{=} 23 + 23$$

$$46 = 46$$

## *Completion Example 6*

Find the three consecutive even integers such that three times the first is 10 more than the sum of the second and third.

### Solution

Let

$n$ = the first even integer,

_____ = the second consecutive even integer, and

_____ = the third consecutive even integer.

Solve the equation:

$3n$ = _____ + _____ + _____

$3n$ = _____ + _____

$3n$ − _______ = _____ − _____

$n$ = _____

The three consecutive even integers are

_____, _____, and _____.

## Completion Example Answers

**2.** Let $x$ = the number.

Translate "fifteen plus a number"to $\mathbf{15 + x}$.

Translate "$\frac{3}{2}$ times the number"to $\mathbf{\frac{3}{2}x}$

Form the equation: $\mathbf{15 + x = \frac{3}{2}x}$

Solve the equation:

$$\mathbf{15 + x = \frac{3}{2}x}$$
$$15 + x - x = \frac{3}{2}x - x$$
$$15 = \frac{1}{2}x$$
$$\mathbf{2} \cdot 15 = \mathbf{2} \cdot \frac{1}{2}x$$
$$\mathbf{30} = x$$

CHECK:

"Fifteen plus the number" translates to 15 + 30 = 45, and

"$\frac{3}{2}$ times the number" translates to $\frac{3}{2}(30) = 45$.

So, **30** is the correct number.

**6.** Let

$n$ = the first even integer,

$\mathbf{n + 2}$ = the second consecutive even integer, and

$\mathbf{n + 4}$ = the third consecutive even integer.

Solve the equation:

$$3n = \mathbf{(n + 2) + (n + 4) + 10}$$
$$3n = \mathbf{2n + 16}$$
$$3n - \mathbf{2n} = \mathbf{2n + 16 - 2n}$$
$$n = \mathbf{16}$$

The three consecutive even integers are **16**, **18**, and **20**.

Name ______________________ Section ____________ Date ____________

ANSWERS

For Exercises 7.4, respond in exercise.

## Exercises 7.4

In Exercises 1-4, translate each of the English phrases indicated in the problem into an algebraic expression, and set up the related equation. Do not solve the equation.

**1.** The difference between a number and ten (**a.**) is equal to twice the number (**b.**).

What is the number?

Let $x$ = the unknown number.

Translation of **a.** : $x - 10$

Translation of **b.** : $2x$

Equation : $x - 10 = 2x$

**2.** The sum of a number and six (**a.**) is equal to six minus the number (**b.**).

What is the number?

Let $x$ = the unknown number.

Translation of **a.** : $x + 6$

Translation of **b.** : $6 - x$

Equation : $x + 6 = 6 - x$

**3.** Three times the sum of a number and two (**a.**) is the same as two times the number (**b.**).

Find the number.

Let $y$ = the unknown number.

Translation of **a.** : $3(y + 2)$

Translation of **b.** : $2y$

Equation : $3(y + 2) = 2y$

4. If the product of a number and one-half is decreased by three-fourths, the result is five-eighths. What is the number?

(a. = "the product of a number and one-half is decreased by three-fourths"; b. = "five-eighths")

Let $y$ = the unknown number.

Translation of **a.** : $\frac{1}{2}y - \frac{3}{4}$

Translation of **b.** : $\frac{5}{8}$

Equation : $\frac{1}{2}y - \frac{3}{4} = \frac{5}{8}$

In Exercises 5-28

**a.** choose a variable and tell what the variable represents,

**b.** translate the related English phrases,

**c.** set up the equation, and

**d.** solve the equation.

**5.** The sum of a number and 32 is 90. What is the number?

a. Let $x$ = the number.
b. $x + 32$; 90
c. $x + 32 = 90$
d. $x = 58$

**6.** The difference between a number and 16 is –50. What is the number?

a. Let $x$ = the number.
b. $x - 16$; $-50$
c. $x - 16 = -50$
d. $x = -34$

**7.** What number should be added to –10 to get a sum of –25?

a. Let x = the number.
b. $-10 + x$; $-25$
c. $-10 + x = -25$
d. $x = -15$

**8.** If the quotient of a number and 4 is decreased by 20, the result is 5. Find the number.

a. Let $n$ = the number.
b. $\frac{n}{4} - 20$; 5
c. $\frac{n}{4} - 20 = 5$
d. $n = 100$

**9.** What number should be subtracted from –10 to get a difference of –25?

a. Let n = the number.
b. $-10 - n$; $-25$
c. $-10 - n = -25$
d. $15 = n$

Name ______________________ Section ____________ Date __________

**ANSWERS**

For Exercises 7.4, respond below exercise.

**10.** If the product of a number and 8 is increased by 24, the result is twice the number. What is the number?

a. Let $y$ = the number.
b. $8y + 24;\ 2y$
c. $8y + 24 = 2y$
d. $y = -4$

**11.** Three times the sum of a number and 4 is –60. What is the number?

a. Let $y$ = the number.
b. $3(y + 4);\ -60$
c. $3(y + 4) = -60$
d. $y = -24$

**12.** Twice a number plus 5 is equal to 20 more than the number. What is the number?

a. Let $x$ = the number.
b. $2x + 5;\ x + 20$
c. $2x + 5 = x + 20$
d. $x = 15$

**13.** If 7 is subtracted from a number and the result is 8 times the number, what is the number?

a. Let $a$ = the number.
b. $a - 7;\ 8a$
c. $a - 7 = 8a$
d. $-1 = a$

**14.** Twenty plus a number is equal to twice the number plus three times the same number. What is the number?

a. Let $x$ = the number.
b. $20 + x;\ 2x + 3x$
c. $20 + x = 2x + 3x$
d. $5 = x$

**15.** Twice the difference between a number and 5 is equal to 6 times the number plus 14. What is the number?

a. Let $y$ = the number.
b. $2(y - 5);\ 6y + 14$
c. $2(y - 5) = 6y + 14$
d. $y = -6$

**16.** The perimeter of ( distance around ) a triangle is 18 centimeters. If two sides are equal in length and the third side is 4 centimeters long, what is the length of each of the other two sides?

a. Let $x$ = length of one of the equal sides.
b. $x + x + 4;\ 18$
c. $x + x + 4 = 18$
d. $x = 7$cm

**17.** A length of wire is bent to form a triangle with two sides equal. If the wire is 30 inches long and the two equal sides are each 8 inches long, how long is the third side?

a. Let $x$ = length of third side.
b. $8 + 8 + x;\ 30$
c. $8 + 8 + x = 30$
d. $x = 14$ in.

**18.** Two sides of a triangle have the same length and the third side is 3 cm less than the sum of the other two sides. If the perimeter of the triangle is 45 cm, what are the lengths of the three sides?

a. Let $x$ = length of one of the equal sides.
b. $x + x + 2x - 3$; 45
c. $x + x + 2x - 3 = 45$
d. 12 cm; 12 cm; 21cm

**19.** The perimeter of a rectangular shaped pool is 168 meters. If the width is 16 meters less than the length, how long are the width and length of the pool?

a. Let $x$ = length of the pool.
b. $2x + 2(x - 16)$; 168
c. $2x + 2(x - 16) = 168$
d. $x = 50$ m; $x - 16 = 34$ m

**20.** A classic car is now selling for $500 more than three times its original price. If the selling price is now $11,000, what was the car's original price?

a. Let $p$ = the original price.
b. $3p + 500$; 11,000
c. $3p + 500 = 11,000$
d. $p$ = $3500

**21.** A real estate agent says that the current value of a home is $80,000 more than twice its value when it was new. If the current value is $260,000, what was the value of the home when it was new? The home is 20 years old.

a. Let $n$ = value of home when new.
b. $2n + 80,000$; 260,000
c. $2n + 80,000 = 260,000$
d. $n$ = $90,000

**22.** The perimeter of ( distance around ) a rectangular parking lot is 400 yards. If the width is 75 yards, what is the length of the parking lot? The concrete in the parking lot is 4 inches thick.

a. Let $x$ = the length of the parking lot.
b. $2x + 2 \cdot 75$; 400
c. $2x + 2 \cdot 75 = 400$
d. $x = 125$ yd

**23.** The sum of two consecutive integers is 67. Find the two integers.

a. Let $n$ = the first integer.
$n + 1$ = the second integer.
b. $n + n + 1$; 67
c. $n + n + 1 = 67$
d. $n = 33$; $n + 1 = 34$

**24.** The sum of three consecutive integers is 351. Find the three integers.

a. Let $n$ = the first integer.
$n + 1$ = the second integer.
$n + 2$ = the third integer.
b. $n + n + 1 + n + 2$; 351
c. $n + n + 1 + n + 2 = 351$
d. $n = 116$
$n + 1 = 117$
$n + 2 = 118$

Name ______________ Section ______________ Date ______________

**ANSWERS**

For Exercises 7.4, respond below exercise.

**25.** If the sum of two consecutive integers is multiplied by 3, the result is –15. What are the two integers?

a. Let $n$ = the first integer
$n + 1$ = the second integer
b. $3(n + n + 1)$; $-15$
c. $3(n + n + 1) = -15$
d. $n = -3$; $n + 1 = -2$

**26.** Find three consecutive odd integers such that 4 times the first is 44 more than the sum of the second and third.

a. Let $n$ = the first odd integer
$n + 2$ = the second odd integer
$n + 4$ = the third odd integer
b. $4n$; $n + 2 + n + 4 + 44$
c. $4n = n + 2 + n + 4 + 44$
d. $n = 25$
$n + 2 = 27$
$n + 4 = 29$

**27.** Find three consecutive odd integers such that the sum of the first and third is equal to 27 less than 3 times the second.

a. Let $n$ = the first odd integer
$n + 2$ = the second odd integer
$n + 4$ = the third odd integer
b. $n + n + 4$; $3(n + 2) - 27$
c. $n + n + 4 = 3(n + 2) - 27$
d. $n = 25$
$n + 2 = 27$
$n + 4 = 29$

**28.** Find three consecutive even integers such that if the first is subtracted from the sum of the second and third, the result is 66.

a. Let $n$ = the first even integer.
$n + 2$ = the second even integer
$n + 4$ = the third even integer
b. $n + 2 + n + 4 - n$; $66$
c. $n + 2 + n + 4 - n = 66$
d. $n = 60$
$n + 2 = 62$
$n + 4 = 64$

**29.** Find three consecutive even integers such that their sum is 168 more than the second.

a. Let $n$ = the first even integer
$n + 2$ = the second even integer
$n + 4$ = the third even integer
b. $n + n + 2 + n + 4$; $n + 2 + 168$
c. $n + n + 2 + n + 4 = n + 2 + 168$
d. $n = 82$
$n + 2 = 84$
$n + 4 = 86$

**30.** Find four consecutive integers whose sum is 90.

a. Let $n$ = the first integer
$n + 1$ = the second integer
$n + 2$ = the third integer
$n + 3$ = the fourth integer
b. $n + n + 1 + n + 2 + n + 3$; $90$
c. $n + n + 1 + n + 2 + n + 3 = 90$
d. $n = 21$
$n + 1 = 22$
$n + 2 = 23$
$n + 3 = 24$

Make up your own word problem that might use the given equation in its solution. Be creative! Then solve the equation and check to see that the answer is reasonable.

**31.** $n + (n + 1) = 31$

**32.** $3(n + 1) = n + 41$

**33.** $2x + 5 = 7$

**34.** $3x - 4 = x + 8$

**35.** $2(x + 6) = 10$

**36.** $\frac{x}{2} + \frac{1}{3} = \frac{3}{4}$

**37.** $n + n + 1 + n + 2 = 33$

**38.** $n + 2(n + 2) = n + 4$

**39.** $x - \frac{2}{3}x = \frac{5}{3}$

**40.** $2(y + 2) - (y + 1) = 2y$

For Exercises 31-40, answers will vary. Some students will be quite imaginative in creating word problems.

## Writing and Thinking About Mathematics

**41.** Discuss, briefly, how you would apply Pòlya's four-step problem-solving process to a problem you have faced today. ( For example, what route to take to school, what to eat for breakfast, where to go to study mathematics, etc. )

**a.** What was the problem?

**b.** What was your plan?

**c.** How did you carry out the plan?

**d.** Did your solution make sense? Could you have solved the problem in a better way?

Discussions will vary. They should follow the pattern of Polya's four-step process in problem solving.

# 7.5 Working with Formulas

## What is a Formula?

A **formula** is a general statement ( usually an equation ) that relates two or more variables. We have already discussed a variety of formulas throughout this text, particularly in terms of geometric figures for perimeter, area, and volume. We have also evaluated formulas and polynomials for given values of the variables. In this section, we will work with formulas in two ways:

1. Evaluate formulas for given values of the variables, and
2. Solve formulas for certain variables in the formulas.

For your interest and understanding, several formulas and their meanings are stated below.

**Objectives**

① Know what a formula is.

② Be able to evaluate a formula with given information.

③ Learn how to use the techniques for solving equations to solve a formula for a particular variable.

1. $I = Prt$ — The simple interest ( $I$ ) earned investing money is equal to the product of the principal ( $P$ ), the rate of interest ( $r$ ), and the time ( $t$ ) in years. ( See Section 6.6 )

2. $C = \pi d$ — The circumference of a circle is equal to the product of pi ( $\pi$ ) and the diameter ( $d$ ). ( See Section 5.3. )

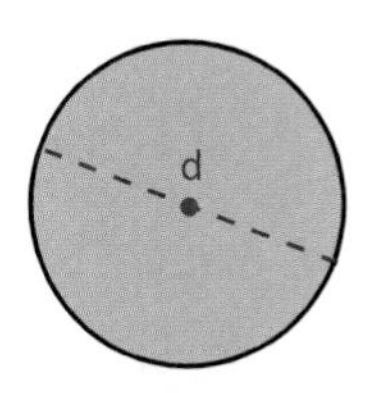


3. $P = 2l + 2w$ — The perimeter ( $P$ ) of a rectangle is equal to twice the length ( $l$ ) plus twice the width ( $w$ ). ( See Chapter 10. )

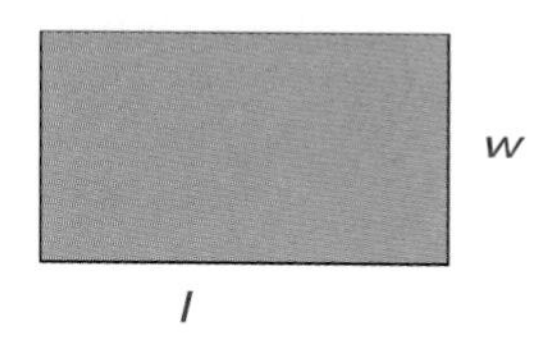


4. $\alpha + \beta + \gamma = 180°$ — The sum of the measures of angles ($\alpha$, $\beta$, and $\gamma$) of a triangle is 180°. ($\alpha$, $\beta$, and $\gamma$ are the Greek letters alpha, beta, and gamma, respectively.) (See Chapter 10.)

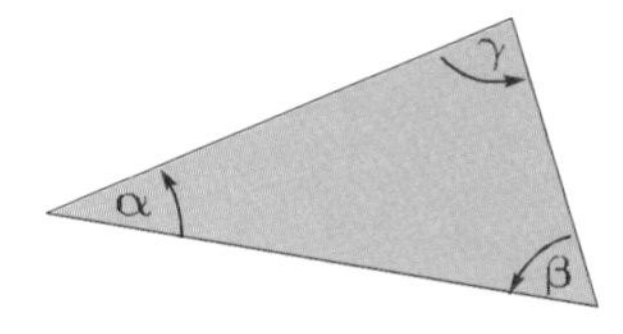


5. $C = \frac{5}{9}(F - 32)$ — Temperature in degrees Celsius ( $C$ ) is equal to $\frac{5}{9}$ of the difference between the Fahrenheit temperature ( $F$ ) and 32. ( See Chapter 10. )

6. $d = rt$ — The distance ( $d$ ) traveled is equal to the product of the rate of speed ( $r$ ) and the time ( $t$ ).

## Evaluating Formulas

If we know values for all but one variable in a formula, then we can substitute these values and solve the equation for the unknown variable. If the equation is first-degree, then we simply use the methods for solving equations discussed in Sections 7.2 and 7.3.

### Example 1

Given the formula $C = \frac{5}{9}(F - 32)$, find the Fahrenheit temperature that corresponds to a Celsius temperature of $100°$.

**Solution**

Substitute 100 for $C$ in the formula and solve the resulting equation for $F$.

$$100 = \frac{5}{9}(F - 32)$$

$$9 \cdot 100 = 9 \cdot \frac{5}{9}(F - 32)$$ Multiply both sides by 9.

$$900 = 5(F - 32)$$ Simplify.

$$900 = 5F - 160$$ Use the distributive property.

$$900 + 160 = 5F - 160 + 160$$ Add 160 to both sides.

$$\frac{1060}{5} = \frac{\cancel{5}F}{\cancel{5}}$$ Divide both sides by 5.

$$212 = F$$

Thus, $100°$C is equal to **212°F**, the boiling point of water at sea level.

### *Completion Example 2*

Two angles of a triangle are measured at $62°$ and $83°$. What is the measure of the third angle?

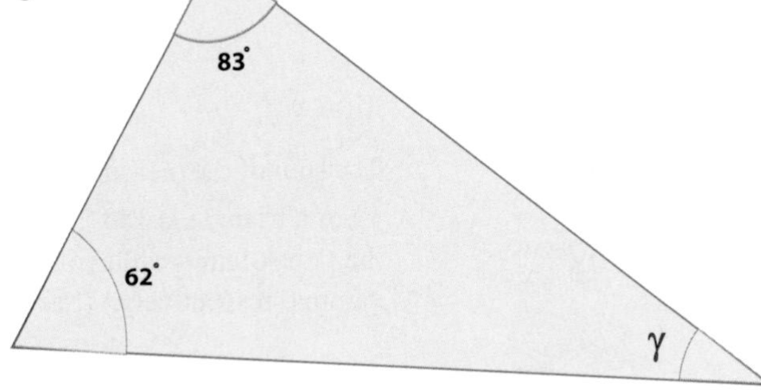


**Solution**

Use the formula $\alpha + \beta + \gamma = 180°$. Substitute $\alpha = 62$ and $\beta = 83$, and solve for $\gamma$.

$$\_\_\_\_ + \_\_\_\_ + \gamma = 180$$

$$\_\_\_\_ + \gamma = 180$$

$$\gamma = 180 - \_\_\_\_$$

$$\gamma = \_\_\_\_$$

The measure of the third angle is _____.

*Now Work Exercises 1 and 2 in the Margin.*

1. Use the formula $I = Prt$ to find the interest earned on a principal of \$5000 at 6% in 9 months.

   \$225

2. Use the formula $d = rt$ to find the time it would take to drive 300 miles at an average rate of 45 mph.

   6 hr 40 mins

## Solving for Any Variable in a Formula

Sometimes a particular formula might be expressed in one form when another form would be more convenient. To change the form of a formula, we need to be able to **solve the formula for any variable in that formula.** In such cases, we solve the equation **treating the variables other than the chosen one as if they were constants.** The following examples illustrate the technique.

The ability to solve equations is one of the most important skills in algebra. Encourage the students to work hard in this section. If they can solve the relatively abstract equations ( formulas ) in this section, they are well on their way to success in algebra.

### Example 3

Given the formula $d = rt$, solve for $t$ in terms of $d$ and $r$.

#### Solution

$d = rt$ Treat $d$ and $r$ as if they were constants.

$\frac{d}{r} = \frac{\not{r}t}{\not{r}}$ Divide both sides by $r$.

$\frac{d}{r} = t$ Simplify.

The formula now indicates that time is equal to distance divided by rate. ( Note that the relationship among distance, rate, and time is not changed. Only the form of the formula is changed. )

### Completion Example 4

Given the equation $3x + 4y = 12$,

**a.** solve for $x$ in terms of $y$, and

**b.** solve for $y$ in terms of $x$. ( We will see in Chapter 8 that such an equation represents the graph of a straight line. )

#### Solution

**a.** Solving for $x$:

$3x + 4y = 12$

$3x + 4y - 4y = 12 - 4y$ Subtract $4y$ from both sides.

$3x = 12 - 4y$ Simplify.

$\frac{3x}{3} = \frac{12 - 4y}{3}$ Divide both sides by 3.

$x = \frac{12 - 4y}{3}$

**b.** Solving for $y$:

$3x + 4y = 12$

$3x + 4y - \_\_\_\_ = 12 - \_\_\_\_$

$\_\_\_\_ = 12 - \_\_\_\_$

$\frac{\_\_\_\_}{4} = \frac{12 - \_\_\_\_}{4}$

$y = \_\_\_\_$

*Now Work Exercises 3 and 4 in the Margin.*

**3.** Solve the equation $2x - y = 7$ for $x$.

$x = \frac{7 + y}{2}$

**4.** Solve the formula $C = \pi d$ for $d$.

$d = \frac{C}{\pi}$

## Completion Example Answers

**2.** Use the formula $\alpha + \beta + \gamma = 180°$. Substitute $\alpha = 62$ and $\beta = 83$, and solve for $\gamma$.

$$\begin{aligned} \mathbf{62} + \mathbf{83} + \gamma &= 80 \\ \mathbf{145} + \gamma &= 180 \\ \gamma &= 180 - \mathbf{145} \\ \gamma &= \mathbf{35} \end{aligned}$$

The measure of the third angle is **35°**.

**4. b.** Solving for $y$:

$$\begin{aligned} 3x + 4y &= 12 \\ 3x + 4y - \mathbf{3x} &= 12 - \mathbf{3x} \\ \mathbf{4y} &= 12 - \mathbf{3x} \\ \frac{\cancel{4}\mathbf{y}}{\cancel{4}} &= \frac{12 - \mathbf{3x}}{4} \\ y &= \frac{\mathbf{12 - 3x}}{\mathbf{4}} \end{aligned}$$

Name ____________ Section ____________ Date ____________

## Exercises 7.5

Substitute the given values into each formula, and find the value of the remaining variable.

**1.** $d = rt$; $r = 60$ mph, $t = 1.5$ hr

**2.** $d = rt$; $r = 50$ mph, $d = 450$ mi

**3.** $C = \pi d$; $d = 3$ ft, $\pi = 3.14$

**4.** $A = \pi r^2$; $r = 6$ cm, $\pi = 3.14$

**5.** $I = Prt$; $I = \$150$, $r = 8\%$

$t = 9$ months $= \dfrac{3}{4}$ yr

**6.** $P = 2l + 2w$; $l = 46$ in., $w = 6$ in.

**7.** $A = P + Prt$; $P = \$1000$, $r = 6\%$, $t = 30$ days $= \dfrac{1}{12}$ yr

**8.** $A = P + Prt$; $P = \$5000$, $r = 5\%$, $t = 6$ months $= \dfrac{1}{2}$ yr

**9.** $\alpha + \beta + \gamma = 180°$; $\beta = 15°$, $\gamma = 90°$

**10.** $C = 2\pi r$; $C = 7.85$ m, $\pi = 3.14$

**ANSWERS**

1. $d = 90$ mi
2. $t = 9$ hr
3. $C = 9.42$ ft
4. $A = 113.04\ \text{cm}^2$
5. $P = \$2500$
6. $P = 104$ in.
7. $A = \$1005$
8. $A = \$5125$
9. $\alpha = 75°$
10. $r = 1.25$ m

11. [ Respond below and in exercise.]

12. [ Respond below and in exercise.]

13. [ Respond below and in exercise.]

14. [ Respond below and in exercise.]

15. [ Respond below and in exercise.]

16. [ Respond below and in exercise.]

17. [ Respond below and in exercise.]

18. [ Respond below and in exercise.]

For each of the following equations,

**a.** solve for $y$ and then

**b.** fill in the corresponding values for $y$ in the table with given $x$ values.

**11.** $3x + y = 14$

a. $y = -3x + 14$

| $x$ | $y$ |
|---|---|
| 0 | 14 |
| −1 | 17 |
| 5 | −1 |

**12.** $5x + y = 10$

a. $y = -5x + 10$

| $x$ | $y$ |
|---|---|
| 0 | 10 |
| 2 | 0 |
| −2 | 20 |

**13.** $4x - y = -6$

a. $y = 4x + 6$

| $x$ | $y$ |
|---|---|
| 2 | 14 |
| 1.5 | 12 |
| −2 | −2 |

**14.** $x - y = -7$

a. $y = x + 7$

| $x$ | $y$ |
|---|---|
| 7 | 14 |
| 4 | 11 |
| −3 | 4 |

**15.** $2x + 2y = 5$

a. $y = \frac{5 - 2x}{2}$

| $x$ | $y$ |
|---|---|
| 0 | 2.5 |
| 2.5 | 0 |
| 3 | −0.5 |

**16.** $3x + 4y = -8$

a. $y = \frac{-3x - 8}{4}$

| $x$ | $y$ |
|---|---|
| 0 | −2 |
| 4 | −5 |
| −2 | −0.5 |

For each of the following equations,

**a.** solve for $x$ and then

**b.** fill in the corresponding values for $x$ in the table with given $y$ values.

**17.** $x - 3y = 15$

a. $x = 3y + 15$

| $x$ | $y$ |
|---|---|
| 30 | 5 |
| 15 | 0 |
| 0 | −5 |

**18.** $x + 4y = 16$

a. $x = -4y + 16$

| $x$ | $y$ |
|---|---|
| 0 | 4 |
| 32 | −4 |
| 16 | 0 |

Name ______________________ Section ______________ Date ____________

**ANSWERS**

**19.** $-x+y=-10$

a. $x=y+10$

| $x$ | $y$ |
|---|---|
| 10 | 0 |
| 0 | −10 |
| 13 | 3 |

**20.** $-x-5y=4$

a. $x=-5y-4$

| $x$ | $y$ |
|---|---|
| −4 | 0 |
| 1 | −1 |
| 0 | −0.8 |

**21.** $2x+2y=5$

a. $x=\dfrac{-2y+5}{2}$

| $x$ | $y$ |
|---|---|
| 0 | 2.5 |
| 2.5 | 0 |
| 0.5 | 2 |

**22.** $3x+4y=-8$

a. $x=\dfrac{-4y-8}{3}$

| $x$ | $y$ |
|---|---|
| 0 | −2 |
| −8 | 4 |
| 8 | −8 |

Solve each of the following formulas for the indicated variable in terms of the other variables. ( See if you can identify any of the formulas from previous experience or previous classes. )

**23.** $f=ma$; solve for $m$.

**24.** $I=Prt$; solve for $P$.

**25.** $C=2\pi r$; solve for $r$.

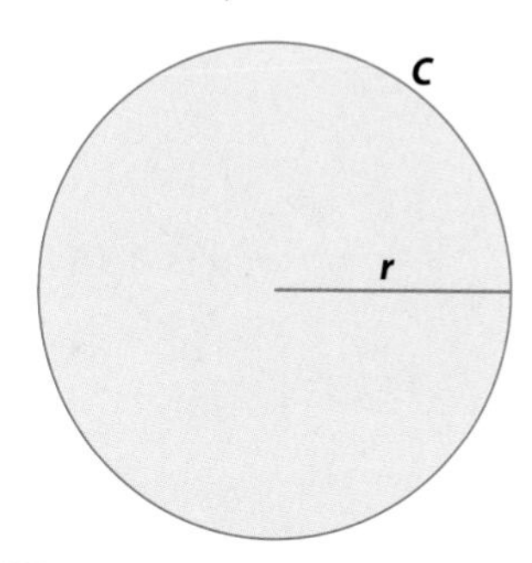


**26.** $V=lwh$; solve for $h$.

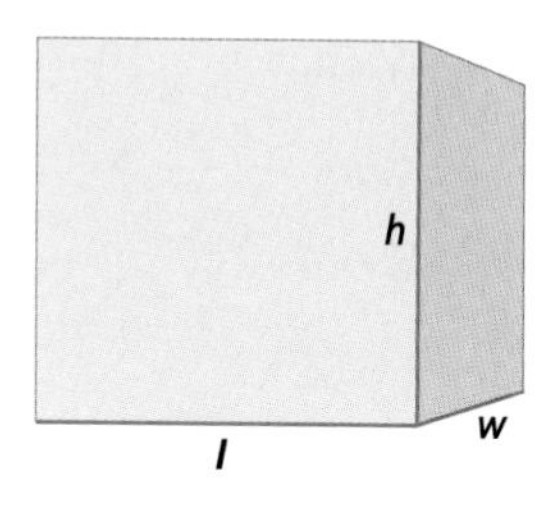


**27.** $P=2l+2w$; solve for $w$.

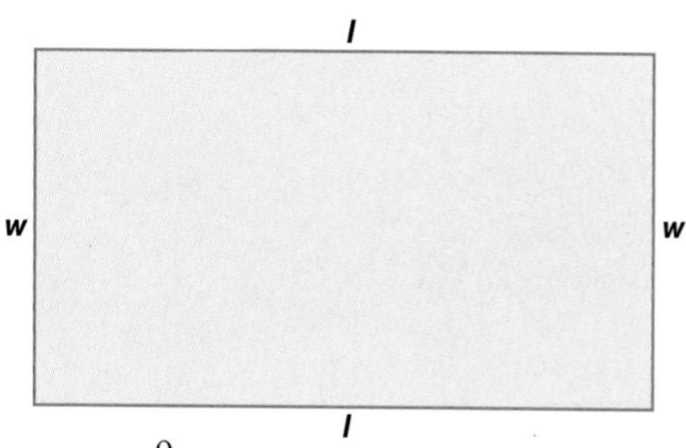


**28.** $A=\dfrac{1}{2}bh$; solve for $b$.

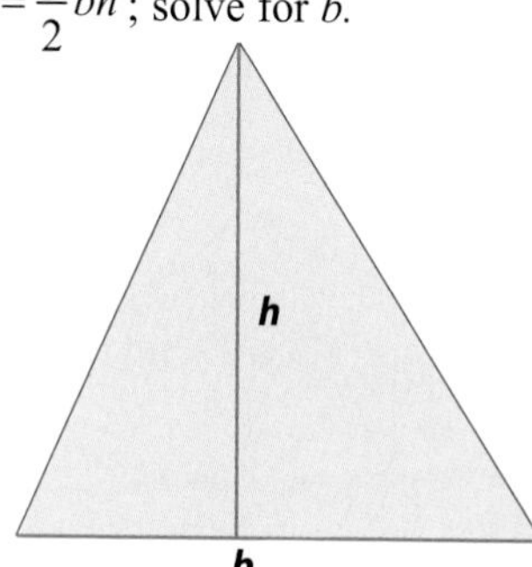


**29.** $F=\dfrac{9}{5}C+32$; solve for $C$.

**30.** $y=mx+b$; solve for $m$.

**19.** [ Respond below and in exercise.]

**20.** [ Respond below and in exercise.]

**21.** [ Respond below and in exercise.]

**22.** [ Respond below and in exercise.]

**23.** $m=\dfrac{f}{a}$

**24.** $P=\dfrac{I}{rt}$

**25.** $r=\dfrac{C}{2\pi}$

**26.** $h=\dfrac{V}{lw}$

**27.** $w=\dfrac{P-2l}{2}$

**28.** $b=\dfrac{2A}{h}$

**29.** $C=\dfrac{5}{9}(F-32)$

**30.** $m=\dfrac{y-b}{x}$

31. a. $z = 0$

b. values less than 60

c. [Respond below exercise.]

**31.** The formula $z = \dfrac{x - \overline{x}}{s}$ is used extensively in statistics. In this formula, $x$ represents one of a set of numbers, $\overline{x}$ represents the average ( or mean ) of those numbers, and $s$ represents a value called the **standard deviation** of the numbers. ( The standard deviation is a positive number and is a measure of how "spread out" the numbers are. ) The values for $z$ are called ***z*-scores**, and they measure the number of standard deviation units a number $x$ is from the mean $\overline{x}$.

**a.** If $\overline{x} = 60$, what will be the Z-score that corresponds to $x = 60$? Does this $z$-score depend on the value of $s$? Explain.

**b.** For what values of $x$ will the corresponding $z$-scores be negative?

**c.** Calculate your $z$-score on the last two exams in this class. ( Your instructor will give you the mean and standard deviation for each test. ) What do these scores tell you about your performance on the two exams?

Answers will vary. z-scores will be a measure of individual performance relative to that of the entire class.

32. a. $z = 1$

b. $z = -1$

c. $z = 1.5$

d. $z = -3$

**32.** Suppose that, for a particular set of exam scores, $\overline{x} = 75$ and $s = 6$. Find the $z$-score that corresponds to a score of

**a.** 81

**b.** 69

**c.** 84

**d.** 57

# 7.6 First-Degree Inequalities $(ax + b < c)$

## A Review of Real Numbers

**Real numbers** ( see Section 4.6 ) include

**integers** such as 5, 0, and –3,

**fractions** such as $\frac{1}{5}, \frac{3}{100}$, and $-\frac{5}{8}$,

**decimals** such as 0.007, 4.56, and 0.33333…. , and

**radicals** such as $\sqrt{2}, \sqrt{15}$, and $\sqrt{10}$.

In fact, real numbers include all those numbers which can be classified as either **rational numbers** or **irrational numbers.** All the numbers that we have discussed in this text are real numbers.

**Rational numbers** are numbers that can be represented either as **terminating decimals** or as **infinite repeating decimals.** The whole numbers, integers, and fractions that we have studied are all rational numbers. Examples of rational numbers are

$$0, \quad -6, \quad \frac{2}{3}, \quad 27.1, \quad -90, \quad \text{and} \quad 19.\overline{6}$$

**Irrational numbers** are numbers that can be represented as infinite nonrepeating decimals. Examples of irrational numbers are

$\pi = 3.1415926535 \ldots$ ( For an interesting discussion of $\pi$, see Table 4 in the back of the text. )

$\sqrt{2} = 1.41423562 \ldots$

$\sqrt{5} = 2.23606797 \ldots$

Because of the limited space on the display screens, most calculators give only 8 to 10-digit accuracy for both rational and irrational numbers. Therefore, you cannot rely on a calculator to distinguish between rational and irrational numbers. Remember, though, that all rational and irrational numbers are classified as real numbers.

The diagram in Figure 7.1 illustrates the relationships among various types of real numbers.

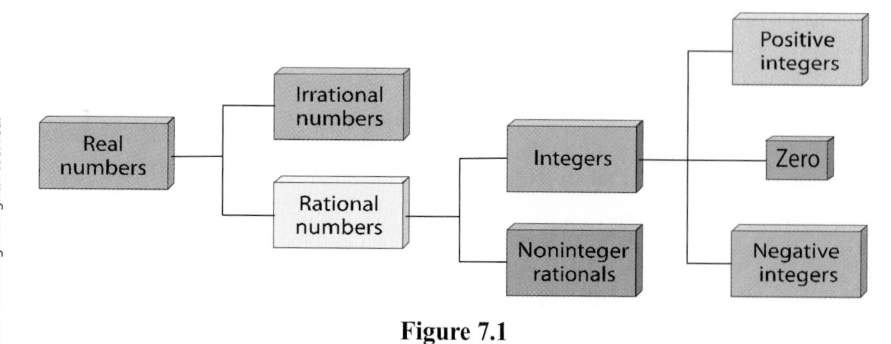


**Figure 7.1**

**Objectives**

① Know that rational numbers and irrational numbers are called real numbers.

② Be able to graph a set of real numbers on a real number line.

③ Know how to solve an inequality and graph the solution on a real number line.

As a point of intevest and discussion and to give the students a better understanding of the concept of an irrational number as an infinite nonrepeating decimal, the value of $\pi$ accurate to over 3500 decimal places is given in Table 4 at the back of the text.

Some students think that $\frac{22}{7} = \pi$. You might discuss the fact that $\frac{22}{7}$ is only an approximation of $\pi$. It is a rational number, and therefore could not be equal to $\pi$.

### Example 1

Given the set of real numbers

$$A = \left\{-5, \ -\pi, \ -1.3, \ 0, \ \sqrt{2}, \ \frac{4}{5}, \ 8\right\}$$

determine which numbers in $A$ are

**a.** Integers,

**b.** Rational numbers, and

**c.** Irrational numbers.

**Solution**

**a.** Integers: $-5, 0, 8$

**b.** Rational numbers: $-5, 0, 8, -1.3, \frac{4}{5}$

Note that each integer is also a rational number.

**c.** Irrational numbers: $-\pi, \sqrt{2}$

## Real Number Lines and Inequalities

Number lines are called **real number lines** because of the following important relationship between real numbers and points on a line.

> **There is a one-to-one correspondence between the real numbers and the points on a line.** That is, each point on a number line corresponds to one real number, and each real number corresponds to one point on a number

You can stress the importance of irrational numbers by discussing the fact that ( by mathematical methods and theory ) there are more irrational numbers than rational numbers. That is, the number of irrational numbers is a higher level of infinity than the number of counting numbers and rational numbers. ( This idea is fascinating for many students. ) Also, this is true even though both types of numbers are dense on the number line. (Between any two rational numbers, there is an infinite number of rational numbers, and between any two irrational numbers, there is an infinite number of rational numbers. )

The locations of some real numbers, including integers, are shown in Figure 7.2. Sometimes a calculator may be used to find the approximate value ( and therefore the approximate location ) of an unfamiliar real number. For example, a calculator will show that

$$\sqrt{6} = 2.449489743...$$

Therefore, $\sqrt{6}$ is located between 2.4 and 2.5 on a real number line.

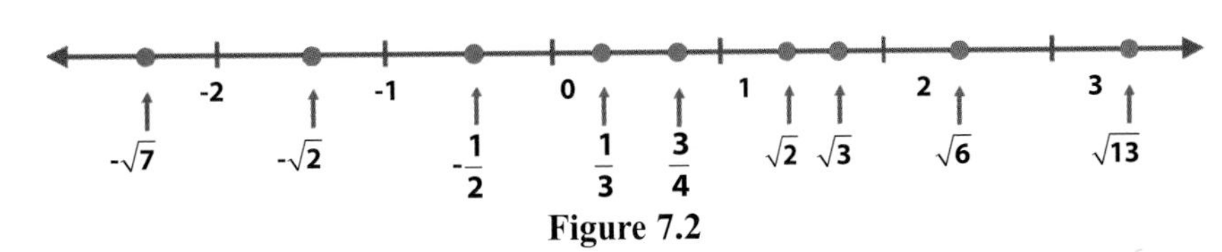


**Figure 7.2**

To compare two real numbers, we use the following symbols of equality and inequality ( reading from **left to right** ):

| | |
|---|---|
| $a = b$ | $a$ is equal to $b$ |
| $a < b$ | $a$ is less than $b$ |
| $a \leq b$ | $a$ is less than or equal to $b$ |
| $a > b$ | $a$ is greater than $b$ |
| $a \geq b$ | $a$ is greater than or equal to $b$ |

These symbols can also be read from **right to left**. For example, we can read

$a < b$ as "$b$ is greater than $a$"

and

$a > b$ as "$b$ is less than $a$"

A slash, /, through a symbol negates that symbol. Thus, for example, $\neq$ is read "is not equal to" and $\nless$ is read "is not less than."

### Example 2

Write the meaning of each of the following inequalities.

**a.** $6 < 7.5$

**b.** $-3 > -10$

**c.** $-14 \neq |-14|$

### Solution

**a.** $6 < 7.5$ "6 is less than 7.5"
or "7.5 is greater than 6"

**b.** $-3 > -10$ "–3 is greater than –10"
or "–10 is less than –3"

**c.** $-14 \neq |-14|$ "–14 is not equal to the absolute value of –14"

On a number line, smaller numbers are to the left of larger numbers. ( Or, larger numbers are to the right of smaller numbers. ) Thus, as shown in figure 7.3 we have

$-1 < 2$ $\quad$ $-2 < -1$ $\quad$ $c > a$ $\quad$ $b > 0$

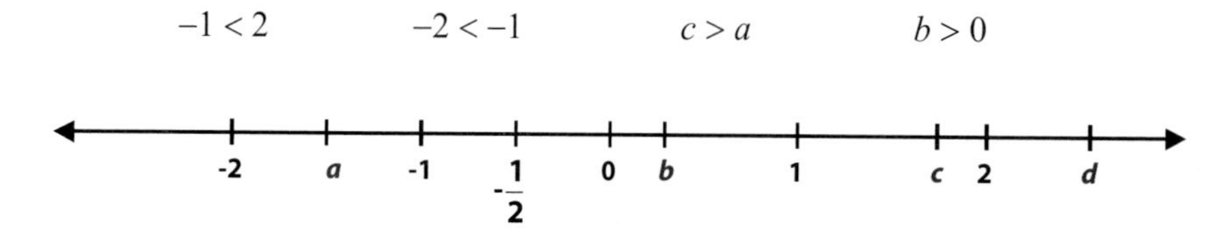


**Figure 7.3**

Suppose that $a$ and $b$ are two real numbers and that $a < b$. The set of all real numbers between $a$ and $b$ is called an **interval of real numbers.** Intervals of real numbers are commonly used in relationship to ideas such as the range of temperatures for plants and animals to live, the length of time to heal a wound, the shelf life of batteries, the stopping distances for a car traveling at a certain speed, and the distances that a camera can be expected to take clear pictures. You can probably think of many other uses of intervals of real numbers. As a more specific example, a headache may last for an interval of 1 to 3 hours after medication has been taken. The following discussion shows that this interval can be indicated in the form

$1 \leq t \leq 3$ where t represents the time of a headache.

Intervals are classified and graphed as indicated in the box that follows. You should keep in mind the following facts as you consider the boxed information:

1. $x$ is understood to represent real numbers.
2. Open dots at endpoints $a$ and $b$ indicate that these points are **not** included in the graph.
3. Solid dots at endpoints $a$ and $b$ indicate that these points **are** included in the graph.

**Intervals of Real Numbers**

| Name | Symbolic Representation | Graph |
|---|---|---|
| Open interval | $a < x < b$ | |
| Closed interval | $a \leq x \leq b$ |  |
| Half-open interval | $a \leq x < b$ |  |
| | $a < x \leq b$ |  |
| Open interval | $x > a$ |  |
| | $x < a$ |  |
| Half-open interval | $x \geq a$ | 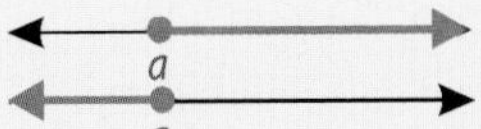 |
| | $x \leq a$ | |

## Example 3

Graph the closed interval $4 \leq x \leq 6$.

### Solution

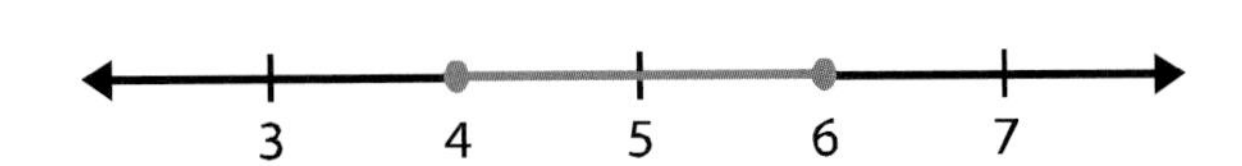


Graph the closed interval $4 \leq x \leq 6$. Note that 4 and 6 are in the interval, as are all the real numbers between 4 and 6. For example, $4\frac{1}{2}$ and 5.99 are in the interval because

$$4 \leq 4\frac{1}{2} \leq 6 \quad \text{and} \quad 4 \leq 5.99 \leq 6.$$

### Example 4

Represent the following graph using algebraic notation, and tell what kind of interval it is.

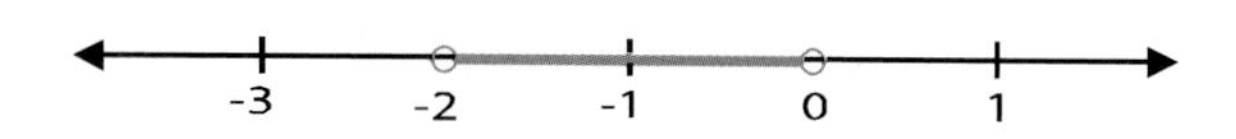


**Solution**

$-2 < y < 0$ is an open interval.

*Now Work Exercises 1- 3 in the Margin.*

## Solving Linear Inequalities

Inequalities of the forms

$ax + b < c$ and $ax + b \leq c$

$ax + b > c$ and $ax + b \geq c$

$c < ax + b < d$ and $c \leq ax + b \leq d$

are called **linear inequalities** or **first-degree inequalities**.

The solutions to linear inequalities are intervals of real numbers, and the methods for solving linear inequalities are similar to those used in solving linear ( or first-degree ) equations. The rules are the same, with one important exception.

**Multiplying or dividing both sides of an inequality by a negative number "reverses the sense" of the inequality.**

Consider the following examples.

We know that $4 < 10$:

| Add 3 | Multiply by 2 | Add – 5 |
|---|---|---|
| $4 < 10$ | $4 < 10$ | $4 < 10$ |
| $4+3 \ ? \ 10+3$ | $2 \cdot 4 \ ? \ 2 \cdot 10$ | $4+(-5) \ ? \ 10+(-5)$ |
| $7 < 13$ | $8 < 20$ | $-1 < 5$ |

In each instance, the sense of the inequality stayed the same, namely <.

Now we see that multiplying or dividing both sides by a negative number reverses the sense of the inequality from < to >:

| Multiply by -6 | Divide by -3 |
|---|---|
| $4 < 10$ | $-6 < 9$ |
| $-6\times4 \ ? \ -6\times10$ | $\frac{-6}{-3} \ ? \ \frac{9}{-3}$ |
| $-24 > -60$ | $2 > -3$ |

In each case, the sense is reversed from < to >.

**1.** Graph the closed interval $-2 \leq x \leq 1$

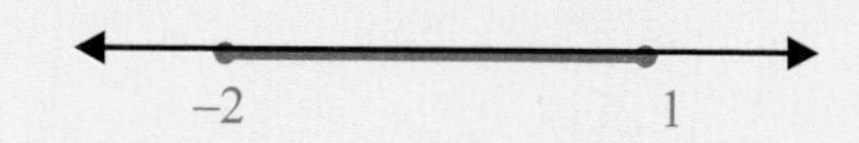


**2.** Graph the open interval $x > 3$

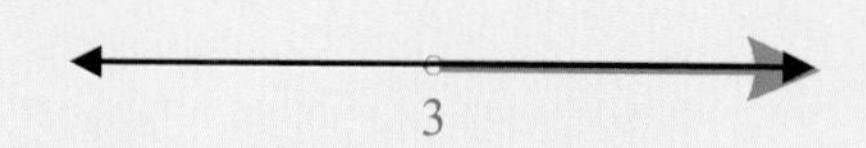


**3.** Represent the graph below using algebraic notation and tell what type of interval it is.

$-3 \leq x < 2$; half-open interval

**As with solving an equation the object of solving an inequality is to find equivalent inequalities that are simpler than the original and to isolate the variable.** The variable may be isolated on one side of an inequality symbol or between two inequality symbols. The difference between the solutions of inequalities and equations is that the solution to an inequality consists of an interval of numbers ( an infinite set of numbers ), whereas the solution to a first-degree equation is a single number.

### Rules for Solving Linear ( or First- Degree ) Inequalities

1. The same number or expression ( positive or negative ) may be added to both sides, and the sense of the inequality will remain the same.
2. Both sides may be multiplied by ( or divided by ) the same **positive** number, and the sense of the inequality will remain the same.
3. Both sides may be multiplied by ( or divided by ) the same **negative** number, but the sense of the inequality must be **reversed.**

### Note About Reading Inequalities

Inequalities can be read from left to right or from right to left. To help in graphing solutions to linear inequalities, read the variable first, regardless of which side the variable is on. Thus, $5 > x$ and $x < 5$ should both be read "$x$ is less than 5." ( See Examples 5 and 6. )

Similarly, in a compound inequality such as $1 < x < 3.5$ the variable $x$ should be read twice: "$x$ is greater than 1 and $x$ is less than 3.5." (See Example 8.)

You are stating the location of the numbers represented by the variable. Read the variable first.

## Example 5

Solve the inequality $x - 3 < 2$, and graph the solution on a number line.

### Solution

$$x - 3 < 2$$
$$x - 3 + 3 < 2 + 3 \quad \text{Add 3 to both sides.}$$
$$x < 5$$

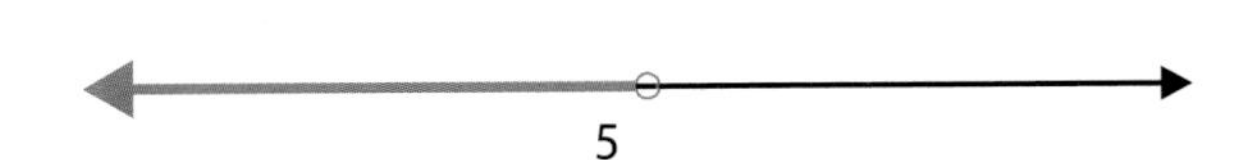

## Example 6

Solve the inequality $-2x + 7 \geq 4$, and graph the solution on a number line.

### Solution

$$-2x + 7 \geq 4$$

$$-2x + 7 - 7 \geq 4 - 7 \qquad \text{Add } -7 \text{ to both sides}$$

$$-2x \geq -3$$

$$\frac{-2x}{-2} \leq \frac{-3}{-2} \qquad \text{Divide both sides by } -2 \text{ and } \textbf{reverse the sense.}$$

$$x \leq \frac{3}{2}$$

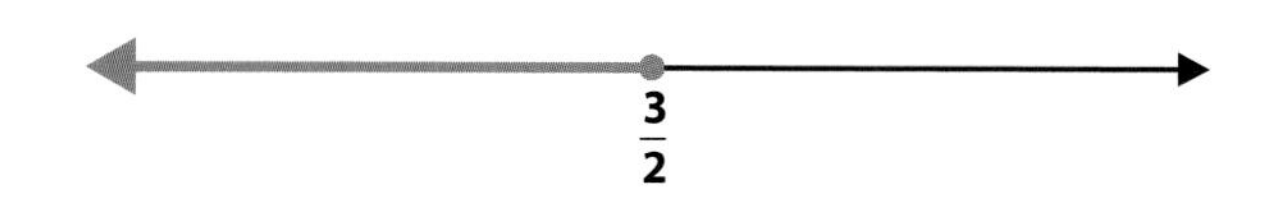


## *Completion Example 7*

Solve the inequality $7y - 8 > y + 10$, and graph the solution on a number line.

### Solution

$$7y - 8 > y + 10$$

$$7y - 8 - y > y + 10 - y \qquad \text{Add } \underline{\qquad} \text{ to both sides.}$$

$$6y - 8 > \underline{\qquad} \qquad \text{Simplify.}$$

$$6y - 8 + \underline{\qquad} > \underline{\qquad} + \underline{\qquad} \qquad \text{Add } \underline{\qquad} \text{ to both sides.}$$

$$6y > \underline{\qquad} \qquad \text{Simplify.}$$

$$\frac{6y}{\underline{\qquad}} > \frac{\underline{\qquad}}{\underline{\qquad}} + \underline{\qquad} \qquad \text{Divide both sides by } \underline{\qquad}.$$

$$y > \underline{\qquad} \qquad \text{Simplify.}$$

Graph.

## Example 8

Find the value of $x$ that satisfy **both** these inequalities:

$$5 < 2x + 3 \quad \textbf{and} \quad 2x + 3 < 10$$

Graph the solution on a number line.

### Solution

Since the variable expression $2x + 3$ is the same in both inequalities and $5 < 10$, we can write the two inequalities in one expression and solve both inequalities at the same time.

| | |
|---|---|
| $5 < 2x + 3 < 10$ | |
| $5 - 3 < 2x + 3 - 3 < 10 - 3$ | Add $-3$ to each part. |
| $2 < 2x < 7$ | Simplify. |
| $\frac{2}{2} < \frac{2x}{2} < \frac{7}{2}$ | Divide each part by 2. |
| $1 < x < 3.5$ | Thus, $x$ is greater than 1 and less than 3.5. |

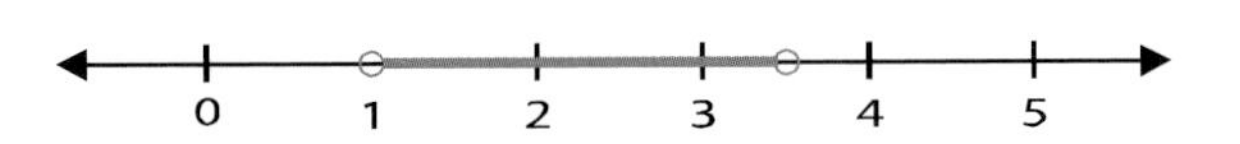


*Now Work Exercises 4- 6 in the Margin.*

Solve each inequality, and graph the solution on a number line.

**4.** $2x + 1 > 7$

$x > 3$

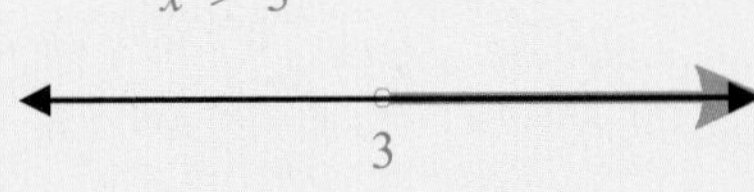


**5.** $3x + 2 \leq 5x + 1$

$x \geq \frac{1}{2}$

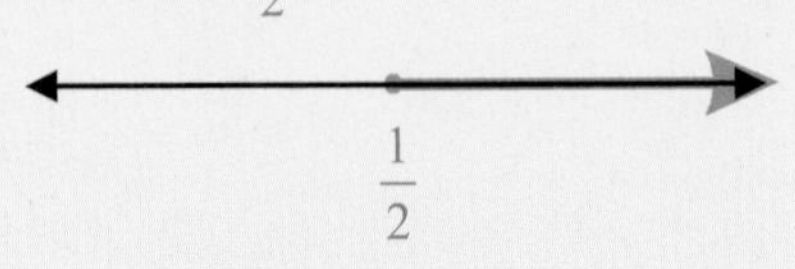


**6.** $-4 < 5y + 1 \leq 11$

$-1 < y \leq 2$

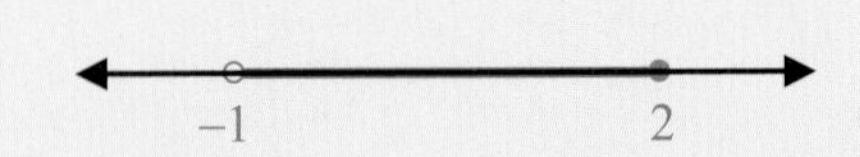


## Completion Example Answer

**7.**

| | |
|---|---|
| $7y - 8 > y + 10$ | |
| $7y - 8 - y > y + 10 - y$ | Add $-y$ to both sides. |
| $6y - 8 > 10$ | Simplify. |
| $6y - 8 + 8 > 10 + 8$ | Add 8 to both sides. |
| $6y > 18$ | Simplify. |
| $\frac{\cancel{6}y}{\cancel{6}} > \frac{18}{6}$ | Divide both sides by 6. |
| $y > 3$ | Simplify. |

Graph:

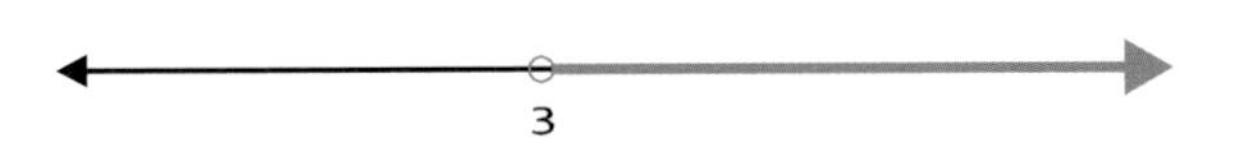

Name ______________________ Section ______________ Date ____________

## Exercises 7.6

Below each exercise, write each of the following inequalities in words and state whether it is true or false. If an inequality is false, rewrite it in a correct form. ( There may be more than one way to correct a false statement. )

**1.** $3 \neq -3$

Three does not equal negative three. True.

**2.** $-5 < -2$

Negative five is less than negative two. True.

**3.** $-13 > -1$

Negative thirteen is greater than negative one. False. Correction: $-13 < -1$

**4.** $|-7| \neq |+7|$

The absolute value of negative seven does not equal the absolute value of positive seven. False. Correction: $|-7| = |+7|$

**5.** $|-7| < +5$

The absolute value of negative seven is less than positive five. False. Correction: $|-7| > +5$

**6.** $|-4| > |+3|$

The absolute value of negative four is greater than the absolute value of positive three. True.

**7.** $-\frac{1}{2} < -\frac{3}{4}$

Negative one-half is less than negative three-fourths. False.

Correction: $-\frac{1}{2} > -\frac{3}{4}$

**8.** $\sqrt{2} > \sqrt{3}$

The square root of two is greater than the square root of three. False.

Correction: $\sqrt{2} < \sqrt{3}$

**9.** $-4 < -6$

Negative four is less than negative 6. False. Correction: $-4 > -6$

**10.** $|-6| > 0$

The absolute value of negative six is greater than 0. True.

Represent each of the following graphs with algebraic notation, and tell what kind of interval it is.

**11.** 

**12.** 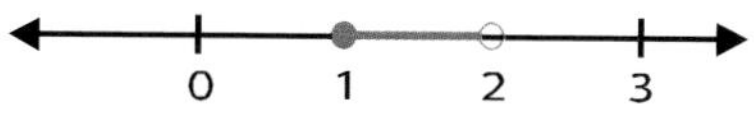

**13.** 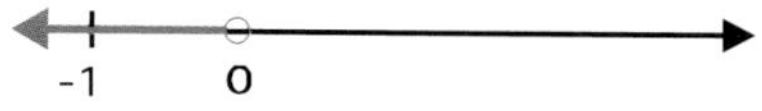

**14.** 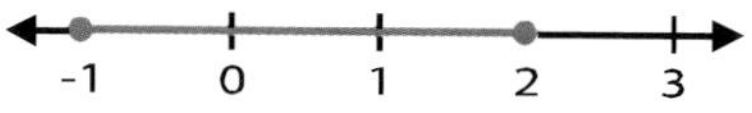

**15.** 

**ANSWERS**

1. [Respond below exercise.]
2. [Respond below exercise.]
3. [Respond below exercise.]
4. [Respond below exercise.]
5. [Respond below exercise.]
6. [Respond below exercise.]
7. [Respond below exercise.]
8. [Respond below exercise.]
9. [Respond below exercise.]
10. [Respond below exercise.]
11. $-1 < x < 2$; open interval
12. $1 \leq x < 2$; half-open interval
13. $x < 0$; open interval
14. $-1 \leq x \leq 2$; closed interval
15. $x \geq -1$; half-open interval

16. open

17. open

18. closed

19. closed

20. half-open

21. half-open

22. closed

23. open

24. open

25. half-open

26. open

27. half-open

In Exercise 16-27, graph the interval on the number line provided. Use the answer blank in the margin to tell what kind of interval each is

**16.** $5 < x < 8$

5 6 7 8

**17.** $-2 < x < 0$

−2 −1 0

**18.** $3 \leq y \leq 6$

3 4 5 6

**19.** $-3 \leq y \leq 5$

−3 0 5

**20.** $-2 < y \leq 1$

−2 −1 0 1

**21.** $4 \leq x < 7$

4 5 6 7

**22.** $47 \leq x \leq 52$

47 48 49 50 51 52

**23.** $-12 < z < -7$

−12 −11 −10 −9 −8 −7

**24.** $x > -5$

−5

**25.** $x \geq 0$

0

**26.** $x < -\sqrt{3}$

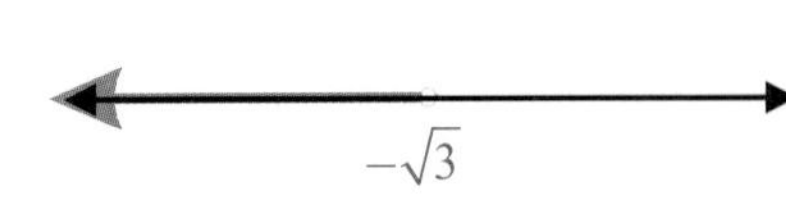


**27.** $x \leq \frac{2}{3}$

$\frac{2}{3}$

Name ______________________ Section ____________ Date __________

In Exercises 28-55, solve each inequality and write the solution in the answer blank in the margin. Then graph the solution on the number line provided.

**28.** $x + 5 < 6$

1

**29.** $x - 6 > -2$

4

**30.** $y - 4 \geq -1$

3

**31.** $y + 5 \leq 2$

$-3$

**32.** $3y \leq 4$

$\frac{4}{3}$

**33.** $5y > -6$

$-\frac{6}{5}$

**34.** $10y + 1 > 5$

$\frac{2}{5}$

**35.** $7x - 2 < 9$

$\frac{11}{7}$

**36.** $x + 3 < 4x + 3$

0

**37.** $x - 4 > 2x + 1$

$-5$

**38.** $3t - 5 \geq t + 5$

**39.** $3t - 8 \leq t + 2$

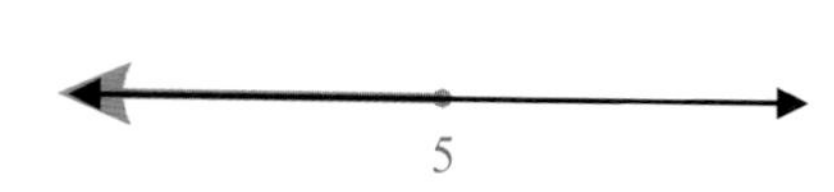


**ANSWERS**

**28.** $x < 1$

**29.** $x > 4$

**30.** $y \geq 3$

**31.** $y \leq -3$

**32.** $y \leq \frac{4}{3}$

**33.** $y > -\frac{6}{5}$

**34.** $y > \frac{2}{5}$

**35.** $x < \frac{11}{7}$

**36.** $x > 0$

**37.** $x < -5$

**38.** $t \geq 5$

**39.** $t \leq 5$

40. $x < 16$

41. $x > -6$

42. $t > 5$

43. $t \leq -1$

44. $t \geq 1$

45. $t > -1$

46. $x \leq 0$

47. $x < -5$

48. $-3 \leq x \leq -2$

49. $1 \leq x \leq 4$

50. $-2 \leq y \leq -\frac{3}{4}$

51. $\frac{1}{3} \leq x \leq 2$

52. $-1 < x < 1$

53. $-3 \leq y \leq 1$

54. $-6 < t \leq -5$

55. $-5 \leq t < -3$

**40.** $\frac{1}{2}x - 2 < 6$

16

**41.** $\frac{1}{3}x + 1 > -1$

−6

**42.** $-t + 4 < -1$

5

**43.** $-t - 5 \geq -4$

−1

**44.** $8t - 2 \geq 5t + 1$

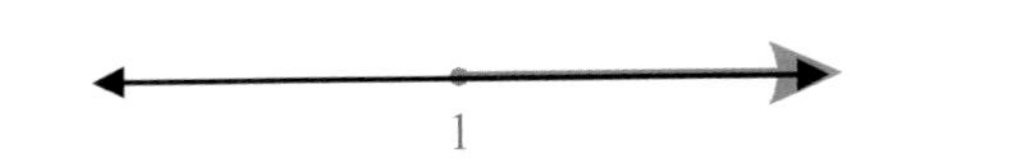

**45.** $6t + 3 > t - 2$

−1

**46.** $-5x - 7 \geq -7 + 2x$

0

**47.** $3x + 15 < x + 5$

−5

**48.** $6 \leq x + 9 \leq 7$

−3 −2

**49.** $-2 \leq x - 3 \leq 1$

1 4

**50.** $-5 \leq 4y + 3 \leq 0$

$-2$ $-\frac{3}{4}$

**51.** $0 \leq 3x - 1 \leq 5$

$\frac{1}{3}$ $2$

**52.** $-2 < 5x + 3 < 8$

−1 1

**53.** $-5 \leq y - 2 \leq -1$

−3 1

**54.** $5 \leq -2t - 5 < 7$

**55.** $14 < -5t - 1 \leq 24$

−5 −3

# Chapter 7 Index of Key Ideas and Terms

List of Key Words in Translating English Phrases — page 562

| Addition | Subtraction | Multiplication | Division |
|---|---|---|---|
| add | subtract ( from ) | multiply | divide |
| sum | difference | product | quotient |
| plus | minus | times | ratio |
| more than | less than | twice, double, | |
| increased by | decreased by | triple, etc. | |
| | | of ( with fractions and percent ) | |

First-Degree Equations — page 571

A **first-degree equation in** $x$ ( or a **linear equation in** $x$ )
is any equation that can be written in the form
$ax + b = c$ where $a$, $b$, and $c$ are constants and $a \neq 0$.

Principles Used in Solving a First-Degree Equation — page 571

1. The **Addition Principle:**

   The equations $A = B$
   and $A + C = B + C$
   have the same solutions.

2. The **Multiplication Principle:**

   For $C \neq 0$, the equations $A = B$
   and $C \cdot A = C \cdot B$

   and $\frac{A}{C} = \frac{B}{C}$

   and $\frac{1}{C} \cdot A = \frac{1}{C} \cdot B$

   have the same solutions.

Understanding How to Solve Equations — page 572

1. Simplify both sides of the equation.
   ( This includes applying the distributive property and/or combining like terms. )
2. If a constant is added to a variable, add the opposite of the constant to both sides of the equation and simplify.
3. If a variable has a constant coefficient other than 1, divide both sides by that coefficient ( that is, in effect, multiply both sides by the reciprocal of that coefficient ).
4. Generally, use the Addition Principle first so that terms with variables are on one side and constant terms are on the other side. Then combine like terms and use the Multiplication Principle.
5. Remember that the object is to isolate the variable ( with coefficient 1 ) on one side of the equation.

Equations with Fractions page 577

Pòlya's Four-Step Process for Solving Word Problems page 581

1. Understand the problem.
2. Devise a plan.
3. Carry out the plan.
4. Look back over the results.

Number Problems page 581

Consecutive Integer Problems page 583

**Consecutive Integers:**

Integers are consecutive if each is 1 more than the previous integer. Three consecutive integers can be represented as $n$, $n + 1$, and $n + 2$

**Consecutive Odd Integers:**

Odd integers are consecutive if each is 2 more than the previous odd integer. Three consecutive odd integers can be represented as $n$, $n + 2$, and $n + 4$ where $n$ is an **odd** integer.

**Consecutive Even Integers:**

Even integers are consecutive if each is 2 more than the previous even integer. Three consecutive even integers can be represented as $n$, $n + 2$, and $n + 4$ where $n$ is an **even** integer.

Formulas

A formula is a general statement ( usually an equation ) that relates two or more variables.

Evaluating Formulas page 596

Solving for Any Variable in a Formula page 597

Real Numbers page 603

Rational Numbers, Irrational Numbers page 603

Real Number Lines page 604

Intervals of Real Numbers page 606

| **Name** | **Symbolic Representation** |
|---|---|
| Open interval | $a < x < b$ |
| Closed interval | $a \leq x \leq b$ |
| Half-open interval | $a \leq x < b$<br>$a < x \leq b$ |
| Open interval | $x > a$<br>$x < a$ |
| Half-open interval | $x \geq a$<br>$x \leq a$ |

Linear ( or First-Degree ) Inequalities ( $ax + b < c$ ) page 607

Rules for Solving

1. The same number ( positive or negative ) may be added to both sides, and the sense of the inequality will remain the same.

2. Both sides may be multiplied by ( or divided by ) the same **positive** number, and the sense of the inequality will remain the same.

3. Both sides may be multiplied by ( or divided by ) the same **negative** number, but the sense of the inequality must be **reversed.**

Name ______________________ Section ____________ Date __________

## Chapter 7 Test

Write an English phrase that indicates the meaning of each algebraic expression.

**1.** $5n + 6$

**2.** $4 - 2n$

**3.** $3(x + 1)$

Write an algebraic expression described by each of the following English phrases.

**4.** The quotient of a number and 7 increased by the product of the number and 6

**5.** Four less than the product of a number and eight

Represent each of the following descriptions with an algebraic expression.

**6.** The number of hours in $x$ days

**7.** The number of days in $x$ hours

**8.** Make up your own word problem that might use the following equation in its solution: $3(x - 5) = 2x + 1$.

Answers will vary.

Solve each of the following equations.

**9.** $x - 2(3 - x) = 30$

**10.** $4y = -3y$

**11.** $5x + 6 = 3x - 4$

**12.** $2(35x + 5) = 4(-60 + 5x)$

**13.** $\frac{1}{2}y + 2 = \frac{3}{4}y - 3$

**14.** The sum of three consecutive integers is equal to 252. Find the integers.

**15.** Twice a number plus seven is equal to three less than four times the number. What is the number?

**ANSWERS**

1. five times a number plus 6
2. four decreased by twice a number
3. three times the sum of a number and 1
4. $\frac{n}{7} + 6n$
5. $8n - 4$
6. $24x$
7. $\frac{x}{24}$
8. [Respond below exercise]
9. $x = 12$
10. $y = 0$
11. $x = -5$
12. $x = -5$
13. $y = 20$
14. 83, 84, 85
15. 5

16. $9000

16. A car owner says that his car is now worth $5000 and that this is $500 more than one-half its original value. What was the original value of his car?

17. 14, 16, 25

17. One side of a triangle is 2 feet more than another side and the third side is 5 feet less than the sum of the other two sides. If the perimeter is 55 feet, what is the length of each side?

18. $2460

18. Given the formula $A = P + Prt$, find the value of $A$ if $P = \$2400$, $r = 10\%$, and $t = 3$ months. ( Remember to express $t$ in years. )

19. $h = \frac{A-2x^2}{4x}$

19. Solve the formula $A = 4hx + 2x^2$ for $h$.

20. a. $y = \frac{18-4x}{3}$

b. [Respond below exercise]

20. a. Solve the formula $4x + 3y = 18$ for $y$, then

b. complete the following table.

| $x$ | $y$ |
|---|---|
| 0 | 6 |
| 1.5 | 4 |
| 2.5 | $\frac{8}{3}$ |
| −3 | 10 |

21. $x > 4$; open

21. Represent the following graph by using algebraic notation, and tell what type of interval it is.

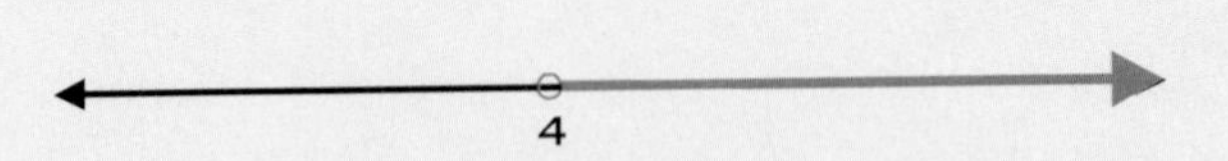

Name ____________ Section ____________ Date ____________

**ANSWERS**

**22.** Graph the interval $0.5 \le x \le 3$, and tell what type of interval it is.

0.5 3

**23.** Solve each inequality, and graph the solution on a real number line.

**a.** $2x - 5 \le 5x - 2$

$-1$

**b.** $7x + 4 < 16 + x$

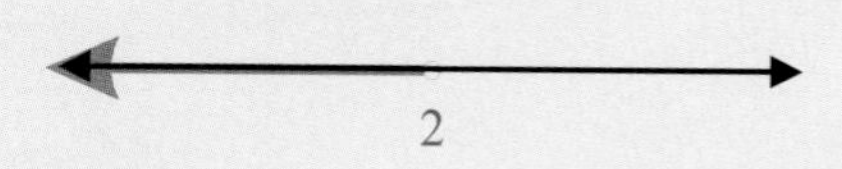


**22.** closed

**23. a.** $-1 \le x$

**b.** $x < 2$

Name ______________________ Section ______________ Date ____________

# Cumulative Review: Chapters 1 - 7

Perform the indicated operations.

**1.** $7000 \cdot 80{,}000$

**2.** $\begin{array}{r} 8.63 \\ 7.25 \\ +\ 16.91 \\ \hline \end{array}$

**3.** $\begin{array}{r} 71.3 \\ \times\ 0.84 \\ \hline \end{array}$

**4.** $10^2(96.377)$

**5.** $\dfrac{14.65}{1000}$

**6.** $0.25\overline{)6.1}$

**7.** $\dfrac{17-32}{-5}$

**8.** $16-(-3)-|10|$

**9.** $(-6)(-1)(-2)(4)$

**10.** **a.** Write the decimal number 83,500 in scientific notation.

**b.** Write the decimal number 0.000671 in scientific notation.

**11.** Find the quotient $(7 \cdot 6 \cdot 5 \cdot 4 \cdot 2) \div 48$.

**12.** Find 68% of 180.

**13.** What percent of 300 is 252?

**14.** The school bookstore bought a textbook from the publisher for $45 and sold the book to the students for $60.

**a.** What amount of profit did the bookstore make on each of these textbooks?

**b.** What was the percent of profit based on cost?

**c.** What was the percent of profit based on selling price?

**ANSWERS**

1. 560,000,000
2. 32.79
3. 59.892
4. 9637.7
5. 0.01465
6. 24.4
7. 3
8. 9
9. −48
10. a. $8.35 \times 10^4$
    b. $6.71 \times 10^{-4}$
11. 35
12. 122.4
13. 84%
14. a. $15
    b. $33.\overline{3}\%$
    c. 25%

Evaluate each of the following expressions. Reduce all fractions to lowest terms.

15. $\frac{5}{8}+\frac{3}{10}$

16. $\frac{1}{3}-\frac{7}{6}$

17. $\left(\frac{5}{6}\right)\left(\frac{4}{15}\right)\left(-\frac{7}{8}\right)$

18. $-\frac{15}{16}\div\frac{32}{35}$

19. $\dfrac{6\frac{1}{2}-2\frac{3}{4}}{1-\frac{7}{10}}$

20. $\frac{2}{3}\div\frac{3}{4}+\frac{3}{10}\left(\frac{1}{3}\right)^2-\frac{1}{6}$

21. $\begin{array}{r} 19\frac{3}{4} \\ +17\frac{7}{8} \\ \hline \end{array}$

22. $\begin{array}{r} 25\frac{1}{10} \\ -30\frac{7}{20} \\ \hline \end{array}$

Solve each of the following equations.

23. $3x + 2x - 17 = 25$

24. $\frac{2}{3}x+10=-\frac{1}{2}$

25. $5(x+2) = 6x - 15$

26. $\frac{5x}{8}-\frac{3}{4}=\frac{2x}{3}+\frac{5}{6}$

27. Find

**a.** the circumference and

**b.** the area of a circle with a radius of 8 feet.

28. Find the length of the hypotenuse of a right triangle with legs of length 10 meters and 20 meters. Write the answer in both simplified radical form and decimal form (accurate to three decimal places ).

29. Find three consecutive odd integers such that the sum of the first and second is 5 less than three times the third.

**15.** $\frac{37}{40}$

**16.** $-\frac{5}{6}$

**17.** $-\frac{7}{36}$

**18.** $-\frac{525}{512}$

**19.** $12\frac{1}{2}$

**20.** $\frac{34}{45}$

**21.** $37\frac{5}{8}$

**22.** $-5\frac{1}{4}$

**23.** $x = 8.4$

**24.** $x = -15.75$

**25.** $x = 25$

**26.** $x = -38$

**27. a.** $16\pi$ ft. (or 50.24 ft.)

**b.** $64\pi$ ft.$^2$ (or 200.96 ft.$^2$)

**28.** $10\sqrt{5}$ m; 22.361 m

**29.** −5, −3, −1

Name ____________ Section ____________ Date ____________

**ANSWERS**

**30.** If the quotient of a number and 7 is decreased by 12, the result is the number plus $(-6)$. What is the number?

**30.** $-7$

**31.** Find the simple interest earned on \$5000 loaned at 18% for 10 months.

**31.** \$750

**32.** What will be the value of \$10,000 invested at 5% and compounded daily for 3 years?

**32.** \$11,618.23

**33.** The purchase price of a new home is \$180,000. The buyer makes a down payment of 20% of the purchase price and pays a loan fee of 1% of the mortgage.

**a.** What is the amount of the mortgage?

**b.** What is the amount of the loan fee?

**33. a.** \$144,000

**b.** \$1440

Solve each of the following inequalities, and graph the solutions on real number lines.

**34.** $5x + 4 < -6$

**35.** $2x + 1 \geq 5x - 14$

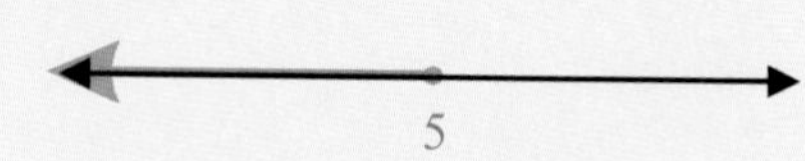


**34.** $x < -2$

**35.** $5 \geq x$

# 8

# ALGEBRAIC TOPICS II

**Chapter 8 Algebraic Topics II**





## WHAT TO EXPECT IN CHAPTER 8

While algebraic concepts have been integrated throughout this text, Chapters 7 and 8 are designed to provide a first look at some slightly more advanced topics from a beginning algebra course. In Chapter 7, the emphasis was on solving equations and working with formulas. In Chapter 8, the topics are related to operating with polynomials and graphing in two dimensions.

Section 8.1 discusses the properties of integer exponents and the use of these properties to simplify expressions that contain both positive and negative integer exponents. These properties of exponents are then applied in Sections 8.2 - 8.4 with the following basic operations: adding, subtracting, multiplying, and factoring with polynomials. While addition and subtraction with polynomials are similar to combining like terms ( and therefore somewhat familiar ), multiplication and factoring with polynomials are new and relatively sophisticated ideas. We leave division with polynomials to the next course in beginning algebra.

Section 8.5 discusses the points in a plane and how they can be represented with ordered pairs of real numbers ( called the *coordinates* of the points ). The plane is separated into four quadrants by two straight lines called *axes*. In Section 8.6, equations of the form $Ax + By = C$ are shown to correspond to graphs of straight lines in a plane.

These topics provide a "jump start" into algebra for those of you who plan to continue your studies in mathematics. Understanding these topics and developing the related skills will help ensure your success in algebra

# Chapter 8 Algebraic Topics II

## Mathematics at Work!

In many real-life applications, the relationship between two variables can be expressed in terms of a formula in which both variables are first-degree. For example, the distance you travel on your bicycle at an average speed of 15 miles per hour can be represented by the formula $d = 15t$, where $t$ is time ( in hours ).

Relationships indicated by first-degree equations and formulas can be "pictured" as straight-line graphs. Once the correct graph has been drawn, this graph can serve in place of the formula in finding a value ( or an approximate value ) of one of the variables corresponding to a known value of the other variable.

The formula $C = \frac{5}{9}(F - 32)$ represents the relationship between degrees Fahrenheit ( $F$ ) and degrees Celsius ( $C$ ). A portion of the graph of this relationship is shown here.

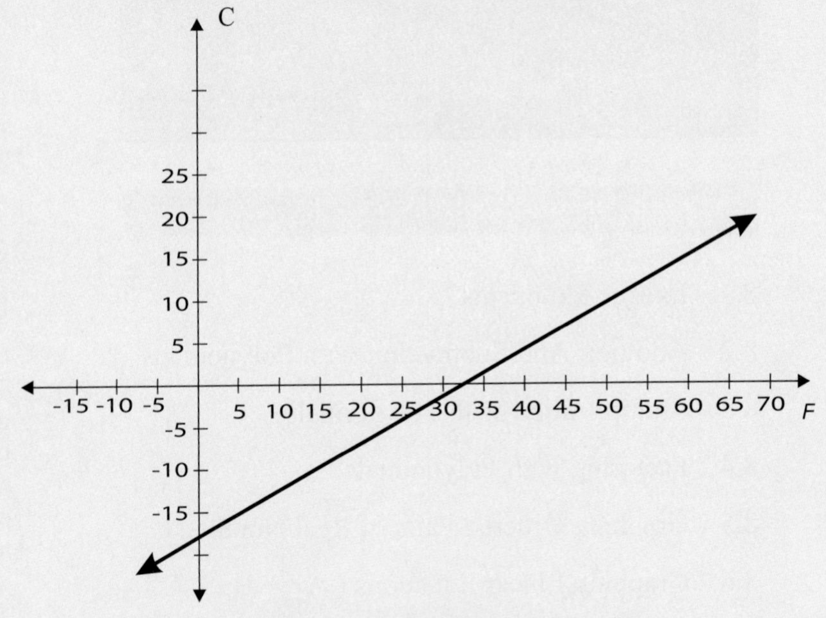


**a.** From the graph, estimate the Celsius temperature corresponding to 32°$F$ ( That is, estimate the value of $C$ given that $F = 32$. )

0°$C$

**b.** From the graph, estimate the Celsius temperature corresponding to 59°$F$ ( That is, estimate the value of $C$ given that $F = 59$. )

15°$C$

# 8.1 Integer Exponents

## Review of Exponents

Exponents have been discussed in several sections and used with various topics throughout this text.

In Section 1.5, we defined and developed the following properties of exponents.

$$\underbrace{a \times a \times a \times \ldots \times a}_{n \text{ factors}} = a^n, \quad a^1 = a, \quad a^0 = 1 \qquad \text{for } a \neq 0$$

and $\quad a^m \cdot a^n = a^{m+n} \qquad$ for whole numbers $m$ and $n$

Expressions with exponents can be evaluated with a calculator that has a key marked $y^x$ ( or $x^y$ ) or with a TI–83 Plus graphing calculator using the keys marked as $x^2$ ( $x^2$ ) and the key marked with a caret ^ ( ^ ).

Other exponent keys are available in the MATH ( MATH ) menu.

In Section 5.4, we noted that calculators use scientific notation for very large and very small decimal numbers, and scientific notation involves both positive and negative integer exponents. For example,

$$3{,}200{,}000 = 3.2 \times 10^6 \qquad \text{and} \qquad 0.000047 = 4.7 \times 10^{-5}$$

In Section 5.6, fractional exponents were used with calculators to find the values of radicals accurate to several decimal places. For example,

$$\sqrt{3} = 3^{\frac{1}{2}} = 3^{0.5} = 1.732050808 \qquad \text{accurate to nine places}$$

In Section 6.6, exponents were used to find compound interest with the formula

$$A = P\left(1 + \frac{r}{n}\right)^{nt}$$

As this discussion indicates, exponents and their properties are an important part of mathematics. In this section, we expand on our previous knowledge of exponents to include expressions that contain both **positive** and **negative** exponents.

**Objectives**

① Know the meanings of 0 and 1 as exponents.

② Learn the following rules for exponents: **Product Rule, Power Rule,** and **Quotient Rule.**

③ Know the meaning of a negative exponent.

④ Learn how to simplify expressions with negative exponents.

## The Product Rule

One of the basic rules ( or properties ) of exponents deals with multiplying powers that have the same base. For example, with 2 as a base,

$$2^2 \cdot 2^4 = \underbrace{(2 \cdot 2)}_{2 \text{ factors}} \cdot \underbrace{(2 \cdot 2 \cdot 2 \cdot 2)}_{4 \text{ factors}} = \underbrace{2 \cdot 2 \cdot 2 \cdot 2 \cdot 2 \cdot 2}_{6 \text{ factors}} = 2^6 = 64$$

More generally, with the variable $a$ as the base,

$$a^2 \cdot a^4 = \underbrace{(a \cdot a)}_{2 \text{ factors}} \cdot \underbrace{(a \cdot a \cdot a \cdot a)}_{4 \text{ factors}} = \underbrace{a \cdot a \cdot a \cdot a \cdot a \cdot a}_{6 \text{ factors}} = a^6$$

or

$$a^2 \cdot a^4 = a^{2+4} = a^6$$

This property is called the **product rule** and is stated here for **integer** exponents.

You might want to warn students to avoid the common error of multiplying the bases when numbers are used as bases. For example, $2^2 \cdot 2^3 \neq 4^5$. You might show that $2^2 \cdot 2^3 = 4 \cdot 8 = 32$ and that $4^5 = 1024$. As another example, $3^2 \cdot 3^2 \neq 9^4$.

**Note:** Although the examples shown in the beginning of this section show only positive integer exponents, the rules of exponents are all valid for **negative integer exponents** as well. Thus, the rules are stated here as they apply to integer exponents. At the end of the section, negative integer exponents are discussed and the rules are summarized in terms of positive and negative integer exponents.

**Product Rule:**

For any real number $a$ and integers $m$ and $n$, $\quad a^m \cdot a^n = a^{m+n}$

( To multiply two powers with the same base, keep the base and add the exponents. )

**Note:** All Rules of Exponents are stated with the understanding that $0^0$ **is undefined**.

### Example 1

Use the Product Rule to simplify each of the following expressions.

**a.** $(-7)^2 \cdot (-7)^3$

**b.** $x^3 \cdot x^5$

**c.** $2y \cdot 5y^9$

#### Solution

In each case the Product Rule is used by adding the exponents.

**a.** $(-7)^2 \cdot (-7)^3 = (-7)^{2+3} = (-7)^5$ ( or $-16{,}807$ )

**b.** $x^3 \cdot x^5 = x^{3+5} = x^8$

**c.** $2y \cdot 5y^9 = (2 \cdot 5)(y^1 \cdot y^9) = 10 \cdot y^{1+9} = 10y^{10}$
( Note that the coefficients are multiplied as usual. )

*Now Work Exercises 1-3 in the Margin.*

Use the Product Rule to simplify each expression.

**1.** $8^2 \cdot 8$

$8^3$ ( or 512 )

**2.** $(-2)^3 \cdot (-2)^4$

$(-2)^7$ ( or $-128$ )

**3.** $-5x^3 \cdot 2x^3$

$-10x^6$

## The Power Rule

An expression such as $(a^2)^3$ is a power raised to a power. To simplify this expression, we could multiply and use the Product Rule as follows:

$$(a^2)^3 = a^2 \cdot a^2 \cdot a^2 = a^{2+2+2} = a^6$$

and

$$(a^2)^4 = a^2 \cdot a^2 \cdot a^2 \cdot a^2 = a^{2+2+2+2} = a^8$$

However, these same results can be obtained by simply **multiplying the exponents.** Thus,

$$(a^2)^3 = a^{2 \cdot 3} = a^6 \qquad \text{and} \qquad (a^2)^4 = a^{2 \cdot 4} = a^8$$

These examples lead to the **Power Rule** for exponents.

> **Power Rule:**
>
> For any real number $a$ and integers $m$ and $n$, $\qquad \mathbf{(a^m)^n = a^{mn}}$
>
> ( To find a power raised to a power, keep the base and multiply the exponents. )

### Example 2

Use the Power Rule to simplify each of the following expressions.

**a.** $(2^3)^3$

**b.** $(3^2)^4$

**c.** $(x^3)^7$

**Solution**

**a.** $(2^3)^3 = 2^{3 \cdot 3} = 2^9$ ( or 512 )

**b.** $(3^2)^4 = 3^{2 \cdot 4} = 3^8$ ( or 6561 )

**c.** $(x^3)^7 = x^{3 \cdot 7} = x^{21}$

*Now Work Exercises 4-6 in the Margin.*

Use the Power Rule to simplify each expression.

**4.** $(3^2)^3$

$3^6$ ( or 729 )

**5.** $(2^2)^6$

$2^{12}$ ( or 4096 )

**6.** $(a^7)^2$

$a^{14}$

## The Quotient Rule

Now consider the fraction $\frac{a^5}{a^3}$, in which the numerator and denominator are powers with the same base. To simplify this expression, we can write

$$\frac{a^5}{a^3} = \frac{\cancel{a} \cdot \cancel{a} \cdot \cancel{a} \cdot a \cdot a}{\underset{1}{\cancel{a} \cdot \cancel{a} \cdot \cancel{a}}} = \frac{a \cdot a}{1} = a^2$$

Similarly,

$$\frac{x^8}{x^5} = \frac{\cancel{x}\cdot\cancel{x}\cdot\cancel{x}\cdot\cancel{x}\cdot\cancel{x}\cdot x\cdot x\cdot x}{\underset{1}{\cancel{x}\cdot\cancel{x}\cdot\cancel{x}\cdot\cancel{x}\cdot\cancel{x}}} = \frac{x\cdot x\cdot x}{1} = x^3$$

These same results can be obtained by **subtracting the exponents.** Thus,

$$\frac{a^5}{a^3} = a^{5-3} = a^2 \quad \text{and} \quad \frac{x^8}{x^5} = x^{8-5} = x^3$$

These examples lead to the **Quotient Rule** for exponents.

You might help the students "discover" at this point how the Quotient Rule leads naturally to negative exponents. Thus, they can anticipate a need for negative exponents and understand why the Product Rule and the Quotient Rule were stated for integer exponents.

**Quotient Rule:**

For any nonzero real number $a$ and integers $m$ and $n$, $\frac{a^m}{a^n} = a^{m-n}$.

[ To divide two powers with the same base, subtract the exponents ( numerator exponent minus denominator exponent ) and keep the base. ]

### Example 3

Use the Quotient Rule to divide and simplify each of the following expressions.

**a.** $\frac{9^5}{9^2}$

**b.** $\frac{p^{10}}{p}$

**c.** $\frac{26x^7}{13x^2}$

**Solution**

**a.** $\frac{9^5}{9^2} = 9^{5-2} = 9^3$ (or 729)

**b.** $\frac{p^{10}}{p} = \frac{p^{10}}{p^1} = p^{10-1} = p^9$

**c.** $\frac{26x^7}{13x^2} = \left(\frac{26}{13}\right)x^{7-2} = 2x^5$

*Now Work Exercises 7-9 in the Margin.*

Use the Quotient Rule to simplify each expression.

**7.** $\frac{3^5}{3^2}$

$3^3$ ( or 27 )

**8.** $\frac{5^3}{5^2}$

5

**9.** $\frac{14t^7}{7t^2}$

$2t^5$

**To Summarize the Three Rules for Exponents:**

1. For the **Product Rule**, terms are being multiplied and the exponents are added.
2. For the **Power Rule,** a term is raised to a power and exponents are multiplied.
3. For the **Quotient Rule**, terms are being divided and the exponents are subtracted.

## The Exponent 0 and Negative Integer Exponents

As previously stated, for any nonzero real number $a$, $a^0 = 1$. Now, this property of 0 as an exponent can be verified by using the Quotient Rule. Consider the fact that any nonzero number divided by itself gives a quotient of 1:

$$\frac{a^3}{a^3} = 1$$

Also, by the Quotient Rule, we have

$$\frac{a^3}{a^3} = a^{3-3} = a^0$$

In general, for any nonzero real number $a$ and any integer $n$ we have the familiar result

$$\frac{a^n}{a^n} = a^{n-n} = a^0 = 1$$

The meaning of negative integer exponents can be developed with this same approach. Consider the expression

$$\frac{5^2}{5^3} = \frac{\overset{1}{\cancel{5}} \cdot \cancel{5}}{\cancel{5} \cdot \cancel{5} \cdot 5} = \frac{1}{5}$$

However, by the Quotient Rule,

$$\frac{5^2}{5^3} = 5^{2-3} = 5^{-1}$$

This means that

$$\frac{1}{5} = 5^{-1}$$

Similarly,

$$\frac{x^2}{x^8} = \frac{\overset{1}{\cancel{x}} \cdot \cancel{x}}{\cancel{x} \cdot \cancel{x} \cdot x \cdot x \cdot x \cdot x \cdot x \cdot x} = \frac{1}{x \cdot x \cdot x \cdot x \cdot x \cdot x} = \frac{1}{x^6}$$

$$\text{and} \quad \frac{x^2}{x^8} = x^{2-8} = x^{-6}$$

Again, these two statements mean that

$$\frac{1}{x^6} = x^{-6}$$

The pattern that can be observed in these examples leads to the following definition of negative exponents.

**Negative Integer Exponents:**

For any nonzero real number $a$ and any integer $n$,

$$a^{-n} = \frac{1}{a^n}$$

( $a^{-n}$ is another form of the reciprocal of $a^n$. )

## Example 4

Write each of the following expressions in an equivalent form with positive exponents and simplify.

**a.** $6^{-3}$

**b.** $(-4)^{-2}$

**c.** $x^{-4}$

**d.** $\frac{1}{x^{-2}}$

### Solution

**a.** $6^{-3} = \frac{1}{6^3} = \frac{1}{216}$

**b.** $(-4)^{-2} = \frac{1}{(-4)^2} = \frac{1}{16}$

**c.** $x^{-4} = \frac{1}{x^4}$

**d.** $\frac{1}{x^{-2}} = \frac{x^0}{x^{-2}} = x^{0-(-2)} = x^2$

Note that $1 = x^0$. Using this form of 1 helps in simplifying the expression.

*Now Work Exercises 10-12 in the Margin.*

Write each expression in an equivalent form with positive exponents and simplify.

**10.** $4^{-2}$

$\frac{1}{4^2} = \frac{1}{16}$

**11.** $(-3)^{-3}$

$\frac{1}{(-3)^3} = -\frac{1}{27}$

**12.** $\frac{13}{x^{-2}}$

$13x^2$

**Summary of Definitions and Rules for Exponents:**

For any real number $a$ and integers $m$ and $n$:

| | | | Example |
|---|---|---|---|
| **The exponent 1:** | $a^1 = a$ | | $7^1 = 7$ |
| **The exponent 0:** | $a^0 = 1$ | $(a \neq 0)$ | $3^0 = 1$ |
| **Product Rule:** | $a^m \times a^n = a^{m+n}$ | | $2^2 \cdot 2^3 = 2^5$ |
| **Power Rule:** | $(a^m)^n = a^{mn}$ | | $(5^2)^3 = 5^6$ |
| **Quotient Rule:** | $\frac{a^m}{a^n} = a^{m-n}$ | $(a \neq 0)$ | $\frac{2^6}{2^3} = 2^{6-3} = 2^3$ |
| **Negative exponents:** | $a^{-n} = \frac{1}{a^n}$ | $(a \neq 0)$ | $6^{-2} = \frac{1}{6^2}$ |

## Example 5

Simplify each of the following expressions so that the answer contains only nonnegative exponents.

**a.** $5^{-4} \cdot 5^2$

**b.** $\frac{3^2}{3^{-3}}$

**c.** $\frac{x}{x^6}$

**d.** $\frac{y^{-5}}{y^{-5}}$

### Solution

**a.** $5^{-4} \cdot 5^2 = 5^{-4+2} = 5^{-2} = \frac{1}{5^2}\left(\text{or } \frac{1}{25}\right)$

**b.** $\frac{3^2}{3^{-3}} = 3^{2-(-3)} = 3^5 \quad (\text{or } 243)$

**c.** $\frac{x}{x^6} = x^{1-6} = x^{-5} = \frac{1}{x^5}$

**d.** $\frac{y^{-5}}{y^{-5}} = y^{-5-(-5)} = y^0 = 1$

*Now Work Exercises 13-16 in the Margin.*

Simplify each expression so that the answer contains only nonnegative exponents.

**13.** $7^{-3} \cdot 7^2$

$\frac{1}{7}$

**14.** $\frac{6^{-2}}{6^{-3}}$

$6$

**15.** $\frac{t^{12}}{t^{-3}}$

$t^{15}$

**16.** $\frac{15x^{-3}}{5x^{-3}}$

$3$

Name ______ Section ______ Date ______

## Exercises 8.1

Use the Product Rule to simplify each of the following expressions. Leave the expressions in base-exponent form.

**1.** $4^2 \cdot 4^2$ **2.** $3^3 \cdot 3^2$ **3.** $x^2 \cdot x^9$

**4.** $y^4 \cdot y^5$ **5.** $3x^2 \cdot 5x^4$ **6.** $2t^5 \cdot 8t^2$

**7.** $(1.5)^3 \cdot (1.5)^4$ **8.** $(2.1)^2 \cdot (2.1)^5$ **9.** $(-6)^4 \cdot (-6)$

**10.** $(-5) \cdot (-5)^3$

Use the Power Rule to simplify each of the following expressions. Leave the expressions in base-exponent form.

**11.** $(x^3)^5$ **12.** $(y^2)^4$ **13.** $(7^2)^2$

**14.** $(6^2)^3$ **15.** $(5.2^3)^4$ **16.** $(3.1^2)^5$

**17.** $[(-3)^4]^2$ **18.** $[(-2)^3]^3$ **19.** $(t^5)^0$

**20.** $(s^5)^0$

Use the Quotient Rule to simplify each of the following expressions. Leave the expressions in base-exponent form.

**21.** $\frac{6^5}{6^3}$ **22.** $\frac{8^3}{8^2}$ **23.** $\frac{x^{10}}{x}$



**ANSWERS**

1. $4^4$
2. $3^5$
3. $x^{11}$
4. $y^9$
5. $15x^6$
6. $16t^7$
7. $1.5^7$
8. $2.1^7$
9. $(-6)^5$
10. $(-5)^4$
11. $x^{15}$
12. $y^8$
13. $7^4$
14. $6^6$
15. $5.2^{12}$
16. $3.1^{10}$
17. $(-3)^8$
18. $(-2)^9$
19. $t^0 = 1$
20. $s^0 = 1$
21. $6^2$
22. $8$
23. $x^9$

24. $y^5$

25. $t^6$

26. $x$

27. $3.5^2$

28. $1.6^2$

29. $(-5)^3$

30. $(-6)^2$

31. $\frac{1}{5^2}$

32. $\frac{1}{8^2}$

33. $\frac{1}{x^5}$

34. $\frac{1}{y^{11}}$

35. $\frac{1}{(-2)^3}$

36. $\frac{1}{(-4)^3}$

37. $\frac{1}{(1.4)^2}$

38. $\frac{1}{(1.2)^3}$

39. $\frac{1}{y^6}$

40. $\frac{1}{s^9}$

41. $4^3$

42. $\frac{1}{3^2}$

43. $2^3$

44. $3^2$

45. $x^3$

46. $y^4$

**24.** $\frac{y^6}{y}$ **25.** $\frac{t^8}{t^2}$ **26.** $\frac{x^7}{x^6}$

**27.** $\frac{3.5^5}{3.5^3}$ **28.** $\frac{1.6^4}{1.6^2}$ **29.** $\frac{(-5)^6}{(-5)^3}$

**30.** $\frac{(-6)^9}{(-6)^7}$

Write each of the following expressions in an equivalent form with positive exponents and simplify. Leave the expressions in base-exponent form.

**31.** $5^{-2}$ **32.** $8^{-2}$ **33.** $x^{-5}$

**34.** $y^{-11}$ **35.** $(-2)^{-3}$ **36.** $(-4)^{-3}$

**37.** $1.4^{-2}$ **38.** $1.2^{-3}$ **39.** $y^{-6}$

**40.** $s^{-9}$

Simplify each of the following expressions so that the answer contains only nonnegative exponents. Leave the expressions in base-exponent form.

**41.** $4^{-2} \cdot 4^5$ **42.** $3^{-5} \cdot 3^3$ **43.** $\frac{1}{2^{-3}}$

**44.** $\frac{1}{3^{-2}}$ **45.** $\frac{1}{x^{-3}}$ **46.** $\frac{1}{y^{-4}}$

Name ______________ Section ______________ Date ______________

**47.** $4x^3 \cdot 5x^2$ **48.** $-3x^5 \cdot 2x^3$ **49.** $(y^3 \cdot y^2)^0$

**50.** $(x^4 \cdot x^5)^0$ **51.** $y^0 \cdot y^{-8}$ **52.** $x^0 \cdot x^{-3}$

**53.** $x^2 \cdot x^3 \cdot x^4$ **54.** $y^3 \cdot y^5 \cdot y$ **55.** $t^{-2} \cdot t^{-3} \cdot t^{-4}$

**56.** $y^{-5} \cdot y^{-1} \cdot y^{-2}$ **57.** $(s^{-3})^{-2}$ **58.** $(t^{-1})^{-3}$

**59.** $(x^{-4})^2$ **60.** $(x^{-5})^3$ **61.** $\frac{4^3}{4^5}$

**62.** $\frac{5^3}{5^6}$ **63.** $x \cdot x^{-1}$ **64.** $y \cdot y^{-1}$

**65.** $3x^2 \cdot 4x^{-2}$ **66.** $2y^3 \cdot 4y^{-3}$ **67.** $\frac{x}{x^{-1}}$

**68.** $\frac{a}{a^{-1}}$ **69.** $\frac{p^{-1}}{p}$ **70.** $\frac{x^{-2}}{x^2}$



**ANSWERS**

**47.** $20x^5$

**48.** $-6x^8$

**49.** $1$

**50.** $1$

**51.** $\frac{1}{y^8}$

**52.** $\frac{1}{x^3}$

**53.** $x^9$

**54.** $y^9$

**55.** $\frac{1}{t^9}$

**56.** $\frac{1}{y^8}$

**57.** $s^6$

**58.** $t^3$

**59.** $\frac{1}{x^8}$

**60.** $\frac{1}{x^{15}}$

**61.** $\frac{1}{4^2}$

**62.** $\frac{1}{5^3}$

**63.** $1$

**64.** $1$

**65.** $12$

**66.** $8$

**67.** $x^2$

**68.** $a^2$

**69.** $\frac{1}{p^2}$

**70.** $\frac{1}{x^4}$

71. a. $\frac{1}{x^6}$

b. $\frac{1}{x}$

c. $x^5$

d. $\frac{1}{x^5}$

72. a. $\frac{1}{a^4}$

b. $a^3$

c. $a^5$

d. $\frac{1}{a^5}$

73. a. $y^{10}$

b. $\frac{1}{y^7}$

c. $y^3$

d. $\frac{1}{y^3}$

74. a. $\frac{1}{x^{12}}$

b. $x$

c. $\frac{1}{x^7}$

d. $x^7$

75. a. $x^{18}$

b. $x^9$

c. $x^3$

d. $\frac{1}{x^3}$

76. [Respond below exercise.]

**71. a.** $(x^2)^{-3}$ **b.** $x^2 \cdot x^{-3}$ **c.** $\frac{x^2}{x^{-3}}$ **d.** $\frac{x^{-3}}{x^2}$

**72. a.** $(a^4)^{-1}$ **b.** $a^4 \cdot a^{-1}$ **c.** $\frac{a^4}{a^{-1}}$ **d.** $\frac{a^{-1}}{a^4}$

**73. a.** $(y^{-2})^{-5}$ **b.** $y^{-2} \cdot y^{-5}$ **c.** $\frac{y^{-2}}{y^{-5}}$ **d.** $\frac{y^{-5}}{y^{-2}}$

**74. a.** $(x^{-3})^4$ **b.** $x^{-3} \cdot x^4$ **c.** $\frac{x^{-3}}{x^4}$ **d.** $\frac{x^4}{x^{-3}}$

**75. a.** $(x^6)^3$ **b.** $x^6 \cdot x^3$ **c.** $\frac{x^6}{x^3}$ **d.** $\frac{x^3}{x^6}$

## Writing and Thinking about Mathematics

**76. a.** Determine whether or not the two expressions $4x^0$ and $(4x)^0$ have the same meaning, and explain your reasoning.

**b.** Evaluate both expressions for $x = 3$

**c.** Evaluate both expressions for $x = -3$

a. $4x^0 = 4 \cdot 1 = 4;\ (4x)^0 = 1$. Explanations will vary.

b. $4 \cdot 3^0 = 4 \cdot 1 = 4;\ (4 \cdot 3)^0 = (12)^0 = 1$

c. $4 \cdot (-3)^0 = 4 \cdot 1 = 4;\ [4 \cdot (-3)]^0 = (-12)^0 = 1$

# 8.2 Addition and Subtraction with Polynomials

## Review of Polynomials and Combining Like Terms

In Section 1.6, we defined polynomials in terms of whole number coefficients and constants. In Section 2.6, we discussed combining like terms and evaluating algebraic expressions. The use of coefficients and constants in polynomials and like terms has progressed from whole numbers (Chapter 1), to integers (Chapter 2), to fractions (Chapter 3), and to decimals ( Chapter 5). In this section, we will review these ideas and the definitions and related concepts that now include real numbers as coefficients and constants. (Remember that real numbers include rational numbers and irrational numbers.)

**Objectives**

① Recognize polynomial expressions

② Know how to classify polynomials.

③ Learn how to add polynomials.

④ Learn how to subtract polynomials.

Each of the following types of algebraic expressions is called a **term**:

1. real numbers ( constants )
2. powers of variables
3. products of real numbers and powers of variables

A number written next to a variable indicates multiplication, and the number is called the **coefficient** of the variable. A term that consists of only a real number, such as 3, $\frac{1}{2}$, 0.86, or $\sqrt{2}$, is called a **constant** or a **constant term**.

**Definition:**

A **monomial in** $x$ is a term of the form

$kx^n$ where $k$ is a real number and $n$ is a whole number.

$n$ is called the **degree** of the monomial, and

$k$ is called the **coefficient**.

( Note that a monomial can be "in" any variable. For example, $3y^2$ is a second degree monomial in $y$. )

Since ( for nonzero $x$ ) $x^0 = 1$, a constant can be multiplied by $x^0$ without changing its value. For example,

$$17 = 17x^0 \quad \text{and} \quad -9.3 = -9.3x^0$$

Thus, any nonzero constant is considered to be a **monomial of degree 0**.

**Definition:**

A **polynomial** is a monomial or the indicated sum or difference of monomials.

The **degree of a polynomial** is the largest of the degrees of its terms.

### Classification of Polynomials

| | | Example |
|---|---|---|
| **Monomial:** | polynomial with one term | $-2x^5$ |
| **Binomial:** | polynomial with two terms | $8x - 32$ |
| **Trinomial:** | polynomial with three terms | $a^2 + 7a + 6$ |

Examples of polynomials are

| | |
|---|---|
| $-14x^5$ | fifth-degree monomial |
| $-\sqrt{2}x^3 - \frac{1}{3}$ | third-degree binomial |
| $5x^2 + 2.5x - 13$ | second-degree trinomial |

**Like terms** ( or **similar terms** ) are terms that contain the same variables ( if any ) raised to the same powers. Whatever power a variable is raised to in one term, it is raised to the same power in other like terms. Constants are like terms.

Example 1 gives several examples that review how to combine like terms and label the resulting polynomials.

### Example 1

Simplify each of the following algebraic expressions by combining like terms whenever possible. If the expression is a polynomial in one variable, state the degree and type of the polynomial **after it is simplified**.

**a.** $4x^2 + 7x^2$

**b.** $x^3 + 8x + 9 - x^3$

**c.** $5x^2 - x^2 - 3x + 4x - 10 + 6$

#### Solution

**a.** $4x^2 + 7x^2 = 11x^2$ second-degree monomial

**b.** $x^3 + 8x + 9 - x^3 = x^3 - x^3 + 8x + 9$

$= 8x + 9$ first-degree binomial

**c.** $5x^2 - x^2 - 3x + 4x - 10 + 6 = 4x^2 + x - 4$ second-degree trinomial

## Addition with Polynomials

The **sum** of two or more polynomials is found by combining like terms. The polynomials may be written horizontally ( as in Example 2 ) or vertically with like terms aligned ( as in Example 3 ).

### Example 2

Add the following three polynomials by combining like terms.

$$(4x^3 - 5x^2 + 15x - 10) + (-7x^2 - 2x + 3) + (4x^2 + 9)$$
$$= 4x^3 + (-5x^2 - 7x^2 + 4x^2) + (15x - 2x) + (-10 + 3 + 9)$$
$$= 4x^3 - 8x^2 + 13x + 2$$

### Example 3

Find the sum of the following two polynomials by combining like terms.

$$\begin{array}{rr} & x^3 - 4x^2 + 2x + 13 \\ + & 2x^3 + 3x^2 - 5x + 16 \\ \hline & 3x^3 - x^2 - 3x + 29 \end{array}$$

### *Completion Example 4*

Write the sum

$$(5x^3 - 8x^2 - 10x + 2) + (3x^3 + 4x^2 - x - 7)$$

in the vertical format and find the sum by combining like terms.

**Solution**

$$5x^3 - 8x^2 - 10x + 2$$

+ ______________

______________

*Now Work Exercises 1 and 2 in the Margin.*

## Subtraction with Polynomials

The opposite of a polynomial can be indicated by writing a negative sign in front of the polynomial. The opposite of an entire polynomial indicates that the sign of every term in the polynomial is changed. For example,

$$-(5x^2 - 6x - 1) = -5x^2 + 6x + 1$$

Another approach is to think of this type of expression as indicating multiplication by $-1$ and use the distributive property as follows:

$$\begin{aligned} -(5x^2 - 6x - 1) &= -1(5x^2 - 6x - 1) \\ &= -1(5x^2) - 1(-6x) - 1(-1) \\ &= -5x^2 + 6x + 1 \end{aligned}$$

The answer is the same either way. In effect, using the distributive property and multiplying every term by $-1$ gives the same result as changing every term in the polynomial.

Find the indicated sums.

1. $(2x^2 - 5x + 3) + (x^2 - 7) + (2x + 10)$

   $3x^2 - 3x + 6$

2. $$\begin{array}{rr} & 3x^4 - 5x^3 + 7x^2 - 2x + 12 \\ + & 3x^4 - 6x^3 - 9x^2 - 2x - 14 \\ \hline & 6x^4 - 11x^3 - 2x^2 - 4x - 2 \end{array}$$

The **difference** of two polynomials can be found by **adding the opposite** of the polynomial being subtracted. The polynomials can be written horizontally ( as in Example 5 ) or vertically with like terms aligned ( as Example 6 ). In either form, the sign of every term in the polynomial being subtracted is changed and then the polynomials are added.

### Example 5

Subtract the polynomials as indicated.

$(9x^3 + 4x^2 - 15) - (-2x^3 + x^2 - 5x - 6)$

$= 9x^3 + 4x^2 - 15 + 2x^3 - x^2 + 5x + 6$ Signs in second polynomial are changed.

$= 11x^3 + 3x^2 + 5x - 9$ Polynomials are added by combining like terms.

When using the vertical alignment format, write a 0 as a placeholder for any missing powers of the variable in order to maintain proper alignment and ensure that like terms will be subtracted ( or added ).

### Example 6

Find the difference: $(8x^4 + 2x^3 - 5x^2 + 0x - 7) - (3x^4 + 5x^3 - x^2 + 6x - 11)$

Writing the polynomials in a vertical format we have:

$$\begin{array}{r} 8x^4 + 2x^3 - 5x^2 + 0x - 7 \\ \underline{-(3x^4 + 5x^3 - x^2 + 6x - 11)} \end{array}$$

Be sure to change the sign of every term in the polynomial being subtracted and then combine like terms.

$$\begin{array}{rl} 8x^4 + 2x^3 - 5x^2 + 0x - 7 & \\ \underline{+\left(-3x^4 - 5x^3 + x^2 - 6x + 11\right)} & \text{signs changed} \\ 5x^4 - 3x^3 - 4x^2 - 6x + 4 & \text{difference} \end{array}$$

*Now Work Exercises 3 and 4 in the Margin.*

### Completion Example Answer

4. $$\begin{array}{r} 5x^3 - 8x^2 - 10x + 2 \\ \underline{+\ 3x^3 + 4x^2 - \ 1x\ - 7} \\ 8x^3 - 4x^2 - 11x - 5 \end{array}$$

Find the indicated differences.

**3.** $$\begin{array}{r} -2x^3 + 5x^2 + 8x - 1 \\ \underline{-(2x^3 - x^2 - 6x + 13)} \\ -4x^3 + 6x^2 + 14x - 14 \end{array}$$

**4.** $$\begin{array}{r} 15x^3 + 10x^2 - 17x - 25 \\ \underline{-(12x^3 + 10x^2 + 3x - 16)} \\ 3x^3 - 20x - 9 \end{array}$$

Name ______________ Section ______________ Date ______________

## Exercises 8.2

Simplify each polynomial and tell what type of polynomial it is and its degree after it is simplified.

1. $5x+6x-10+3$
2. $8x-9x+14-5$
3. $3x^2-x^2+7x-x+2$
4. $4x^2+3x^2-x+2x+18$
5. $a^3+4a^2+a^2-a^3$
6. $y^4-2y^3+3y^3-y^4$
7. $-2y^2-y^2+10y-3y+2+5$
8. $-5a^2+2a^2-4a-2a+4+1$
9. $5x^3+2x-8x+17+3x$
10. $-4x^3+5x^2-3x^2+12x-x$

Add or subtract as indicated, and simplify if possible.

11. $(3x-5)+(2x-5)$
12. $(7x+8)+(-3x+8)$
13. $(x^2+4x-6)+(x^2-4x+2)$
14. $(x^2-3x-10)+(2x^2-3x-10)$



**ANSWERS**

1. $11x-7$; first- degree binomial
2. $-x+9$; first-degree binomial
3. $2x^2+6x+2$; second-degree trinomial
4. $7x^2+x+18$; second-degree trinomial
5. $5a^2$; second-degree monomial
6. $y^3$; third-degree monomial
7. $-3y^2+7y+7$; second-degree trinomial
8. $-3a^2-6a+5$; second-degree trinomial
9. $5x^3-3x+17$; third-degree trinomial
10. $-4x^3+2x^2+11x$; third-degree trinomial
11. $5x-10$
12. $4x+16$
13. $2x^2-4$
14. $3x^2-6x-20$

15. $4y^3 + 3y^2 + 2y - 9$

16. $-2y^3 + 4y^2 - 10$

17. $x - 2$

18. $-x - 11$

19. $a^2 + 5a$

20. $-a^2 - 10a + 6$

21. $12x^3 + x^2 - 6x$

22. $5x^3 - 11x + 10$

23. [Respond below exercise.]

24. [Respond below exercise.]

25. [Respond below exercise.]

26. [Respond below exercise.]

**15.** $(4y^3 + 2y - 7) + (3y^2 - 2)$

**16.** $(-2y^3 + y^2 - 4) + (3y^2 - 6)$

**17.** $(8x + 3) - (7x + 5)$

**18.** $(4x - 9) - (5x + 2)$

**19.** $(2a^2 + 3a - 1) - (a^2 - 2a - 1)$

**20.** $(a^2 - 5a + 3) - (2a^2 + 5a - 3)$

**21.** $(9x^3 + x^2 - x) - (-3x^3 + 5x)$

**22.** $(4x^3 - 9x + 11) - (-x^3 + 2x + 1)$

Find the sums of the polynomials in Exercises 23–26.

**23.**
$$\begin{array}{r} x^2 + 5x - 7 \\ -3x^2 + 2x - 1 \\ \hline -2x^2 + 7x - 8 \end{array}$$

**24.**
$$\begin{array}{r} 2x^2 + 4x - 6 \\ -3x^2 - 9x + 2 \\ \hline -x^2 - 5x - 4 \end{array}$$

**25.**
$$\begin{array}{r} x^3 + 2x^2 + x \\ -2x^3 - 2x^2 - 2x + 6 \\ \hline -x^3 - x + 6 \end{array}$$

**26.**
$$\begin{array}{r} x^3 + 6x^2 + 7x - 8 \\ 7x^2 + 2x + 1 \\ \hline x^3 + 13x^2 + 9x - 7 \end{array}$$

Name ______________ Section ______________ Date ______________

Subtract in Exercises 27–30.

**27.**
$$\begin{array}{r} 9x^2+3x-2 \\ -(4x^2+5x+3) \\ \hline 5x^2-2x-5 \end{array}$$

**28.**
$$\begin{array}{r} -3x^2+6x-7 \\ -(2x^2-x+7) \\ \hline -5x^2+7x-14 \end{array}$$

**29.**
$$\begin{array}{r} 4x^3+0x^2+10x-15 \\ -(x^3+5x^2-3x-9) \\ \hline 3x^3-5x^2+13x-6 \end{array}$$

**30.**
$$\begin{array}{r} x^3-8x^2+11x+6 \\ -(-3x^3+8x^2-2x+6) \\ \hline 4x^3-16x^2+13x \end{array}$$

**ANSWERS**

**27.** [Respond below exercise.]

**28.** [Respond below exercise.]

**29.** [Respond below exercise.]

**30.** [Respond below exercise.]

**31.** [Respond below exercise.]

## WRITING AND THINKING ABOUT MATHEMATICS

**31. a.** Find the sum of these two polynomials:

$$\begin{array}{r} 2x^3-4x^2+3x+20 \\ 3x^3+x^2-5x+10 \\ \hline 5x^3-3x^2-2x+30 \end{array}$$

**b.** Now, substitute 1 for $x$ in each of the polynomials in part **a.**, including the sum. Does the sum of the first two values equal the value of the sum?

$2(1)^3-4(1)^2+3(1)+20=2-4+3+20=21$

$3(1)^3+(1)^2-5(1)+10=3+1-5+10=9$

$5(1)^3-3(1)^2-2(1)+30=5-3-2+30=30$

$21+9\stackrel{?}{=}30$

$30=30$

**c.** Repeat the process in part **b.** using $x = 3$.

$2(3)^3-4(3)^2+3(3)+20=54-36+9+20=47$

$3(3)^3+(3)^2-5(3)+10=81+9-15+10=85$

$5(3)^3-3(3)^2-2(3)+30=135-27-6+30=132$

$47+85\stackrel{?}{=}132$

$132=132$

**d.** Substituting a value of $x$ in each polynomial seems to provide a method for checking answers. Do you think that this method will catch all errors? Would substituting $x = 0$ be a good idea? Briefly discuss your reasoning.

Answers may vary. This does provide a legitimate check, though it is not necessarily fool-proof. Using x = 0 is not a good idea, since it will merely cause all variable terms to have value 0.

[Respond below
**32.** exercise.]

**32. a.** Find the difference of these two polynomials:

$$\begin{array}{r} 2x^3 - 4x^2 + 3x + 20 \\ -(3x^3 + x^2 - 5x + 10) \\ \hline -x^3 - 5x^2 + 8x + 10 \end{array}$$

**b.** Now, substitute 2 for $x$ in each of the polynomials in part **a.**, including the difference. Does the sum of the first two values equal the value of the difference?

$2(2)^3 - 4(2)^2 + 3(2) + 20 = 16 - 16 + 6 + 20 = 26$
$3(2)^3 + (2)^2 - 5(2) + 10 = 24 + 4 - 10 + 10 = 28$
$-(2)^3 - 5(2)^2 + 8(2) + 10 = -8 - 20 + 16 + 10 = -2$

$26 - 28 \stackrel{?}{=} -2$
$-2 = -2$

**c.** Repeat the process in part **b.** using $x = 3$.

$2(3)^3 - 4(3)^2 + 3(3) + 20 = 54 - 36 + 9 + 20 = 47$
$3(3)^3 + (3)^2 - 5(3) + 10 = 81 + 9 - 15 + 10 = 85$
$-(3)^3 - 5(3)^2 + 8(3) + 10 = -27 - 45 + 24 + 10 = -38$

$47 - 85 \stackrel{?}{=} -38$

$-38 = -38$

**d.** Substituting a value of $x$ in each polynomial seems to provide a method for checking answers. Do you think that this method will catch all errors? Would substituting $x = 0$ be a good idea? Briefly discuss your reasoning.

Answers will vary. See Exercise 31, part **d**.

# 8.3 Multiplication with Polynomials

## Multiplying a Monomial and a Polynomial

The Product Rule for exponents discussed in Section 1.6 and again in Section 8.1 is needed for multiplying two ( or more ) polynomials. Multiplication with more than two polynomials is not discussed at this time.

**Product Rule for Exponents:**

For any nonzero real number $a$ and integers $m$ and $n$,

$$a^m \times a^n = a^{m+n}$$

**Objectives**

1. Learn how to multiply a monomial and a polynomial by using the distributive property.
2. Learn how to multiply two binomials by using the distributive property.
3. Be able to use the FOIL method to multiply two binomials.
4. Know how to use a vertical format to multiply two polynomials.

**To multiply a monomial times a polynomial with two or more terms,** the distributive property can be used in the form

$$a(b+c) = ab + ac$$

For example

$$6x(3x+5) = 6x \cdot 3x + 6x \cdot 5 = 18x^2 + 30x$$

and

$$\begin{aligned} 4x^2(3x^2 - 5x + 2) &= 4x^2(3x^2) + 4x^2(-5x) + 4x^2(+2) \\ &= 12x^4 - 20x^3 + 8x^2 \end{aligned}$$

Note that in the second example above each term being multiplied by $4x^2$ was placed in parentheses. This use of parentheses helps in getting the correct signs in the product ( particularly if terms have negative coefficients ).

### Example 1

Find each product.

**a.** $3a(2a^2 + 3a - 4)$

**b.** $-2x^3(x^3 - 5x)$

**c.** $y^3(y^2 + 7y + 6)$

**Solution**

**a.** $\begin{aligned} 3a(2a^2 + 3a - 4) &= 3a(2a^2) + 3a(3a) + 3a(-4) \\ &= 6a^3 + 9a^2 - 12a \end{aligned}$

**b.** $\begin{aligned} -2x^3(x^3 - 5x) &= -2x^3(x^3) - 2x^3(-5x) \\ &= -2x^6 + 10x^4 \end{aligned}$

**c.** $\begin{aligned} y^3(y^2 + 7y + 6) &= y^3 \cdot y^2 + y^3 \cdot 7y + y^3 \cdot 6 \\ &= y^5 + 7y^4 + 6y^3 \end{aligned}$

## Multiplying Two Binomials

**To multiply two binomials,** we can apply the distributive property in the form

$(b + c)a = ba + ca,$ where another binomial takes the place of $a$.

For example, to find the product $(x + 5x)(x + 8)$, we treat the binomial $(x + 8)$ as a single term and multiply on the right as follows:

$$
\begin{aligned}
(b+c)a &= ba + ca \\
(x+5)\overbrace{(x+8)} &= x\overbrace{(x+8)} + 5\overbrace{(x+8)} \\
&= x \cdot x + x \cdot 8 + 5 \cdot x + 5 \cdot 8 \\
&= x^2 + 8x + 5x + 40 \\
&= x^2 + 13x + 40
\end{aligned}
$$

Apply the distributive property twice more.

### Example 2

Find each product.

**a.** $(x + 5)(x - 10)$

**b.** $(2x - 3)(3x - 7)$

#### Solution

**a.** $(x + 5)(x - 10) = x(x - 10) + 5(x - 10)$
$= x^2 - 10x + 5x - 50$
$= x^2 - 5x - 50$

**b.** $(2x - 3)(3x - 7) = 2x(3x - 7) - 3(3x - 7)$
$= 6x^2 - 14x - 9x + 21$
$= 6x^2 - 23x + 21$

*Now Work Exercises 1-4 in the Margin.*

Find each product and simplify.

**1.** $5a(3a^2 + 9a - 2)$
$15a^3 + 45a^2 - 10a$

**2.** $7x^2(-2x^3 + 8x^2 + 3x - 5)$
$-14x^5 + 56x^4 + 21x^3 - 35x^2$

**3.** $(x + 4)(x + 6)$
$x^2 + 10x + 24$

**4.** $(x - 9)(2x + 1)$
$2x^2 - 17x - 9$

## Using the Foil Method to Multiply Two Binomials

The process of using the distributive property to multiply two binomials can be shortened considerably by using the **F–O–I–L method**. The word **FOIL** is a memory device developed as follows:

F L

$(2x + 5)(3x + 1) = 2x(3x + 1) + 5(3x + 1)$

I

$= \underbrace{2x \cdot 3x} + \underbrace{2x \cdot 1} + \underbrace{5 \cdot 3x} + \underbrace{5 \cdot 1}$

O

| F | O | I | L |
|---|---|---|---|
| First terms | Outside terms | Inside terms | Last terms |

With practice, this method can be done mentally in the following manner:

$$F = 2x \cdot 3x \quad L = 5 \cdot 1$$

$$(2x+5)(3x+1) = \overset{F}{6x^2} + \overset{O}{2x} + \overset{I}{15x} + \overset{L}{5}$$

$$I = 5 \cdot 3x \qquad = 6x^2 + 17x + 5$$

$$O = 2x \cdot 1$$

### Example 3

Use the **FOIL** method to find each of the following products.

**a.** $(x+4)(x-7)$

**b.** $(3x-1)(2x-9)$

**c.** $(a+6)(a+5)$

#### Solution

**a.** $(x+4)(x-7) = x^2 - 7x + 4x - 28 = x^2 - 3x - 28$

**b.** $(3x-1)(2x-9) = 6x^2 - 27x - 2x + 9 = 6x^2 - 29x + 9$

**c.** $(a+6)(a+5) = a^2 + 5a + 6a + 30 = a^2 + 11a + 30$

*Now Work Exercises 5 and 6 in the Margin.*

Use the FOIL method to find each of the following products.

**5.** $(x-3)(x-10)$

$x^2 - 13x + 30$

**6.** $(4x+7)(x-8)$

$4x^2 - 25x - 56$

## The Difference of Two Squares: $(x+a)(x-a) = x^2 - a^2$

Consider the product of two binomials that are the sum and difference of the same two terms. For example,

$$(x+6)(x-6) \qquad \text{and} \qquad (2x+5)(2x-5)$$

Using the FOIL method, we find

$$(x+6)(x-6) = \overset{F}{x^2} \overset{O}{- 6x} \overset{I}{+ 6x} \overset{L}{- 36} = x^2 - 36$$

and

$$(2x+5)(2x-5) = \overset{F}{4x^2} \overset{O}{- 10x} \overset{I}{+ 10x} \overset{L}{- 25} = 4x^2 - 25$$

In each case, the two middle terms, $-6x$ and $+6x$ and $-10x$ and $+10x$, are opposites of each other and their sum is 0. Therefore, the resulting product has only two terms, and these two terms represent **the difference of two squares.** With this fact about the middle terms, we can move directly to the product when multiplying the sum and difference of the same two terms.

### The Difference of Two Squares

The product of the sum and difference of the same two terms will always be the difference of the squares of the terms:

$$(x+a)(x-a) = x^2 - a^2$$

In Examples 4-8, we have multiplied from left to right, because this allows for easy placement of the + and – signs and alignment of like terms. Also, we usually read polynomials from left to right. To illustrate these advantages, you might have the students try multiplying two polynomials from right to left. Also, to help them remember the technique for multiplying polynomials, you might contrast this method with that for multiplying two whole numbers.

### Example 4

Find each product.

**a.** $(x+7)(x-7)$

**b.** $(4y+3)(4y-3)$

**c.** $(2x+1)(2x-1)$

#### Solution

Since, in each case, the two binomials are the sum and difference of the same two terms, the product will be the difference of the squares of the terms.

**a.** $(x+7)(x-7) = x^2-49$

**b.** $(4y+3)(4y-3) = 16y^2-9$

**c.** $(2x+1)(2x-1) = 4x^2-1$

## Perfect Square Trinomials: $(x+a)^2 = x^2+2ax+a^2$
## $(x-a)^2 = x^2-2ax+a^2$

Now consider the situation in which the two binomials being multiplied are the same. That is, we want to **square a binomial.** As the following examples illustrate, there is a pattern that, after some practice, allows us to go directly to the product.

$$\begin{aligned}(x+6)^2 &= (x+6)(x+6)\\ &= x^2+6x+6x+36\\ &= x^2+2\cdot 6x+36\\ &= x^2+12x+36\end{aligned}$$

$$\begin{aligned}(x+7)^2 &= (x+7)(x+7)\\ &= x^2+7x+7x+49\\ &= x^2+2\cdot 7x+49\\ &= x^2+14x+49\end{aligned}$$

And the basic pattern is, $\boldsymbol{(x+a)^2 = x^2+2ax+a^2}$

The expression $\boldsymbol{x^2+2ax+a^2}$ is a trinomial that is the result of squaring a binomial, and this trinomial is called a **perfect square trinomial**.

An interesting device for remembering the result of squaring a binomial is the square shown here where the total area is the sum of the shaded areas.

The sides of the square shown in the figure are of length $(x+a)$. The area of the square is $\boldsymbol{(x+a)^2}$. The sum of the indicated areas in the figure is $\boldsymbol{x^2+2ax+a^2}$.

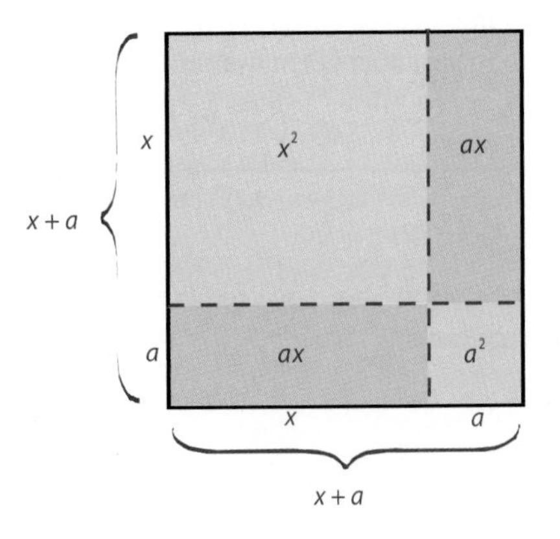

Another **perfect square trinomial** results if a binomial of the form $(x-a)$ is squared. The sign of the middle term in the trinomial will be –, instead of +.

$$\begin{aligned}(x-10)^2 &= (x-10)(x-10)\\ &= x^2-10x-10x+100\\ &= x^2-2\cdot 10x+100\\ &= x^2-20x+100\end{aligned}$$

In general, $\mathbf{(x-a)^2 = x^2-2ax+a^2}$

### Example 5

Find the following products.

**a.** $(y+8)^2$

**b.** $(y-8)^2$

**c.** $(3x+1)^2$

#### Solution

**a.** $$\begin{aligned}(y+8)^2 &= y^2+2\cdot 8y+64\\ &= y^2+16y+64\end{aligned}$$

**b.** $$\begin{aligned}(y-8)^2 &= y^2-2\cdot 8y+64\\ &= y^2-16y+64\end{aligned}$$

**c.** $$\begin{aligned}(3x+1)^2 &= 9x^2+2\cdot 3x\cdot 1+1\\ &= 9x^2+6x+1\end{aligned}$$

**Perfect Square Trinomials**

The square of a binomial gives a perfect square trinomial.

$$\begin{aligned}(x+a)^2 &= x^2+2ax+a^2\\ (x-a)^2 &= x^2-2ax+a^2\end{aligned}$$

## Multiplying Any Two Polynomials

Another method for finding the product of two polynomials is to arrange the polynomials in a vertical format and multiply by applying the distributive property. This method is particularly helpful when at least one of the polynomials has more than two terms because of the way the terms are organized.

**Multiplying Polynomials Vertically**

1. Arrange the polynomials in a vertical format with one polynomial directly below the other.
2. Multiply each term of the top polynomial by each term of the bottom polynomial. Be sure to align like terms.
3. Combine like terms.

When the original polynomials are written, there is no concern about lining up like terms. However, as terms are multiplied, like terms must be aligned vertically. As you multiply from left to right, the powers of the variable decrease. If a power is missing in the order of decreasing exponents, leave a space in the product so that like terms are aligned. Example 7 illustrates this situation.

### Example 6

Find the product $(2x + 3)(3x^2 + 4x - 5)$.

**Solution**

Arrange the polynomials in a vertical format, and multiply each term in the top polynomial by $2x$.

$$\begin{array}{l} 3x^2 + 4x - 5 \\ \underline{2x + 3 \qquad\quad} \\ 6x^3 + 8x^2 - 10x \end{array}$$

Next, multiply each term in the top polynomial by +3, and align like terms.

$$\begin{array}{rrrr} & 3x^2 & +4x & -5 \\ & & 2x & +3 \\ \hline 6x^3 & +8x^2 & -10x & \\ & 9x^2 & +12x & -15 \\ \hline \end{array}$$

Now combine like terms vertically to find the product.

$$\begin{array}{rrrr} & 3x^2 & +4x & -5 \\ & & 2x & +3 \\ \hline 6x^3 & +8x^2 & -10x & \\ & 9x^2 & +12x & -15 \\ \hline 6x^3 & +17x^2 & +2x & -15 \end{array}$$

Product

### Example 7

Multiply $(3x - 1)(x^3 + 2x + 6)$.

**Solution**

$$\begin{array}{rrrrr} & x^3 & & +2x & +6 \\ & & & 3x & -1 \\ \hline 3x^4 & & +6x^2 & +18x & \\ & -x^3 & & -2x & -6 \\ \hline 3x^4 & -x^3 & +6x^2 & +16x & -6 \end{array}$$

Note the spaces left so that like terms can be added vertically.

Product

### Completion Example 8

Find the product $( x^2 + x + 2 )( x^2 + 3x - 4 )$.

**Solution**

$x^2 + 3x - 4$
$x^2 + x + 2$
$x^4 + 3x^3 - 4x^2$ Multiply by $x^2$.
________ Multiply by $x$.
________ Multiply by 2.
________
________ Combine like terms.

*Now Work Exercises 7 and 8 in the Margin.*

**Common Error:**

Many beginning algebra students make the following error:

$(x + a)^2 = x^2 + a^2$ **INCORRECT**

$(x + 5)^2 = x^2 + 25$ **INCORRECT**

Avoid this error by remembering that the square of a binomial is a trinomial.

$(x + 5)^2 = x^2 + 2 \cdot 5x + 25$

$= x^2 + 10x + 25$ **CORRECT**

### Completion Example Answer

**8.**
$x^2 + 3x - 4$
$x^2 + x + 2$
$x^4 + 3x^3 - 4x^2$
$\mathbf{x^3 + 3x^2 - 4x}$
$\mathbf{2x^2 + 6x - 8}$
$\mathbf{x^4 + 4x^3 + x^2 + 2x - 8}$

Multiply as indicated and simplify.

**7.** $( 2x + 5 )( x^2 - 8x + 1 )$
$2x^3 - 11x^2 - 38x + 5$

**8.** ( **Note:** Be sure to align like terms as you multiply. )

$2x^3 + 4x^2 - 7$
$5x + 6$
$10x^4 + 32x^3 + 24x^2 - 35x - 42$

Name ______________________ Section ____________ Date __________

## Exercises 8.3

Find each indicated product, and simplify if possible. Tell which answers, if any, are the difference of two squares and which answers, if any, are perfect square trinomials.

**1.** $-4x^2(-3x^2)$

**2.** $(-5x^3)(-2x^2)$

**3.** $9a^2(2a)$

**4.** $-7a^2(2a^4)$

**5.** $4y(2y^2+y+2)$

**6.** $5y(3y^2-2y+1)$

**7.** $-1(4x^3-2x^2+3x-5)$

**8.** $-1(7x^3+3x^2-4x-2)$

**9.** $7x^2(-2x^2+3x-12)$

**10.** $-3x^2(x^2-x+13)$

**11.** $(x+2)(x+5)$

**12.** $(x+3)(x+10)$

**13.** $(x-4)(x-3)$

**14.** $(x-7)(x-2)$

**ANSWERS**

1. $12x^4$

2. $10x^5$

3. $18a^3$

4. $-14a^6$

5. $8y^3+4y^2+8y$

6. $15y^3-10y^2+5y$

7. $-4x^3+2x^2-3x+5$

8. $-7x^3-3x^2+4x+2$

9. $-14x^4+21x^3-84x^2$

10. $-3x^4+3x^3-39x^2$

11. $x^2+7x+10$

12. $x^2+13x+30$

13. $x^2-7x+12$

14. $x^2-9x+14$

15. $a^2+4a-12$

16. $a^2-2a-35$

17. $7x^2-20x-3$

18. $15x^2-8x-12$

19. $x^2-81$
difference of 2 squares

20. $y^2-4$
difference of 2 squares

21. $x^2-9$
difference of 2 squares

22. $y^2-25$
difference of 2 squares

23. $x^2-2.25$
difference of 2 squares

24. $x^2-6.25$
difference of 2 squares

25. $9x^2-49$
difference of 2 squares

26. $25x^2-1$
difference of 2 squares

27. $4y^2-81$
difference of 2 squares

28. $x^2-2x+1$
perfect square trinomial

**15.** $(a+6)(a-2)$

**16.** $(a-7)(a+5)$

**17.** $(7x+1)(x-3)$

**18.** $(5x-6)(3x+2)$

**19.** $(x+9)(x-9)$

**20.** $(y+2)(y-2)$

**21.** $(x+3)(x-3)$

**22.** $(y+5)(y-5)$

**23.** $(x+1.5)(x-1.5)$

**24.** $(x+2.5)(x-2.5)$

**25.** $(3x+7)(3x-7)$

**26.** $(5x+1)(5x-1)$

**27.** $(2y+9)(2y-9)$

**28.** $(x-1)^2$

Name ______________________ Section ______________ Date ____________

**ANSWERS**

**29.** $(y+1)^2$

**30.** $(x+11)^2$

**31.** $(2x+3)^2$

**32.** $(6x+5)^2$

**33.** $(2y-1)^2$

**34.** $(3x-10)^2$

**35.** $(5x+6)^2$

First, write the polynomials in a vertical format, and then find the indicated products. Be sure to leave spaces if some degrees are missing.

**36.** $(x+4)(x^2+5x+3)$

$$\begin{array}{r} x^2 + 5x + 3 \\ x + 4 \\ \hline x^3 + 9x^2 + 23x + 12 \end{array}$$

**37.** $(x+5)(x^2-2x+1)$

$$\begin{array}{r} x^2 - 2x + 1 \\ x + 5 \\ \hline x^3 + 3x^2 - 9x + 5 \end{array}$$

**38.** $(x-2)(2x^3+x^2-7x)$

$$\begin{array}{r} 2x^3 + x^2 - 7x \\ x - 2 \\ \hline 2x^4 - 3x^3 - 9x^2 + 14x \end{array}$$

**39.** $(y-6)(4y^3-y^2-7y)$

$$\begin{array}{r} 4y^3 - y^2 - 7y \\ y - 6 \\ \hline 4y^4 - 25y^3 - y^2 + 42y \end{array}$$

**40.** $(3y+1)(y^3-6y^2-10)$

$$\begin{array}{r} y^3 - 6y^2 - 10 \\ 3y + 1 \\ \hline 3y^4 - 17y^3 - 6y^2 - 30y - 10 \end{array}$$

**41.** $(2y+9)(y^3-5y^2+11)$

$$\begin{array}{r} y^3 - 5y^2 + 11 \\ 2y + 9 \\ \hline 2y^4 - y^3 - 45y^2 + 22y + 99 \end{array}$$

**29.** $y^2+2y+1$
perfect square trinomial

**30.** $x^2+22x+121$
perfect square trinomial

**31.** $4x^2+12x+9$
perfect square trinomial

**32.** $36x^2+60x+25$
perfect square trinomial

**33.** $4y^2-4y+1$
perfect square trinomial

**34.** $9x^2-60x+100$
perfect square trinomial

**35.** $25x^2+60x+36$
perfect square trinomial

**36.** [Respond below exercise.]

**37.** [Respond below exercise.]

**38.** [Respond below exercise.]

**39.** [Respond below exercise.]

**40.** [Respond below exercise.]

**41.** [Respond below exercise.]

[Respond below exercise.]
42. ________

[Respond below exercise.]
43. ________

[Respond below exercise.]
44. ________

[Respond below exercise.]
45. ________

[Respond below exercise.]
46. ________

[Respond below exercise.]
47. ________

[Respond below exercise.]
48. ________

[Respond below exercise.]
49. ________

[Respond below exercise.]
50. ________

Find the indicated products. Be sure to align like terms as you are multiplying and to leave spaces if some degrees are missing.

**42.** $x^4 + 2x^3 - 5x^2 + 3$
$\underline{x^2 + x - 1}$
$x^6 + 3x^5 - 4x^4 - 7x^3 + 8x^2 + 3x - 3$

**43.** $x^4 - 3x^2 + 4x - 1$
$\underline{x^2 - 2x + 1}$
$x^6 - 2x^5 - 2x^4 + 10x^3 - 11x^2 + 6x - 1$

**44.** $x^3 - 7x + 2$
$\underline{x^2 + 2x + 3}$
$x^5 + 2x^4 - 4x^3 - 12x^2 - 17x + 6$

**45.** $y^3 + 3y - 4$
$\underline{2y^2 - y + 3}$
$2y^5 - y^4 + 9y^3 - 11y^2 + 13y - 12$

**46.** $y^4 + 3y^2 - 5$
$\underline{y^4 - y^2 + 1}$
$y^8 + 2y^6 - 7y^4 + 8y^2 - 5$

**47.** $a^4 - 4a^2 + 7$
$\underline{a^4 + a^2 - 1}$
$a^8 - 3a^6 + 2a^4 + 11a^2 - 7$

**48.** $a^3 + a^2 + a + 1$
$\underline{a^3 + a^2 + a + 1}$
$a^6 + 2a^5 + 3a^4 + 4a^3 + 3a^2 + 2a + 1$

**49.** $x^3 - x^2 + x - 1$
$\underline{x^3 - x^2 + x - 1}$
$x^6 - 2x^5 + 3x^4 - 4x^3 + 3x^2 - 2x + 1$

**50.** $x^3 + 2x - 1$
$\underline{x^2 + 2x + 1}$
$x^5 + 2x^4 + 3x^3 + 3x^2 - 1$

Name ______________________ Section ______________ Date __________

## WRITING AND THINKING ABOUT MATHEMATICS

[Respond below
51. exercise.]

**51. a.** Find the product of these two polynomials:

$$\begin{array}{r} 2x^3 - 4x^2 - 5x + 6 \\ 5x - 2 \\ \hline 10x^4 - 24x^3 - 17x^2 + 40x - 12 \end{array}$$

**b.** Now, substitute 2 for $x$ in each of the polynomials in part **a.**, including the product. Does the product of the first two values equal the value of the product?

$2(2)^3 - 4(2)^2 - 5(2) + 6 = 16 - 16 - 10 + 6 = -4$

$5(2) - 2 = 10 - 2 = 8$

$10(2)^4 - 24(2)^3 - 17(2)^2 + 40(2) - 12 = 160 - 192 - 68 + 80 - 12 = -32$

$-4 \cdot 8 \stackrel{?}{=} -32$

$-32 = -32$

**c.** Repeat the process in part **b.** using $x = -2$.

$2(-2)^3 - 4(-2)^2 - 5(-2) + 6 = -16 - 16 + 10 + 6 = -16$

$5(-2) - 2 = -10 - 2 = -12$

$10(-2)^4 - 24(-2)^3 - 17(-2)^2 + 40(-2) - 12 = 160 + 192 - 68 - 80 - 12 = 192$

$-12(-16) \stackrel{?}{=} 192$

$192 = 192$

**d.** Substituting a value of $x$ in each polynomial seems to provide a method for checking answers. Do you think that this method will catch all errors? Would substituting $x = 0$ be a good idea? Briefly discuss your reasoning.

Answers may vary. This does provide a legitimate check, though it is not necessarily fool-proof. Using x = 0 is not a good idea, since it will merely cause all variable terms to have value 0.

# 8.4 Introduction to Factoring Polynomials

## Greatest Common Factor ( GCF )

Factoring is the reverse of multiplication. Thus, factoring polynomials involves remembering how to multiply polynomials. In this way, factoring is built on previous knowledge and skills with multiplication. In Section 8.3, we developed techniques using the distributive property and the **FOIL** method for multiplying polynomials. The result of multiplication is called the **product**, and the expressions and/or numbers being multiplied are called **factors** of the product. For example, by using the distributive property,

$$\underbrace{3x}_{\text{factor}}\underbrace{(x+5)}_{\text{factor}} = 3x \cdot x + 3x \cdot 5 = \underbrace{3x^2 + 15x}_{\text{product}}$$

**Objectives**

① Know that GCF stands for Greatest Common Factor.

② Learn how to factor out the GCF in a polynomial.

③ Learn to recognize and factor the difference of two squares and perfect square trinomials.

Now, we want to reverse this process. That is, given the product, we want to find the factors. This is called **factoring**. ( As you will see in a course in algebra, factoring is very useful in solving equations and simplifying algebraic expressions. )

Just as in arithmetic where multiplication and division are related, factoring with polynomials is closely related to division. From our work with exponents,

we know the Quotient Rule: for $a \neq 0$, $\dfrac{a^m}{a^n} = a^{m-n}$.

You may want to teach ( or just make reference to ) the material on greatest common factor ( GCF ) in Appendix IV. In case you choose to do this, I have included some algebraic terms there for the students to relate to this section.

This rule can be used when dividing two monomials. For example,

$$\frac{45x^7}{5x^3} = 9x^{7-3} = 9x^4 \quad \text{and} \quad \frac{21a^6}{-3a} = -7a^{6-1} = -7a^5$$

To divide a polynomial by a monomial, a procedure discussed in detail in Appendix V, we divide **each term** in the polynomial by the monomial. For example,

$$\frac{10x^3 - 14x^2 + 6x}{2x} = \frac{10x^3}{2x} - \frac{14x^2}{2x} + \frac{6x}{2x} = 5x^2 - 7x + 3$$

This concept of dividing each term by a monomial is part of finding a monomial factor of a polynomial. Finding the **Greatest Common Monomial Factor ( GCF )** in a polynomial means to **find the monomial with the highest degree and the largest integer coefficient that will divide into each term of the polynomial.** ( In later courses you will find that the GCF can itself be a polynomial. )

Thus, two of the factors of a polynomial will be the GCF and the sum of the various terms found by dividing by the GCF. For example, consider factoring

$$10x^3 - 14x^2 + 6x$$

We have already seen that $2x$ will divide into each term and yield **integer** coefficients. ( Note that we could divide by just 2 or divide by just $x$. However, $2x$ is the GCF. ) We say that $2x$ is **factored out** as follows:

$$\begin{aligned} 10x^3 - 14x^2 + 6x &= 2x \cdot 5x^2 + 2x \cdot (-7x) + 2x \cdot 3 \\ &= 2x(5x^2 - 7x + 3) \end{aligned}$$

and $2x$ and ( $5x^2 - 7x + 3$ ) are factors. After some practice, finding the polynomial factor $5x^2 - 7x + 3$ should be done by mentally dividing each term by $2x$.

Now, factor out the greatest common monomial factor ( GCF ) from the polynomial

$$18x^6 - 12x^4 - 6x^3$$

Each term is divisible by several monomials. For example, 6, $2x$, $3x^2$, $2x^3$ , and $6x^3$ will all divide into each term. However, $6x^3$ is the GCF because it has the largest coefficient and the largest exponent of all the common factors. We can factor out $6x^3$ as follows:

$$18x^6 - 12x^4 - 6x^3 = 6x^3 \cdot 3x^3 + 6x^3 \cdot (-2x) + 6x^3 \cdot (-1)$$

$$= 6x^3(3x^3 - 2x - 1)$$

**Remember that in factoring, the answer can be easily checked by multiplying the factors to see if the product is the original polynomial.**

If all the terms in the original polynomial have negative coefficients or if just the leading term has a negative coefficient, we choose the GCF to be negative. In this way, the leading term in the polynomial factor will be positive. For example,

$$-2x + 14 = -2 \cdot x - 2 \cdot (-7)$$

$$= -2(x - 7)$$ Here $-2$ is factored out.

$$-15x^2 - 25x = -5x \cdot 3x - 5x \cdot 5$$

$$= -5x(3x + 5)$$ Here $-5x$ is factored out.

**To factor completely** means to find factors of the polynomial so that none of the factors are themselves factorable. For example,

$4x + 12 = 2(2x + 6)$ is not factored completely because 2 is a factor of $2x + 6$.

$4x + 12 = 4(x + 3)$ is factored completely.

Also, note that not every polynomial has a common monomial factor. In this case, we just rewrite the original polynomial or, as we will see later, try some other factoring technique.

## Example 1

Factor each polynomial by finding and factoring out the GCF.

**a.** $x^3 + 9x$

**b.** $5x^3 - 15x^2$

**c.** $x^4 + 6x^2 + 1$

**d.** $-4a^5 + 20a^4 - 24a^2$

**Solution**

**a.** $x^3 + 9x = x \cdot x^2 + x \cdot 9$
$= x(x^2 + 9)$

Note that $x^2 + 9$ is the **sum** of two squares. The sum of two squares is not factorable.

**b.** $5x^3 - 15x^2 = 5x^2 \cdot x + 5x^2 \cdot (-3)$
$= 5x^2(x - 3)$

**c.** $x^4 + 6x^2 + 1$

There is no common monomial factor other than 1.

**d.** Since the leading term is negative, $-4a^2$ is factored out.

$$-4a^5 + 20a^4 - 24a^2 = -4a^2 \cdot (a^3) - 4a^2 \cdot (-5a^2) - 4a^2 \cdot (+6)$$
$$= -4a^2(a^3 - 5a^2 + 6)$$

*Now Work Exercises 1-3 in the Margin.*

Factor each polynomial by finding and factoring out the GCF.

**1.** $x^4 + 5x^3$

$x^3(x + 5)$

**2.** $-10x^2 - 20x - 15$

$-5(2x^2 + 4x + 3)$

**3.** $6x^3 + 18x^2$

$6x^2(x + 3)$

## Factoring Special Products

In Section 8.3 we discussed multiplication that resulted in the difference of two squares and perfect square trinomials. For example,

$(x + 3)(x - 3) = x^2 - 9$ — The difference of two squares

$(x + 3)^2 = x^2 + 2 \cdot 3x + 9$
$= x^2 + 6x + 9$ — A perfect square trinomial

$(x - 3)^2 = x^2 - 2 \cdot 3x + 9$
$= x^2 - 6x + 9$ — A perfect square trinomial

Now, in factoring, we can use these ideas in reverse. They are so common and useful that we write them here as formulas.

**Factoring Special Products**

**I.** $x^2 - a^2 = (x + a)(x - a)$ — **Difference of two squares**

**II.** $x^2 + 2ax + a^2 = (x + a)^2$ — **Perfect square trinomial** (The + signs correspond.)

**III.** $x^2 - 2ax + a^2 = (x - a)^2$ — **Perfect square trinomial** (The – signs correspond.)

As you can tell, I did not go deeply into factoring polynomials here. My idea is that giving the students an early look at basic forms of the difference of squares and perfect square trinomials may help them overcome some difficulties in handling more complicated expressions in Beginning Algebra. To tie these concepts more closely with those in Section 8.3, you might choose to discuss some elementary factoring of trinomials by using the FOIL method.

In general, when factoring polynomials, look for a common monomial factor first. Then, after factoring out the GCF, see if the remaining polynomial is the difference of two squares or a perfect square trinomial. In a beginning algebra course, you will also learn how to use the FOIL method in reverse to factor polynomials.

### Example 2

Factor each of the following polynomials completely.

**a.** $y^2 - 25$

**b.** $16x^2 - 49$

**c.** $x^2 - 14x + 49$

**d.** $2y^2 + 12y + 18$

**e.** $5y^2 + 180$

#### Solution

**a.** Difference of two squares:
$y^2 - 25 = (y+5)(y-5)$

**b.** Difference of two squares:
$16x^2 - 49 = (4x+7)(4x-7)$

**c.** Perfect square trinomial:
$x^2 - 14x + 49 = (x-7)^2$

Note: $49 = 7^2$ and $-14x = -2 \times 7 \times x$

**d.** $2y^2 + 12y + 18 = 2(y^2 + 6y + 9)$
$= 2(y+3)^2$

First factor out the GCF 2. Note that $y^2 + 6y + 9$ is a perfect square trinomial. ($9 = 3^2$ and $6y = 2 \cdot 3 \cdot y$)

**e.** $5y^2 + 180 = 5(y^2 + 36)$

Note that $y^2 + 36$ is the **sum** of two squares. The sum of two squares is **not** factorable.

*Now Work Exercises 4–8 in the Margin.*

Factor each of the following polynomials completely.

**4.** $4x^2 + 100$
$4(x^2 + 25)$

**5.** $4x^2 - 100$
$4(x+5)(x-5)$

**6.** $4x^2 - 80x + 400$
$4(x - 10)^2$

**7.** $5x^2 - 15x - 15$
$5(x^2 - 3x - 3)$

**8.** $16x^2 + 8x + 8$
$8(2x^2 + x + 1)$

Name ____________ Section ____________ Date ____________

## Exercises 8.4

Factor each of the following polynomials by factoring out the GCF.

1. $2x+14$
2. $3x+15$
3. $5x-10$
4. $6x-18$
5. $3x^2-5x$
6. $4x^2+3x$
7. $5x^2+10x$
8. $9x^2-3x$
9. $2x^2-2x$
10. $8x^2+10x$
11. $5a^3+5a^2+5a$
12. $12a^3-4a^2+2a$
13. $6y^3-12y^2-6y$
14. $10y^3+20y^2+40y$
15. $15x^4+10x^3$
16. $16x^4-20x^2$
17. $72x^4-27x^3+9x^2$
18. $25a^5+25a^3-50a^2$
19. $30a^6+18a^4+12a^3+12a^2$
20. $28y^6-20y^5+20y^3-4y^2$

Factor each of the following special products.

21. $x^2-1$
22. $x^2-25$
23. $x^2-169$
24. $x^2-196$

**ANSWERS**

1. $2(x+7)$
2. $3(x+5)$
3. $5(x-2)$
4. $6(x-3)$
5. $x(3x-5)$
6. $x(4x+3)$
7. $5x(x+2)$
8. $3x(3x-1)$
9. $2x(x-1)$
10. $2x(4x+5)$
11. $5a(a^2+a+1)$
12. $2a(6a^2-2a+1)$
13. $6y(y^2-2y-1)$
14. $10y(y^2+2y+4)$
15. $5x^3(3x+2)$
16. $4x^2(4x^2-5)$
17. $9x^2(8x^2-3x+1)$
18. $25a^2(a^3+a-2)$
19. $6a^2(5a^4+3a^2+2a+2)$
20. $4y^2(7y^4-5y^3+5y-1)$
21. $(x+1)(x-1)$
22. $(x+5)(x-5)$
23. $(x+13)(x-13)$
24. $(x+14)(x-14)$

25. $(y+15)(y-15)$

26. $(y+8)(y-8)$

27. $(6y+1)(6y-1)$

28. $(5z+2)(5z-2)$

29. $(4z+9)(4z-9)$

30. $(3z+4)(3z-4)$

31. $(x+1)(x+1)$

32. $(x-1)(x-1)$

33. $(y-5)(y-5)$

34. $(y+5)(y+5)$

35. $(z+6)(z+6)$

36. $(z-7)(z-7)$

37. $(x-9)(x-9)$

38. $(x-15)(x-15)$

39. $(x+20)(x+20)$

40. $(x+14)(x+14)$

41. $2(x+5)(x+5)$

42. $2(x^2+6x+36)$

43. $4(x-1)(x-1)$

44. $5(3y+4)(3y-4)$

45. $2(y+5)(y-5)$

46. $50(y+1)(y-1)$

**25.** $y^2-225$

**26.** $y^2-64$

**27.** $36y^2-1$

**28.** $25z^2-4$

**29.** $16z^2-81$

**30.** $9z^2-16$

**31.** $x^2+2x+1$

**32.** $x^2-2x+1$

**33.** $y^2-10y+25$

**34.** $y^2+10y+25$

**35.** $z^2+12z+36$

**36.** $z^2-14z+49$

**37.** $x^2-18x+81$

**38.** $x^2-30x+225$

**39.** $x^2+40x+400$

**40.** $x^2+28x+196$

Factor each of the following polynomials completely. If there are no factors other than 1, write *not factorable.*

**41.** $2x^2+20x+50$

**42.** $2x^2+12x+72$

**43.** $4x^2-8x+4$

**44.** $45y^2-80$

**45.** $2y^2-50$

**46.** $50y^2-50$

Name ______________________ Section ______________ Date __________

**47.** $6x^3 - 12x^2 + 6x$

**48.** $8x^3 + 32x^2 + 32x$

**49.** $9y^3 - 36y^2 + 36y$

**50.** $7y^4 + 42y^3 + 63y^2$

**51.** $x^2 - 5x + 10$

**52.** $x^2 + 4x - 2$

**53.** $5x^2 - 5x + 5$

**54.** $10x^2 + 50x + 20$

**55.** $4x^2 + 36x - 28$

**56.** $6x^2 + 6$

**57.** $7x^2 + 28$

**58.** $5x^3 + 125x$

**59.** $4y^3 + 100y$

**60.** $8y^4 + 72y^2$

**ANSWERS**

**47.** $6x(x-1)(x-1)$

**48.** $8x(x+2)(x+2)$

**49.** $9y(y-2)(y-2)$

**50.** $7y^2(y+3)(y+3)$

**51.** not factorable

**52.** not factorable

**53.** $5(x^2 - x + 1)$

**54.** $10(x^2 + 5x + 2)$

**55.** $4(x^2 + 9x - 7)$

**56.** $6(x^2 + 1)$

**57.** $7(x^2 + 4)$

**58.** $5x(x^2 + 25)$

**59.** $4y(y^2 + 25)$

**60.** $8y^2(y^2 + 9)$

# 8.5 Graphing Ordered Pairs of Real Numbers

## Equations in Two Variables

Equations such as

$$d = 40t, \qquad I = 1.18P, \qquad \text{and} \qquad y = 2x - 5$$

represent relationships between pairs of variables. These equations are said to be **equations in two variables**. The first equation, $d = 40t$, can be interpreted as follows: The distance $d$ traveled in time $t$ at a rate of 40 miles per hour is found by multiplying 40 by $t$ ( where $t$ is measured in hours ). Thus, if $t = 3$ hours, then $d = 40(3) = 120$ miles. The pair ( 3, 120 ) is called an **ordered pair** and is in the form ( $t$, $d$ ).

We say that the ordered pair ( 3, 120 ) **is a solution of** ( or **satisfies** ) the equation $d = 40t$. Similarly, ( 5, 200 ) represents $t = 5$ and $d = 200$ and satisfies the equation $d = 40t$. In the same way, ( 100, 18 ) satisfies the equation $I = 0.18P$, where $P = 100$ and $I = 0.18(100) = 18$. In this equation, the interest $I$ is equal to 18% ( or 0.18 ) times the principal $P$, and the ordered pair (100, 18) is in the form ( $P$, $I$ ).

For the equation $y = 2x - 5$, ordered pairs are in the form ( $x$, $y$ ), and ( 3, 1 ) satisfies the equation: If $x = 3$, then $y = 2(3) - 5 = 1$. In the ordered pair ( $x$, $y$ ), $x$ is called the **first coordinate** ( or **first component** ), and $y$ is called the **second coordinate** ( or **second component** ). To find ordered pairs that satisfy an equation in two variables, we can **choose any value** for one variable and find the corresponding value for the other variable by substituting into the equation. For example,

For the equation $\mathbf{y = 2x - 5}$

| For $x$: | Substitution: | Ordered Pair: |
|---|---|---|
| $x = 2$ | $y = 2(2) - 5 = 4 - 5 = -1$ | ( 2, −1 ) |
| $x = 0$ | $y = 2(0) - 5 = 0 - 5 = -5$ | ( 0, −5 ) |
| $x = 6$ | $y = 2(6) - 5 = 12 - 5 = 7$ | ( 6, 7 ) |

All the ordered pairs ( 2, −1 ), ( 0, −5 ), and ( 6, 7 ) satisfy the equation $y = 2x - 5$. There are an infinite number of such ordered pairs.

Since the equation $y = 2x - 5$ is solved for $y$, we say that the value $y$ "depends" on the choice of $x$. Thus, in an ordered pair of the form ( $x$, $y$ ), the second coordinate $y$ is called the **dependent variable,** and the first coordinate $x$ is called the **independent variable**.

Some examples of ordered pairs for each of the three equations discussed are shown on the following page in table form. Remember that the choices for the value of the independent variable are arbitrary; other values could have been chosen.

**Objectives**

① Learn the terminology related to graphs in two dimensions, such as **ordered pairs, first coordinate, second coordinate, $x$-axis, $y$-axis,** and **quadrant**.

② Be able to list the set of ordered pairs represented on a graph.

③ Know how to graph a set of ordered pairs of real numbers.

| $d = 40t$ | | | $I = 0.18P$ | | | $y = 2x - 5$ | |
|---|---|---|---|---|---|---|---|
| $t$ | $d$ | | $P$ | $I$ | | $x$ | $y$ |
| 1 | 40 | | 100 | 18 | | –2 | –9 |
| 2 | 80 | | 200 | 36 | | 0 | –5 |
| 3 | 120 | | 1000 | 180 | | 1 | –3 |
| 4 | 160 | | 5000 | 900 | | 5 | 5 |

## Graphing Ordered Pairs

Ordered pairs of real numbers can be graphed as points in a plane by using the **Cartesian coordinate system** [ named after the famous French mathematician René Descartes ( 1596 – 1650 ) ]. In this system, the plane is separated into four **quadrants** by two number lines that are perpendicular to each other. The lines intersect at a point called the **origin,** represented by the ordered pair ( 0, 0 ). The horizontal number line represents the independent variable and is called the **horizontal axis** ( or the ***x*–axis** ). The vertical number line represents the dependent variable and is called the **vertical axis** ( or the ***y*–axis** ). ( See Figure 8.1. )

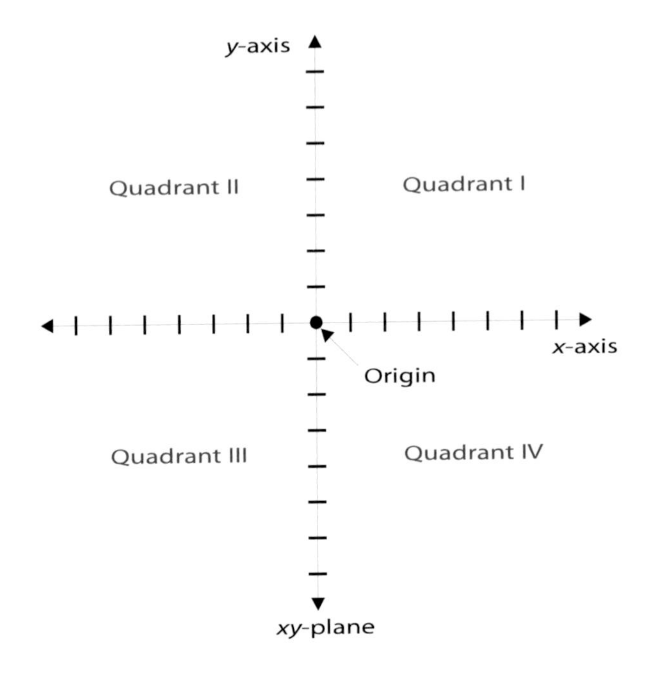


**Figure 8.1**

On the $x$-axis:

positive values of $x$ are indicated to the right of the origin and

negative values of $x$ are to the left of the origin.

On the $y$-axis:

positive values of $y$ are indicated above the origin and

negative values of $y$ are below the origin.

The relationship between ordered pairs of real numbers and the points in a plane is similar to the correspondence between real numbers and points on a number line and is expressed in the following statement:

> **There is a one-to-one correspondence between the points in a plane and ordered pairs of real numbers.** That is, each point in a plane corresponds to one ordered pair of real numbers, and each ordered pair of real numbers corresponds to one point in a plane. ( See Figure 8.2. )

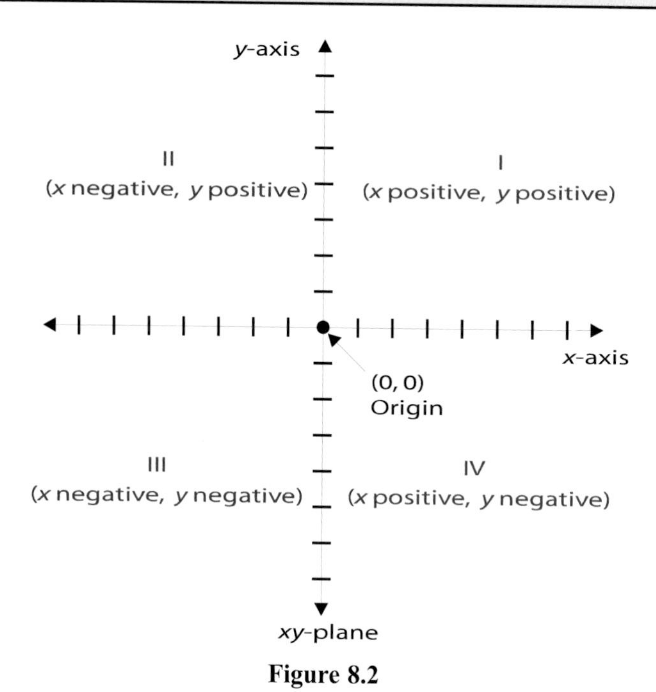


**Figure 8.2**

The graphs of the points $A(3, 1)$, $B(-2, 3)$, $C(-3, -1)$, $D(1, -2)$, and $E(2, 0)$ are shown in Figure 8.3. The point $E(2, 0)$ is on an axis and not in any quadrant. Each ordered pair is called the **coordinates** of the corresponding point. For example, the coordinates of point $A$ are given by the ordered pair $(3, 1)$.

| POINT | QUADRANT |
|---|---|
| $A(3,1)$ | I |
| $B(-2,3)$ | II |
| $C(-3,-1)$ | III |
| $D(1,-2)$ | IV |
| $E(2,0)$ | $x$-axis |

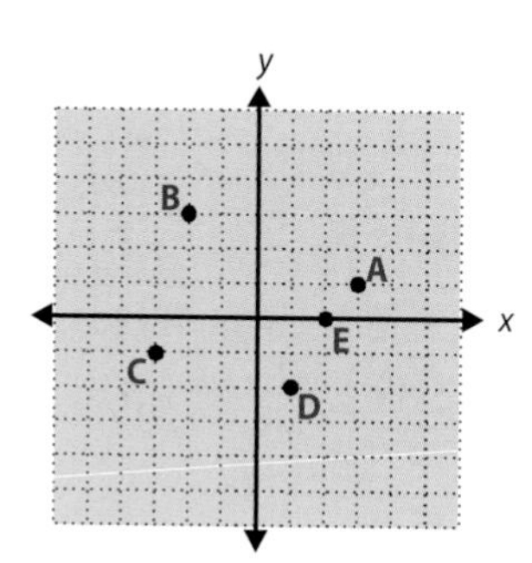


**Figure 8.3**

> **Note:** Unless a scale is labeled on the $x$-axis or on the $y$-axis, the grid lines are assumed to be one unit apart in both the horizontal and vertical directions.

## Example 1

Graph the following set of ordered pairs.

$\{(-2, 1), (0, 3), (1, 2), (2, -2)\}$

**Solution**

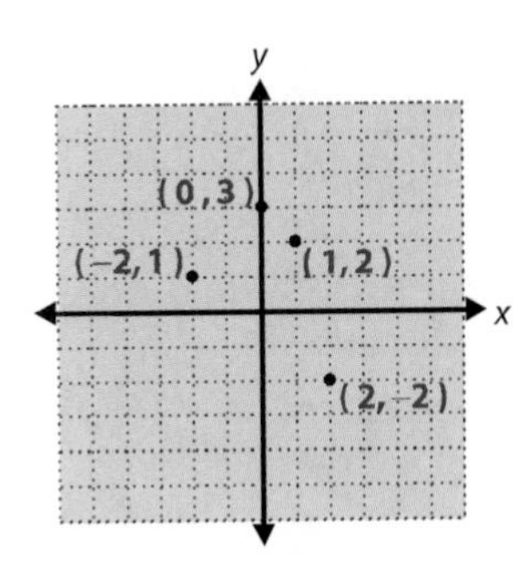


## Example 2

Graph the following set of ordered pairs.

$\{(-3, -5), (-2, -3), (-1, -1), (0, 1), (1, 3)\}$

**Solution**

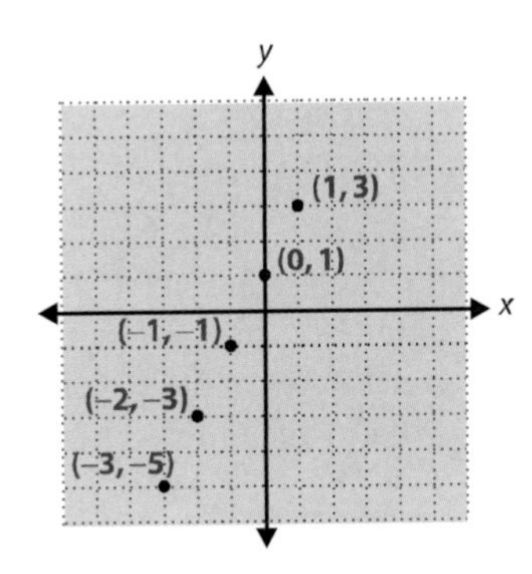


## Example 3

The graph of a set of points is given. List the set of ordered pairs that correspond to the points in the graph.

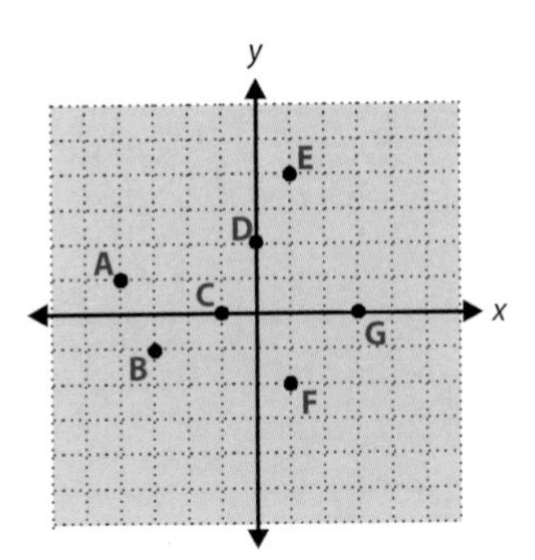


**Solution**

$\{A(-4, 1), B(-3, -1), C(-1, 0), D(0, 2), E(1, 4), F(1, -2), G(3, 0)\}$

The section is entitled, "Graphing Ordered Pairs of *Real Numbers*," but you may have noticed that in the discussion and examples, only ordered pairs of **integers** have been used. This is so because ordered pairs of integers are relatively easy to locate on a graph and relatively easy to read from a graph. These points, with integer scales indicated, are at the intersections of the horizontal and vertical lines on the graph paper.

Ordered pairs with fractions, decimals, or radicals must be located by estimating the positions of the points. And even after graphing such points, a person might not be able to read the precise coordinates intended because relatively large dots must be used so that the graphed points can be seen. For example, distinguishing between the two points $\left(\frac{1}{2},\frac{7}{12}\right)$ and $\left(\frac{1}{2},\frac{8}{12}\right)$ would be almost impossible with a normal scale.

Example 4 illustrates the difficulties encountered in graphing ordered pairs with fractions and decimals. **Even with these difficulties, you should understand that these points do exist but that graphing them involves estimating positions.**

In preparation for Section 8.6 and graphing in general, you might want to develop an early class discussion on ordered pairs of real numbers. For example, you could point out that for the equation y = 2x - 5, if $x = \sqrt{2}$ then $y = 2\sqrt{2} - 5$ and the corresponding ordered pair is $\left(\sqrt{2}, 2\sqrt{2} - 5\right)$. Also, locating and plotting the graph of this point involves approximating ( or estimating ) even though this point is just as "valid" as an ordered pair of integers.

## Example 4

Graph the following set of ordered pairs.

$$\left\{(-2,1.3)\left(0,-\frac{7}{8}\right),\left(\frac{1}{2},\frac{3}{4}\right),(1.5,2.6),(\pi,0)\right\}.$$

**Solution**

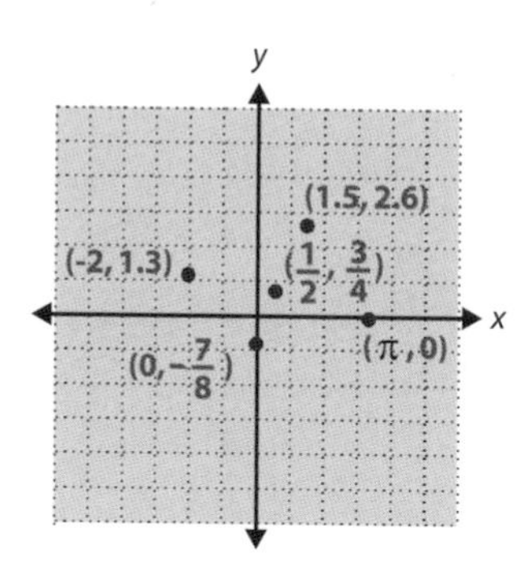

Name ________________ Section ________ Date ________

## Exercises 8.5

Below each graph in Exercises 1-10, list the set of ordered pairs that correspond to the points on the graph.

**1.**

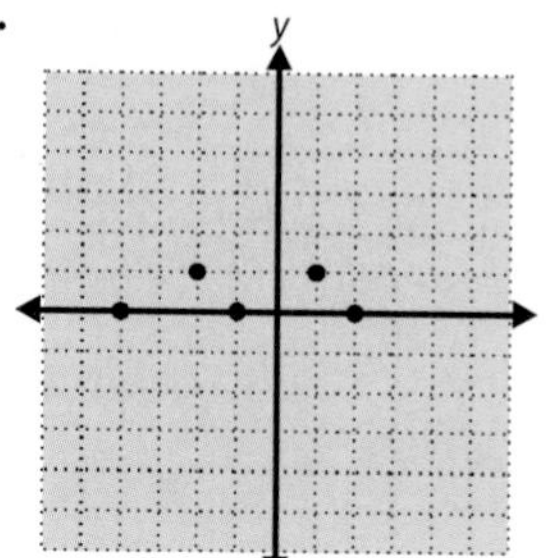


**2.**

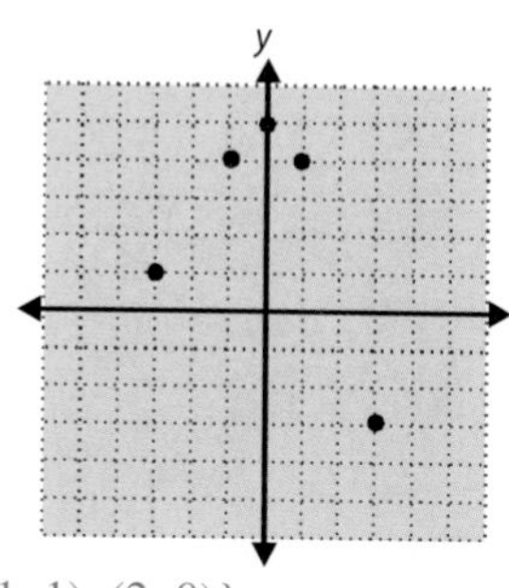


**3.**

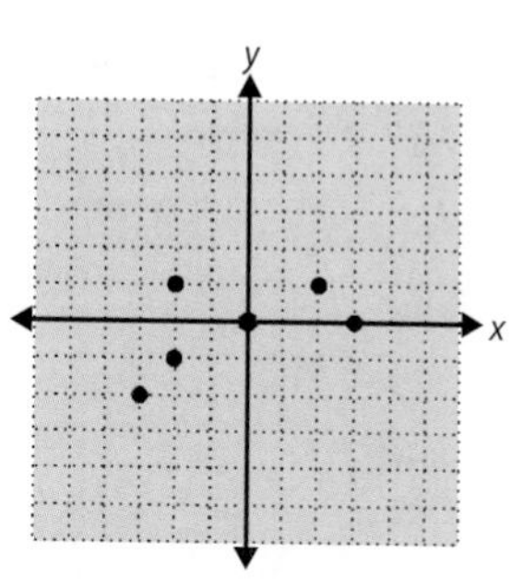


1. {(–4, 0), (–2, 1), (–1, 0), (1, 1), (2, 0)}
2. {(–3, 1), (–1, 4), (0, 5), (1, 4), (3, –3)}
3. {(–3, –2), (–2, –1), (–2, 1), (0, 0), (2, 1), (3, 0)}

**4.**

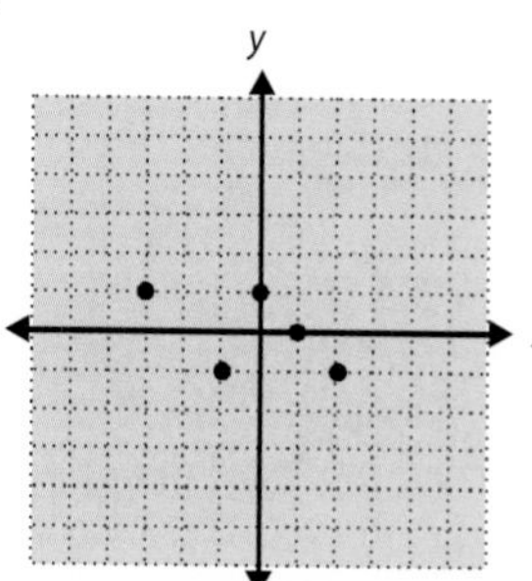


**5.**

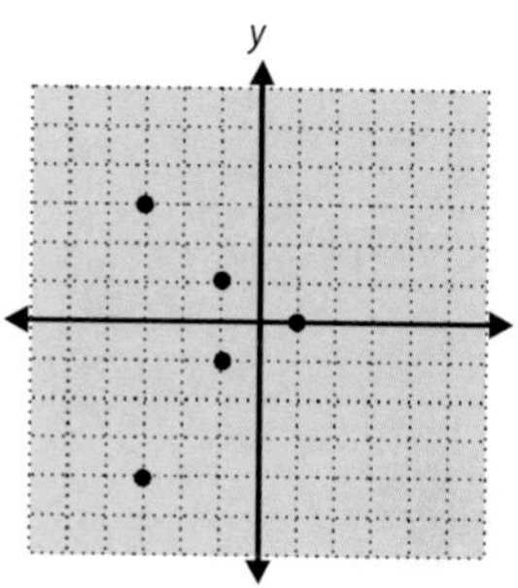


**6.**

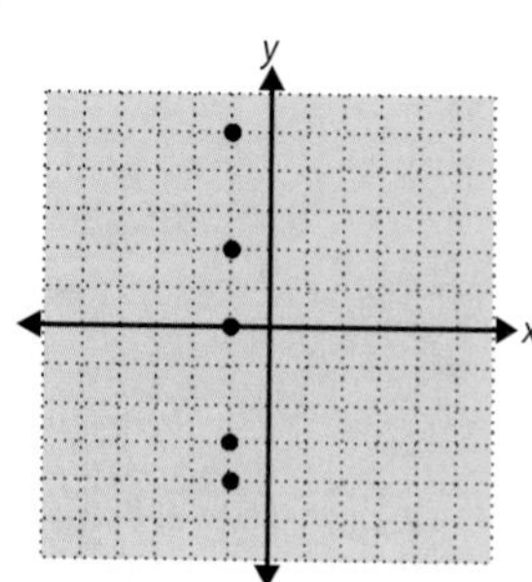


4. {(–3, 1), (–1, –1), (0, 1), (1, 0), (2, –1)}
5. {(–3, –4), (–3, 3), (–1, –1), (–1, 1), (1, 0)}
6. {(–1, 5), (–1, 2), (–1, 0), (–1, –3), (–1, –4)}

**7.**

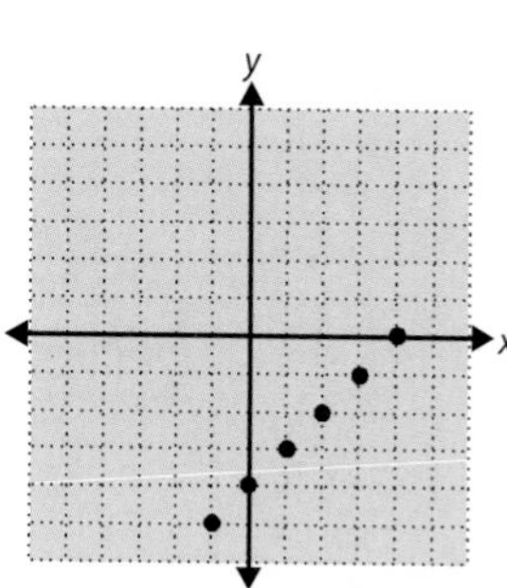


**8.**

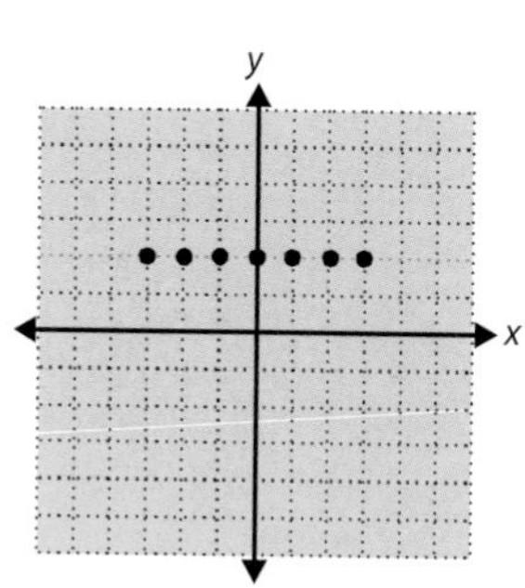


**9.**

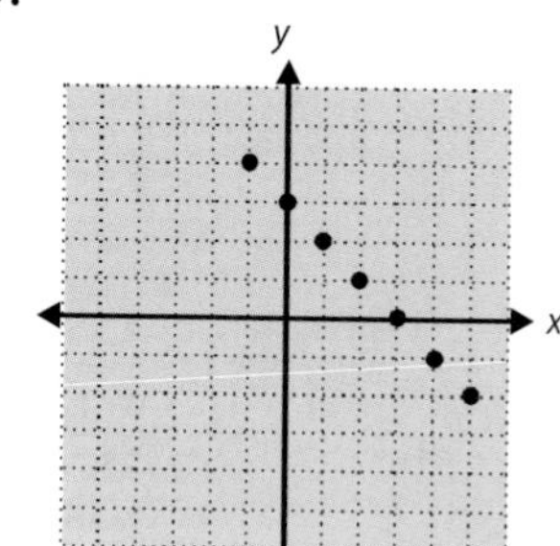


7. {(–1, –5), (0, –4), (1, –3), (2, –2), (3, –1), (4, 0)}
8. {(–3, 2), (–2, 2), (–1, 2), (0, 2), (1, 2), (2, 2), (3, 2)}
9. {(–1, 4), (0, 3), (1, 2), (2, 1), (3, 0), (4, –1), (5, –2)}

**10.**

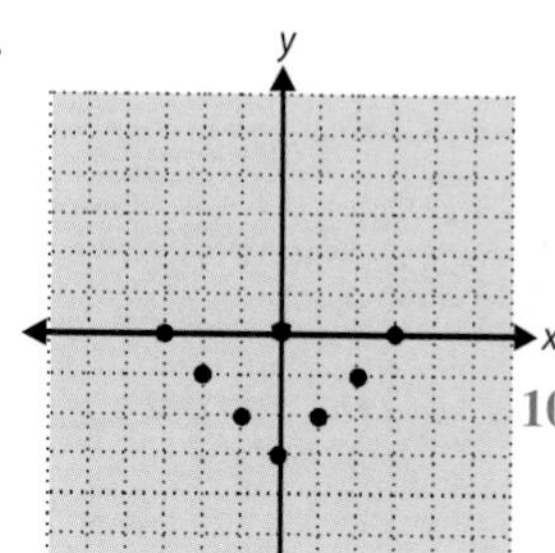


10. {(–3, 0), (–2, –1), (–1, –2), (0, 0), (0, –3), (1, –2), (2, –1), (3, 0)}

Graph each of the following sets of ordered pairs.

**11.** $\{(-2, 4), (-1, 3), (0, 1), (1, -2), (1, 3)\}$

**12.** $\{(-5, 1), (-3, \ ), (-2, -1), (0, 2), (2, -1)\}$

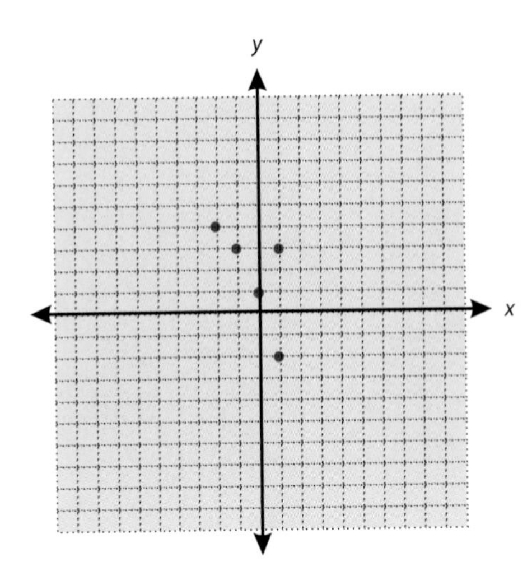


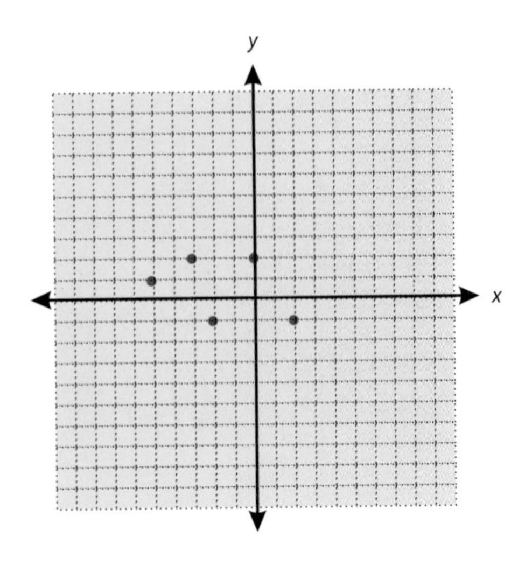

**13.** $\{(-1, 2), (1, 3), (2, -2), (3, 4), (4, -2)\}$

**14.** $\{(-2, 3), (-1, 0), (0, -3), (2, 3), (4, -1)\}$

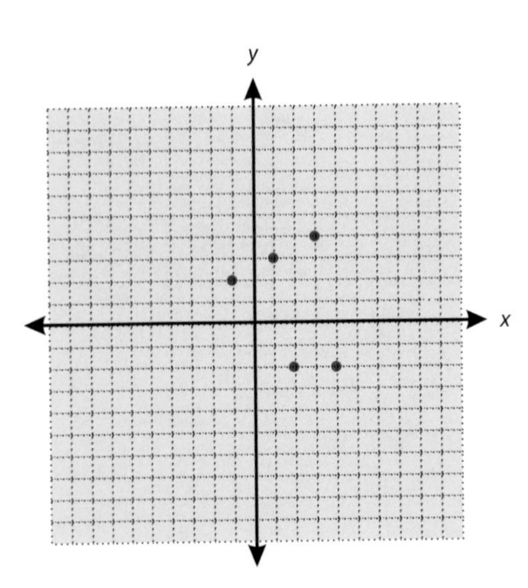


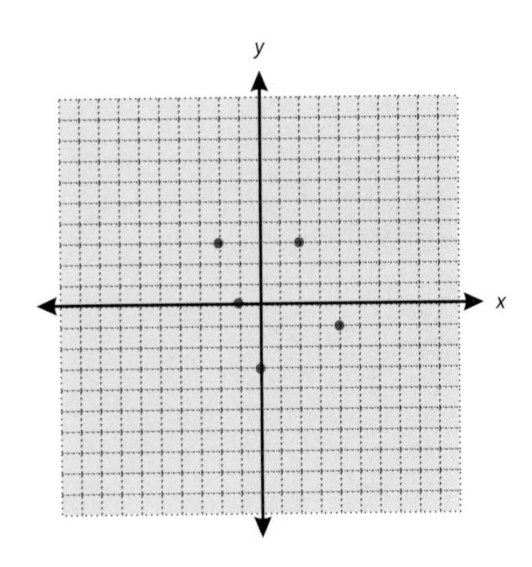


**15.** $\{(0, -3), (1, -1), (2, 1), (3, 3), (4, 5)\}$

**16.** $\{(-3, 3), (-2, 2), (-1, 1), (0, 0), (1, -1)\}$

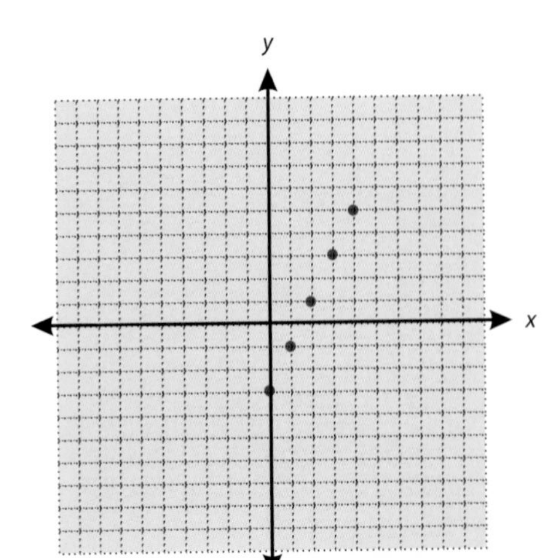


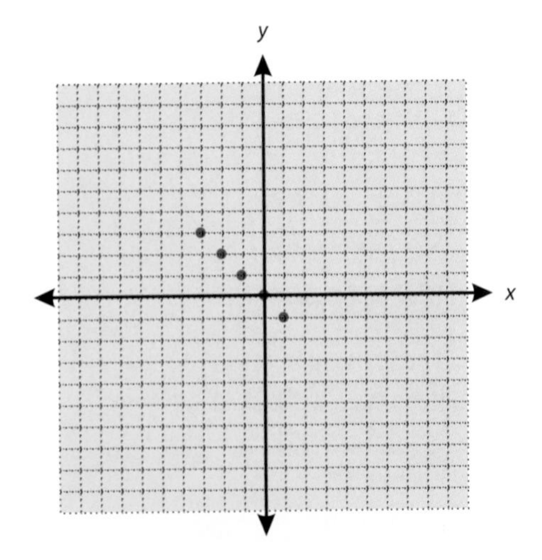

Name ______________________ Section ______________ Date ___________

**17.** $\{(-2, -1), (0, -1), (2, -1), (4, -1), (6, -1)\}$

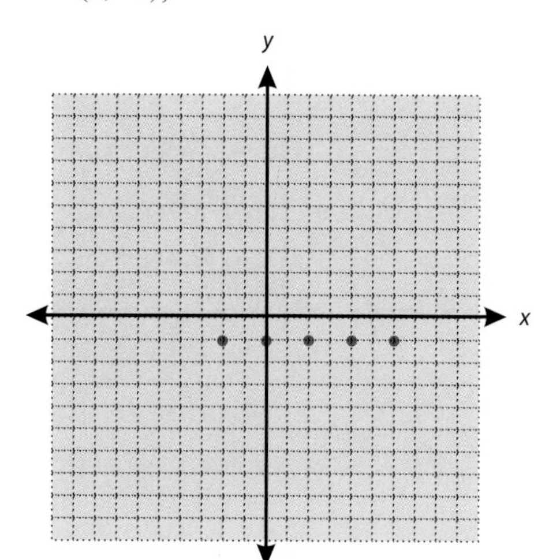

**18.** $\{(-3, 1), (-2, 1), (-1, 1), (0, 1), (1, 1)\}$

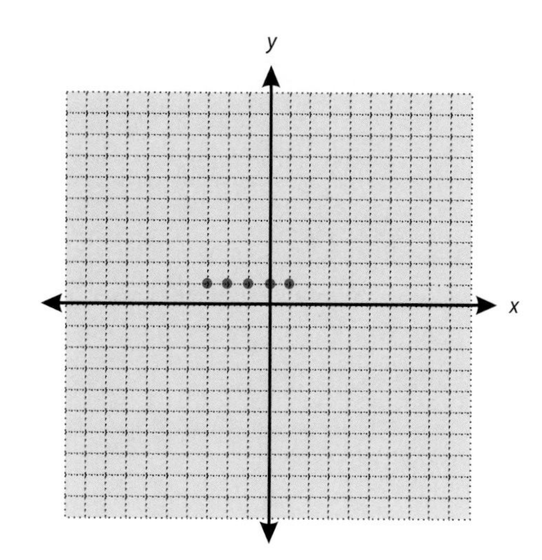

**19.** $\{(-3, 3), (-1, 1), (0, 0), (1, 1), (3, 3)\}$

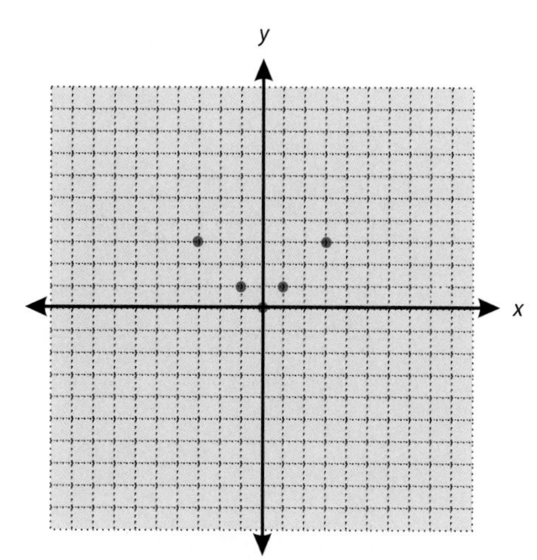

**20.** $\{(-2,-2), (-1,-1), (0, 0), (1,-1), (2,-2)\}$

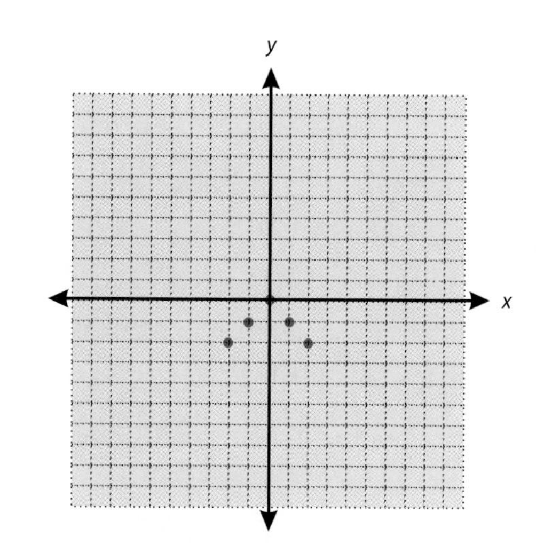

**21.** $\{(-3, 9), (-2, 4), (-1, 1), (1, 1), (2, 4), (3, 9)\}$

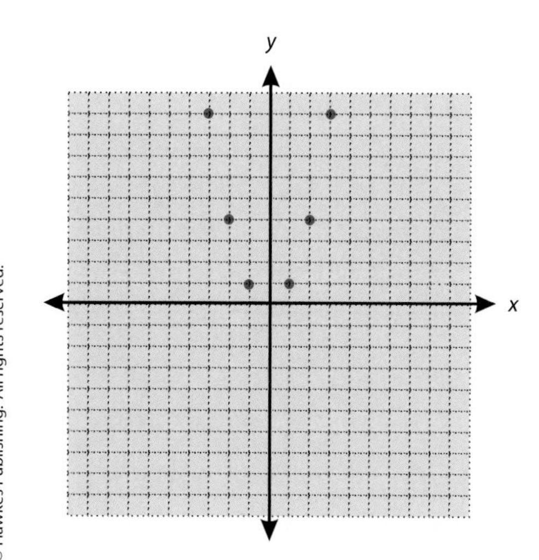

**22.** $\{(-3, -9), (-2, -4), (-1, -1), (1, -1), (2, -4), (3, -9)\}$

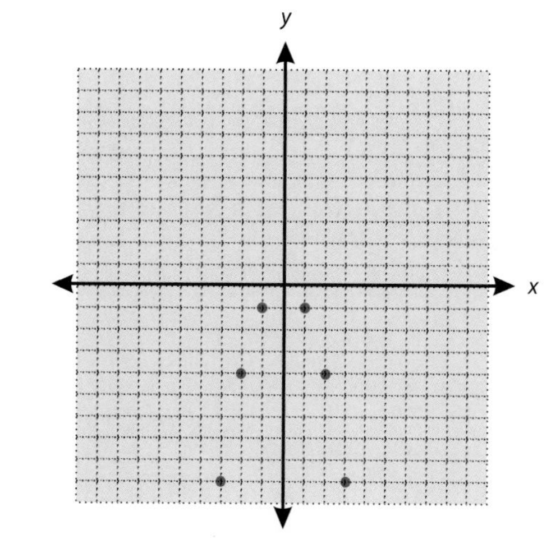

Graph each of the following sets of ordered pairs.

**23.** $\{(-4, 0), (-2, 0), (0, 0), (2, 0), (4,0)\}$

**24.** $\{(0, -3), (0, -1), (0, 0), (0, 1), (0, 3)\}$

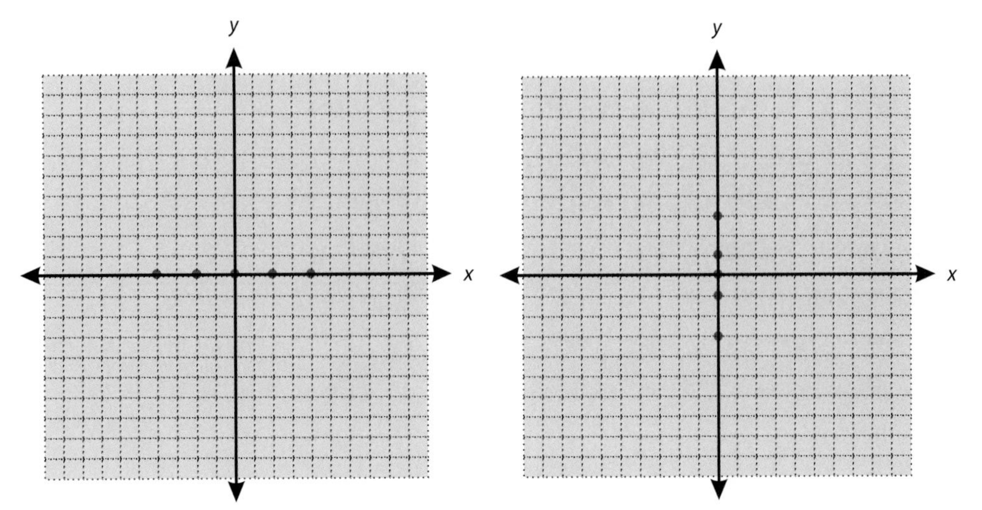


**25.** $\{(-2, 1), (1, 4), (2, 5), (3, 6), (4, 7)\}$

**26.** $\{(-1, -5), (0, -2), (1, 1), (2, 4), (3, 7)\}$

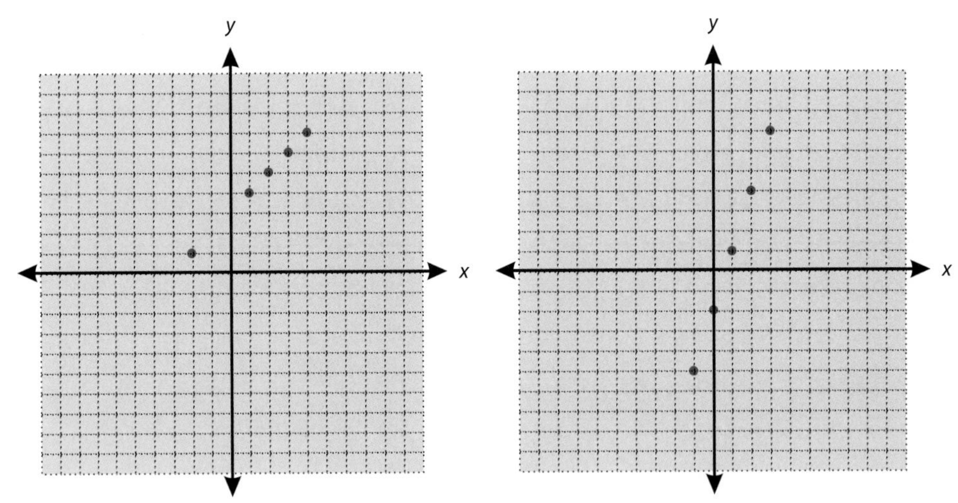


**27.** $\{(-2, -7), (-1, -5), (2, 1), (3, 3), (4, 5)\}$

**28.** $\{(0, 1), (1, -1), (2, -3), (3, -5), (5, -9)\}$

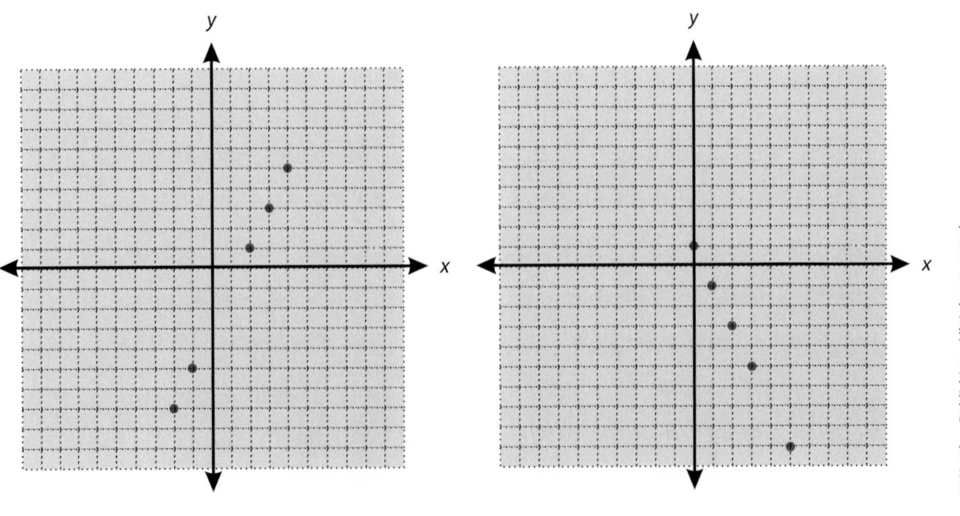

Name ______________________ Section ______________ Date ____________

**29.** $\{(-2, 8), (0, 2), (1, -1), (2, -4), (3, -7)\}$

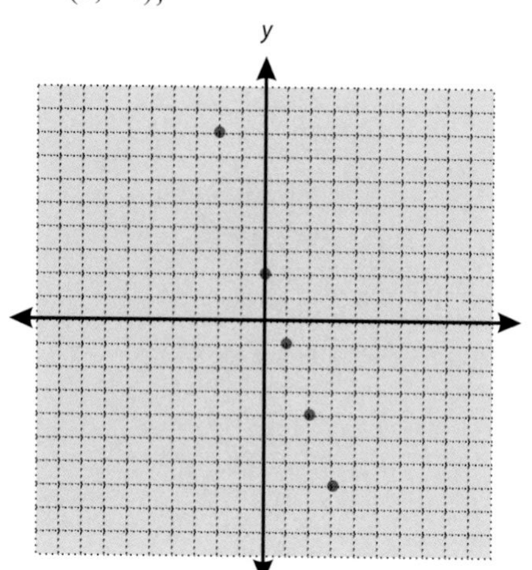


**30.** $\{(-2, -1), (-1, 1), (1, 5), (2, 7), (3, 9)\}$

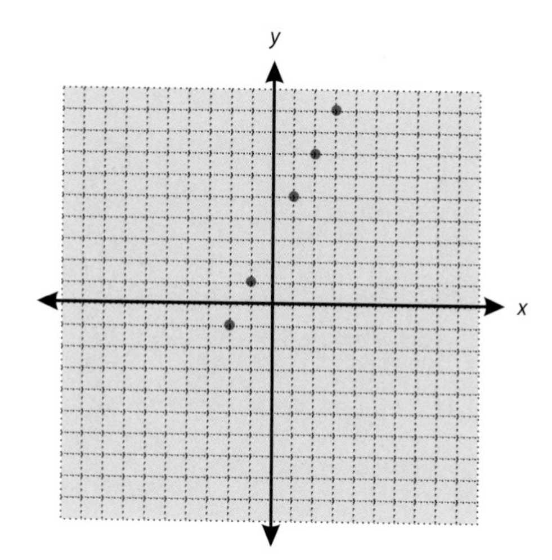


Graph the following sets of ordered pairs by estimating the positions of the points as best as you can.

**31.** $\left\{\left(-3, \frac{1}{2}\right), \left(-2, \frac{1}{4}\right), \left(-1, \frac{1}{8}\right), (0, 0)\right\}$

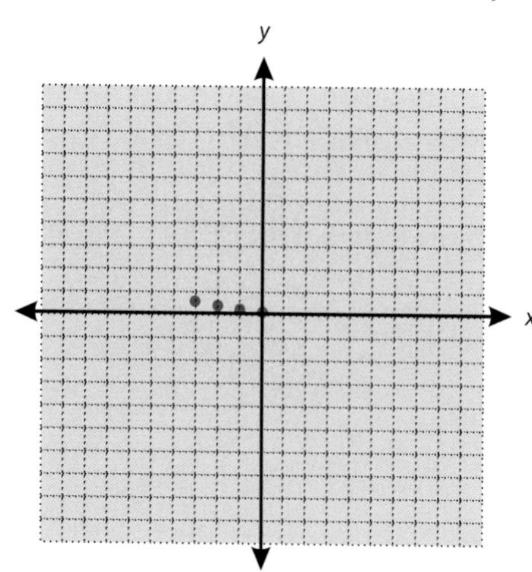


**32.** $\left\{(-1, -1), \left(0, -\frac{1}{2}\right), \left(\frac{1}{2}, 0\right), \left(1.5, -\frac{3}{8}\right)\right\}$

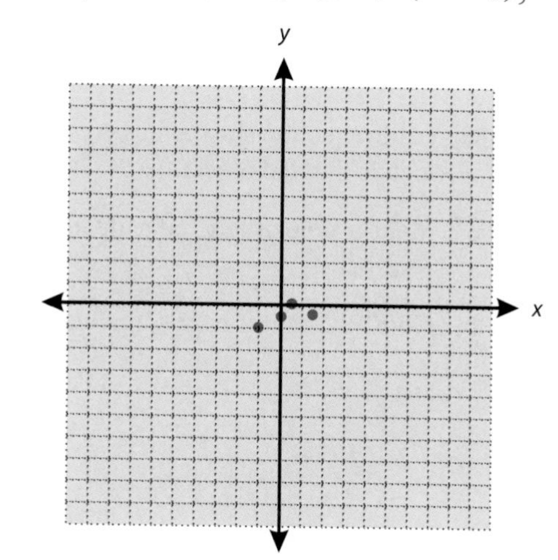


**33.** $\{(-2.6, -2.7), (-1.4, 0.75), (0.2, -2.8), (1.3, 1.7), (2.8, -0.9)\}$

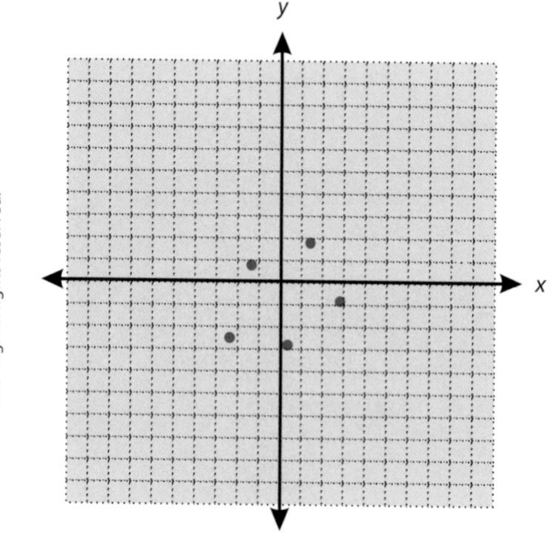


**34.** $\left\{\left(-1.5, -1\frac{1}{2}\right), \left(\frac{5}{6}, \frac{5}{6}\right), (\sqrt{5}, \sqrt{5}), (\pi, \pi)\right\}$

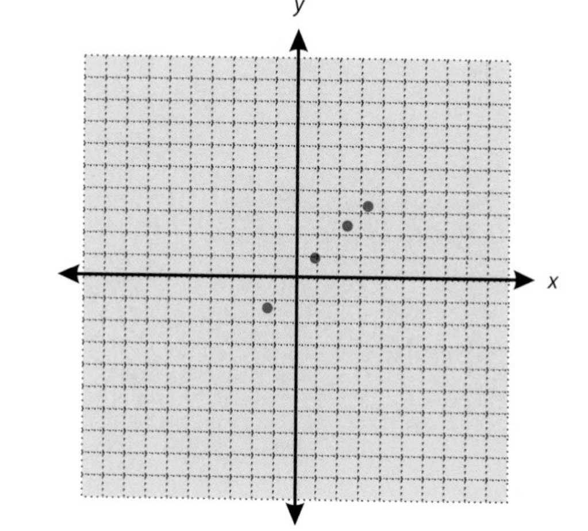

**35.** Given the equation $I = 0.12P$, where $I$ is the interest earned on a principal $P$ at a rate of 12%:

**a.** Make a table of ordered pairs for the values of $P$ and $I$ if $P$ has the values \$100, \$200, \$300, \$400, and \$500.

**b.** Graph the points corresponding to the ordered pairs.

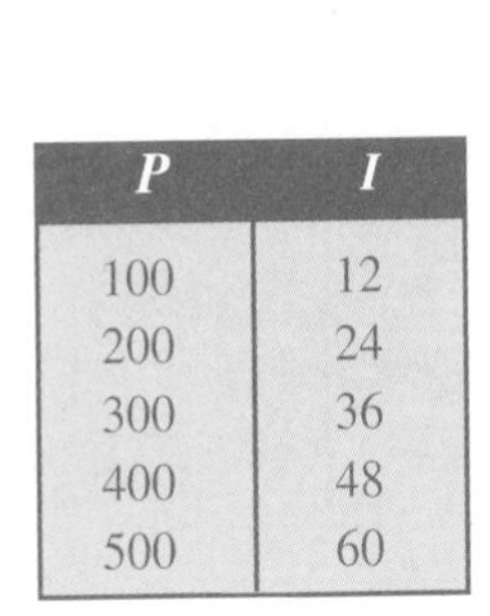

| $P$ | $I$ |
|---|---|
| 100 | 12 |
| 200 | 24 |
| 300 | 36 |
| 400 | 48 |
| 500 | 60 |

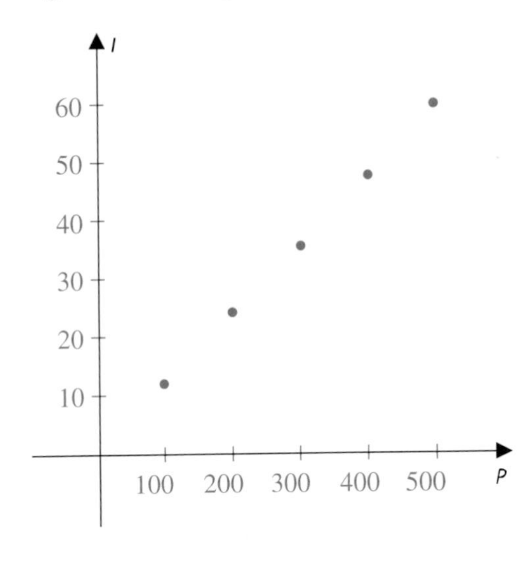


**36.** Given the equation $d = 16t^2$, where $d$ is the distance an object falls in feet and $t$ is the time in seconds that the object falls:

**a.** Make a table of ordered pairs for the values of $t$ and $d$ if $t$ has the values 1, 2, 3, 4, and 5 seconds.

**b.** Graph the points corresponding to the ordered pairs.

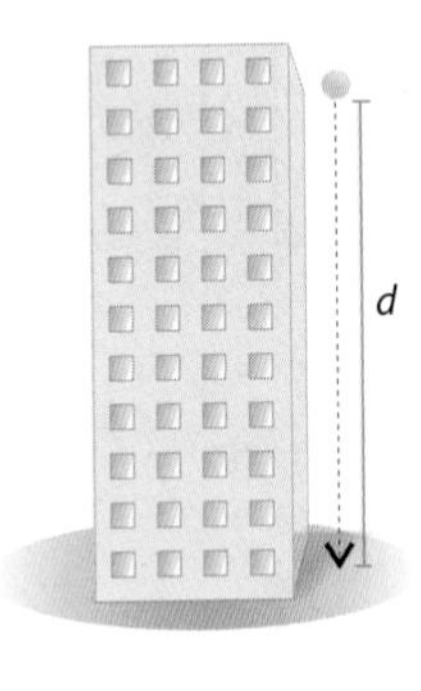


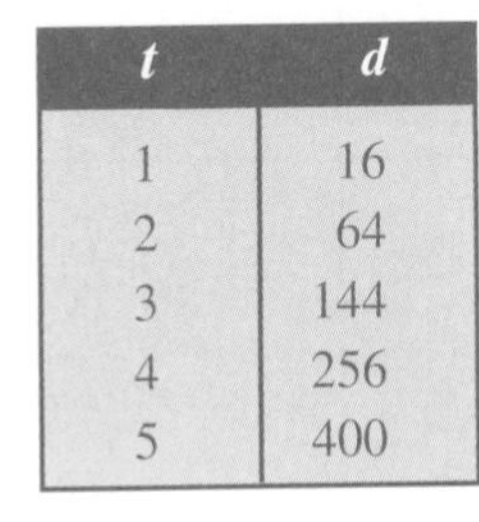

| $t$ | $d$ |
|---|---|
| 1 | 16 |
| 2 | 64 |
| 3 | 144 |
| 4 | 256 |
| 5 | 400 |

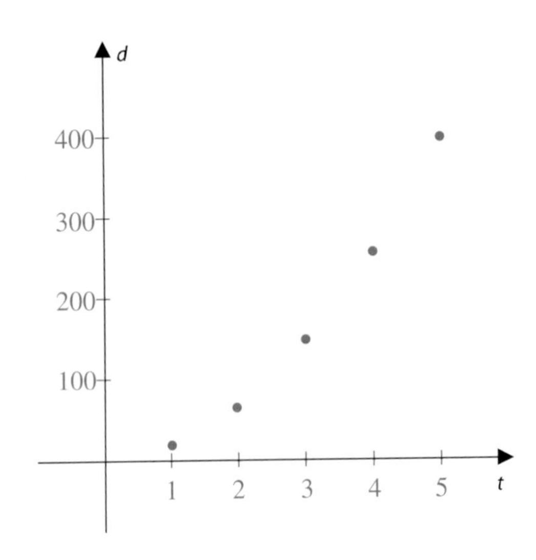

Name ______________________ Section ______________ Date ____________

**37.** Given the equation $F = \frac{9}{5}C + 32$ where $C$ is temperature in degrees Celsius and $F$ is the corresponding temperature in degrees Fahrenheit:

**a.** Make a table of ordered pairs for the values of $C$ and $F$ if $C$ has the values $-20°$, $-15°$, $-10°$, $-5°$, $0°$, $5°$, and $10°$.

**b.** Graph the points corresponding to the ordered pairs.

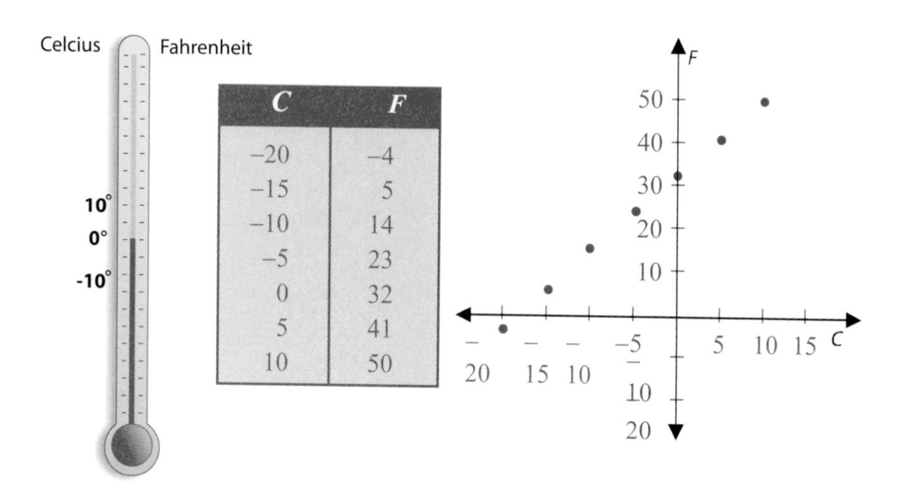


| C | F |
|---|---|
| −20 | −4 |
| −15 | 5 |
| −10 | 14 |
| −5 | 23 |
| 0 | 32 |
| 5 | 41 |
| 10 | 50 |

**38.** Given the equation $V = 25h$, where $V$ is the volume ( in $cm^3$ ) of a box with height $h$ in centimeters and a fixed base of area 25 $cm^2$.

**a.** Make a table of ordered pairs for the values of $h$ and $V$ if $h$ has the values 3 cm, 5 cm, 6 cm, 8 cm, and 10 cm.

**b.** Graph the points corresponding to the ordered pairs.

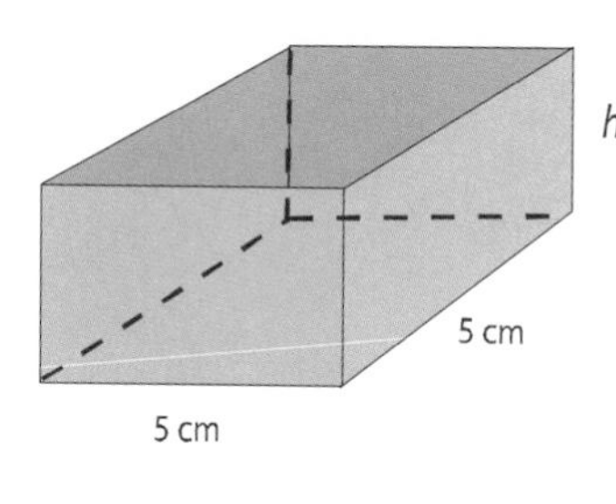


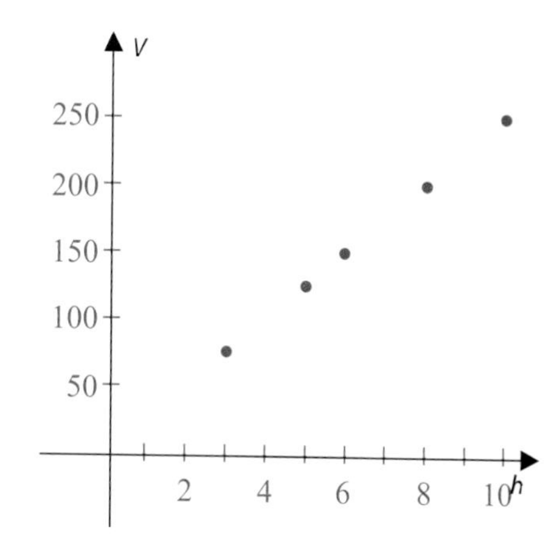


| h | V |
|---|---|
| 3 | 75 |
| 5 | 125 |
| 6 | 150 |
| 8 | 200 |
| 10 | 250 |

# 8.6 Graphing Linear Equations ( Ax + By = C )

## Linear Equations in Standard Form

There are an infinite number of ordered pairs of real numbers that satisfy the equation $y = 3x + 1$. In Section 8.5, we substituted integer and fraction values of $x$ in introducing the concepts of ordered pairs and graphing ordered pairs. However, the discussion was based on ordered pairs of real numbers, and this means that we can also substitute values for $x$ that are irrational numbers. That is, radicals can be substituted just as well as integers and fractions. The corresponding $y$ values may be expressions with integers, fractions, or radicals. For example,

**Objectives**

① Be familiar with the standard form for the equation of a line: $Ax + By = C$.

② Know how to find the $y$-intercept and the $x$-intercept of a line.

③ Know how to plot points that satisfy a linear equation and draw the graph of the corresponding line.

Equation $\mathbf{y = 3x + 1}$:

| For $x$: | Substitution: | Ordered Pair: |
|---|---|---|
| $x = \frac{1}{3}$ | $y = 3\cdot\frac{1}{3} + 1 = 1 + 1 = 2$ | $\left(\frac{1}{3}, 2\right)$ |
| $x = -\frac{3}{4}$ | $y = 3\left(-\frac{3}{4}\right) + 1 = -\frac{9}{4} + \frac{4}{4} = -\frac{5}{4}$ | $\left(-\frac{3}{4}, -\frac{5}{4}\right)$ |
| $x = \sqrt{2}$ | $y = 3\sqrt{2} + 1$ | $\left(\sqrt{2}, 3\sqrt{2} + 1\right)$ |

The important idea, here, is that even though we cannot actually substitute all real numbers for $x$ ( we do not have enough time or paper ), there is a corresponding real value for $y$ for any real value of $x$ we choose. In Figure 8.4, a few points that satisfy the equation $y = 3x + 1$ have been graphed so that you can observe a pattern.

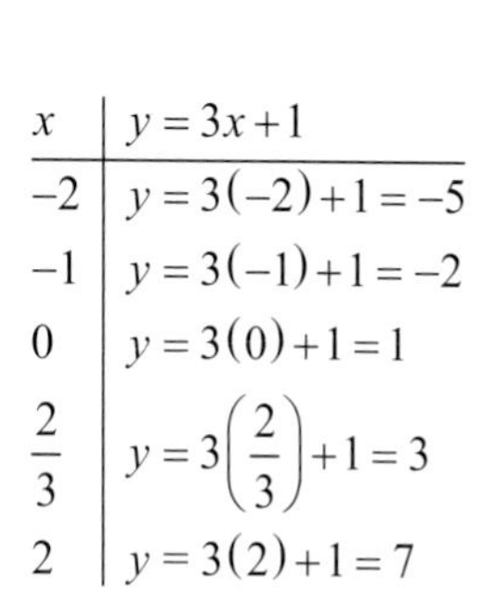

| $x$ | $y = 3x + 1$ |
|---|---|
| $-2$ | $y = 3(-2) + 1 = -5$ |
| $-1$ | $y = 3(-1) + 1 = -2$ |
| $0$ | $y = 3(0) + 1 = 1$ |
| $\frac{2}{3}$ | $y = 3\left(\frac{2}{3}\right) + 1 = 3$ |
| $2$ | $y = 3(2) + 1 = 7$ |

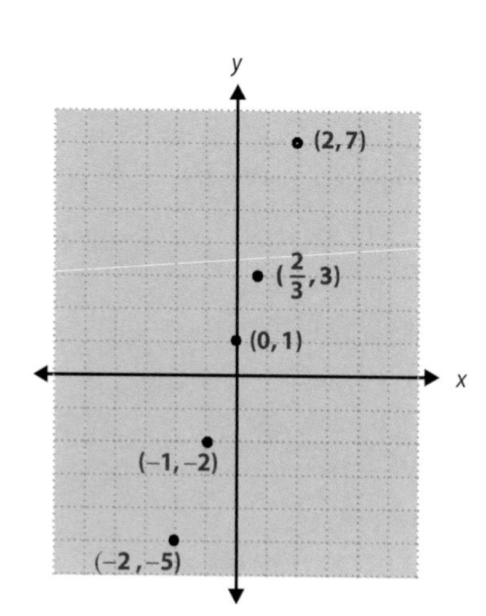


**Figure 8.4**

The points in Figure 8.4 appear to lie on a straight line, and in fact, they do. We can draw a straight line through all the points, as shown in Figure 8.5, and **any point that lies on the line will satisfy the equation.**

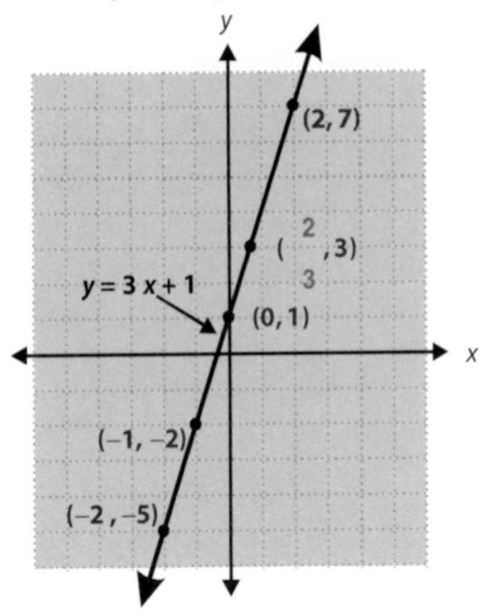


**Figure 8.5**

> The points ( ordered pairs of real numbers ) that satisfy any equation of the form
>
> $\mathbf{A}x + \mathbf{B}y = \mathbf{C}$,
>
> where **A** and **B** are not both 0, will lie on a straight line. The equation is called a **linear equation in two variables** and is in the **standard form** for the equation of a line.

The linear equation $y = 3x + 1$ is solved for $y$ and is not in standard form. However, it can be written in the standard form by adding or subtracting terms to both sides of the equation as

$-3x + y = 1$ or $3x - y = -1$

All three forms are acceptable and correct.

## Example 1

Write the linear equation $y = -2x + 5$ in standard form.

### Solution

By adding $2x$ to both sides, we get the standard form:

$$2x + y = 5$$

## Graphing Linear Equations

**Since we know that the graph of a linear equation is a straight line,** we need only to graph two points ( because two points determine a line ), and then we can draw the line through these two points: A good check against possible error is to locate three points instead of only two. Also, the values chosen for $x$ or $y$ should be such that the points are not too close together.

You must understand the following ideas about linear equations and the corresponding graphs of lines:

1. Choose **any** value of $x$ ( or $y$ ) that you like.
2. Substitute this value into the equation, and solve for the corresponding value of $y$ ( or $x$ ).
3. This will give you one point that must lie on the graph of the line.
4. No matter what points you locate, they will all be on the same line.

### Example 2

Draw the graph of the linear equation $x + 2y = 6$.

#### Solution

We find and plot three ordered pairs that satisfy the equation and then sketch the graph. ( **Note:** The values chosen of $x$ are purely arbitrary. Other values could have been chosen. The result would still be the same graph. That is, locating different points on the graph will not change the position of the line. )

**For $x = -2$**

$$\begin{aligned} x + 2y &= 6 \\ -2 + 2y &= 6 \\ 2y &= 8 \\ y &= 4 \end{aligned}$$

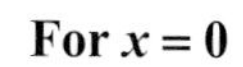

**For $x = 0$**

$$\begin{aligned} x + 2y &= 6 \\ 0 + 2y &= 6 \\ 2y &= 6 \\ y &= 3 \end{aligned}$$

**For $x = 2$**

$$\begin{aligned} x + 2y &= 6 \\ 2 + 2y &= 6 \\ 2y &= 4 \\ y &= 2 \end{aligned}$$

| $x$ | $y$ |
|---|---|
| −2 | 4 |
| 0 | 3 |
| 2 | 2 |

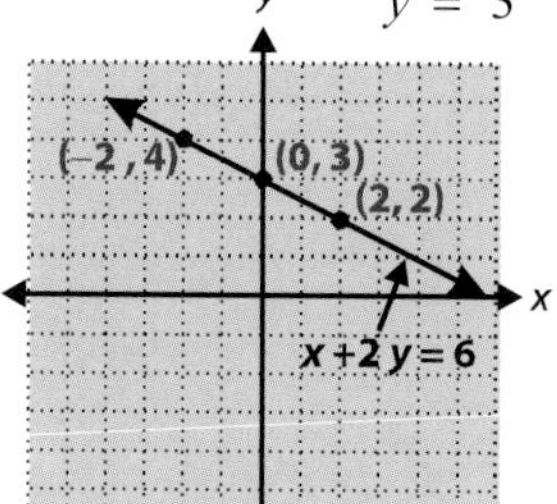


( Locating three points helps in avoiding errors. Avoid choosing points close together. )

### Example 3

Draw the graph of the linear equation $x + 2y = 4$.

#### Solution

The process of substituting values for $x$ and finding the corresponding values for $y$ can be made easier by first solving the equation for $y$ as follows:

$$\begin{aligned} x + 2y &= 4 \\ 2y &= 4 - x \\ y &= \frac{4 - x}{2} \end{aligned}$$

We can now substitute values for $x$ in this form of the equation.

| **For $x = -2$** | **For $x = 0$** | **For $x = 4$** |
|---|---|---|
| $y = \dfrac{4-(-2)}{2}$ | $y = \dfrac{4-0}{2}$ | $y = \dfrac{4-4}{2}$ |
| $y = \dfrac{6}{2} = 3$ | $y = \dfrac{4}{2} = 2$ | $y = \dfrac{0}{2} = 0$ |

| $x$ | $y$ |
|---|---|
| –2 | 3 |
| 0 | 2 |
| 4 | 0 |

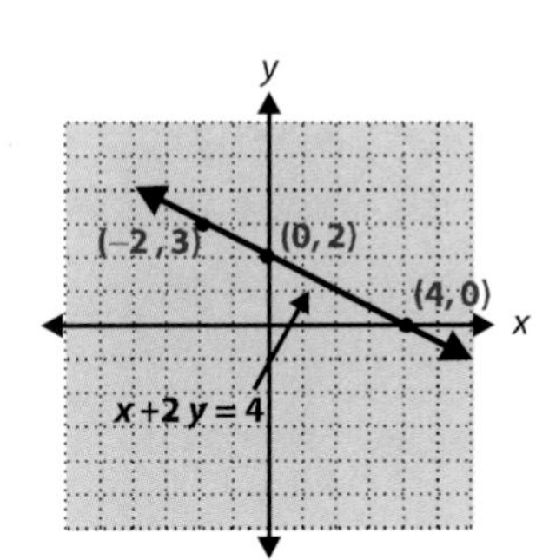


The ***y*-intercept** of a line is the point where the line crosses the $y$-axis. This point can be located by letting $x = 0$. Similarly, the ***x*-intercept** is the point where the line crosses the $x$-axis and is found by letting $y = 0$. When the line is not vertical or horizontal, the $y$-intercept and the $x$-intercept are generally easy to locate and are frequently used for drawing the graph of a linear equation.

## Example 4

Draw the graph of the linear equation $x - 2y = 8$ by locating the $y$-intercept and the $x$-intercept.

### Solution

**Find the *y*-intercept**

$$x = 0$$
$$0 - 2y = 8$$
$$y = -4$$

**Find the *x*-intercept**

$$y = 0$$
$$x - 2 \cdot 0 = 8$$
$$x = 8$$

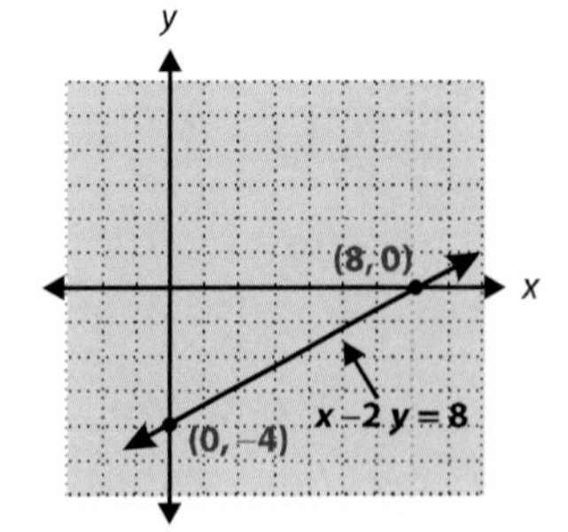


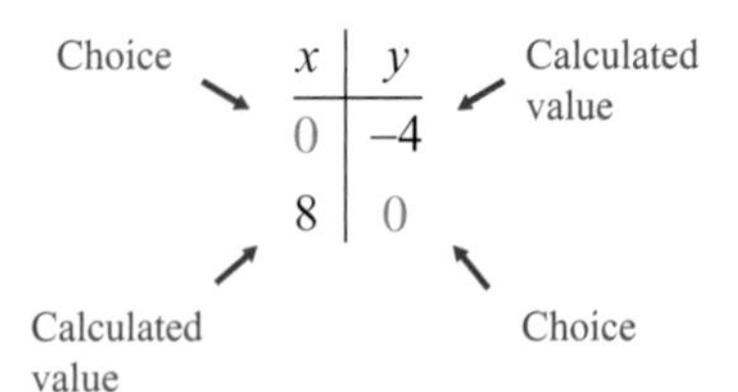


| $x$ | $y$ |
|---|---|
| 0 | –4 |
| 8 | 0 |

### Example 5

Locate the $y$-intercept and the $x$-intercept, and draw the graph of the linear equation $3x - y = 3$.

**Solution**

**Find the $y$-intercept**

$$x = 0$$
$$3 \cdot 0 - y = 3$$
$$y = -3$$

**Find the $x$-intercept**

$$y = 0$$
$$3x - 0 = 3$$
$$x = 1$$

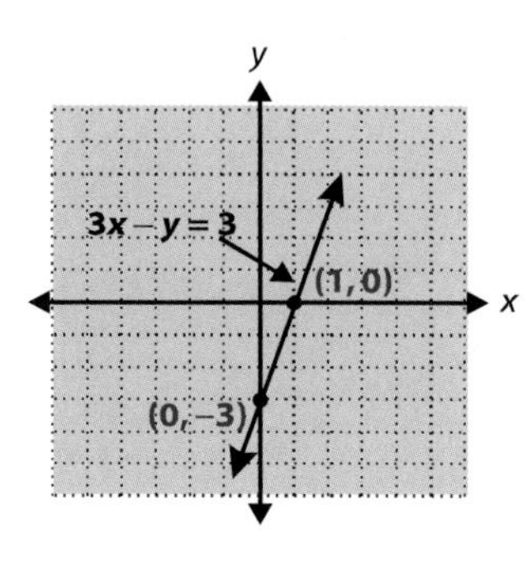


| $x$ | $y$ |
|---|---|
| 0 | −3 |
| 1 | 0 |

If A = 0 and B ≠ 0 in the standard form $Ax + By = C$, the equation takes the form $By = C$ and can be solved for $y$ as $y = \frac{C}{B}$. For example, we can write $0x + 3y = 6$ as $y = 2$. Thus, no matter what value $x$ has, the value of $y$ is 2. The graph of the equation $y = 2$ is a **horizontal line,** as shown in Figure 8.6. For horizontal lines ( other than the $x$-axis itself ), there is no $x$-intercept.

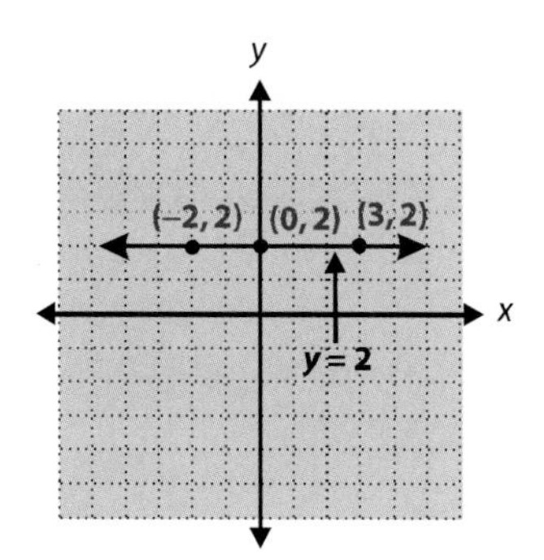


The $y$-coordinate is 2 for all points on the line $y = 2$.

| $x$ | $y$ |
|---|---|
| −2 | 2 |
| 0 | 2 |
| 3 | 2 |

**Figure 8.6**

If B = 0 and A ≠ 0 in the standard form $Ax + By = C$, the equation takes the form $Ax = C$ or $x = \frac{C}{A}$. For example, we can write $5x + 0y = -5$ as $x = -1$. Thus, no matter what value $y$ has, the value of $x$ is $-1$. The graph of the equation $x = -1$ is a **vertical line,** as shown in Figure 8.7. For vertical lines (other than the $y$-axis itself), there is no $y$-intercept.

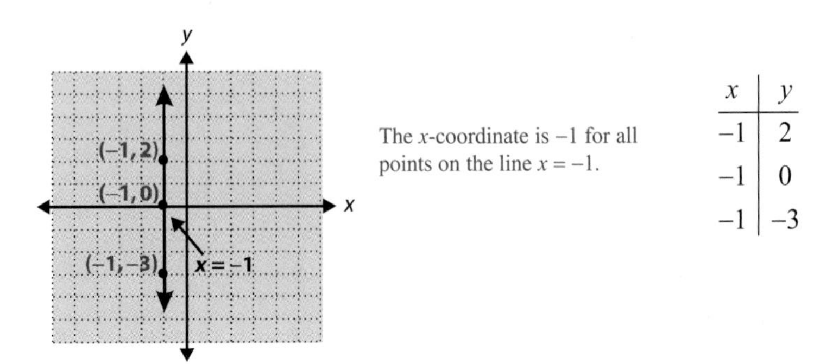


| $x$ | $y$ |
|---|---|
| −1 | 2 |
| −1 | 0 |
| −1 | −3 |

**Figure 8.7**

## Example 6

Graph the horizontal line $y = -1$ and the vertical line $x = 3$ on the same coordinate system.

### Solution

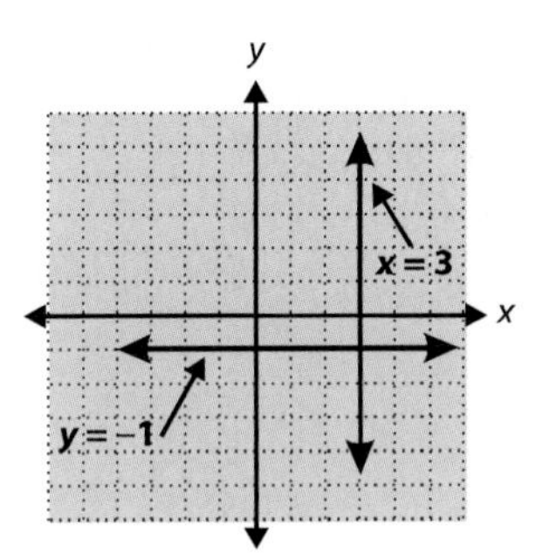

Name ______________ Section ______________ Date ______________

## Exercises 8.6

For each linear equation, fill in the table so that the corresponding points will lie on the graph of the equation. Do not graph the equation.

**1.** $y = 2x + 3$

| $x$ | $y$ |
|---|---|
| 2 | 7 |
| −1 | 1 |
| 0 | 3 |

**2.** $y = 3x - 1$

| $x$ | $y$ |
|---|---|
| 0 | −1 |
| 1 | 2 |
| 2 | 5 |

**3.** $y = 3x$

| $x$ | $y$ |
|---|---|
| −1 | −3 |
| 3 | 9 |
| 0 | 0 |

**4.** $y = -4x$

| $x$ | $y$ |
|---|---|
| 2 | −8 |
| −2 | 8 |
| 3 | −12 |

**5.** $2x + 3y = 6$

| $x$ | $y$ |
|---|---|
| 0 | 2 |
| 3 | 0 |
| 6 | −2 |

**6.** $x - 3y = 9$

| $x$ | $y$ |
|---|---|
| 6 | −1 |
| 0 | −3 |
| 15 | 2 |

**7.** $x = -2$

| $x$ | $y$ |
|---|---|
| −2 | 5 |
| −2 | −1 |
| −2 | 7 |

**8.** $y = 5$

| $x$ | $y$ |
|---|---|
| −5 | 5 |
| 1 | 5 |
| 3 | 5 |

**9.** $2x - 3y = 12$

| $x$ | $y$ |
|---|---|
| 0 | −4 |
| 6 | 0 |
| 12 | 4 |

**10.** $x + 3y = -9$

| $x$ | $y$ |
|---|---|
| −6 | −1 |
| 0 | −3 |
| −3 | −2 |

**ANSWERS**

1. [Respond in exercise.]
2. [Respond in exercise.]
3. [Respond in exercise.]
4. [Respond in exercise.]
5. [Respond in exercise.]
6. [Respond in exercise.]
7. [Respond in exercise.]
8. [Respond in exercise.]
9. [Respond in exercise.]
10. [Respond in exercise.]

11. (2, −12) ______

______

12. (2, −12) ______

(−2, −16) ______

13. (2, 1), (3, −1) ______

(5, −5) ______

14. (0, −2), (−4, −18) ______

______

15. (3, 1) ______

______

16. (−2, 0), (0, −3) ______

______

17. (0, 0), (−3, 6), ______

(−1, 2) ______

18. (1.5, −0.75), (0, 0) ______

______

19. (−3, 15), (−3, −1.2), ______

(−3, 3) ______

20. (−2.3, 5), (6, 5) ______

______

Determine which of the ordered pairs in each table satisfy the given equation. (**Hint:** Substitute the values into the equation to see if you get a true statement or not.)

**11.** $x + y = -10$

| $x$ | 5 | 0 | 2 | −1 |
|---|---|---|---|---|
| $y$ | 5 | 10 | −12 | −11 |

**12.** $x - y = 14$

| $x$ | 2 | −2 | 0 | 7 |
|---|---|---|---|---|
| $y$ | −12 | −16 | 14 | 7 |

**13.** $y = -2x + 5$

| $x$ | 2 | 3 | 4 | 5 |
|---|---|---|---|---|
| $y$ | 1 | −1 | 3 | −5 |

**14.** $y = 4x - 2$

| $x$ | −1 | 0 | 4 | −4 |
|---|---|---|---|---|
| $y$ | −4 | −2 | −12 | −18 |

**15.** $2x + y = 7$

| $x$ | 4 | 3 | 2 | 1 |
|---|---|---|---|---|
| $y$ | 1 | 1 | 1 | 1 |

**16.** $3x + 2y = -6$

| $x$ | −2 | −2 | 2 | 0 |
|---|---|---|---|---|
| $y$ | 0 | −6 | 0 | −3 |

**17.** $y = -2x$

| $x$ | −5 | 0 | −3 | −1 |
|---|---|---|---|---|
| $y$ | −10 | 0 | 6 | 2 |

**18.** $x = -2y$

| $x$ | 0.5 | 1.5 | −0.5 | 0 |
|---|---|---|---|---|
| $y$ | −0.5 | −0.75 | −1 | 0 |

**19.** $x = -3$

| $x$ | −3 | 4 | −3 | −3 |
|---|---|---|---|---|
| $y$ | 15 | 8 | −1.2 | 3 |

**20.** $y = 5$

| $x$ | 4 | −2.3 | 6 | 0 |
|---|---|---|---|---|
| $y$ | 0 | 5 | 5 | −5 |

In Exercises 21 and 22, four equations and the graphs of four lines are given. By substituting ordered pairs of numbers that you see on the lines into the equations, match each equation with its graph. Respond below exercises.

**21. a.** $2x + y = 6$ **b.** $x - 2y = 10$ **c.** $2x + 2y = 6$ **d.** $2x - y = 4$

**(i)**

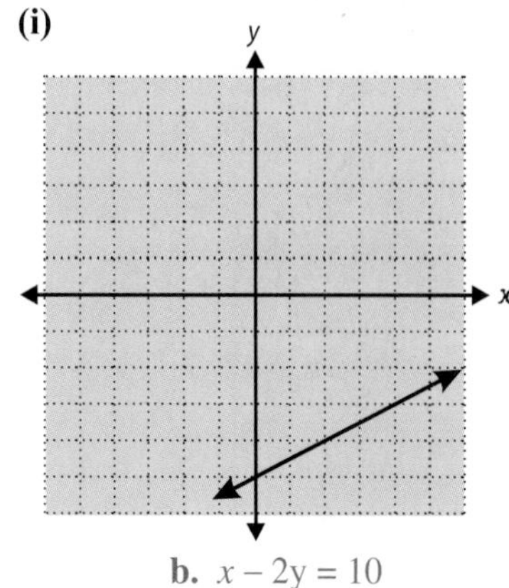


b. $x - 2y = 10$

**(ii)**

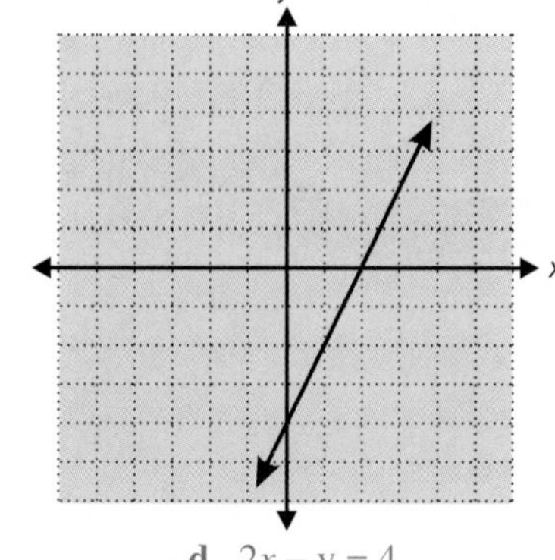


d. $2x - y = 4$

Name ______________________ Section ____________ Date __________

(iii)

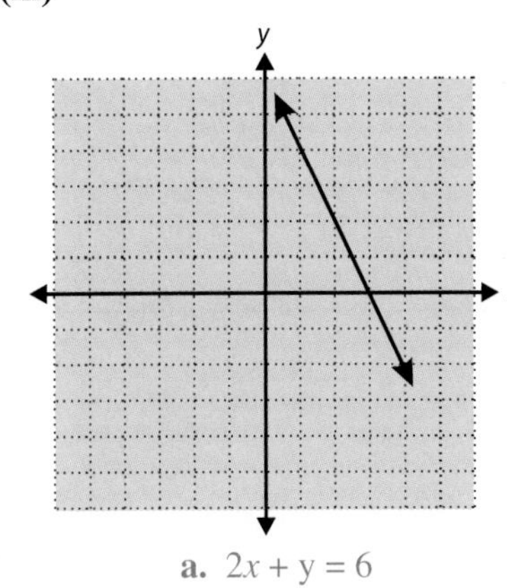

a. $2x + y = 6$

(iv)

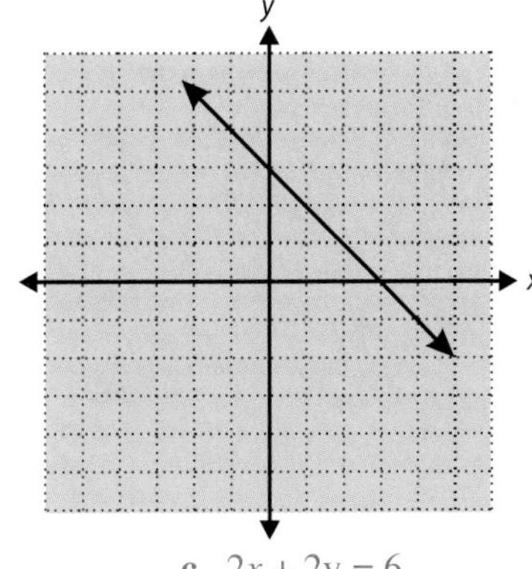

c. $2x + 2y = 6$

22. a. $x + y = 5$ b. $x + 2y = 8$ c. $x + y = -1$ d. $2x + 3y = -2$

(i)

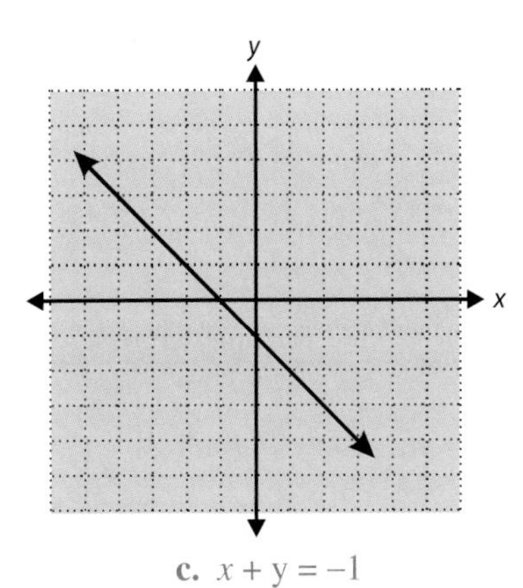

c. $x + y = -1$

(ii)

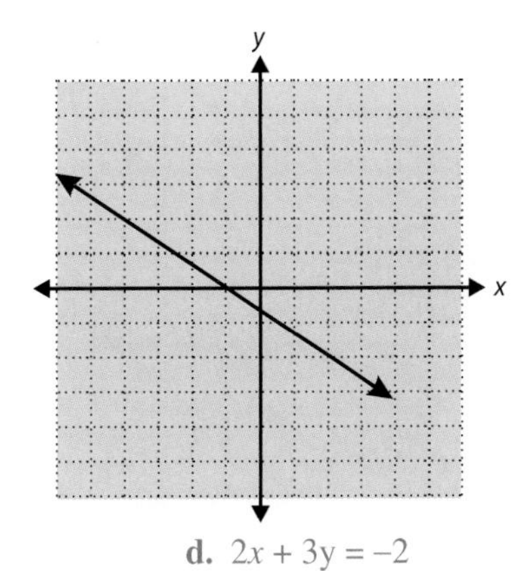

d. $2x + 3y = -2$

(iii)

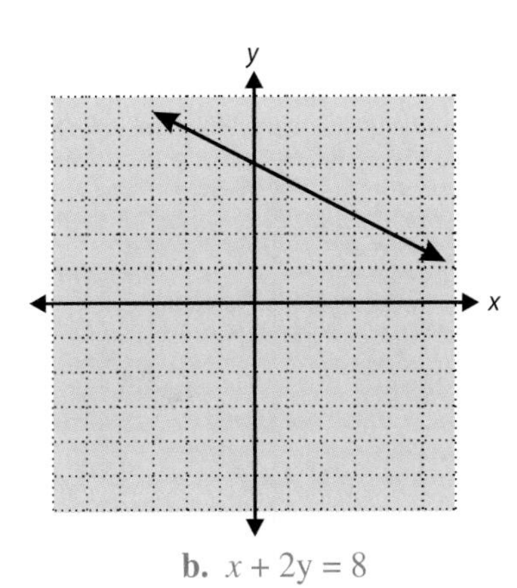

b. $x + 2y = 8$

(iv)

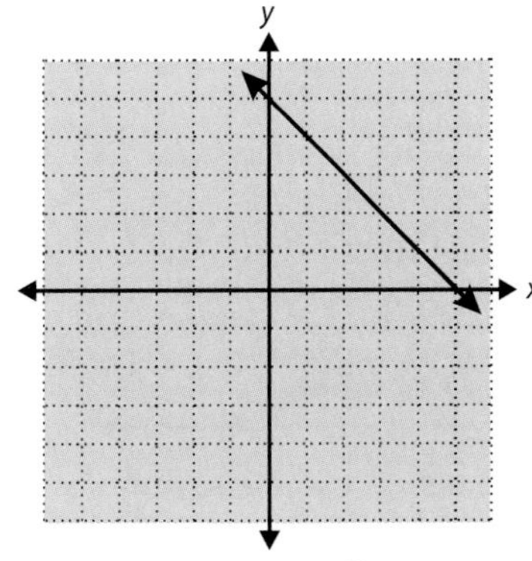

a. $x + y = 5$

Graph the following linear equations. Label at least three points on each line.

**23.** $y = x + 1$

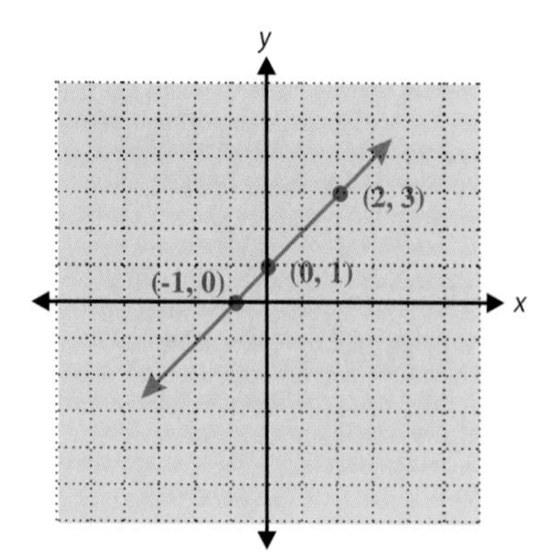


**24.** $y = x + 3$

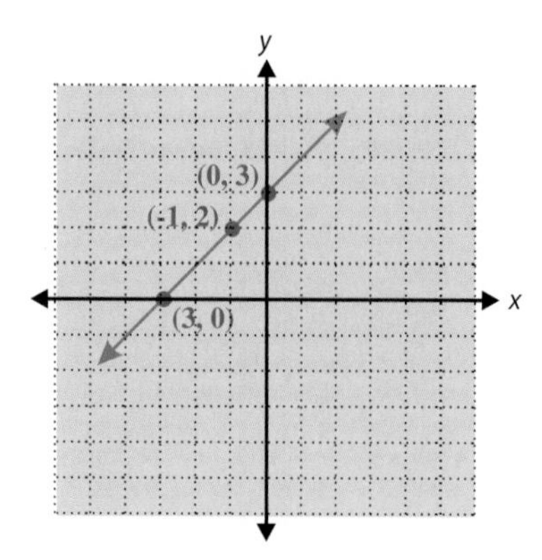


**25.** $y = x - 6$

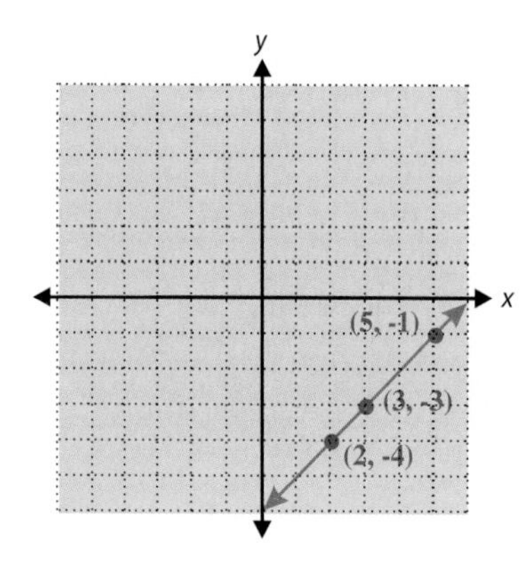


**26.** $y = x - 2$

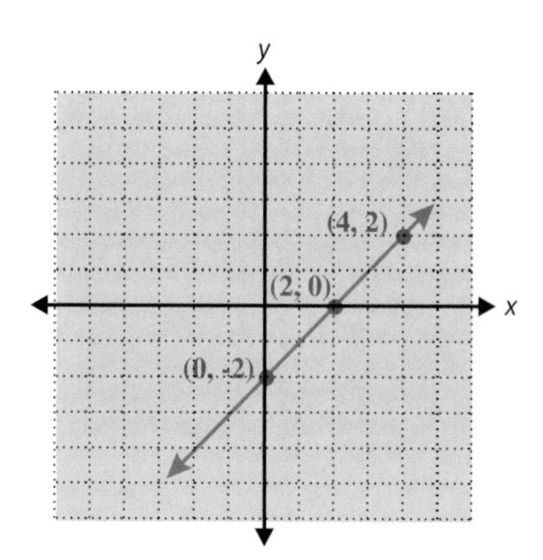


**27.** $y = 2x$

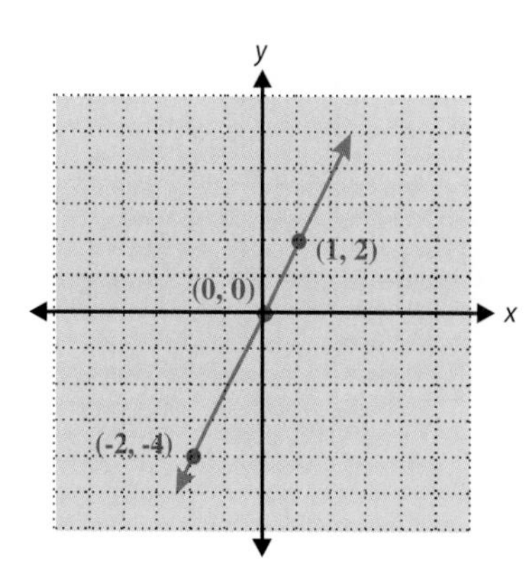


**28.** $y = 4x$

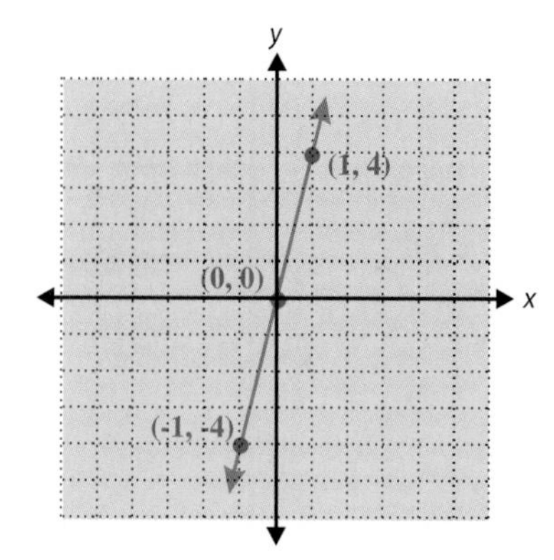

Name ______________________ Section ______________ Date ____________

**29.** $y = -x$

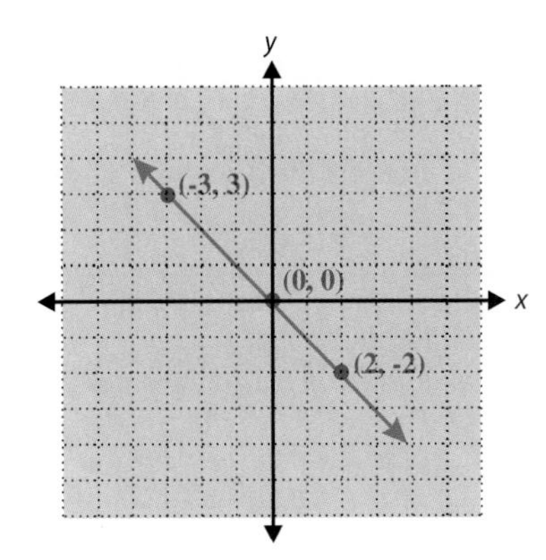


**30.** $y = -3x$

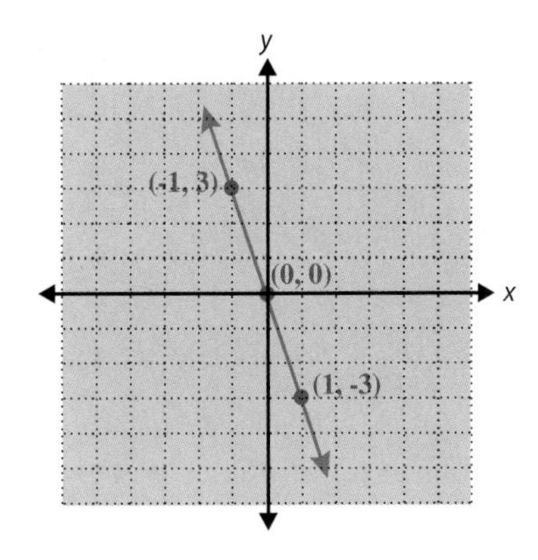


**31.** $y = x$

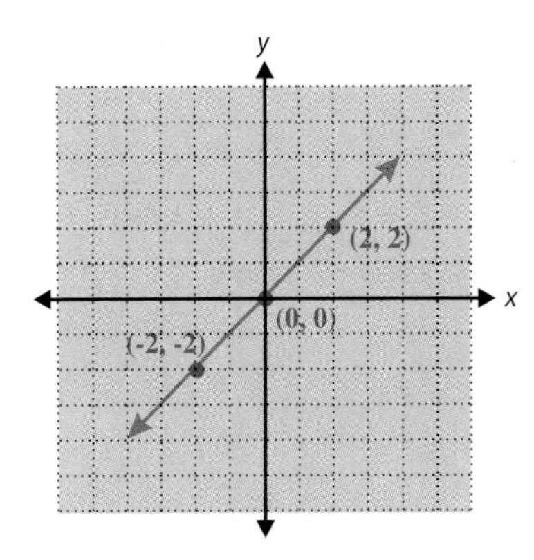


**32.** $y = 2 - x$

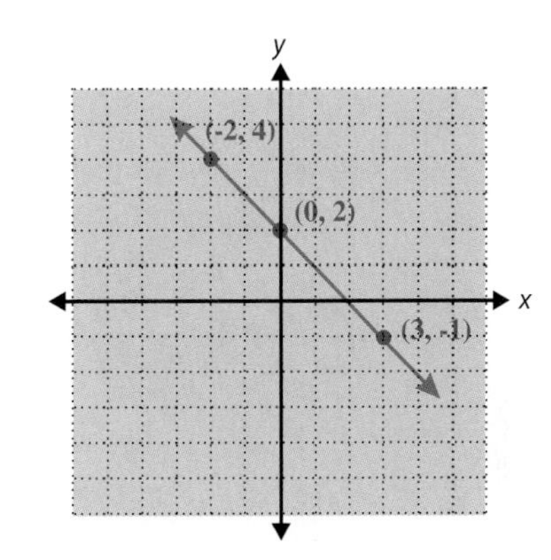


**33.** $y = 4 - x$

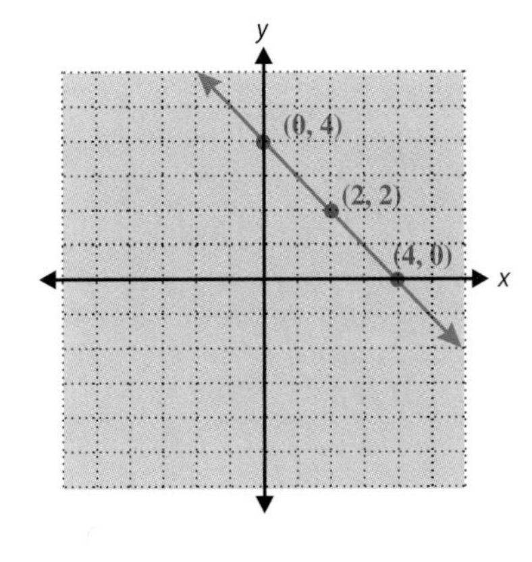


**34.** $y = 7 - x$

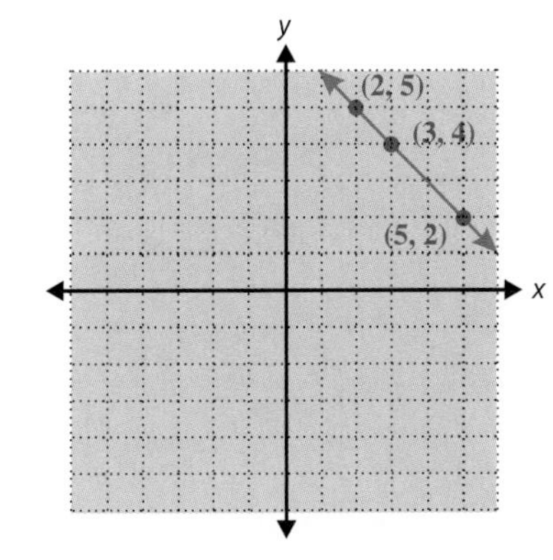

**35.** $y = 2x + 1$

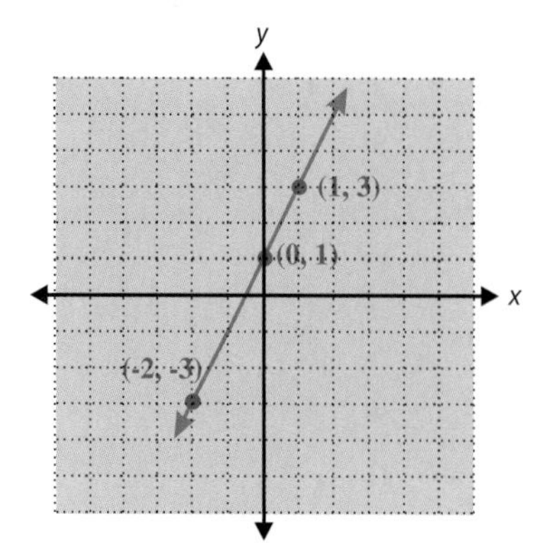


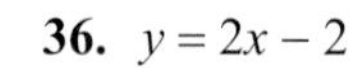

**36.** $y = 2x - 2$

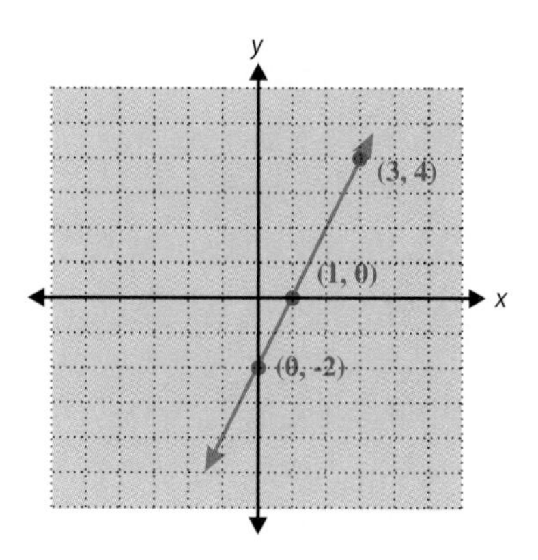


**37.** $y = 2x - 5$

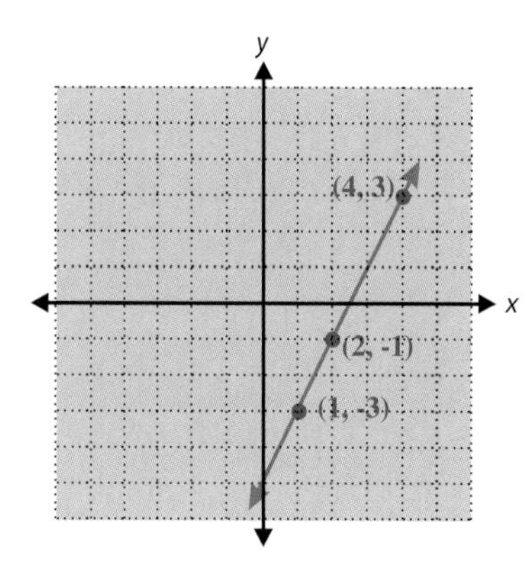


**38.** $y = 2x + 3$

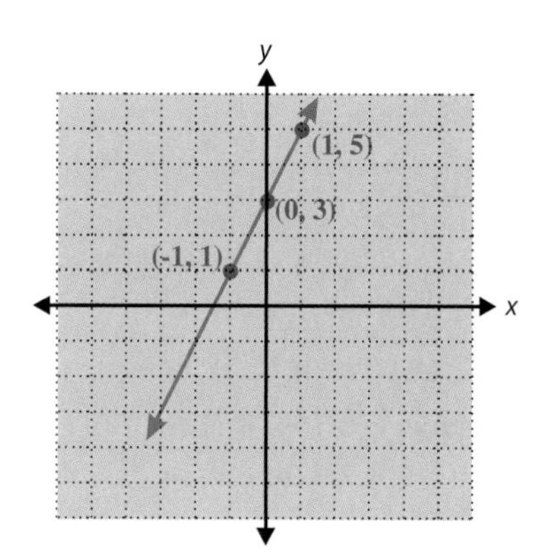


**39.** $y = -2x + 4$

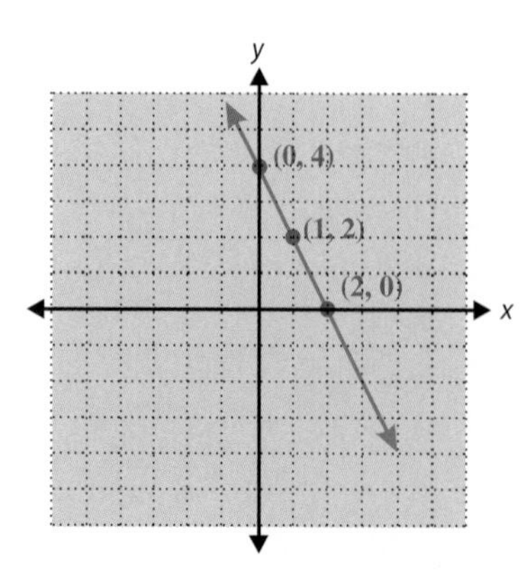


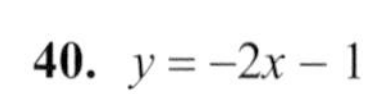

**40.** $y = -2x - 1$

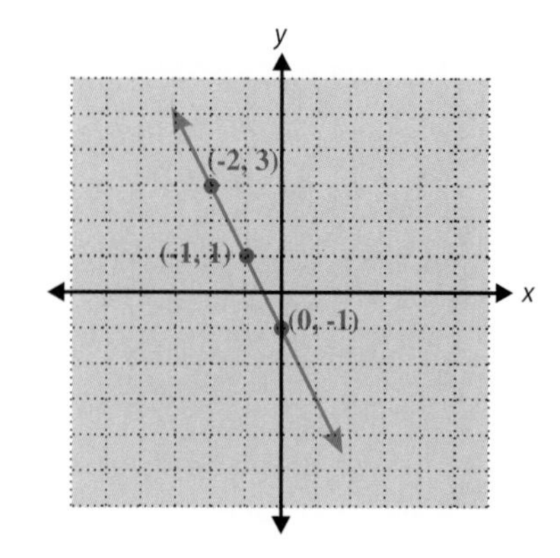

Name ______________________ Section ______________ Date ____________

**41.** $y = -3x + 3$

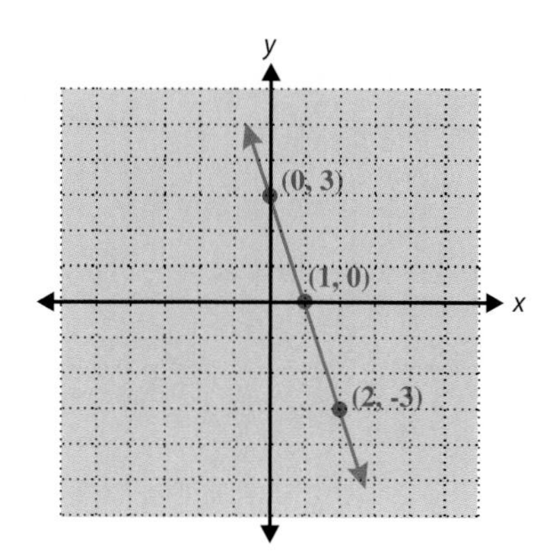


**42.** $y = -3x - 5$

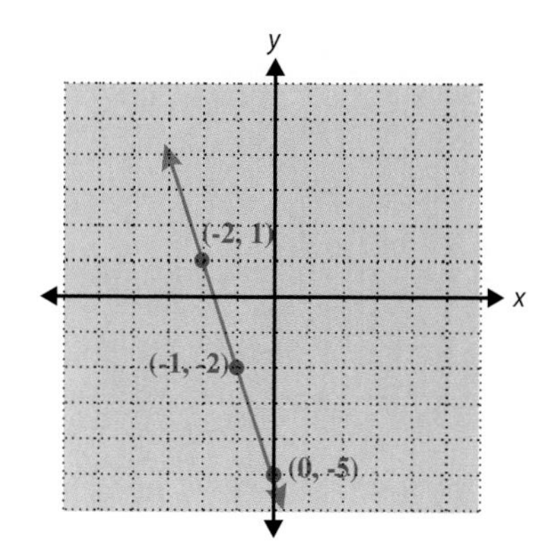


**43.** $x - 2y = 4$

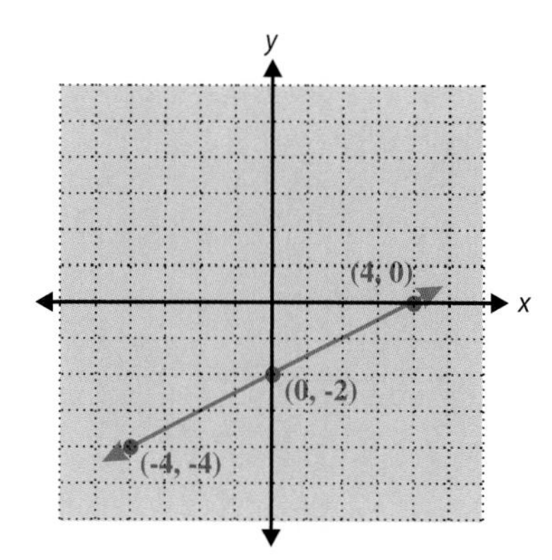


**44.** $x - 4y = 12$

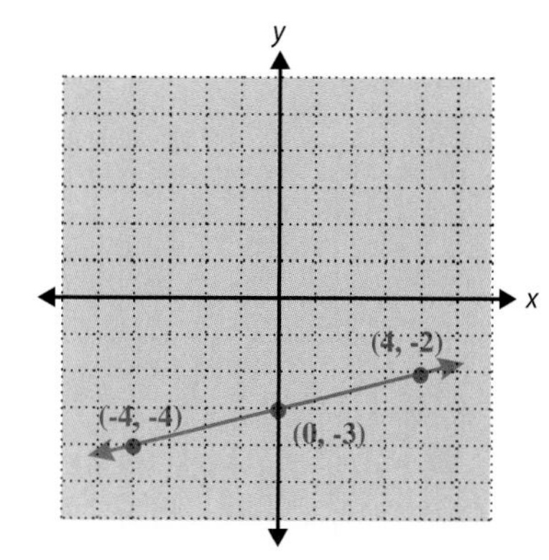


**45.** $-2x + 3y = 6$

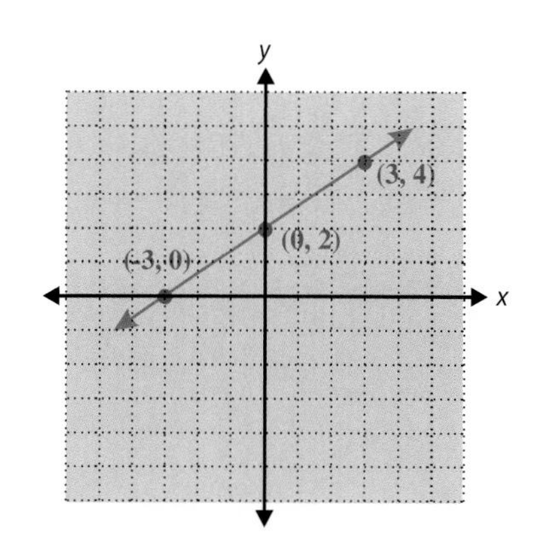


**46.** $2x - 5y = 10$

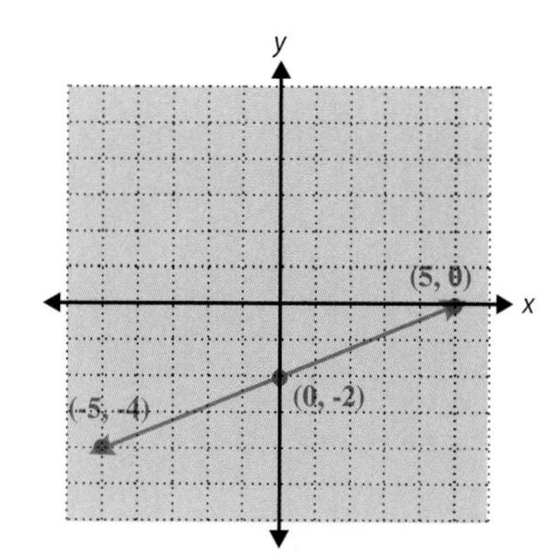

**47.** $-2x + y = 4$

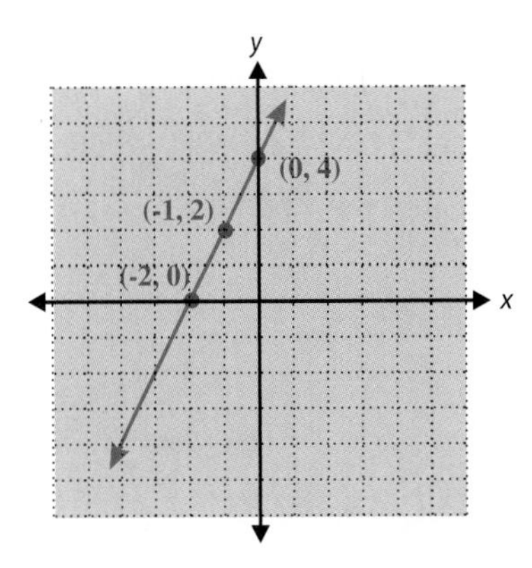


**48.** $3x + 4y = 12$

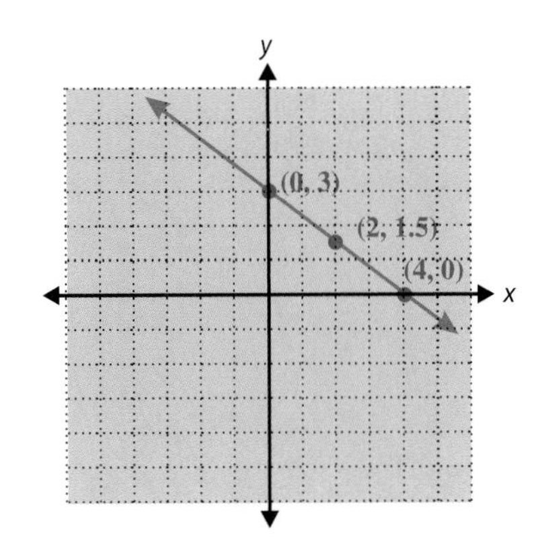


**49.** $3x + 5y = 15$

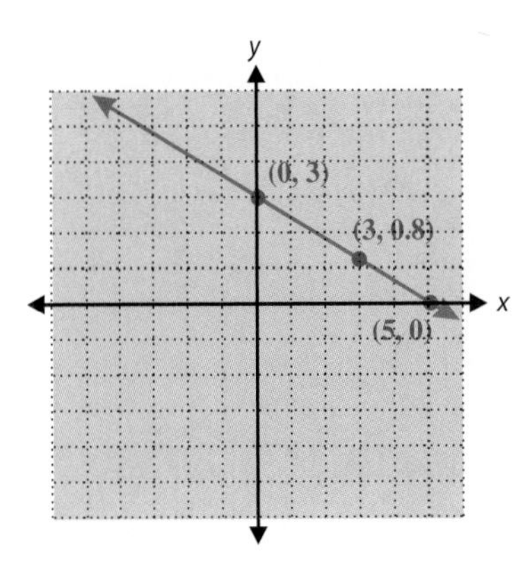


**50.** $4x + y = 8$

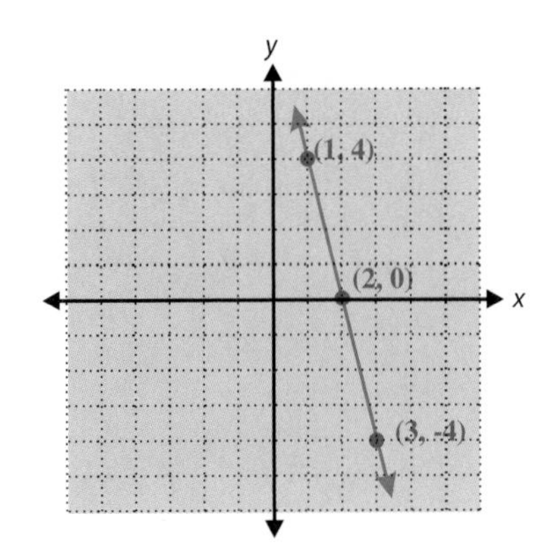


**51.** $x + 4y = 8$

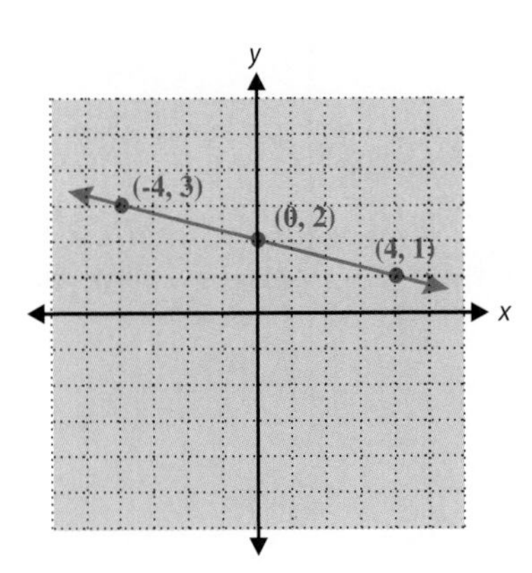


**52.** $2x - 3y = 6$

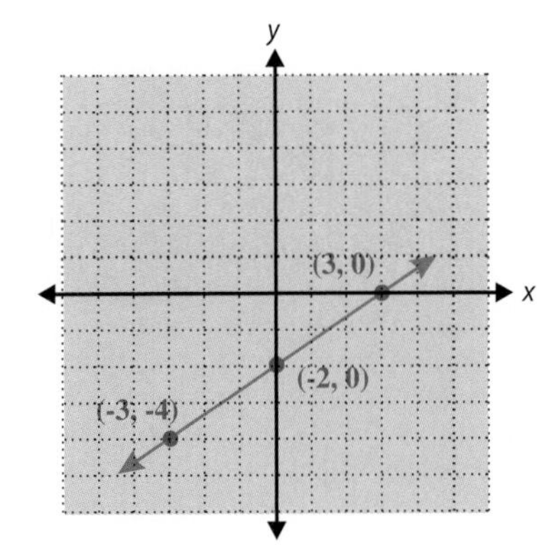

Name ______________________ Section ______________ Date ___________

**53.** $x = 3$

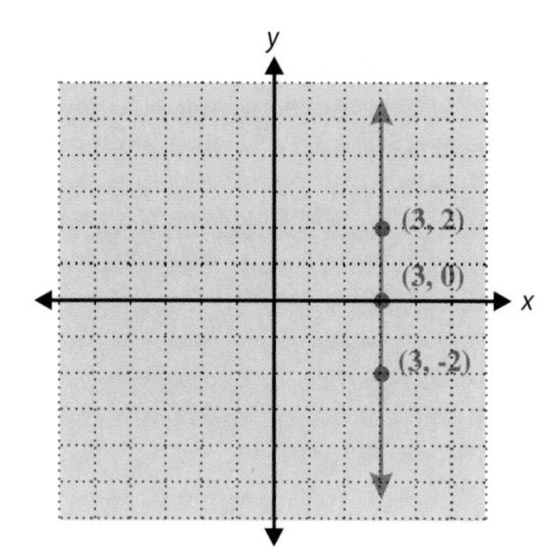


**54.** $y = 4$

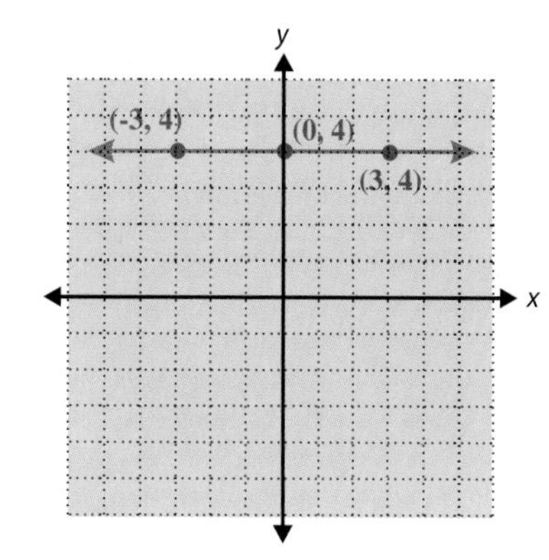


**55.** $y = -3$

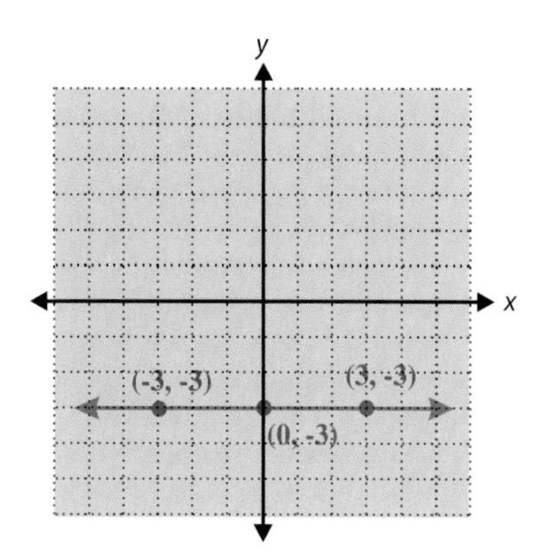


**56.** $x = -6$

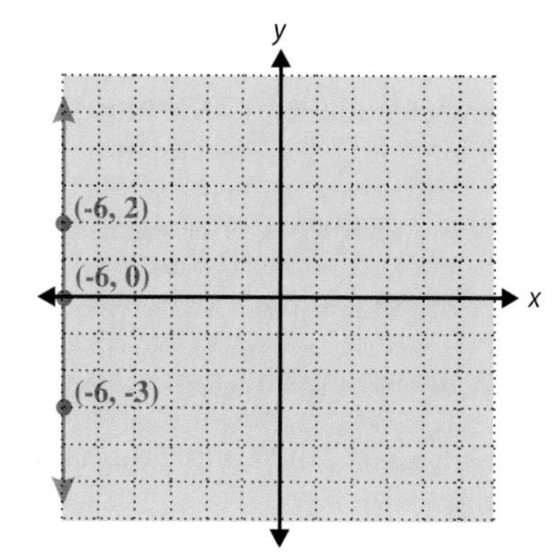


**57.** $x - 4 = 0;\ y + 2 = 0$

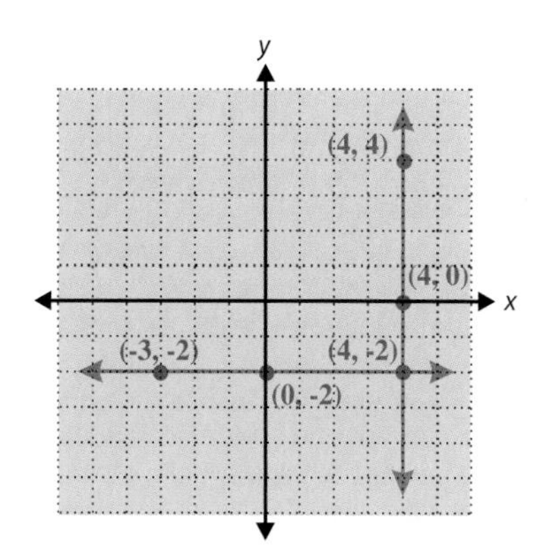


**58.** $y - 5 = 0;\ 3x = -9$

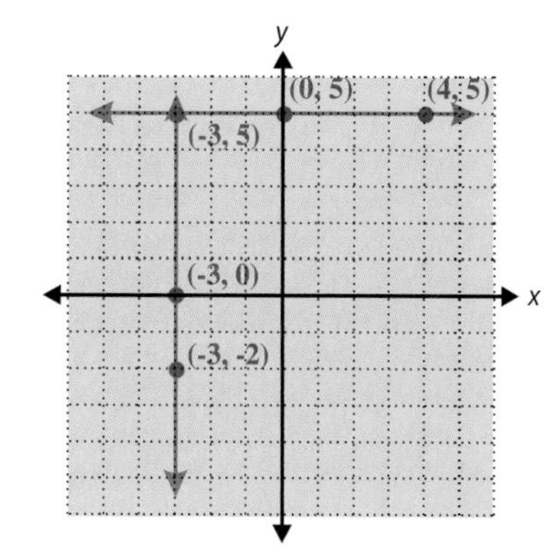

**59.** $4x = 1.6;\ 3y = -1.2$

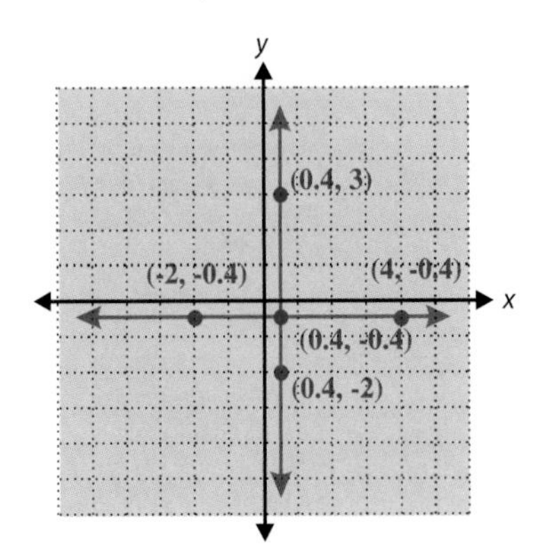


**60.** $y = \frac{3}{4};\quad x = -\frac{5}{4}$

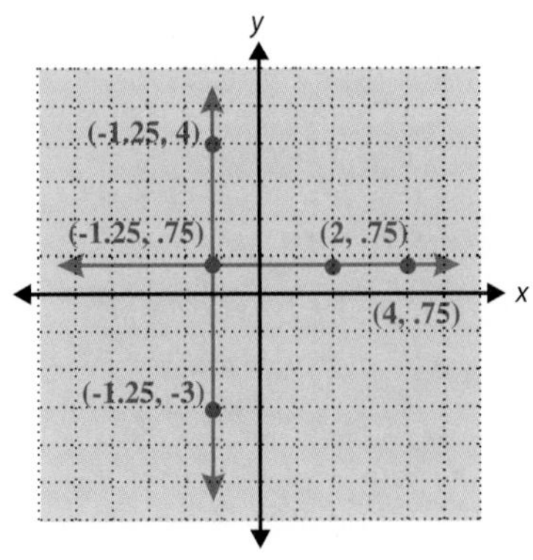


## COLLABORATIVE LEARNING EXERCISE ( with a graphing calculator )

**61.** With the class separated into teams of two to four students, each team is to use a graphing calculator ( TI-83 Plus ) to graph each pair of lines with the calculator and find the point of intersection. Each team should proceed as follows:

1. Press the Y= key in the upper left corner of the keyboard.
2. Enter the first equation after $y_1 =$ . ( Use the X key to enter the variable $x$. )
3. Enter the second equation after $y_2 =$ .
4. Press ZOOM then 6 to get the Standard Window. Both graphs will appear on the display.
5. Press TRACE and use the left and right arrows to move the cursor to the point of intersection.
6. Check the point of intersection in both equations to see if the point is correct. ( Decimal values may be only approximate. )

**a.** $y = 2x + 3$
$y = -x + 9$

**b.** $y = -x + 5$
$y = 3x + 1$

**c.** $y = -2x + 1$
$y = -0.5x + 2.5$

**61. a.** ( 2, 7 )

**b.** ( 1, 4 )

**c.** ( −1, 3 )

## Chapter 8 Index of Key Ideas and Terms

Name ______________ Section ______________ Date ______________

## Chapter 8 Test

Use the definitions and rules for exponents to simplify each of the following expressions so that the answer contains only positive exponents.

**1.** $12x^2 \cdot 5x^3$

**2.** $5^{-5} \cdot 5^3$

**3.** $\dfrac{y^7}{y^8}$

**4.** $(x^{-3})^2$

**5.** $(y^{-5})^{-4}$

**6.** $y^{-6} \cdot y^6$

**7.** $(-3)^2(-3)^3(-3)$

**8.** $\dfrac{x^{-3}}{x^{-5}}$

**9.** $\dfrac{a^{-4}}{a}$

Simplify each of the following polynomials, and state its type and degree.

**10.** $5x^3 - x^3 + 4x^2 + 5x - 12x$

**11.** $3x^2 + 7x - x - x^2 + 3 - 2x^2 + 1$

**12.** Evaluate the polynomial $2x^3 - 8x^2 + 7x + 10$ for $x = -2$.

Perform the indicated operations, and simplify if possible.

**13.** $(4y^3 + 7y^2 - 8y - 14) + (-5y^3 - 2y^2 + 6y - 4)$

**14.** $(-5y^2 + 9y - 1) + (-3y^2 + 9y - 3) + (6y^2 + y + 20)$

**ANSWERS**

1. $60x^5$
2. $\dfrac{1}{5^2} = \dfrac{1}{25}$
3. $\dfrac{1}{y}$
4. $\dfrac{1}{x^6}$
5. $y^{20}$
6. $1$
7. $(-3)^6 = 729$
8. $x^2$
9. $\dfrac{1}{a^5}$
10. $4x^3 + 4x^2 - 7x$; third-degree trinomial
11. $6x + 4$; first-degree binomial
12. $-52$
13. $-y^3 + 5y^2 - 2y - 18$
14. $-2y^2 + 19y + 16$

15. $4x+27$

16. $-2a^2-4a$

17. $15x^3+10x^2+20.5x$

18. $8a^2-10a-33$

19. $10x^4-23x^3+30x^2$ $-20x+8$

20. $9x^4-9x^3+6x^2$ $+15x+7$

21. a. $3(x-9)$

b. $2(x^2+25)$

c. $3x^2(x^2+9x-3)$

22. a. $(a-8)^2$

b. $(a+12)^2$

c. $(5a+1)(5a-1)$

23. a. $2(y+8)^2$

b. $3(y+5)(y-5)$

c. not factorable

**15.** $(8x+17)-(4x-10)$

**16.** $(2a^2+5a-3)-(4a^2+9a-3)$

Find each of the following products.

**17.** $5x(3x^2+2x+4.1)$

**18.** $(4a-11)(2a+3)$

**19.** $\begin{array}{r} 2x^2-3x+2 \\ \underline{5x^2-4x+4} \end{array}$

**20.** $\begin{array}{r} 3x^2-6x+7 \\ \underline{3x^2+3x+1} \end{array}$

Factor each polynomial completely.

**21. a.** $3x-27$ **b.** $2x^2+50$ **c.** $3x^4+27x^3-9x^2$

**22. a.** $a^2-16a+64$ **b.** $a^2+24a+144$ **c.** $25a^2-1$

**23. a.** $2y^2+32y+128$ **b.** $3y^2-75$ **c.** $y^2+6y+10$

Name ______________________ Section ______________ Date ___________

**ANSWERS**

24. {(–4, 1), (–3, 0), (–2, –1), (0, 0), (1, 1), (2, 0)}

25. {(–1, 1), (0, 1), (1, 1), (2, 1)}

List the set of ordered pairs that correspond to the points in each graph.

**24.**

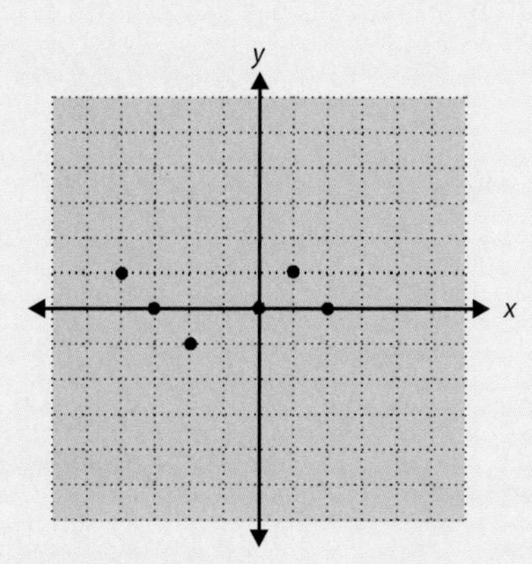

**25.**

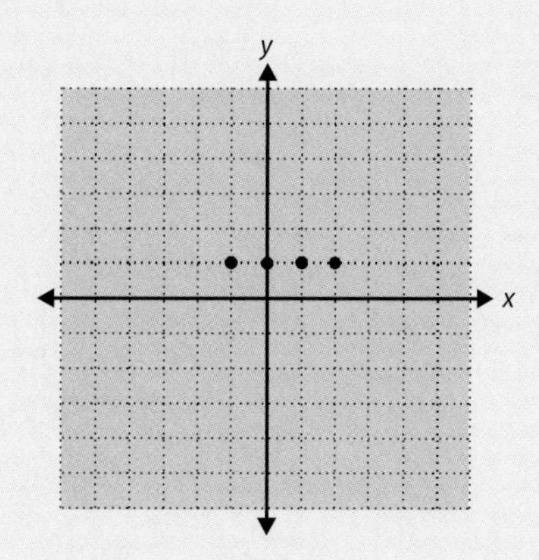

Graph the following sets of ordered pairs.

**26.** $\{(-2, 5), (-1, 3), (0, 1), (1, -1)\}$

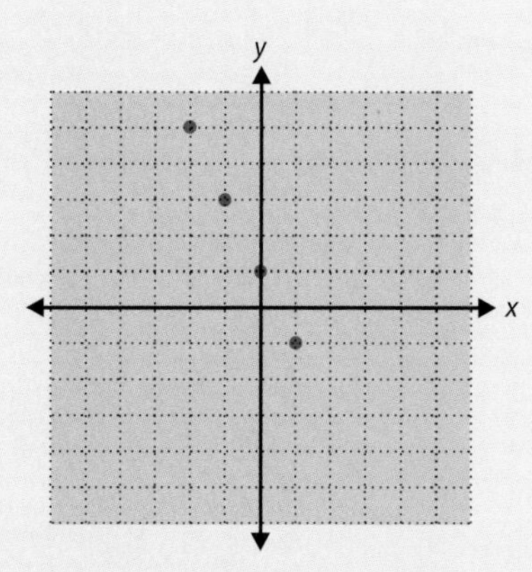

**27.** $\left\{\left(-3,\frac{1}{2}\right),(-0.5,0.3),\left(\frac{3}{4},\frac{7}{4}\right),(2,0)\right\}$

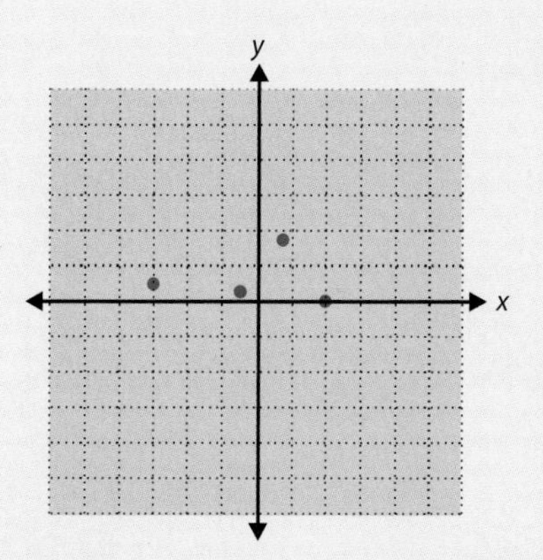

Graph the following linear equations.

**28.** $y = -2x + 1$

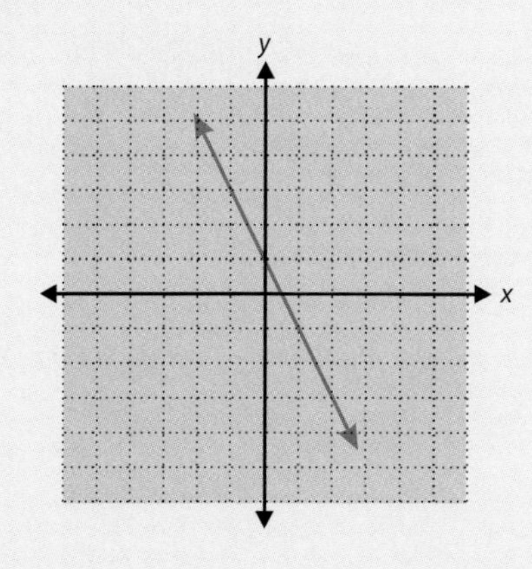

**29.** $2x - 3y = 6$

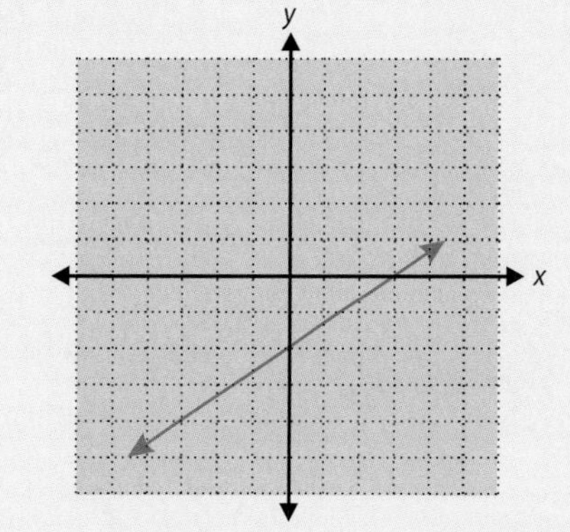

**30.** Graph the following pair of linear equations on the same set of axes.

**a.** $x = 2.5$

**b.** $y = 3x - 2$

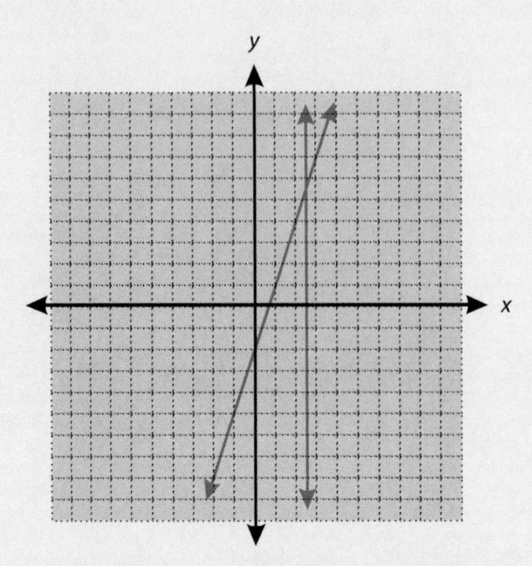

Name ______________ Section ______________ Date ______________

## Cumulative Review: Chapters 1 – 8

All answers should be in simplest form. Reduce all fractions to lowest terms, and express all improper fractions as mixed numbers.

Find each of the following sums.

**1.** $\frac{2}{3}+\frac{5}{6}+\frac{2}{9}$

**2.** $17\frac{5}{8}+12\frac{7}{10}$

**3.** $6.09 + 10.6 + 7$

**4.** $(-2.03)+(16.7)+(-5.602)$

Find each difference.

**5.** $\frac{3}{4}-\frac{5}{9}$

**6.** $4\frac{1}{6}-2\frac{2}{3}$

**7.** $142.01 - 67.135$

**8.** Subtract $-4.5$ from $-0.03$

Find each of the following products.

**9.** $2400 \cdot 30{,}000$

**10.** $\frac{5}{12}\cdot\frac{3}{5}\cdot\frac{4}{7}$

**11.** $\left(3\frac{2}{3}\right)\left(4\frac{4}{11}\right)$

**12.** $(-7.03)(0.28)$

Find each quotient.

**13.** $\frac{4}{9}\div\frac{8}{15}$

**14.** $5\frac{1}{4}\div\frac{7}{10}$

**ANSWERS**

1. $1\frac{13}{18}$
2. $30\frac{13}{40}$
3. 23.69
4. 9.068
5. $\frac{7}{36}$
6. $1\frac{1}{2}$
7. 74.875
8. 4.47
9. 72,000,000
10. $\frac{1}{7}$
11. 16
12. −1.9684
13. $\frac{5}{6}$
14. $7\frac{1}{2}$

15. 0.608

16. undefined

17. $2^3 \cdot 3^2 \cdot 11$

18. 30

19. 5990

20. 723.1

21. 22

22. $\frac{5}{36}$

23. $3\sqrt{5} + 20\sqrt{3}$

24. [ Respond in exercise. ]

25. $2(n+6) - 5$

26. $x = -12$

27. $x = 18$

**15.** $1.0336 \div 1.7$

**16.** $-3.7 \div 0$

**17.** Find the prime factorization of 792.

**18.** What is the quotient if $(6 \cdot 5 \cdot 4 \cdot 3 \cdot 2) \div 24$? Find this quotient by using prime factors and show your work.

Round off each of the following numbers as indicated.

**19.** 5987.04 ( nearest ten )

**20.** 723.149 ( nearest tenth )

Evaluate each of the following expressions.

**21.** $(2^2 \cdot 3 \div 4 + 2) \cdot 5 - 3$

**22.** $\left(\frac{1}{2}\right)^2 - \left(\frac{1}{3}\right)^2$

**23.** Simplify $3\sqrt{20} + 4\sqrt{75} - \sqrt{45}$

**24.** Fill in the blank with the appropriate symbol: <, >, or =.
$-|-7|$ __<__ $-(-7)$

**25.** Write an algebraic expression for the following English phrase:
"5 less than twice the sum of a number and 6."

Solve each of the following equations.

**26.** $-2(3 - x) = 3(x + 2)$

**27.** $\frac{5}{6}x - 1 = \frac{2}{3}x + 2$

Name ______________________ Section ______________ Date __________

**ANSWERS**

**28.** Solve $y = mx + b$ for $x$ in terms of the other variables.

**29.** Solve $2 < 3 - 4x \le 5$, and graph the solution on the number line.

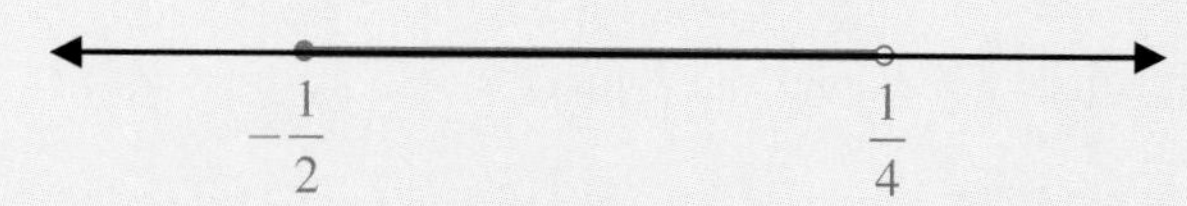


**30.** Find the length of the hypotenuse of a right triangle if one of its legs is 5 inches long and the other leg is 10 inches long.

**31.** List the set of ordered pairs that correspond to the points in the graph.

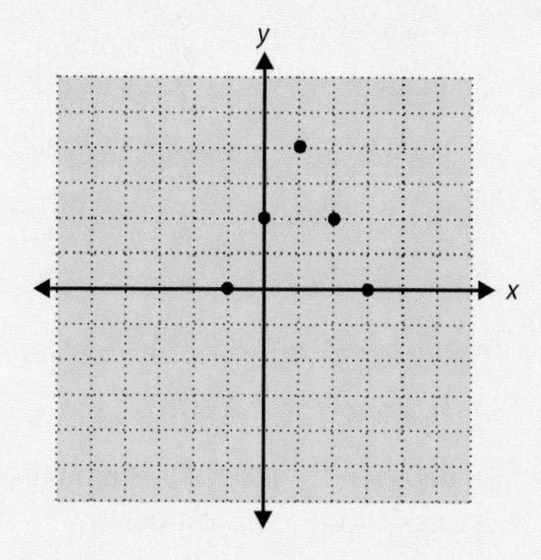


Graph each of the following equations.

**32.** $3x + 4y = 0$

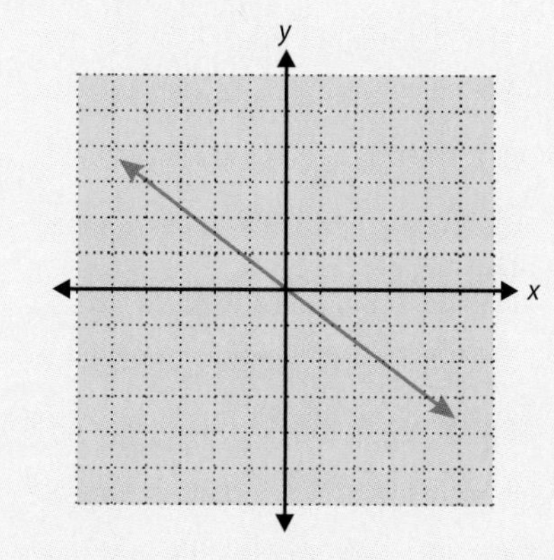


**33.** $y = \frac{1}{3}x + 4$

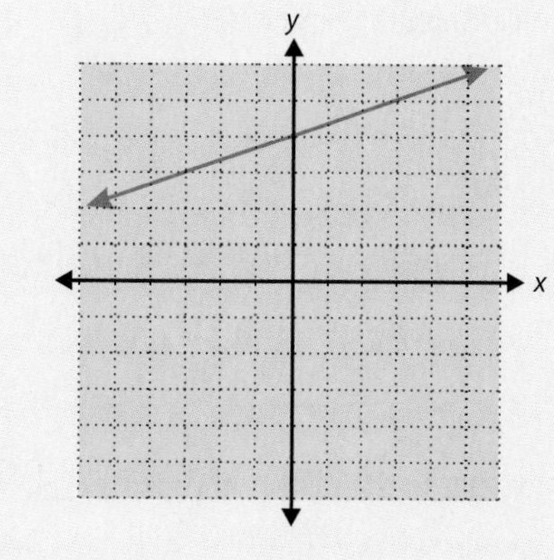


**28.** $x = \frac{y-b}{m}$

**29.** $-\frac{1}{2} \le x < \frac{1}{4}$

**30.** $5\sqrt{5}$ in.

( or 11.18 in. )

**31.** {(-1, 0), (0, 2),

(1, 4), (2, 2), (3, 0)}

**32.** [Use grid provided]

**33.** [Use grid provided]

34. 36

35. 210%

36. a. $P = 17.8$ ft

b. $A = 15.6$ ft$^2$

37. a. $C = 125.6$ in.

b. $A = 1256$ in.$^2$

38. 225 mi

39. a. $1969

b. $419

40. $-15, -14$

41. 21

42. $3.45 \times 10^6$

**34.** 16% of what number is 5.76?

**35.** What percent of 2.1 is 4.41?

**36.** Find

**a.** the perimeter and

**b.** the area of a rectangle that is 6.5 feet long and 2.4 feet wide.

**37.** Find

**a.** the circumference and

**b.** the area of a circle that has a radius of 20 inches. ( Use $\pi = 3.14$. )

**38.** Two towns are shown 13.5 centimeters apart on a map that has a scale of 3 centimeters to 50 miles. What is the actual distance between the towns ( in miles )?

**39. a.** If $1550 is deposited in an account paying 8% compounded monthly, what will be the total amount in the account at the end of 3 years ( to the nearest dollar )?

**b.** How much interest will be earned?

**40.** Find two consecutive integers such that 3 more than twice the smaller is equal to 13 less than the larger.

**41.** Evaluate the polynomial $3x^3 - 10x^2 - 30x + 25$ for $x = -2$.

**42.** Write the number 3,450,000 in scientific notation.

Name ______ Section ______ Date ______

**ANSWERS**

Use the properties of exponents to simplify each expression so that it has only nonnegative exponents.

**43.** $7^{-1} \cdot 7^{3}$

**44.** $(a^{5})^{-2}$

**45.** $\frac{x^{-4}}{x^{6}}$

**46.** $6y^{2} \cdot 3y^{-4}$

**47.** Find the sum $(-3x^{3} + 4x^{2} - 10x + 33) + (2x^{3} + 5x^{2} - x - 10)$.

**48.** Find the difference $(8x^{2} - 14x - 12) - (3x^{2} - 10x - 30)$.

**49.** Find the product $(5x + 3)(2x - 3)$.

**50.** Find the product:

$$\begin{array}{r} x^{2} - 3x - 4 \\ x^{2} + x + 2 \\ \hline \end{array}$$

**51.** Factor each of the following polynomials completely.

**a.** $3x^{2} - 108$ **b.** $5x^{2} + 10x + 35$ **c.** $6x^{4} + 36x^{3} - 12x^{2}$

**52.** Factor each of the following polynomials completely.

**a.** $25a^{2} - 50a + 25$ **b.** $2a^{2} + 24a + 72$ **c.** $a^{3} + 18a^{2} + 81a$

**43.** $7^{2}$

**44.** $\frac{1}{a^{10}}$

**45.** $\frac{1}{x^{10}}$

**46.** $\frac{18}{y^{2}}$

**47.** $-x^{3} + 9x^{2} - 11x + 23$

**48.** $5x^{2} - 4x + 18$

**49.** $10x^{2} - 9x - 9$

**50.** $x^{4} - 2x^{3} - 5x^{2}$

$-10x - 8$

**51. a.** $3(x+6)(x-6)$

**b.** $5(x^{2} + 2x + 7)$

**c.** $6x^{2}(x^{2} + 6x - 2)$

**52. a.** $25(a-1)^{2}$

**b.** $2(a+6)^{2}$

**c.** $a(a+9)^{2}$

# 9

# STATISTICS

**Chapter 9 Statistics**



## WHAT TO EXPECT IN CHAPTER 9

Chapter 9 deals with some basic statistical concepts that appear regularly in daily newspapers, magazines, corporate reports, sports, insurance studies, and so on. In Section 9.1, the statistical terms **mean, median, mode,** and **range** are defined, and the student is to perform calculations on given data to find these **statistics**.

The knowledge and skills related to reading and understanding graphs are discussed in Section 9.2 and 9.3. The types of graphs discussed (bar graphs, circle graphs, pictographs, line graphs, histograms, and frequency polygons) can be found in almost every newspaper and magazine. These graphs provide the reader with considerable information (such as trends in the stock market, relative amounts of success in medical treatments, and percentages of income spent in categories of a budget) at a glance.

In Sections 9.2 and 9.3, the graphs are given and the students are asked questions concerning these graphs. In Section 9.4, only the data (called **databases**) are given, and the students are to create their own graphs (bar graphs and circle graphs). In this way, the students can develop a thorough understanding of fundamental statistical and graphing concepts.

**Mathematics at Work!**

Graphs ( circle graphs, pictographs, bar graphs, and line graphs ) appear almost daily in newspapers, magazines, and journals and are a basic part of many textbooks. *USA Today* has a regular feature on the front page entitled "USA SNAPSHOTS" ( a look at statistics that shape the nation ), which uses graphs, so that information can be communicated with very few words.

Even though graphs are convenient and effective, you ( as a consumer and citizen ) should be aware of how some graphs can imply ideas that are not true. In reading graphs, you should learn to look past the graphics to the actual numerical scales and other information stated in the graph. Ask yourself such questions:

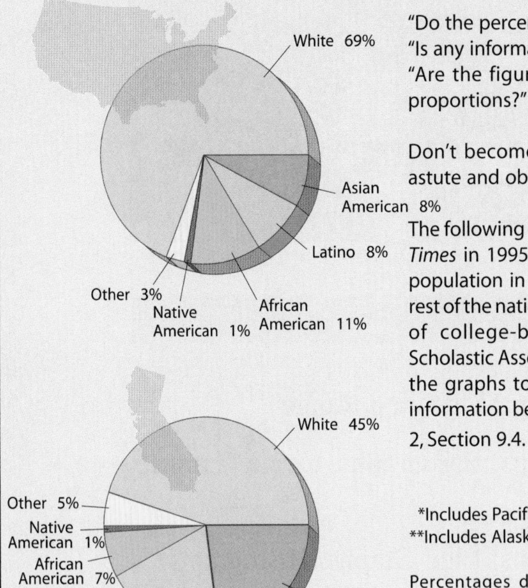


"Do the percents total 100?"
"Is any information missing or misleading?"
"Are the figures or diagrams used in proper proportions?"

Don't become a complete pessimist; just be astute and observant.

The following graphs were published in the *L.A. Times* in 1995 to illustrate the diversity of the population in California as compared with the rest of the nation. The graphs show the ethnicity of college-bound seniors that took the Scholastic Assessment Test ( SAT ) in 1995. Study the graphs to see if you can understand the information being illustrated. ( Also see Example 2, Section 9.4. )

*Includes Pacific islanders
**Includes Alaskan natives

Percentages don't add up to 100% because of rounding.

# 9.1 Mean, Median, Mode, and Range

## Statistical Terms

*Statistics* is the study of how to gather, organize, analyze, and interpret information. A **statistic** is a particular measure or characteristic of a sample ( part ) of a larger collection of items. The larger collection of items ( people, animals, or objects ) is called the **population** of interest.

In this text, we will study only four numerical statistics that are easily found or calculated: **mean, median, mode,** and **range**. The mean, median, and mode are measures that describe the "average" or "middle" of a set of data. The range is a measure that describes how "spread out" the data is. Other measures that you might read about in newspapers and magazines or study in a statistics course are

**standard deviation** and **variance** ( both measures of how "spread out" data are )

***z*-score** ( a measure that compares numbers that are expressed in different units — for example, your test score on a mathematics exam and your performance in a physical exercise )

**correlation coefficient** ( a measure of how two different types of data might be related — for example, the relationship between a person's income and the number of years of schooling they have )

**Objectives**

① Learn the meanings of the terms **mean, median, mode,** and **range**.

② Know how to find the mean, median, mode, and range of a set of data items.

You will need a semester or two of algebra in order to be able to study and understand these and other statistics, so keep working hard. The following terms and their definitions are necessary for understanding the ideas and problems in this section. They are listed here for easy reference.

The formula for *z*-scores is given in Section 7.5, Exercises #31 and #32. You might calculate the mean and standard deviation for 2 exams in your class and have the students calculate their own *z*-scores for each exam. This could lead to an interesting discussion of negative numbers and how the same raw score on different exams can yield different z-scores (a nice application of algebraic concepts).

**Terms Used in the Study of Statistics**

**Data:** Value( s ) measuring some information of interest
( We will consider only numerical data. )

**Statistic:** A single number describing some characteristic of the data

**Mean:** The arithmetic average of the data
( Find the sum of all the data, and divide by the number of data items. )

**Median:** The middle of the data after the data have been arranged in order
( The median may or may not be one of the data items. )

**Mode:** The single data item that appears most frequently
( A set of data may have more than one mode. The data sets in this text, however, will have either one mode or no mode. )

**Range:** The difference between the largest and smallest data items

## Finding the Mean, Median, Mode, and Range of a Set of Data

Two sets of data, Group A and Group B, are shown here and used in the discussion and calculations in Examples 1 - 3.

| Group A: Body Temperature ( in Farenheit degrees ) of 8 People | | | | | | | |
|---|---|---|---|---|---|---|---|
| 96.4° | 98.6° | 98.7° | 99.8° | 99.2° | 101.2° | 98.6° | 97.1° |

| Group B: The Time ( in minutes ) of 11 Movies | | | | | |
|---|---|---|---|---|---|
| 100 min | 90 min | 113 min | 110 min | 88 min | 90 min |
| 155 min | 88 min | 105 min | 93 min | 90 min | |

### Example 1

Find the mean temperature for the 8 people in Group A.

#### Solution

The **mean** is the arithmetic average of the data. Therefore, we add the 8 temperatures and divide the sum by 8. ( You can, of course, perform these calculations with a calculator. )

Fahrenheit

$$\begin{array}{r} 96.4 \\ 98.6 \\ 98.7 \\ 99.8 \\ 99.2 \\ 101.2 \\ 98.6 \\ +\ 97.1 \\ \hline 789.6 \end{array} \qquad \begin{array}{r} 98.7 \\ 8\overline{)789.6} \\ 72\phantom{.6} \\ \hline 69\phantom{.} \\ 64\phantom{.} \\ \hline 56 \\ 56 \\ \hline 0 \end{array}$$

mean temperature

98.7°

The mean temperature is **98.7°F**.

The **median** is another type of "average." In a set of **ranked data** ( data arranged in order, smallest to largest or largest to smallest ), the median is the middle value. As we will see, the determination of this value depends on whether there is an odd number of data items or an even number of data items.

**To Find the Median:**

1. Rank the data. ( Arrange the data in order, either from smallest to largest or largest to smallest. )
2. The median can be found by counting from the top down ( or from the bottom up ) to the position $\frac{n+1}{2}$ where $n$ represents the number of data items.
   a. If there is an **odd** number of items, the median is the middle item.
   b. If there is an **even** number of items, the median is the value found by averaging the two middle items. ( **Note:** This value may or may not be in the data. )

## Example 2

Find the median temperature for the 8 people in Group A and the median time for the movies in Group B.

### Solution

First, we rank both sets of data in order from smallest to largest.

| Group A ( Temperatures ) | | Group B ( Movie Times ) | |
|---|---|---|---|
| 1. 96.4 | | 1. 88 | |
| 2. 97.1 | | 2. 88 | |
| 3. 98.6 | | 3. 90 | |
| 4. 98.6 | ↙ median | 4. 90 | |
| 5. 98.7 | ↖ | 5. 90 | |
| 6. 99.2 | | 6. 93 | ← median |
| 7. 99.8 | | 7. 100 | |
| 8. 101.2 | | 8. 105 | |
| | | 9. 110 | |
| | | 10. 113 | |
| | | 11. 155 | |

Group A has 8 items. With the formula for the position $\frac{n+1}{2} = \frac{8+1}{2} = \frac{9}{2} = 4.5$, the median is the average of the items in the fourth and fifth positions.

$$\text{Median temperature} = \frac{98.6 + 98.7}{2} = \frac{197.3}{2} = 98.65^\circ \text{F}$$

( Note that 8 is an **even** number and the median is the average of the two middle items. )

Group B has 11 items. With the formula for the position

$\frac{n+1}{2} = \frac{11+1}{2} = \frac{12}{2} = 6$, the median is the item in the sixth position.

Median movie time = 93 minutes

( Note that 11 is an **odd** number and the median is the middle item. )

( **Comment:** Note that in Group A the median is not one of the data items, while in Group B the median is one of the data items. )

Once the data have been ranked ( as was done in Example 2 ), the mode ( if there is one ) and the range are easily determined. Remember that the mode is the item that occurs most frequently, and the range is the difference between the largest and smallest items in each set of data.

### Example 3

For both Group A and Group B, find

**a.** the mode and

**b.** the range

#### Solution

**a.** From the ranked data in Example 2, we can see the most frequent item in each group:

For Group A, the mode is 98.6°. ( 98.6° occurs twice and no other item occurs more than once. )

For Group B, the mode is 90 minutes. ( 90 min. occurs three times and no other item occurs more than twice. )

**b.** Again referring to the ranked data in Example 2, we can calculate each range as follows:

Range = ( Largest value ) – ( Smallest value )

Group A range = 101.2° – 96.4° = 4.8°

Group B range = 155 – 88 = 67 minutes

### *Completion Example 4*

Suppose that your grade in this class is based on 5 exam scores: 4 sectional exams and 1 comprehensive final exam, with each exam scored on a basis of 100 points. On the first 4 exams you have scores of 85, 78, 82, and 70. What is the lowest score you could get on the final exam and still earn a grade of B in the course? ( Assume that to get a B you must average between 80 and 89. )

#### Solution

We solve the problem in two ways: first by using arithmetic and the total number of points possible and second by using an algebraic inequality.

ARITHMETIC APPROACH

**a.** Find the total number of points needed for a B. Since 5 exams are to average 80 ( or more ), then the total number of points must be at least

$$\begin{array}{r} 80 \\ \times\ 5 \\ \hline \underline{\qquad} \end{array}$$ Total points ( or more ) needed for a B

**b.** On the first 4 exams you have accumulated

$$\begin{array}{r} 85 \\ 78 \\ 82 \\ +\ 70 \\ \hline \underline{\qquad} \end{array}$$ Points accumulated

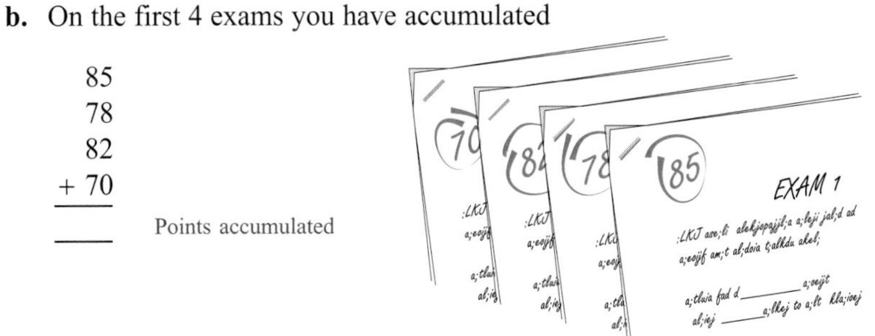


**c.** To average 80 ( or more ) on your 5 exams, you need

$$\begin{array}{r} 400 \\ -\ \underline{\qquad} \\ \underline{\qquad} \end{array}$$

400 Total Points needed
− ____ Points accumulated
____ Points needed on the final exam

ALGEBRAIC APPROACH

**a.** Let $x$ = the score you need on the final exam.

**b.** The $x$ must satisfy the following inequality:

$$\frac{85+78+82+70+x}{5} \geq 80$$

$$\frac{\underline{\qquad}+x}{5} \geq 80$$

$$315+x \geq \underline{\qquad}(80)$$

$$x \geq 400-\underline{\qquad}$$

$$x \geq \underline{\qquad}$$

By either method, you need at least a score of ____ on your final exam to earn a B for the course.

**Note:** Of the four statistics mentioned in this section, the mean and median are most commonly used. Many people who use statistics professionally believe that the mean ( or arithmetic average ) is relied on too much in reporting central tendencies for data such as income, housing costs, and taxes, because a few very high or very low items can distort the picture of a central tendency. For example, the median of 93 minutes for the movies in Group B is probably more representative than the mean, which is 102 minutes. This is so because the one high time of 155 minutes raises the mean considerably, whereas the median is not affected by this one extreme outer value. When you read an article in a magazine or newspaper that reports means or medians, you should now have a better understanding of the implications of these statistical measures.

## Completion Example Answer

**4.** ARITHMETIC APPROACH

**a.** Find the total number of points needed for a B. Since 5 exams are to average 80 ( or more ), then the total number of points must be at least

$$\begin{array}{r} 80 \\ \times\ 5 \\ \hline \mathbf{400} \end{array} \quad \text{Total points ( or more ) needed for a B}$$

**b.** On the first 4 exams you have accumulated

$$\begin{array}{r} 85 \\ 78 \\ 82 \\ +\ 70 \\ \hline \mathbf{315} \end{array} \quad \text{Points accumulated}$$

**c.** To average 80 (or more) on your 5 exams, you need

$$\begin{array}{rl} 400 & \text{Total Points needed} \\ -\ \mathbf{315} & \text{Points accumulated} \\ \hline \mathbf{85} & \text{Points needed on the final exam} \end{array}$$

ALGEBRAIC APPROACH

**a.** Let $x =$ the score you need on the final exam.

**b.** The $x$ must satisfy the following inequality:

$$\frac{85+78+82+70+x}{5} \geq 80$$
$$\frac{\mathbf{315}+x}{5} \geq 80$$
$$315+x \geq \mathbf{5}(80)$$
$$x \geq 400-\mathbf{315}$$
$$x \geq \mathbf{85}$$

By either method, you need at least a score of **85** on your final exam to earn a B for the course.

Name ______________________ Section ______________ Date ____________

## Exercises 9.1

In Exercises 1-12, find the mean, median, mode ( if any ), and range of the given data.

**1.** Nine test scores selected at random from Dr. Wright's course in statistics:

95 82 85 71 65 85 62 77 98

**a.** Mean = 80 **b.** Median = 82

**c.** Mode = 85 **d.** Range = 36

**2.** The ages of the first five U.S. presidents on the date of their inaugurations: ( The presidents were Washington, Adams, Jefferson, Madison, and Monroe. Who was the sixth president? ) (Adams)

57 61 57 57 58

**a.** Mean = 58 **b.** Median = 57

**c.** Mode = 57 **d.** Range = 4

**3.** Family incomes in a survey of eight students:

$35,000 $28,000 $42,000 $71,000

$63,000 $36,000 $51,000 $63,000

**a.** Mean = $48,625 **b.** Median = $46,500

**c.** Mode = $63,000 **d.** Range = $43,000

**4.** Resident tuition charged by 10 colleges in North Carolina in 1999:

$1970 $2100 $1770 $8820 $8930

$1960 $2260 $2410 $2140 $1800

**a.** Mean = $3416 **b.** Median = $2120

**c.** Mode = none **d.** Range = $7160

**ANSWERS**

1. [Respond in exercise.]

2. [Respond in exercise.]

3. [Respond in exercise.]

4. [Respond in exercise.]

5. [Respond in exercise.]

6. [Respond in exercise.]

7. [Respond in exercise.]

8. [Respond in exercise.]

**5.** Nonresident tuition charged by the same 10 colleges in North Carolina in 1992:

| | | | | |
|---|---|---|---|---|
| $9070 | $8580 | $11,430 | $11,580 | $10,540 |
| $9240 | $9320 | $8900 | $8930 | $9420 |

**a.** Mean = $9701 **b.** Median = $9280

**c.** Mode = none **d.** Range = $3000

**6.** Ages of 20 students surveyed in a chemistry class:

| | | | | | | | | | |
|---|---|---|---|---|---|---|---|---|---|
| 18 | 23 | 23 | 23 | 22 | 18 | 21 | 20 | 18 | 20 |
| 19 | 20 | 21 | 19 | 23 | 36 | 35 | 26 | 17 | 24 |

**a.** Mean = 22.3 **b.** Median = 21

**c.** Mode = 23 **d.** Range = 19

**7.** The distances from Chicago to

Boston: 980 miles
Dallas: 930 miles
Detroit: 280 miles
Los Angeles: 2110 miles
New Orleans: 950 miles
Seattle: 2050 miles
Cleveland: 345 miles
Denver: 1050 miles
Indianapolis: 190 miles
San Francisco: 2210 miles
Miami: 1390 miles

**a.** Mean = 1135 miles **b.** Median = 980 miles

**c.** Mode = none **d.** Range = 2020 miles

**8.** Passengers ( to the nearest thousand in 1999 ) in the world's 10 busiest airports:

Chicago, O'Hare: 72,568,000
London, Heathrow: 62,264,000
Atlanta,Hartsfield: 77,940,000
San Francisco: 40,387,000
Frankfurt - Main: 45,858,000
Dallas/Ft. Worth: 60,000,000
Los Angeles: 63,877,000
Tokyo, Haneda: 54,338,000
Denver: 38,034,000
Paris, Charles de Gaulle: 43,597,000

**a.** Mean = 55,886,300 **b.** Median = 57,169,000

**c.** Mode = none **d.** Range = 39,906,000

Name ______________________ Section ____________ Date __________

ANSWERS

9. [Respond in exercise.]

10. [Respond in exercise.]

11. [Respond in exercise.]

12. [Respond in exercise.]

**9.** Capacity ( in thousands of cubic meters ) of the world's 5 largest dams:

New Cornelia Tailings, Arizona: 209,500
Pati, Argentina: 238,200
Tarbela, Pakistan: 121,720
Syncrude Tailings, Canada: 540,000
Chapeton, Argentina: 396,200

**a.** Mean = 301,124,000 $m^3$ **b.** Median = 238,200,000 $m^3$

**c.** Mode = none **d.** Range = 418,280,000 $m^3$

**10.** Volumes ( to nearest thousand ) in top college libraries in United States:

Harvard: 13,617,000
U. of Illinois: 9,024,000
U of Texas: 7,495,000
UCLA: 7,010,000
Yale: 9,932,000
UC Berkeley: 8,628,000
U of Michigan: 6,973,000
Columbia: 6,906,000

**a.** Mean = 8,698,125 volumes **b.** Median = 8,061,500 volumes

**c.** Mode = none **d.** Range = 6,711,000 volumes

**11.** The winning margin for each of the first 34 Super Bowls:

25 19 9 16 3 21 7 17 10 4 18 17 4 12

17 5 10 29 22 36 19 32 4 45 1 13 35 17

23 10 14 7 15 7

**a.** Mean = 16.0 **b.** Median = 15.5

**c.** Mode = 17 **d.** Range = 44

**12.** The Medal of Honor is the nation's highest military award for uncommon valor by men and women in battle. The medals awarded by war are:

Civil War: 1520
Korean Expedition ( 1871 ): 15
Philippines/Samoa ( 1899 - 1913 ): 91
Dominican Republic ( 1904 ): 3
Mexico ( Veracruz ) ( 1914 ): 55
Miscellaneous ( 1861 - 1920 ): 166
Haitian Action ( 1919 - 20 ): 2
World War II: 433
Vietnam War: 238
Indian Wars ( 1861 - 1898 ): 428
Spanish-American War: 109
Boxer Rebellion ( 1900 ): 59
Nicaragua ( 1911 ): 2
Haiti ( 1915 ): 6
World War I: 124
Misc. ( 1920 - 1940 ): 18
Korean War: 131
Somalia: 2

**a.** Mean = 189 **b.** Median = 75

**c.** Mode = 2 **d.** Range = 1518

13. 79

14. 72

15. a. 3.3

b. 2.4

c. answers vary

16. [Respond in exercise.]

**13.** Suppose that you are to take four hourly exams and a final exam in your chemistry class. Each exam has a maximum of 100 points, and you must average between 75 and 82 points to receive a passing grade of C. If you have scores of 83, 65, 70, and 78 on the hourly exams, what is the minimum score you can make on the final exam and receive a grade of C? ( First explain your strategy in solving this problem. Then solve the problem. )

**14.** Suppose that the instructor in the class in Exercise 13 has informed the class that the lowest of your hourly exam scores will be replaced by your score on the final exam provided that your final exam score is higher. ( That is, the final exam score may be counted twice. ) Now, what is the minimum score that you can make on the final exam and still receive a grade of C? ( First explain your strategy in solving this problem. Then solve the problem. )

## Writing and Thinking about Mathematics

**15.** You grade point average ( GPA ) is a form of a **weighted average**. That is, 4 units of A counts more than 4 units of B. The most common weight for grades is A — 4 points, B—3 points, C—2 points, D—1 point, and F—0 points. To find a GPA,

1. Multiply the points for each grade by the number of units for the course.
2. Find the sum of these products.
3. Divide this sum by the total number of units taken.

Find the GPA ( to the nearest tenth ) for each of the following situations:

**a.** 3 units of A in astronomy, 4 units of B in geometry, 3 units of C in sociology, and 4 units of A in biology

**b.** 5 units of B in history, 4 units of C in calculus, 3 units of A in computer science, and 4 units of D in geology

**c.** Your own GPA for the last semester ( or your anticipated GPA for this semester ).

## Collaborative Learning Exercise

**16.** With the class separated into teams of 2 to 4 students, each team is to go on campus and survey 50 students and ask each student how many minutes it takes him or her to drive to school from home.

**a.** Each team is to find the mean, median, mode, and range for the 50 responses.

**b.** Each team is to bring all the data to class, and the class is to pool the information and find the mean, median, mode, and range for the pooled data.

**c.** The class is to discuss the results of the individual teams and the pooled data and what use such information might have for the administration of the college.

Answers will vary. The students should become aware of the idea that statistics based on samples give results more representative of the population as the sample sizes become larger.

# 9.2 Reading Graphs: Bar Graphs, Circle Graphs, Pictographs

## Introduction to Graphs

Graphs are **pictures** of numerical information. Well-drawn graphs can organize and communicate information accurately, effectively, and fast. Most computers have sophisticated graphing capabilities, and anyone whose work involves a computer in any way will probably be expected to understand graphs and, at some time, create graphs. [ Computer programs called spreadsheets, such as Excel, are particularly useful in creating and analyzing graphs. ]

There are many different types of graphs, each particularly well suited to the display and clarification of certain types of information. In this section, we will discuss three types of graphs: bar graphs, circle graphs (or pie charts), and pictographs. Their basic uses are listed below.

1. **Bar Graphs:**
   To emphasize comparative amounts
2. **Circle Graphs** ( or **Pie Charts** ):
   To help in understanding percents or parts of a whole
3. **Pictographs:**
   To emphasize the topic being related as well as quantities

A common characteristic of all graphs is that they are intended to communicate information about numerical data quickly and easily. With this in mind, note the following three properties of all graphs:

1. They should be clearly labeled.
2. They should be easy to read.
3. They should have appropriate titles.

**Objectives**

① Understand how to read and interpret the information shown in **bar graphs, circle graphs,** and **pictographs**.

② Be able to perform appropriate operations related to the data shown in a bar graph, circle graph, or pictograph.

As a class project (or individual project), you might have students collect examples of graphs that they find in newspapers, journals, and magazines. You could have them discuss (either orally or in writing) the good features and any potentially misleading features, such as breaks in scales on bar graphs or pictographs that are out of proportion, relative to their implied use.

## Reading Bar Graphs

### *Completion Example 1* ( Bar Graph )

Figure 9.1 shows a bar graph. Note that the scale on the left and the data along bottom ( tribe names ) are clearly labeled and the graph itself has a title.

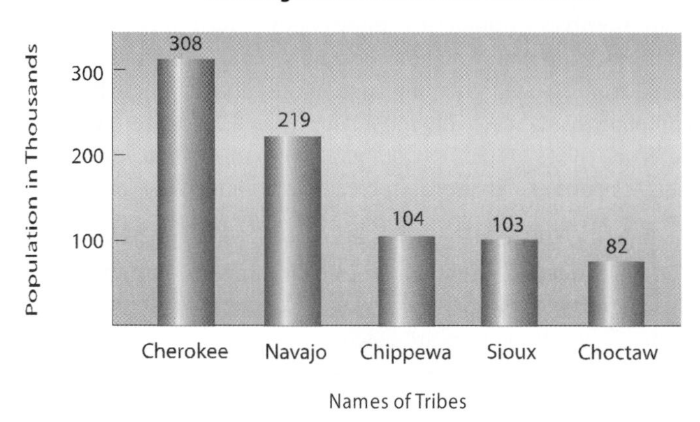


**Figure 9.1**

**a.** Which tribe had the most number of people? Cherokee

**b.** How many? ________

**c.** Which of these five tribes had the least number of people? ________

**d.** How many more Native Americans were Navajo than Sioux? ________

**e.** What was the mean number of people per tribe? ________

## Reading Circle Graphs ( or Pie Charts )

### *Completion Example 2* ( Circle Graph )

According to the *Information Please Almanac*, in 2000 there were 557,000,000 computers in use in the world. The circle graph in Figure 9.2 shows the approximate distribution of these computers by country.

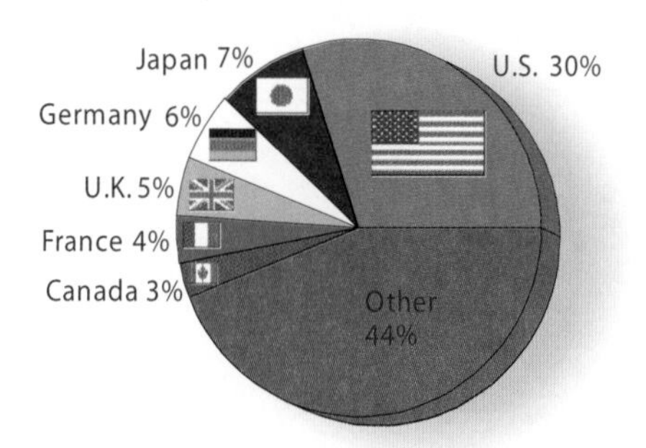


**Figure 9.2**

Using a calculator and the percentages shown in the graph, find the missing numbers of computers.

| Country(s) | Number of Computers in 2000 |
|---|---|
| **a.** U.S. | $0.30 \times 557{,}000{,}000 =$ 167,100,000 |
| **b.** Japan | $0.07 \times 557{,}000{,}000 =$ ________ |
| **c.** Germany | $0.06 \times 557{,}000{,}000 =$ ________ |
| **d.** U.K. | $0.05 \times 557{,}000{,}000 =$ ________ |
| **e.** France | ____________ = ________ |
| **f.** Canada | ____________ = ________ |
| **g.** Others | ____________ = ________ |

## Reading Pictographs

### *Completion Example 3* ( Pictograph )

Figure 9.3 shows the number of subscribers ( in millions ) for six magazines in 1999: *Reader's Digest*, *TV Guide*, *National Geographic*, *Better Homes and Gardens*, *Good Housekeeping*, and *Sports Illustrated.*

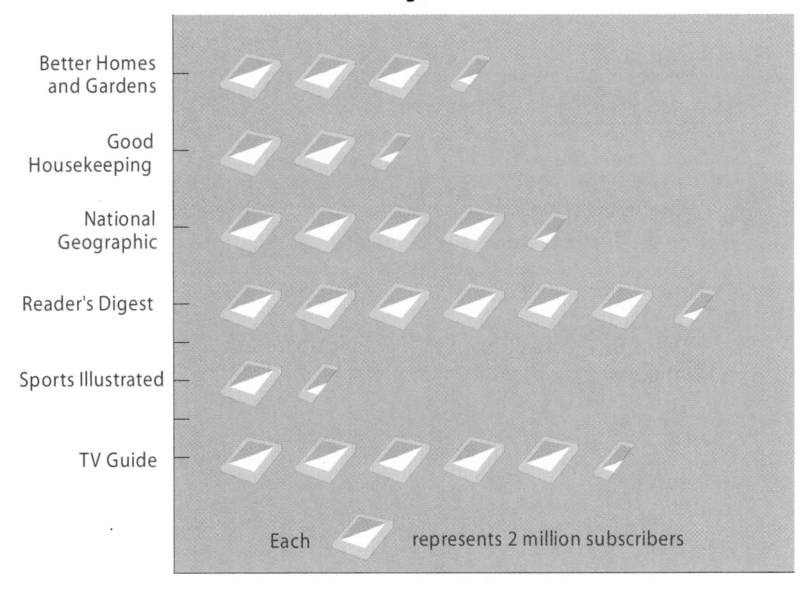


**Figure 9.3**

Of the magazines shown here:

**a.** Which magazine had the most number of subscribers in 1999?
*Reader's Digest*

**b.** How many subscribers? _____

**c.** Which magazine had the least number of subscribers? _____

**d.** How many subscribers? _____

**e.** What was the mean number of subscribers for the six magazines?
_____

## Completion Example Answers

1. **a.** Which tribe had the most number of people? Cherokee

   **b.** How many? **308,000**

   **c.** Which of these five tribes had the least number of people? **Choctaw**

   **d.** How many more Native Americans were Navajo than Sioux? **116,000**

   **e.** What was the mean number of people per tribe? **163,200**

2. | **Country( s )** | **Number of Computers in 1992** |
   |---|---|
   | **a.** U.S. | $0.30 \times 557{,}000{,}000 = 167{,}100{,}000$ |
   | **b.** Japan | $0.07 \times 557{,}000{,}000 =$ **38,990,000** |
   | **c.** Germany | $0.06 \times 557{,}000{,}000 =$ **33,420,000** |
   | **d.** U.K. | $0.05 \times 557{,}000{,}000 =$ **27,850,000** |
   | **e.** France | **0.04 × 557,000,000** = **22,280,000** |
   | **f.** Canada | **0.03 × 557,000,000** = **16,710,000** |
   | **g.** Others | **0.44 × 557,000,000** = **245,080,000** |

3. **a.** Which magazine had the most number of subscribers in 1993? *Reader's Digest*

   **b.** How many subscribers? **13,000,000**

   **c.** Which magazine had the least number of subscribers? ***Sports Illustrated***

   **d.** How many subscribers? **3,000,000**

   **e.** What was the mean number of subscribers for the six magazines? **8,000,000**

Name ______________________ Section ____________ Date __________

## Exercises 9.2

Answer the questions related to each of the graphs. Some questions can be answered directly from the graphs; others may require some calculation.

1. The bar graph shows the approximate amounts of time per week spent watching TV for six groups ( by age and sex ) of people 18 years of age and older.

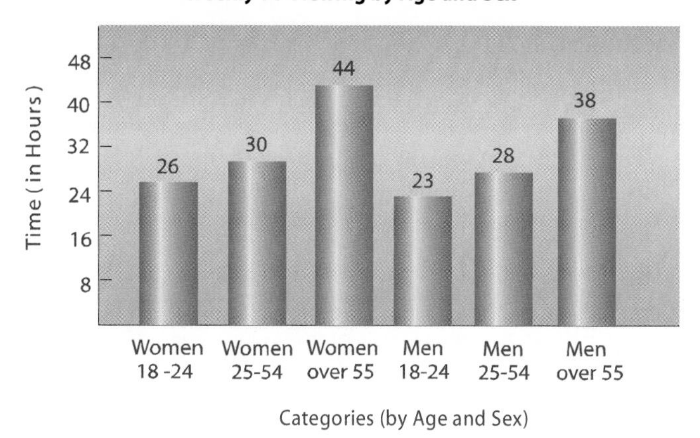


   a. Which group watched TV the most each week?

   b. Which group watched TV the least?

   c. Which age category ( regardless of sex ) watched TV the most? (*Hint*: Add the amounts for each age category.)

   d. How much time was this?

**ANSWERS**

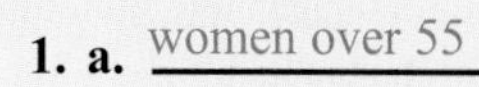

**1. a.** women over 55

**b.** men 18-24

**c.** over 55

**d.** 82 hours

2. a. Cornell

b. Dartmouth

c. about 700

d. none

2. The following bar graph shows the enrollment ( to the nearest hundred ) of the eight "Ivy League" universities in 1999.

   a. Which university had the largest enrollment in 1999?

   b. Which university had the least enrollment in 1999?

   c. What was the difference between the enrollment at Brown and the enrollment at Yale?

   d. Which universities had the same enrollment?

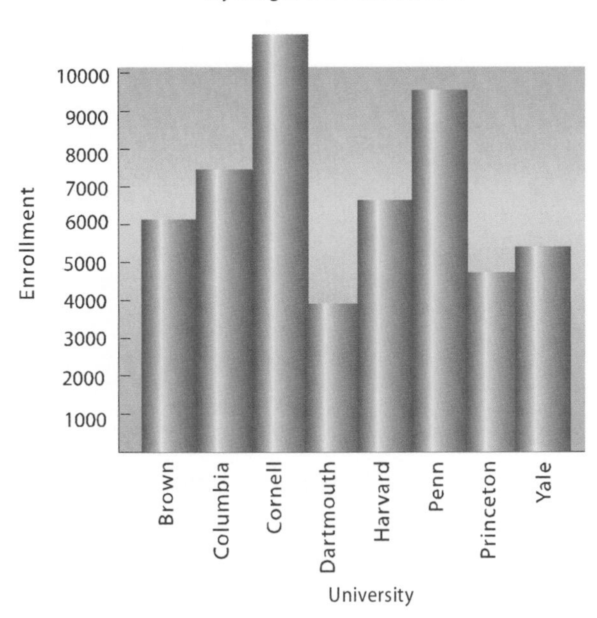

Name ______________________ Section ______________ Date ____________

**ANSWERS**

3. a. agriculture/biological sciences

b. math

c. about 1800

d. about 21%

**3.** The bar graph shows the number of doctorates ( to the nearest hundred ) conferred in selected science and engineering fields in 1997.

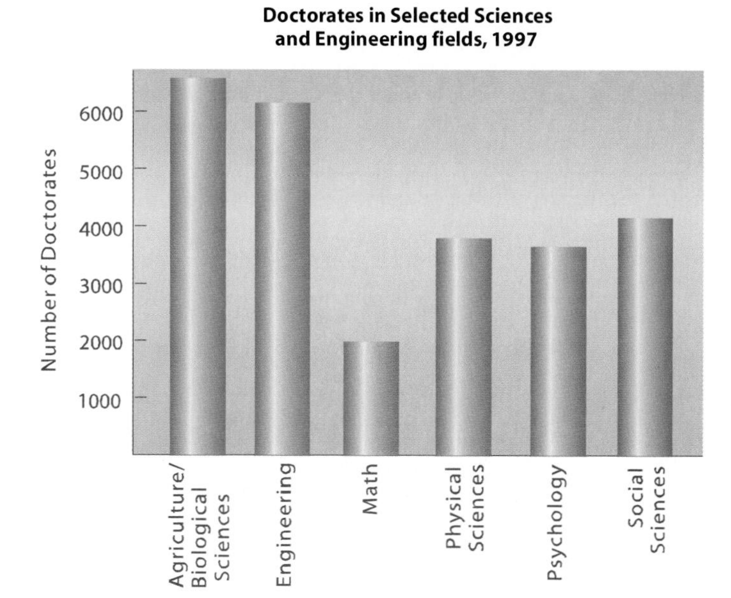


**a.** Which field of study had the most number of doctorates conferred?

**b.** Which field of study had the least number of doctorates conferred?

**c.** What was the difference between the number of doctorates conferred in mathematics and physical science?

**d.** Of the doctorates represented in this graph, what percent were conferred in agriculture/ biological science?

4. a. Lake Superior

b. about 24,500 square miles

c. Lake Huron and Lake Michigan

d. about 18,800 square miles

e. 20,000 square miles (Lake Michigan)

**4.** Lake Superior, with an area of about 32,000 square miles, is the world's largest freshwater lake. ( **Note:** Historically, Lake Huron and Lake Michigan are considered two lakes.Technically, Lake Huron and Lake Michigan could be considered one lake, since they are at the same elevation and are connected by the 120-foot-deep Mackinac Strait. This strait could be considered just a narrowing of one lake instead of a separation of two lakes. )

**a.** Which of the Great Lakes is the largest?

**b.** What is the difference in size between the largest and smallest of the lakes?

**c.** Which two lakes are closest in size?

**d.** What is the mean size of these lakes?

**e.** What is the median size of these lakes?

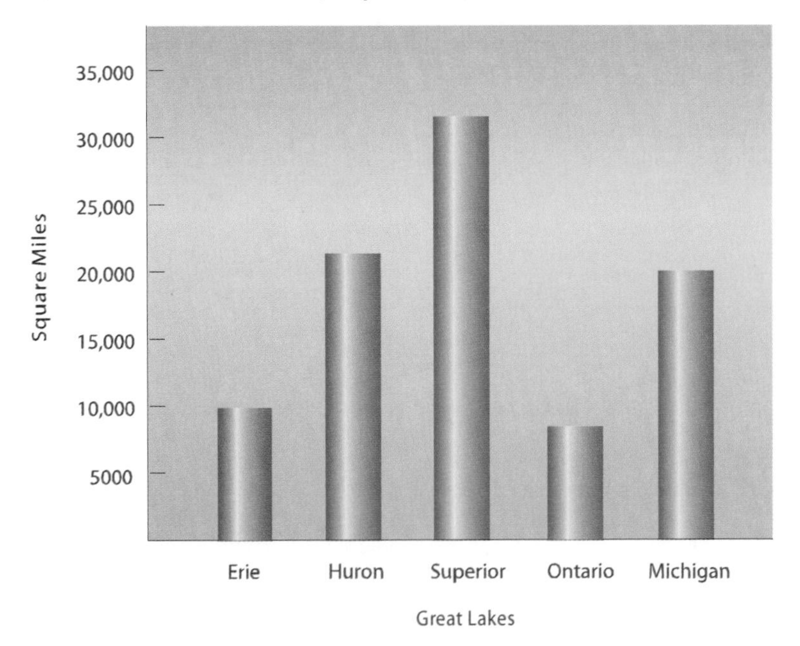

Name ______________________ Section ____________ Date __________

**ANSWERS**

5. stocks: \$125,000; bonds: \$200,000; cash: \$100,000; real estate: \$75,000

6. Democrat: 450; Republican: 350; other: 200

7. pepperoni: 900; sausage: 450; vegetarian: 180; other: 270

**5.** A retired person has accumulated \$500,000 in savings and wishes to invest this money prudently. The types of investments and corresponding percentages recommended by a financial advisor are shown in the following circle graph. What amount of money should this person invest in each category?

**Recommended Categories of Investment**

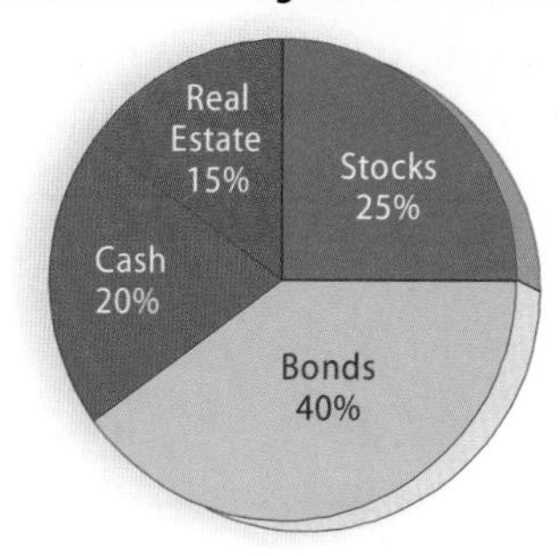


**6.** The political party affiliation of 1000 students interviewed on a college campus is shown in the following circle graph. How many students were affiliated with each party?

**Political Party Affiliation of 1000 College Students**

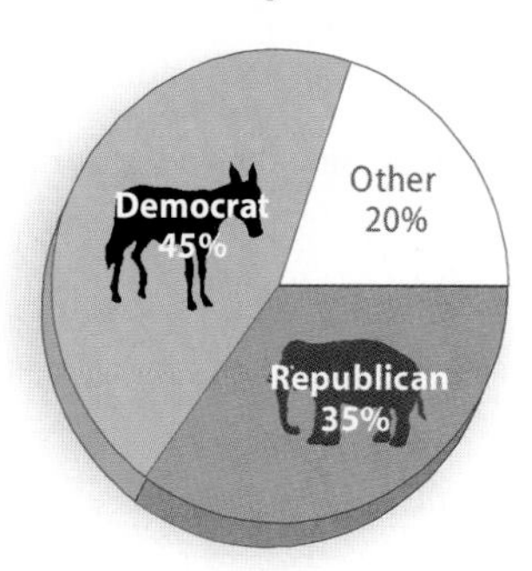


**7.** The Spaghetti and Pizza Haus sells about 1800 pizzas each month. Use the circle graph shown here to find the number of pizzas with each kind of topping that the Haus sells each month.

**Toppings for Pizza at the Spaghetti and Pizza Haus**

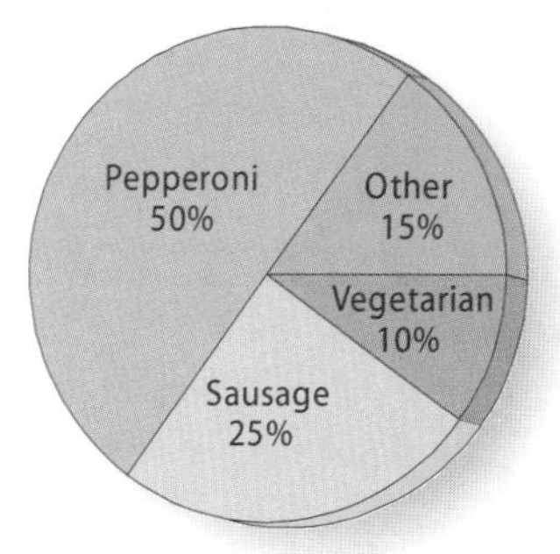

8. bad tracks: 36;

human error: 20;

other causes: 9.6;

faulty equipment: 14.4

9. a. 1992

b. about 2700

c. about 1050

d. about 2720

**8.** The following circle graph shows various reasons and percents related to train derailments. Use the graph to determine how many out of a reported 80 train derailments were caused by each reason.

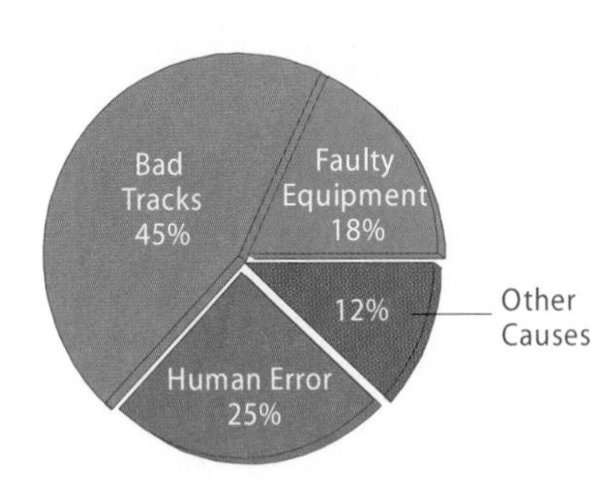


**9.** The following pictograph shows the sales of homes in Orange County, California, for the years 1991-1995.

**a.** During which year were the sales lowest?

**b.** What were the sales in 1993?

**c.** What was the difference in sales in 1991 and 1992?

**d.** What was the mean number of sales per year for these 5 years?

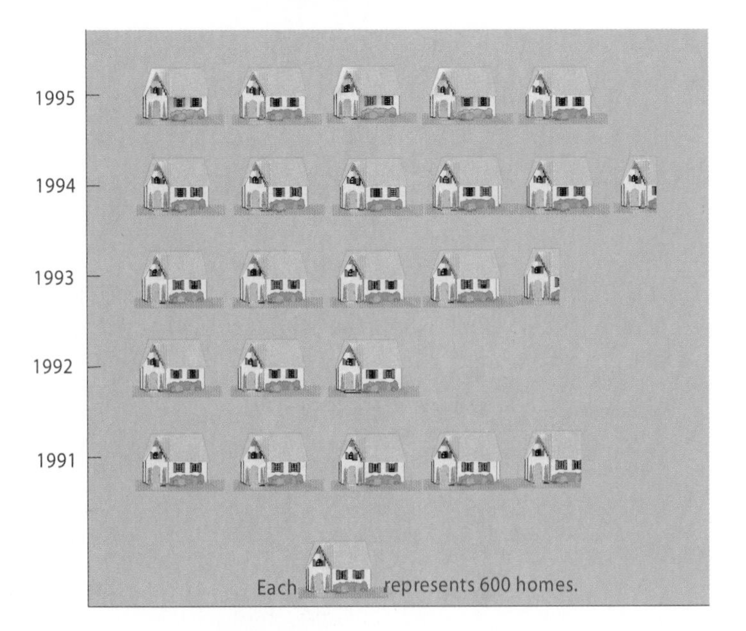

Name ______________________ Section ______________ Date ____________

**ANSWERS**

**10.** The following pictograph shows the numbers of various types of vehicles sold in 1 month at Meyer's Auto Center.

**a.** How many compact cars were sold during that month?

**b.** How many vehicles of all types were sold during that month?

**c.** What was the most number of any one type sold?

**d.** What type was this?

**Types of Vehicles Sold in One Month**

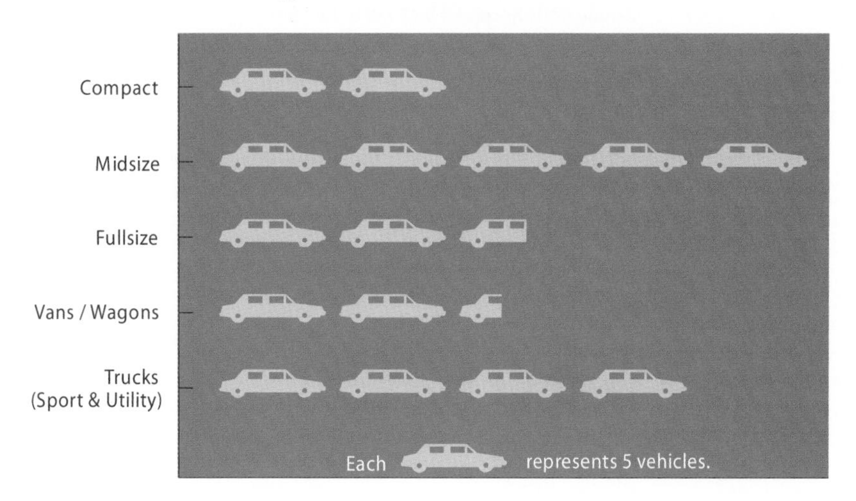


**10. a.** 10

**b.** about 80

**c.** 25

**d.** midsize

**11.** The following pictograph show the number of bicycles sold during 1 week at the Sea Side Bicycle shop.

**a.** How many bicycles were sold on Monday?

**b.** How many bicycles were sold on Friday?

**c.** On which day were the fewest bicycles sold?

**d.** What was the total number of bicycles sold that week?

**Sales of Bicycles in one week**

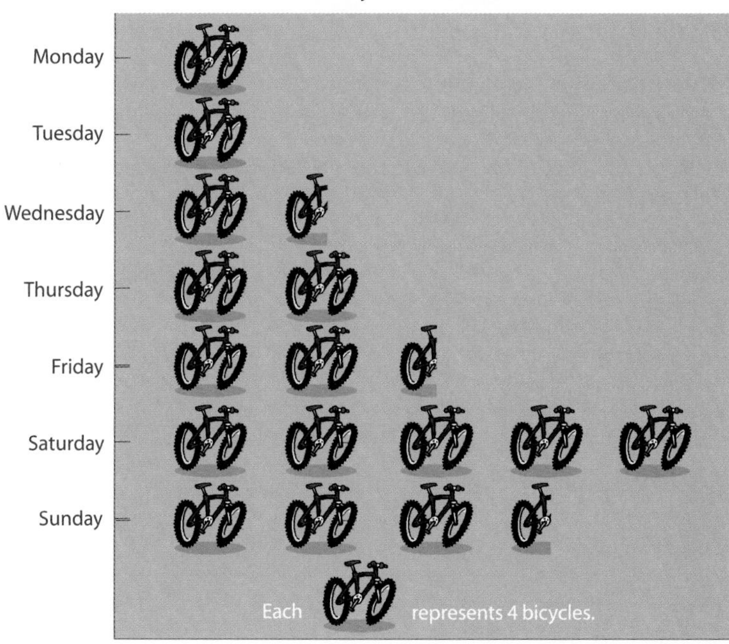


**11. a.** 4

**b.** 10

**c.** Monday and Tuesday

**d.** 66

12. a. 10

b. 65

c. 65

d. 42

13. a. no

b. 11%

c. 28,600,000 ;

2,600,000

d. 6,650,000 ;

15,750,000

**12.** The following pictograph shows the number of phone calls received by a 1-hour radio talk show during 1 week in Seattle. ( Not all calls actually get on the air. )

**a.** What was the difference between the number of calls on Monday and the number of calls on Tuesday?

**b.** How many calls were received on Friday's show?

**c.** What was the most number of calls on the one show?

**d.** What was the mean number of calls per show received that week?

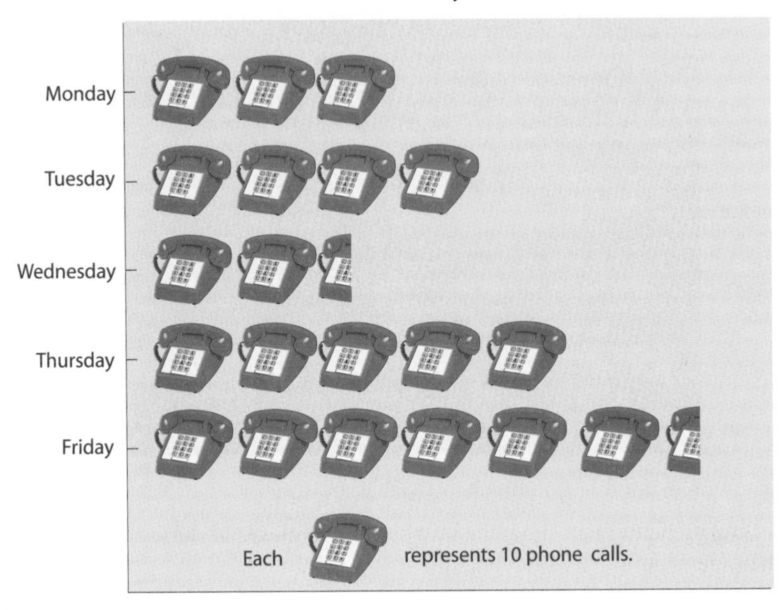


**13.** Refer to the two circle graphs in Mathematics at Work! at the beginning of this chapter.

**a.** Do the percents total 100% in both graphs?

**b.** What is the difference in percents of Latinos in California and the Nation?

**c.** If the population in the nation was about 260,000,000 in 1995, about how many citizens were African American? About how many were Native American?

**d.** If the population in California was about 35,000,000 in 1995, about how many citizens of California were Latino? About how many were White?

# 9.3 Reading Graphs: Line Graphs, Histograms, Frequency Polygons

In this section, we will discuss three more types of graphs: line graphs, histograms, and frequency polygons. The general purpose or nature of each of these types of graphs is described below.

1. **Line Graph:** To indicate tendencies ( or trends ) over a period of time.
2. **Histogram:** A special type of bar graph with intervals of numbers indicated on the baseline
3. **Frequency Polygon:** A type of line graph formed by connecting the midpoints of the tops of the rectangles of a histogram.

Remember that all graphs should have the following three basic properties:

1. They should be clearly labeled.
2. They should be easy to read.
3. They should have appropriate titles.

**Objectives**

① Understand how to read and interpret the information shown in **line graphs, histograms,** and **frequency polygons**.

② Be able to perform appropriate operations related to the data shown in a line graph, histogram, or frequency polygon.

## Reading Line Graphs

### *Completion Example 1* ( Line Graphs)

Environmental engineers are interested in the "elevation" of water ( the number of feet above sea level ) in water tables under the ground. The readings of water levels can tell how the flow of possible contaminants such as gasoline ( which contains the carcinogen benzene ) might affect the water supply for a community. Figure 9.4 on the following page illustrates the trend in water elevations for one particular well drilled for an oil company in Los Angeles.

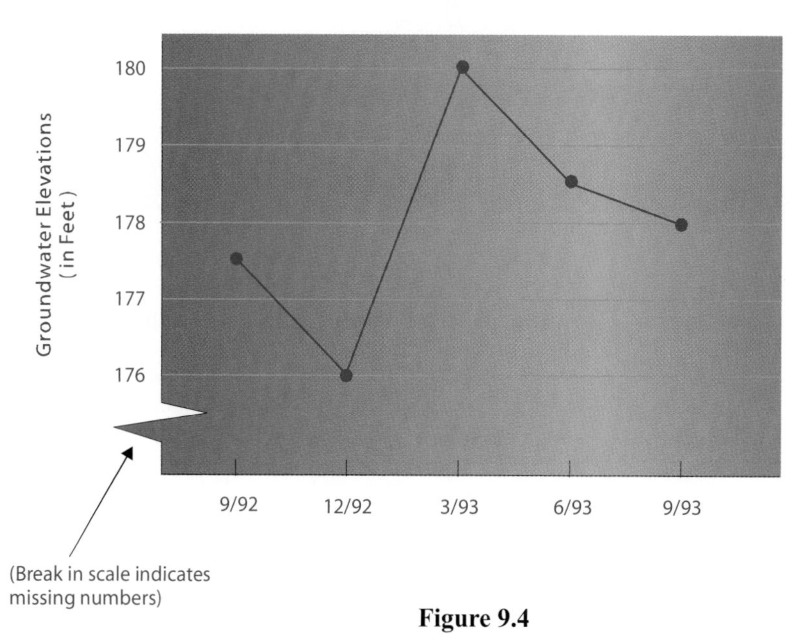


**Figure 9.4**

**a.** At what time was the water level the highest? March 1993

**b.** What was the water level then? _____

**c.** Can you think of any reason for the water level to be so high at this particular time in Los Angeles? _____

**d.** What was the mean elevation of water for this year at this well? _____

## Reading Histograms

A histogram looks very much like a bar graph, but there are some important distinctions that must be understood. In a histogram, the baseline is labeled with numbers that indicate the boundaries of a range of numbers called a **class.** In effect, the baseline is part of a real number line. The bars are not spaced apart. They are placed next to each other with no spaces between them. The following terminology is related to histograms ( and frequency polygons ).

### Terms Related to Histograms

**Class:** A range ( or interval ) of numbers that contains data items

**Lower Class Limit:** The smallest number that belongs to a class

**Upper Class Limit:** The largest number that belongs to a class

**Class Boundaries:** Numbers that are halfway between the upper limit of one class and the lower limit of the next class.

**Class Width:** The difference between the class boundaries of any one class ( the width of each bar )

**Frequency:** The number of data items in a class

### Example 2 ( Histogram )

Major League Baseball was halted on August 12, 1994, by a players' strike. Among the final statistics for the 1994 season is the total number of runs scored by each of the 28 teams. Those totals have been grouped as shown in table 9.1, and Figure 9.5 shows the related histogram that can be drawn based on this data.

**Table 9.1 A Frequency Distribution of Total Runs Scored per Team in Major League Baseball in 1994**

| Class Number | Class Limits ( Range of Total Runs ) | Frequency ( Number of Teams in Each Class ) |
|---|---|---|
| 1 | 461-500 | 4 |
| 2 | 501-540 | 5 |
| 3 | 541-580 | 9 |
| 4 | 581-620 | 5 |
| 5 | 621-660 | 3 |
| 6 | 661-700 | 2 |

**Figure 9.5 Histogram for Table 9.1**

The following statements describe various characteristics and facts related to the histogram in Figure 9.5.

**a.** There are 6 classes represented.

**b.** For the first class, 461 is the lower class limit and 500 is the upper class limit.

**c.** There are 9 teams in the third class. That is, for these 9 teams, the total number of runs that each scored was between 541 and 580, inclusive.

**d.** The class boundaries for the second class are 500.5 and 540.5.

**e.** The width of each class is 40( $540.5 - 500.5 = 40$ ). ( Note that this is **not** the difference between the class limits, since $540 - 501 = 39$. )

**f.** The percent of teams in the third class is found by dividing the frequency of the class ( 9 ) by the total number of teams ( 28 ):

$$\frac{9}{28} \approx 0.3214 \approx 32\%$$

Therefore, about 32% of the data are in the third class. Or about 32% of the teams scored between 541 and 580 runs, inclusive.

## Reading Frequency Polygons

The midpoint of each class is called the **class mark** of that class. For example, the class mark of the first class in Table 9.1 is found by averaging the two class boundaries ( or by averaging the two class limits ):

$$\frac{460.5 + 500.5}{2} = \frac{961}{2} = 480.5 \qquad \text{class mark of first class}$$

A variation on both the histogram and the line graph can be created by connecting the midpoints of the tops of the frequency rectangles with straight line segments. This is, in effect, a line graph where each point represent the frequency of the class mark of each class. The resulting graph is called a **frequency polygon.** Two additional line segments may be drawn to connect the points at each end to the baseline. ( See Figures 9.6 and 9.7. ) This is done to indicate the fact that the frequency is 0 for any data outside the classes indicated by the frequency polygon ( or related histogram ).

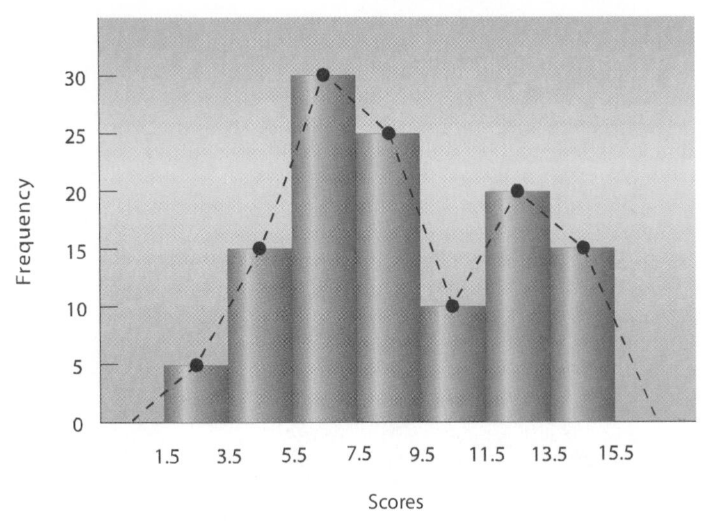


**Figure 9.6**

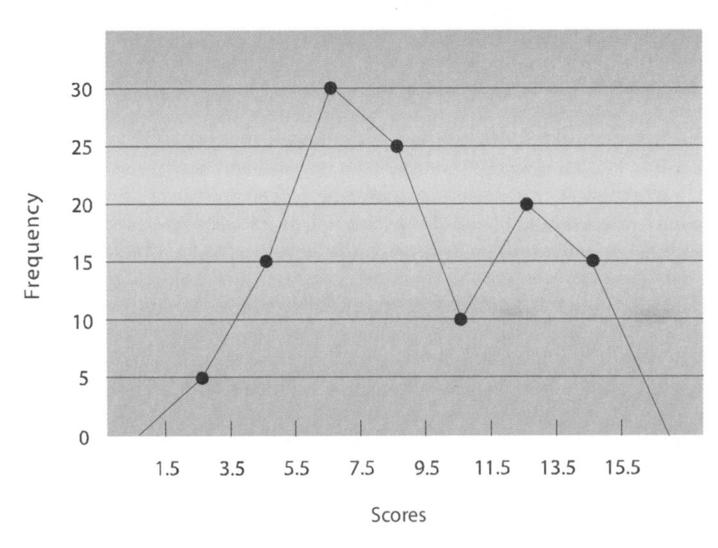


**Figure 9.7**

### *Completion Example 3*

Answer the following questions by referring to the frequency polygon in Figure 9.7.

**a.** Which class contains the highest number of test scores? the third class

**b.** How many scores are between 3.5 and 5.5? ______

**c.** How many students took the test? ______

**d.** How many students scored less than 7.5? ______

**e.** How many students scored more than 11? ______

**f.** Approximately what percent of the students scored between 5.5 and 11.5? ______

**g.** What is the class mark for the fourth class? ______

## Completion Example Answers

**1.** **a.** At what time was the water level the highest? **March 1993**

**b.** What was the water level then? **180 feet**

**c.** Can you think of any reason for the water level to be so high at this particular time in Los Angeles? **Probably the rainy season**

**d.** What was the mean elevation of water for this year at this well? **178 feet**

**3.** **a.** Which class contains the highest number of test scores? **the third class**

**b.** How many scores are between 3.5 and 5.5? **15**

**c.** How many students took the test? **120**

**d.** How many students scored less than 7.5? **50**

**e.** How many students scored more than 11? **35**

**f.** Approximately what percent of the students scored between 5.5 and 11.5? **about 54%**

**g.** What is the class mark for the fourth class? **8.5**

Name ______________________ Section ______________ Date ____________

ANSWERS

1. a. February and August

b. $190,000

c. October

d. $208,000

e. $205,000

f. [Respond below exercise.]

## Exercises 9.3

1. The following line graph shows the median prices of homes sold in Orange County California for each month for 1 year.

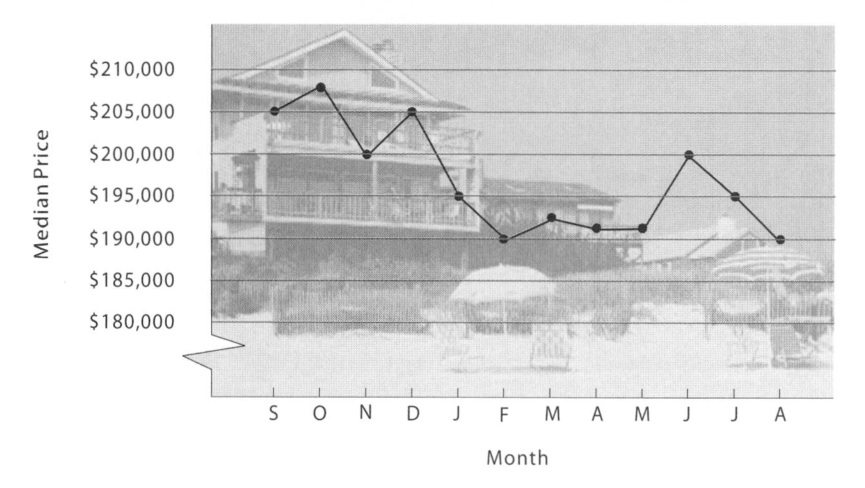


**a.** During which month was the median price the lowest?

**b.** What was this price?

**c.** During which month was the median price the highest?

**d.** What was this price?

**e.** What was the median price in December?

**f.** Describe the general trend in the median price over this year?
The general trend is a decline.

2. a. March and

September 1993

b. about 228 feet

c. December 1990

d. 214 feet

e. about 223.4 feet

2. The following line graph shows the "elevation" of water ( the number of feet above sea level ) in a well drilled by environmental engineers in Ventura, California over a 3-year period.

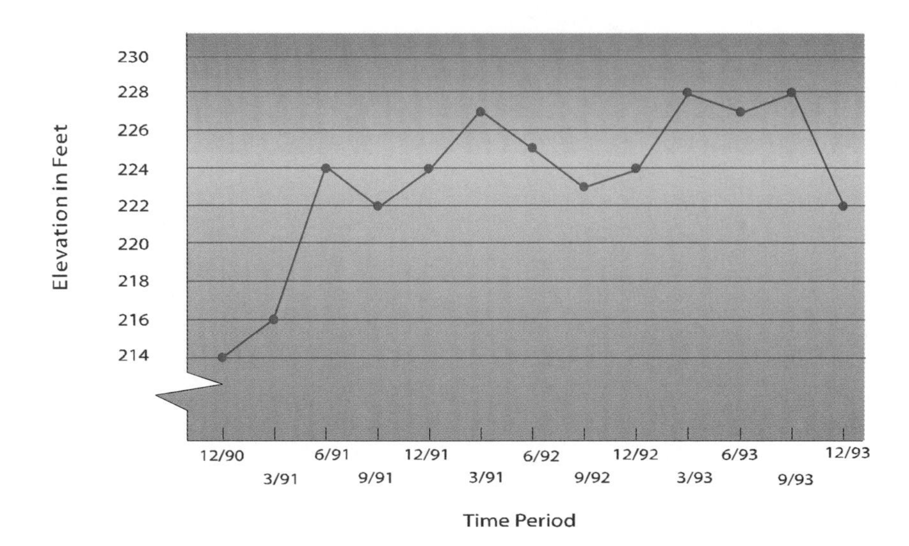


**a.** At what time was the water level the highest?

**b.** What was the water level then?

**c.** At what time was the water level the lowest?

**d.** What was the water level then?

**e.** What was the mean elevation for these 3 years at this well?

Name ______________________ Section __________ Date __________

**ANSWERS**

**3. a.** central cities

**b.** rural

**c.** rural; suburbs

**d.** 1970

**3.** The following three line graphs show how the demographics of the population in the United States have changed over a 50-year period. Three regions are represented: rural, central cities, and suburbs.

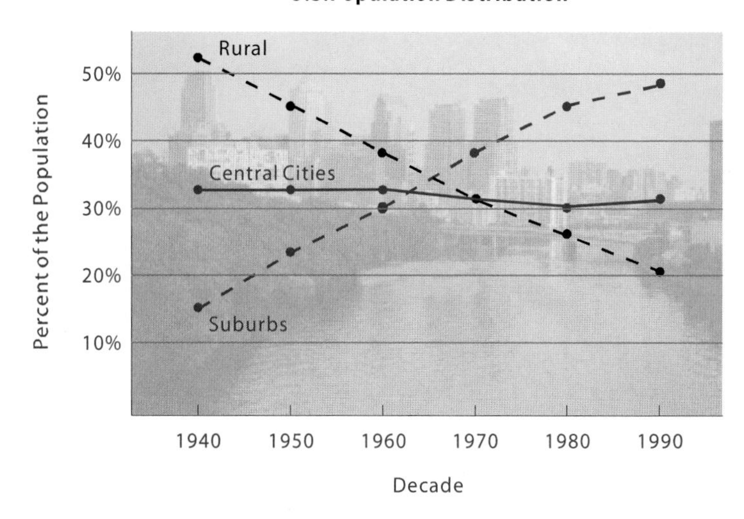


**a.** Which region has maintained a rather steady percent of the population from 1940 to 1990?

**b.** Which region has shown a decline in the percent of the population from 1940 to 1990?

**c.** Which region had the largest percent of the population in 1950? in 1980?

**d.** At about what time was the percent of the population living in rural areas about the same as the percent of the population living in central cities?

4. a. 21,000,000 ; 25,000,000

b. 40,000,000 ; 34,000,000

c. 49,000,000 ; 45,000,000

d. 4,000,000

e. [Respond below exercise.]

**4.** The following two line graphs show the numbers of children living in the United States from 1940 until today and projected numbers to 2040. Two age groups are represented: children 9 years old or younger and children aged 10 to 19 years.

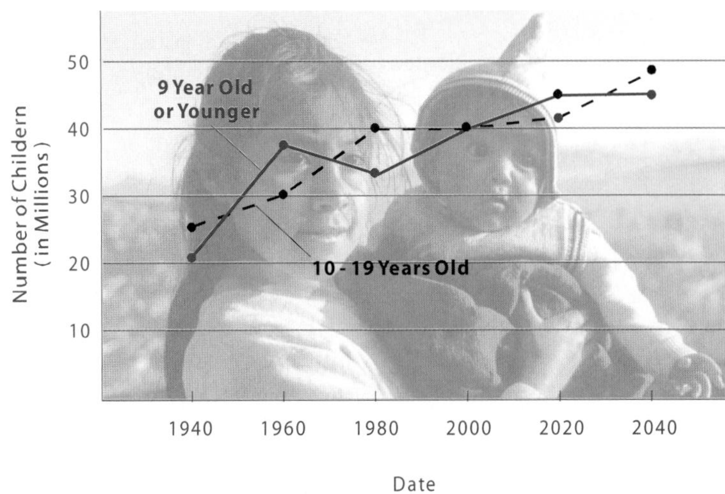


**a.** What were the approximate numbers of children in each category of 1940?

**b.** In 1980?

**c.** What are the projected numbers of children in each category for 2040?

**d.** What is the projected growth in the number of children 9 years old or younger from the year 2000 to the year 2040?

**e.** In your opinion, what do the number represented in this graph indicate for education and schools in the next 40 years?

These projections indicate that there will be a need for more schools and more teachers.

Name ______________________ Section ______________ Date ___________

5. The histogram for the following data is shown below.

**A Frequency Distribution of Per Capita Income for the 50 States in 1998**

| Class Number | Class Limits (Range of Income) | Frequency (Number of States in Each Class) | Class Marks |
|---|---|---|---|
| 1 | $18,001 - $21,000 | 5 | $19,500.50 |
| 2 | $21,001 - $24,000 | 13 | $22,500.50 |
| 3 | $24,001 - $27,000 | 18 | $25,500.50 |
| 4 | $27,001 - $30,000 | 10 | $28,500.50 |
| 5 | $30,001 - $33,000 | 3 | $31,500.50 |
| 6 | $33,001 - $36,000 | 0 | $34,500.50 |
| 7 | $36,001 - $39,000 | 1 | $37,500.50 |

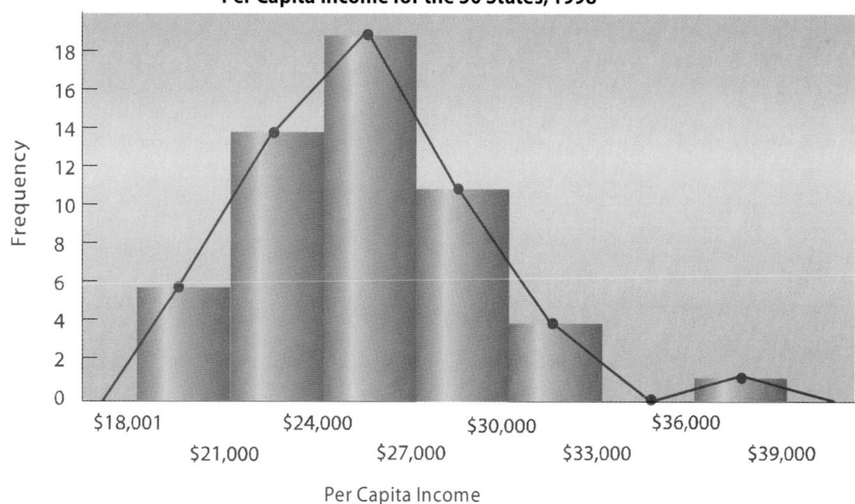


**a.** Which class had the most number of states?

**b.** How many states had per capita income between $18,001 and $27,000?

**c.** How many states had per capita income of more than $33,000?

**d.** What percent of the states had per capita income of more than $33,000?

**e.** Find the class mark for each class.

**f.** Draw the corresponding frequency polygon over the top of the histogram shown.

**ANSWERS**

**5. a.** the third class

**b.** 36

**c.** 1

**d.** 2%

**e.** [Respond in table]

**f.** [Draw on the figure]

6. a. the third class

b. 19

c. 8

d. 28%

e. [Respond in table]

f. [Draw on the figure]

**6.** The histogram for the following data is shown below.

**A Frequency Distribution of 25 Exam Scores in Dr. Wright's Statistics Course**

| Class Number | Class Limits | Frequency | Class Marks |
|---|---|---|---|
| 1 | 50 - 59 | 3 | 54.5 |
| 2 | 60 - 69 | 5 | 64.5 |
| 3 | 70 - 79 | 10 | 74.5 |
| 4 | 80 - 89 | 4 | 84.5 |
| 5 | 90 - 99 | 3 | 94.5 |

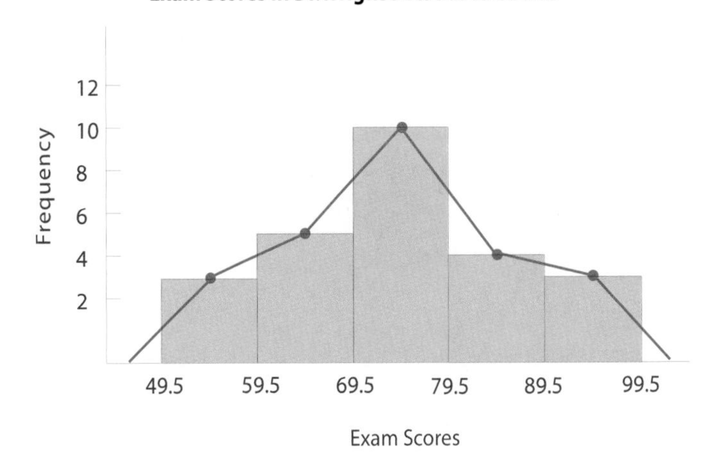


**a.** Which class had the most number of students?

**b.** How many students scored between 59.5 and 89.5?

**c.** How many students scored below 70?

**d.** What percent of the students scored 80 or better?

**e.** Find the class mark for each class.

**f.** Draw the corresponding frequency polygon over the top of the histogram shown.

Name ______________ Section ______________ Date ______________

7. The histogram for the following data is shown below.

**A Frequency Distribution of the Ages of 40 Blood Donors**

| Class Number | Class Limits | Frequency | Class Marks |
|---|---|---|---|
| 1 | 25 - 29 | 7 | 27 |
| 2 | 30 - 34 | 5 | 32 |
| 3 | 35 - 39 | 11 | 37 |
| 4 | 40 - 44 | 8 | 42 |
| 5 | 45 - 49 | 3 | 47 |
| 6 | 50 - 54 | 2 | 52 |
| 7 | 55 - 59 | 3 | 57 |
| 8 | 60 - 64 | 1 | 62 |

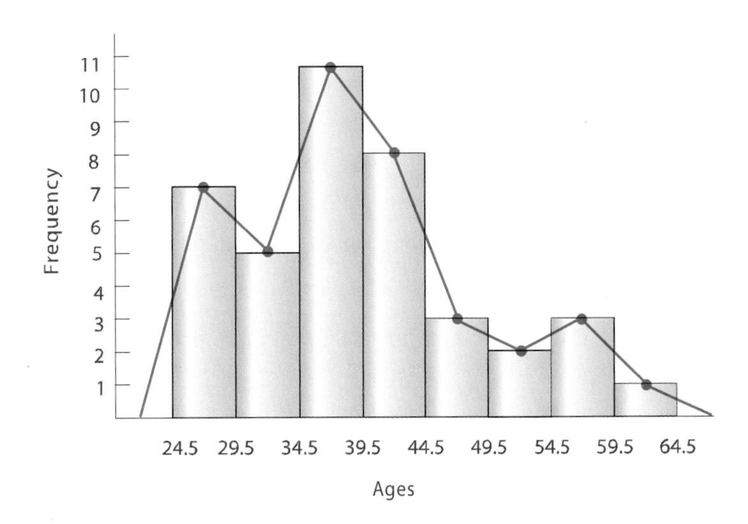


a. Which age group had 7 donors?

b. What are the class boundaries of the third class?

c. How many donors were younger than 40?

d. What percent of the donors were younger than 40?

e. Find the class mark for each class.

f. Draw the corresponding frequency polygon over the top of the histogram shown.

**ANSWERS**

7. a. the first class

b. 34.5 and 39.5

c. 23

d. 57.5%

e. [Respond in table]

f. [Draw on the figure]

8. a. 13

b. $34,500 and $39,500

c. 37

d. 74%

e. [Respond in table]

f. [Draw on the figure]

**8.** The histogram for the following data is shown below.

**A Frequency Distribution of Starting Salaries for 50 Graduating Doctors of Veterinary Medicine, 1994**

| Class Number | Class Limits | Frequency | Class Marks |
|---|---|---|---|
| 1 | \$20,000 - \$24,000 | 6 | \$22,000 |
| 2 | \$25,000 - \$29,000 | 12 | \$27,000 |
| 3 | \$30,000 - \$34,000 | 19 | \$32,000 |
| 4 | \$35,000 - \$39,000 | 10 | \$37,000 |
| 5 | \$40,000 - \$44,000 | 3 | \$42,000 |

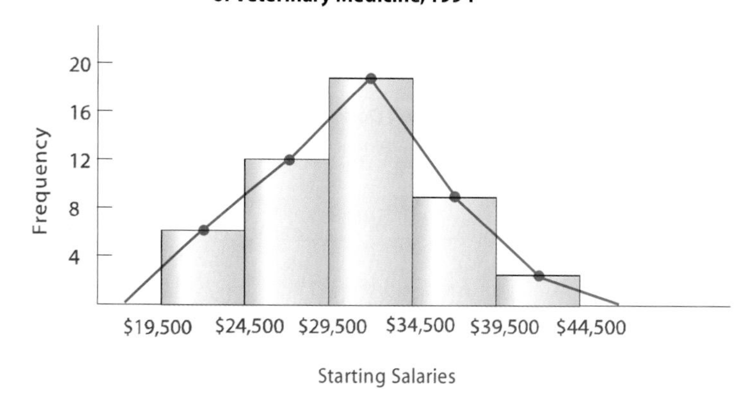


**a.** How many doctors ( DVM ) had a starting salary of more than \$34,000?

**b.** What are the boundaries of the fourth class?

**c.** How many of the doctors earned between \$19,500 and \$34,500?

**d.** What percent of the doctors had a starting salary of less than \$35,000?

**e.** Find the class mark for each class.

**f.** Draw the corresponding frequency polygon over the top of the histogram shown.

# 9.4 Constructing Graphs from Databases

In this section, you will be given a table of data and asked to construct your own graph, either a bar graph or a circle graph.

> [ **Important Note:** All of the graphs discussed here can be easily done with a computer and a spreadsheet program such as Excel. If you have access to a computer, your instructor may choose to have you work the problems in this section with a spreadsheet program. ]

**Objectives**

① Be able to organize and represent given data in the appropriate form of a bar graph or a circle graph.

## Constructing a Bar Graph

A bar graph may have either horizontal or vertical bars. In both cases, the length of each bar represents the frequency of the data in a category being graphed. ( See Completion Example 1 of Section 9.2. ) For consistency and to simplify the directions, we will discuss the construction of vertical bar graphs only.

**Steps to Follow in Constructing a Vertical Bar Graph**

1. Draw a vertical axis and a horizontal axis.
2. Mark an appropriate scale on the vertical axis to represent the frequency of each category. ( The scale must be uniform. That is, the distance between consecutive marks must represent the same amount. )
3. Mark the categories of data along the horizontal axis.
4. Draw the vertical bar for each category so that the height of the bar reaches the frequency of the data in that category.

If you want your students to find up-to-date data and/or create their own graphs, you might encourage them to use the internet. Some example sources are http://www.infoplease.com and http://www.cia.gov/cia/publications/factbook/index.html

### Example 1

Construct a bar graph that represents the following data.

| City | Population |
|---|---|
| 1. New York, NY | 7,311,966 |
| 2. Los Angeles, CA | 3,489,779 |
| 3. Chicago, IL | 2,768,483 |
| 4. Houston, TX | 1,690,180 |
| 5. Philadelphia, PA | 1,552,572 |
| 6. San Diego, CA | 1,148,851 |
| 7. Dallas, TX | 1,022,497 |
| 8. Phoenix, AZ | 1,012,230 |

( Based on the 1994 edition of **County and City Data Book** by the US Census Bureau )

**Steps 1 and 2:** Draw the vertical axis and horizontal axis and mark a scale on the vertical axis that will encompass the numbers from 0 to 7.3 million people. ( On this graph, we have chosen to mark the numbers from 1.0 to 8.0 in a scale of 1 unit. )

**Steps 3 and 4:** The horizontal axis marks are labeled with the names of the cities represented. The height of each vertical bar corresponds to the population ( in millions ) of each city as given.

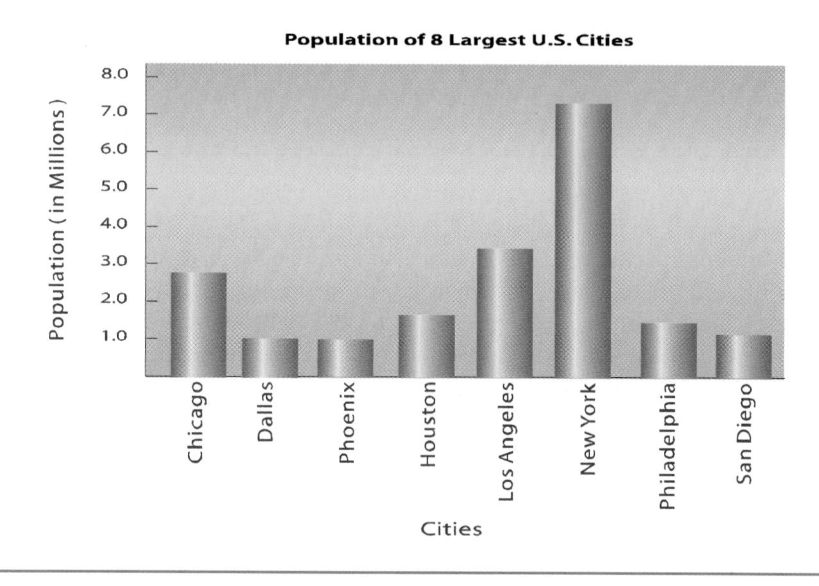


## Constructing a Circle Graph ( or Pie Chart )

A **circle graph** ( or **pie chart** ) is a circle that is marked in sectors ( pie-shaped wedges ) that correspond to percentages of data in each category represented. ( See Example 2 in Section 9.2. ) Refer to Section 4.6 for information on angles and protractors and Section 5.5 for information on circles. Section 10.1 discusses angles in greater detail.

If a radius of a circle is rotated completely around the circle ( like the motion of the minute hand of a clock in 1 hour ), the radius is said to rotate 360°. Therefore, to find the central angle needed in the graph to represent a percent of the data, multiply that percent ( in decimal form ) by 360°.

### Steps to Follow in Constructing a Circle Graph ( or Pie Chart )

1. Find the central angle (angle at the center of the circle) for each category by multiplying the corresponding percent (in decimal form) times 360°.
2. Draw a circle.
3. Draw each central angle ( use a protractor ), and label each sector with the name and corresponding percent of each category.

## Example 2

Construct a circle graph ( pie chart ) that represents the following data.

| Ethnic Breakdown of Students Who Took the SAT ( Scholastic Assessment Test ), Nationwide, 1995 | |
|---|---|
| **Ethnicity** | **Percent** |
| African-American | 11% |
| Asian-American | 8% |
| Latino | 8% |
| Native American | 1% |
| White | 69% |
| Other | 3% |

**Step 1:** Find each percent of 360°.

**a.** 11% of 360°= $0.11 \times 360° = 39.6°$

**b.** 8% of 360° = $0.08 \times 360° = 28.8°$

**c.** 8% of 360° = $0.08 \times 360° = 28.8°$

**d.** 1% of 360° = $0.01 \times 360° = 3.6°$

**e.** 69% of 360°= $0.69 \times 360° = 248.4°$

**f.** 3% of 360° = $0.03 \times 360° = 10.8°$

**Steps 2 and 3:** Draw a circle, mark the central angles as close to the actual degrees as is practical, and label each sector.

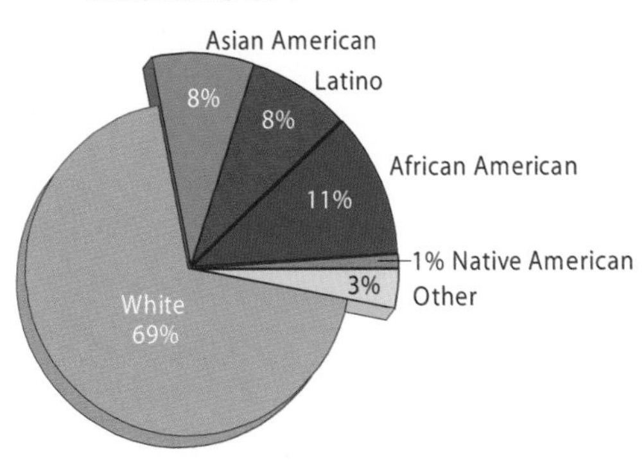

Name ______________________ Section ______________ Date __________

## Exercises 9.4

**1.** Construct a bar graph that represents the following data.

**Largest Islands of the World**

| Island | Area in Square Miles (nearest ten thousand) |
|---|---|
| Greenland | 840,000 |
| New Guinea | 320,000 |
| Borneo | 290,000 |
| Madagascar | 230,000 |
| Baffin | 180,000 |
| Sumatra | 180,000 |
| Honshu | 90,000 |
| Great Britain | 90,000 |

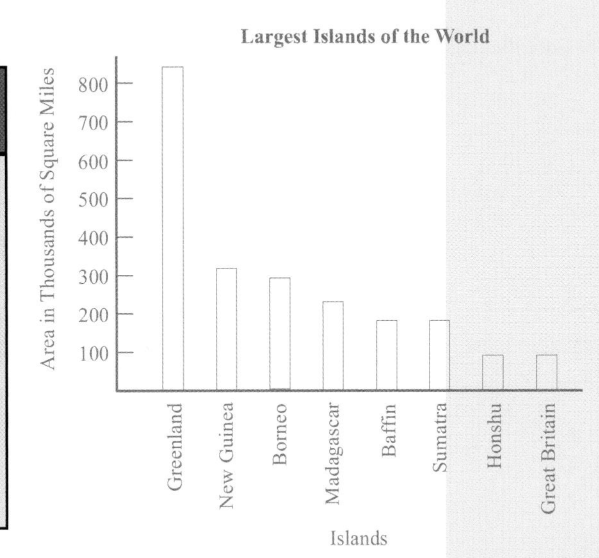


**2.** Construct a bar graph that represents the following data.

**Top 10 Box Office Grossing Motion Pictures Worldwide, 1999**

| Motion Picture | Gross Income (in millions of dollars) |
|---|---|
| Star Wars: The Phantom Menace | $922.6 |
| The Sixth Sense | $660.7 |
| Toy Story 2 | $485.7 |
| The Matrix | $456.4 |
| Tarzan | $435.3 |
| The Mummy | $401.7 |
| Notting Hill | $354.8 |
| The World is Not Enough | $352.0 |
| American Beauty | $330.2 |
| Austin Powers: The Spy Who Shagged Me | $310.3 |

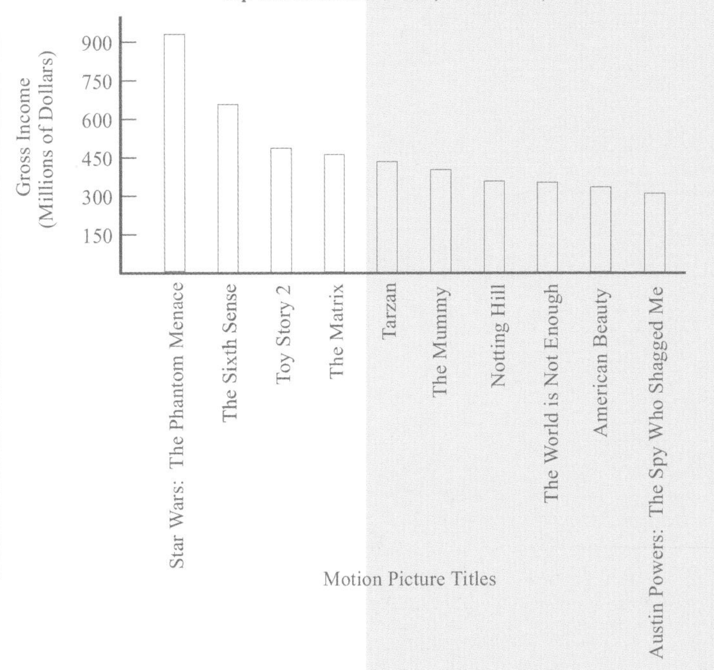

**3.** Construct a bar graph that represents the following data.

**Well-Known U.S. Skyscrapers**

| Building and City | Height (in Stories) |
|---|---|
| AMOCO ( Chicago ) | 80 |
| Chrysler ( NY ) | 77 |
| Empire State ( NY ) | 102 |
| First Interstate World Center ( LA ) | 73 |
| John Hancock Center ( Chicago ) | 100 |
| Sears Tower ( Chicago ) | 110 |
| Texas ( Houston ) | 75 |
| World Trade Center ( NY ) | 110 |

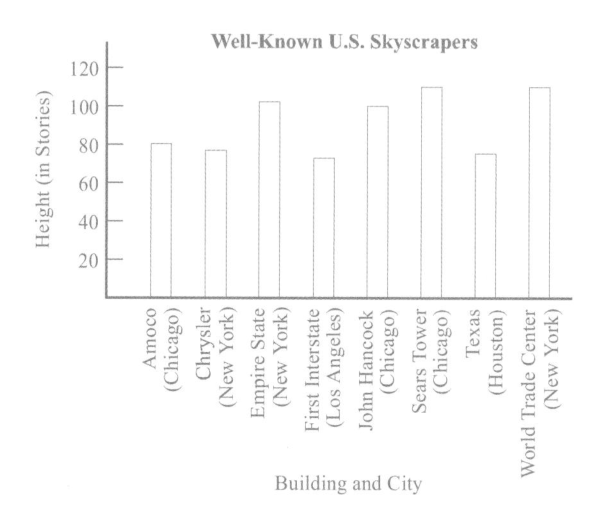


**4.** Construct a circle graph that represents the following data.

**Percentage of Population with Particular Blood Types**

| Type of Blood | Percent of Population |
|---|---|
| O positive ( $O^+$ ) | 38% |
| O negative ( $O^-$ ) | 7% |
| A positive ( $A^+$ ) | 34% |
| A negative ( $A^-$ ) | 6% |
| B positive ( $B^+$ ) | 9% |
| B negative ( $B^-$ ) | 2% |
| AB positive ( $AB^+$ ) | 3% |
| AB negative ( $AB^-$ ) | 1% |

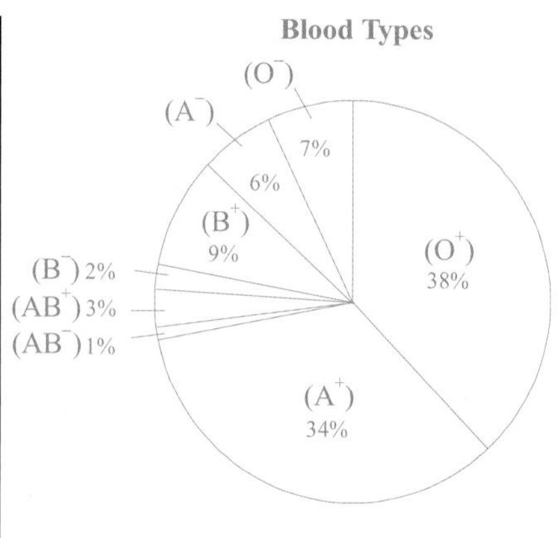


( http://www.aabb.org/All_About_Blood/FAQs/aabb_faqs.htm )

Name ________________ Section ____________ Date __________

**5.** Construct a bar graph that represents the following data.

**Population of Countries that Begin with the Letter L [ mid 1999 estimates ]**

| Country | Population |
|---|---|
| Laos | 5,497,459 |
| Latvia | 2,404,926 |
| Lebanon | 3,578,036 |
| Lesotho | 2,143,141 |
| Liberia | 3,164,156 |
| Libya | 5,115,450 |
| Liechtenstein | 32,204 |
| Lithuania | 3,620,756 |
| Luxembourg | 437,389 |

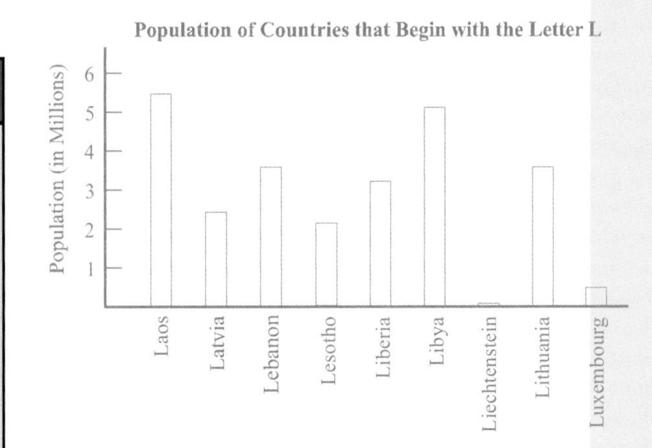


**6.** Construct a bar graph that represents the following data.

**Notable Modern Bridges**

| Bridge and Location | Span (in Meters) |
|---|---|
| Brooklyn ( New York ) | 500 |
| Bosporus ( Istanbul ) | 1100 |
| Fourth Road ( Scotland ) | 1000 |
| George Washington ( New York ) | 1100 |
| Golden Gate ( San Francisco) | 1300 |
| Humber ( Hull, Britain ) | 1400 |
| Mackinac Straits ( Michigan ) | 1200 |
| Newport ( Rhode Island ) | 500 |
| Ponte 25 de Abril ( Lisbon ) | 1000 |
| Verazano-Narrows ( New York ) | 1300 |

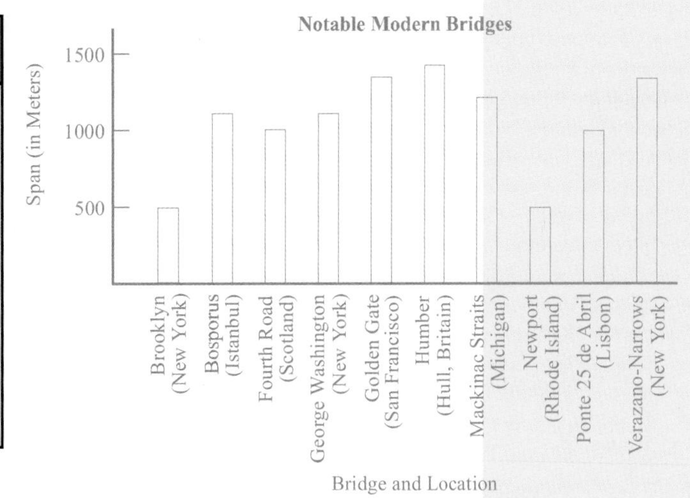

**7.** Construct a circle graph that represents the following data.

**World ( not U.S. ) Sources of Energy for 1998**

| Source of Energy | Percent |
|---|---|
| Oil | 40% |
| Coal | 23% |
| Natural Gas | 22% |
| Water | 7% |
| Nuclear | 7% |
| Other | 1% |

( http://www.infoplease.com )

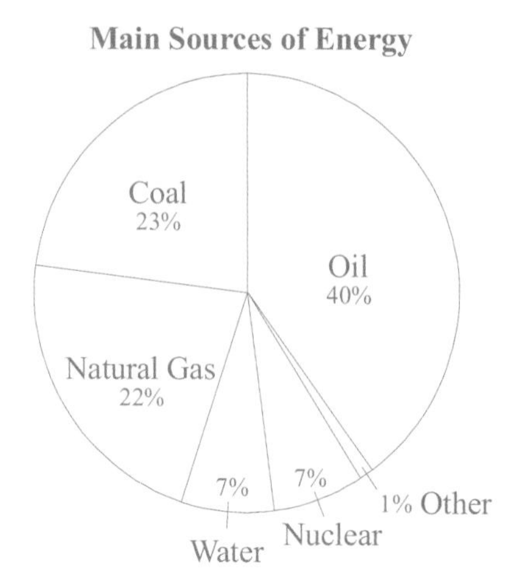


**8.** Construct a circle graph that represents the following data.

**Total Medals Awarded by Country, 1999 Winter Olympic Games ( Nagano )**

| Country | Number of Medals |
|---|---|
| Germany | 29 |
| Norway | 25 |
| Russia | 18 |
| Austria | 17 |
| Canada | 15 |
| United States | 13 |
| Others | 88 |
| | 205 Total |

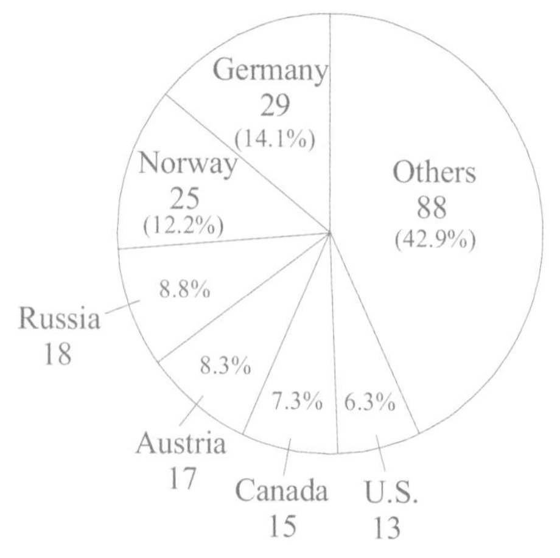

Name ______________________ Section ______________ Date __________

**9.** Construct a circle graph that represents the levels of cholesterol from a sample of 100 students reporting for a physical examination.

**Cholesterol Levels of 100 Students**

| Cholesterol Level | Number of Students |
|---|---|
| Recommended | 35 |
| Borderline | 15 |
| Moderate Risk | 40 |
| High Risk | 10 |

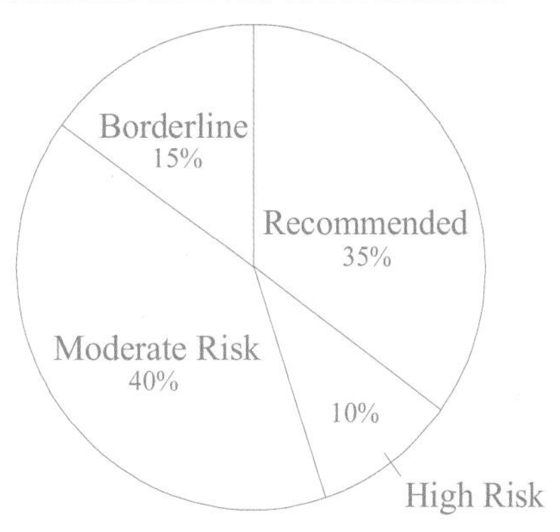


**10.** Construct a circle graph that represents the following data.

**1999 Distribution of Employment by Occupational Group**

| Occupation | Percent |
|---|---|
| Managerial | 30% |
| Sales & technician | 29% |
| Service | 13% |
| Production & repair | 11% |
| Operators & laborers | 14% |
| Farming & forestry | 3% |

( http://www.infoplease.com )

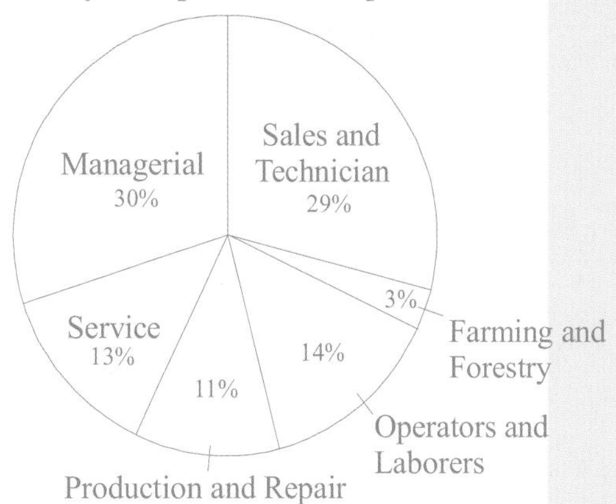

**11.** Construct a circle graph that represents the following data.

**World On-Line Households by Region, 2000**

| Region | Percent |
|---|---|
| North America | 57% |
| Europe | 25% |
| Asia / Pacific Rim | 15% |
| Other | 3% |

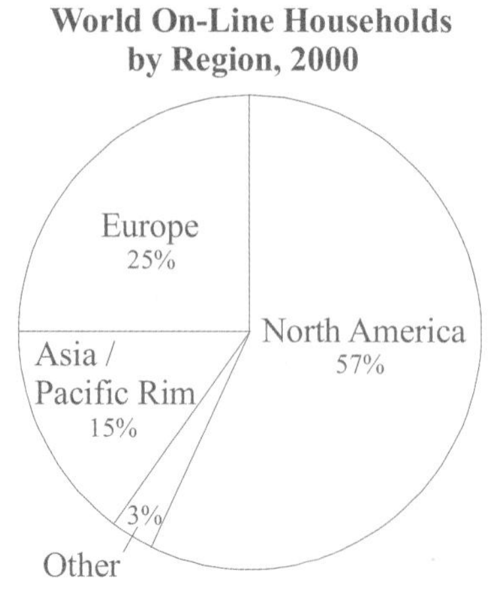


SOURCE: U.S. Dept. of Commerce, National Telecommunications and Information Administration. *Falling Through the Net: Defining the Digital Divide*

**12.** Construct a bar graph that represents the following data.

**World Population by Decade**

| Year | Total world population ( mid-year figures) |
|---|---|
| 1950 | 2,556,000,053 |
| 1960 | 3,039,451,023 |
| 1970 | 3,706,618,163 |
| 1980 | 4,453,831,714 |
| 1990 | 5,278,639,789 |
| 2000 | 6,082,966,429 |

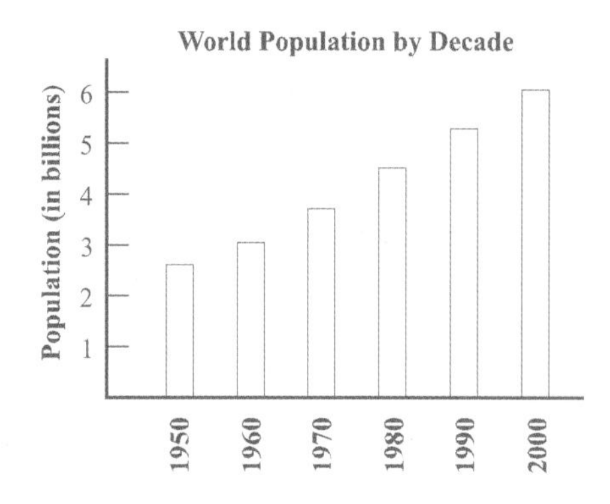


SOURCE: U.S. Bureau of the Census, International Database

Name ______________________ Section ______________ Date __________

**13.** Construct a bar graph that represents the following data.

**World's 10 Most Populous Cities**

| Name | Country | Est. Population ( in millions ) |
|---|---|---|
| Tokyo | Japan | 34.8 |
| New York | USA | 20.2 |
| Seoul | South Korea | 19.9 |
| Mexico City | Mexico | 19.8 |
| São Paulo | Brazil | 17.9 |
| Bombay ( Mumbai ) | India | 17.9 |
| Osaka | Japan | 17.9 |
| Los Angeles | USA | 16.2 |
| Cairo | Egypt | 14.4 |
| Manila | Philippines | 13.5 |

Source: Thomas Brinkhoff, *Principal Agglomerations and Cities of the World*

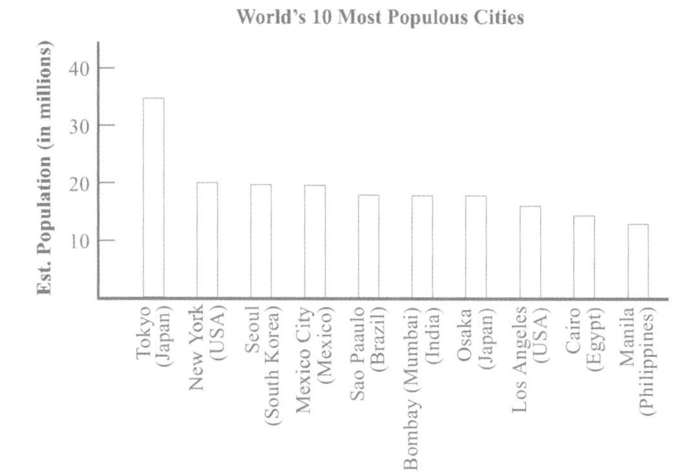

## COLLABORATIVE LEARNING EXERCISE

**14.** With the class separated into teams of two to four students, find access to a computer and a computer spreadsheet program, such as Excel, and enter the following data related to livestock on farms. Have the program find the mean and median for each category of livestock for the years listed here. Also, have the program draw a bar graph for each category of livestock by using the years listed here on the base line. Each team leader is to discuss how the team learned to use the program and bring printouts of the graphs. A general classroom discussion related to the power and efficiency of the computer should follow.

**Livestock on Farms ( in thousands )**

| Year | Cattle | Dairy Cows | Sheep | Swine | Chickens | Turkeys |
|---|---|---|---|---|---|---|
| **1998** | 99,501 | 9,191 | 7,616 | 60,915 | 403,495 | n.a. |
| **1997** | 101,460 | 9,309 | 7,937 | 56,141 | 386,974 | 300,620 |
| **1996** | 103,487 | 9,416 | 8,461 | 58,264 | 384,622 | 302,708 |
| **1995** | 102,755 | 9,487 | 8,886 | 59,990 | 383,829 | 292,856 |
| **1994** | 100,988 | 9,528 | 9,714 | 57,904 | 379,640 | 286,605 |
| **1990** | 95,816 | 10,015 | 11,358 | 53,788 | 357,241 | 282,445 |
| **1985** | 109,582 | 10,311 | 10,716 | 64,462 | 347,443 | 185,427 |
| **1980** | 111,242 | 10,758 | 12,699 | 67,318 | 400,585 | 165,243 |
| **1975** | 132,028 | 11,220 | 14,515 | 54,693 | 384,101 | 124,165 |
| **1970** | 112,369 | 13,303 | 20,423 | 57,046 | 422,096 | 116,139 |
| **1965** | 109,000 | 16,981 | 25,127 | 56,106 | 401,813 | 105,914 |

SOURCE: U.S. Department of Agriculture, National Agricultural Statistics Service

## Chapter 9 Index of Key Ideas and Terms

Statistical Terms page 715

Data: Value( s ) measuring some information of interest

Statistic: A single number describing some characteristic of the data

Mean: The arithmetic average of the data
( Find the sum of all the data, and divide by the number of data items. )

Median: The middle of the data after the data have been arranged in order

( The position of the median of $n$ items is $\frac{n+1}{2}$. )

Mode: The single data item that appears most frequently

Range: The difference between the largest and smallest data items

Properties of all Graphs page 725

1. Be clearly labeled
2. Be easy to read
3. Have a title

Types of Graphs ( Section 9.2 ) page 725

Bar Graphs ( To emphasize comparative amounts )

Circle Graphs( To help in understanding percents or parts of a whole )

Pictographs ( To emphasize the topic being related as well as quantities )

Types of Graphs ( Section 9.3 ) page 737

Line Graphs ( To indicate tendencies or trends over a period of time )

Histograms ( A special type of bar graph with intervals of numbers indicated on the baseline )

Frequency Polygons
( A type of line graph formed by connecting the midpoints of the tops of the rectangles of a histogram )

Constructing Bar Graphs page 751

1. Draw a vertical axis and a horizontal axis.
2. Mark an appropriate scale on the vertical axis to represent the frequency of each category.
3. Mark the categories of data along the horizontal axis.
4. Draw the vertical bar for each category so that the height of the bar reaches the frequency of the data in that category.

Constructing Circle Graphs page 752

1. Find the central angle for each category by multiplying the corresponding percent ( in decimal form ) times 360°.
2. Draw a circle.
3. Draw each central angle, and label each sector with the name and corresponding percent of each category.

Name ______________________ Section ______________ Date __________

# Chapter 9 Test

In Exercises 1 and 2, find

**a.** The mean,

**b.** The median,

**c.** The mode ( if any ), and

**d.** The range for the given date.

**1.** The ages of 20 students in a geometry class:

| 18 | 22 | 25 | 18 | 17 | 20 | 20 | 19 | 22 | 22 |
|---|---|---|---|---|---|---|---|---|---|
| 19 | 21 | 20 | 23 | 31 | 29 | 20 | 18 | 20 | 24 |

**2.** The number of inches of snow at a ski resort for 5 months:

25 38 60 62 48

Refer to the circle graph ( pie chart ) shown here for Exercises 3-5.

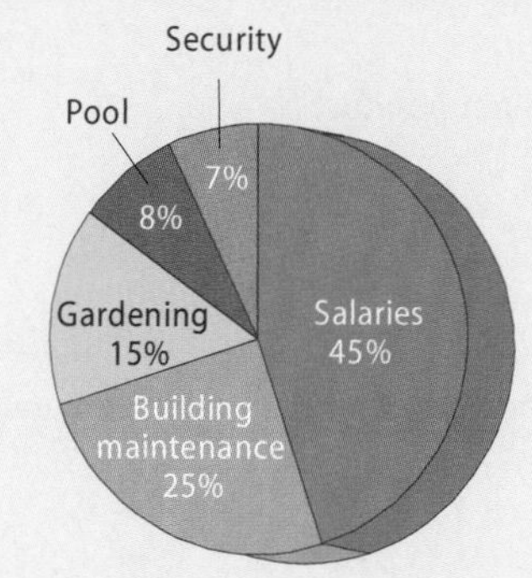


**3.** The budget for an apartment complex is shown in the graph. What percent of the budget is spent for salaries and security combined?

**4.** How much will be spent for building maintenance in 1 year if the budget is \$250,000?

**5.** How much will be spent for salaries and security combined in 6 months if the total annual budget is \$250,000?

**ANSWERS**

**1. a.** mean = 21.4

**b.** median = 20

**c.** mode = 20

**d.** range = 14

**2. a.** mean = 46.6 in.

**b.** median = 48 in.

**c.** none

**d.** range = 37 in.

**3.** 52%

**4.** \$62,500

**5.** \$65,000

6. a. 58°

b. Saturday

7. a. 61°

b. Friday

8. $\frac{68}{7} \approx 10°$

Refer to the line graph shown here for Exercises 6-8.

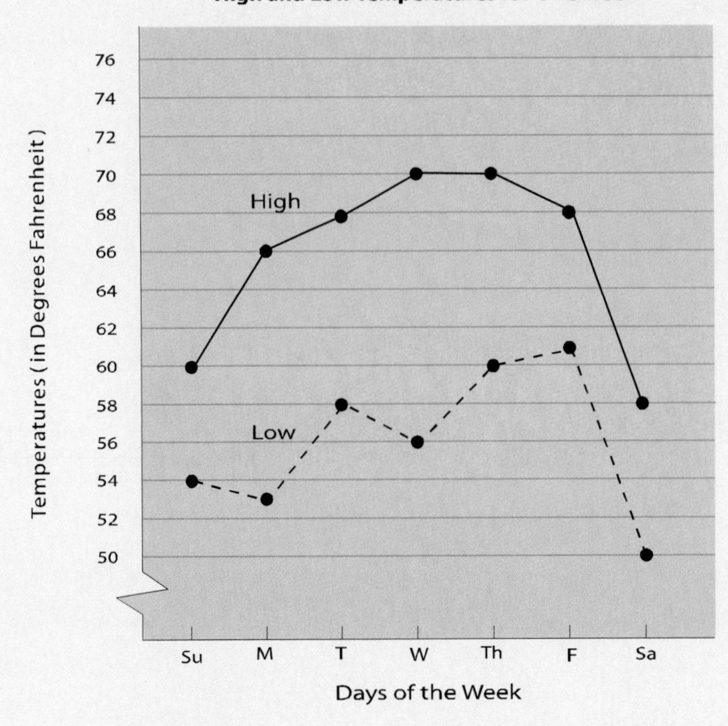


**6. a.** What was the lowest high temperature?

**b.** On what day did this occur?

**7. a.** What was the highest low temperature?

**b.** On what day did this occur?

**8.** Find the mean of the differences between the daily high and low temperatures for the week shown.

In Exercises 9-11, refer to the histogram below.

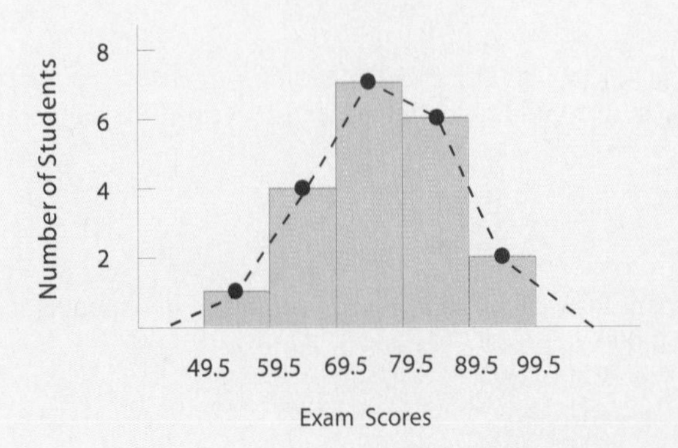

Name ______________ Section ______________ Date ______________

**ANSWERS**

**9. a.** How many classes are represented?

**b.** What is the width of each class?

**c.** What is the frequency of the fourth class?

**9. a.** 5

**b.** 10

**c.** 6

**10. a.** What percent of the scores were less than 80?

**b.** What percent of the scores were between 59.5 and 89.5?

**10. a.** 60%

**b.** 85%

**11. a.** Find the class mark for each class.

**b.** Draw the corresponding frequency polygon over the top of the histogram.

**11. a.** 54.5; 64.5; 74.5;

**b.** 84.5; 94.5

**12.** [Respond below exercise.]

**12.** Construct a bar graph that represents the following data.

**Survey of Student Majors for 150 Students**

| Major | Number of Students |
|---|---|
| Biology | 10 |
| Business | 35 |
| Computer Science | 5 |
| Engineering | 15 |
| Fine Arts | 15 |
| Liberal Arts | 20 |
| Physical Science | 5 |
| Social Science | 45 |

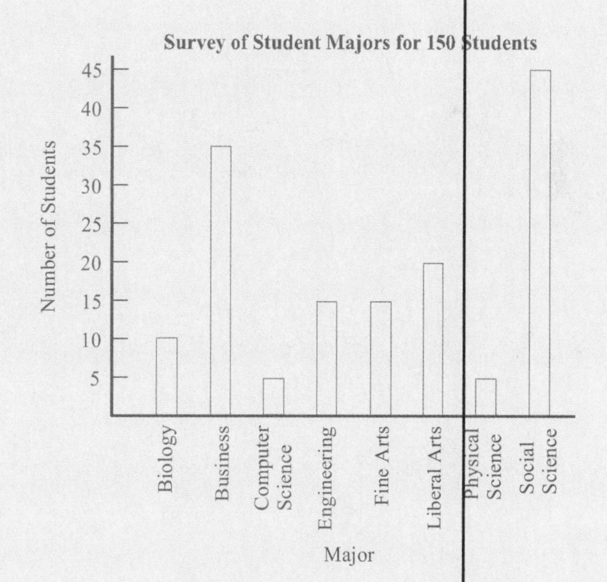

[Respond below exercise.]

13. ________

**13.** Construct a circle graph ( pie chart ) that represents the following data.

**Monthly Expenses for a Family Budget**

| Category | Percent |
|---|---|
| Housing | 30% |
| Food | 25% |
| Savings | 10% |
| Clothing | 9% |
| Entertainment | 7% |
| Insurance | 4% |
| Auto | 13% |
| Miscellaneous | 2% |

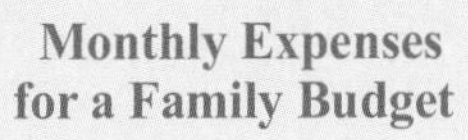


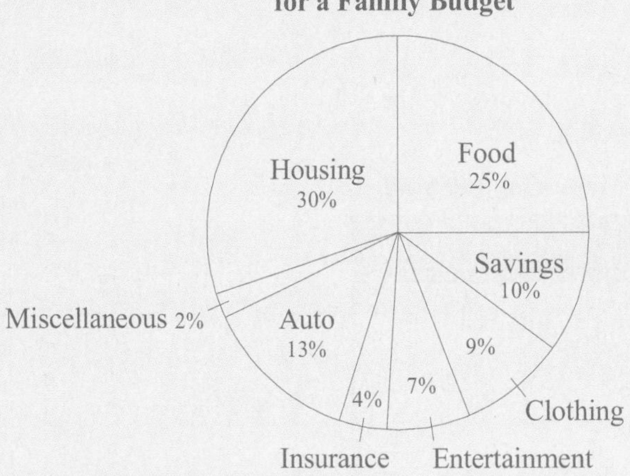

Name ______________________ Section ______________ Date __________

## Cumulative Review: Chapters 1- 9

All fractions should be reduced to lowest terms

1. Write the number two hundred thousand, sixteen and four hundredths in decimal notation.

2. Round off 17.986 to the nearest hundredth.

3. Find the decimal equivalent of $\frac{14}{35}$.

4. Write $\frac{9}{5}$ as a percent.

5. Write $1\frac{1}{2}\%$ in decimal form.

Find the value of each expression by performing the indicated operations.

6. $\frac{2}{15}+\frac{11}{15}+\frac{7}{15}$

7. $4-\frac{10}{11}$

8. $\left(\frac{2}{5}\right)\left(-\frac{5}{6}\right)\left(\frac{4}{7}\right)$

9. $2\frac{4}{15}+3\frac{1}{6}+4\frac{7}{10}$

10. $70\frac{1}{4}-23\frac{5}{6}$

11. $\left(-4\frac{5}{7}\right)\left(-2\frac{6}{11}\right)$

**ANSWERS**

1. 200,016.04
2. 17.99
3. 0.4
4. 180%
5. 0.015
6. $\frac{4}{3}$
7. $3\frac{1}{11}$ $\left(\text{or } \frac{34}{11}\right)$
8. $-\frac{4}{21}$
9. $10\frac{2}{15}$
10. $46\frac{5}{12}$
11. 12

12. $1\frac{4}{5}$

13. $-560,000$

14. $-26.9$

15. $75.74$

16. $0.00049923$

17. $-80.6$

18. $7$

19. $-11.11$

20. $-2.75$

21. $2^2 \cdot 3^2 \cdot 11$

22. $210$

23. $3.2 \times 10^7$

24. a. yes

b. 336

25. [Respond below exercise.]

**12.** $6 \div 3\frac{1}{3}$

**13.** $(700)(-800)$

**14.** $40.3 - 67.2$

**15.** $71 + |-0.35| + 4.39$

**16.** $(0.27)(0.043)^2$

**17.** $27.404 \div (-0.34)$

Use the rules for order of operation to evaluate each expression.

**18.** $\left(36 \div 3^2 \cdot 2\right) + 12 \div 4 - 2^2$

**19.** $1.7^2 - 3\frac{1}{2} \div \frac{1}{4}$

**20.** Evaluate $3x^2 - 5x - 17$ for $x = -1.5$.

**21.** Find the prime factorization of 396.

**22.** Find the LCM of 14, 21, and 30.

**23.** Write 32,000,000 in scientific notation.

**24.** **a.** Does 15 divide the product $(7 \cdot 6 \cdot 5 \cdot 4 \cdot 3 \cdot 2)$?

**b.** If so, what is the quotient?

**25.** Division by ___0___ is undefined.

Name ______________________ Section ____________ Date __________

**ANSWERS**

**26.** 15% of <u>50</u> is 7.5.

**27.** $9\frac{1}{4}\%$ of 200 is <u>18.5</u>.

**28.** Solve for $x$: $\frac{1\frac{2}{3}}{x} = \frac{10}{2.25}$

**29.** Solve for $x$: $7x + 21 = 3(x - 4) - 3$

**30.** Solve for $l$ in terms of the other variables in the formula.

$P = 2l + 2w$

**31.** An investment pays 6% compounded monthly.

**a.** What will be the value of $10,000 in 5 years?

**b.** How much interest will have been earned?

**32.** Find

**a.** The mean,

**b.** The median, and

**c.** The range of the following set of numbers:

27 36 45 72 63 36 27 18 36 90

**33.** In a certain company, 3 of every 5 employees are female. How many male employees are there out of the 490 people working for this company?

**34.** The sum of two numbers is 521. If one of the numbers is 196, what is the other number?

**35.** A customer received a 2% discount for paying cash for her new computer. If she paid $2009 in cash, what was the amount of the discount?

**26.** [Respond in exercise.]

**27.** [Respond in exercise.]

**28.** $x = 0.375$

**29.** $x = -9$

**30.** $l = \frac{P - 2w}{2}$

**31. a.** $13,488.51

**b.** $3488.51

**32. a.** mean = 45

**b.** median = 36

**c.** range = 72

**33.** 196

**34.** 325

**35.** $41

36. a. $C = 62.8$ ft

b. $A = 314$ ft²

37. a. $x < 6$

b. $x \geq \frac{1}{3}$

38. [Use grids provided.]

39. a. $8x^2 - 2x - 7$

b. $-5x^3 + 4x^2 + x + 11$

40. a. $28x^4 + 12x^3 - 20x^2$

b. $2x^3 - 9x^2 - 24x - 9$

41. a. $a^{18}$

b. $\frac{1}{6^2} = \frac{1}{36}$

c. $12x^7$

**36.** Find

**a.** The circumference and

**b.** The area of a circle with a radius of 10 feet. ( Use $\pi = 3.14$. )

**37.** Solve each inequality, and graph the solution on a number line.

**a.** $2x + 3 < 15$

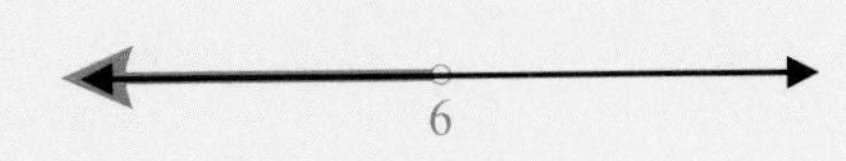


**b.** $5 - 3x \leq 4$

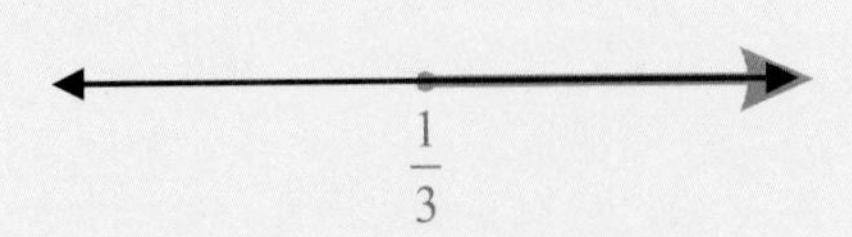


**38.** Graph each of the following equations.

**a.** $y = 2x$

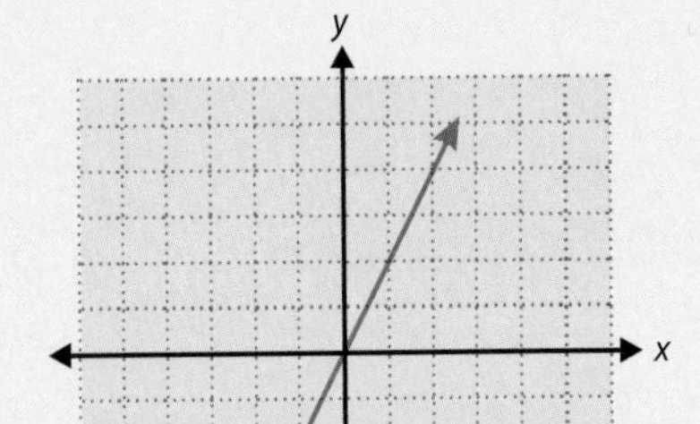


**b.** $2y - x = 6$

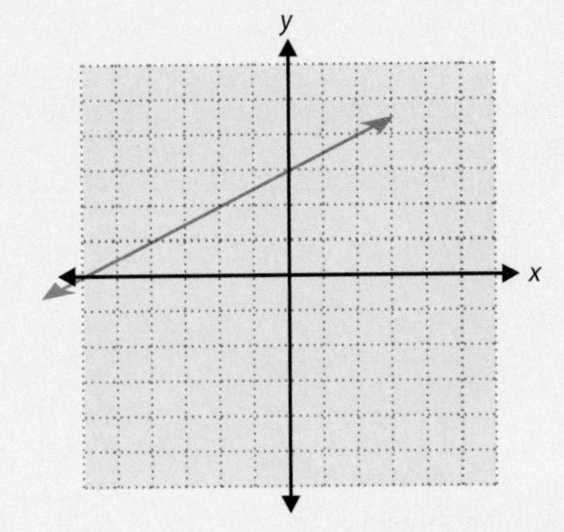


**39.** Find the sum or difference as indicated.

**a.** $( 3x^2 + 5x - 8 ) + ( 5x^2 - 7x + 1 )$

**b.** $( 2x^3 + 4x^2 - 3x + 9 ) - ( 7x^3 - 4x - 2 )$

**40.** Find each product.

**a.** $4x^2( 7x^2 + 3x - 5 )$

**b.** $( 2x + 3 )( x^2 - 6x - 3 )$

**41.** Simplify each expression by using the properties of exponents.

**a.** $( a^2 )^9$

**b.** $\frac{6^5}{6^7}$

**c.** $4x^4 \cdot 3x^3$

Name ______________________ Section ______________ Date __________

**ANSWERS**

**42.** Completely factor each of the following polynomials

**a.** $4x^3 - 8x^2 + 12x$

**b.** $5x^4 - 125x^2$

**c.** $y^2 - 16y + 64$

**d.** $y^2 + 60y + 900$

**42. a.** $4x\left(x^2 - 2x + 3\right)$

**b.** $5x^2(x+5)(x-5)$

**c.** $(y-8)(y-8)$

**d.** $(y+30)(y+30)$

# 10

# GEOMETRY AND MEASUREMENT

## Chapter 10 Geometry and Measurement





## WHAT TO EXPECT IN CHAPTER 10

Chapter 10 develops a formal approach to geometry and shows how the two measurement systems, the metric system and the U.S. customary system, can be used independently and interchangeably. Section 10.1 starts with the undefined terms **point, line,** and **plane** and expands to include definitions and properties of angles and triangles. Angles are classified by their measure; triangles are classified by properties of their sides and angles.

Sections 10.2 - 10.4 are presented in the same general pattern. First, units of measure in the metric system are discussed; second, equivalent units of measure in the metric system and the U.S. customary system are related; and third, geometric formulas are discussed in detail with a variety of applications and figures. The topics are length and perimeter ( Section 10.2 ), area ( Section 10.3 ), and volume ( Section 10.4 ). In each section, tables of units of measure in both the metric system and the U.S. customary system are given, and techniques for changing units of measure are shown.

# Chapter 10 Geometry and Measurement

## Mathematics at Work!

Geometric figures and their properties are part and parcel of our daily lives. **Geometry** is an integral part of art, employing concepts such as a vanishing point and ways of creating three-dimensional perspective on a two-dimensional canvas. Architects and engineers use geometric shapes in designing buildings and other infrastructures for beauty and structural strength. Geometric patterns are evident throughout nature in snow flakes, leaves, pine cones, butterflies, trees, crystal formations, camouflage for deer, fish, and tigers, and so on.

As an example of geometry in the human-created world, consider sewer covers in streets that you drive over every day. Why are these covers circular in shape? Why are they not in the shape of a square or a rectangle? One reason might be that a circle cannot "fall through" a slightly smaller circle, whereas a square or rectangle can be turned so that either will "fall through" a slightly smaller version of itself. Thus, circular sewer covers are much safer than other shapes because they avoid the risk of the cover accidentally falling down into the hole.

To understand this idea, draw a square, a rectangle, a triangle, and a hexagon ( six-sided figure ) on a piece of paper. Cut out these figures, and show how easily they can be made to pass through the holes made by the corresponding cutouts. Then follow the same procedure with a circle. The difference will be clear.

# 10.1 Angles and Triangles

## Introduction to Geometry

Several geometry concepts have been discussed throughout this text. For example,

| | |
|---|---|
| Perimeter ( Section 1.1 ) | Area ( Section 1.1 ) |
| Volume ( Section 4.3 ) | Angles ( Section 4.6 ) |
| Triangles ( Section 4.6 ) | Circles ( Section 5.5 ) |

These discussions were brief and were based on generally known ideas or specific directions given in a problem rather than detailed definitions and formulas. They were designed to show applications of the topics in the corresponding chapters ( such as decimals, proportions, and square roots ) and to ensure that some basic geometric concepts are known before a more in-depth course in algebra is attempted.

This section provides a more formal introduction to geometry and introduces special terminology related to angles and triangles. The remaining sections of this chapter emphasize formulas for perimeter, area, and volume for geometric figures as well as metric and U.S. customary measurements.

**Plane Geometry** is the study of the properties of figures in a plane. The most basic ideas in plane geometry are **point, line,** and **plane**. These terms are considered so fundamental that any attempt to define their meaning would use terms more complicated and more difficult to understand than the terms themselves. Thus, they are simply not defined and are called **undefined terms.** These undefined terms provide the foundation for the study of geometry and are used in the definitions of higher-level ideas such as rays, angles, triangles, circles, and so on.

| Undefined Term | Representation | Discussion |
|---|---|---|
| **1. Point** | $A$ • point $A$ | A dot represents a point. Points are labeled with capital letters. |
| **2. Line** | line $\ell$ or line $\overleftrightarrow{AB}$ | A line has no beginning or end. Lines are labeled with small letters or by two points on the line. |
| **3. Plane** | plane P | Flat surfaces, such as table tops or walls, represent planes. Planes are labeled with capital letters. |

### Objectives

1. Recognize the terms **point, line,** and **plane,** and know that they are undefined terms.
2. Know the definition of an **angle.**
3. Learn how to classify an angle by its measure as **acute, right, obtuse,** or **straight.**
4. Know the meanings of the terms **complementary angles, supplementary angles, vertical angles** and **adjacent angles.**
5. Learn how to classify a triangle by its sides: **scalene, isosceles,** or **obtuse.**
6. Learn how to classify a triangle by its angles: **acute, right,** or **obtuse.**
7. Know that for any triangle:
   - **a.** The sum of the measures of its angles is 180°.
   - **b.** The sum of the lengths of any two sides must be greater than the length of the third side.

## Angles

Angles and the measurement of angles with a protractor were discussed briefly in Section 4.6. Here we give a formal definition of an angle by first defining a **ray** by using the two undefined terms **point** and **line.**

**Definitions:**

A **ray** consists of a point ( called the **endpoint** ) and all the points on a line on one side of that point.

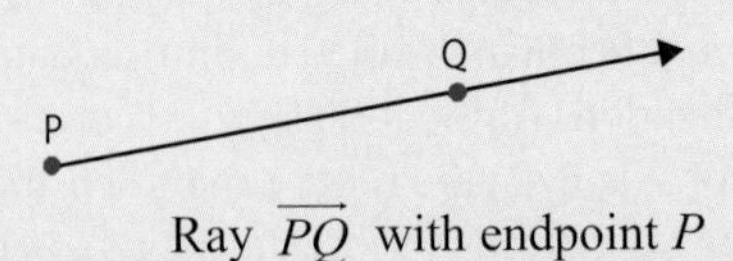


Ray $\overrightarrow{PQ}$ with endpoint $P$

An **angle** consists of two rays with a common endpoint. The common endpoint is called the **vertex** of the angle.

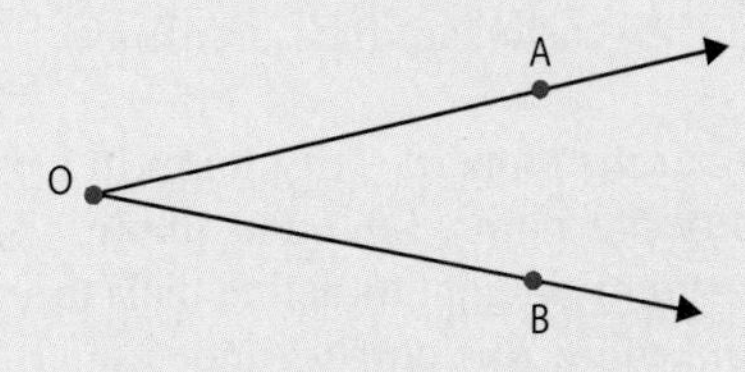


$\angle AOB$ with vertex $O$

In an angle, the two rays are called the **sides** of the angle. Thus, as shown in the figure, $\angle AOB$ has sides $\overrightarrow{OA}$ and $\overrightarrow{OB}$ .

Every angle has a **measure** ( or **measurement** ) associated with it. If a circle is divided into 360 equal arcs and two rays are drawn from the center of the circle through two successive points of division on the circle, then that angle is said to **measure one degree** ( symbolized 1° ).

Figure 10.1 illustrates the use of a device ( called a *protractor* ) that shows the **measure** of $\angle AOB = 60°$. We write m $\angle AOB = 60°$.

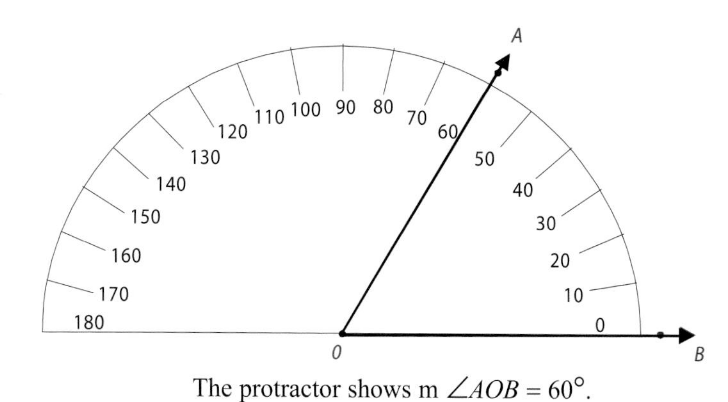


The protractor shows m $\angle AOB = 60°$.

**Figure 10.1**

To measure an angle with a protractor, lay the bottom edge of the protractor along one side of the angle with the vertex at the marked center point. Then read the measure from the protractor where the other side of the angle crosses it. You may need to extend one side of the angle for it to intersect the protractor.

Angles can be classified ( or named ) according to their measures.

**Three common ways of labeling angles:**

**1.** Using three capital letters with the vertex as the middle letter.

$\angle AOB$

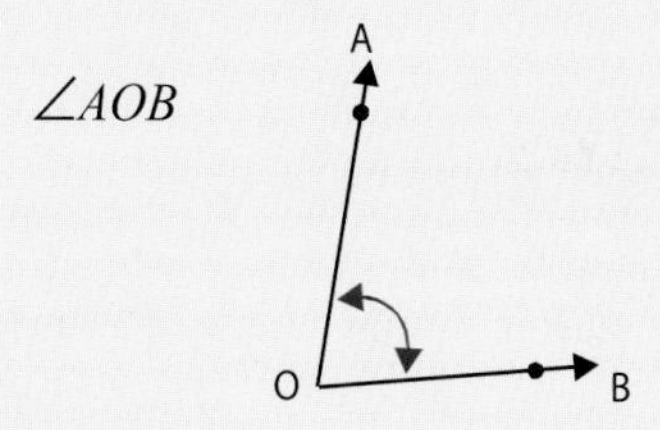


**2.** Using single numbers such as 1, 2, 3.

$\angle 1$

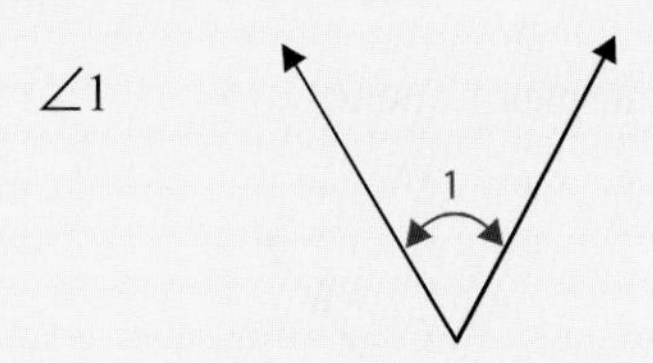


**3.** Using the single capital letter at the vertex when the meaning is clear.

$\angle B$

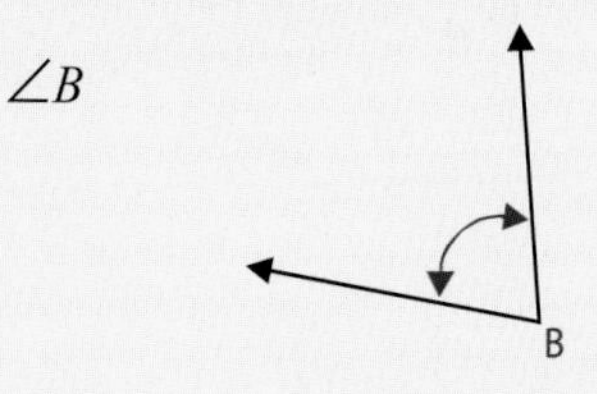


**Figure 10.2**

You might mention that a fourth way to label angles is to use Greek letters such as $\alpha$, $\beta$, and $\gamma$. These letters are not used in this section, but they were used in Section 7.5 in the formula for the sum of the measures of the angles of a triangle.

**Types of Angles**

| Name | Measure | Illustration |
|---|---|---|
| **1. Acute** | $0^0 < m\angle A < 90^0$ | $\angle A$ is an acute angle. |
| **2. Right** | $m\angle A = 90^0$ | $\angle A$ is a right angle. The rays are perpendicular to each other. |
| **3. Obtuse** | $90^0 < m\angle A < 180^0$ | $\angle A$ is an obtuse angle. |
| **4. Straight** | $m\angle A = 180^0$ | $\angle A$ is a straight angle. The rays are in opposite directions. |

The following figure is used for Examples 1 and 2.

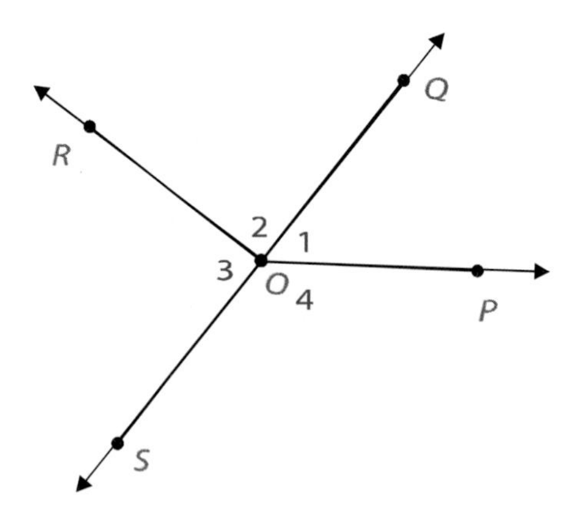


**Figure for Examples 1 and 2**

## Example 1

Use a protractor to check that the measures of the angles are as follows.

**a.** $m\angle 1 = 45°$ **b.** $m\angle 2 = 90°$

**c.** $m\angle 3 = 90°$ **d.** $m\angle 4 = 135°$

## Example 2

Tell whether each of the following angles is acute, right, obtuse, or straight.

**a.** $\angle 1$

**b.** $\angle 2$

**c.** $\angle POR$

### Solution

**a.** $\angle 1$ is acute since $0° < m\angle 1 < 90°$.

**b.** $\angle 2$ is a right angle since $m\angle 2 = 90°$.

**c.** $\angle POR$ is obtuse since $m\angle POR = 45° + 90° = 135° > 90°$.

**Definitions:**

1. Two angles are **complementary** if the sum of their measures is 90°.
2. Two angles are **supplementary** if the sum of their measures is 180°.
3. Two angles are **equal** if they have the same measure.

## Example 3

In the figure,

**a.** $\angle 1$ and $\angle 2$ are complementary because $m\angle 1 + m\angle 2 = 90°$.

**b.** $\angle COD$ and $\angle COA$ are supplementary because

$m\angle COD + m\angle COA = 70° + 110° = 180°$.

**c.** $\angle AOD$ is a straight angle because $m\angle AOD = 180°$.

**d.** $\angle BOA$ and $\angle BOD$ are supplementary, and in this case $m\angle BOA = m\angle BOD = 90°$.

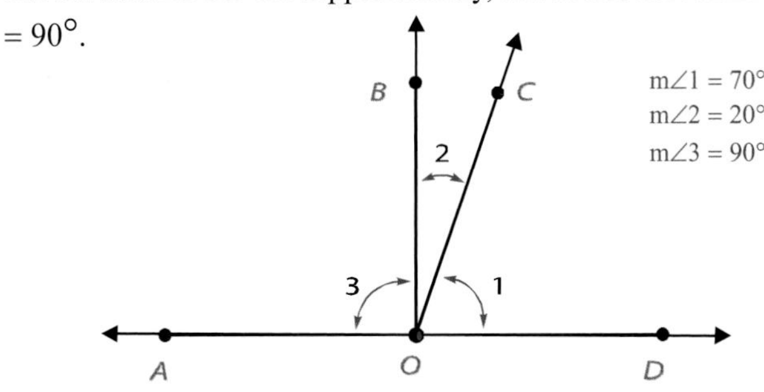

## Example 4

In the figure below, $\overleftrightarrow{PS}$ is a straight line and $m\angle QOP = 30°$ and $m\angle QOR = 30°$. Find the measures of

**a.** $\angle QOS$ and

**b.** $\angle SOP$.

**c.** Are any pairs complementary?

**d.** Are any pairs supplementary?

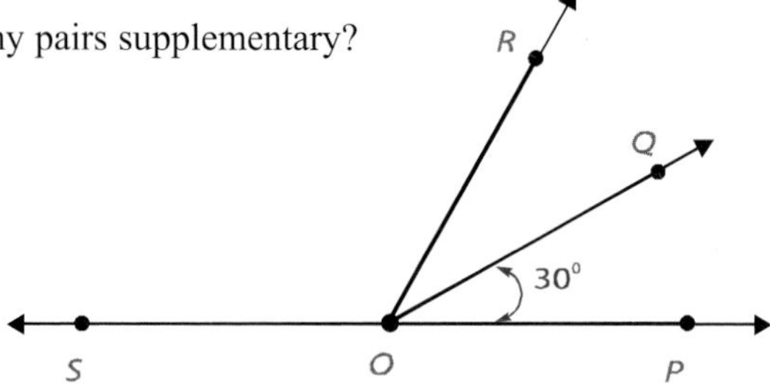


### Solution

**a.** $m\angle QOS = 150°$.

**b.** $m\angle SOP = 180°$.

**c.** No pairs are complementary. No two angles have a total measure of 90°.

**d.** Yes. $\angle QOP$ and $\angle QOS$ are supplementary, and $\angle ROP$ and $\angle ROS$ are supplementary. The sum of the measures of each pair is 180°.

If two lines intersect, then two pairs of **vertical angles** are formed. **Vertical angles** are also called **opposite angles.** ( See Figure 10.3. )

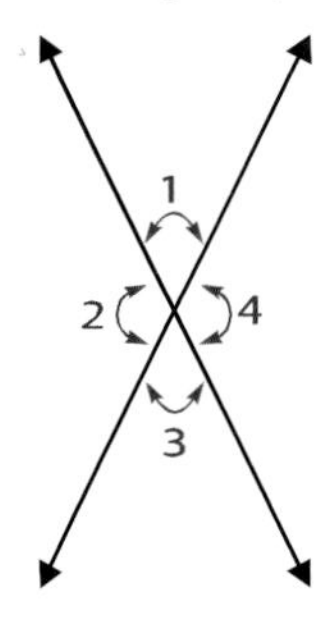


$\angle 1$ and $\angle 3$ are vertical angles.

$\angle 2$ and $\angle 4$ are vertical angles.

**Figure 10.3**

**Vertical angles are equal.** That is, **vertical angles have the same measure.** ( In Figure 10.3, $m\angle 1 = m\angle 3$ and $m\angle 2 = m\angle 4$. )

Two angles are **adjacent** if they have a common side. ( See Figure 10.4. )

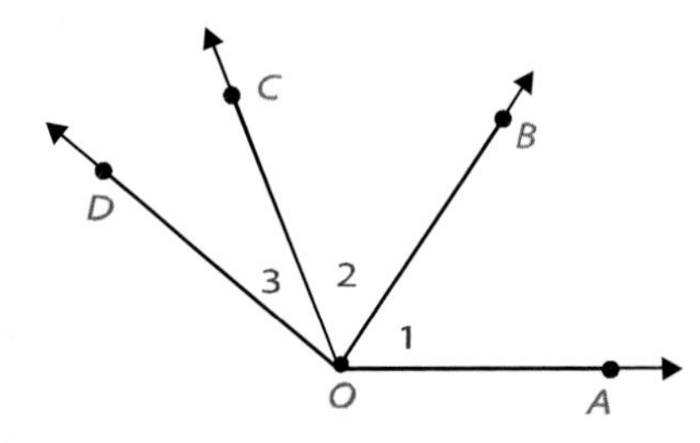


$\angle 1$ and $\angle 2$ are adjacent. They have the common side $\overrightarrow{OB}$.

$\angle 2$ and $\angle 3$ are adjacent. They have the common side $\overrightarrow{OC}$.

**Figure 10.4**

## Example 5

In the figure below, $\overleftrightarrow{AC}$ and $\overleftrightarrow{BD}$ are straight lines.

**a.** Name an angle adjacent to $\angle EOD$.

**b.** What is $m\angle AOD$?

### Solution

**a.** Four angles adjacent to $\angle EOD$ are $\angle AOE$, $\angle BOE$, $\angle BOD$, and $\angle COD$.

**b.** Since $\angle BOC$ and $\angle AOD$ are vertical angles, they have the same measure. So $m\angle AOD = 60°$.

## Triangles

A **triangle** consists of three line segments which join three points that do not lie on a straight line. The line segments are called the **sides** of the triangle, and the points are called the **vertices** of the triangle. If the points are labeled $A$, $B$, and $C$, the triangle is symbolized $\Delta ABC$. ( See Figure 10.5. )

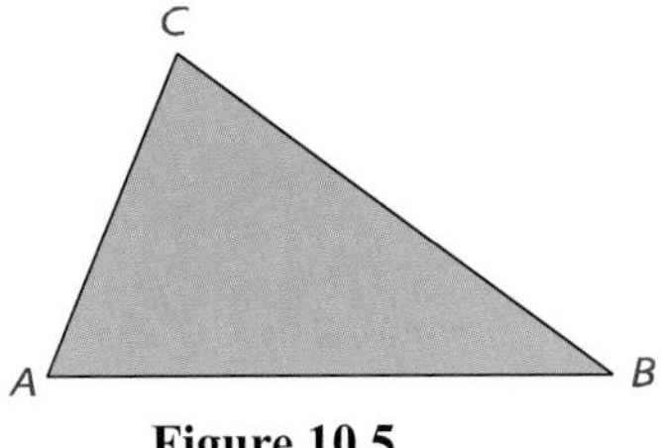


$\Delta ABC$ with vertices $A$, $B$, and $C$ and sides $\overline{AB}, \overline{BC}$, and $\overline{AC}$.

**Figure 10.5**

The sides of a triangle are said to determine three angles, and these angles are labeled by the vertices. Thus, the angles of $\Delta ABC$ are $\angle A$, $\angle B$, and $\angle C$. ( Since the definition of an angle involves rays, we can think of the sides of the triangle extended as rays that form these angles. )

**Triangles are classified in two ways:**

1. According to the lengths of their sides, and
2. According to the measures of their angles.

The corresponding names and properties are listed in the following tables.

**Special Note:** The line segment with endpoints $A$ and $B$ is indicated by placing a bar over the letters, as in $\overline{AB}$. The length of the segment is indicated by writing only the letters, as in $AB$.

**Triangles Classified by Sides**

| Name | Property | Illustration |
|---|---|---|
| **1. Scalene** | No two sides are equal. 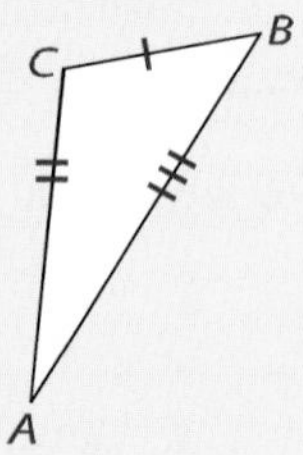 | $\Delta ABC$ is scalene since no two sides are equal. |
| **2. Isosceles** | Two sides are equal. 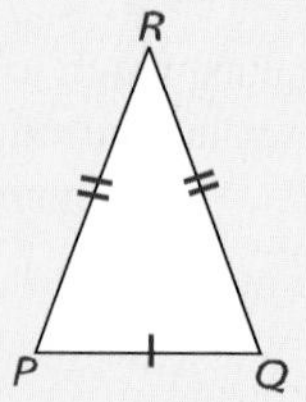 | $\Delta PQR$ is isosceles since $PR = QR$. |
| **3. Equilateral** | All three sides are equal. 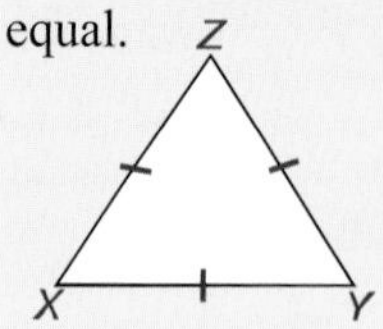 | $\Delta XYZ$ is equilateral since $XY = XZ = YZ$. |

C B A

R P Q

Z X Y

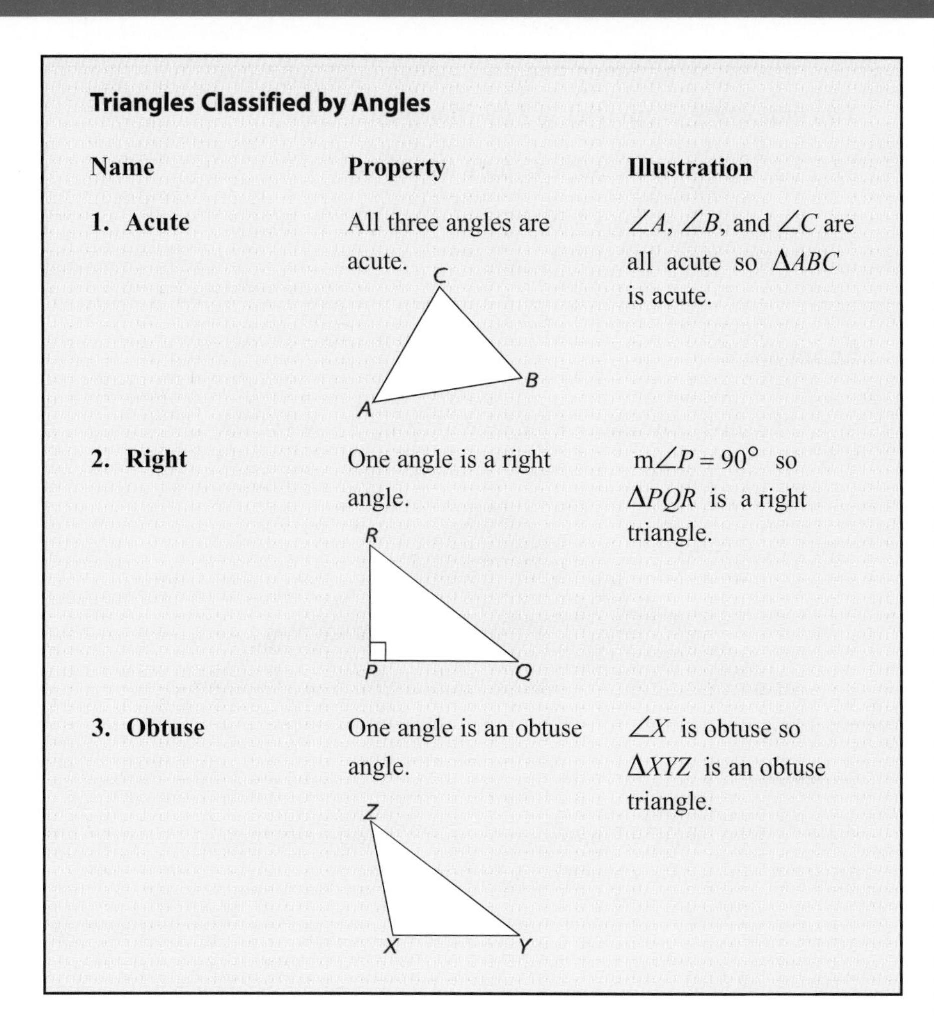

**Triangles Classified by Angles**

| Name | Property | Illustration |
|---|---|---|
| **1. Acute** | All three angles are acute. | $\angle A$, $\angle B$, and $\angle C$ are all acute so $\Delta ABC$ is acute. |
| **2. Right** | One angle is a right angle. | $m\angle P = 90°$ so $\Delta PQR$ is a right triangle. |
| **3. Obtuse** | One angle is an obtuse angle. | $\angle X$ is obtuse so $\Delta XYZ$ is an obtuse triangle. |



Every triangle is said to have six parts -- namely, three angles and three sides. Two sides of a triangle are said to **include** the angle at their common endpoint or vertex. The third side is said to be **opposite** this angle. In a triangle, **equal sides are opposite equal angles**.

The sides in a right triangle have special names. The longest side, opposite the right angle, is called the **hypotenuse,** and the other two sides are called **legs**. ( See Figure 10.6. )

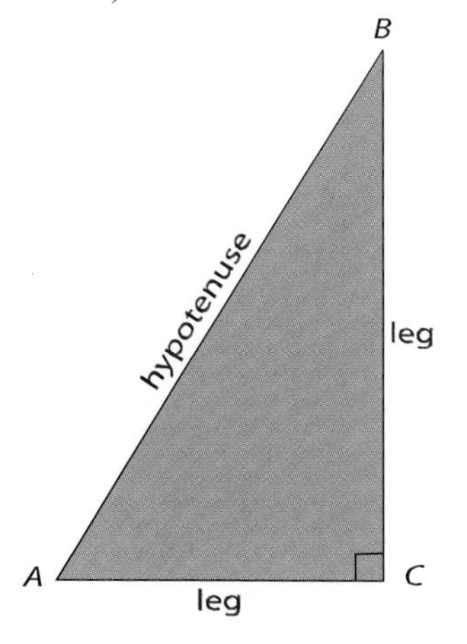



$\Delta ABC$ is a right triangle.
$m\angle C = 90°$. $\overline{AB}$ is opposite $\angle C$.

**Figure 10.6**

**Two Important Properties of Any Triangle**

1. The sum of the measures of the angles is 180°.
2. The sum of the lengths of any two sides must be greater than the length of the third side.

## Example 6

In $\Delta ABC$ below, $AB = AC$. What kind of triangle is $\Delta ABC$?

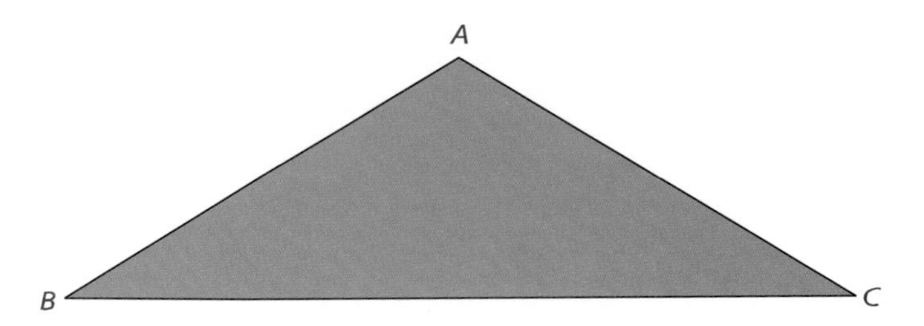


### Solution

$\Delta ABC$ can be classified in two ways. $\Delta ABC$ is **isosceles** because two sides are equal. $\Delta ABC$ is also **obtuse** because $m\angle A > 90°$.

## Example 7

Suppose the lengths of the sides of $\Delta PQR$ are as shown in the figure below. Is this possible?

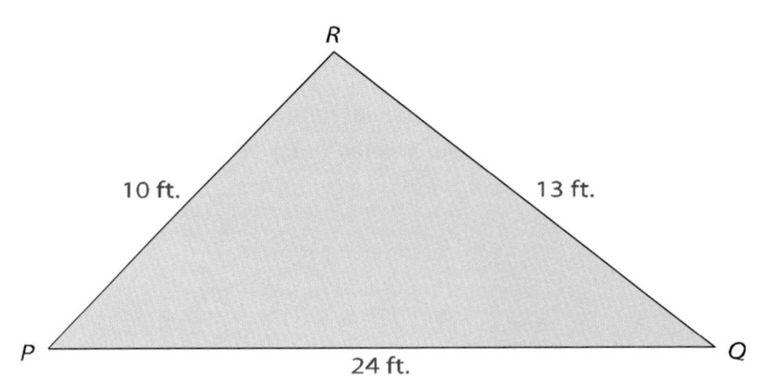


### Solution

This is **not** possible because $PR + QR = 10$ ft. + 13 ft. = 23 ft. and $PQ = 24$ ft., which is longer than the sum of the other two sides. In a triangle, the sum of the lengths of any two sides must be greater than the length of the third side.

## Example 8

In $\Delta BOR$ below, $m\angle B = 50°$ and $m\angle O = 70°$.

**a.** What is $m\angle R$?

**b.** What kind of triangle is $\Delta BOR$?

**c.** Which side is opposite $\angle R$?

**d.** Which sides include $\angle R$?

**e.** Is $\Delta BOR$ a right triangle? Why or why not?

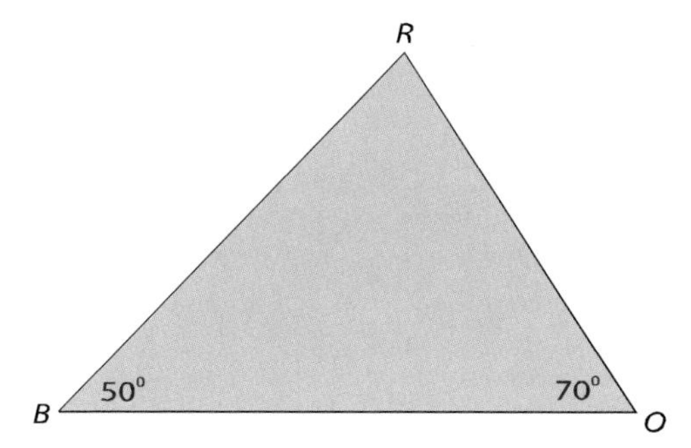


### Solution

**a.** The sum of the measure of the angles must be 180°. Since $50°+70° = 120°$, $m\angle R = 180° - 120° = 60°$.

**b.** $\Delta BOR$ is an acute triangle because all the angles are acute. Also, $\Delta BOR$ is a scalene because no two sides are equal.

**c.** $\overline{BO}$ is opposite of $\angle R$.

**d.** $\overline{RB}$ and $\overline{RO}$ include $\angle R$.

**e.** $\Delta BOR$ is not a right triangle because none of the angles are right angles.

Name ______________________ Section ______________ Date ___________

## Exercises 10.1

1. **a.** If $\angle 1$ and $\angle 2$ are complementary and $m\angle 1 = 15°$, what is $m\angle 2$?

   **b.** If $m\angle 1 = 3°$, what is $m\angle 2$?

   **c.** If $m\angle 1 = 45°$, what is $m\angle 2$?

   **d.** If $m\angle 1 = 75°$, what is $m\angle 2$?

2. **a.** If $\angle 3$ and $\angle 4$ are supplementary and $m\angle 3 = 45°$, what is $m\angle 4$?

   **b.** If $m\angle 3 = 90°$, what is $m\angle 4$?

   **c.** If $m\angle 3 = 110°$, what is $m\angle 4$?

   **d.** If $m\angle 3 = 135°$, what is $m\angle 4$?

3. The supplement of an acute angle is an obtuse angle.

   **a.** What is the supplement of a right angle?

   **b.** What is the supplement of an obtuse angle?

   **c.** What is the complement of an acute angle?

4. In the figure shown below $\overleftrightarrow{DC}$ is a line and $m\angle BOA = 90°$.

   **a.** What type of angle is $\angle AOC$?

   **b.** What type of angle is $\angle BOC$?

   **c.** What type of angle is $\angle BOA$?

   **d.** Name a pair of complementary angles.

   **e.** Name two pairs of supplementary angles.

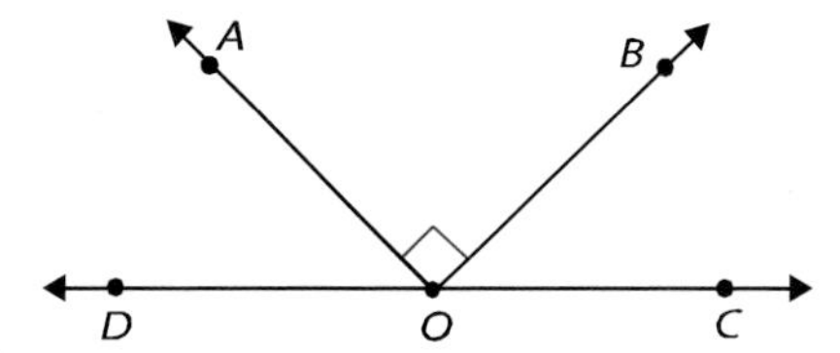


5. An **angle bisector** is a ray that divides an angle into two angles with equal measures. If $\overrightarrow{OX}$ bisects $\angle COD$ and $m\angle COD = 50°$, what is the measure of each of the equal angles formed?

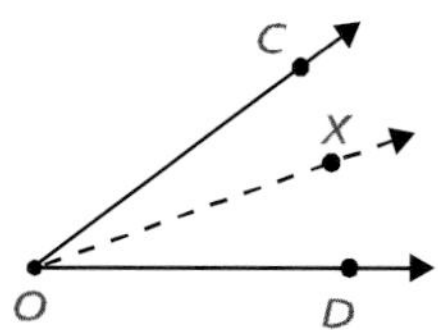


**ANSWERS**

1. **a.** $m\angle 2 = 75°$
   **b.** $m\angle 2 = 87°$
   **c.** $m\angle 2 = 45°$
   **d.** $m\angle 2 = 15°$

2. **a.** $m\angle 4 = 135°$
   **b.** $m\angle 4 = 90°$
   **c.** $m\angle 4 = 70°$
   **d.** $m\angle 4 = 45°$

3. **a.** a right angle
   **b.** an acute angle
   **c.** an acute angle

4. **a.** an obtuse angle
   **b.** an acute angle
   **c.** a right angle
   **d.** $\angle AOD$, $\angle BOC$
   **e.** $\angle AOD$ and $\angle AOC$, $\angle BOC$ and $\angle BOD$

5. 25°

6. 15°

7. 40°

8. 95°

9. 55°

10. 70°

11. 110°

12. [ Respond in exercise. ]

13. m∠A = 55°; m∠B = 51°; m∠C = 74°

14. m∠P = 65°; m∠Q = 115°; m∠R = 105°; m∠S = 75°

15. m∠L = 104°; m∠M = 109°; m∠N = 105°; m∠O = 106°; m∠P = 116°

16. m∠D = 100°; m∠E = 45°; m∠F = 35°

In Exercises 6–11, given that m∠*AOB* = 30° and m∠*BOC* = 80° and $\overrightarrow{OX}$ and $\overrightarrow{OY}$ are angle bisectors, find the measures of the following angles.

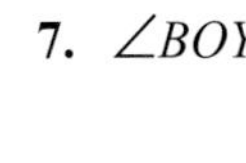


6. ∠*AOX*

7. ∠*BOY*

8. ∠*COX*

9. ∠*YOX*

10. ∠*AOY*

11. ∠*AOC*

**12.** In the figure shown below:

**a.** Name all the pairs of supplementary angles.

**b.** Name all the pairs of complementary angles.

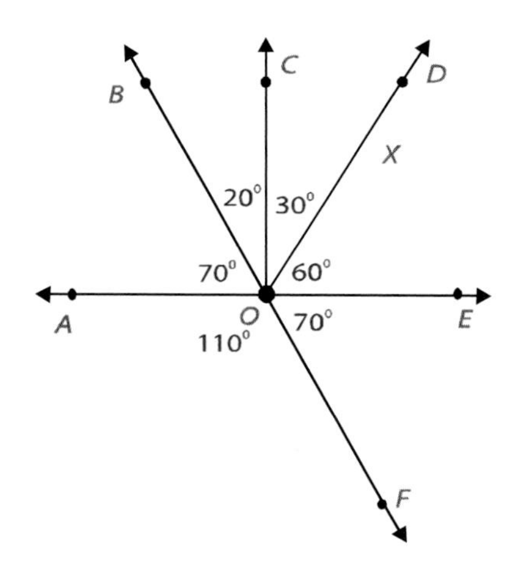


a. ∠*BOA* and ∠*BOE*; ∠*COA* and ∠*COE*; ∠*DOA* and ∠*DOE*; ∠*COB* and ∠*COF*; ∠*DOB* and ∠*DOF*; ∠*EOB* and ∠*EOF*; ∠*AOF* and ∠*FOE*; ∠*BOA* and ∠*AOF*;

b. ∠*BOA* and ∠*BOC*; ∠*COD* and ∠*DOE*; ∠*BOC* and ∠*EOF*

In Exercises 13–16, use a protractor to measure all the angles in each figure. Each line segment may be extended as a ray to form the side of an angle.

**13.**

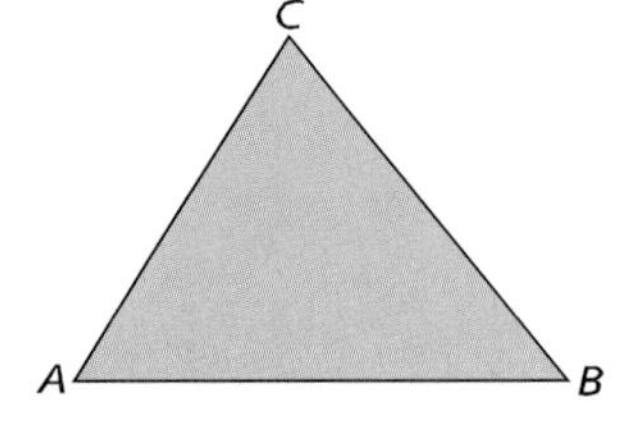


**14.**

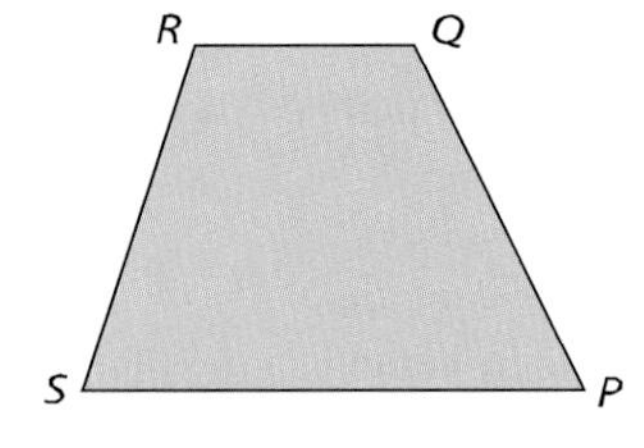


**15.**

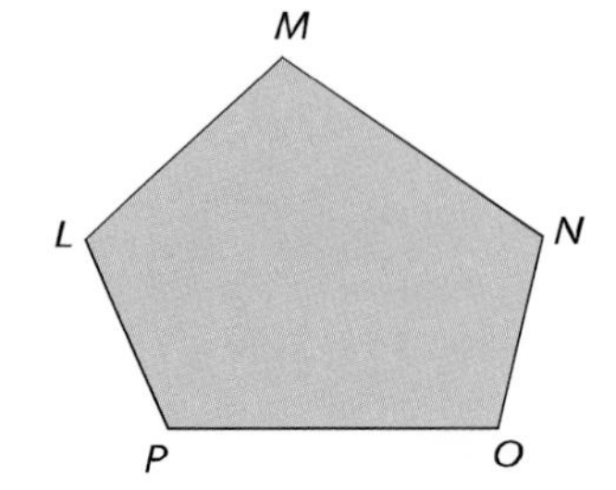


**16.**

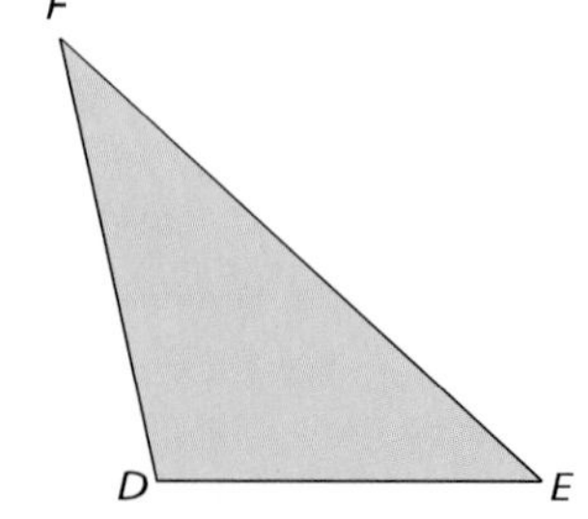

Name ______________________ Section ____________ Date __________

**17.** Name the type of angle formed by the hands on a clock.

**a.** at six o'clock **b.** at three o'clock

**c.** at one o' clock **d.** at five o'clock

a. 

b. 

c. 

d. 

**18.** What is the measure of each angle formed by the hands of the clock in Exercise 17?

**19.** The figure at right shows two intersecting lines.

**a.** If $m\angle 1 = 30°$, what is $m\angle 2$?

**b.** Is $m\angle 3 = 30°$? Give a reason for your answer other than the fact that $\angle 1$ and $\angle 3$ are vertical angles.

**c.** Name four pairs of adjacent angles.

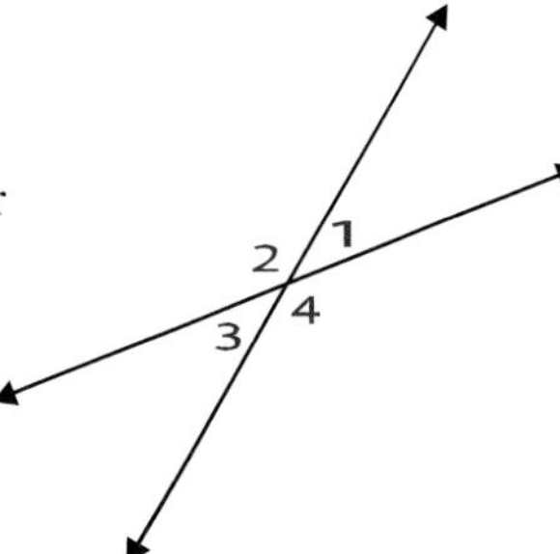


**20.** In the figure here, $\overleftrightarrow{AB}$ is a line.

**a.** Name two pairs of adjacent angles.

**b.** Name two vertical angles if there are any.

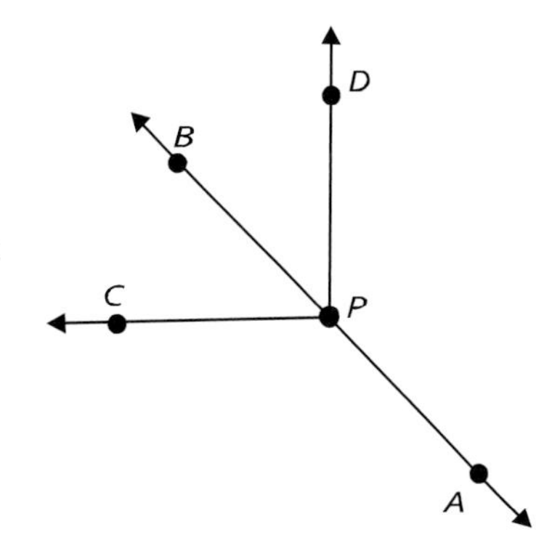


**ANSWERS**

**17. a.** straight

**b.** right

**c.** acute

**d.** obtuse

**18. a.** 180°

**b.** 90°

**c.** 30°

**d.** 150°

**19. a.** $m\angle 2 = 150°$

**b.** yes; $\angle 2$ and $\angle 3$ are supplementary

**c.** $\angle 1$ and $\angle 2$; $\angle 1$ and $\angle 4$; $\angle 2$ and $\angle 3$; $\angle 3$ and $\angle 4$

**20. a.** $\angle CPB$ and $\angle BPD$, $\angle CPB$ and $\angle CPA$. There are others.

**b.** There are none.

21. $m\angle 2 = 150°$;

$m\angle 3 = 30°$;

$m\angle 4 = 150°$

22. a. $m\angle 2 = 70°$;

$m\angle 3 = 90°$;

$m\angle 4 = 20°$;

$m\angle 5 = 70°$

b. $\angle 3$

c. $\angle 2$ and $\angle 5$

23. $m\angle 1 = m\angle 4 = 140°$;

$m\angle 4 = m\angle 6 = 140°$;

$m\angle 1 = m\angle 6 = 140°$;

$m\angle 3 = m\angle 5 = 40°$;

$m\angle 2 = m\angle 5 = 40°$

24. scalene, obtuse

25. equilateral, acute

26. right, scalene

27. obtuse, scalene

**21.** Given that $m\angle 1 = 30°$ in the figure shown here, find the measures of the other three angles.

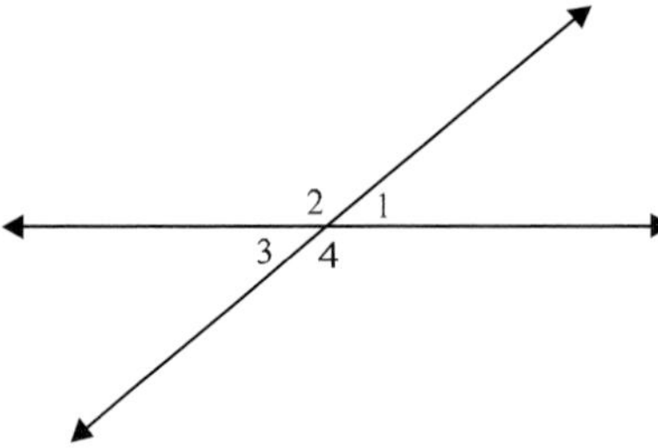


**22.** In the figure shown below, *l*, *m*, and *n* are straight lines with $m\angle 1 = 20°$ and $m\angle 6 = 90°$.

**a.** Find the measures of the other four angles.

**b.** Which angle is supplementary to $\angle 6$?

**c.** Which angles are complementary to $\angle 1$?

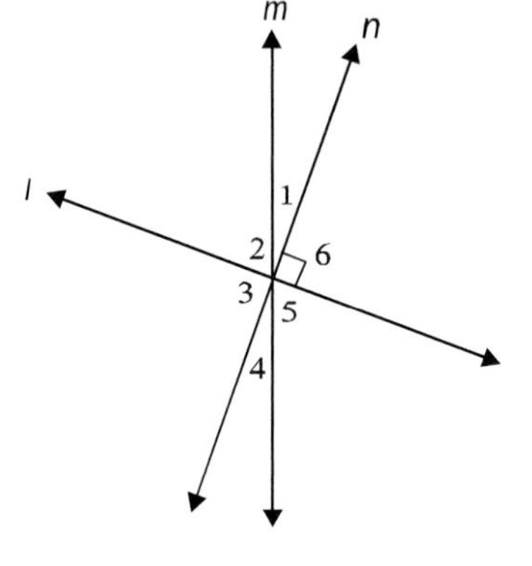


**23.** In the figure shown below, $m\angle 2 = m\angle 3 = 40°$. Find all other pairs of angles that have equal measures.

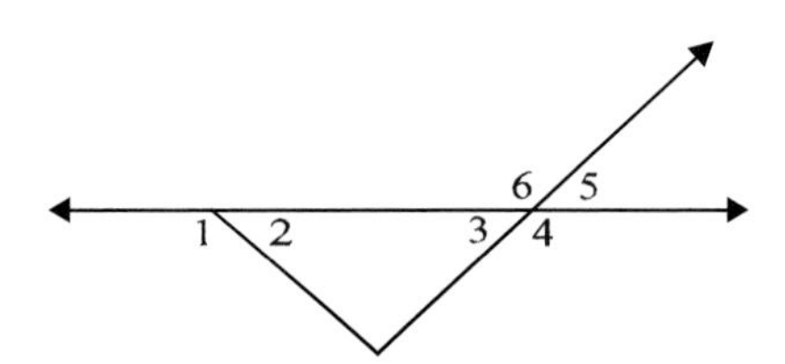


Name each of the following triangles in two ways.

**24.**

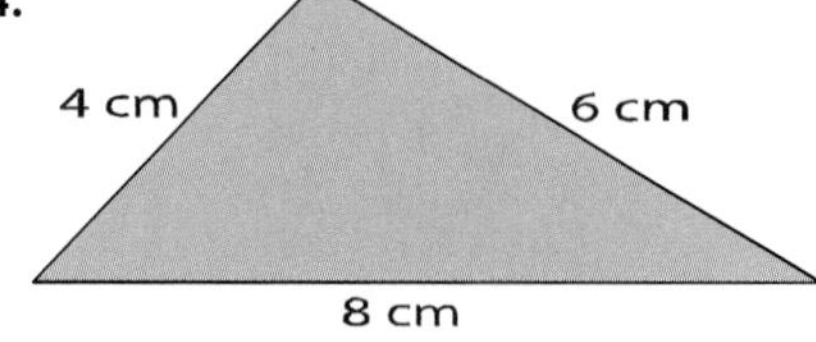


**25.**

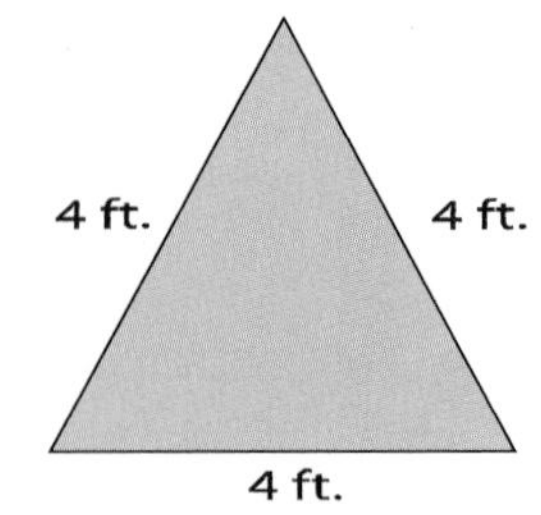


**26.**

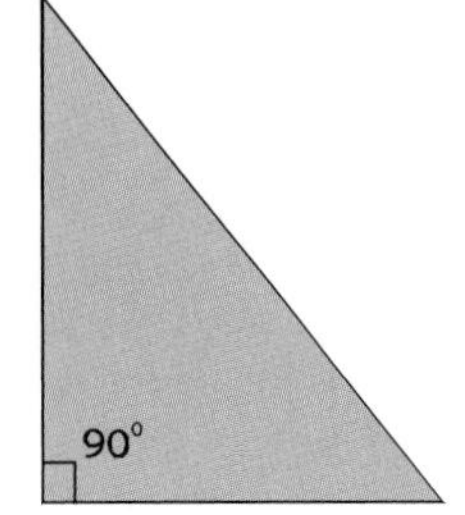


**27.**

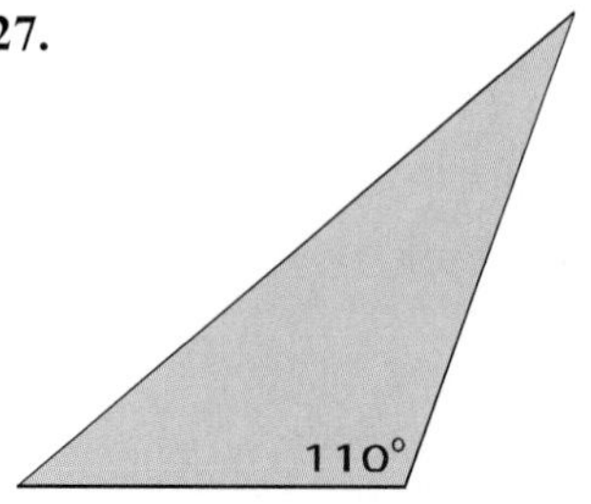

Name ______________________ Section ____________ Date __________

**28.**

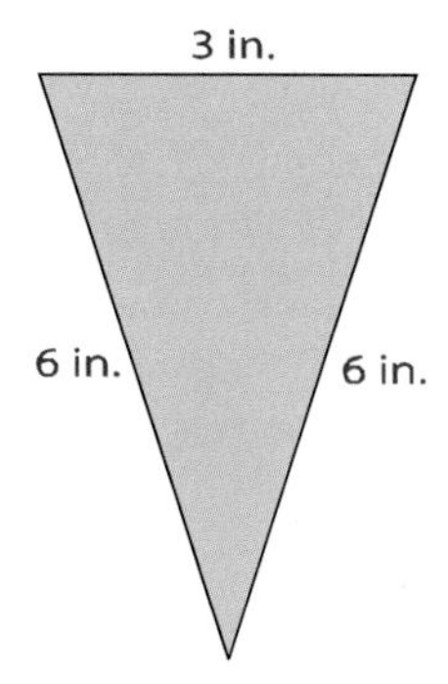


**29.**

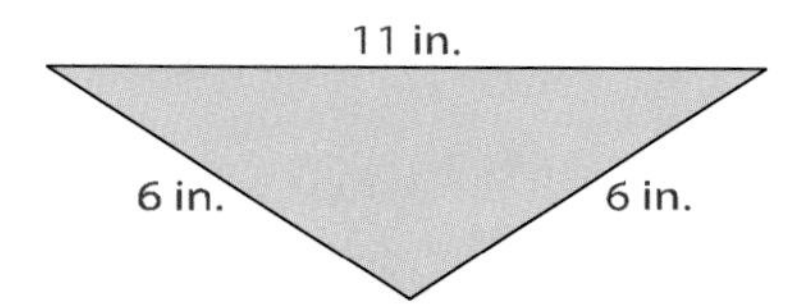


**30.**

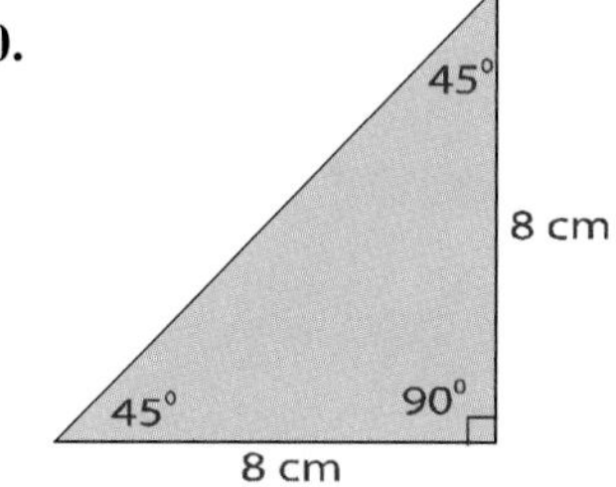


**31.**

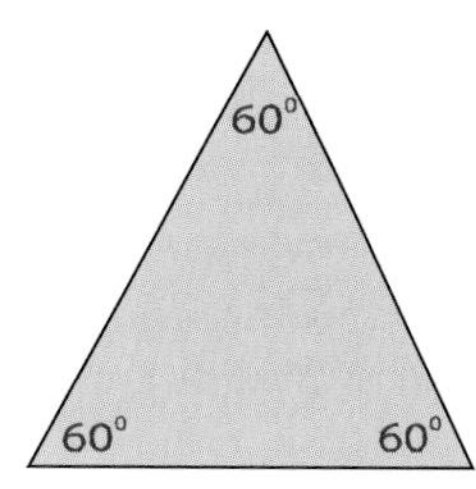


**32.**

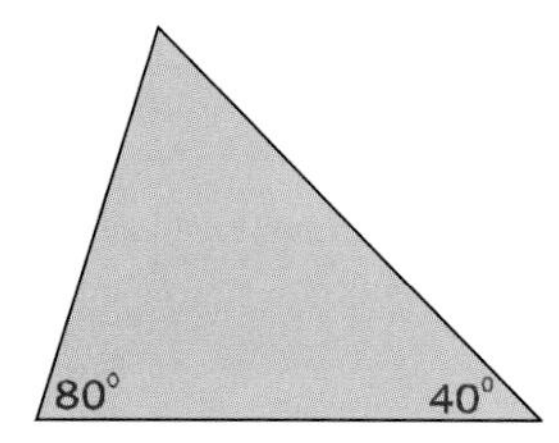


**33.** Suppose the lengths of the sides of $\Delta ABC$ are as shown in the figure below. Is this possible?

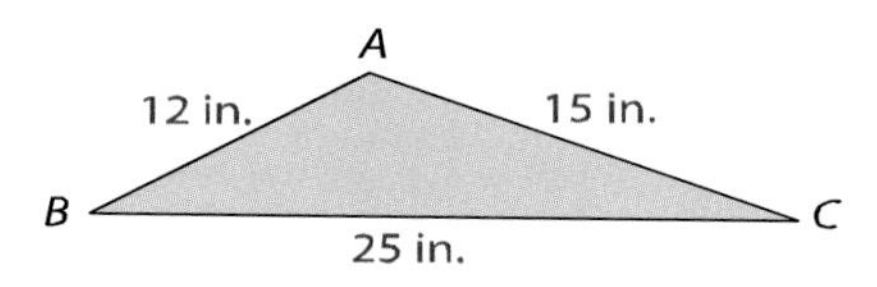


**34.** Suppose the lengths of the sides of $\Delta DEF$ are as shown in the figure below. Is this possible?

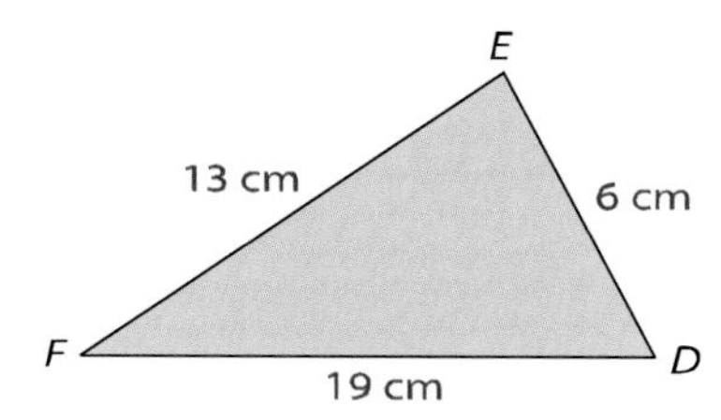


**ANSWERS**

**28.** isosceles, acute

**29.** isosceles, obtuse

**30.** right, isosceles

**31.** acute, equilateral

**32.** acute, scalene

**33.** yes, since $25 < 12 + 15$

**34.** no, since $19 = 13 + 6$

**35.** In $\Delta XYZ$ below, $m\angle X = 30°$ and $m\angle Y = 70°$.

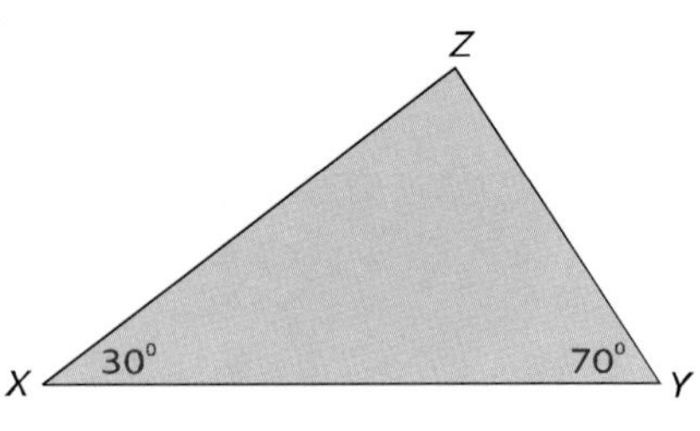


**a.** What is $m\angle Z$?

**b.** What kind of triangle is $\Delta XYZ$?

**c.** Which side is opposite $\angle X$?

**d.** Which sides include $\angle X$?

**e.** Is $\Delta XYZ$ a right triangle? Why or why not?

**35. a.** $m\angle Z = 80°$

**b.** acute, scalene

**c.** $\overleftrightarrow{YZ}$

**d.** $\overleftrightarrow{XZ}$ and $\overleftrightarrow{XY}$

**e.** no; no $\angle = 90°$

**36.** In ΔSTU below, $m\angle T = 50°$ and $m\angle U = 40°$.

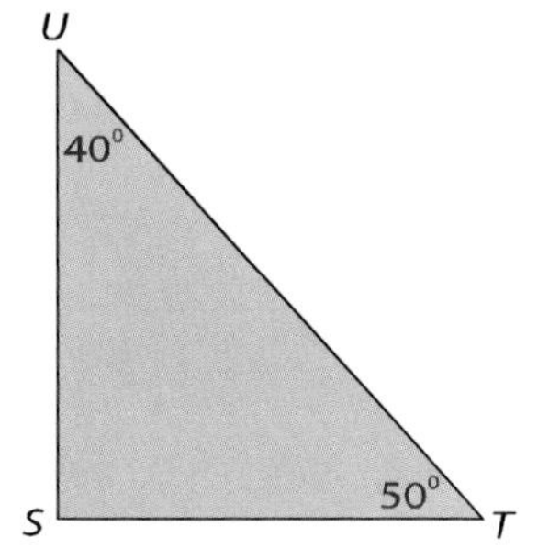


**a.** What is $m\angle S$?

**b.** What kind of triangle is $\Delta STU$?

**c.** Which side is opposite $\angle$T ?

**d.** Which sides include $\angle T$?

**e.** Is $\Delta STU$ a right triangle? Why or why not?

**36. a.** $m\angle S = 90°$

**b.** right, scalene

**c.** $\overleftrightarrow{US}$

**d.** $\overleftrightarrow{ST}$ and $\overleftrightarrow{UT}$

**e.** yes, $m\angle S = 90°$

# 10.2 Length and Perimeter

## Metric Units of Length

The metric system of measurement is used by about 90% of the people in the world. The United States is the only major industrialized country still committed to the U.S. customary system ( formerly called the *English system* ). Even in the United States, the metric system is used in many fields of study and business, such as medicine, science, the computer industry, and the military. Industries involved in international trade must be familiar with the metric system.

The **meter** is the basic unit of length in the metric system. Smaller and larger units are named by putting a prefix in front of the basic unit: for example, **centi**meter and **kilo**meter. The prefixes we will use are shown in boldface print in Table 10.1. Other prefixes that indicate extremely small units are *micro-*, *nano-*, *pico-*, *femto-*and *atto-*. Prefixes that indicate extremely large units are *mega-*, *giga-*, and *tera-*.

| Metric Measures of Length | | |
|---|---|---|
| 1 **milli**meter ( mm ) | = 0.001 meter | 1 m = 10 dm |
| 1 **centi**meter ( cm ) | = 0.01 meter | 1 m = 100 cm |
| 1 **deci**meter ( dm ) | = 0.1 meter | 1 m = 1000 mm |
| 1 meter ( m ) | = 1.0 meter ( the basic unit ) | |
| 1 **deca**meter ( dam ) | = 10 meters | |
| 1 **hecto**meter ( hm ) | = 100 meters | |
| 1 **kilo**meter ( km ) | = 1000 meters | |

**Table 10.1**

As indicated in Table 10.1, the metric units of length are related to each other by powers of 10. That is, simply multiply by 10 to get the equivalent measure expressed in the next lower ( or smaller ) unit. Thus, you will have **more** of a **smaller** unit. For example,

**1 m = 10 dm = 100 cm = 1000 mm**

Conversely, divide by 10 to get the equivalent measure expressed in the next higher ( or larger ) unit. Thus, you will have **less** of a **larger** unit. For example,

**1 mm = 0.1 cm = 0.01 dm = 0.001m**

### Objectives

① Recognize measures of length in the metric system.

② Be able to change metric measures of length within the metric system.

③ Know how to use tables to convert units of length between the metric system and U.S. customary system.

④ Know the formulas for the perimeter of several geometric figures.

In the United States, a serious effort to convert to the metric system was made in the 1970's. Road signs included kilometers (as well as miles) and gasoline pumps indicated liters (in place of gallons). People did not react favorably and we reverted to the English system in our daily lives. However, many industries (particularly those that deal with imports) have converted to the metric system.

Emphasize the importance of memorizing the prefixes in their proper order, so that conversions can be easily made.

You might introduce and discuss the meanings of "new" measures such as millidollar, centidollar, decidollar, dekadollar, hectodollar, and kilodollar. Ask the students to relate the meanings of the words such as cent, percent, centennial, century, decagon, decathlon, and kilowatt to the prefixes in the metric system.

Point out that this ease of conversion between units in the metric system is one of its principal advantages over the U.S. customary system. There is no systematic method of conversion in the U.S. system.

**Changing Metric Measures of Length**

| Procedure | Example |
|---|---|
| To change to a measure of length that is<br>one unit smaller, multiply by 10,<br>two units smaller, multiply by 100,<br>three units smaller, multiply by 1000,<br>and so on. | **Larger to Smaller**<br>6 cm = 60 mm<br>3 m = 300 cm<br>12 m = 12 000 mm |
| To change to a measure of length that is<br>one unit larger, divide by 10,<br>two units larger, divide by 100,<br>three units larger, divide by 1000,<br>and so on. | **Smaller to Larger**<br>35 mm = 3.5 cm<br>250 cm = 2.5 m<br>32 m = 0.032 km |

## Example 1

The following examples illustrate how to change from larger to smaller units of length in the metric system.

**a.** 5.6 m = 100( 5.6 ) cm = 560 cm

**b.** 5.6 m = 1000( 5.6 ) mm= 5600 mm

**c.** 23.5 cm= 10( 23.5 ) mm = 235 mm

**d.** 1.42 km= 1000( 1.42 ) m= 1420 m

## Example 2

The following examples illustrate how to change from smaller to larger units of length in the metric system.

**a.** $375 \text{ cm} = \left(\frac{375}{100}\right)\text{m} = 3.75\text{m}$

**b.** $375 \text{ mm} = \left(\frac{375}{1000}\right)\text{m} = 0.375\text{m}$

**c.** $1055 \text{ m} = \left(\frac{1055}{1000}\right)\text{km} = 1.055\text{km}$

Another technique for changing units in the metric system can be illustrated with the concept of a number line with the metric prefixes below in order.

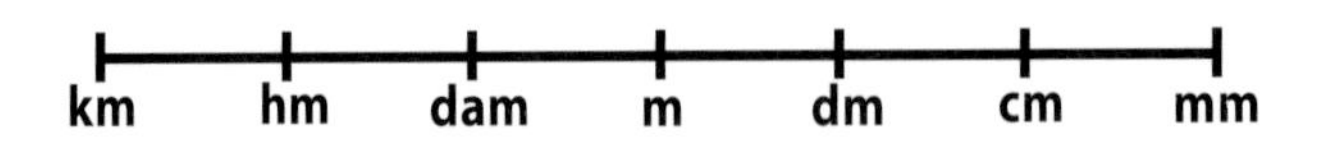


Simply move the decimal point in the direction of change.

## Example 3

Change 56 cm to the equivalent measure in meters.

### Solution

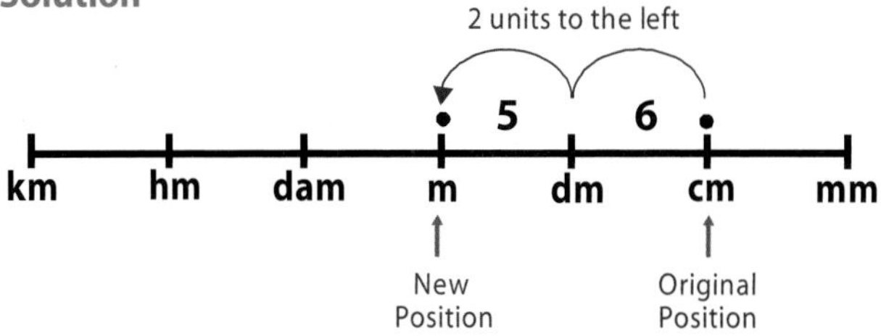


Thus, 56 cm = 0.56 m

## Example 4

Change 13.5 m to an equivalent measure in millimeters.

### Solution

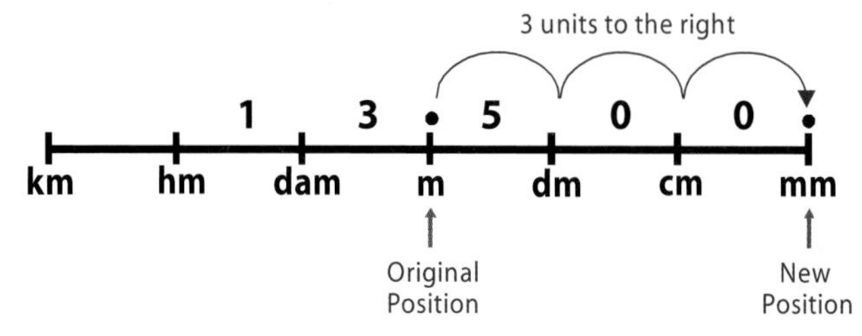


Thus, 13.5 m = 13 500 mm

**Note:** In the metric system,

1. A 0 is written to the left of the decimal point if there is no whole number part ( 0.287 m ).
2. No commas are used in writing numbers. If a number has more than four digits ( left or right of the decimal point ), the digits are grouped in threes from the decimal point with a space between the groups ( 25 000 m or 0.000 34 m ).

*Now Work Exercises 1-4 in the Margin.*

Change the following units as indicated.

1. 35 m = 3500 cm
2. 6.4 cm = 64 mm
3. 5.9 m = 0.0059 km
4. 320 mm = 0.32 m

To help students become familiar with both systems, you might have them buy rulers marked in metric and U.S. customary units and measure objects in the room (such as books, table tops, their heights, etc.) and discuss the units they chose to use for various measures and why.

## U.S. Customary and Metric Equivalent Units of Length

In the U.S. customary system, the units are not systematically related as are the units in the metric system. Historically, some of the units were associated with parts of the body, which would vary from person to person. For example, a foot was the length of a person's foot and a yard was the distance from the tip of one's nose to the tip of one's fingers with the arm outstretched.

There is considerably more consistency now because the official weights and measures are monitored by the government. Table 10.2 shows some of the basic units of length in the U.S. customary system.

| U.S. Customary Units of Length |
|---|
| 1 foot ( ft. ) = 12 inches (in.) |
| 1 yard ( yd. ) = 3 ft. |
| 1 mile ( mi. ) = 5280 ft. |

**Table 10.2**

We will use the equivalent measures ( rounded off ) in Table 10.3 to convert units of length from the U.S. customary system to the metric system, and vice versa. Examples 5 and 6 illustrate how to make conversions using these equivalents and multiplying by the number of given units.

| U.S. - Metric Length Equivalents | |
|---|---|
| **U.S. to Metric** | **Metric to U.S.** |
| 1 in. = 2.54 cm ( exact ) | 1 cm = 0.394 in. |
| 1 ft. = 0.305 m | 1 m = 3.28 ft. |
| 1 yd. = 0.914 m | 1 m = 1.09 yd. |
| 1 mi. = 1.61 km | 1 km = 0.62 mi. |

**Table 10.3**

**Important Note About Conversions between U.S. and Metric Equivalents**

Most of the conversion units are not exact and slightly different answers are possible, depending on the conversion units used. Generally, the conversions will be accurate to at least the first three digits. Therefore, we will give conversions rounded off to the place of the first three digits only.

### Example 5

**a.** 6 ft. = _____ cm

**b.** How many kilometers is the same as 25 miles?

#### Solution

**a.** 6 ft. = 6(12 in.) = 72 in. = 72 (1 in.) = 72(2.54 cm) = 183 cm (rounded off)

**b.** 25 mi. = 25 (1 mi.) = 25(1.61 km) = 40.25 km = 40.3 km (rounded off )

### Example 6

**a.** How many feet are there in 30 m?

**b.** Convert 10 km to miles.

#### Solution

**a.** 30 m = 30 ( 1 m ) = 30( 3.28 ft. ) = 98.4 ft.

**b.** 10 km = 10 ( 1 km ) = 10( 0.62 mi. ) = 6.2 mi.

*Now Work Exercises 5-8 in the Margin.*

## Geometry: Formulas for Perimeter

The geometric figures and formulas discussed in this section illustrate applications of measures of length. The formulas are independent of any measurement system, and the units used are clearly labeled in each figure. Some of the exercises ask that the answers be given in both U.S. customary units and metric units. For these exercises, refer to Table 10.3. ( **Note:** For easy reference, some of the ideas related to circles that were discussed in Section 5.5 are listed again. )

**Important Terms**

| | |
|---|---|
| **Perimeter:** | Total distance around a plane geometric figure |
| **Circumference:** | Perimeter of a circle |
| **Radius:** | Distance from the center of a circle to a point on the circle |
| **Diameter:** | Distance from one point on a circle to another point on the circle measured through the center |

**Note:** We will use $\pi = 3.14$, but you should remember that 3.14 is only an approximation to $\pi$. ( See the discussion preceding Table 4 in the appendices for more information about $\pi$. )

Use Table 10.3 as a reference to convert the following measures of length as indicated.

**5.** 2 ft. = ___61___ cm

**6.** 5 yd. = ___4.57___ m

**7.** How many inches are there in 50 cm?
19.7 in.

**8.** Convert 15 km to feet.
49,200 ft.

### Geometric Figures and the Formulas for Finding Their Perimeters

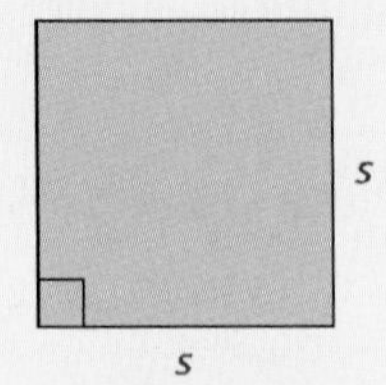


**Square**

$P = 4s$

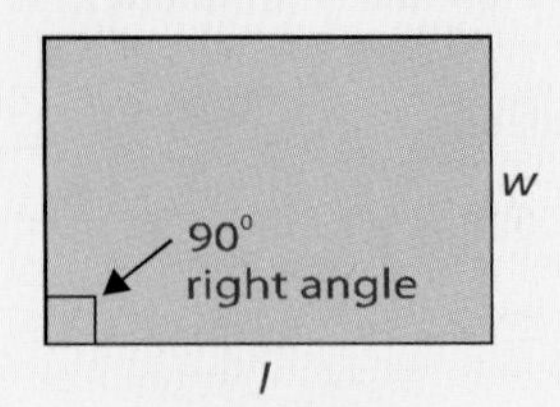


**Rectangle**

$P = 2l + 2w$

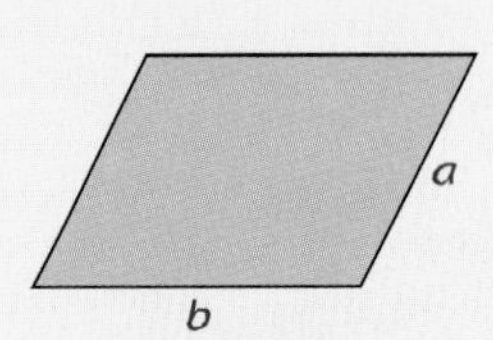


**Parallelogram**

$P = 2b + 2a$

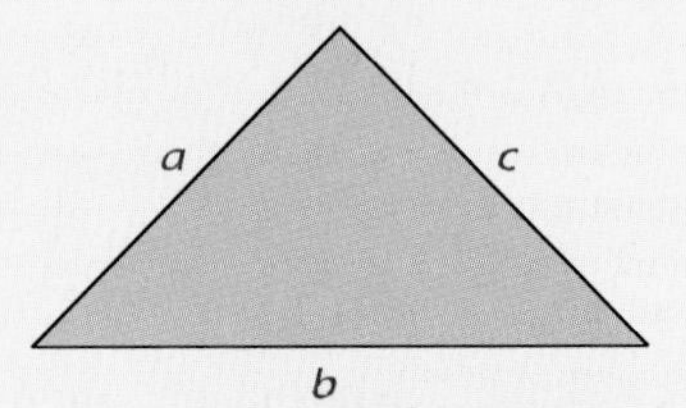


**Triangle**

$P = a + b + c$

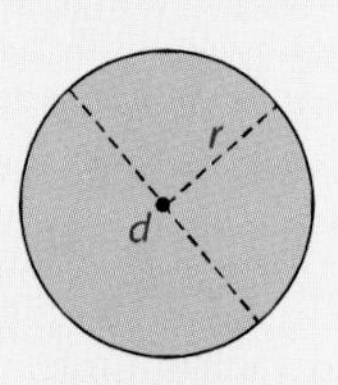


**Circle**

$C = 2\pi r$

$C = \pi d$

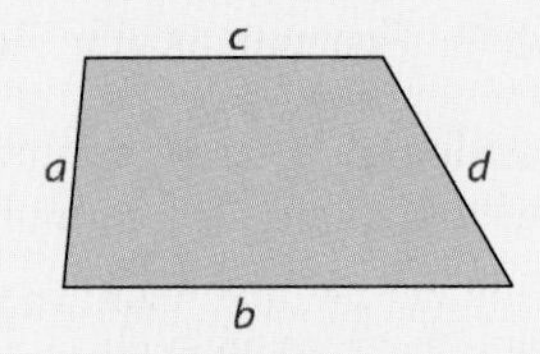


**Trapezoid**

$P = a + b + c + d$

## Example 7

Find the perimeter of a rectangle with length 20 inches and width 12 inches. Write the answer in both inches and centimeters.

### Solution

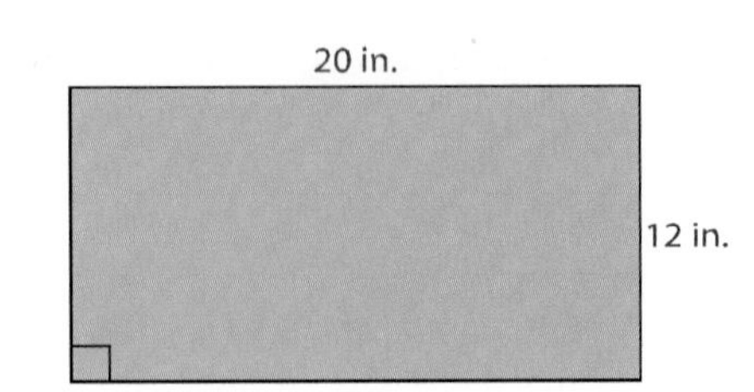


$P = 2l + 2w$

$P = 2 \cdot 20 + 2 \cdot 12$

$P = 40 + 24 = 64$ in.

64 in. = 64 ( 1 in. ) = 64( 2.54 cm ) = 162.56 cm

The perimeter is 64 in., or 163 cm (rounded off).

## Example 8

Find the perimeter of a triangle with sides of 4 cm, 80 mm, and 0.6 dm. Find the perimeter in both millimeters and centimeters.

### Solution

Sketch the figure carefully so that your drawing shows how the lengths of the sides are related. ( That is, the longest side must be labeled as 80 mm. ) Change all the units to the same unit of measure. In this case, any of the three units of measure will do. Millimeters have been chosen.

4 cm = 40 mm, 80 mm = 80 mm, 0.6 dm = 60 mm

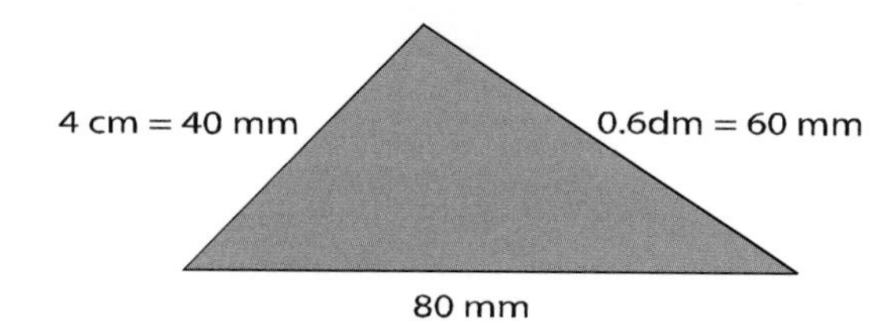


$P = a + b + c$

$P = 40 + 60 + 80 = 180$ mm

$180 \text{ mm} = \frac{180}{10} = 18 \text{ cm}$

So, the perimeter is 180 mm, or 18 cm.

## Example 9

Find the circumference of a circle with a diameter of 1.5 meters. Write the answer in both meters and feet.

### Solution

Sketch the circle and label a diameter.

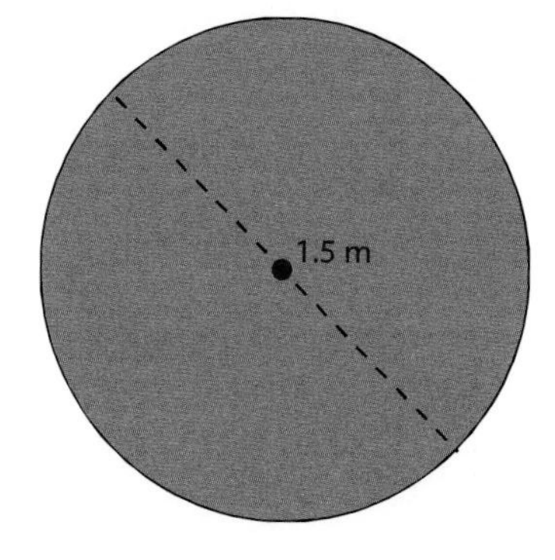


$C = \pi d$

$C = 3.14 \cdot 1.5 = 4.71$ m

4.71m = 4.71 ( 1 m ) = 4.71( 3.28 ft. ) =15.4 ft. ( rounded off )

So, the circumference is 4.71 m, or 15.4 ft.

### Example 10

Find the perimeter of a figure that is a semicircle ( half of a circle ) with a diameter of 100 mm. Find the perimeter in both millimeters and meters.

**Solution**

A sketch of the figure will help.

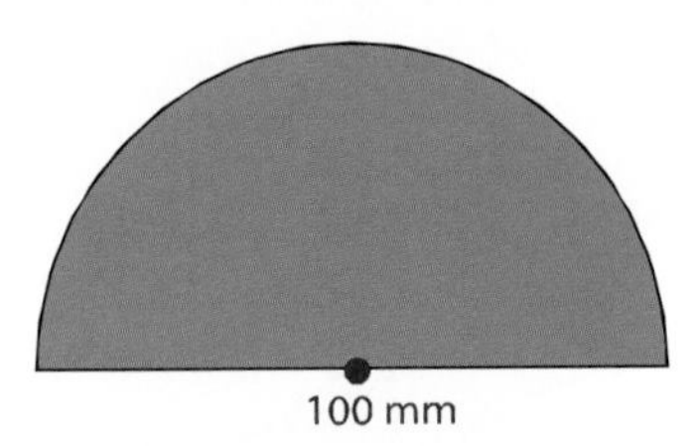


Find the perimeter of the figure by adding the length of the diameter (100 mm) to the length of the arc of the semicircle.

$$\text{Length of arc of semicircle} = \frac{1}{2}\pi d = \frac{1}{2}(3.14)\cdot 100 = 157 \text{ mm}$$

$$\text{Perimeter of figure} = 157 + 100 = 257 \text{ mm}$$

$$257 \text{ mm} = \frac{257}{1000} \text{ m} = 0.257 \text{ m}$$

So, the perimeter of the figure is 257 mm, or 0.257 m.

Name ______________________ Section ______________ Date ____________

ANSWERS

For Exercises 1-16, respond in exercise.

## Exercises 10.2

Change the following units of length in the metric system as indicated.

**1.** 3 m = 300 cm

**2.** 0.8 m = 80 cm

**3.** 1.5 m = 1500 mm

**4.** 1.9 cm = 19 mm

**5.** 45 cm = 450 mm

**6.** 27 dm = 270 cm

**7.** 13.6 km = 13 600 m

**8.** 4.38 km = 4380 m

**9.** 18.25 m = 182.5 dm

**10.** 60 m = 600 dm

**11.** 4.8 mm = 0.48 cm

**12.** 36 mm = 3.6 cm

**13.** 6.5 cm = 0.065 m

**14.** 82 cm = 0.82 m

**15.** 1300 mm = 1.3 m

**16.** 750 mm = 0.75 m

17. [ Respond in exercise. ]

18. [ Respond in exercise. ]

19. [ Respond in exercise. ]

20. [ Respond in exercise. ]

21. 0.245 m

22. 0.23 m

23. 10 km

24. 1500 m

25. 200 m

26. 473 cm

27. 0.32 cm

28. 0.087 m

29. 17 350 mm

30. 140 km

**17.** 5.25 cm = 0.0525 m

**18.** 185 m = 0.185 km

**19.** 5500 m = 5.5 km

**20.** 140 cm = 14 dm

**21.** Change 245 mm to meters.

**22.** Convert 23 cm to meters.

**23.** How many kilometers are in 10 000 m?

**24.** How many meters are in 1.5 km?

**25.** What number of meters is equivalent to 20 000 cm?

**26.** Express 4.73 m in centimeters.

**27.** How many centimeters are in 3.2 mm?

**28.** Change 87 mm to meters.

**29.** Express 17.35 m in millimeters.

**30.** What number of kilometers is equivalent to 140 000 m?

Name ______________________ Section ______________ Date ____________

Use Table 10.3 as a reference to convert the following measures of length as indicated. (**Note:** Remember that use of the approximate equivalents in the table may lead to more than one approximate conversion.) Use 3-digit accuracy in your answers.

**31.** 3 in. = 7.62 cm

**32.** 3 ft. = 91.5 cm

**33.** 4 yd. = 3.66 m

**34.** 30 mi. = 48.3 km

**35.** 100 cm = 39.4 in.

**36.** 150 cm = 4.93 ft.

**37.** 67 m = 220 ft.

**38.** 8 km = 4.96 mi

**39.** 20 km = 12.4 mi.

**40.** 5 m = 197 in.

**41.** How many inches are in 6 m?

**42.** How many centimeters are in 6 ft.?

**43.** Convert 300 cm to feet.

**44.** Convert 300 km to miles.

**45.** Change 7.6 m to inches.

**46.** Change 1.5 ft. to cm.

**47.** Match each formula for perimeter to its corresponding geometric figure.

B **a.** square — A. $P = 2l + 2w$

C **b.** parallelogram — B. $P = 4s$

D **c.** circle — C. $P = 2b + 2a$

A **d.** rectangle — D. $C = 2\pi r$

F **e.** trapezoid — E. $P = a + b + c$

E **f.** triangle — F. $P = a + b + c + d$

**ANSWERS**

For Exercises 31-40, respond in exercise.

**41.** 236 in.

**42.** 183 cm

**43.** 9.84 ft. (or 9.85 ft.)

**44.** 186 mi.

**45.** 299 in.

**46.** 45.8 cm

**47.** [ Respond in exercise. ]

48. $P$ = 18 cm

$P$ = 7.09 in.

49. $P$ = 104 mm

$P$ = 4.10 in.

50. $C$ = 31.4 ft.

$C$ = 9.58 m

51. $C$ = 18.8 yd.

$C$ = 17.2 m

52. $P$ = 0.12 m

$P$ = 12 cm

$P$ = 120 mm

**48.** Find the perimeter of a triangle with sides of 4 cm, 8 cm, and 6 cm, in both centimeters and inches.

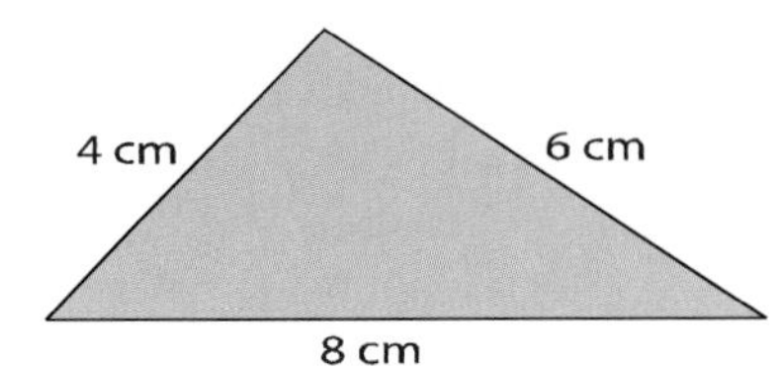


**49.** Find the perimeter of a rectangle with length 35 mm and width 17 mm, in both millimeters and inches.

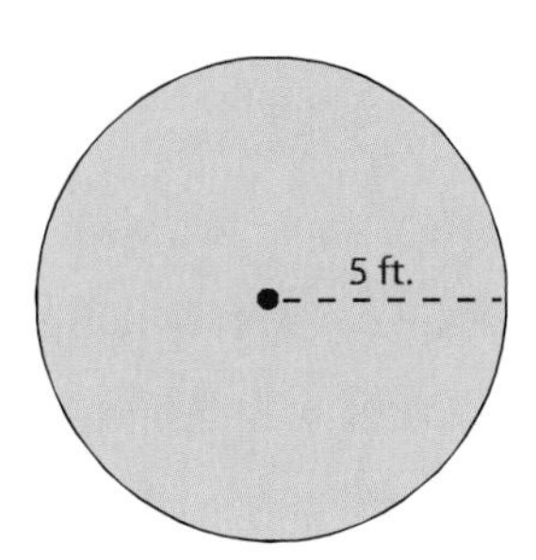


**50.** Find the circumference of a circle with a radius 5 ft., in both feet and meters.

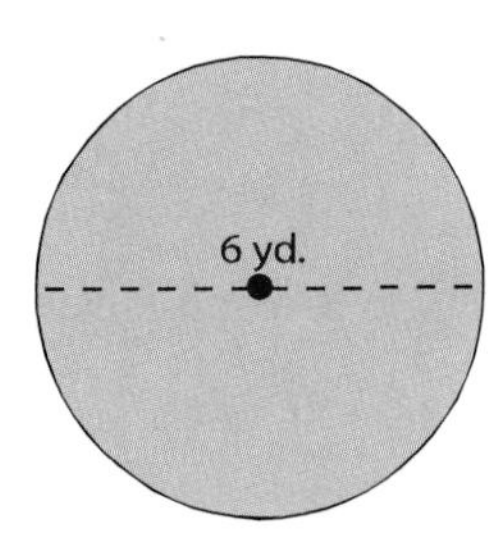


**51.** Find the circumference of a circle with diameter 6 yd., in both yards and meters.

6 yd.

**52.** Find the perimeter of a triangle with sides of 5 cm, 40 mm, and 0.03 m. Write the answer in equivalent measures in meters, centimeters, and millimeters.

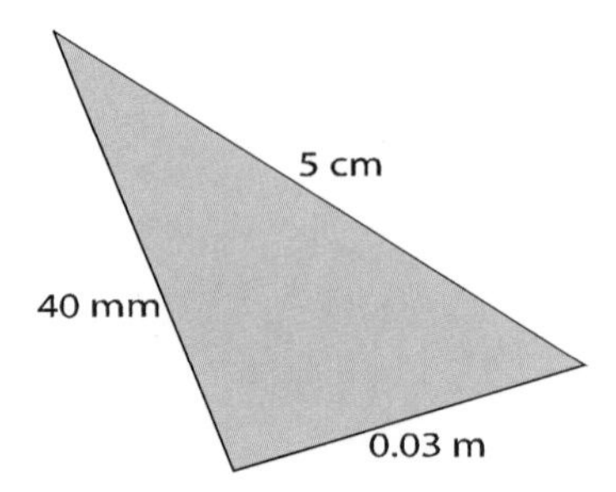

Name ____________________ Section ____________ Date ____________

Find the perimeter of each of the following figures with indicated dimensions. Write the answers in equivalent measures in centimeters, millimeters, and inches regardless of the given units.

**53.**

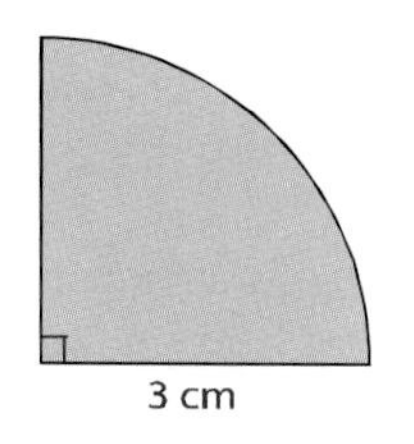


**54.**

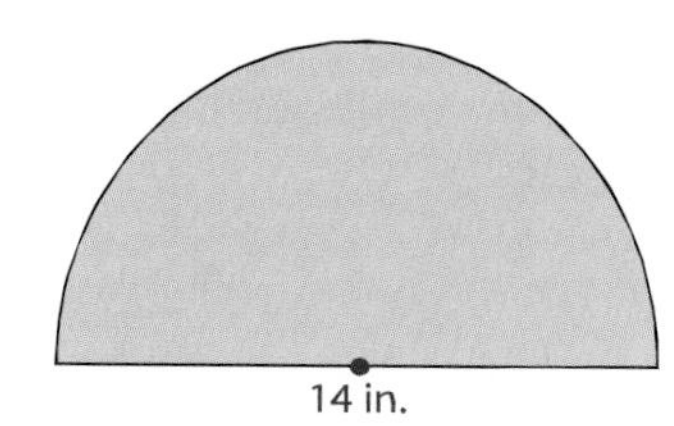


**55.**

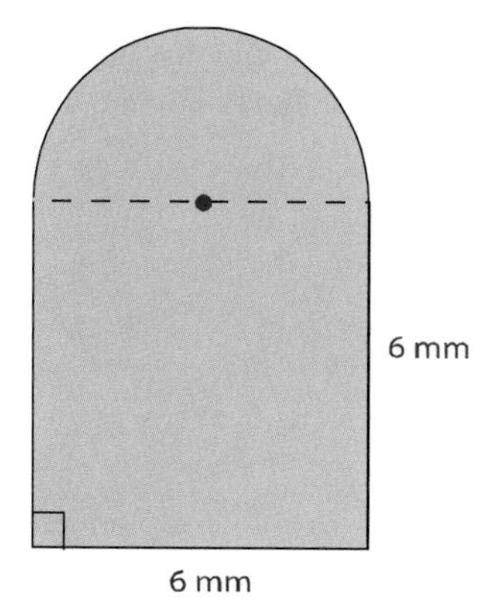


**56.**

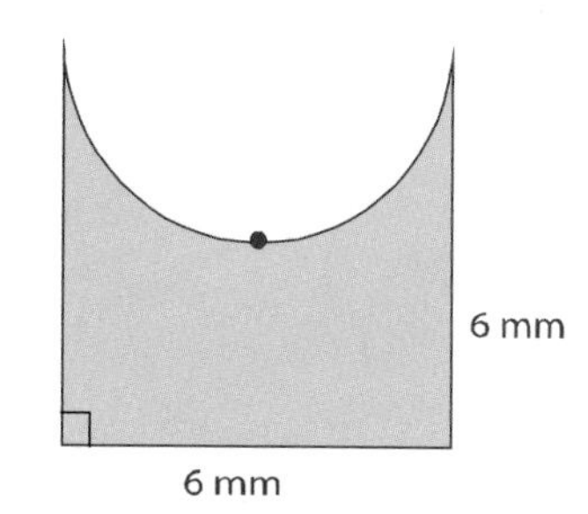


**57.**

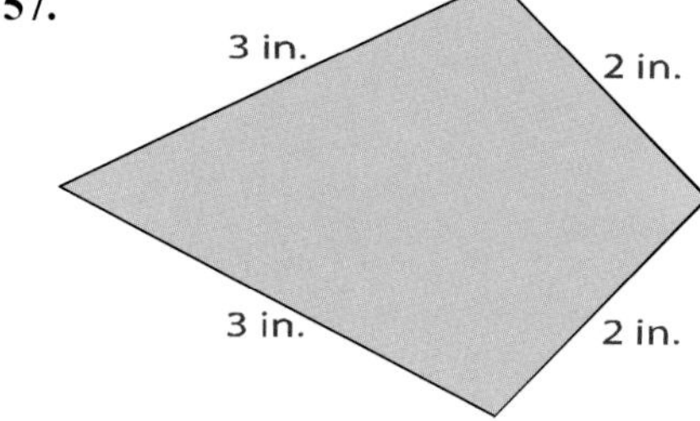


**58.**

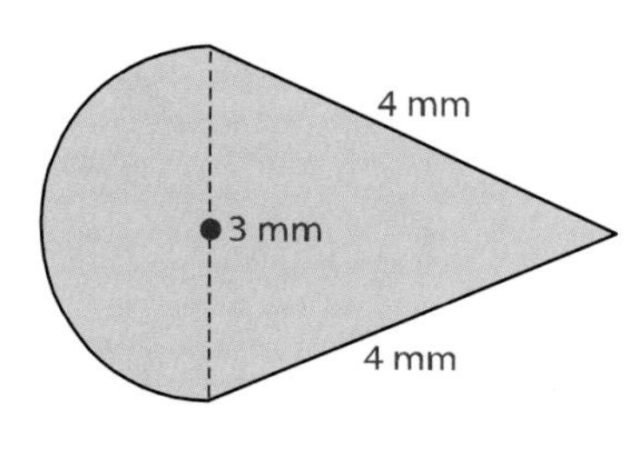


**59.**

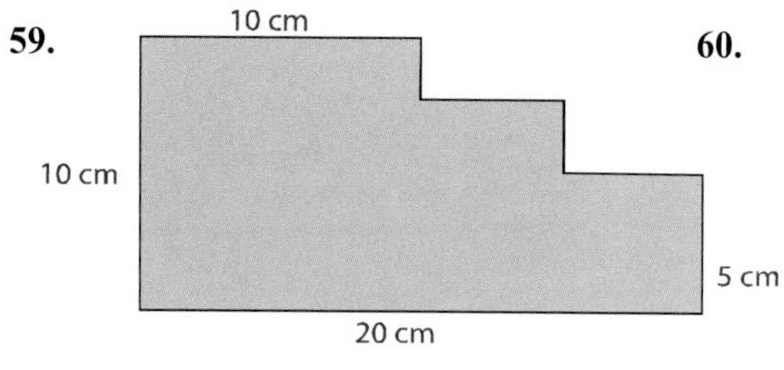


**60.**

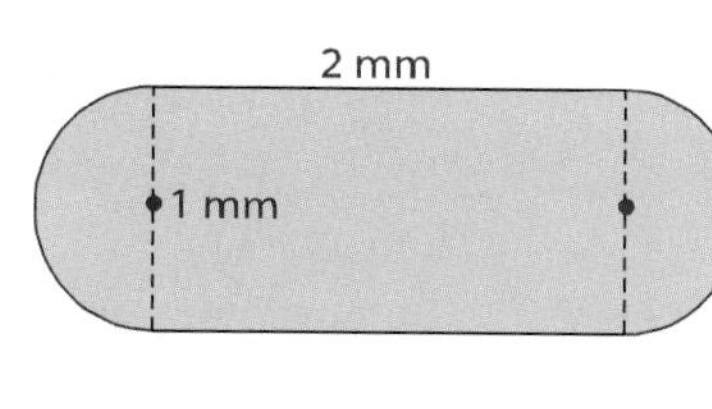




**ANSWERS**

**53.** $P = 107$ mm

$P = 10.7$ cm

$P = 4.22$ in.

**54.** $P = 914$ mm

$P = 91.4$ cm

$P = 36.0$ in.

**55.** $P = 27.4$ mm

$P = 2.74$ cm

$P = 1.08$ in.

**56.** $P = 27.4$ mm

$P = 2.74$ cm

$P = 1.08$ in.

**57.** $P = 254$ mm

$P = 25.4$ cm

$P = 10$ in.

**58.** $P = 12.7$ mm

$P = 1.27$ cm

$P = 0.500$ in.

**59.** $P = 600$ mm

$P = 60$ cm

$P = 23.6$ in.

**60.** $P = 7.14$ mm

$P = 0.714$ cm

$P = 0.281$ in.

61. [ Respond in exercise. ]

## Writing and Thinking about Mathematics

**61.** First draw a circle with a diameter of 10 cm ( radius 5 cm ). Next, draw a square outside the circle so that each side of the square just touches the circle. ( The square is said to be *circumscribed* about the circle, and the circle is said to be *inscribed* in the square. )

**a.** Divide the perimeter of the square by the diameter of the circle. What number did you get?

Next ( on the same figure ) draw an octagon ( 8-sided figure ) that is circumscribed about the circle. Measure the perimeter of the octagon as accurately as you can.

**b.** Divide the perimeter of the octagon by the diameter of the circle. What number did you get?

**c.** Consider drawing circumscribed figures with more sides ( such as a 16-sided figure, a 32-sided figure, and so on ) about the circle. Describe the numbers you think you would get by dividing the perimeters of these figures by the diameter of the circle.

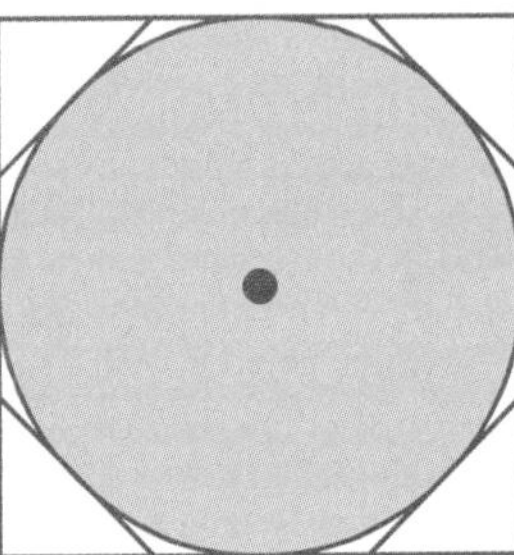

a. 4

b. Answers will vary somewhat, but they should be about 3.3.

c. Answers will vary. The numbers will become closer to pi as the number of sides increases.

# 10.3 Area

## Metric Units of Area

Recall from Section 1.1 that **area** is a measure of the interior, or enclosure, of a surface. For example, the two rectangles in Figure 10.7 have different areas because they have different amounts of interior space. That is, different amounts of space are enclosed by the sides of the figures.

These two rectangles have different **areas**.

**Figure 10.7**

Also recall that area is measured in **square units.** In metric units, a square that is 1 centimeter long on each side is said to have an area of 1 square centimeter ( or **1 $cm^2$**). Figure 10.8 illustrates the concept of area with a rectangle of area of 28 $cm^2$.

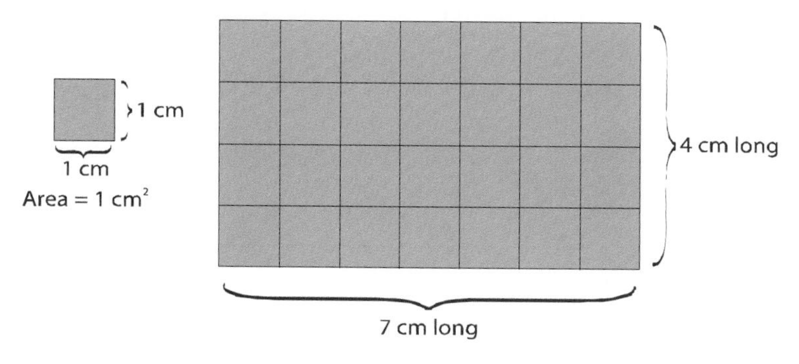


Area = 4cm × 7cm = 28$cm^2$

These are 28 squares that are each 1$cm^2$ in the large rectangle.

**Figure 10.8**

**Objectives**

1. Recognize measures of area in the metric system.
2. Be able to change metric measures of area within the metric system.
3. Know how to use tables to convert units of area between the metric and U.S. customary systems.
4. Know the formulas for the area of several geometric figures.

You might want to discuss the difference between certain terms of measure such as 5 square centimeters (5 $cm^2$) and a 5-centimeter square. A 5-centimeter square contains 25 $cm^2$.

Table 10.4 shows metric area measures useful for relatively small areas. For example, the area of a tennis court might be measured in square meters, while the area of this page of paper might be measured in square centimeters. Other measures, listed in Table 10.5, are used for measuring land.

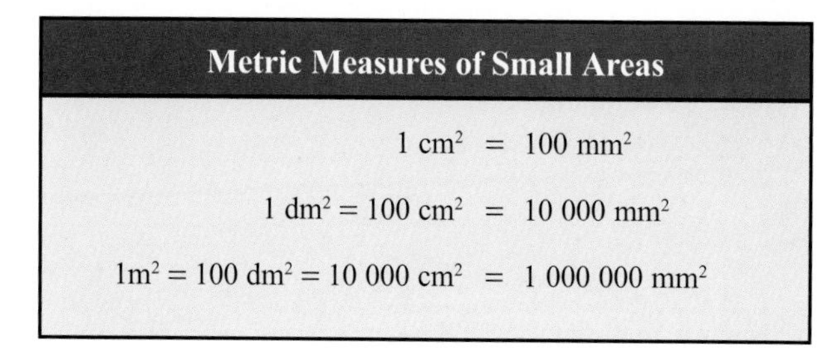

| Metric Measures of Small Areas |
|---|
| 1 $cm^2$ = 100 $mm^2$ |
| 1 $dm^2$ = 100 $cm^2$ = 10 000 $mm^2$ |
| 1$m^2$ = 100 $dm^2$ = 10 000 $cm^2$ = 1 000 000 $mm^2$ |

**Table 10.4**

Note that each unit of area in the metric system is 100 times the next smaller unit of area – not just 10 times, as it is with length. Also, remember, to multiply by 100, move the decimal point two places **to the right**; to multiply by 10 000, move the decimal point four places **to the right**, and so on. Remember that to divide by 100 move the decimal point two places **to the left**; to divide by 10 000, move the decimal point four places **to the left**; and so on.

**Changing Metric Measures of Area**

| Procedure | Example |
|---|---|
| To change to a measure of area that is | |
| one unit smaller, multiply by 100, | 8.2 $cm^2$ = 820 $mm^2$ |
| two units smaller,multiply by 10 000, | 3.56 $m^2$ = 35 600 $cm^2$ |
| three units smaller, multiply by 1 000 000, | 43 $m^2$ = 43 000 000$m^2$ |
| and so on. | |
| To change to a measure of area that is | |
| one unit larger, divide by 100, | 75 $mm^2$ = 0.75 $cm^2$ |
| two units larger, divide by 10 000, | 52 $cm^2$ = 0.0052 $m^2$ |
| three units larger, divide by 1 000 000, | 6500 $m^2$ = 0.0065 $km^2$ |
| and so on. | |

### Example 1

**a.** A square 1 centimeter on each side encloses 100 square millimeters:
$1 \text{ cm}^2 = 100 \text{ mm}^2$

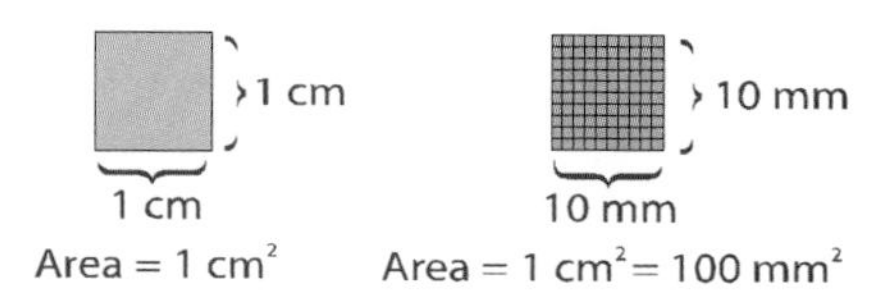


**b.** A square 1 decimeter (10 cm) on a side encloses 1 square decimeter:
$1 \text{ dm}^2 = 100 \text{ cm}^2 = 10\ 000 \text{ mm}^2$

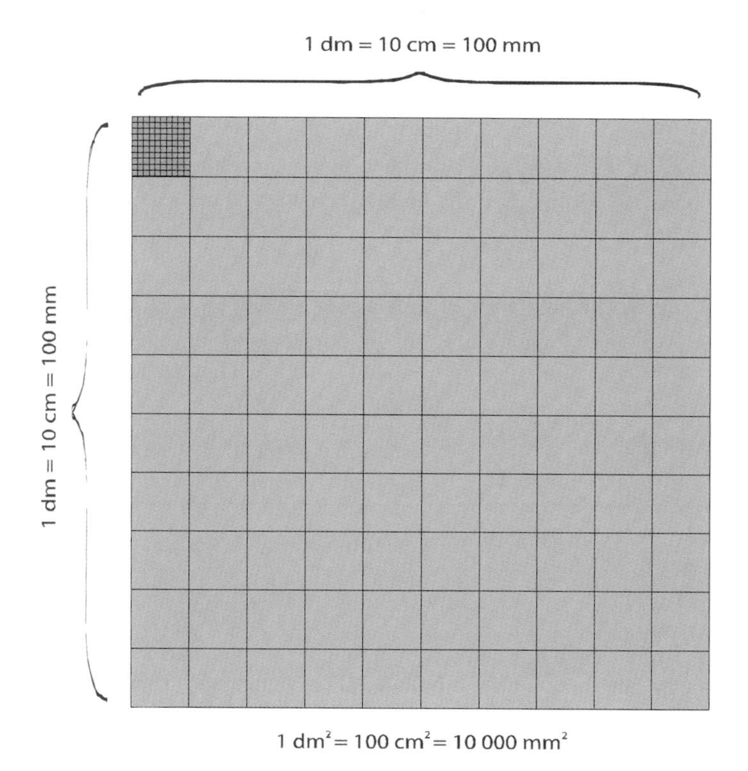


As a class activity, you might have several students come to the board and draw what they think is a square with area 1 m² and include a square decimeter and a square centimeter. Then have them use a meter stick to draw the squares with more accuracy.

### Example 2

The following examples illustrate how to change from larger to smaller units of area in the metric system.

**a.** $5 \text{ cm}^2 = 5(1 \text{ cm}^2) = 5(100 \text{ mm}^2) = 500 \text{ mm}^2$
(Note: mm² is one unit smaller than cm².)

**b.** $3 \text{ dm}^2 = 3(1 \text{ dm}^2) = 3(10\ 000 \text{ mm}^2) = 30\ 000 \text{ mm}^2$
(Note: mm² is two units smaller than dm².)

**c.** $9.8 \text{ km}^2 = 9.8(1 \text{ km}^2) = 9.8(1\ 000\ 000 \text{ m}^2) = 9\ 800\ 000 \text{ m}^2$
(Note: m² is three units smaller than km².)

## Example 3

The following examples illustrate how to change from smaller to larger units of area in the metric system.

**a.** $1.4 \text{ m}^2 = \dfrac{1.4}{100} \text{ dam}^2 = 0.014 \text{ dam}^2$

( Note: $\text{dam}^2$ is one unit larger than $\text{m}^2$ . )

**b.** $3500 \text{ mm}^2 = \dfrac{3500}{100} \text{ cm}^2 = 35 \text{ cm}^2$

( Note: $\text{cm}^2$ is one unit larger than $\text{mm}^2$ . )

**c.** $3500 \text{ mm}^2 = \dfrac{3500}{1\,000\,000} \text{ m}^2 = 0.0035 \text{ m}^2$

( Note: $\text{m}^2$ is three units larger than $\text{mm}^2$ . )

A square with each side 10 meters long encloses an area of 1 **are** ( **a** ). A **hectare** ( **ha** ) is 100 ares. The are and hectare are used to measure land area in the metric system.

| Metric Measurements of Land Area |
| --- |
| $1 \text{ a} = 100 \text{ m}^2$ |
| $1 \text{ ha} = 100\text{a} = 10\,000 \text{ m}^2$ |

**Table 10.5**

## Example 4

How many ares are in 1 $\text{km}^2$? ( **Note:** For comparison, 1 km is about 0.6 mile, so 1 $\text{km}^2$ is about 0.6 mi. × 0.6 mi. = 0.36 $\text{mi.}^2$. )

### Solution

Remember that 1 km = 1000m, so

$$\begin{aligned} 1 \text{ km}^2 &= 1000 \text{ m} \times 1000 \text{ m} \\ &= 1\,000\,000 \text{ m}^2 \\ &= 10\,000 \text{ a} \qquad \text{Divide m}^2 \text{ by 100 to get ares.} \end{aligned}$$

*Now Work Exercises 1 - 4 in the Margin.*

Change the units as indicated.

**1.** $23 \text{ cm}^2 =$ 2300 $\text{mm}^2$

**2.** $5200 \text{ mm}^2 =$ 52 $\text{cm}^2$

= 0.52 $\text{dm}^2$

**3.** $3.6 \text{ a} =$ 360 $\text{m}^2$

**4.** $49\,500 \text{ m}^2 =$ 495 a

= 4.95 ha

## U.S. Customary and Metric Equivalent Units of Area

As with units of length, the units of area in the U.S. customary system are not systematically related, as are the units of area in the metric system. Table 10.6 shows some of the basic units of area in the U.S. customary system.

| U.S. Customary Units of Area |
|---|
| 1 ft.$^2$ = 144 in.$^2$ |
| 1 yd.$^2$ = 9 ft.$^2$ |
| 1 acre = 4840 yd.$^2$ = 43,560 ft.$^2$ |

**Table 10.6**

We will use the equivalent measures ( rounded off ) in Table 10.7 to convert units of area from the U.S. customary system to the metric system, and vice versa. Examples 5 and 6 illustrate how to make conversions using these equivalents and multiplying by the given number of units.

| U.S. Customary and Metric Area Equivalents | |
|---|---|
| **U.S. to Metric** | **Metric to U.S.** |
| 1 in.$^2$ = 6.45 cm$^2$ | 1 cm$^2$ = 0.155 in.$^2$ |
| 1 ft.$^2$ = 0.093 m$^2$ | 1 m.$^2$ = 10.764 ft.$^2$ |
| 1 yd.$^2$ = 0.836 m$^2$ | 1 m$^2$ = 1.196 yd.$^2$ |
| 1 acre = 0.405 ha | 1 ha = 2.47 acres |

**Table 10.7**

**Important Note About Conversions between U.S. and Metric Equivalents**

As stated in Section 10.2, most of the conversion units are not exact and slightly different answers are possible, depending on the conversion units used. We will give area conversions rounded off to the place of the first four digits only.

### Example 5

**a.** 40 yd.$^2$ = _____ m$^2$

**b.** How many hectares are there in 5 acres?

#### Solution

**a.** 40 yd.$^2$ = 40( 1 yd.$^2$ ) = 40( 0.836 m$^2$ ) = 33.44 m$^2$

**b.** 5 acres = 5( 1 acre ) = 5( 0.405 ha ) = 2.025 ha

Convert the measures as indicated.

**5.** $6 \text{ in.}^2 = \underline{38.70}$ $\text{cm}^2$

**6.** $625 \text{ ft}^2. = \underline{58.13}$ $\text{m}^2$

**7.** $50 \text{ m}^2 = \underline{538.2}$ $\text{ft.}^2$

**8.** $3 \text{ ha} = \underline{7.410}$ acres

### Example 6

**a.** Convert 5 hectares to acres.

**b.** Change 100 $\text{cm}^2$ to square inches.

#### Solution

**a.** $5 \text{ ha} = 5(\ 1 \text{ ha}\ ) = 5(\ 2.47 \text{ acres}\ ) = 12.35 \text{ acres}$

**b.** $100 \text{ cm}^2 = 100(\ 1 \text{ cm}^2\ ) = 100(\ 0.155 \text{ in.}^2) = 15.5 \text{ in.}^2$

*Now Work Exercises 5-8 in the Margin.*

## Geometry: Formulas for Area

The following formulas for the areas of the geometric figures shown are valid regardless of the system of measurement used. **Be able to label your answers with the correct units.** Labeling the units is a good way, particularly in word problems, to remind yourself whether you are dealing with length, area, or volume.

**Note:** In triangles and other figures, the letter $h$ is used to represent the **height** of the figure. The height is also called the **altitude.**

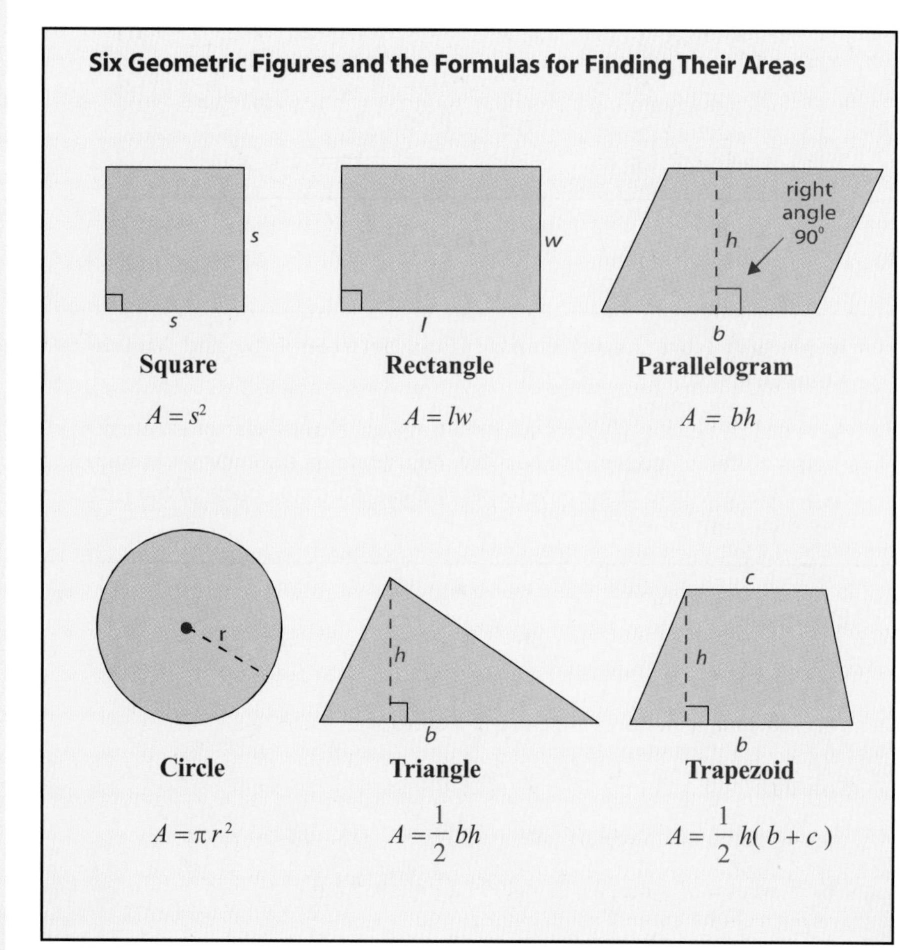

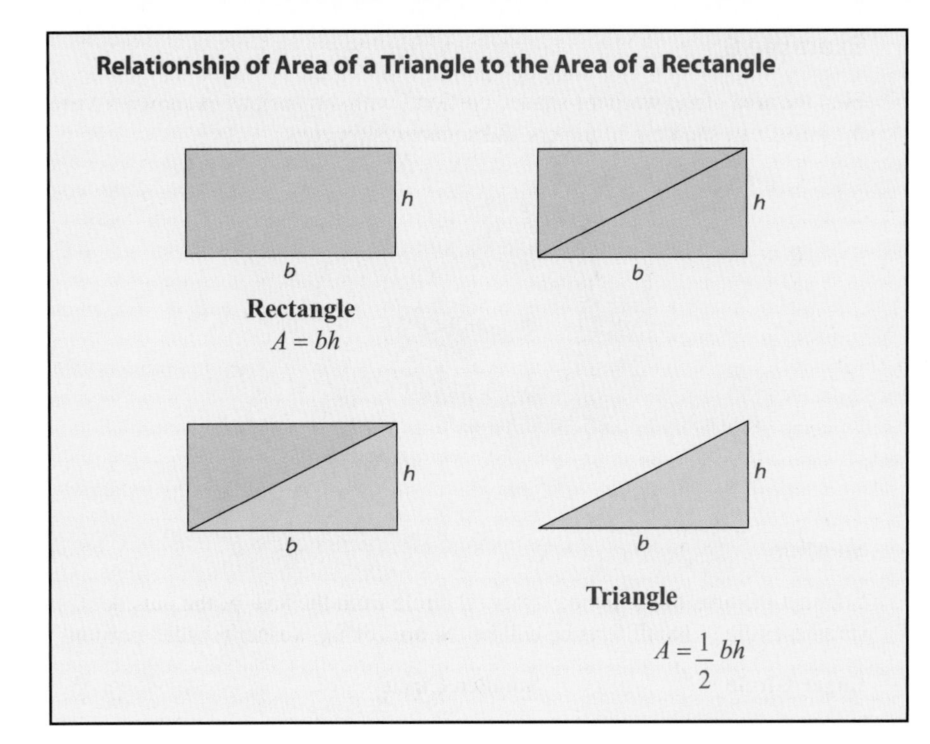


To help the students understand the formula of the triangle to the area of a rectangle, draw several rectangles of height $h$ and length $b$. Show with dotted lines that the altitude is always the same (the length of the side labeled $h$) and that the area is one-half of the area of the rectangle. You can have them "cut out" the triangle in a piece of paper and lay the remainder pieces of the rectangle onto the triangle to show the area concepts.

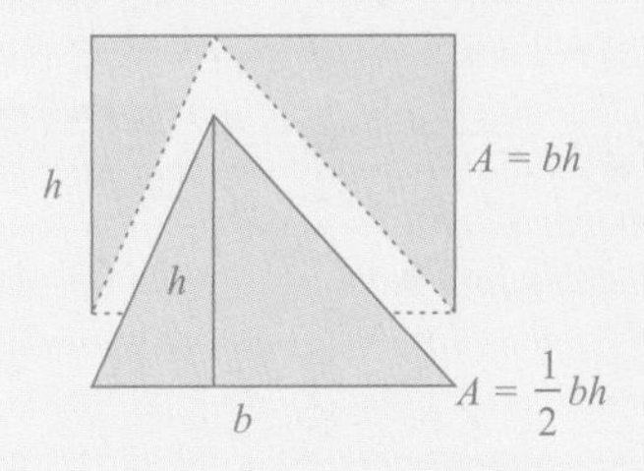


**OR**

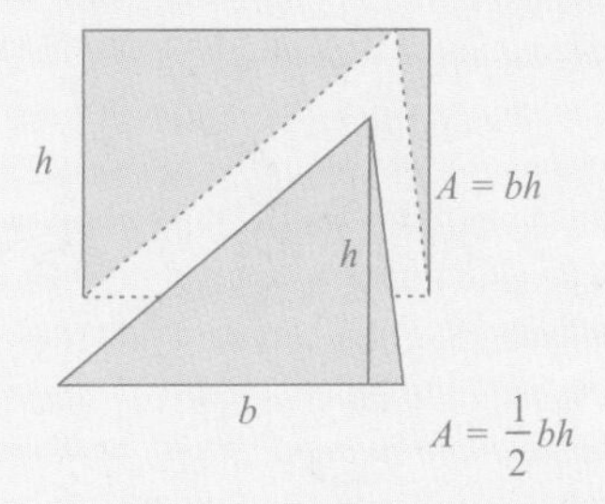


## Example 7

Find the area of the figure shown here with the indicated dimensions. Write the answer in both square centimeters and square inches.

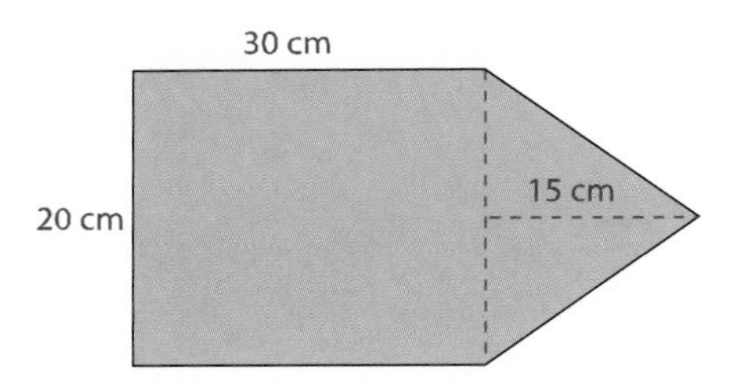


### Solution

Find the area of the rectangle and the area of the triangle, and then add the two areas.

**Rectangle**

$A = lw$

$A = 20 \cdot 30$

$= 600 \text{ cm}^2$

**Triangle**

$A = \frac{1}{2} bh$

$A = \frac{1}{2} \cdot 20 \cdot 15$

$= 150 \text{ cm}^2$

Total area $= 600 \text{ cm}^2 + 150 \text{ cm}^2 = 750 \text{ cm}^2$

In square inches:

$750 \text{ cm}^2 = 750(\ 1 \text{ cm}^2\ ) = 750(\ 0.155 \text{ in.}^2) = 116.25 \text{ in.}^2$

$= 116.3 \text{ in.}^2$ ( rounded off )

## Example 8

Find the area of the washer ( shaded portion ) with dimensions as shown. Write the answer in square millimeters and square centimeters.

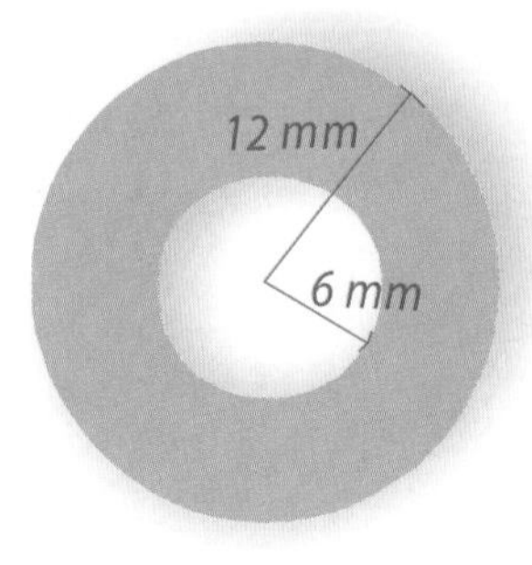


### Solution

Subtract the area of the inside (smaller) circle from the area of the outside (larger) circle. This difference will be the area of the washer (shaded portion).

| **Lager Circle** | **Smaller Circle** |
| --- | --- |
| $A = \pi r^2$ | $A = \pi r^2$ |
| $A = 3.14(12)^2$ | $A = 3.14(6)^2$ |
| $= 3.14(144)$ | $= 3.14(36)$ |
| $= 452.16 \text{ mm}^2$ | $= 113.04 \text{ mm}^2$ |

Area of shaded portion = $452.16 \text{ mm}^2 - 113.04 \text{ mm}^2 = 339.12 \text{ mm}^2$

In square centimeters:

$$339.12 \text{ mm}^2 = \frac{339.12}{100} \text{ cm}^2 = 3.3912 \text{ cm}^2 = 3.391 \text{ cm}^2 \text{ ( rounded off )}$$

An alternate approach is to leave $\pi$ in the first answers and substitute later as follows:

$$A = \pi(12)^2 = 144\pi \text{ mm}^2 \quad \text{and} \quad A = \pi(6)^2 = 36\pi \text{ mm}^2$$

$$\text{Total area} = (144\pi - 36\pi)\text{mm}^2 = 108\pi \text{ mm}^2$$
$$= 108(3.14) \text{ mm}^2$$
$$= 339.12 \text{ mm}^2 = 339.1 \text{ mm}^2 \text{ ( rounded off )}$$

This approach is more algebraic in nature.

### Example 9

Find the area enclosed by a semicircle and a diameter if the diameter is 12 feet long. Write the answer in both square feet and square meters.

#### Solution

First, sketch the figure. ( A semicircle is half of a circle. )

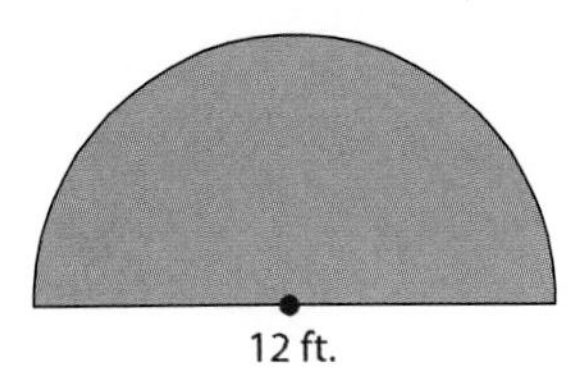


For a circle, $A = \pi r^2$. So, for a semicircle, $A = \frac{1}{2}\pi r^2$

For this semicircle, $d = 12$ ft., so $r = 6$ ft.

$$A = \frac{1}{2}\pi \cdot 6^2 = \frac{1}{2}(3.14)(36) = 56.52 \text{ ft.}^2$$

In square meters:

$$56.52 \text{ ft.}^2 = 56.52(0.093 \text{ m}^2) = 5.256 \text{ m}^2 \text{ ( rounded off )}$$

Name ______ Section ______ Date ______

ANSWERS

For Exercises 1-14, respond in exercise.

## Exercises 10.3

Change the following units of area in the metric system as indicated.

**1.** 13 $cm^2$ = 1300 $mm^2$

**2.** 6.5 $cm^2$ = 650 $mm^2$

**3.** 9.6 $cm^2$ = 960 $mm^2$

**4.** 4.52 $cm^2$ = 452 $mm^2$

**5.** 500 $mm^2$ = 5 $cm^2$

**6.** 39 $mm^2$ = 0.39 $cm^2$

**7.** 14 $dm^2$ = 1400 $cm^2$

**8.** 19.5 $dm^2$ = 1950 $cm^2$

**9.** 0.5 $dm^2$ = 5000 $mm^2$

**10.** 3 $dm^2$ = 30 000 $mm^2$

**11.** 13 $dm^2$ = 1300 $cm^2$ = 130 000 $mm^2$

**12.** 6.4 $dm^2$ = 640 $cm^2$ = 64 000 $mm^2$

**13.** 11.5 $m^2$ = 1150 $dm^2$ = 115 000 $cm^2$ = 11 500 000 $mm^2$

**14.** 3.6 $m^2$ = 360 $dm^2$ = 36 000 $cm^2$ = 3 600 000 $mm^2$

15. [ Respond in exercise. ]

16. [ Respond in exercise. ]

17. [ Respond in exercise. ]

18. [ Respond in exercise. ]

19. [ Respond in exercise. ]

20. [ Respond in exercise. ]

21. 0.02 ha

22. 1.5 ha

23. 575 ha

24. 40 ha

25. 75 000 a

26. [ Respond in exercise. ]

27. [ Respond in exercise. ]

**15.** $0.04\ m^2$ = 4 $dm^2$ = 400 $cm^2$ = 40 000 $mm^2$

**16.** $0.6\ m^2$ = 60 $dm^2$ = 6000 $cm^2$ = 600 000 $mm^2$

**17.** 6.7 a = 670 $m^2$

**18.** 0.45 a = 45 $m^2$

**19.** 200 a = 20 000 $m^2$

**20.** 0.8 a = 80 $m^2$

**21.** How many hectares are in 2 a?

**22.** How many hectares are in 150 a?

**23.** Change 5.75 $km^2$ to hectares.

**24.** Change 0.4 $km^2$ to hectares.

**25.** How many ares are in 750 ha?

**26.** 9.56 ha = 956 a = 95 600 $m^2$

**27.** 0.27 ha = 27 a = 2700 $m^2$

Name ______________________ Section ____________ Date __________

**ANSWERS**

For Exercises 28-42, respond in exercise.

43. 494 acres

44. 988 acres

45. 269.1 ft.$^2$

46. 4.65 in.$^2$

**28.** 35 km$^2$ = 350 000 a = 3500 ha

**29.** 6.25 km$^2$ = 62 500 a = 625 ha

**30.** 15 km$^2$ = 15 000 000 m$^2$

Use Table 10.7 as a reference to convert the following measures of area as indicated.

**31.** 4 in.$^2$ = 25.8 cm$^2$

**32.** 18 in.$^2$ = 116.1 cm$^2$

**33.** 500 ft.$^2$ = 46.5 m$^2$

**34.** 200 ft.$^2$ = 18.6 m$^2$

**35.** 10 yd.$^2$ = 8.36 m$^2$

**36.** 230 yd.$^2$ = 192.3 m$^2$

**37.** 15 m$^2$ = 161.5 ft.$^2$

**38.** 2.6 m$^2$ = 27.99 ft.$^2$

**39.** 14 cm$^2$ = 2.17 in.$^2$

**40.** 3.25 cm$^2$ = 0.5038 in.$^2$

**41.** 2000 acres = 810 ha

**42.** 350 acres = 141.8 ha

**43.** How many acres are in 200 ha?

**44.** How many acres are in 400 ha?

**45.** Change 25 m$^2$ to square feet.

**46.** Change 30 cm$^2$ to square inches.

47. [ Respond in exercise. ]

48. $A = 875 \text{ in.}^2$

$A = 5644 \text{ cm}^2$

49. $A = 16 \text{ cm}^2$

$A = 2.48 \text{ in.}^2$

50. $A = 78.5 \text{ yd}^2$

$A = 65.63 \text{ m}^2$

51. $A = 38.5 \text{ mm}^2$

$A = 0.385 \text{ cm}^2$

**47.** Match each formula for or to its corresponding geometric figure.

| | | |
|---|---|---|
| C **a.** square | | A. $A = lw$ |
| B **b.** parallelogram | | B. $A = bh$ |
| D **c.** circle | | C. $A = s^2$ |
| A **d.** rectangle | | D. $A = \pi r^2$ |
| F **e.** trapezoid | | E. $A = \frac{1}{2}bh$ |
| E **f.** triangle | | F. $A = \frac{1}{2}h(b+c)$ |

**48.** Find the area of a rectangle 35 inches long and 25 inches wide, in both square inches and square centimeters.

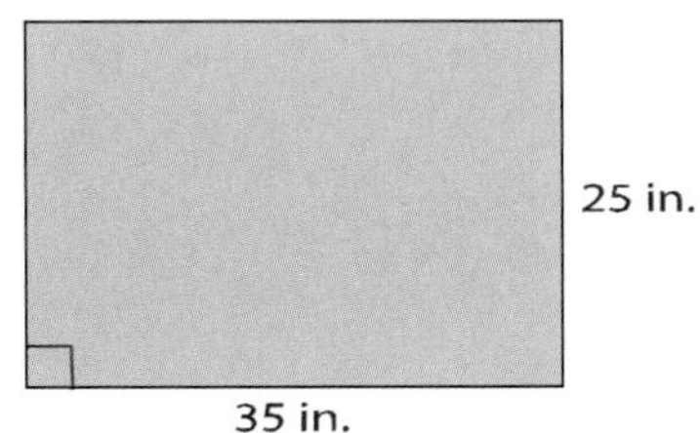


**49.** Find the area of a triangle with base 4 centimeters and altitude 8 centimeters, in both square centimeters and square inches.

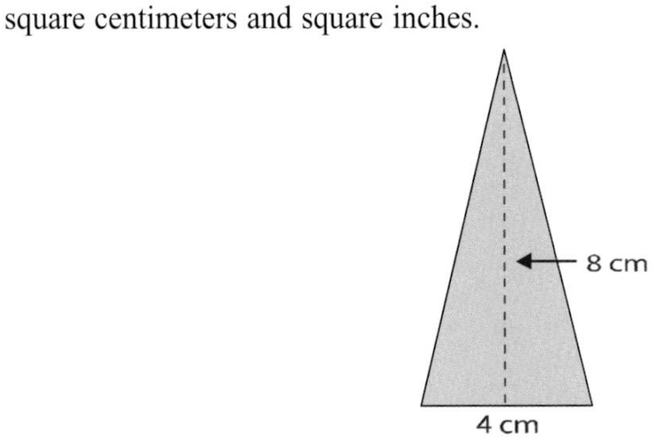


**50.** Find the area of a circle with radius 5 yards, in both square yards and square meters.

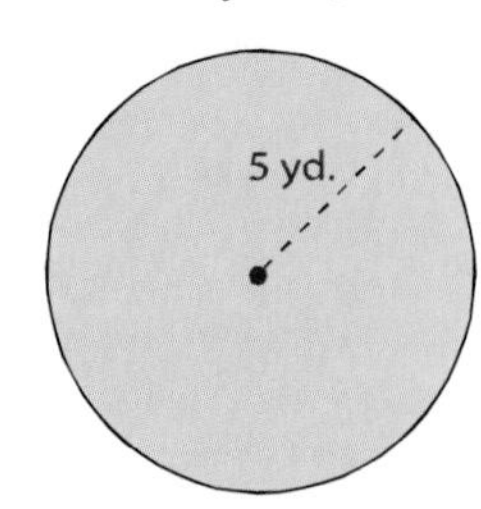


**51.** Find the area of a trapezoid with parallel sides of 3.5 mm and 4.2 mm and altitude of 10 mm, in both square millimeters and square centimeters.

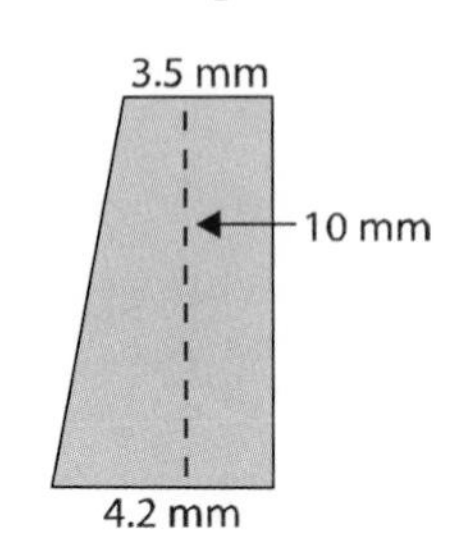

Name ______________________ Section ______________ Date ____________

**52.** Find the area of a parallelogram with height 1 m and base 50 cm, in both square meters and square centimeters.

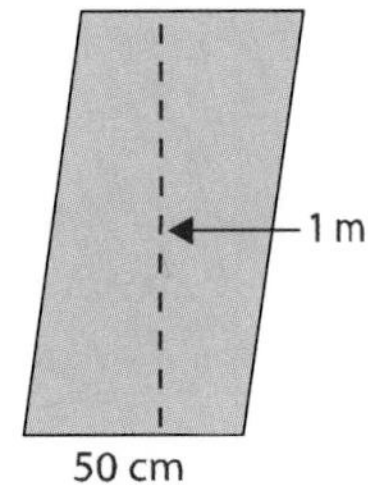


Find the area of each of the following figures with the indicated dimensions. Write the answers in equivalent measures in square centimeters, square millimeters, and square inches, regardless of the given units.

**53.**

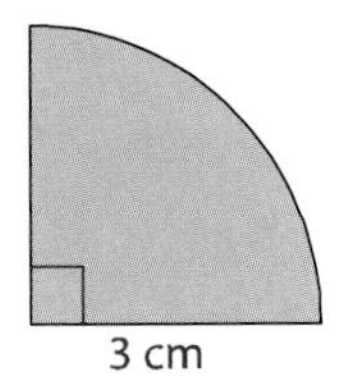


**54.**

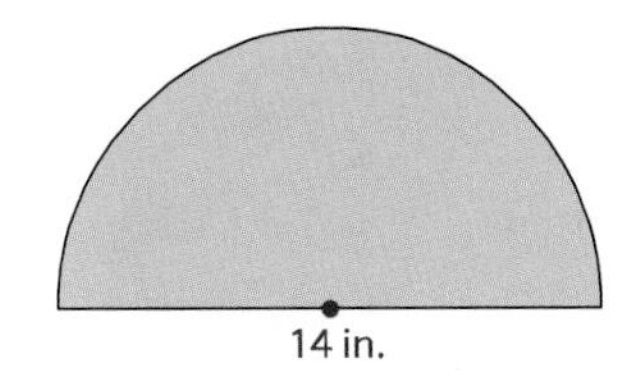


**55.**

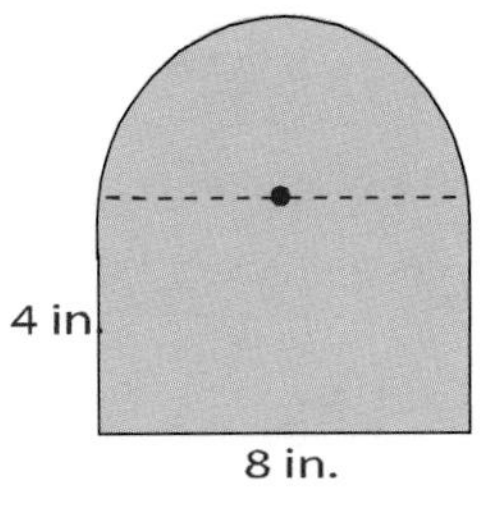


**56.**

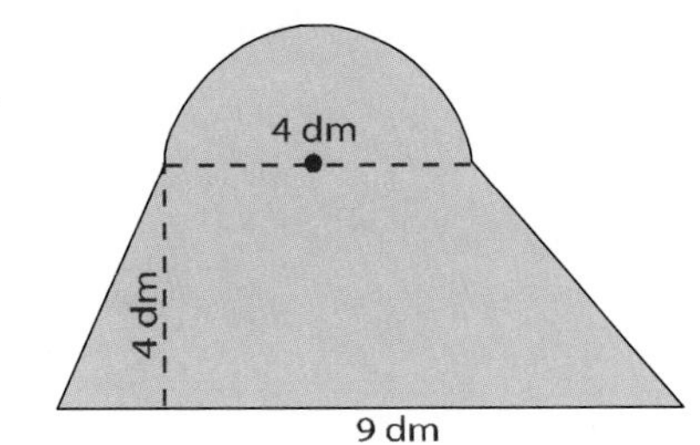


**57.**

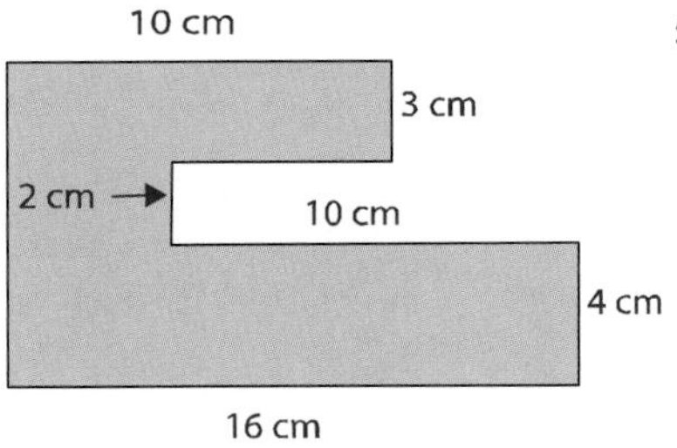


**58.**

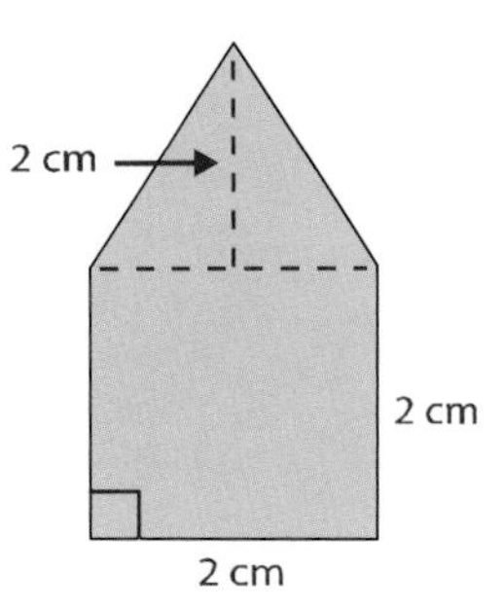


**59.**

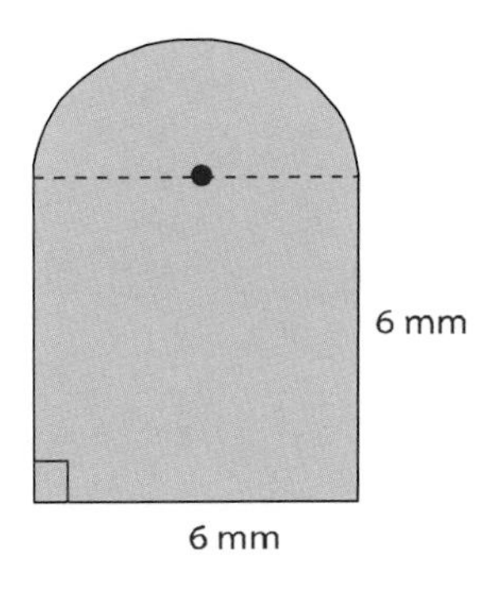


**60.**

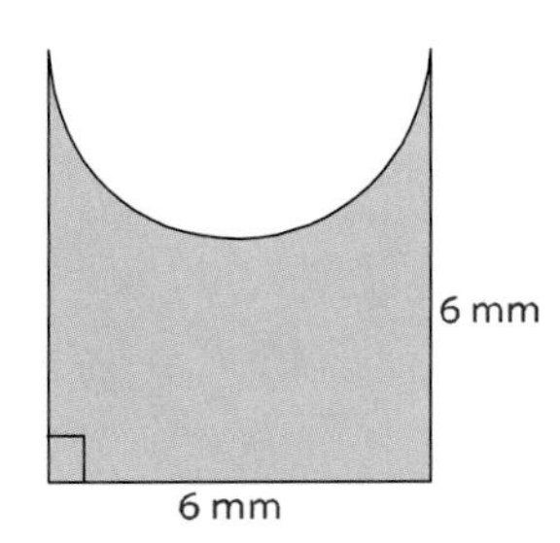


**ANSWERS**

**52.** $A = 0.5 \text{ m}^2$
$A = 5000 \text{ cm}^2$

**53.** $A = 706.5 \text{ mm}^2$
$A = 7.065 \text{ cm}^2$
$A = 1.095 \text{ in.}^2$

**54.** $A = 49\,620 \text{ mm}^2$
$A = 496.2 \text{ cm}^2$
$A = 76.93 \text{ in.}^2$

**55.** $A = 36\,840 \text{ mm}^2$
$A = 368.4 \text{ cm}^2$
$A = 57.12 \text{ in.}^2$

**56.** $A = 322\,800 \text{ mm}^2$
$A = 3228 \text{ cm}^2$
$A = 500.3 \text{ in.}^2$

**57.** $A = 10\,600 \text{ mm}^2$
$A = 106 \text{ cm}^2$
$A = 16.43 \text{ in.}^2$

**58.** $A = 600 \text{ mm}^2$
$A = 6 \text{ cm}^2$
$A = 0.93 \text{ in.}^2$

**59.** $A = 50.13 \text{ mm}^2$
$A = 0.5013 \text{ cm}^2$
$A = 0.07770 \text{ in.}^2$

**60.** $A = 21.87 \text{ mm}^2$
$A = 0.2187 \text{ cm}^2$
$A = 0.03390 \text{ in.}^2$

61. $A = 304\ 440\ \text{mm}^2$

$A = 3044.4\ \text{cm}^2$

$A = 472\ \text{in.}^2$

62. $A = 7\ 536\ 000\ 000\ \text{mm}^2$

$A = 75\ 360\ 000\ \text{cm}^2$

$A = 11\ 680\ 800\ \text{in.}^2$

Find the areas of the shaded regions. Write the answers in equivalent measures in square centimeters, square millimeters, and square inches, regardless of the given units.

**61.**

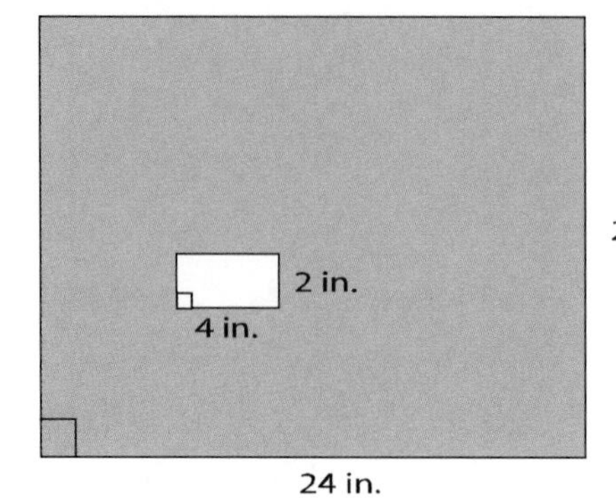


**62.**

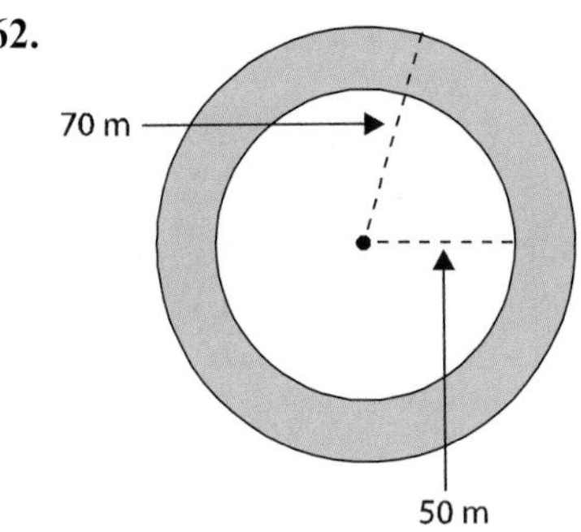

# 10.4 Volume

## Metric Units of Volume

Recall from Section 4.3 that **volume** is a measure of the space enclosed by a three-dimensional figure and is measured in **cubic units.** The volume of space contained within a cube that is 1 centimeter on each edge is **one cubic centimeter** ( or **1 cm³** ), as shown in Figure 10.9. A cubic centimeter is about the size of a sugar cube.

### Objectives

① Recognize measures of volume in the metric system.

② Be able to change metric measures of volume within the metric system.

③ Know how to use tables to convert units of volume between the metric and U.S. customary systems.

④ Know the formulas for the volume of several geometric figures.

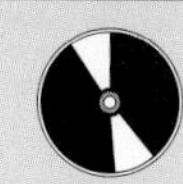

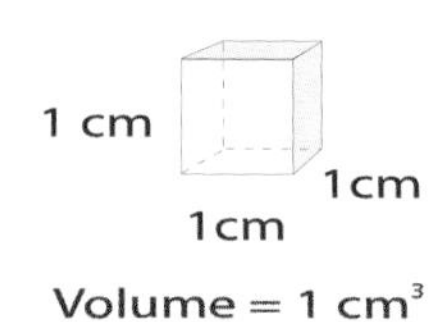


**Figure 10.9**

Figure 10.10 illustrates the concept of volume with a rectangular solid that has edges of length 3 cm, 2 cm, and 5 cm and a volume of 30 cm³.

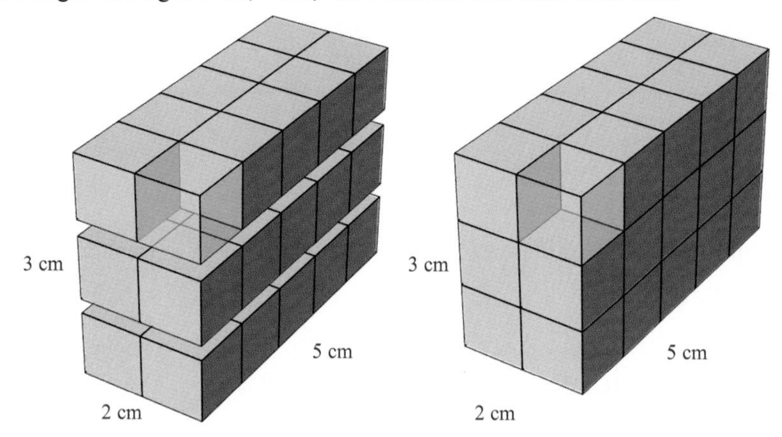


Volume = 30 cm³

**Figure 10.10**

The size of a cubic meter can be surprising to the students. As a model, you might hold three mutually perpendicular meter sticks in the corner of the room. You can point out that this volume filled with water would weigh about 1 metric ton.

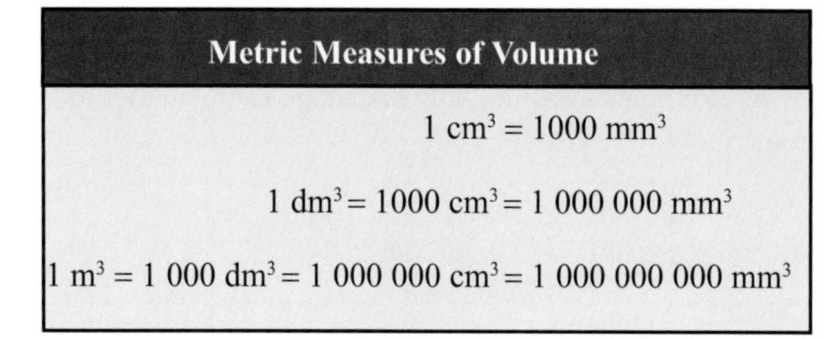

| Metric Measures of Volume |
| --- |
| $1 \text{ cm}^3 = 1000 \text{ mm}^3$ |
| $1 \text{ dm}^3 = 1000 \text{ cm}^3 = 1\,000\,000 \text{ mm}^3$ |
| $1 \text{ m}^3 = 1\,000 \text{ dm}^3 = 1\,000\,000 \text{ cm}^3 = 1\,000\,000\,000 \text{ mm}^3$ |

**Table 10.8**

### Changing Metric Measures of Volume

| Procedure | Example |
|---|---|
| To change to a measure of volume | |
| that is one unit smaller, multiply by 1000, | $6 \text{ cm}^3 = 6000 \text{ mm}^3$ |
| two units smaller, multiply by 1000 000, | $3 \text{ m}^3 = 3\,000\,000 \text{ cm}^3$ |
| three units smaller, multiply by 1 000 000 000, | $12 \text{ m}^3 = 12\,000\,000\,000 \text{mm}^3$ |
| and so on. | |
| To change to a measure of area | |
| that is one unit larger, divide by 1000, | $35 \text{ mm}^3 = 0.035 \text{ cm}^3$ |
| two units larger, divide by 1 000 000, | $250 \text{ cm}^3 = 0.00025 \text{ m}^3$ |
| three units larger, divide by 1 000 000 000, | $32 \text{ m}^3 = 0.000\,000\,032 \text{ km}^3$ |
| and so on. | |

### Example 1

The following examples illustrate how to change from larger to smaller units of volume in the metric system.

**a.** $15 \text{ cm}^3 = 15(1 \text{ cm}^3) = 15(1000 \text{ mm}^3) = 15\,000 \text{ mm}^3$

**b.** $4.1 \text{ dm}^3 = 4.1(1 \text{ dm}^3) = 4.1(1000 \text{ cm}^3) = 4100 \text{ cm}^3$

**c.** $22.6 \text{ m}^3 = 22.6(1 \text{ m}^3) = 22.6(1\,000\,000 \text{ cm}^3) = 22\,600\,000 \text{ cm}^3$

**d.** $4 \text{ m}^3 = 4(1 \text{ m}^3) = 4(1000 \text{ dm}^3) = 4000 \text{ dm}^3 = 4000(1000 \text{ cm}^3)$

$= 4\,000\,000 \text{ cm}^3$

### Example 2

The following examples illustrate how to change from smaller to larger units of volume in the metric system.

**a.** $8 \text{ cm}^3 = \dfrac{8}{1000} \text{ dm}^3 = 0.008 \text{ dm}^3$

**b.** $0.8 \text{ mm}^3 = \dfrac{0.8}{1000} \text{ cm}^3 = 0.0008 \text{ cm}^3$

*Now Work Exercises 1-4 in the Margin.*

Change the units as indicated.

1. $9 \text{ cm}^3 = $ 9000 $\text{mm}^3$
2. $7.5 \text{ m}^3 = $ 7 500 000 $\text{cm}^3$
3. $0.54 \text{ mm}^3 = $ 0.000 54 $\text{cm}^3$
4. $5982 \text{ cm}^3 = $ 0.005 982 $\text{m}^3$

## U.S. Customary and Metric Equivalent Units of Volume

Of particular interest is the volume of a cube that is 1 decimeter along each edge. The volume of such a cube (see Figure 10.11) is **1 cubic decimeter** (or **1 dm³**). In terms of cubic centimeters, this same cube has volume.

$$\mathbf{10\text{ cm} \times 10\text{ cm} \times 10\text{ cm} = 1000\text{ cm}^3}$$

Thus

$$\mathbf{1\text{ dm}^3 = 1000\text{ cm}^3}$$

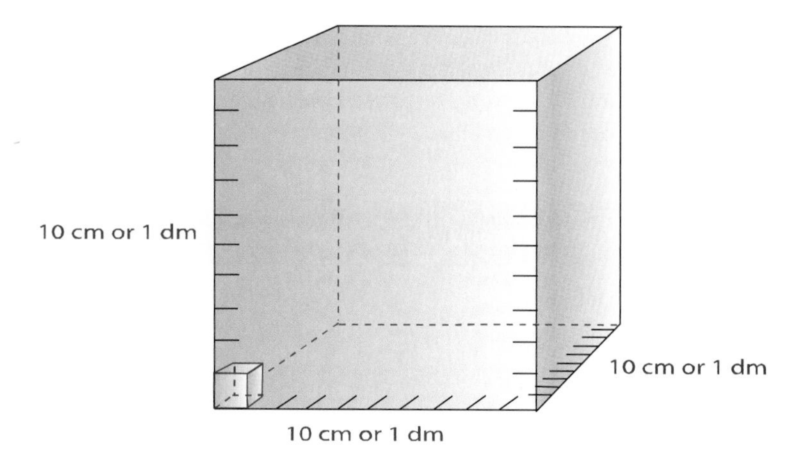


**Figure 10.11**

The **liquid volume** contained in a cube that is 10 cm along each edge ( as illustrated in Figure 10.11 ) is called a **liter** ( **L** ). That is,

$\mathbf{1L = 1000\text{ cm}^3}$ and $\mathbf{1mL = 1\text{ cm}^3}$.

You are probably familiar with 1-liter and 2-liter bottles of soda on your grocer's shelf. Table 10.9 indicates measures of liquid volume and equivalents in the metric system, and Table 10.10 indicates measures of liquid volume in the U.S. customary system.

The fact that 1 mL = 1 cm³ may be confusing to some students. You might have them bring in containers (for example, 1-liter and 2-liter soft drink bottles) and discuss the volume in both milliliters and cubic centimeters. You might show that one liter contains the same volume as a cube that measures 10 cm on each edge.

| Metric Units of Liquid Volume and Equivalents | |
|---|---|
| **Metric Units** | **Equivalents** |
| 1 **milli**liter ( mL )= 0.001 liter | 1 L = 1000 mL |
| 1 liter ( L ) = 1.0 liter | 1 kL = 1000 L |
| 1 **hecto**liter ( hL )= 100 liters | 1 kL = 10 hL |
| 1 **kilo**liter ( kL ) = 1000 liters | 1 L = 1 dm³ |
| | 1 kL = 1 m³ |
| | 1 mL= 1 cm³ |

**Table 10.9**

| U.S. Customary Units of Liquid Volume | | |
|---|---|---|
| 1 pint ( pt ) | = | 16 fluid ounces ( fl oz ) |
| 1 quart ( qt ) | = | 2 pt = 32 fl oz |
| 1 gallon ( gal ) | = | 4 qt |

**Table 10.10**

We will use the equivalent measures ( rounded off ) in table 10.11 to convert units of volume from the U.S. customary system to the metric system, and vice versa. Examples 3 and 4 illustrate how to make conversions using these equivalents and multiplying by the number of given units.

| U.S. Customary and Metric Volume Equivalents | |
|---|---|
| **U.S. to Metric** | **Metric to U.S.** |
| 1 in.$^3$ = 16.387 cm$^3$ | 1cm$^3$ = 0.06 in.$^3$ |
| 1 ft.$^3$ = 0.028 m$^3$ | 1m$^3$ = 35.315 ft.$^3$ |
| 1 qt = 0.946 L | 1L = 1.06 qt |
| 1 gal = 3.785 L | 1L = 0.264 gal |

**Table 10.11**

**Important Note About Conversions between U.S. Customary and Metric**

As stated in Sections 10.2 and 10.3, most of the conversion units are not exact and slightly different answers are possible, depending on the conversion units used. We will give volume conversions rounded off to the place of the first four digits only.

## Example 3

**a.** 20 gal = _____ L

**b.** How many liters are there in 6 qt?

### Solution

**a.** 20 gal = 20( 1 gal ) = 20( 3.785 L ) = 75.7 L

**b.** 6 qt = 6( 1 qt ) = 6( 0.946 L ) = 5.676 L

### Example 4

**a.** Change 42 L to gallons.

**b.** Express 10 cm³ in cubic inches.

#### Solution

**a.** $42\text{ L} = 42(1\text{ L}) = 42(0.264\text{ gal}) = 11.088\text{ gal} = 11.09\text{ gal}$ ( rounded off )

**b.** $10\text{ cm}^3 = 10(1\text{ cm}^3) = 10(0.06\text{ in.}^3) = 0.6\text{ in.}^3$

*Now Work Exercises 5-8 in the Margin.*

Change the units as indicated.

**5.** 340 mL = 0.34 L

**6.** 700 mL = 700 cm³

**7.** 18 L = 4.752 gal

**8.** 10 qt = 9.46 L

## Geometry: Formulas for Volume

The following formulas for the volumes of common geometric solids shown are valid regardless of the measurement system used. Always be sure to label your answers with the correct units of measure.

**Five Geometric Solids and the Formulas for their Volumes**

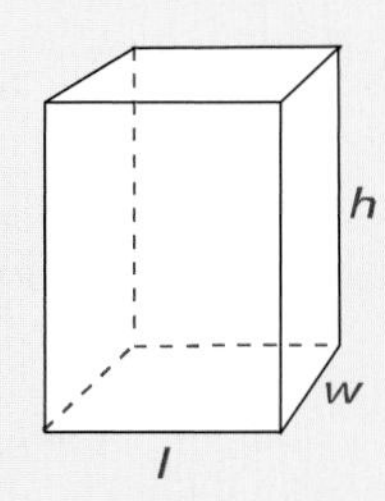


**Rectangle Solid**

$V = lwh$

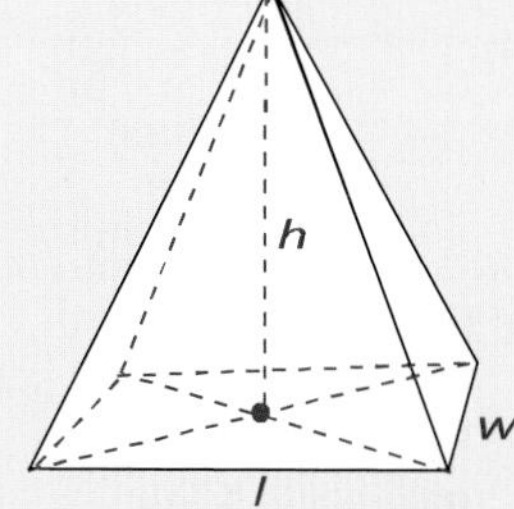


**Rectangular Pyramid**

$V = \frac{1}{3}lwh$

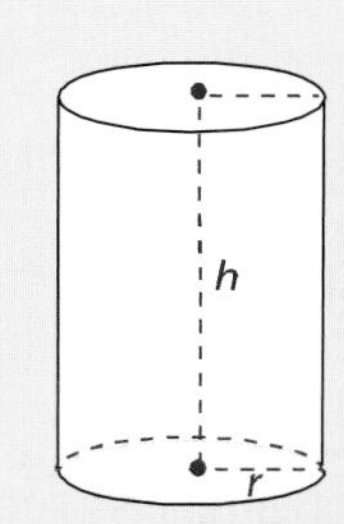


**Right Circular Cylinder**

$V = \pi r^2 h$

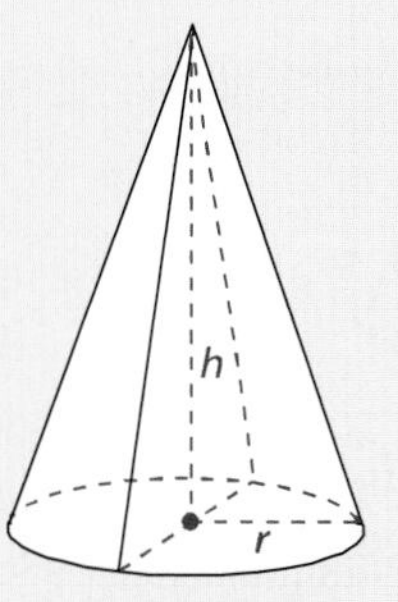


**Right Circular Cone**

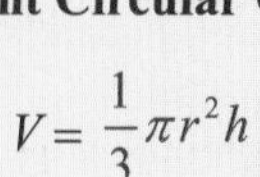

$V = \frac{1}{3}\pi r^2 h$

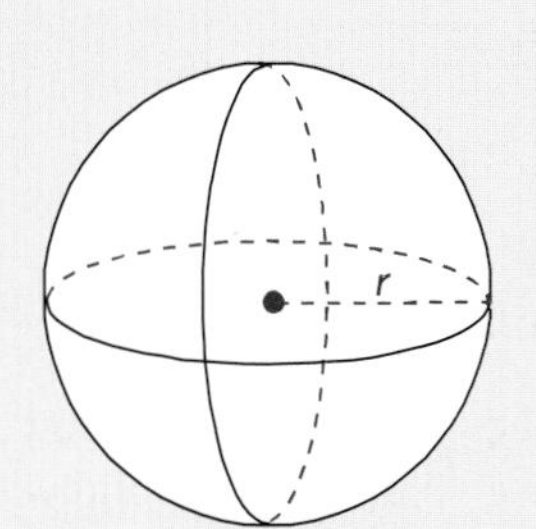


**Sphere**

$V = \frac{4}{3}\pi r^3$

## Example 5

Find the volume of the rectangular solid with length 8 cm, width 4 cm, and height 12 cm. Write the answer in both cubic centimeters and milliliters.

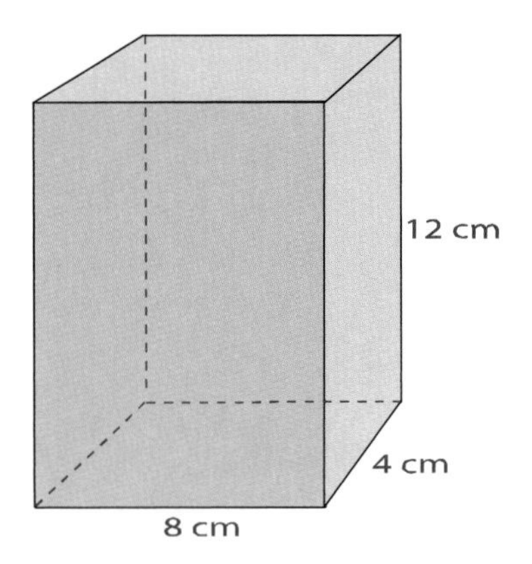


**Solution**

$V = lwh$

$V = 8 \cdot 4 \cdot 12$

$= 384 \text{ cm}^3$

Since $1 \text{ cm}^3 = 1$ mL, $384 \text{ cm}^3 = 384$ mL.

## Example 6

Find the volume of the solid with the dimensions indicated. Write the answer in both cubic centimeters and liters.

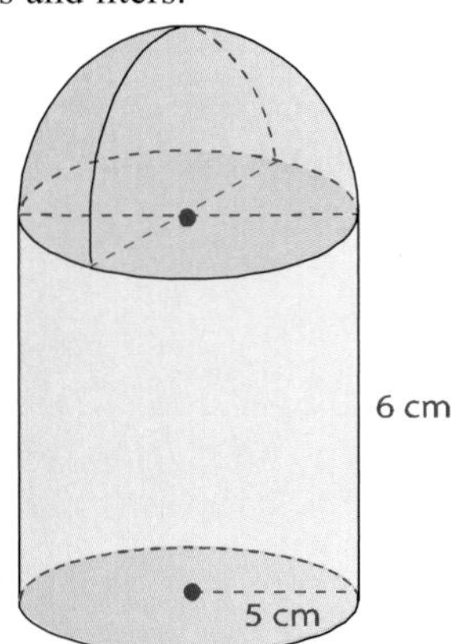


**Solution**

From the illustration, the solid is a hemisphere resting on top of a cylinder. Thus, the volume of the solid will be sum of the volumes of these two figures.

| **Cylinder** | **Hemisphere** | **Total Volume** | |
|---|---|---|---|
| $V = \pi r^2 h$ | $V = \frac{1}{2} \cdot \frac{4}{3} \pi r^3$ | $471.00 \text{ cm}^3$ | cylinder |
| | | $+\ 261.67 \text{ cm}^3$ | hemisphere |
| $V = 3.14(5^2)(6)$ | $V = \frac{2}{3}(3.14)(5^3)$ | $732.67 \text{ cm}^3$ | total volume |
| $= 471 \text{ cm}^3$ | $= 261.67 \text{ cm}^3$ | | |

Since $1 \text{ cm}^3 = 1 \text{ mL} = 0.001$ L, the total volume in liters is $732.67 \text{ cm}^3 = 0.73267$ L or 0.733 L ( rounded off ).

Name ______________________ Section ______________ Date ____________

ANSWERS

For Exercises 1-12, respond in exercise.

## Exercises 10.4

**1.** $1 cm^3 =$ 1000 $mm^3$

$1 dm^3 =$ 1000 $cm^3$

$1 m^3 =$ 1000 $dm^3$

$1 km^3 =$ 1 000 000 000 $m^3$

**2.** 1 dm = 10 cm

1 dm = 100 mm

$1 dm^2 =$ 100 $cm^2$

$1 dm^2 =$ 10 000 $mm^2$

$1 dm^3 =$ 1000 $cm^3$

$1 dm^3 =$ 1 000 000 $mm^3$

**3.** 1 m = 10 dm

1 m = 100 cm

$1 m^2 =$ 100 $dm^2$

$1 m^2 =$ 10 000 $cm^2$

$1 m^3 =$ 1000 $dm^3$

$1 m^3 =$ 1 000 000 $cm^3$

**4.** 1km = 1000 m

$1 km^2 =$ 1 000 000 $m^2$

$1 km^3 =$ 1 000 000 000 $m^3$

1km = 10 000 dm

$1 km^2 =$ 100 ha

$1 km^3 =$ 1 000 000 000 kL

Change the following units of volume in the metric system as indicated.

**5.** $85 m^3 =$ 85 000 $dm^3$

**6.** $0.7 m^3 =$ 700 $dm^3$

**7.** $325 cm^3 =$ 325 000 $mm^3$

**8.** $10.6 cm^3 =$ 10 600 $mm^3$

**9.** $9.15 m^3 =$ 9 150 000 $cm^3$

**10.** $0.672 m^3 =$ 672 000 $cm^3$

**11.** $75 dm^3 =$ 0.075 $m^3$

**12.** $8.38 dm^3 =$ 0.008 38 $m^3$

13. [ Respond in exercise. ]

14. [ Respond in exercise. ]

15. 0.035 $cm^3$

16. 4200 $mm^3$

17. 0.000 029 $dm^3$

18. 15 000 $mm^3$

19. 12 000 $cm^3$

20. 30 mL

21. [ Respond in exercise. ]

22. [ Respond in exercise. ]

23. [ Respond in exercise. ]

24. [ Respond in exercise. ]

25. 6700 mL

26. 0.0158 L

27. 0.3 L

28. 6000 $cm^3$

29. [ Respond in exercise. ]

30. [ Respond in exercise. ]

**13.** 135 $mm^3$ = 0.135 $cm^3$

**14.** 0.6 $mm^3$ = 0.0006 $cm^3$

**15.** How many $cm^3$ are in 35 $mm^3$?

**16.** How many $mm^3$ are in 4.2 $cm^3$?

**17.** Change 29 $mm^3$ to $dm^3$.

**18.** Change 15 $cm^3$ to $mm^3$.

**19.** Convert 12 $dm^3$ to $cm^3$.

**20.** Convert 30 $cm^3$ to mL.

**21.** 4.6 kL = 4 600 L

**22.** 65 000 L = 65 kL

**23.** 380 L = 0.38 kL

**24.** 560 mL = 0.56 L

**25.** Change 6.7 L to milliliters.

**26.** Change 15.8 mL to liters.

**27.** Change 300 $cm^3$ to liters.

**28.** Convert 6L to cubic centimeters.

**29.** 95L = 95 000 mL = 95 000 $cm^3$

**30.** 7.53 L = 7530 mL = 7530 $cm^3$

Name ____________ Section ____________ Date ____________

**ANSWERS**

For Exercises 31-47, respond in exercise.

Use Table 10.11 as a reference to convert the following measures of volume as indicated.

**31.** 18 in.$^3$ = 295.0 cm$^3$

**32.** 2.5 in.$^3$ = 41.0 cm$^3$

**33.** 100 cm$^3$ = 6 in.$^3$

**34.** 1000 cm$^3$ = 60 in.$^3$

**35.** 27 ft.$^3$ = 0.756 m$^3$

**36.** 1.5 ft.$^3$ = 0.042 m$^3$

**37.** 1000 m$^3$ = 35 300 ft.$^3$

**38.** 8 m$^3$ = 283.0 ft.$^3$

**39.** 10 qt = 9.46 L

**40.** 20 qt = 18.9 L

**41.** 20 L = 21.2 qt

**42.** 45 L = 47.7 qt

**43.** 52 L = 13.7 gal

**44.** 500 L = 132 gal

**45.** 20 gal = 75.7 L

**46.** 30 gal = 114.0 L

**47.** Match each formula for volume to its corresponding geometric figure.

C **a.** rectangular solid

E **b.** rectangular pyramid

D **c.** right circular cylinder

B **d.** right circular cone

A **e.** sphere

A. $V = \frac{4}{3}\pi r^3$

B. $V = \frac{1}{3}\pi r^2 h$

C. $V = lwh$

D. $V = \pi r^2 h$

E. $V = \frac{1}{3}lwh$

48. $V = 70 \text{ in.}^3$

$V = 1147 \text{ cm}^3$

49. $V = 1.178 \text{ ft.}^3$

$V = 0.033 \text{ m}^3$

50. $V = 381.5 \text{ cm}^3$

$V = 22.89 \text{ in.}^3$

51. $V = 12.56 \text{ dm}^3$

$V = 0.012\ 56 \text{ m}^3$

52. $V = 180 \text{ cm}^3$

$V = 180\ 000 \text{ mm}^3$

53. 96 mL

**48.** Find the volume of a rectangular solid with length 5 in., width 2 in., and height 7 in., in both cubic inches and cubic centimeters.

**49.** Find the volume of right circular cylinder 1.5 ft. height and 1 ft. in diameter, in both cubic feet and cubic meters.

**50.** Find the volume of a Christmas tree ornament in the shape of a sphere with radius 4.5 cm, in both cubic centimeters and cubic inches.

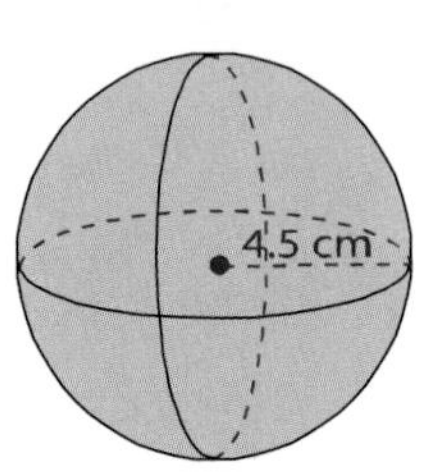


**51.** Find the volume of a right circular cone 3 dm high with a 2 dm radius, in both cubic decimeters and cubic meters.

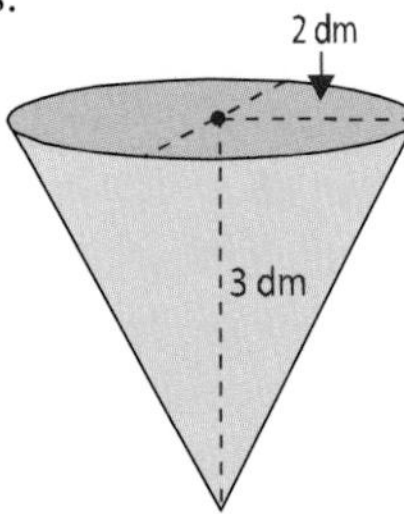


**52.** Find the volume of a rectangular pyramid with length 18 cm, width 10 cm, and altitude 3 cm, in both cubic centimeters and cubic millimeters.

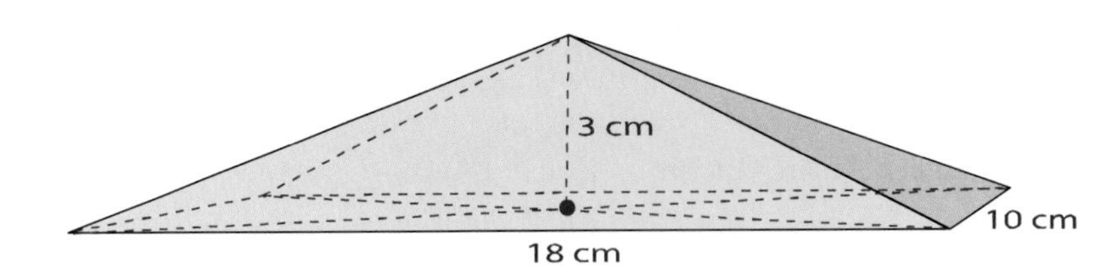


**53.** Disposable paper drinking cups like those used at water coolers are often cone-shaped. Find the volume of such a cup that is 9 cm high with a 3.2 cm radius. Express the answer to the nearest milliliter.

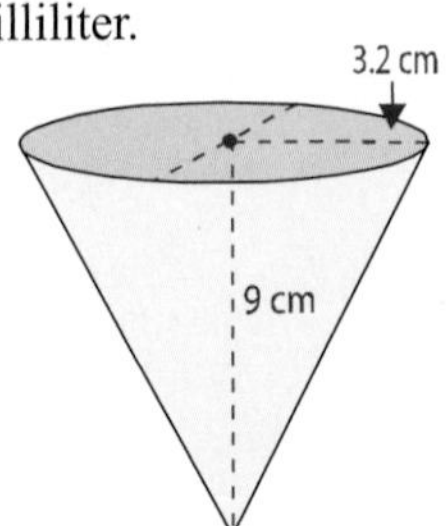

Name ______________________ Section ____________ Date __________

**54.** A manufacturer is to design a can in the shape of a right circular cylinder to hold 0.5 L of juice concentrate. If the can must have a diameter of 7 cm, how tall will the can be ( to the nearest centimeter )?

Find the volume of each of the following figures with the indicated dimensions. Write the answers in equivalent measures in cubic centimeters, cubic millimeters, and cubic inches.

**55.**

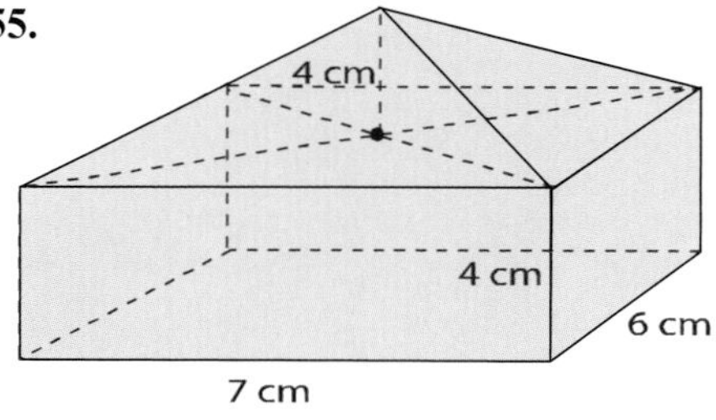


**56.**

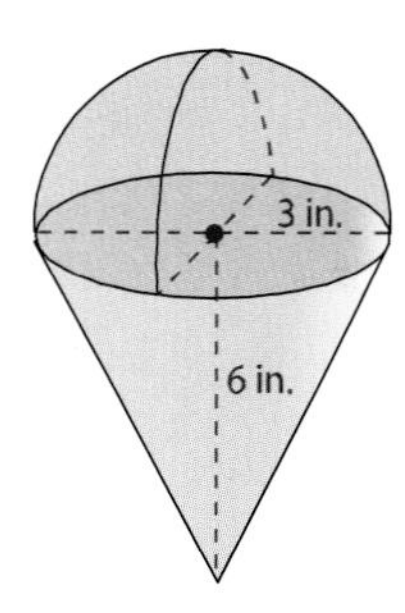


**57.**

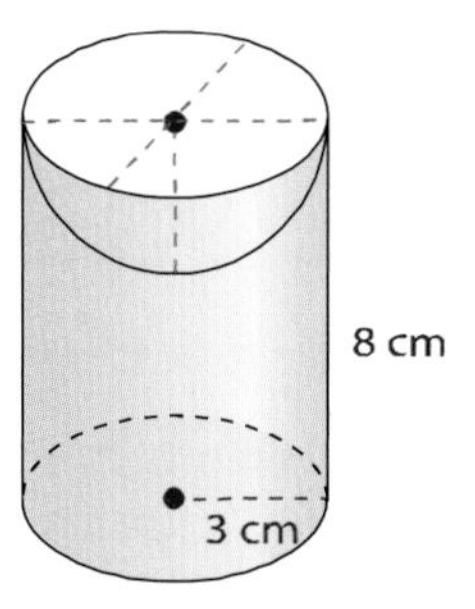


**58.**

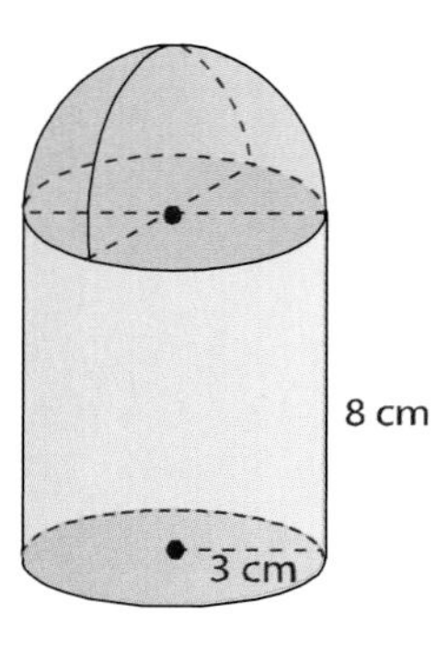


**59.**

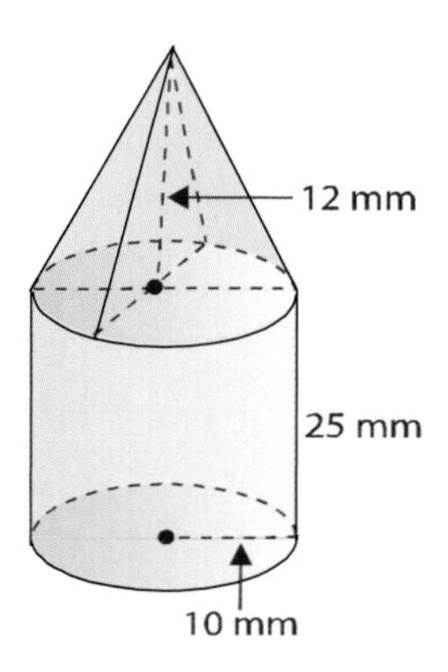


**60.**

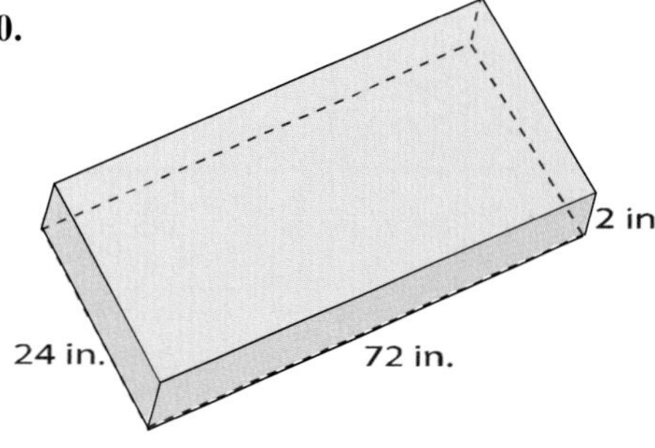


**ANSWERS**

**54.** 13.0 cm

**55.** $V = 224\ 000\ \text{mm}^3$

$V = 224\ \text{cm}^3$

$V = 13.44\ \text{in.}^3$

**56.** $V = 1\ 852\ 000\ \text{mm}^3$

$V = 1852\ \text{cm}^3$

$V = 113.0\ \text{in.}^3$

**57.** $V = 169\ 600\ \text{mm}^3$

$V = 169.6\ \text{cm}^3$

$V = 10.1736\ \text{in.}^3$

**58.** $V = 282\ 600\ \text{mm}^3$

$V = 282.6\ \text{cm}^3$

$V = 16.956\ \text{in.}^3$

**59.** $V = 9106\ \text{mm}^3$

$V = 9.106\ \text{cm}^3$

$V = 0.5464\ \text{in.}^3$

**60.** $V = 56\ 630\ 000\ \text{mm}^3$

$V = 56\ 633\ \text{cm}^3$

$V = 3456\ \text{in.}^3$

# Chapter 10 Index of Key Ideas and Terms

Undefined Terms ( in Geometry ) page 777

Point: A dot represents a point. Points are labeled with capital letters. page 777

Line: A line has no beginning or end. Lines are labeled with small letters or by two points on the line. page 777

Plane: Flat surfaces, such as table tops or walls, represent planes. Planes are labeled with capital letters. page 777

Ray page 778

A ray consists of a point ( called the **endpoint** ) and all the points on a line on one side of that point.

Angle page 778

An angle consists of two rays with a common endpoint. The common endpoint is called the **vertex** of the angle.

Types of Angles page 780

Defined by Measure

Acute: $0° < m\angle A < 90°$

Right: $m\angle A = 90°$

Obtuse: $90° < m\angle A < 180°$

Straight: $m\angle A = 180°$

For Two Angles: page 781

Two angles are **complementary** if the sum of their measures = 90°.

Two angles are **supplementary** if the sum of their measures =180°.

Two angles are **equal** if they have the same measure.

Vertical Angles page 782

Two angles are **vertical** if they are formed by two intersecting straight lines and are opposite to each other.

For Triangles:

Classified by Sides page 784

Scalene: No two sides are equal.

Isosceles: Two sides are equal.

Equilateral: All three sides are equal.

Classified by Angles page 785

Acute: All three angles are acute.

Right: One angle is a right angle.

Obtuse: One angle is an obtuse angle.

Two Properties of All Triangles page 786

1. The sum of the measures of the angles is 180°.
2. The sum of the lengths of any two sides must be greater than the length of the third side.

Prefixes for the Metric System page 795

Milli, centi, deci, deka, hecto, kilo

Length

Metric Measures of Length page 795

Changing Metric Measures of Length page 796

U.S. Customary Units of Length page 798

U.S. - Metric Length Equivalents page 798

Perimeter: Total distance around a plane geometric figure page 799

Circumference: Perimeter of a circle page 799

Radius: Distance from the center of a circle to a point on the circle page 799

Diameter: Distance from one point on a circle to another point on the circle measured through the center page 799

Geometric Formulas for Perimeter page 800

Square: $P = 4s$

Rectangle: $P = 2l + 2w$

Parallelogram: $P = 2b + 2a$

Triangle: $P = a + b + c$

Circle: $C = 2\pi r$ or $C = \pi d$

Trapezoid: $P = a + b + c + d$

Area

Metric Measures of Small Area page 809

Changing Metric Measures of Area page 810

Metric Measures of Land Area page 812

U.S. Customary Units of Area page 813

U.S. -- Metric Area Equivalents page 813

Name ______________________ Section ____________ Date __________

# Chapter 10 Test

1. **a.** If $\angle 1$ and $\angle 2$ are complementary and $m\angle 1 = 35°$, what is $m\angle 2$?

   **b.** If $\angle 3$ and $\angle 4$ are supplementary and $m\angle 3 = 15°$, what is $m\angle 4$?

2. In the figure shown here, $\overleftrightarrow{AB}$ is a line with $m\angle DOC = 120°$.

   **a.** What type of angle is $\angle BOD$?

   **b.** What type of angle is $\angle AOB$?

   **c.** Name two pairs of supplementary angles.

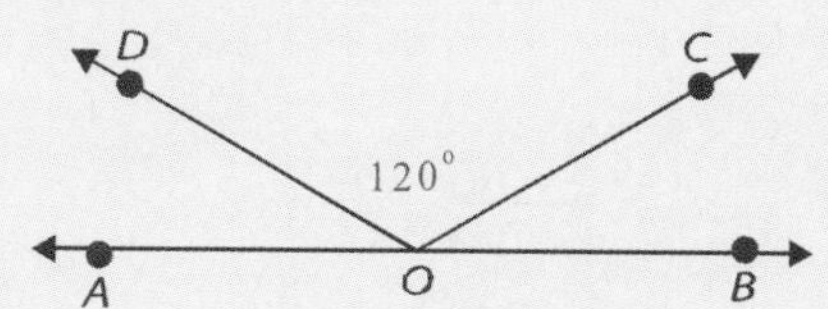


3. Name the type of triangle in each figure two ways ( in terms of angles and sides ).

**a.**

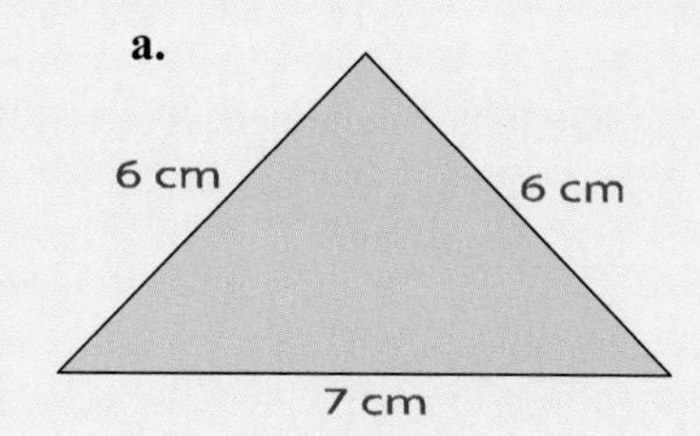


**b.**

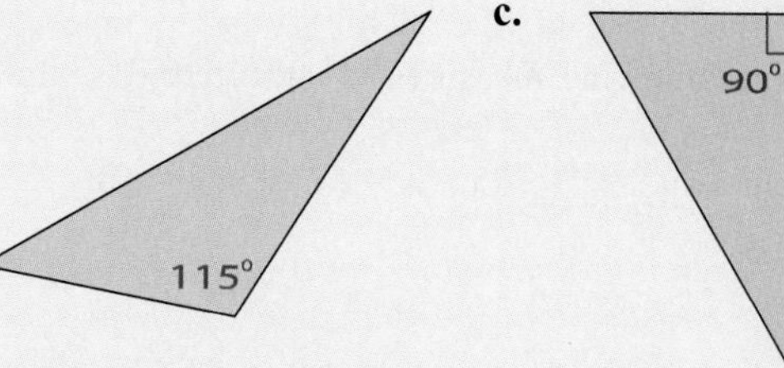


**c.** 90°

4. **a.** Find the measure of $\angle x$.

   **b.** What kind of triangle is $\Delta RST$?

   **c.** Which side is opposite $\angle S$?

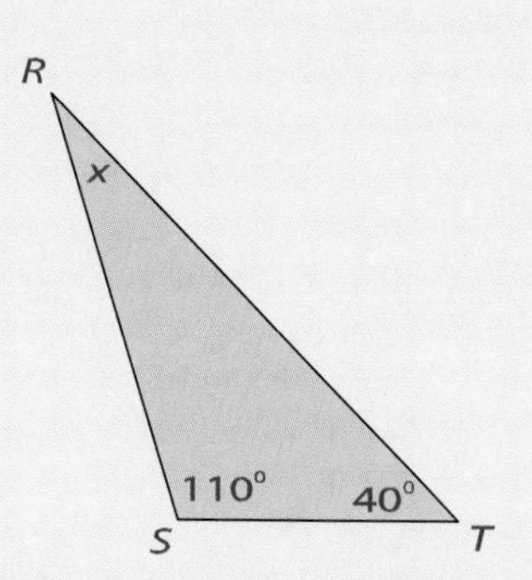


5. Which is longer, 10 mm or 10 cm? How much longer?

6. Which has the greater volume, 10 mL or 10 $cm^3$? How much greater?

**ANSWERS**

1. a. $m\angle 2 = 55°$
   b. $m\angle 4 = 165°$

2. a. obtuse
   b. straight
   c. $\angle AOD$ and $\angle DOB$; $\angle AOC$ and $\angle COB$

3. a. isosceles, acute
   b. obtuse, scalene
   c. right, scalene

4. a. $m\angle x = 30°$
   b. obtuse, scalene
   c. $\overline{RT}$

5. 10 cm is longer; 90 mm longer

6. The volume is the same. 10 mL = 10 $cm^3$

For Exercises 7-22, respond in exercise.

**23.a.** 94.2 cm

**b.** 706.5 cm²

Change the following units of measure in the metric system as indicated.

**7.** 56 cm = 0.56 m

**8.** 33 m = 3300 cm

**9.** 11 000 mm = 11 m

**10.** 83.5 mL = 0.0835 L

**11.** 960 $mm^2$ = 9.6 $cm^2$

**12.** 15 $m^2$ = 150 000 $cm^2$

**13.** 75 a = 0.75 ha

**14.** 75 ha = 7500 a

**15.** 140 $cm^3$ = 140 000 $mm^3$

**16.** 32 000 $cm^3$ = 0.032 $m^3$

Convert the following measures as indicated. Use the appropriate table where necessary.

**17.** 50 cm = 19.7 in.

**18.** 25 L = 6.6 gal

**19.** 200 km = 124 mi.

**20.** 7.8 $in.^2$ = 50.31 $cm^2$

**21.** 30 qt = 28.38 L

**22.** 39 $ft^3$ = 1.092 $m^3$

**23.** Find

**a.** the circumference and

**b.** the area of a circle with diameter 30 cm.

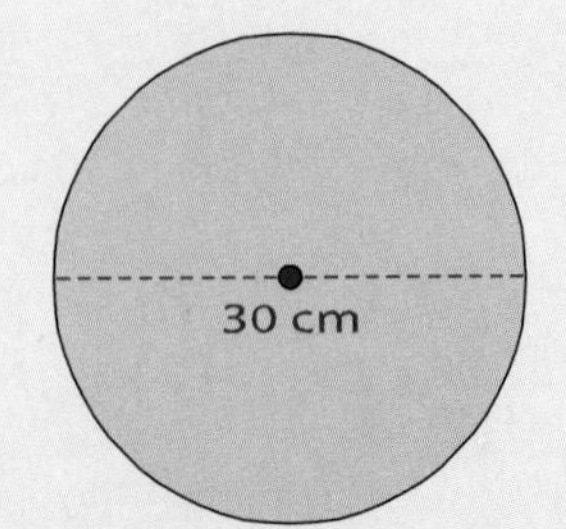

Name ______________________ Section ____________ Date __________

**ANSWERS**

**24.** Find

**a.** the perimeter and

**b.** the area of a rectangle that is 8.5 in. wide and 11.6 in. long.

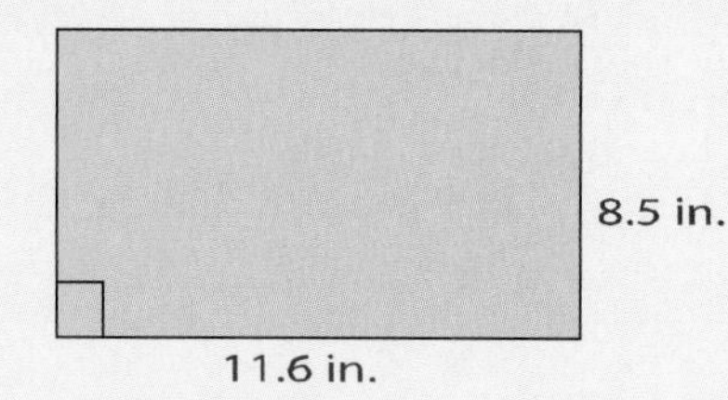


**25.** Find the volume of a sphere with diameter 12 in., in both cubic inches and cubic centimeters.

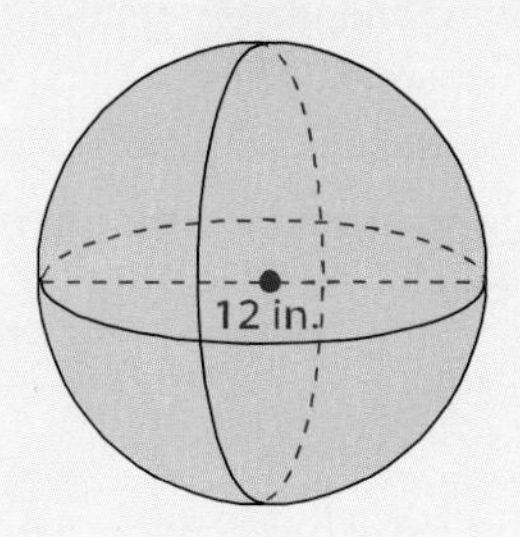


**26.** Find

**a.** the perimeter and

**b.** the area of the triangle with dimensions as shown here.

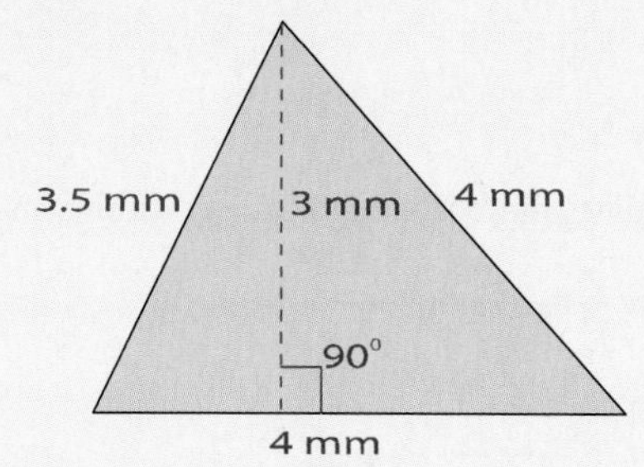


**27.** Find

**a.** the perimeter and

**b.** the area of the figure with dimensions as shown here.

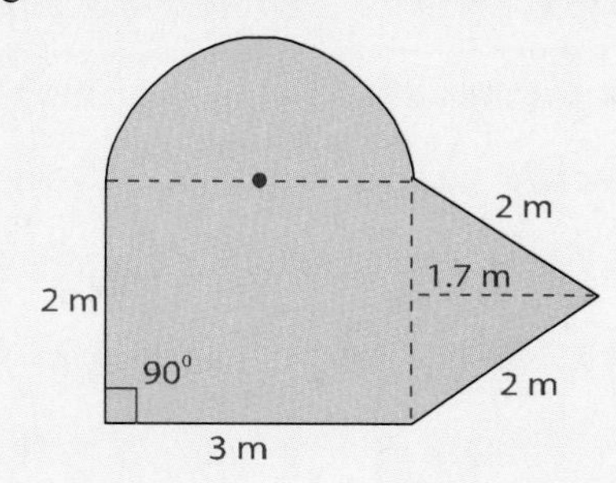


**24. a.** 40.2 in.

**b.** 98.6 in.$^2$

**25.** 904.32 in.$^3$

= 14 800 cm$^3$

**26. a.** 11.5 mm

**b.** 6 mm$^2$

**27. a.** 13.71 m

**b.** 11.2 m$^2$

28. a. 30 ft.

b. 30 ft.$^2$

29. 486 m$^2$

30. 90 cm$^3$ = 5.4 in.$^3$

**28.** Find

**a.** the perimeter and

**b.** the area of a right triangle with sides of length 5 ft., 12 ft., and 13 ft. Sketch the figure.

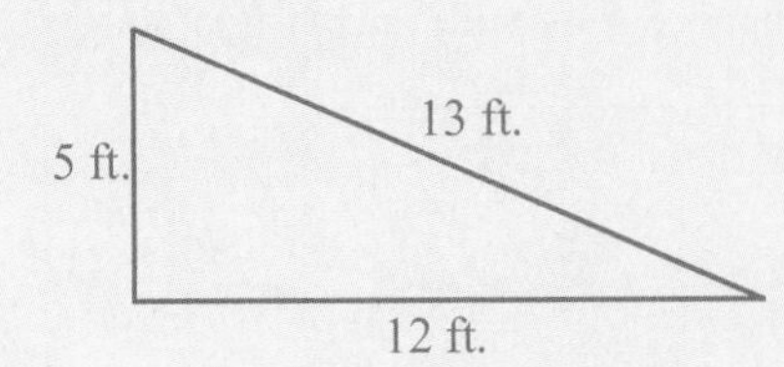


**29.** A rectangular swimming pool is surrounded by a concrete border. If the swimming pool is 50 meters long and 25 meters wide and the concrete border is 3 meters wide all around the pool, what is the area of the concrete border?

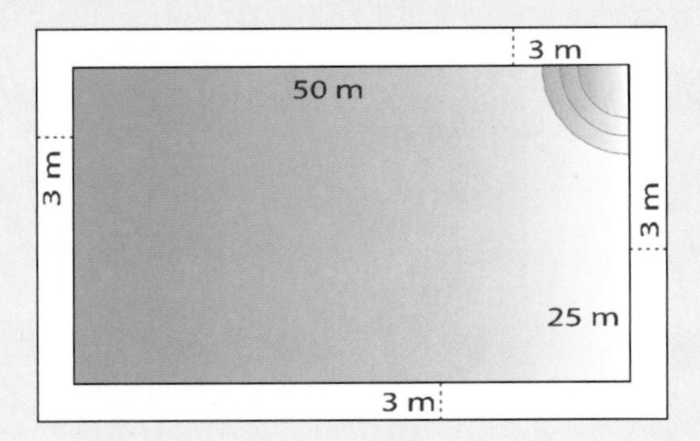


**30.** Draw a sketch of rectangular solid with length 5 cm, width 3 cm, and height 6 cm. Find the volume of this solid in both cubic centimeters and cubic inches.

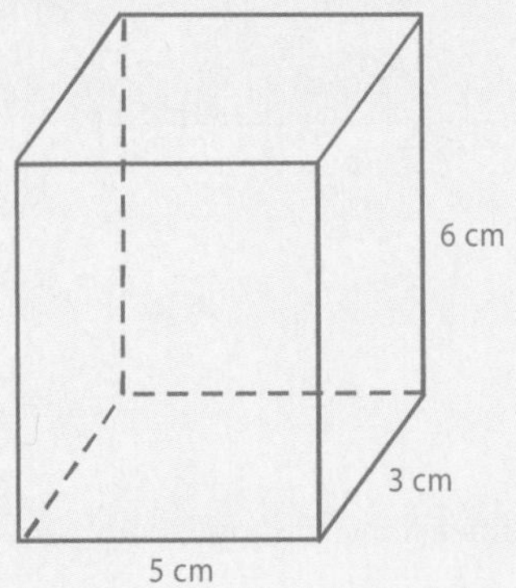

Name ______________________ Section ______________ Date ____________

## Cumulative Review: Chapters 1 - 10

1. The number 0 is called the additive identity. The number ___1___ is called the multiplicative identity.

2. Write the number five hundred twenty and sixty-five thousandths in decimal form.

3. Find the quotient of 89.5 and 17.6 to the nearest tenth.

4. Simplify the expression: $\sqrt{72}+2\sqrt{50}$

5. Use a calculator to find the value of $\sqrt{63}$ to the nearest thousandth.

6. Find the LCM of 35, 49, and 105.

7. Evaluate the expression: $-16+(12\cdot 3+2^3)\div 4-20$

8. Evaluate the expression: $(0.5)^2+2\frac{1}{5}\div\frac{11}{6}-0.3$

9. Evaluate each expression.

   **a.** $(8-10)(15-10)$

   **b.** $(1.2)^2-(1.5)^2$

   **c.** $\left(-\frac{2}{3}\right)\div\left(-\frac{7}{15}\right)\div(-2)$

**ANSWERS**

1. [ Respond in exercise.]
2. 520.065
3. 5.1
4. $16\sqrt{2}$
5. 7.937
6. 735
7. −25
8. 1.15
9. a. −10
   b. −0.81
   c. $-\frac{5}{7}$

10. $8950.46

11. $2000

12. [Respond below exercise.]

13. [Respond in exercise.]

14. [Respond in exercise.]

15. [Respond in exercise.]

16. [Respond in exercise.]

17. a. $17.50

b. $45

18. $x = -1.05$

19. $x = -32.5$

20. $y = \frac{10-5x}{3}$

21. 17

22. a. mean = 45

b. median = 46

c. mode = 41

d. range = 35

**10.** If an investment pays 10% interest compounded daily, what will be the value of $6000 in 4 years?

**11.** If a savings account earned $75 simple interest at 5% for 9 months, what was the original principal in the account?

**12.** The circle graph shows a family budget for one year. What amount will be spent in each category if the family income is $45,000?

Food: $9000
Housing: $11,250
Transportation: $6750
Miscellaneous: $6750
Savings: $2250
Education: $4500
Taxes: $4500

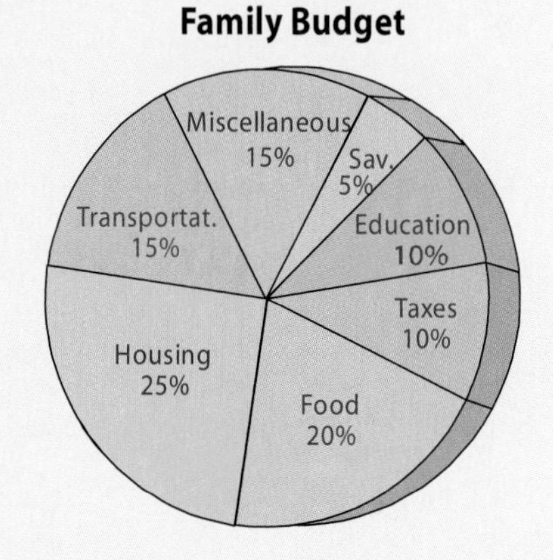


**13.** 18 is 20 % of 90

**14.** 70% of 130 is 91.

**15.** 37.5% of 103.4 is 38.775.

**16.** 86% of 14,000 is 12,040.

**17.** Graphing calculators are on sale at a discount of 28%, and the original price was $62.50.

**a.** What is the approximate amount of the discount?

**b.** What is the sale price?

**18.** Solve for $x$: $4x - 10 = 16x + 2.6$

**19.** Solve for $x$: $3(x - 5) = 5(x + 10)$

**20.** Solve for $y$: $5x + 3y = 10$

**21.** If a number is increased by twenty-two, the result is three times the difference between the number and four. What is the number?

**22.** Find

**a.** the mean,

**b.** the median,

**c.** the mode, and

**d.** the range for the following data:

26 51 49 61 41 46 41

Name ______________________ Section ______________ Date ____________

**ANSWERS**

**23.** Write each of the following numbers in scientific notation.

**a.** 193,000,000

**b.** 0.000386

**24.** Find the value of $x$ and the value of $y$ if the two triangles shown are similar triangles.

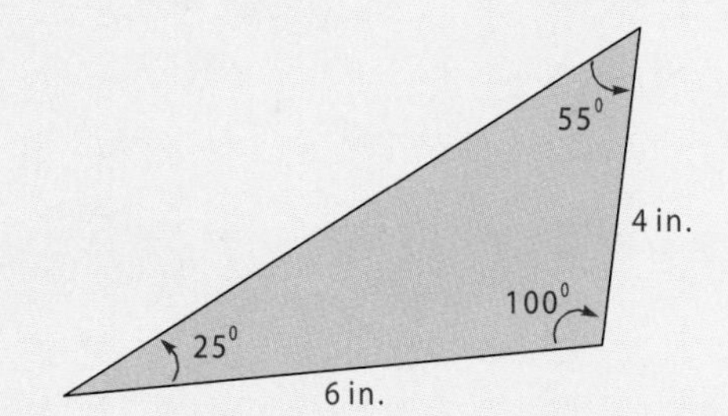


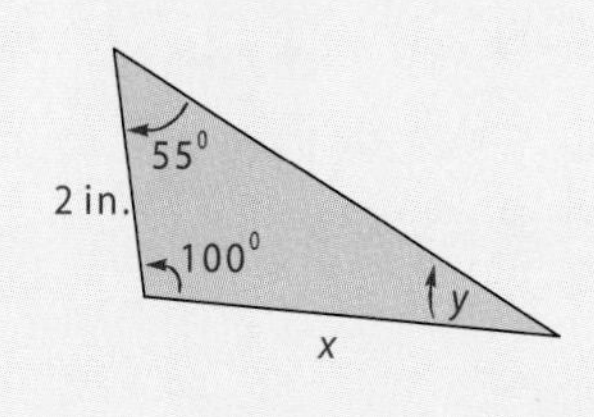


**25. a.** Find the length of the hypotenuse ( in radical notation ) for the right triangle shown here.

**b.** Use a calculator to find this value to the nearest hundredth.

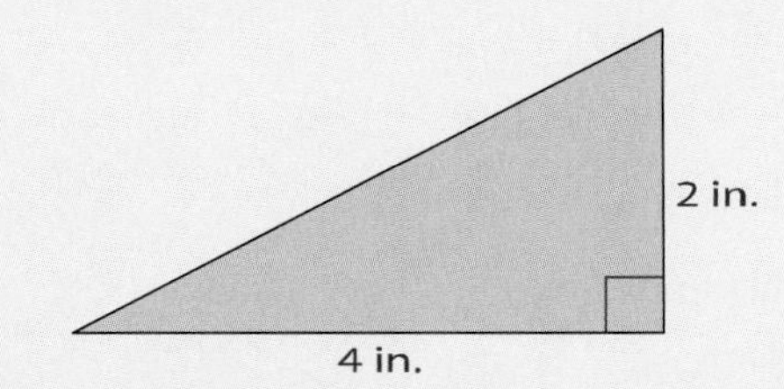


**26.** Solve the following inequality, and graph the solution on a real number line.
$2x + 3x - 5 > x + 7$

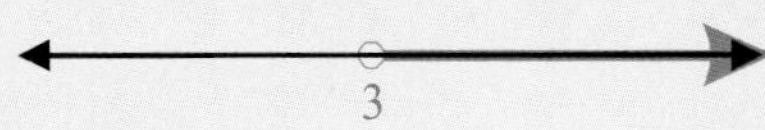


**27.** Graph each of the following linear equations.

**a.** $y = -2$

**b.** $x + 2y = 4$

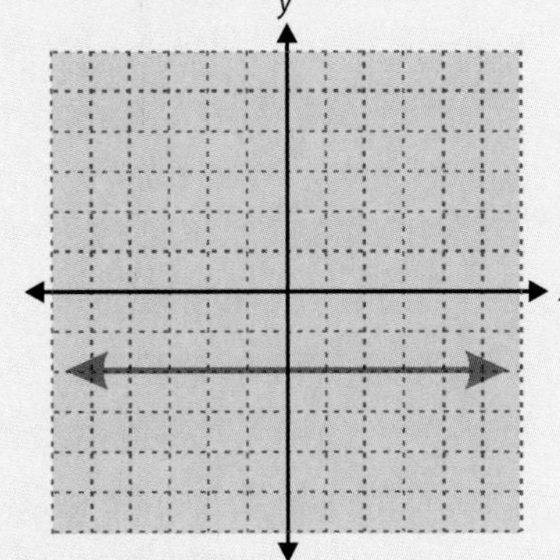


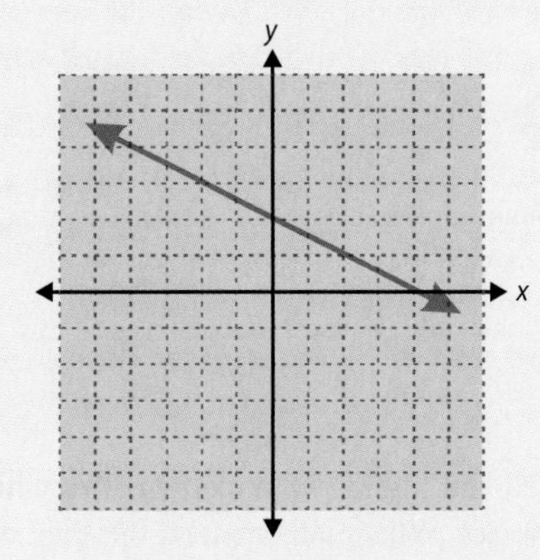


**23. a.** $1.93 \times 10^8$

**b.** $3.86 \times 10^{-4}$

**24.** $x = 3$ in.; $y = 25°$

**25. a.** $\sqrt{20}$ in. $= 2\sqrt{5}$ in.

**b.** 4.47 in.

**26.** $x > 3$

**27.** [ Use grids provided.]

28. a. $\frac{1}{a^6}$

b. $\frac{1}{x^2}$

c. $1$

d. $-10y^7$

29. a. $9x^2 - 7$; second-degree binomial

b. $3x^3 + 2x^2 - 4x$; third-degree trinomial

c. $6x^5 + 12x^3 - 6x^2$; fifth-degree trinomial

d. $6x^2 + x - 15$; second-degree trinomial

30. a. $3x(x + 5)$

b. $5x(x - 2)(x + 2)$

c. $2(y + 5)^2$

31. [ Respond below exercise.]

32. [ Respond below exercise.]

33. [ Respond below exercise.]

34. [ Respond below exercise.]

**28.** Simplify each of the following expressions with no negative exponents in the answer.

**a.** $(a^2)^{-3}$ **b.** $x^3 \cdot x^{-5}$

**c.** $(2x^7)^0$ **d.** $-5y^4 \cdot 2y^3$

**29.** Perform the indicated operations and simplify. Tell the degree and type of each polynomial answer.

**a.** $(5x^2 + 7x - 10) + (4x^2 - 7x + 3)$

**b.** $(-2x^3 + x^2 - 8x + 2) - (-5x^3 - x^2 - 4x + 2)$

**c.** $6x^2(x^3 + 2x - 1)$

**d.** $(3x + 5)(2x - 3)$

**30.** Completely factor each of the following polynomials.

**a.** $3x^2 + 15x$

**b.** $5x^3 - 20x$

**c.** $2y^2 + 20y + 50$

## WRITING AND THINKING ABOUT MATHEMATICS

**31.** Given the equation $5(x + 10) = 3x$, make up your own word problem that can be solved by using this equation. Does the solution of the equation make sense as an answer to your problem? Why or why not?

Answers will vary. For example: The sum of a number and ten is multiplied by five, and this product equals three times the number. What is the number? Since the solution is $x = -25$, any word problem must be such that a negative solution is sensible.

**32.** Make up a word problem that involves a circle with a diameter or 1 meter and a square with a side of 100 centimeters.

Answers will vary. For example: A tile patio uses a simple pattern of large circles inscribed in large squares. The squares are 100 cm on a side and the circle inside each square is 1 m in diameter. How much area enclosed in each square is outside the inscribed circle?
Answer: $100^2 - \pi(50)^2 = 10{,}000 - 2500(3.14) = 10{,}000 - 7850 = 2150 \text{ cm}^2$.

**33.** Make up your own word problem that uses the numbers 72, 80, and 97 and has 83 as the answer.

Answers will vary. The problem should be to find the mean of these values. For example: Find Elliot's test average in European Literature if he has scored 72, 80, and 97 on three exams.

**34.** Explain briefly what mathematics you used today. What mathematics did you use this week? What mathematical thinking did you use to solve some problem this week?

Answers will vary.

# Appendix I

## Practice with Basic Addition Facts

Find all of the following sums mentally as fast as you can. Check your answers. Redo **all** of the exercises until you get **all** of the answers right and improve your time.

| | | | | |
|---|---|---|---|---|
| **1.** $\begin{array}{r} 2 \\ +6 \\ \hline \end{array}$ | **2.** $\begin{array}{r} 0 \\ +3 \\ \hline \end{array}$ | **3.** $\begin{array}{r} 5 \\ +0 \\ \hline \end{array}$ | **4.** $\begin{array}{r} 7 \\ +2 \\ \hline \end{array}$ | **5.** $\begin{array}{r} 3 \\ +8 \\ \hline \end{array}$ |
| **6.** $\begin{array}{r} 2 \\ +8 \\ \hline \end{array}$ | **7.** $\begin{array}{r} 0 \\ +7 \\ \hline \end{array}$ | **8.** $\begin{array}{r} 6 \\ +6 \\ \hline \end{array}$ | **9.** $\begin{array}{r} 4 \\ +7 \\ \hline \end{array}$ | **10.** $\begin{array}{r} 2 \\ +1 \\ \hline \end{array}$ |
| **11.** $\begin{array}{r} 8 \\ +4 \\ \hline \end{array}$ | **12.** $\begin{array}{r} 9 \\ +0 \\ \hline \end{array}$ | **13.** $\begin{array}{r} 2 \\ +2 \\ \hline \end{array}$ | **14.** $\begin{array}{r} 1 \\ +4 \\ \hline \end{array}$ | **15.** $\begin{array}{r} 1 \\ +1 \\ \hline \end{array}$ |
| **16.** $\begin{array}{r} 4 \\ +9 \\ \hline \end{array}$ | **17.** $\begin{array}{r} 9 \\ +8 \\ \hline \end{array}$ | **18.** $\begin{array}{r} 6 \\ +7 \\ \hline \end{array}$ | **19.** $\begin{array}{r} 7 \\ +8 \\ \hline \end{array}$ | **20.** $\begin{array}{r} 1 \\ +5 \\ \hline \end{array}$ |
| **21.** $\begin{array}{r} 3 \\ +6 \\ \hline \end{array}$ | **22.** $\begin{array}{r} 7 \\ +3 \\ \hline \end{array}$ | **23.** $\begin{array}{r} 1 \\ +7 \\ \hline \end{array}$ | **24.** $\begin{array}{r} 7 \\ +9 \\ \hline \end{array}$ | **25.** $\begin{array}{r} 0 \\ +2 \\ \hline \end{array}$ |
| **26.** $\begin{array}{r} 2 \\ +3 \\ \hline \end{array}$ | **27.** $\begin{array}{r} 8 \\ +6 \\ \hline \end{array}$ | **28.** $\begin{array}{r} 2 \\ +0 \\ \hline \end{array}$ | **29.** $\begin{array}{r} 2 \\ +7 \\ \hline \end{array}$ | **30.** $\begin{array}{r} 8 \\ +8 \\ \hline \end{array}$ |
| **31.** $\begin{array}{r} 9 \\ +9 \\ \hline \end{array}$ | **32.** $\begin{array}{r} 5 \\ +5 \\ \hline \end{array}$ | **33.** $\begin{array}{r} 5 \\ +6 \\ \hline \end{array}$ | **34.** $\begin{array}{r} 0 \\ +1 \\ \hline \end{array}$ | **35.** $\begin{array}{r} 3 \\ +0 \\ \hline \end{array}$ |
| **36.** $\begin{array}{r} 3 \\ +4 \\ \hline \end{array}$ | **37.** $\begin{array}{r} 3 \\ +7 \\ \hline \end{array}$ | **38.** $\begin{array}{r} 0 \\ +8 \\ \hline \end{array}$ | **39.** $\begin{array}{r} 0 \\ +0 \\ \hline \end{array}$ | **40.** $\begin{array}{r} 2 \\ +9 \\ \hline \end{array}$ |
| **41.** $\begin{array}{r} 4 \\ +4 \\ \hline \end{array}$ | **42.** $\begin{array}{r} 9 \\ +6 \\ \hline \end{array}$ | **43.** $\begin{array}{r} 8 \\ +1 \\ \hline \end{array}$ | **44.** $\begin{array}{r} 5 \\ +8 \\ \hline \end{array}$ | **45.** $\begin{array}{r} 9 \\ +7 \\ \hline \end{array}$ |
| **46.** $\begin{array}{r} 9 \\ +5 \\ \hline \end{array}$ | **47.** $\begin{array}{r} 5 \\ +7 \\ \hline \end{array}$ | **48.** $\begin{array}{r} 9 \\ +2 \\ \hline \end{array}$ | **49.** $\begin{array}{r} 6 \\ +8 \\ \hline \end{array}$ | **50.** $\begin{array}{r} 6 \\ +9 \\ \hline \end{array}$ |

| | | | | |
|---|---|---|---|---|
| **51.** $5 + 3$ | **52.** $0 + 4$ | **53.** $7 + 7$ | **54.** $3 + 3$ | **55.** $6 + 1$ |
| **56.** $4 + 8$ | **57.** $4 + 6$ | **58.** $5 + 4$ | **59.** $1 + 2$ | **60.** $6 + 2$ |
| **61.** $9 + 1$ | **62.** $3 + 9$ | **63.** $8 + 7$ | **64.** $4 + 0$ | **65.** $1 + 9$ |
| **66.** $7 + 4$ | **67.** $9 + 4$ | **68.** $0 + 9$ | **69.** $1 + 0$ | **70.** $5 + 2$ |
| **71.** $6 + 0$ | **72.** $0 + 5$ | **73.** $8 + 9$ | **74.** $1 + 6$ | **75.** $1 + 8$ |
| **76.** $6 + 3$ | **77.** $4 + 2$ | **78.** $6 + 4$ | **79.** $1 + 3$ | **80.** $3 + 5$ |

## Appendix I Answers

| | | | | | |
|---|---|---|---|---|---|
| **1.** 8 | **2.** 3 | **3.** 5 | **4.** 9 | **5.** 11 | **6.** 10 |
| **7.** 7 | **8.** 12 | **9.** 11 | **10.** 3 | **11.** 12 | **12.** 9 |
| **13.** 4 | **14.** 5 | **15.** 2 | **16.** 13 | **17.** 17 | **18.** 13 |
| **19.** 15 | **20.** 6 | **21.** 9 | **22.** 10 | **23.** 8 | **24.** 16 |
| **25.** 2 | **26.** 5 | **27.** 14 | **28.** 2 | **29.** 9 | **30.** 16 |
| **31.** 18 | **32.** 10 | **33.** 11 | **34.** 1 | **35.** 3 | **36.** 7 |
| **37.** 10 | **38.** 8 | **39.** 0 | **40.** 11 | **41.** 8 | **42.** 15 |
| **43.** 9 | **44.** 13 | **45.** 16 | **46.** 14 | **47.** 12 | **48.** 11 |
| **49.** 14 | **50.** 15 | **51.** 8 | **52.** 4 | **53.** 14 | **54.** 6 |
| **55.** 7 | **56.** 12 | **57.** 10 | **58.** 9 | **59.** 3 | **60.** 8 |
| **61.** 10 | **62.** 12 | **63.** 15 | **64.** 4 | **65.** 10 | **66.** 11 |
| **67.** 13 | **68.** 9 | **69.** 1 | **70.** 7 | **71.** 6 | **72.** 5 |
| **73.** 17 | **74.** 7 | **75.** 9 | **76.** 9 | **77.** 6 | **78.** 10 |
| **79.** 4 | **80.** 8 | | | | |

# Appendix II

## Practice with Basic Multiplication Facts

Find all the following products mentally as fast as you can. Check your answers. Redo **all** the exercises until you get **all** the answers right and improve your time.

| | | | | |
|---|---|---|---|---|
| **1.** $2 \times 6$ | **2.** $3 \times 4$ | **3.** $5 \times 0$ | **4.** $7 \times 2$ | **5.** $3 \times 8$ |
| **6.** $5 \times 7$ | **7.** $9 \times 2$ | **8.** $6 \times 6$ | **9.** $4 \times 7$ | **10.** $5 \times 2$ |
| **11.** $8 \times 4$ | **12.** $9 \times 0$ | **13.** $2 \times 2$ | **14.** $1 \times 4$ | **15.** $1 \times 1$ |
| **16.** $4 \times 9$ | **17.** $9 \times 8$ | **18.** $6 \times 7$ | **19.** $7 \times 8$ | **20.** $6 \times 9$ |
| **21.** $3 \times 6$ | **22.** $7 \times 3$ | **23.** $8 \times 2$ | **24.** $7 \times 9$ | **25.** $1 \times 4$ |
| **26.** $2 \times 3$ | **27.** $8 \times 6$ | **28.** $2 \times 0$ | **29.** $7 \times 2$ | **30.** $8 \times 8$ |
| **31.** $9 \times 9$ | **32.** $5 \times 5$ | **33.** $5 \times 6$ | **34.** $4 \times 7$ | **35.** $3 \times 0$ |
| **36.** $5 \times 2$ | **37.** $3 \times 7$ | **38.** $2 \times 3$ | **39.** $7 \times 2$ | **40.** $9 \times 2$ |
| **41.** $4 \times 4$ | **42.** $9 \times 6$ | **43.** $8 \times 1$ | **44.** $5 \times 8$ | **45.** $9 \times 7$ |
| **46.** $9 \times 5$ | **47.** $5 \times 4$ | **48.** $0 \times 7$ | **49.** $6 \times 8$ | **50.** $1 \times 5$ |

| | | | | |
|---|---|---|---|---|
| **51.** $\begin{array}{r} 5 \\ \times 3 \\ \hline \end{array}$ | **52.** $\begin{array}{r} 9 \\ \times 8 \\ \hline \end{array}$ | **53.** $\begin{array}{r} 7 \\ \times 7 \\ \hline \end{array}$ | **54.** $\begin{array}{r} 3 \\ \times 3 \\ \hline \end{array}$ | **55.** $\begin{array}{r} 6 \\ \times 1 \\ \hline \end{array}$ |
| **56.** $\begin{array}{r} 4 \\ \times 8 \\ \hline \end{array}$ | **57.** $\begin{array}{r} 6 \\ \times 9 \\ \hline \end{array}$ | **58.** $\begin{array}{r} 2 \\ \times 8 \\ \hline \end{array}$ | **59.** $\begin{array}{r} 4 \\ \times 4 \\ \hline \end{array}$ | **60.** $\begin{array}{r} 1 \\ \times 7 \\ \hline \end{array}$ |
| **61.** $\begin{array}{r} 9 \\ \times 1 \\ \hline \end{array}$ | **62.** $\begin{array}{r} 3 \\ \times 9 \\ \hline \end{array}$ | **63.** $\begin{array}{r} 8 \\ \times 7 \\ \hline \end{array}$ | **64.** $\begin{array}{r} 4 \\ \times 0 \\ \hline \end{array}$ | **65.** $\begin{array}{r} 6 \\ \times 8 \\ \hline \end{array}$ |
| **66.** $\begin{array}{r} 7 \\ \times 4 \\ \hline \end{array}$ | **67.** $\begin{array}{r} 9 \\ \times 4 \\ \hline \end{array}$ | **68.** $\begin{array}{r} 5 \\ \times 6 \\ \hline \end{array}$ | **69.** $\begin{array}{r} 3 \\ \times 7 \\ \hline \end{array}$ | **70.** $\begin{array}{r} 2 \\ \times 1 \\ \hline \end{array}$ |
| **71.** $\begin{array}{r} 6 \\ \times 0 \\ \hline \end{array}$ | **72.** $\begin{array}{r} 0 \\ \times 5 \\ \hline \end{array}$ | **73.** $\begin{array}{r} 8 \\ \times 9 \\ \hline \end{array}$ | **74.** $\begin{array}{r} 9 \\ \times 7 \\ \hline \end{array}$ | **75.** $\begin{array}{r} 5 \\ \times 4 \\ \hline \end{array}$ |
| **76.** $\begin{array}{r} 6 \\ \times 3 \\ \hline \end{array}$ | **77.** $\begin{array}{r} 4 \\ \times 2 \\ \hline \end{array}$ | **78.** $\begin{array}{r} 6 \\ \times 4 \\ \hline \end{array}$ | **79.** $\begin{array}{r} 8 \\ \times 1 \\ \hline \end{array}$ | **80.** $\begin{array}{r} 3 \\ \times 5 \\ \hline \end{array}$ |

## Appendix II Answers

| | | | | | |
|---|---|---|---|---|---|
| **1.** 12 | **2.** 12 | **3.** 0 | **4.** 14 | **5.** 24 | **6.** 35 |
| **7.** 18 | **8.** 36 | **9.** 28 | **10.** 10 | **11.** 32 | **12.** 0 |
| **13.** 4 | **14.** 4 | **15.** 1 | **16.** 36 | **17.** 72 | **18.** 42 |
| **19.** 56 | **20.** 54 | **21.** 18 | **22.** 21 | **23.** 16 | **24.** 63 |
| **25.** 4 | **26.** 6 | **27.** 48 | **28.** 0 | **29.** 14 | **30.** 64 |
| **31.** 81 | **32.** 25 | **33.** 30 | **34.** 28 | **35.** 0 | **36.** 10 |
| **37.** 21 | **38.** 6 | **39.** 14 | **40.** 18 | **41.** 16 | **42.** 54 |
| **43.** 8 | **44.** 40 | **45.** 63 | **46.** 45 | **47.** 20 | **48.** 0 |
| **49.** 48 | **50.** 5 | **51.** 15 | **52.** 72 | **53.** 49 | **54.** 9 |
| **55.** 6 | **56.** 32 | **57.** 54 | **58.** 16 | **59.** 16 | **60.** 7 |
| **61.** 9 | **62.** 27 | **63.** 56 | **64.** 0 | **65.** 48 | **66.** 28 |
| **67.** 36 | **68.** 30 | **69.** 21 | **70.** 2 | **71.** 0 | **72.** 0 |
| **73.** 72 | **74.** 63 | **75.** 20 | **76.** 18 | **77.** 8 | **78.** 24 |
| **79.** 8 | **80.** 15 | | | | |

# Appendix III Ancient Numeration Systems

## III.1 Egyptian, Myan, Attic Greek, and Roman Systems

The number systems used by ancient peoples are interesting from a historical point of view, but from a mathematical point of view, they are difficult to work with. One of the many things that determine the progress of any civilization is its system of numeration. Humankind has made its most rapid progress since the invention of the zero and the place value system (which we will discuss in the next section) by the Hindu-Arabic peoples about A.D. 800.

### Egyptian Numerals (Hieroglyphics)

The ancient Egyptians used a set of symbols called **hieroglyphics** as early as 3500 B.C. (See Table III.1.) To write the numeral for a number, the Egyptians wrote the symbols next to each other from left to right, and the number represented was the sum of the values of the symbols. The most times any symbol was used was nine. Instead of using a symbol ten times, they used the symbol for the next higher number. They also grouped the symbols in threes or fours.

**Table III.1 Egyptian Hieroglyphic Numerals**

| Symbol | Name | | Value |
|---|---|---|---|
| 𓏺 | Staff (vertical stroke) | 1 | one |
| 𓎆 | Heel bone (arch) | 10 | ten |
| 𓍢 | Coil of rope (scroll) | 100 | one hundred |
| 𓆼 | Lotus flower | 1000 | one thousand |
| 𓂭 | Pointing finger | 10,000 | ten thousand |
| 𓆐 | Bourbot (tadpole) | 100,000 | one hundred thousand |
| 𓁨 | Astonished man | 1,000,000 | one million |

**Example 1**

𓆼 𓍢𓍢𓍢𓍢𓍢𓍢 𓎆𓎆 𓏺𓏺𓏺𓏺𓏺𓏺𓏺 represents the number one thousand six hundred twenty-seven, or 1000 + 600 + 20 + 7 = 1627

## The Mayan System

The Mayans used a system of dots and bars (for numbers from 1 to 19) combined with a place value system. A dot represented one and a bar represented five. They had a symbol, [zero symbol], for zero and based their system, with one exception, on twenty. (See Table III.2.) The symbols were arranged vertically, smaller values starting at the bottom. The value of the third place up was 360 (18 times the value of the second place), but all other places were 20 times the value of the previous place.

| Table III.2 | Mayan Numerals |
|---|---|
| **Symbol** | **Value** |
| • | 1 one |
| — | 5 five |
| [zero symbol] | 0 zero |

### Example 2

| | |
|---|---|
| ••• over one bar | ( 3 + 5 = 8 ) |

### Example 3

| | |
|---|---|
| •••• over three bars | ( 3 × 5+ 4 = 19 ) |

### Example 4

| | |
|---|---|
| • • • | 3 20's |
| [zero symbol] | 0 units |

( 3 · 20 + 0 = 60 )

### Example 5

| | |
|---|---|
| • • | 2 7200's |
| [zero symbol] | 0 360's |
| • over one bar | 6 20's |
| •• over one bar | 7 units |

( 2 × 7200 + 0 × 360 + 6 × 20 + 7 = 14,527)

**(Note:** [zero symbol] is used as a place holder.)

## Attic Greek System

The Greeks used two numeration systems, the Attic (see Table III.3) and the Alexandrian (see Section III.2). For information on the Alexandrian system). In the Attic system, no numeral was used more than four times. When a symbol was needed five or more times, the symbol for five was used, as shown in the examples.

| Table III.3 | Attic Greek Numerals | |
|---|---|---|
| **Symbol** | **Value** | |
| I | 1 | one |
| Γ | 5 | five |
| Δ | 10 | ten |
| H | 100 | one hundred |
| X | 1000 | one thousand |
| M | 10,000 | ten thousand |

### Example 6

X X 𐅅 H H 𐅄 I I I I  ( $2 \times 1000 + 7 \times 100 + 5 \times 10 + 4 = 2754$ )

### Example 7

𐅆 H H H H Δ Δ Γ  ( $5 \times 1000 + 4 \times 100 + 2 \times 10 + 5 = 5425$ )

## Roman System

The Romans used a system (Table III.4) that we still see in evidence as hours on clocks and dates on buildings.

| Table III.4 | Roman Numerals | |
|---|---|---|
| **Symbol** | **Value** | |
| I | 1 | one |
| V | 5 | five |
| X | 10 | ten |
| L | 50 | fifty |
| C | 100 | one hundred |
| D | 500 | five hundred |
| M | 1000 | one thousand |

The symbols were written largest to smallest, from left to right. The value of the numeral was the sum of the values of the individual symbols. Each symbol was used as many times as necessary, with the following exceptions: When the Romans got to 4, 9, 40, 90, 400, or 900, they used a system of subtraction.

I V = 5 − 1 = 4  X C = 100 − 10 = 90
I X = 10 − 1 = 9  C D = 500 − 100 = 400
X L = 50 − 10 = 40  C M = 1000 − 100 = 900

### Example 8

V I I represents 7

### Example 9

D X L I V represents 544

### Example 10

M C C C X X V I I I represents 1328

Name ______________________ Section ______________ Date ____________

## Exercises III.1

Find the values of the following ancient numbers.

1. 𓍢𓍢𓎆𓎆𓎆𓎆𓎆𓏺𓏺𓏺

2. 𓆐𓂭𓂭𓂭𓂭𓂭𓂭𓆼𓆼𓆼𓎆𓎆𓎆𓎆𓏺

3. 𓂭𓂭𓆼𓏺𓏺𓏺𓏺𓏺𓏺𓏺𓍢𓍢𓍢𓎆𓎆

4. 𝋭

5. 𝋧<br>𝋠

6. 𝋬<br>𝋰

7. 𐅄III

8. 𐅅HHΔΔΔ𐅄I

9. XXHHH𐅄ΔII

10. XCVII

11. DCCXLIV

12. MMMCDLXV

13. CMLXXVIII

14. Write 64

a. as an Egyptian numeral

𓎆𓎆𓎆𓎆𓎆𓎆𓏺𓏺𓏺𓏺

b. as a Mayan numeral

𝋣<br>𝋤

c. as an Attic Greek numeral

𐅄ΔIIII

d. as a Roman numeral

LXIV

**ANSWERS**

1. 253
2. 163,041
3. 21,327
4. 13
5. 140
6. 256
7. 53
8. 781
9. 2362
10. 97
11. 744
12. 3465
13. 978
14. [Respond below exercise]

15. [Respond below exercise]

16. [Respond below exercise]

17. [Respond below exercise]

**15.** Follow the instructions for Exercise 14, using 532 in place of 64.

**a.** 9 9 9 9 9 ∩∩∩ I I

**b.**
•
• • •
—
• •
—
—

**c.** $\Gamma^{H}$Δ Δ Δ I I

**d.** D X X X I I

**16.** Follow the same instructions, using 1969.

**a.** ⚘ 9 9 9 9 9 9 9 9 9 ∩∩∩∩∩∩ IIIII IIII

**b.**
—
• • •
—
• • • •
—

**c.** X $\Gamma^{H}$ H H H H $\Gamma^{\Delta}$ Δ Γ I I I I

**d.** M C M L X I X

**17.** Follow the same instructions, using 846.

**a.** 9 9 9 9 9 9 9 9 ∩∩∩∩ III III

**b.**
• • •
—
•
—

**c.** $\Gamma^{H}$ H H H Δ Δ Δ Δ Γ I

**d.** D C C C X L V I

# III.2 BABYLONIAN, ALEXANDRIAN GREEK, AND CHINESE-JAPANESE SYSTEMS

## Babylonian System (Cuneiform Numerials)

The Babylonians (about 3500 B.C.) used a place value system based on the number sixty, called a **sexagesimal system**. They had only two symbols, ∨ and <. (See table III.5.) These wedge shapes are called **cuneiform numerals**, since **cuneus** means "wedge" in Latin.

| Table III.5 | Cuneiform | |
|---|---|---|
| **Symbol** | **Value** | |
| ∨ | 1 | one |
| < | 10 | ten |

The symbol for one was used as many as nine times, and the symbol for ten as many as five time; however, since there was no symbol for zero, many Babylonian numbers could be read several ways. For our purposes, we will group the symbols to avoid some of the ambiguities inherent in the system.

### Example 1

∨∨∨ << ∨∨∨/∨∨ <<< ∨∨

(3 × 60²) + (25 × 60¹) + (32 × 1)
= (3 × 3600) + (25 × 60) + 32
= 10,800 + 1500 + 32
= 12,332

### Example 2

∨ <<<< ∨∨∨/∨∨∨ <∨∨∨∨/∨∨∨

$(1 \times 60^2) + (46 \times 60^1) + (17 \times 1) = 3600 + 2760 + 17 = 6377$

## Alexandrian Greek System

The Greeks used two numeration systems, the Attic and the Alexandrian. We discussed the Attic Greek system in Section III.1.

In the Alexandrian system (Table III.6), the letters were written next to each other, largest to smallest, from left to right. Since the numerals were also part of the Greek alphabet, an accent mark or bar was sometimes used above a letter to indicate that it represented a number. Multiples of 1000 were indicated by strikes (/) in front of the unit symbols, and multiples of 10,000 were indicated by placing the unit symbols above the symbol M.

| Table III.6 | Alexandrian Greek Symbols | | | | | | |
|---|---|---|---|---|---|---|---|
| **Symbol** | **Name** | **Value** | | **Symbol** | **Name** | **Value** | |
| Α | Alpha | 1 | one | Ν | Nu | 50 | fifty |
| Β | Beta | 2 | two | Ξ | Xi | 60 | sixty |
| Γ | Gamma | 3 | three | Ο | Omicron | 70 | seventy |
| Δ | Delta | 4 | four | Π | Pi | 80 | eighty |
| Ε | Epsilon | 5 | five | C | Koppa | 90 | ninety |
| F | Digamma (or Vau) | 6 | six | Ρ | Ro | 100 | one hundred |
| | | | | Σ | Sigma | 200 | two hundred |
| Ζ | Zeta | 7 | seven | Τ | Tau | 300 | three hundred |
| Η | Eta | 8 | eight | Υ | Upsilon | 400 | four hundred |
| Θ | Theta | 9 | nine | Φ | Phi | 500 | five hundred |
| Ι | Iota | 10 | ten | Χ | Chi | 600 | six hundred |
| Κ | Kappa | 20 | twenty | Ψ | Psi | 700 | seven hundred |
| Λ | Lambda | 30 | thirty | Ω | Omega | 800 | eight hundred |
| Μ | Mu | 40 | forty | ͲͲ | Sampi | 900 | nine hundred |

### Example 3

$\overline{\Phi\,\Xi\,Z}$ (500 + 60 + 7 = 567)

### Example 4

$\overline{\text{B}}$
Μ Τ Ν Δ (20,000 + 300 + 50 + 4 = 20,354)

## Chinese - Japanese System

The Chinese-Japanese system (Table III.7) uses a different numeral for each of the digits up to ten, then a symbol for each power of ten. A digit written above a power of ten is to be multiplied by that power, and all such results are to be added to find the value of the numeral.

| Table III.7 | Chinese-Japanese Numerals | | |
|---|---|---|---|
| **Symbol** | **Value** | **Symbol** | **Value** |
| 一 | 1 one | 七 | 7 seven |
| 二 | 2 two | 八 | 8 eight |
| 三 | 3 three | 九 | 9 nine |
| 四 | 4 four | 十 | 10 ten |
| 五 | 5 five | 百 | 100 one hundred |
| 六 | 6 six | 千 | 1000 one thousand |

## Example 5

三 十 — 30

九 — 9

(30 + 9 =39)

## Example 6

5000

四 百 — 400

八 十 — 80

二 — 2

(5000 + 400 + 80 + 2 = 5482)

Name ____________ Section ____________ Date ____________

## Exercises III.2

Find the value of each of the following ancient numerals.

**1.** ∨ <<∨∨∨ ∨∨∨

**2.** <<<∨∨

**3.** <<< ∨∨

**4.** $\overline{\text{Y N E}}$

**5.** $\overline{\Delta\text{/Z}}$
M /ZΠ

**6.** $\overline{\Sigma\text{ K B}}$

**7.** 四
十
六

**8.** 五
千
一
百
八

**9.** 九
百
九
十
九

Write the following numbers as (a) Babylonian numerals, (b) Alexandrian Greek numerals, and (c) Chinese-Japanese numerals.

**10.** 472

**a.** ∨∨∨∨ <<< ∨∨
∨∨∨ <<

**b.** $\overline{\text{Y O B}}$

**c.** 四
百
七
十
二

**11.** 596

**a.** ∨∨∨∨∨ <<< ∨∨∨
∨∨∨∨ << ∨∨∨

**b.** $\overline{\Phi\text{ C F}}$

**c.** 五
百
九
十
六

**1.** 4,983
**2.** 32
**3.** 1,802
**4.** 455
**5.** 47,900
**6.** 222
**7.** 46
**8.** 5,108
**9.** 999
**10.** [ Respond below exercise]
**11.** [ Respond below exercise]

12. [ Respond below and in exercise.]

13. [ Respond below and in exercise.]

14. [ Respond below and in exercise.]

15. [ Respond below and in exercise.]

**12.** 5,047

**a.** ∨∨ ∨∨∨∨
∨<< ∨∨ ∨∨∨

**b.** $\overline{/\mathrm{EMZ}}$

**c.** 五
千
四
十
七

**13.** 3,665

**a.** ∨∨∨
∨∨ ∨∨

**b.** $\overline{/\Gamma \mathrm{X} \Xi \mathrm{C} \mathrm{E}}$

**c.** 三
千
六
百
六
十
五

**14.** 7,293

**a.** ∨∨ ∨ <<< ∨∨∨

**b.** $\overline{/\mathrm{Z}\Sigma\mathrm{C}\Gamma}$

**c.** 七
千
二
百
九
十
三

**15.** 10,852

**a.** ∨∨∨ <<< ∨∨
<<

**b.** $\overline{\mathrm{A}}$
MΩNB

**c.** 十
千
八
百
五
十
二

# Appendix IV Greatest Common Divisor

## Greatest Common Divisor (GCD)

Consider the two numbers 12 and 18. Is there a number (or numbers) that will divide into **both** 12 and 18? To help answer this question, the divisors for 12 and 18 are listed below.

Set of divisors for 12: ( 1, 2, 3, 4, 6, 12 )
Set of divisors for 18: ( 1, 2, 3, 6, 9, 18 )

The **common divisors** for 12 and 18 are 1, 2, 3, and 6. The **greatest common divisor (GCD)** for 12 and 18 is 6; that is, of all the common divisors of 12 and 18, 6 is the largest divisor.

### Example 1

List the divisors of each number in the set { 36, 24, 48 } and find the greatest common divisor (GCD).

Set of divisors for 36: { **1, 2, 3, 4, 6,** 9, **12,** 18, 36 }
Set of divisors for 24: { **1, 2, 3, 4, 6,** 8, **12,** 24 }
Set of divisors for 48: { **1, 2, 3, 4, 6,** 8, **12,** 16, 24, 48 }

**Solution**

The common divisors are **1, 2, 3, 4, 6,** and **12. GCD = 12.**

**Definition:**

The **greatest common divisor (GCD)*** of a set of natural numbers is the largest natural number that will divide into all the numbers in the set.

As Example 1 illustrates, listing all the divisors of each number before finding the GCD can be tedious and difficult. **The use of prime factorizations leads to a simple technique for finding the GCD.**

**Technique for Finding the GCD of a Set of Natural Numbers**

1. Find the prime factorization of each number.
2. Find the prime factors common to all factorizations.
3. Form the product of these primes, using each prime the number of times it is common to **all** factorizations.
4. This product is the GCD. If there are no primes common to all factorizations, the GCD is 1.

*The largest common divisor is, of course, the largest common factor, and the GCD could be called the **greatest common factor** and be abbreviated **GCF**.

### Example 2

Find the GCD for { 36, 24, 48 }.

**Solution**

$$\left.\begin{aligned} 36 &= 2\cdot2\cdot3\cdot3 \\ 24 &= 2\cdot2\cdot2\cdot3 \\ 48 &= 2\cdot2\cdot2\cdot2\cdot3 \end{aligned}\right\}\text{GCD} = 2\cdot2\cdot3 = 12$$

The factor 2 appears twice, and the factor 3 appears once in **all** the prime factorizations.

### Example 3

Find the GCD for { 360, 75, 30 }.

**Solution**

$$\left.\begin{aligned} 360 &= 36\cdot10 = 4\cdot9\cdot2\cdot5 = 2\cdot2\cdot2\cdot3\cdot3\cdot5 \\ 75 &= 3\cdot25 = 3\cdot5\cdot5 \\ 30 &= 6\cdot5 = 2\cdot3\cdot5 \end{aligned}\right\}\text{GCD} = 3\cdot5 = 15$$

Each of the factors 3 and 5 appears only once in **all** the prime factorizations.

### Example 4

Find the GCD for { 168, 420, 504 }.

**Solution**

$$\left.\begin{aligned} 168 &= 8\cdot21 = 2\cdot2\cdot2\cdot3\cdot7 \\ 420 &= 10\cdot42 = 2\cdot5\cdot6\cdot7 = 2\cdot2\cdot3\cdot5\cdot7 \\ 504 &= 4\cdot126 = 2\cdot2\cdot6\cdot21 = 2\cdot2\cdot2\cdot3\cdot3\cdot7 \end{aligned}\right\}\text{GCD} = 2\cdot2\cdot3\cdot7 = 84$$

In **all** the prime factorizations, 2 appears twice, 3once, and 7once.

If the GCD of two numbers is 1 (that is, they have no common prime factors), then the two numbers are said to be **relatively prime.** The numbers themselves may be prime or they may be composite.

### Example 5

Find the GCD for { 15, 18 }.

**Solution**

$$\left.\begin{aligned} 15 &= 3\cdot5 \\ 8 &= 2\cdot2\cdot2 \end{aligned}\right\}\text{GCD} = 1$$

8 and 15 are relatively prime.

### Example 6

Find the GCD for { 20, 21 }.

**Solution**

$$\left.\begin{aligned} 20 &= 2\cdot2\cdot5 \\ 21 &= 3\cdot7 \end{aligned}\right\}\text{GCD} = 1$$

8 and 15 are relatively prime.

Name ______________ Section ______________ Date ______________

## Exercises IV

Find the GCD for each of the following sets of numbers.

**1.** { 12, 8 } **2.** { 16, 28 } **3.** { 85, 51 }

**4.** { 20, 75 } **5.** { 20, 30 } **6.** { 42, 48 }

**7.** { 15, 21 } **8.** { 27, 18 } **9.** { 18, 24 }

**10.** { 77, 66 } **11.** { 182, 184 } **12.** { 110, 66 }

**13.** { 8, 16, 64 } **14.** { 121, 44 } **15.** { 28, 52, 56 }

**16.** { 98, 147 } **17.** { 60, 24, 96 } **18.** { 33,, 55, 77 }

**19.** { 25, 50, 75 } **20.** { 30, 78, 60 } **21.** { 17, 15, 21 }

**ANSWERS**

**1.** 4
**2.** 4
**3.** 17
**4.** 5
**5.** 10
**6.** 6
**7.** 3
**8.** 9
**9.** 6
**10.** 11
**11.** 2
**12.** 22
**13.** 8
**14.** 11
**15.** 4
**16.** 49
**17.** 12
**18.** 11
**19.** 25
**20.** 6
**21.** 1

22. 20

23. 1

24. 21

25. 35

26. relatively prime

27. relatively prime

28. 2

29. relatively prime

30. 7

31. relatively prime

32. relatively prime

33. relatively prime

34. 22

35. relatively prime

**22.** { 520, 220 }

**23.** { 14, 55 }

**24.** { 210, 231, 84 }

**25.** { 140, 245, 420 }

State whether the following pairs of numbers are relatively prime. If they are not relatively prime, state their GCD.

**26.** { 35,24 }

**27.** { 11, 23 }

**28.** { 14, 36 }

**29.** { 72, 35 }

**30.** { 42, 77 }

**31.** { 16, 51 }

**32.** { 20, 21 }

**33.** { 8, 15 }

**34.** { 66, 22 }

**35.** { 10, 27 }

# Appendix V Division with Polynomials

## Division of a Polynomial by a Monomial

When two or more fractions with the same denominator are added ( or subtracted ), the numerators are added ( or subtracted ) and the common denominator is used. For example,

$$\frac{x}{2}+\frac{1}{2}=\frac{x+1}{2} \quad \text{and} \quad \frac{5x}{3}-\frac{4}{3}=\frac{5x-4}{3}$$

In each case, the numerator is a polynomial and the denominator is a monomial.

The reverse of this process is called **division by a monomial**.

**Objectives**

① Learn how to divide a polynomial by a monomial, term by term.

② Know how to apply the division algorithm to division with polynomials.

**To Divide a Polynomial by a Monomial**

1. Divide **each term** in the numerator by the denominator.
2. Simplify each resulting fraction.

### Example 1

**a.** $\dfrac{6x^2+12x-3}{3}$

**b.** $\dfrac{12x^3-2x^2+14x+5}{2x^2}$

**Solution**

**a.** $\dfrac{6x^2+12x-3}{3}=\dfrac{6x^2}{3}+\dfrac{12x}{3}-\dfrac{3}{3}$

$=2x^2+4x-1$

**b.** $\dfrac{12x^3-2x^2+14x+5}{2x^2}=\dfrac{12x^3}{2x^2}-\dfrac{2x^2}{2x^2}+\dfrac{14x}{2x^2}+\dfrac{5}{2x^2}$

$=6x-1+\dfrac{7}{x}+\dfrac{5}{2x^2}$

## Division of a Polynomial by a Polynomial

**Long division** in arithmetic is a process of repeated steps that is called the **division algorithm**. For example, to divide 382 by 15, we multiply and subtract and multiply and subtract until the remainder is less than 15:

$$\begin{array}{r} 25 \\ 15\overline{)382} \\ \underline{30} \\ 82 \\ \underline{75} \\ 7 \end{array}$$

← quotient; divisor ↗; ↖ dividend; ← remainder

A similar process, also called the **division algorithm**, is used to divide a polynomial by a polynomial. That is, **we multiply and subtract and multiply and subtract until the remainder is of smaller degree than the divisor.** This division algorithm is illustrated in the following examples. Study the process carefully by reading the

## Example 2

Find the quotient and remainder for

$$\frac{5x^2+17x+10}{x+3}$$

### Solution

**Step 1:** Write both polynomials in order of descending powers.

$$x+3\overline{)5x^2+17x+10}$$

**Step 2:** Trial divide ( mentally ) $5x^2$ by $x$. $\frac{5x^2}{x}=5x$. Write $5x$ in the quotient above $5x^2$.

$$\begin{array}{r} 5x \phantom{+17x+10} \\ x+3\overline{)5x^2+17x+10} \end{array}$$

**Step 3:** Multiply $5x$ times ( $x + 3$ ), and write the terms below the like terms in the dividend.

$$\begin{array}{r} 5x \phantom{+17x+10} \\ x+3\overline{)5x^2+17x+10} \\ \underline{5x^2+15x} \phantom{+10} \end{array}$$

**Step 4:** Subtract by changing signs and adding.

$$\begin{array}{r} 5x \phantom{+17x+10} \\ x+3\overline{)5x^2+17x+10} \\ \underline{\pm 5x^2 \pm 15x} \phantom{+10} \\ 2x \phantom{+10} \end{array}$$

**Step 5:** Bring down the next term, 10.

$$\begin{array}{r} 5x \phantom{+17x+10} \\ x+3\overline{)5x^2+17x+10} \\ \underline{\pm 5x^2 \pm 15x} \phantom{+10} \\ 2x+10 \end{array}$$

Continue the same pattern as steps 2 – 4

**Step 6:** Divide $2x$ by $x$: $\dfrac{2x}{x} = +2$. Write the result in the quotient.

$$\begin{array}{r} 5x+2 \\ x+3 \overline{)\ 5x^2+17x+10} \\ \underline{\pm 5x^2 \pm 15x} \\ 2x+10 \end{array}$$

**Step 7:** Multiply 2 times ( $x + 3$ ), and write the terms below the like terms in the new dividend

$$\begin{array}{r} 5x+2 \\ x+3 \overline{)\ 5x^2+17x+10} \\ \underline{\pm 5x^2 \pm 15x} \\ 2x+10 \\ \underline{2x+\ 6} \end{array}$$

**Step 8:** Subtract by changing signs and adding.

$$\begin{array}{r} 5x+2 \\ x+3 \overline{)\ 5x^2+17x+10} \\ \underline{\pm 5x^2 \pm 15x} \\ 2x+10 \\ \underline{\pm 2x \pm\ 6} \\ 4 \end{array}$$

The quotient is $5x + 2$, and the remainder is 4.

CHECK: Multiply the divisor and quotient, and add the remainder. The result must be the original dividend.

$$\begin{aligned} (x+3)(5x+2)+4 &= 5x^2+15x+2x+6+4 \\ &= 5x^2+17x+10 \end{aligned}$$

## Example 3

Find the quotient and remainder for

$$\frac{6x^2-7x+11}{2x+1}$$

### Solution

**Step 1:** Write both polynomials in order of descending powers.

$$2x+1 \overline{)6x^2-7x+11}$$

**Step 2:** Trial divide (mentally) $6x^2$ by $2x$. $\dfrac{6x^2}{2x} = 3x$. Write $3x$ in the quotient above $6x^2$.

$$\begin{array}{r} 3x\phantom{-7x+11} \\ 2x+1 \overline{)\ 6x^2-7x+11} \end{array}$$

**Step 3:** Multiply $3x$ times $(2x+1)$, and write the terms below the like terms in the dividend.

$$\begin{array}{r} 3x \phantom{-7x+11} \\ 2x+1\overline{\big)\, 6x^2 - 7x + 11} \\ \underline{6x^2 + 3x} \phantom{+11} \end{array}$$

**Step 4:** Subtract by changing signs and adding.

$$\begin{array}{r} 3x \phantom{-7x+11} \\ 2x+1\overline{\big)\, 6x^2 - 7x + 11} \\ \underline{\pm 6x^2 \pm 3x} \phantom{+11} \\ -10x \phantom{+11} \end{array}$$

**Step 5:** Bring down the next term, 11.

$$\begin{array}{r} 3x \phantom{-7x+11} \\ 2x+1\overline{\big)\, 6x^2 - 7x + 11} \\ \underline{\pm 6x^2 \pm 3x} \phantom{+11} \\ -10x + 11 \end{array}$$

Continue the same pattern as steps 2 – 4

**Step 6:** Divide $-10x$ by $2x$: $\frac{-10x}{2x} = -5$. Write the result in the quotient.

$$\begin{array}{r} 3x - 5 \phantom{+11} \\ 2x+1\overline{\big)\, 6x^2 - 7x + 11} \\ \underline{\pm 6x^2 \pm 3x} \phantom{+11} \\ -10x + 11 \end{array}$$

**Step 7:** Multiply $-5$ times $(2x+1)$, and write the terms below the like terms in the new dividend.

$$\begin{array}{r} 3x - 5 \phantom{+11} \\ 2x+1\overline{\big)\, 6x^2 - 7x + 11} \\ \underline{\pm 6x^2 \pm 3x} \phantom{+11} \\ -10x + 11 \\ \underline{-10x - \phantom{1}5} \end{array}$$

**Step 8:** Subtract by changing signs and adding.

$$\begin{array}{r} 3x - 5 \phantom{+11} \\ 2x+1\overline{\big)\, 6x^2 - 7x + 11} \\ \underline{\pm 6x^2 \pm 3x} \phantom{+11} \\ -10x + 11 \\ \underline{\pm 10x \pm \phantom{1}5} \\ 16 \end{array}$$

The quotient is $3x - 5$, and the remainder is 16.

CHECK: Multiply the divisor and quotient, and add the remainder. The result must be the original dividend.

$$\begin{aligned} (2x+1)(3x-5) + 16 &= 6x^2 + 3x - 10x - 5 + 16 \\ &= 6x^2 - 7x + 11 \end{aligned}$$

Name ______________ Section ______________ Date ______________

## Exercises V

Divide each polynomial by the monomial denominator, and simplify each resulting fraction.

1. $\dfrac{3x^2+9x+6}{3}$

2. $\dfrac{8x^2+4x-12}{4}$

3. $\dfrac{14x^2-8x+5}{2}$

4. $\dfrac{15x^2-10x-20}{-5}$

5. $\dfrac{4x^2-5x}{x}$

6. $\dfrac{7x^2+9x}{x}$

7. $\dfrac{6x^3+12x^2-2x+6}{3x^2}$

8. $\dfrac{5x^3+4x^2+2x-18}{2x^2}$

9. $\dfrac{14x^3-28x-1}{7x^2}$

10. $\dfrac{12x^3-12x^2-6}{12x^2}$

11. $\dfrac{5x^3+11x^2-2x}{x^3}$

12. $\dfrac{x^3+3x^2+x}{x^3}$

13. $\dfrac{-16x^3-12x^2-2x+8}{4x^2}$

14. $\dfrac{-6x^3+3x+15}{3x^3}$

Use the division algorithm ( as illustrated in Examples 2 and 3 ) to find the quotient and remainder in each of the following problems.

15. $\dfrac{2x^2+5x+6}{x+2}$

16. $\dfrac{3x^2+4x+2}{x+1}$

**ANSWERS**

1. $x^2+3x+2$
2. $2x^2+x-3$
3. $7x^2-4x+\dfrac{5}{2}$
4. $-3x^2+2x+4$
5. $4x-5$
6. $7x+9$
7. $2x+4-\dfrac{2}{3x}+\dfrac{2}{x^2}$
8. $\dfrac{5x}{2}+2+\dfrac{1}{x}-\dfrac{9}{x^2}$
9. $2x-\dfrac{4}{x}-\dfrac{1}{7x^2}$
10. $x-1-\dfrac{1}{2x^2}$
11. $5+\dfrac{11}{x}-\dfrac{2}{x^2}$
12. $1+\dfrac{3}{x}+\dfrac{1}{x^2}$
13. $-4x-3-\dfrac{1}{2x}+\dfrac{2}{x^2}$
14. $-2+\dfrac{1}{x^2}+\dfrac{5}{x^3}$
15. $2x+1$; R 4
16. $3x+1$; R 1

17. $y-19$; R 80

18. $4y-1$; R $-6$

19. $4y-7$; R 12

20. $6y-6$; R 1

21. $10x+3$; R 6

22. $5x+15$; R 2

23. $x^2+3x-20$; R 42

24. $y^2-2y+2$; R $-7$

17. $\dfrac{y^2-15y+4}{y+4}$

18. $\dfrac{4y^2+3y-7}{y+1}$

19. $\dfrac{8y^2-10y+5}{2y+1}$

20. $\dfrac{12y^2+6y-17}{2y+3}$

21. $\dfrac{20x^2-24x-3}{2x-3}$

22. $\dfrac{15x^2+40x-13}{3x-1}$

23. $\dfrac{x^3+5x^2-14x+2}{x+2}$

24. $\dfrac{y^3-3y^2+4y-9}{y-1}$

Name ______________________ Section ____________ Date __________

In long division, if the remainder is 0, then the quotient and divisor are **factors** of the dividend. Show in each of the following division problems that the denominator is a factor of the numerator, and find the corresponding factor ( that is, the quotient ).

**25.** $\dfrac{2x^2+5x-12}{x+4}$

**26.** $\dfrac{3x^2+8x+5}{x+1}$

**27.** $\dfrac{4x^2-25x+25}{x-5}$

**28.** $\dfrac{5x^2-9x-2}{x-2}$

**29.** $\dfrac{4y^2+8y+3}{2y+1}$

**30.** $\dfrac{10y^2+9y-9}{2y+3}$

**31.** $\dfrac{x^3-5x^2-22x+6}{x+3}$

**32.** $\dfrac{x^3+7x^2+7x-6}{x+2}$

**33 .** $\dfrac{y^3-2y^2-11y-20}{y-5}$

**34.** $\dfrac{y^3-y^2-15y+20}{y+4}$

**25.** $2x-3$

**26.** $3x+5$

**27.** $4x-5$

**28.** $5x+1$

**29.** $2y+3$

**30.** $5y-3$

**31.** $x^2-8x+2$

**32.** $x^2+5x-3$

**33.** $y^2+3y+4$

**34.** $y^2-5y+5$

# Tables

## Table 1 U.S. Customary and Metric Equivalents

In the following tables, the equivalents are rounded off.

| U.S. to Metric | Metric to U.S. |
|---|---|
| *Length Equivalents* | |
| 1 in. = 2.54 cm (exact) | 1 cm = .394 in. |
| 1 ft = 0.305 m | 1 m = 3.28 ft |
| 1 yd = 0.914 m | 1 m = 1.09 yd |
| 1 mi = 1.61 km | 1 km = 0.62 mi |
| *Area Equivalents* | |
| 1 in.$^2$ = 6.45 cm$^2$ | 1 cm$^2$ = 0.155 in.$^2$ |
| 1 ft$^2$ = 0.093 m$^2$ | 1 m$^2$ = 10.764 ft$^2$ |
| 1 yd$^2$ = 0.836 m$^2$ | 1 m$^2$ = 1.196 yd$^2$ |
| 1 acre = 0.405 ha | 1 ha = 2.47 acres |
| *Volume Equivalents* | |
| 1 in.$^3$ = 16.387 cm$^3$ | 1 cm$^3$ = .394 in.$^3$ |
| 1 ft$^3$ = 0..28 m$^3$ | 1 m$^3$ = 35.315 ft$^3$ |
| 1 qt = 0.946 L | 1 L = 1.06 qt |
| 1 gal = 3.785 L | 1 L = 0.264 gal |
| *Mass Equivalents* | |
| 1 oz = 28.35g | 1 g = 0.035 oz |
| 1 lb = 0.454 kg | 1 kg = 2.205 lb |

## Table 2 Celsius and Fahrenheit Equivalents

Conversion formulas: $F = \frac{9}{5}C + 32$ $C = \frac{5}{9}(F - 32)$

| Celsius | Fahrenheit | |
|---|---|---|
| 100° | 212° | ← water boils at sea level |
| 95° | 203° | |
| 90° | 194° | |
| 85° | 185° | |
| 80° | 176° | |
| 75° | 167° | |
| 70° | 158° | |
| 65° | 149° | |
| 60° | 140° | |
| 55° | 131° | |
| 50° | 122° | |
| 45° | 113° | |
| 40° | 104° | |
| 35° | 95° | |
| 30° (comfort range) | 86° (comfort range) | |
| 25° (comfort range) | 77° (comfort range) | |
| 20° (comfort range) | 68° (comfort range) | |
| 15° | 59° | |
| 10° | 50° | |
| 5° | 41° | |
| 0° | 32° | ← water freezes at sea level |

## Table 3 Powers, Roots, and Prime Factorizations

| $x$ | $x^2$ | $\sqrt{x}$ | $x^3$ | $\sqrt[3]{x}$ | Prime Factorization |
|---|---|---|---|---|---|
| 1 | 1 | 1.0000 | 1 | 1.0000 | — |
| 2 | 4 | 1.4142 | 8 | 1.2599 | prime |
| 3 | 9 | 1.7321 | 27 | 1.4423 | prime |
| 4 | 16 | 2.0000 | 64 | 1.5874 | $2 \cdot 2$ |
| 5 | 25 | 2.2361 | 125 | 1.7100 | prime |
| 6 | 36 | 2.4495 | 216 | 1.8171 | $2 \cdot 3$ |
| 7 | 49 | 2.6458 | 343 | 1.9129 | prime |
| 8 | 64 | 2.8284 | 512 | 2.0000 | $2 \cdot 2 \cdot 2$ |
| 9 | 81 | 3.0000 | 729 | 2.0801 | $3 \cdot 3$ |
| 10 | 100 | 3.1623 | 1000 | 2.1544 | $2 \cdot 5$ |
| 11 | 121 | 3.3166 | 1331 | 2.2240 | prime |
| 12 | 144 | 3.4641 | 1728 | 2.2894 | $2 \cdot 2 \cdot 3$ |
| 13 | 169 | 3.6056 | 2197 | 2.3513 | prime |
| 14 | 196 | 3.7417 | 2744 | 2.4101 | $2 \cdot 7$ |
| 15 | 225 | 3.8730 | 3375 | 2.4662 | $3 \cdot 5$ |
| 16 | 256 | 4.0000 | 4096 | 2.5198 | $2 \cdot 2 \cdot 2 \cdot 2$ |
| 17 | 289 | 4.1231 | 4913 | 2.5713 | prime |
| 18 | 324 | 4.2426 | 5832 | 2.6207 | $2 \cdot 3 \cdot 3$ |
| 19 | 361 | 4.3589 | 6859 | 2.6684 | prime |
| 20 | 400 | 4.4721 | 8000 | 2.7144 | $2 \cdot 2 \cdot 5$ |
| 21 | 441 | 4.5826 | 9261 | 2.7589 | $3 \cdot 7$ |
| 22 | 484 | 4.6904 | 10,648 | 2.8020 | $2 \cdot 11$ |
| 23 | 529 | 4.7958 | 12,167 | 2.8439 | prime |
| 24 | 576 | 4.8990 | 13,824 | 2.8845 | $2 \cdot 2 \cdot 2 \cdot 3$ |
| 25 | 625 | 5.0000 | 15,625 | 2.9240 | $5 \cdot 5$ |
| 26 | 676 | 5.0990 | 17,576 | 2.9625 | $2 \cdot 13$ |
| 27 | 729 | 5.1962 | 19,683 | 3.0000 | $3 \cdot 3 \cdot 3$ |
| 28 | 784 | 5.2915 | 21,952 | 3.0366 | $2 \cdot 2 \cdot 7$ |
| 29 | 841 | 5.3852 | 24,389 | 3.0723 | prime |
| 30 | 900 | 5.4772 | 27,000 | 3.1072 | $2 \cdot 3 \cdot 5$ |
| 31 | 961 | 5.5678 | 29,791 | 3.1414 | prime |
| 32 | 1024 | 5.6569 | 32,768 | 3.1748 | $2 \cdot 2 \cdot 2 \cdot 2 \cdot 2$ |
| 33 | 1089 | 5.7446 | 35,937 | 3.2075 | $3 \cdot 11$ |
| 34 | 1156 | 5.8310 | 39,304 | 3.2396 | $2 \cdot 17$ |
| 35 | 1225 | 5.9161 | 42,875 | 3.2711 | $5 \cdot 7$ |
| 36 | 1296 | 6.0000 | 46,656 | 3.3019 | $2 \cdot 2 \cdot 3 \cdot 3$ |
| 37 | 1369 | 6.0828 | 50,653 | 3.3322 | prime |
| 38 | 1444 | 6.1644 | 54,872 | 3.3620 | $2 \cdot 19$ |
| 39 | 1521 | 6.2450 | 59,319 | 3.3912 | $3 \cdot 13$ |
| 40 | 1600 | 6.3246 | 64,000 | 3.4200 | $2 \cdot 2 \cdot 2 \cdot 5$ |
| 41 | 1681 | 6.4031 | 68,921 | 3.4482 | prime |
| 42 | 1764 | 6.4807 | 74,088 | 3.4760 | $2 \cdot 3 \cdot 7$ |
| 43 | 1849 | 6.5574 | 79,507 | 3.5034 | prime |
| 44 | 1936 | 6.6333 | 85,184 | 3.5303 | $2 \cdot 2 \cdot 11$ |
| 45 | 2025 | 6.7082 | 91,125 | 3.5569 | $3 \cdot 3 \cdot 5$ |
| 46 | 2116 | 6.7823 | 97,336 | 3.5830 | $2 \cdot 23$ |
| 47 | 2209 | 6.8557 | 103,823 | 3.6088 | prime |
| 48 | 2304 | 6.9282 | 110,592 | 3.6342 | $2 \cdot 2 \cdot 2 \cdot 2 \cdot 3$ |
| 49 | 2401 | 7.0000 | 117,649 | 3.6593 | $7 \cdot 7$ |
| 50 | 2500 | 7.0711 | 125,000 | 3.6840 | $2 \cdot 5 \cdot 5$ |

## Table 3 (continued)

| $x$ | $x^2$ | $\sqrt{x}$ | $x^3$ | $\sqrt[3]{x}$ | Prime Factorization |
|---|---|---|---|---|---|
| 51 | 2601 | 7.1414 | 132,651 | 3.7084 | $3 \cdot 17$ |
| 52 | 2704 | 7.2111 | 140,608 | 3.7325 | $2 \cdot 2 \cdot 13$ |
| 53 | 2809 | 7.2801 | 148,877 | 3.7563 | prime |
| 54 | 2916 | 7.3485 | 157,464 | 3.7798 | $2 \cdot 3 \cdot 3 \cdot 3$ |
| 55 | 3025 | 7.4162 | 166,375 | 3.8030 | $5 \cdot 11$ |
| 56 | 3136 | 7.4833 | 175,616 | 3.8259 | $2 \cdot 2 \cdot 2 \cdot 7$ |
| 57 | 3249 | 7.5498 | 185,193 | 3.8485 | $3 \cdot 19$ |
| 58 | 3364 | 7.6158 | 195,112 | 3.8709 | $2 \cdot 29$ |
| 59 | 3481 | 7.6811 | 205,379 | 3.8930 | prime |
| 60 | 3600 | 7.7460 | 216,000 | 3.9149 | $2 \cdot 2 \cdot 3 \cdot 5$ |
| 61 | 3721 | 7.8103 | 226,981 | 3.9365 | prime |
| 62 | 3844 | 7.8740 | 238,328 | 3.9579 | $2 \cdot 31$ |
| 63 | 3969 | 7.9373 | 250,047 | 3.9791 | $3 \cdot 3 \cdot 7$ |
| 64 | 4096 | 8.0000 | 262,144 | 4.0000 | $2 \cdot 2 \cdot 2 \cdot 2 \cdot 2 \cdot 2$ |
| 65 | 4225 | 8.0623 | 274,625 | 4.0207 | $5 \cdot 13$ |
| 66 | 4356 | 8.1240 | 287,496 | 4.0412 | $2 \cdot 3 \cdot 11$ |
| 67 | 4489 | 8.1854 | 300,763 | 4.0615 | prime |
| 68 | 4624 | 8.2462 | 314,432 | 4.0817 | $2 \cdot 2 \cdot 17$ |
| 69 | 4761 | 8.3066 | 328,509 | 4.1016 | $3 \cdot 23$ |
| 70 | 4900 | 8.3666 | 343,000 | 4.1213 | $2 \cdot 5 \cdot 7$ |
| 71 | 5041 | 8.4262 | 357,911 | 4.408 | prime |
| 72 | 5184 | 8.4853 | 373,248 | 4.1602 | $2 \cdot 2 \cdot 2 \cdot 3 \cdot 3$ |
| 73 | 5329 | 8.5440 | 389,017 | 4.1793 | prime |
| 74 | 5476 | 8.6023 | 405,224 | 4.1983 | $2 \cdot 37$ |
| 75 | 5625 | 8.6603 | 421,875 | 4.2172 | $3 \cdot 5 \cdot 5$ |
| 76 | 5776 | 8.7178 | 438,976 | 4.2358 | $2 \cdot 2 \cdot 19$ |
| 77 | 5929 | 8.7750 | 456,533 | 4.2543 | $7 \cdot 11$ |
| 78 | 6084 | 8.8318 | 474,552 | 4.2727 | $2 \cdot 3 \cdot 13$ |
| 79 | 6241 | 8.8882 | 493,039 | 4.2908 | prime |
| 80 | 6400 | 8.9443 | 512,000 | 4.3089 | $2 \cdot 2 \cdot 2 \cdot 2 \cdot 5$ |
| 81 | 6561 | 9.0000 | 531,441 | 4.3267 | $3 \cdot 3 \cdot 3 \cdot 3$ |
| 82 | 6724 | 9.0554 | 551,368 | 4.3445 | $2 \cdot 41$ |
| 83 | 6889 | 9.1104 | 571,787 | 4.3621 | prime |
| 84 | 7056 | 9.1652 | 592,704 | 4.3795 | $2 \cdot 2 \cdot 3 \cdot 7$ |
| 85 | 7225 | 9.2195 | 614,125 | 4.3968 | $5 \cdot 17$ |
| 86 | 7396 | 9.2736 | 636,056 | 4.4140 | $2 \cdot 43$ |
| 87 | 7569 | 9.3274 | 658,503 | 4.4310 | $3 \cdot 29$ |
| 88 | 7744 | 9.3808 | 681,472 | 4.4480 | $2 \cdot 2 \cdot 2 \cdot 11$ |
| 89 | 7921 | 9.4340 | 704,969 | 4.4647 | prime |
| 90 | 8100 | 9.4868 | 729,000 | 4.4814 | $2 \cdot 3 \cdot 3 \cdot 5$ |
| 91 | 8281 | 9.5394 | 753,571 | 4.4979 | $7 \cdot 13$ |
| 92 | 8464 | 9.5917 | 778,688 | 4.5144 | $2 \cdot 2 \cdot 23$ |
| 93 | 8649 | 9.6437 | 804,357 | 4.5307 | $3 \cdot 31$ |
| 94 | 8836 | 9.6954 | 830,584 | 4.5468 | $2 \cdot 47$ |
| 95 | 9025 | 9.7468 | 857,375 | 4.5629 | $5 \cdot 19$ |
| 96 | 9216 | 9.7980 | 884,736 | 4.5789 | $2 \cdot 2 \cdot 2 \cdot 2 \cdot 2 \cdot 3$ |
| 97 | 9409 | 9.8489 | 912,673 | 4.5947 | prime |
| 98 | 9604 | 9.8995 | 941,192 | 4.6104 | $2 \cdot 7 \cdot 7$ |
| 99 | 9801 | 9.9499 | 970,299 | 4.6261 | $3 \cdot 3 \cdot 11$ |
| 100 | 10,000 | 10.0000 | 1,000,000 | 4.6416 | $2 \cdot 2 \cdot 5 \cdot 5$ |

## About Table 4 and the Value of $\pi$

As discussed in the text on page 423, $\pi$ is an irrational number, and so the decimal form of $\pi$ is an infinite nonrepeating decimal. Mathematicians even in ancient times realized that $\pi$ is a constant value obtained form the ration of a circle's circumference to its diameter, but they had no sense that it might be an irrational number. As early as about 1800 B.C. the Babylonians gave $\pi$ a value of 3, and around 1600 B.C. the ancient Egyptians were using the approximation of 256/81, what would be decimal value of about 3.1605. In the third century B.C. the Greek mathematician Archimedes used polygons approximating a circle to determine that the value of $\pi$ must lie between 223/71(=3.1408) and 22/7(=3.1429). He was thus accurate to two decimal places. About seven hundred years later, in the fourth century A.D. Chinese mathematician Tsu Chung-Chi refined Archimedes' method and expressed the constant as 355/113, which was correct to six decimal places. By 1610, Ludolph van Ceulen of Germany had also used a polygon method to find $\pi$ accurate to 35 decimal places.

Knowing that the decimal expression of $\pi$ would not terminate, mathematicians still sought a repeating pattern in its digits. Such a pattern would mean that $\pi$ was a rational number and that there would be some ration of two whole numbers that would produce the correct decimal representations. Finally, in 1767, Johann Heinrich Lambert provided a proof to show that $\pi$ is indeed irrational and thus is nonrepeating as well as nonterminating.

Since Lambert's proof, mathematicians have still made an exercise of calculating $\pi$ to more and more decimal places. The advent of the computer age in this century has made that work immeasurably easier, and on occasion you will still see newspaper articles pronouncing that mathematics researchers have reached a new high in the number of decimal places in their approximations. In 1988 that number was 201,326,000 decimal places. Within 1 year that record was more than doubled, and most recent approximations of $\pi$ now reach beyond one billion decimal places! For your understanding, appreciation and interest, the value of $\pi$ is given in table 4 to a mere 3742 decimal places as calculated by a computer program. To show $\pi$ calculated to one billion decimal places would take ever page of nearly 400 copies of this text!

## Table 4 The Value of π

$\pi =$

3.1415926535897932384626433832795028841971693993751058209749445923
0781640628620899862803482534211706798214808651328230664709384460955
0582231725359408128481117450284102701938521105559644622948954930
3819644288109756659334461284756482337867831652712019091456485669234
603486104543266482133936072602491412737245870066063155881748815209
209628292540917153643678925903600113305305488204665213841469519415
116094330572703657595919530921861173819326117931051185480744623799
627495673518857527248912279381830119491298336733624406566430860213
949463952247371907021798609437027705392171762931767523846748184676
694051320005681271452635608277857713427577896091736371787214684409
012249534301465495853710507922796892589235420199561121290219608640
344181598136297747713099605187072113499999983729780499510597317328
160963185950244594553469083026425223082533446850352619311881710100
031378387528865875332083814206171776691473035982534904287554687311
595628638823537875937519577818577805321712268066130019278766111959
092164201989317350758794071254473111718097804727231772406421600638
937727837523652530717563316184124192197145015070106730948374303859
610721200087334751567600265277892904773130242681534696284150129696
714915377086549795883517465228938852415140310252843311394291058644
710915270067784892420072080733973760384335797111714553221342295836
620431489509816238414392975605608197094221363929920082606229159466
477754588305514321322836567140385864939655863689079064969741046673
075208583606042171529614202734994650778359711717363164894771857114
570743308004222507738836233705306681612070123714050507026212411300
647559557360598523722815408353566611009339695157603402317872340019
790386224683907778449548800382697970231006962844514704751209177828
258636542096403012819764657388665022638462718330870450854154019618
298340183221655279065438237729291704421640534235925410924840890930
152162693031397494699346259607437878619267556387265161568060041143
140899173995326655630799672282642800940604493951546607152474616534
793053369697647973976556261575586490108182005989151753240451344039
317232346294735519549231102263481825013036020216254046422826903789
282162094818878750627630781950694661387267352089159104677525503987
301878458088319442698601695345867642724174667272234720901594284590
592087524538076996619427484670568786175224024366239286476427637504
673994734823425345407968523681771127258388347311357920672104430735
288054242495016523293327551101332627799352597961338816380461712815
933840823069732501351461261493058653361041266373683628162339378199
013632978463543627876673949558739448879563500163945581307581125736
757511911200366402289038290668060851167782897846348347966511551582
343737275403130269393364793315565209044549903280543067242914136317
996653686199919336633264065936692639423051718170025487254294329713
769901019599359167188297053106843984864332672009991008954179841877
142858458640104858010304617865143915766652874668596080925979961635
785916354698851328383628468283296541170485547527486495724719246971
031023759535620602198063155121581530342702948239049785083895970007
973714392390151829067561159597462435580357262845484392931730235523
727975619003494065082978556445523531466186293489598270364818585933
280059893621511226709953155751203391883595324889873898931290262140
398247972412513331991543262715293218871103744067004299020049496567
755539616804071168718167740653262164694290740955669500514006664363
428590091366035578510435629034189151350508262131532639240365217307
607794302772631156500108988657228002959802770180828202840134549273
910227275180401346573231584699932273329353205227111223347119817008
509537342760918825588910014173620245824880097814309974128027317539
076821356337385505389049053689889051607793797319066978706268395520
785757350272286778742824446205994968924323596628…

# Answer Key

## CHAPTER 1

### Margin Exercises 1.1

**1.** 124,923 **2.** 2395 **3.** 17; Associative Property of Addition **4.** 23; Commutative Property of Addition **5.** $n = 0$; Additive Identity **6.** $x = 12$; Commutative Property of Addition **7.** 387 **8.** 4777 **9.** 2044 **10.** Commutative Property of Multiplication **11.** Associative Property of Multiplication **12.** $x = 25$; Multiplicative Identity **13.** $n = 14$; Commutative Property of Multiplication **14.** 63,000 **15.** 150,000 **16.** 774 **17.** 4725 **18.** 7936 **19.** 3796 square yards **20.** 819 R 0 **21.** 504 R 9 **22.** 2308 R 4

### Exercises 1.1, page 17

**1.** Zero **3.** Answers will vary **5.** Commutative Property of Addition **7.** Associative Property of Addition **9.** 150 **11.** 2127 **13.** 596 **15.** 4,301,692 **17.** $18,527 **19.** $1750 **21.** 12 ft. **23.** 24 m **25.** 12 ft. **27.** 250 yds. **29.** Answers will vary. A factor of a number is a divisor of that number. **31.** $y = 7$; Associative Property of Multiplication **33.** $n = 1$; Multiplicative Identity **35.** $x = 8$; Commutative Property of Multiplication **37.** 64 **39.** 49 **41.** 140 **43.** 3813 **45.** 32,640 **47.** 6 **49.** 8 **51.** 1 **53.** 4 **55.** 0 **57.** Undefined **59.** 0 **61.** Undefined **63.** 8 R 1 **65.** 11 R 9 **67.** 9 R 0 **69.** 45 or 35 **71.** 39 or 56 **73. a.** $21,600 **b.** $46,800 **c.** $144,000 **75.** 37,380 sq. ft. **77.** 600 pages **79.** 30 sq. ft. **81.** 384 sq. in.

### Margin Exercises 1.2

**1.** 9700 **2.** 10,000 **3.** Estimate: 360; Sum: 314 **4.** Estimate: 2200; Difference: 1838 **5.** Estimate: 400; Product: 432 **6.** Estimate:4000; Product: 5031 **7.** Estimate: 15; Quotient: 21 R.8 **8.** Estimate: 500; Quotient: 602 R.0 **9.** 1867 **10.** 4,235,868 **11.** 9900 R 12 **12.** 4442 R 116

### Exercises 1.2, page 39

**1.** 870 **3.** 90 **5.** 400 **7.** 970 **9.** 4300 **11.** 600 **13.** 200 **15.** 75,200 **17.** 8000 **19.** 7000 **21.** 14,000 **23.** 63,000 **25.** 80,000 **27.** 260,000 **29.** 120,000 **31.** 280,000 **33.** 230; 224 **35.** 2400; 2281 **37.** 8000; 7576 **39.** 1000; 1687 **41. a.** 400 **b.** 480 **c.** 300 **d.** 720 **43. a.** 100 **b.** 6 **c.** 400 **d.** 200 **45.** 280; 264 **47.** 1000; 1176 **49.** 3200; 2916 **51.** 80,000; 85,680 **53.** 10; 12 R 0 **55.** 20; 20 R 13 **57.** 4; 4 R 192 **59.** 200; 201 R 62 **61.** n = 1854 **63.** $x = 677{,}543$ **65.** $y = 6{,}580{,}030$ **67.** $a = 780$ **69.** $3000; $2548 **71.** 40,000; 36,750 sq. ft. **73.** 237 R 13 **75.** 728 R 3 **77.** 750 R 17 **79.** 13,799 R 83

### Exercises 1.3, page 47

**1.** 785 **3.** 1741 **5.** 65 **7.** $220 per month **9.** $1296 **11.** $150 **13.** $538 **15.** $7717 **17 a.** 24ft. **b.** 24 sq.ft. **19.** 174 cm **21.** 380 sq. in. **23.** 111 sq. in. **25.** 299 sq. ft. **27.** Octagon; stop sign **29 a.** Error **b.** 0 **c.** 0 **d.** Error

### Margin Exercises 1.4

**1.** $x = 19$ **2.** $y = 4$ **3.** $x = 6$ **4.** $20 = y$

### Exercises 1.4, page 61

**1.** 16 **3.** 18 **5.** 15 **7.** 3 **9.** $x = 4$ **11.** $13 = x$ **13.** $y = 0$ **15.** $y = 18$ **17.** $10 = n$ **19.** $125 = n$ **21.** $x = 4$ **23.** $y = 4$ **25.** $4 = n$ **27.** $x = 6$ **29.** $47 = y$ **31.** $x = 7$ **33.** $x = 5$ **35.** $x = 12$ **37.** $3 = y$ **39.** $9 = x$ **41.** $n = 3$ **43.** $0 = x$ **45.** $0 = x$ **47.** $6 = x$ **49.** $n = 2$

### Margin Exercises 1.5

**1. a.** 8 **b.** 2 **c.** 64 **2. a.** 15 **b.** 0 **c.** 1 **3.** $9^3$ **4.** $2^3 \cdot 5^2$ **5.** $14^2$ **6.** 32,768 **7.** 175,616 **8.** 146,410,000 **9.** 62,748,517 **10.** 23 **11.** 16 **12.** 19 **13.** 29 **14.** 36

### Exercises 1.5, page 73

**1. a.** 2 **b.** 3 **c.** 8 **3. a.** 4 **b.** 2 **c.** 16 **5. a.** 9 **b.** 2 **c.** 81 **7. a.** 11 **b.** 2 **c.** 121 **9. a.** 3 **b.** 5 **c.** 243 **11. a.** 19 **b.** 0 **c.** 1 **13. a.** 1 b. 6 c. 1 **15. a.** 24 **b.** 1 **c.** 24 **17. a.** 20 **b.** 3 **c.** 8000

**19.** **a.** 30 **b.** 2 **c.** 900 **21.** $2^4$ or $4^2$ **23.** $5^2$ **25.** $7^2$ **27.** $6^2$ **29.** $10^3$ **31.** $2^3$ **33.** $6^3$
**35.** $3^5$ **37.** $10^5$ **39.** $3^3$ **41.** $6^5$ **43.** $11^3$ **45.** $2^3 \cdot 3^2$ **47.** $2 \cdot 3^2 \cdot 11^2$ **49.** $3^3 \cdot 7^3$
**51.** 64 **53.** 121 **55.** 400 **57.** 169 **59.** 900 **61.** 2704 **63.** 390,625 **65.** 1,953,125
**67.** 78,12 **69.** 21 **71.** 19 **73.** 28 **75.** 2 **77.** 18 **79.** 0 **81.** 0 **83.** 46
**85.** 9 **87.** 74 **89.** 20 **91.** 132 **93.** 176 **95.** 61 **97.** $x = 10$ **99.** $x = 6$
**101.** Error; Answers will vary.

### Exercises 1.6, page 83

**1.** 2; Trinomial **3.** 2; Binomial **5.** 12; Monomial **7.** 4; Trinomial **9.** 5; Trinomial
**11.** 20 **13.** 8 **15.** 53,248 **17.** 9 **19.** 81 **21.** 0 **23.** 3 **25.** 0 **27.** 5
**29.** 245 **31.** 74 **33.** 706 **35.** 4, 10, 18, 28, 40, 54 **37.** 22 **39.** 29 **41.** 5
**43.** 224,939 **45.** 300 **47.** 3650 **49.** Answers will vary.

### Chapter 1 Test, Page 89

**1.** 680 **2.** 14,000 **3.** 16; Commutative Property of Addition **4.** 3; Associative Property of Multiplication **5.** 0; Additive Identity **6.** 4; Associative Property of Addition
**7.** Estimate : 12,300; Sum : 12,007 **8.** Estimate : 850,000; Sum : 870,526 **9.** Estimate : 2000; Difference : 2484 **10.** Estimate : 50,000; Difference : 52,170 **11.** Estimate : 1500; Product : 1175
**12.** Estimate : 270,000; Product : 222,998 **13.** 170 R 240 **14.** 58,000 **15.** **a.** 150 m **b.** 174 m **c.** 486 sq.m **16.** \$39,080 **17.** $x = 42$ **18.** $x = 35$ **19.** $3 = y$ **20.** **a.** 4 **b.** 3 **c.** 64
**21.** **a.** 32,768 **b.** 6859 **c.** 1 **22.** 8 **23.** 47 **24.** 12 **25.** 17
**26.** Second-degree Binomial **27.** Fourth-degree Trinomial **28.** 24 **29.** 81 **30.** **a.** Yes; $10^2 = 100$ **b.** No; 80 is not a perfect square **c.** Yes; $15^2 = 225$ **31.** Answers may vary.

## CHAPTER 2

### Margin Exercises 2.1

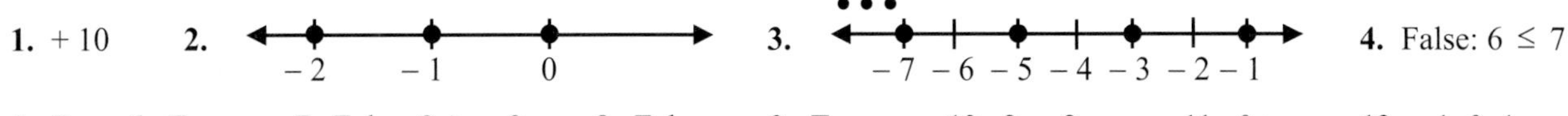


**5.** True **6.** True **7.** False: $2 \geq -2$ **8.** False **9.** True **10.** 2, −2 **11.** 0 **12.** −1, 0, 1

### Exercises 2.1, page103

**1.** > **3.** > **5.** < **7.** < **9.** = **11.** > **13.** True **15.** True **17.** False: $|2| > -2$ **19.** True
**21.** False: $|-3| = 3$ **23.** False: $-4 < |-4|$

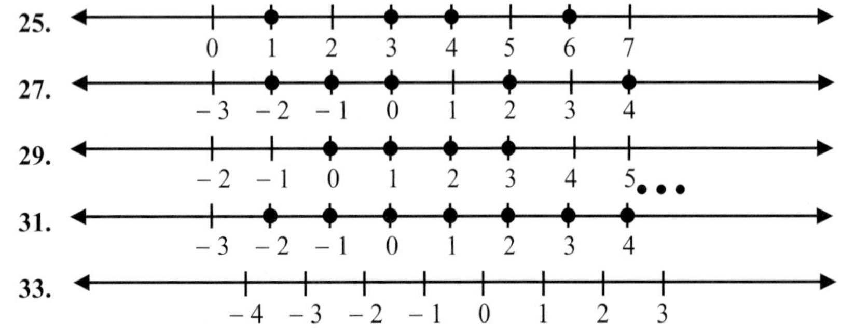


**35.** 5, −5 **37.** 2, −2 **39.** No Solution **41.** 23, −23

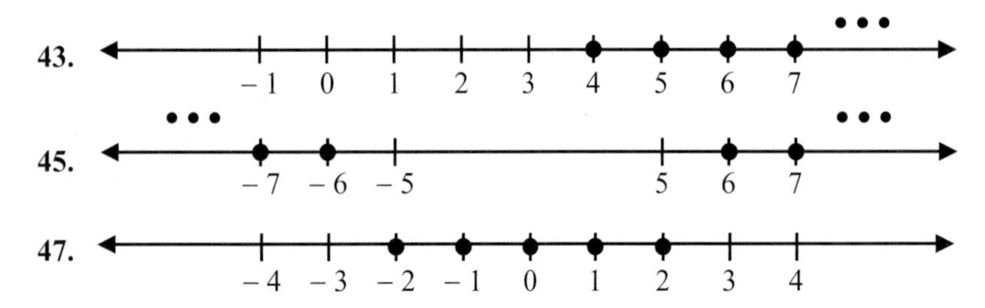

**49.** Sometimes **51.** Sometimes **53.** Answers will vary. If $y$ represents a negative number, then $-y$ will represent a positive number. **55.** Answers will vary. **a.** The absolute value of an integer is never negative. **b.** $|x| \geq 0$

### Margin Exercises 2.2

**1.** 8 **2.** 9 **3.** $-10$ **4.** $-10$ **5.** 0 **6.** $-28$ **7.** $-47$ **8.** 23 **9.** Yes **10.** Yes

### Exercises 2.2, page 113

**1.** $-15$ **3.** 40 **5.** 9 **7.** $-2$ **9.** $-8$ **11.** $-17$ **13.** 0 **15.** 0 **17.** $-18$ **19.** $-45$ **21.** 0 **23.** $-32$ **25.** 3 **27.** $-19$ **29.** $-10$ **31.** $-24$ **33.** 31 **35.** 76 **37.** $-310$ **39.** 33 **41.** 0 **43.** 18 **45.** 14 **47.** 35 **49.** No **51.** Yes **53.** Yes **55.** Yes **57.** No **59.** Sometimes **61.** Sometimes **63.** Never **65.** Never **67.** If they are opposites **69.** $-3405$ **71.** $-2575$ **73.** $-23{,}094$ **75.** 845

### Margin Exercises 2.3

**1.** 0 **2.** $-11$ **3.** 5 **4.** 19 **5.** 9 **6.** $-17$ **7.** $-107$ **8.** 0 **9.** 84 **10.** 37

### Exercises 2.3, page 121

**1. a.** $-8$ **b.** $-20$ **3. a.** $-29$ **b.** $-9$ **5. a.** $-8$ **b.** $-32$ **7.** 6 **9.** 5 **11.** $-20$ **13.** $-34$ **15.** 4 **17.** $-4$ **19.** 6 **21.** $-15$ **23.** $-22$ **25.** $-19$ **27.** $-7$ **29.** $-21$ **31.** $-2$ **33.** 0 **35.** 0 **37.** $-28$ **39.** $-45$ **41.** $-1$ **43.** $-19$ **45.** $-8$ **47.** $-11$ **49.** 0 **51.** 127 **53.** $>$ **55.** $<$ **57.** $>$ **59.** $=$ **61.** Yes **63.** Yes **65.** Yes **67.** No **69.** Yes **71.** $3°$ **73.** 4238 pts. **75.** 85 yds. **77. a.** -12,533 **b.** 57,456 **79. a.** -618,000 **b.** 1,618,000 **81. a.** -5 b. 19 **83.** $-2$
**85.** Answers will vary. Think of $-x$ as the "opposite of $x$." If x is positive, then $-x$ is negative. If $x$ is negative, then $-x$ is positive. Also, note that 0 is its own opposite.

### Recycle Bin, page 127

**1.** 4 **2.** 11 **3.** 25 **4.** 1 **5.** 92 **6. a.** 1296 **b.** 5832 **c.** 9,765,625 **7.a.** 1 **b.** 1. **c.** undefined **8. a.** 10 **b.** 14 **c.** 0

### Margin Exercises 2.4

**1.** $-15$ **2.** $-32$ **3.** 24 **4.** 0 **5.** $-60$ **6.** 126 **7.** $-3$ **8.** $-4$ **9.** 4 **10.** 4 **11.** $-17$ **12.** $-3$ **13.** 23 **14.** 13 **15.** $-12$

### Exercises 2.4, page 137

**1.** $-24$ **3.** $-24$ **5.** 28 **7.** $-18$ **9.** $-50$ **11.** 0 **13.** $-14$ **15.** 105 **17.** $-32$ **19.** 0 **21.** $-1$ **23.** $-50$ **25.** 60 **27.** 0 **29. a.** 1 **b.** $-1$ **c.** 1 **d.** 1 **31.** $-4$ **33.** $-2$ **35.** $-6$ **37.** $-4$ **39.** $-14$ **41.** $-3$ **43.** 5 **45.** 8 **47.** 0 **49.** Undefined **51.** $-70$ **53. a.** $-17$ **b.** $-27$ **55.** $-14$ **57.** $-22$ **59.** $-9$ **61.** 27 **63.** 31 **65.** 45 **67.** 1026 **69.** $-9$ **71.** 4624 **73.** $-105$ **75.** $x \div y = 0$ if $x = 0$ and $y \neq 0$.

### Exercises 2.5, page 143

**1.** 72 years old **3.** $-18{,}500$ ft **5.** 16,300ft **7.** $-105$ points **9.** 0 **11.** $-6$ **13.** \$78 per share **15.** 85 **17.** 8 degrees **19.** 92 **21.a.** $-33$ **b.** 17 **23.** 100; 80 **25.** $-16{,}250$;$-18{,}152$ **27.** $-1000$; $-80$

### Margin Exercises 2.6

**1.** $16x$ **2.** $6x - 8a$ **3.** $4x^2 - 3b$ **4.** $-x^2$ **5.** $8xy - x^2$; $-52$ **6.** $6x$; $-12$ **7.** $6x + y$; $-9$ **8. a.** $-16$ **b.** 16 **c.** $-8$ **d.** $-8$ **9. a.** 49 **b.** $-49$ **c.** $-343$ **d.** $-343$ **10. a.** $-1$ **b.** 1 **c.** $-1$ **d.** $-1$

### Exercises 2.6, page 153

**1.** $6x + 18$ **3.** $-7x - 35$ **5.** $-11n - 33$ **7.** $5x$ **9.** $2y$ **11.** $-2n$ **13.** 0 **15.** $-3y$ **17.** $9x - 24$ **19.** $15x^2 + 5x - 1$ **21.** $4x^2 - 4x - 3$ **23.** $5x - 1$ **25.** $4x - 28$ **27.** $-xy + 11x - y$ **29.** $2x^2y + 2$ **31.** $8a^2b + 7ab^2 + 3ab$ **33.** $5x - a - 5$ **35.** $-2xyz - 6$ **37.** $4x^2 - 4x + 5$; 29 **39.** $2y^2 - 6y - 5$; 3 **41.** $-2x^3 + x^2 - x$; 22 **43.** $7y^3 + 5y^2 - 6$; $-8$ **45.** $3y^2 - 16y - 5$; 14 **47.** $-a^3 - 2a - 7$; $-4$ **49.** $3ab - 3a + 3b$; 3 **51.** $9a - b - 15$; $-22$ **53.** $6abc - 5ab + 2bc$; 14 **55.** $a^2b^2c^2 + 4ab^2 + bc^2$; 2 **57.** $12a + 8b - 3c - 64$; $-101$ **59.** 0; 0 **61.a.** $-16$ **b.** 16 **c.** $-64$ **d.** $-64$ **63.a.** $-216$ **b.** $-1$ **c.** $-343$ **d.** $-32$ **65. a.** $-2$ **b.** 3 **c.** $-10$ **d.** 21

### Margin Exercises 2.7

**1.** $x = -26$ **2.** $n = -8$ **3.** $x = 19$ **4.** $n = 1$

### Exercises 2.7, page 161

**1.** $-12$ **3.** $-25$ **5.** $-15$ **7.** $-3$ **9.** $x = -14$ **11.** $x = -37$ **13.** $y = -18$ **15.** $y = -9$
**17.** $n = 0$ **19.** $n = -105$ **21.** $x = 4$ **23.** $y = -5$ **25.** $n = -21$ **27.** $x = -21$
**29.** $y = -96$ **31.** $x = 43$ **33.** $x = 5$ **35.** $x = -18$ **37.** $y = 8$ **39.** $x = 0$ **41.** $x = 0$
**43.** $y = 0$ **45.** $x = 6$ **47.** $n = -42$ **49.** $n = -11$

### Exercises Chapter 2 Test, page 169

**1.** −2 −1 0 1 2 3 4 **2.** $-2, -1, 0, 1, 2$

**3.** 0 1 2 3 4 **4.** $-3$ **5.** $-165$ **6.** $-40$

**7.** $-2$ **8.** $-420$ **9.** 0 **10.** 480 **11.** $-8$ **12.** 0 **13.** 4 **14.** $-15$ **15.** $-35$
**16.** 43 **17.** $-27,000$ ft **18.** $-60$ **19.** 5 yd gain **20.** 64 mph **21.** 0 **22.** 76 **23.** $5x$
**24.** $-2x^2 + 8x + 15$ **25.** $-3y^3 + 3y^2 - 4y$ **26.** $5y^2 - 6$ **27.** $-51$ **28.** 96 **29.** $x = -2$
**30.** $y = 3$ **31.** $x = -28$ **32.** $y = 1$

### Exercises Chapter 2 Cumulative Review, page 171

**1.** 6200 **2.** 30,000 **3.** 760 **4.** 23,000 **5.** $n = 27$; Commutative Property of Addition
**6.** $n = 2$; Associative Property of Multiplication **7. a.** 0 **b.** 1 **c.** Undefined **8.** Estimate: 9000; Sum: 9671
**9.** Estimate: 880,000; Sum: 881,879 **10.** Estimate: 3000; Difference: 2715 **11.** Estimate: 40,000; Difference: 37,579 **12.** Estimate: 2400; Product: 1995 **13.** Estimate: 240,000; Product: 215,992
**14.** 487 R 0 **15.** 106 R 438

**16.** 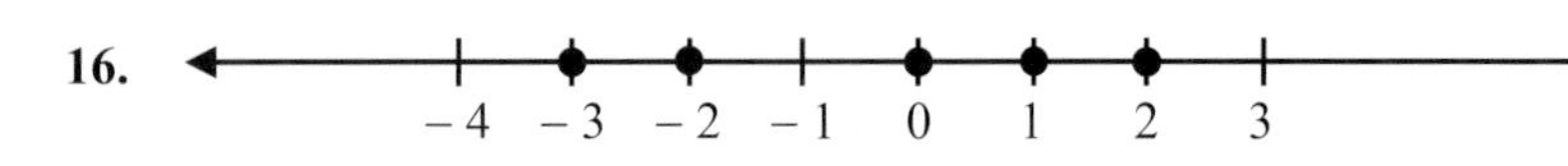


**17.** $5, -5$

**18.** −4 −3 −2 −1 0 1 2 3 **19.** $-42$ **20.** $-205$ **21.** $-17$

**22.** 0 **23.** $-180$ **24.** $-12$ **25.** 0 **26.** Undefined **27.** 57 **28.** \$788 **29.** 63 in.
**30. a.** 290 ft. **b.** 4500 sq. ft. **31.** 1° **32.** $-21,000$ ft. **33.** $n = 25,458$ **34.** 570,240 **35.** 21,356
**36.** $-7^2 = -1 \cdot 7^2 = -49$; $(-7)^2 = 49$ **37. a.** 15 **b.** 2 **c.** 225 **38.** 27 **39.** $-110$ **40.** $7x^3 + 8x - 6$; trinomial; third-degree **41.** $14y^2 + 2y - 1$; trinomial; second-degree **42.** $37x^2 + 10$; binomial; second-degree
**43.** $-8$ **44.** 22 **45.** $-8$ **46.** 9 **47.** 6 **48.** $-1$ **49.** 23 **50.** 754 **51.** Answers will vary.
**52.** Answers will vary. Perimeter is linear measure, and area is a square measure. Perimeter is a measure of a distance, and area is a measure of an interior.

## CHAPTER 3

### Margin Exercises 3.1

**1.** Yes, 6 divides 8034 because the units digit is even and the sum of the digits, 15, is divisible by 3. **2.** No, 3 does not divide 206 because 3 does not divide the sum of the digits: $2 + 0 + 6 = 8$. **3.** The sum of the digits is 15, so 3 divides the number, but 9 does not. **4.** 12,375 is divisible by 5, because the units digit is 5. 12,375 is not divisible by 10, because the units digit is not 0. **5.** All the numbers 2, 5, and 6 are factors of 2400. The units digit is 0, so the number is divisible by 5. Since the number is even and the sum of the digits is 6, which is divisible by 3, it is divisible by 6.
**6.** 1742 is divisible by 2. **7.** 8020 is divisible by 2, 5, and 10. **8.** 33,031 is not divisible by any of these numbers.
**9.** 39 divides this product because $39 = 3 \cdot 13$ and 3 and 13 are both factors of this product. The quotient is $5 \cdot 8 = 40$.
**10.** 35 does not divide this product because $35 = 5 \cdot 7$ and 7 is not a factor of this product. **11.** 100 divides this product because $100 = 2 \cdot 5 \cdot 10$. Since 4 can be written as $2 \cdot 2$, the product can be written $6 \cdot 5 \cdot 2 \cdot 2 \cdot 10 \cdot 7$. So, 2, 5, and 10 are all factors of the product. The quotient is $6 \cdot 2 \cdot 7 = 84$.

## Exercises 3.1, page 183

**1.** Answers will vary. **3.** Answers will vary **5.** 2 **7.** 3, 9 **9.** 3, 9 **11.** 2 **13.** 5 **15.** 3, 9 **17.** 2, 5, 10 **19.** None **21.** 2, 3, 6 **23.** 2, 3, 5, 6, 10 **25.** 3 **27.** 2, 3, 5, 6, 9, 10 **29.** 2 **31.** 2, 3, 6, 9 **33.** 2 **35.** 2, 3, 6 **37.** None **39.** 2 **41.** 2, 5, 10 **43.** None **45.** 3 **47.** 2, 3, 6 **49.** 2 **51. a. i.** 1 **ii.** 7 **iii.** 7 **iv.** 8 **b. i.** 1, 4, 7 **ii.** 4 **iii.** None **iv.** 2, 5, 8 **53.** Answers will vary. Any integer that ends in 5 is divisible by 5 and not divisible by 10. For, example, 25 and 35. **55.** Answers will vary. Not necessarily. To be divisible by 27, an integer must have 3 as a factor three times. Two counterexamples are 9 and 18. **57.** $2 \cdot 3 \cdot 3 \cdot 7 = 6 \cdot 21$. The product is 126, and 6 divides the product 21 times. **59.** $2 \cdot 3 \cdot 5 \cdot 9 = 10 \cdot 27$. The product is 270, and 10 divides the product 27 times. **61.** $3 \cdot 5 \cdot 7 \cdot 10 = 25 \cdot 42$. The product is 1050, and 25 divides the product 42 times. **63.** $2 \cdot 3 \cdot 3 \cdot 7 \cdot 15 = 45 \cdot 42$. The product is 1890, and 45 divides the product 42 times. **65.** $2 \cdot 3 \cdot 5 \cdot 7 \cdot 12 = 40 \cdot 63$. The product is 2520 and 40 divide is the product 63 times. **67. a.** $744 = 700 + 40 + 4 = 7 \cdot 100 + 4 \cdot 10 + 4 = 7(99+1) + 4(9+1) + 4 = 7 \cdot 99 + 4 \cdot 9 + (7+4+4)$. We see that 3 is a factor of both 99 and 9. Therefore, if 3 is a factor of 744, then 3 must also be a factor of $7 + 4 + 4$, which is the sum of the digits of the number 744. **b.** $abcd = a000 + b00 + c0 + d = a \cdot 1000 + b \cdot 100 + c \cdot 10 + d = a(999+1) + b(99+1) + c(9+1) + d = a \cdot 999 + b \cdot 99 + c \cdot 9 + (a+b+c+d)$. We see that 3 is a factor of 999, 99, and 9. Therefore, if 3 is a factor of abcd, then 3 must also be a factor of $a + b + c + d$, which is the sum of the digits of the number abcd. **69.** $3 \cdot 5 \cdot 12 = 180$, $4 \cdot 5 \cdot 10 = 200$, $3 \cdot 10 \cdot 12 = 360$; $4 \cdot 5 \cdot 12 = 240$, $4 \cdot 10 \cdot 12 = 480$, $5 \cdot 10 \cdot 12 = 600$, $3 \cdot 4 \cdot 5 \cdot 10 = 600$, $3 \cdot 4 \cdot 5 \cdot 12 = 720$, $3 \cdot 4 \cdot 10 \cdot 12 = 1440$, $3 \cdot 5 \cdot 10 \cdot 12 = 1800$, $4 \cdot 5 \cdot 10 \cdot 12 = 2400$ **71.** Answers will vary. The question is designed to help the students become familiar with their calculators and to provide an early introduction to scientific notation.

## Margin Exercises 3.2

**1.** Composite **2.** Prime **3.** Prime **4.** Prime **5.** Prime **6.** Composite **7.** Division by prime numbers shows that $221 = 13 \cdot 17$. Therefore, 221 is composite. **8.** Division by prime numbers up to 17 shows no divisions. When 239 is divided by 17, the quotient is less than 17. Therefore, 239 is prime, and its only divisors are 1 and 239.

## Exercises 3.2, page 195

**1.** 3, 6, 9, 12, 15, ... **3.** 8, 16, 24, 32, 40, ... **5.** 11, 22, 33, 44, 55, ... **7.** 20, 40, 60, 80, 100, ... **9.** 41, 82, 123, 164, 205, ... **11.**

| 1 | **2** | **3** | 4 | **5** | 6 | **7** | 8 | 9 | 10 |
|---|---|---|---|---|---|---|---|---|---|
| **11** | 12 | **13** | 14 | 15 | 16 | **17** | 18 | **19** | 20 |
| 21 | 22 | **23** | 24 | 25 | 26 | 27 | 28 | **29** | 30 |
| **31** | 32 | 33 | 34 | 35 | 36 | **37** | 38 | 39 | 40 |
| **41** | 42 | **43** | 44 | 45 | 46 | **47** | 48 | 49 | 50 |
| 51 | 52 | **53** | 54 | 55 | 56 | 57 | 58 | **59** | 60 |
| **61** | 62 | 63 | 64 | 65 | 66 | **67** | 68 | 69 | 70 |
| **71** | 72 | **73** | 74 | 75 | 76 | 77 | 78 | **79** | 80 |
| 81 | 82 | **83** | 84 | 85 | 86 | 87 | 88 | **89** | 90 |
| 91 | 92 | 93 | 94 | 95 | 96 | **97** | 98 | 99 | 100 |

**13.** Prime **15.** Prime

**17.** Composite; 1, 65; 5, 13 **19.** Prime **21.** Composite; 1, 502; 2, 251 **23.** Composite; 1, 517; 11, 47 **25.** Composite; 1, 1073; 29, 37 **27.** Composite; 1, 961; 31, 31 **29.** 3 and 4 **31.** 2 and 12 **33.** 4 and 5 **35.** 3 and 12 **37.** 3 and 19 **39.** 4 and 4 **41.** 4 and 13 **43.** 6 and 11 **45.** 5 and 13 **47.** 5 and 12 **49.** $-5$ and $-15$: $(-5)(-15) = 75$ and $(-5) + (-15) = -20$. Answers will vary as to reasoning. **51.** 3 and $-10$: $3(-10) = -30$ and $3 + (-10) = -7$. Answers will vary as to reasoning. **53. a.** Not true. The number 2 is prime and even. **b.** Not true. Some counterexamples are 9, 15, and 21. **55.** N = 2: $2^2 - 1 = 4 - 1 = 3$; 3 is prime. N = 3: $2^3 - 1 = 8 - 1 = 7$; 7 is prime. N = 5: $2^5 - 1 = 32 - 1 = 31$ 31 is prime. N = 7: $2^7 - 1 = 128 - 1 = 127$ 127 is prime. N = 13: $2^{13} - 1 = 8192 - 1 = 8191$ 8191 is prime. N = 11: $2^{11} - 1 = 2048 - 1 = 2047$ 2047 is composite. The factors of 2047 are: 1, 23, 89, 2047.



## Margin Exercises 3.3

**1.** $2 \cdot 37$ **2.** $2^2 \cdot 13$ **3.** $2^2 \cdot 5 \cdot 23$ **4.** $2^3 \cdot 7 \cdot 11$ **5.** 1, 3, 7, 9, 21, 63 **6.** 1, 2, 3, 6, 9, 18, 27, 54 **7.** 1, 2, 4, 8, 11, 22, 44, 88

## Exercises 3.3, page 203

**1.** $20 = 2^2 \cdot 5$ **3.** $32 = 2^5$ **5.** $50 = 2 \cdot 5^2$ **7.** $70 = 2 \cdot 5 \cdot 7$ **9.** $51 = 3 \cdot 17$ **11.** $62 = 2 \cdot 31$ **13.** $99 = 3^2 \cdot 11$ **15.** 37 is prime. **17.** $120 = 2^3 \cdot 3 \cdot 5$ **19.** $196 = 2^2 \cdot 7^2$ **21.** $361 = 19^2$ **23.** $65 = 5 \cdot 13$ **25.** $1000 = 2^3 \cdot 5^3$ **27.** $100{,}000 = 2^5 \cdot 5^5$ **29.** $600 = 2^3 \cdot 3 \cdot 5^2$ **31.** 107 is prime.

**33.** $309 = 3 \cdot 103$ **35.** $165 = 3 \cdot 5 \cdot 11$ **37.** $675 = 3^3 \cdot 5^2$ **39.** $216 = 2^3 \cdot 3^3$ **41.** $1692 = 2^2 \cdot 3^2 \cdot 47$
**43.** $676 = 2^2 \cdot 13^2$ **45. a.** $42 = 2 \cdot 3 \cdot 7$ **b.** 1, 2, 3, 6, 7, 14, 21, 42 **47. a.** $300 = 2^2 \cdot 3 \cdot 5^2$
**b.** 1, 2, 3, 4, 5, 6, 10, 12, 15, 20, 25, 30, 50, 60, 75, 100, 150, 300 **49. a.** $66 = 2 \cdot 3 \cdot 11$ **b.** 1, 2, 3, 6, 11, 22, 33, 66
**51. a.** $78 = 2 \cdot 3 \cdot 13$ **b.** 1, 2, 3, 6, 13, 26, 39, 78 **53. a.** $150 = 2 \cdot 3 \cdot 5^2$ **b.** 1, 2, 3, 5, 6, 10, 15, 25, 30, 50, 75, 150
**55. a.** $90 = 2 \cdot 3^2 \cdot 5$ **b.** 1, 2, 3, 5, 6, 9, 10, 15, 18, 30, 45, 90 **57. a.** $1001 = 7 \cdot 11 \cdot 13$
**b.** 1, 7, 11, 13, 77, 91, 143, 1001 **59. a.** $585 = 3^2 \cdot 5 \cdot 13$ **b.** 1, 3, 5, 9, 13, 15, 39, 45, 65, 117, 195, 585
**61. a.** $700 = 2^2 \cdot 5^2 \cdot 7$; $(2+1)(1+1)(2+1) = 3 \cdot 2 \cdot 3 = 18$ There are 18 factors of 700. These factors are: 1, 2, 4, 5, 7, 10, 14, 20, 25, 28, 35, 50, 70, 100, 140, 175, 350, and 700. **b.** $660 = 2^2 \cdot 3^1 \cdot 5^1 \cdot 11^1$ $(2+1)(1+1)(1+1)(1+1) = 3 \cdot 2 \cdot 2 \cdot 2 = 24$. There are 24 factors of 660. These factors are: 1, 2, 3, 4, 5, 6, 10, 11, 12, 15, 20, 22, 30, 33, 44, 55, 60, 66, 110, 132, 165, 220, 330, 660. **c.** $450 = 2^1 \cdot 3^2 \cdot 5^2$ $(1+1)(2+1)(2+1) = 2 \times 3 \times 3 = 18$. There are 18 factors of 450. These factors are: 1, 2, 3, 5, 6, 9, 10, 15, 18, 25, 30, 45, 50, 75, 90, 150, 225, 450. **d.** $148{,}225 = 5^2 \times 7^2 \times 11^2$ $(2+1)(2+1)(2+1) = 3 \times 3 \times 3 = 27$. There are 27 factors of 148,225. These factors are: 1, 5, 7, 11, 25, 35, 49, 55, 77, 121, 175, 245, 275, 385, 539, 605, 847, 1225, 1925, 2695, 3025, 4235, 5929, 13,475, 21,175, 29,645, and 148,225.

### The Recycle Bin, page 206

**1.** −1 **2.** −27 **3.** −1 **4.** −56 **5.** 36 **6.** −27 **7.** −10 **8.** 5

### Margin Exercises 3.4

**1.** $2^3 \cdot 3 \cdot 5 \cdot 7 = 840$ **2.** $2^3 \cdot 3 \cdot 5 \cdot 7 = 840$ **3. a.** LCM = 140 **b.** $140 = 20 \cdot 7 = 35 \cdot 4 = 70 \cdot 2$
**4. a.** LCM = 840 **b.** $840 = 24 \cdot 35 = 35 \cdot 24 = 42 \cdot 20$ **5.** $350xy^2$ **6.** $60a^3b^3$

### Exercises 3.4, page 213

**1.** 105 **3.** 30 **5.** 66 **7.** 140 **9.** 150 **11.** 180 **13.** 196 **15.** 60 **17.** 484
**19.** 56 **21.** 120 **23.** 1,192,464 **25.** 7,485,696 **27. a.** LCM = 120 **b.** $120 = 6 \cdot 20 = 24 \cdot 5 = 30 \cdot 4$
**29. a.** LCM = 108 **b.** $108 = 12 \cdot 9 = 18 \cdot 6 = 27 \cdot 4$ **31. a.** LCM = 14,157 **b.** $14{,}157 = 99 \cdot 143 = 121 \cdot 117 = 143 \cdot 99$
**33. a.** LCM = 7840 **b.** $7840 = 40 \cdot 196 = 56 \cdot 140 = 160 \cdot 49 = 196 \cdot 40$ **35. a.** LCM = 194,040
**b.** $194{,}040 = 45 \cdot 4312 = 56 \cdot 3465 = 98 \cdot 1980 = 99 \cdot 1960$ **37.** $600xy^2z$ **39.** $112a^3bc$ **41.** $210a^2b^3$
**43.** $84ab^2c^2$ **45.** $180m^2np^2$ **47.** $108x^2y^2z^2$ **49.** $726ab^3$ **51.** $120y^4$ **53.** $1140c^4$ **55.** $4590a^2bc^2x^2y$
**57. a.** 60 minutes **b.** 4, 3, and 2 trips, respecitvely **59. a.** Every 24 days **b.** Every 24 days **61.** Every 180 days
**63.** Explanations will vary. A multiple of a number is always divisible by the number. By definition, the LCM of a set of numbers is a multiple of each of those numbers. **65. a.** 300 **b.** $735a^3b^5c^4$ **c.** $756x^2y^2z^3$

### The Recycle Bin, page 218

**1.** −15 **2.** 56 **3.** −300 **4.** 16 **5.** −104 **6.** 6 **7.** −139 **8.** −14

### Margin Exercises 3.5

**1.** $\frac{3}{35}$ **2.** 0 **3.** $-\frac{1}{4}$ **4.** $\frac{16}{33}$ **5.** $\frac{15}{32}$ **6.** $\frac{27}{4}$; commutative property of multiplication

**7.** $\frac{1}{24}$; associative property of multiplication **8.** $-\frac{35}{48}$; commutative property of multiplication **9.** $\frac{3}{5}$ **10.** $\frac{3x}{2y}$

**11.** $-\frac{7}{15}$ **12.** $\frac{35}{46}$ **13.** $\frac{3}{7}$ **14.** $\frac{xy}{6}$ **15.** $\frac{4}{15}$

### Exercises 3.5, page 229

**1.a.** 0 **b.** 0 **c.** Undefined **d.** Undefined **3.** $\frac{3}{25}$ **5.** $\frac{4}{9}$ **7.** $-\frac{18}{35}$ **9.** $\frac{1}{4}$ **11.** $\frac{9}{16}$ **13.** $\frac{4}{27}$

**15.** 0 **17.** $-\frac{12}{5}$ **19.** $\frac{1}{150}$ **21.** $\frac{125}{512}$ **23.** Commutative Property of Multiplication **25.** Associative Property of Multiplication **27.** $\frac{3}{4} = \frac{3}{4} \cdot \frac{3}{3} = \frac{9}{12}$ **29.** $\frac{6}{7} = \frac{6}{7} \cdot \frac{2}{2} = \frac{12}{14}$ **31.** $\frac{3n}{-8} = \frac{3n}{-8} \cdot \frac{-5}{-5} = \frac{-15n}{40}$

**33.** $\frac{-5x}{13}=\frac{-5x}{13}\cdot\frac{3y}{3y}=\frac{-15xy}{39y}$ **35.** $\frac{4a^2}{17b}=\frac{4a^2}{17b}\cdot\frac{2a}{2a}=\frac{8a^3}{34ab}$ **37.** $\frac{-9x^2}{10y^2}=\frac{-9x^2}{10y^2}\cdot\frac{10y}{10y}=\frac{-90x^2y}{100y^3}$ **39.** $\frac{4}{5}$

**41.** $\frac{22}{65}$ **43.** $-\frac{2}{3}$ **45.** $\frac{-27}{56x^2}$ **47.** $-\frac{2}{3x}$ **49.** $\frac{17x}{2}$ **51.** $\frac{6a}{b}$ **53.** $\frac{11y}{14}$ **55.** $-\frac{1}{6}$ **57.** $\frac{21}{8}$

**59.** $\frac{-273a}{80}$ **61.** $\frac{21x^2}{2}$ **63.** $-\frac{77}{4}$ **65.** $\frac{7}{10}$; $\frac{3}{10}$ **67.** $\frac{9}{4}$ inches **69.** $\frac{4}{5}$; 60 miles

### Margin Exercises 3.6

**1.** $\frac{18}{11}$ **2.** $\frac{5}{-1}$ or $-5$ **3.** $\frac{7x}{2}$ **4.** 1 **5.** $\frac{x}{4}$ **6.** $-\frac{4}{5}$

### Exercises 3.6, page 241

**1.** $\frac{25}{13}$ **3.** 0 **5.** $10\div 2=5$, while $10\div\frac{1}{2}=10\cdot 2=20$ **7.** Explanations will vary. Refer to the discussion on page 15 of Section 1.1. **9.** $\frac{25}{24}$ **11.** $\frac{14}{11}$ **13.** 1 **15.** $\frac{9}{16}$ **17.** $-\frac{1}{4}$ **19.** $\frac{1}{16}$ **21.** Undefined **23.** 0 **25.** $-\frac{8}{5}$ **27.** $\frac{9}{20}$ **29.** $\frac{4}{3a}$ **31.** $\frac{4x^2}{3y^2}$ **33.** $98x^2$ **35.** $-300a$ **37.** $\frac{29}{155}$ **39.** $-\frac{3x}{8}$ **41. a.** $\frac{2}{5}$ **b.** $\frac{12}{25}$

**43. a.** larger **b.** 315 **45. a.** more **b.** less **c.** 75 **d.** $\frac{108}{5}$ **47.** $\frac{896}{9}$ **49. a.** Yes, it would pass. **b.** Pass by 3 votes. **51. a.** 1000 **b.** 2500 **53. a.** more **b.** less **c.** 8000 **55. a.** more **b.** less **c.** 900

### The Recycle Bin, page 246

**1.** $-2x-6$ **2.** $-y^2+2y+6$ **3.** $-11x^2-7x-6$ **4.** 0 **5. a.** $8x^2-10x+1$ **b.** 53 **6. a.** $-3y^3-4y^2+20y-100$ **b.** $-157$ **7. a.** $4y^2+4xy-5x$ **b.** 22

### Margin Exercises 3.7

**1.** $x=-1$ **2.** $x=6$ **3.** $x=\frac{9}{10}$ **4.** $x=\frac{15}{26}$

### Exercises 3.7, page 255

**1.** Write the equation. Simplify. Divide both sides by 2. Simplify. **3.** Write the equation. Multiply each term by 10. Simplify. Add –2 to both sides. Simplify. Divide both sides by 5. Simplify. **5.** $x=12$ **7.** $y=55$ **9.** $x=5$

**11.** $n=7$ **13.** $x=-11$ **15.** $x=-6$ **17.** $y=-4$ **19.** $n=-\frac{14}{5}$ **21.** $y=-\frac{1}{2}$ **23.** $x=\frac{1}{3}$

**25.** $x=-\frac{25}{2}$ **27.** $x=5$ **29.** $x=-\frac{2}{3}$ **31.** $y=39$ **33.** $x=-210$ **35.** $x=\frac{70}{3}$ **37.** $y=-\frac{10}{3}$

**39.** $y=\frac{14}{15}$ **41.** $x=-2$ **43.** $n=2$ **45.** $x=\frac{55}{8}$ **47.** $x=2$ **49.** $n=-\frac{3}{2}$ **51.** $x=-\frac{1}{10}$

**53.** $n=-\frac{8}{9}$ **55.** $y=-10$ **57.** $x=64$ **59.** $y=-45$ **61.** $x=-\frac{3}{4}$ **63.** $n=-\frac{5}{6}$ **65.** $x=81$

## Chapter 3 Test, page 263

**1.** 2, 3, 6, 9 **2.** 2, 5, 10 **3.** None **4.** 11, 22, 33, 44, 55, 66, 77,88, 99 **5.** 2 **6. a.** False. The number 5 is not a factor twice of $5 \cdot 4 \cdot 3 \cdot 2 \cdot 1$. **b.** True. $2 \cdot 5 \cdot 6 \cdot 7 \cdot 9 = 2 \cdot 5 \cdot 2 \cdot 3 \cdot 7 \cdot 9 = 27 \cdot 140$. **7.** 2, 11, 13, 37 **8.** $2^4 \cdot 5$ **9.** $2^2 \cdot 3^2 \cdot 5$ **10.** $3^2 \cdot 5^2$ **11.** 1, 2, 3, 5, 6, 9, 10, 15,18, 30, 45, 90 **12.** 840 **13.** $360x^3y^2$ **14.** $168a^4$ **15. a.** 378 seconds **b.** 7 laps and 6 laps, respectively **c.** The faster runner must pass the slower runner at some point before a full lap is gained. **16.** integers, zero **17.** 0, undefined

**18.** Any two examples of the form $\frac{a}{b} \cdot \frac{c}{d} = \frac{c}{d} \cdot \frac{a}{b}$ **19.** $\frac{6}{5}$ **20.** $-\frac{2}{9x}$ **21.** $\frac{5a^2}{4b}$ **22.** $\frac{1}{2}$ **23.** $\frac{x^2}{2y}$

**24.** $405x^2$ **25.** $-\frac{12a}{35b}$ **26.** $x = -\frac{9}{5}$ **27.** $x = 3$ **28.** $y = -6$ **29.** $n = \frac{2}{15}$ **30.** 1413 kilobytes

## Cumulative Review Chapter 1 - 3, page 265

**1.** 63,000 **2. a.** 7 **b.** 2 **c.** 49 **3.** $n = 0$; Additive Identity **4.** $n = 14$; Comm. Property of Add. **5.** 1561 **6.** 4545 **7.** 2625 **8.** 150 estimate; 200 R 10 **9.** −39 **10.** 2, 3, 5, 6, 9, 10 **11.** 84 **12. a.** 117,649 **b.** 59,049 **13.** −58 **14. a.** 5, 7, 11, 13 **b.** $5^3 = 125$; $7^3 = 343$; $11^3 = 1331$; $13^3 = 2197$ **15.** $3 \cdot 5^3$ **16.** $2^2 \cdot 3 \cdot 13$ **17.** $2^3 \cdot 5^2$ **18.** 1, 2, 3, 4, 5,6, 10, 12,15, 20, 30, 60 **19.** 900 **20.** $2100x^2y^3z^4$ **21.** −18

**22. a.** $2x^2 - 2x + 3$ **b.** 7 **23. a.** $-2y^3 + 6y - 2$ **b.** −6 **24.** $\frac{36}{5}$ **25.** $\frac{8b}{35a}$ **26.** $-\frac{5}{x}$ **27.** −1

**28.** $x = 8$ **29.** $y = -\frac{31}{3}$ **30.** $x = -\frac{9}{10}$ **31.** $n = 14$ **32.** $x = 1$ **33.** $y = -\frac{1}{8}$ **34.** $x = 1$

**35.** $x = -\frac{3}{22}$ **36.** $246 **37.** every 180 days **38. a.** 1040 sq. inches **b.** 132 inches

## CHAPTER 4

### Margin Exercises 4.1

**1.** $\frac{7}{15}$ **2.** $\frac{53}{40}$ **3.** $\frac{537}{100}$ **4.** $\frac{21}{20}$ **5.** $\frac{2}{3}$ **6.** $\frac{17}{30}$ **7.** $-\frac{25}{48}$ **8.** $\frac{2}{13}$ **9.** $\frac{14}{a}$ **10.** $\frac{16+3x}{8x}$

**11.** $\frac{4y-5}{4}$ **12.** $\frac{3n+7}{6}$

### Exercises 4.1, page 279

**1. a.** $\frac{1}{9}$ **b.** $\frac{2}{3}$ **3. a.** $\frac{4}{49}$ **b.** $\frac{4}{7}$ **5. a.** $\frac{25}{16}$ **b.** $\frac{5}{2}$ **7. a.** $\frac{64}{81}$ **b.** $\frac{16}{9}$ **9. a.** $\frac{9}{25}$ **b.** $\frac{6}{5}$ **11.** $\frac{3}{7}$

**13.** $\frac{4}{3}$ **15.** $\frac{8}{25}$ **17.** $-\frac{1}{5}$ **19.** $\frac{1}{16}$ **21.** $\frac{17}{21}$ **23.** $\frac{1}{3}$ **25.** $-\frac{15}{14}$ **27.** 0 **29.** $-\frac{2}{25}$ **31.** $\frac{7}{12}$

**33.** $-\frac{34}{945}$ **35.** $\frac{1}{3}$ **37.** $-\frac{63}{50}$ **39.** $\frac{2}{15}$ **41.** $\frac{3}{16}$ **43.** $\frac{3}{8}$ **45.** $\frac{19}{10}$ **47.** $\frac{21}{16}$ **49.** $\frac{5x+3}{15}$

**51.** $\frac{y-2}{6}$ **53.** $\frac{3+4x}{3x}$ **55.** $\frac{8a-7}{8}$ **57.** $\frac{10a+3}{100}$ **59.** $\frac{4x+13}{20}$ **61.** $\frac{3x+35}{5x}$ **63.** $\frac{2n-9}{3n}$

**65.** $\frac{n-4}{5}$ **67.** $\frac{35+4x}{5x}$ **69.** $\frac{9}{25}$ **71. a.** $\frac{13}{30}$ **b.** $1040 **73.** $\frac{19}{10}$ oz.

### The Recycle Bin, 284

**1.** $\frac{3}{5}$ **2.** $\frac{1}{7}$ **3.** $\frac{x}{2}$ **4.** $\frac{3n}{8}$ **5.** $\frac{5}{3a}$

## Margin Exercises 4.2

**1.** $\frac{21}{4}$ **2.** $\frac{32}{3}$ **3.** $\frac{65}{9}$ **4.** $\frac{101}{32}$ **5.** $1\frac{9}{11}$ **6.** $3\frac{3}{5}$ **7.** $17\frac{1}{2}$ **8.** $24\frac{5}{6}$

## Exercises 4.2, page 291

**1.** $\frac{14}{5}$ **3.** $\frac{3}{2}$ **5.** $\frac{5}{4}$ **7.** $\frac{6}{5}$ **9.** $\frac{3}{2}$ **11.** $\frac{19}{4}$ **13.** $\frac{17}{15}$ **15.** $\frac{9}{4}$ **17.** $\frac{32}{3}$ **19.** $\frac{15}{2}$

**21.** $\frac{28}{3}$ **23.** $\frac{31}{10}$ **25.** $\frac{123}{20}$ **27.** $\frac{4257}{1000}$ **29.** $\frac{263}{100}$ **31.** $2\frac{1}{3}$ **33.** $1\frac{1}{4}$ **35.** $1\frac{1}{5}$ **37.** $1\frac{1}{2}$

**39.** $1\frac{1}{2}$ **41.** 5 **43.** $\$1\frac{1}{4}$ **45.** $2\frac{13}{24}$ ft **47.** $\frac{751}{45}$ **49.** $\frac{1338}{41}$ **51.** $\frac{2547}{35}$ **53.** $\frac{103,334}{501}$

## The Recycle Bin, page 294

**1.** $\frac{9}{4}$ **2.** $\frac{9}{10}$ **3.** $x^2$ **4.** $\frac{9a^2}{7}$ **5.** $\frac{3x}{10}$ **6.** $\frac{8}{35a}$

## Margin Exercises 4.3

**1.** 16 **2.** $1\frac{1}{5}$ **3.** $-21\frac{2}{3}$ **4.** 45 **5.** $-2\frac{1}{2}$ **6.** $5\frac{1}{4}$ **7.** $-1$ **8.** 9

## Exercises 4.3, page 303

**1.** $3\frac{1}{5}$ **3.** $13\frac{1}{3}$ **5.** 35 **7.** 24 **9.** $12\frac{1}{4}$ **11.** $32\frac{32}{35}$ **13.** $40\frac{1}{4}$ **15.** $-72\frac{9}{20}$

**17.** $-15\frac{99}{160}$ **19.** $1\frac{3}{10}$ **21.** $851\frac{1}{30}$ **23.** $-1291\frac{5}{42}$ **25.** $640\frac{1}{16}$ **27.** 60 **29.** 60 **31.** $1\frac{7}{8}$

**33.** 4 **35.** $3\frac{1}{7}$ **37.** $-1\frac{1}{2}$ **39.** $-1\frac{5}{6}$ **41.** $-\frac{9}{32}$ **43.** $2\frac{2}{7}$ **45. a.** $\frac{11}{16}$ **b.** above ground **c.** 20 ft.

**d.** 44 ft. **47. a.** $\frac{2}{5}$ **b.** 180 pages **c.** 15 hours **49. a.** $20\frac{3}{4}$ m **b.** $26\frac{7}{16}$ $m^2$ **51.** 10 $cm^2$

**53.** $2907\frac{44}{125}$ $cm^3$ **55. a.** less **b.** \$320 **57. a.** more **b.** 60 passengers **59. a.** 20 gallons **b.** $\$27\frac{1}{2}$ ( or \$27.50 )

**61.** Answers for a, b, and c will vary. The product is smaller if the other number is positive, the same if the other number is 0, and larger if the other number is negative.

## The Recycle Bin, page 309

**1.** $\frac{3}{4}$ **2.** $-\frac{29}{60}$ **3.** $-\frac{5}{6}$ **4.** $\frac{9}{14}$ **5.** $\frac{x-5}{12}$ **6.** $\frac{24+7x}{8x}$

## Margin Exercises 4.4

**1.** $8\frac{7}{10}$ **2.** $29\frac{5}{12}$ **3.** $15\frac{3}{11}$ **4.** $6\frac{7}{10}$ **5.** 7 **6.** $3\frac{7}{8}$

## Exercises 4.4, page 319

**1.** 10 **3.** $11\frac{3}{10}$ **5.** $10\frac{11}{12}$ **7.** $17\frac{9}{28}$ **9.** $21\frac{1}{16}$ **11.** $13\frac{3}{4}$ **13.** $5\frac{5}{8}$ **15.** $30\frac{37}{60}$ **17.** $8\frac{2}{5}$

**19.** $8\frac{1}{6}$ **21.** 2 **23.** $3\frac{3}{10}$ **25.** $7\frac{1}{6}$ **27.** $2\frac{16}{21}$ **29.** $20\frac{19}{24}$ **31.** $53\frac{41}{60}$ **33.** $15\frac{4}{7}$ **35.** $\frac{11}{16}$

**37.** $1\frac{2}{5}$ **39.** $7\frac{1}{4}$ **41.** $-12\frac{1}{4}$ **43.** $-5\frac{3}{8}$ **45.** $-6\frac{19}{24}$ **47.** $13\frac{13}{30}$ **49.** $-3\frac{8}{15}$ **51.** $-50\frac{29}{30}$ **53.** $-45\frac{3}{20}$

**55.** $5\frac{3}{4}$ ft. **57.** $10\frac{13}{24}$ inches **59.** $29\frac{1}{10}$ cm.

## The Recycle Bin, page 324

**1.** $x = \frac{1}{2}$ **2.** $x = -\frac{5}{2}$ or $x = -2\frac{1}{2}$ **3.** $y=1$ **4.** $n = -5$ **5.** $x = \frac{3}{11}$

## Margin Exercises 4.5

**1.** $\frac{15}{8}\left(\text{or } 1\frac{7}{8}\right)$ **2.** 4 **3.** $-\frac{1}{10}$ **4.** $2\frac{1}{2}$ **5.** $\frac{3x+2}{3}$ **6.** $1\frac{13}{20}$

## Exercises 4.5, page 329

**1.** $\frac{3}{28}$ **3.** $-\frac{3}{10}$ **5.** $\frac{3}{2}\left(\text{or } 1\frac{1}{2}\right)$ **7.** $\frac{5}{3}\left(\text{or } 1\frac{2}{3}\right)$ **9.** $\frac{26}{135}$ **11.** $-2$ **13.** $\frac{21}{22}$ **15.** $\frac{235}{4}\left(\text{or } 58\frac{3}{4}\right)$

**17.** $24y$ **19.** $\frac{1}{6}$ **21.** $\frac{7}{60}$ **23.** $\frac{31}{28}\left(\text{or } 1\frac{3}{28}\right)$ **25.** $\frac{379}{60}\left(\text{or } 6\frac{10}{60}\right)$ **27.** $\frac{341}{30}\left(\text{or } 11\frac{11}{30}\right)$ **29.** $-\frac{4}{21}$

**31.** $\frac{15x-13}{15}$ **33.** $\frac{42y+13}{42}$ **35.** $\frac{14a+9}{35}$ **37.** $\frac{3-2x}{7x}$ **39.** $4\frac{7}{16}$ **41.** $\frac{128}{135}$ **43.** $-\frac{481}{1600}$

**45. a.** $\frac{\frac{4}{5}+\frac{2}{15}}{2\frac{1}{4}-\frac{7}{8}}$ **b.** $\left(\frac{4}{5}+\frac{2}{15}\right)\div\left(2\frac{1}{4}-\frac{7}{8}\right)$ **47.** The square of any number between 0 and 1 is always smaller than the original number. Explanations will vary.

## Margin Exercises 4.6

**1.** $\frac{3}{4}$ **2.** $\frac{3}{5}$ **3.** $\frac{25 \text{ washers}}{2 \text{ bolts}}$ **4.** True **5.** False **6.** True **7.** $x = 20$ **8.** $y = 600$

## Exercises 4.6, page 345

**1.** $\frac{2}{5}$ **3.** $\frac{1}{15}$ **5.** $\frac{6}{5}$ **7.** $\frac{19 \text{ mi.}}{1 \text{ gal.}}$ (or 19 mpg.) **9.** $\frac{1 \text{ hit}}{4 \text{ at-bats}}$ **11.** $\frac{\$3 \text{ profit}}{\$10 \text{ invested}}$ **13.** True **15.** False

**17.** True **19.** $x = 12$ **21.** $w = 6$ **23.** $x = 12\frac{1}{2}$ **25.** $B = 900$ **27.** $A = 180$ **29.** $500

**31.** 3 in. for the width, $7\frac{1}{2}$ in. for the length **33.** 60 ft. **35.** 7 in. by 10 in. **37.** $x = 7\frac{1}{2}, y = 15$ **39.** $x = 7, y = 5$ **41.** $x = 50°, y = 60°$ **43.** $x = 50°, y = 50°$ **45.** 400 ft. **47. a.** 55 lb. **b.** 6 bags **c.** $72 **49. a.** The statement is misleading because the numbers 4 and 5 are not in the same units. **b.** The ratio of 4 quarters to 5 dollars is 1: 5.

## Chapter 4 Test, page 355

**1.** A proportion is a statement that two ratios are equal. **2.** $\frac{3}{7}$ **3.** $6\frac{3}{4}$ **4.** $3\frac{21}{50}$ **5.** $-\frac{501}{100}$ **6.** $\frac{38}{13}$

**7.** $2\frac{1}{6}$ **8.** $4\frac{47}{180}$ **9.** $\frac{4}{5}$ **10.** $-\frac{13}{36}$ **11.** $-\frac{11}{30}$ **12.** $8\frac{5}{12}$ **13.** $-\frac{8}{11}$ **14.** $-\frac{1}{6}$ **15.** $7\frac{11}{42}$

**16.** $14\frac{19}{30}$ **17.** $-16$ **18.** $11\frac{3}{5}$ **19.** $\frac{x-1}{9}$ **20.** $-\frac{1}{3}$ **21.** $A = 750$ **22.** $x = \frac{1}{96}$ **23. a.** $27\frac{1}{2}$ in. **b.** $46\frac{7}{8}$ in.$^2$ **24. a.** larger **b.** $19\frac{1}{2}$ **25.** $1633\frac{73}{200}$ cm.$^3$ **26. a.** 5000 per hour **b.** 600,000 per week

**27.** $\$7\frac{1}{2}$ **28.** $x = 3\frac{3}{4}$, $y = 2\frac{1}{2}$

## Cumulative Review: Chapters 1- 4, page 359

**1. a.** 5 **b.** 3 **c.** 125 **2.** 180,000 **3. a.** B **b.** A **c.** D **d.** C **4.** 630,000 **5.** – 125
**6. a.** 2, 3, 5, 7, 11, 13, 17, 19, 23 **b.** 4, 9, 25, 49, 121, 169, 289, 361, 529 **7.** 1575 **8.** 0, undefined **9.** 44
**10.** 14,275 **11.** 588 **12.** 28,538 **13.** 204 **14.** $-\frac{1}{40}$ **15.** $\frac{1}{63}$ **16.** $\frac{1}{24}$ **17.** $-\frac{1}{4}$ **18.** $68\frac{23}{60}$
**19.** $-6\frac{21}{40}$ **20.** $-\frac{121}{81}$ **21.** $\frac{7}{4}\left(\text{or } 1\frac{3}{4}\right)$ **22.** $3\frac{9}{38}$ **23.** $-\frac{47}{6}\left(\text{or } -7\frac{5}{6}\right)$ **24.** $\frac{x-2}{15}$ **25.** $x = -27$
**26.** $y = 7$ **27.** $x = -\frac{1}{2}$ **28.** $n = 13$ **29. a.** more **b.** \$815 **30. a.** 15 gal. **b.** Yes **31. a.** $16\frac{3}{40}$ m.
**b.** $52\frac{363}{500}$ ft. **32. a.** $116\frac{1}{2}$ ft. **b.** $807\frac{5}{8}$ $\text{ft.}^2$ **33.** $3\frac{1}{5}$ in. **34. a.** $x = 3$ **b.** $y = 6\frac{3}{4}$ **35.** Responses will vary.
**36.** Responses will vary.

## CHAPTER 5

### Margin Exercises 5.1

**1.** Thirty and eight tenths **2.** seven and six hundredths **3.** eighteen and five hundred sixty-two thousandths
**4.** three and seven ten-thousandths **5.** 10.004 **6.** 600.5 **7.** 700.007 **8.** 0.707 **9.** 8.6 **10.** 5.04
**11.** 0.018 **12.** 240

### Exercises 5.1, page 371

**1.** 6.5 **3.** 19.075 **5.** 62.547 **7.** $13\frac{2}{100}$ **9.** $200\frac{6}{10}$ **11.** 0.4 **13.** 0.23 **15.** 5.028 **17.** 600.66
**19.** 3495.342 **21.** nine tenths **23.** six and five hundredths **25.** fifty and seven thousandths
**27.** eight hundred and nine thousandths **29.** five thousand and five thousandths **31. a.** 7 **b.** 8 **c.** 8,7,8,8 **d.** 34.8
**33.** 89.0 **35.** 18.1 **37.** 14.3 **39.** 0.39 **41.** 8.00 **43.** 0.08 **45.** 0.057 **47.** 0.002 **49.** 32.458
**51.** 479 **53.** 164 **55.** 300 **57.** 5200 **59.** 76,500 **61.** 500 **63.** 62,000 **65.** 103,000 **67.** 7,305,000
**69.** 0.00076 **71.** Nine hundred fourteen thousandths; one and nine hundredths; thirty-nine and thirty-seven hundredths; three and thirty-seven hundredths **73.** Two and eight hundred twenty-five ten thousandths **75.** three and fourteen thousand, one hundred fifty-nine hundred thousandths **77.** Thirty-five and eight tenths; twenty-six and nine tenths; eighteen and nine tenths; twelve and three tenths; seven and two tenths **79.** one hundred one and seventy-five hundredths

### Recycle Bin, page 376

**1. a.** 11,000 **b.** 10,891 **2. a.** 1900 **b.** 1864 **3. a.** 770,000 **b.** 784,791 **4. a.** 0 **b.** 149 **5. a.** 6000
**b.** 6488 **6. a.** 75,000 **b.** 74,784

### Margin Exercises 5.2

**1.** 59.804 **2.** $50.78x$ **3.** – 13.04 **4.** – 157.9 **5.** – 35.18 **6.** $x = -40.05$

### Exercises 5.2, page 383

**1.** 3.3 **3.** 5.65 **5.** 4.7925 **7.** 117.385 **9.** 718.15 **11.** 1.55 **13.** 15.89 **15.** 54.946
**17.** 4.888 **19.** 30.464 **21.** $-13.4x$ **23.** $45.5y$ **25.** $-0.7x$ **27.** $6.3t$ **29.** $10.3x - 4.1y$
**31.** $x = -107.3$ **33.** $y = -1.7$ **35.** $z = 18.39$ **37.** $w = 3.84$ **39.** \$6950 **41.** 12 in.
**43.** 467.846 million **45.** 15.85 in.; 64.9 in. **47.** 74.96 **49.** 9.9209 **51.** 89.9484 **53.** 0.660932
**55.** 1406.8 **57.** 510.989 **59.** – 34.35

### Recycle Bin, page 387

**1.** 4,800,000 **2.** – 3915 **3.** –10,472 **4.** 2,100,000 ( estimate ); 1,854,804 ( product ) **5.** – 3 **6.** 96
**7.** – 519 **8.** – 35

### Margin Exercises 5.3

**1.** 0.27 **2.** – 0.046 **3.** 297.28916 **4.** 0.003752 **5.** 83.6 **6.** – 97.35 **7.** 14,820 **8.** 1872
**9.** 1.14 **10.** 5.6 **11.** 14.9 **12.** 5.0 **13.** 0.732 **14.** 1.6 **15.** 0.08346

## Exercises 5.3, page 401

**1. a.** 1.0 **b.** 3.0 **c.** 6.0 **d.** 0.06 **e.** 5.0 **3. a.** 2 **b.** 2000 **c.** 75 **d.** 5 **e.** 0.02 **5.** 0.35 **7.** 10.8 **9.** 0.0004 **11.** – 0.112 **13.** – 1 **15.** 0.00429 **17.** 2.036 **19.** 0.002028 **21.** 1.632204 **23.** 3.24 **25.** – 0.79 **27.** – 0.006 **29.** 0.7 **31.** 21.3 **33.** 5.04 **35.** 2.290 **37.** 209.167 **39.** 376 **41.** 950 **43.** 730,000 **45.** 0.026483 **47.** 0.00178 **49. a.** 53.6 in. **b.** 179.56 in.$^2$ **51.** \$310.10 **53. a.** about 20 mpg **b.** 22 mpg **55.** \$825 **57.** about 6.45 hrs **59.** 4.4 yards per carry **61.** $x = -7711.772$ **63.** $a = -11{,}052.278$ **65.** $t = -28.808$ **67.** $y = -233.198$ **69.** Answers will vary

## Margin Exercises 5.4

**1.** $\frac{7}{20}$ **2.** $\frac{1}{8}$ **3.** $\frac{12}{5}$ **4.** 0.14 **5.** – 0.56 **6.** $12.96 = 12\frac{24}{25}$ **7.** $15.84 = 15\frac{21}{25}$ **8.** $-\frac{3}{8}$

**9.** –18.75 **10.** $-6.625 \left(\text{or } -\frac{53}{8} \text{ or } -6\frac{5}{8}\right)$ **11.** $8.47 \times 10^3$ **12.** $2.6 \times 10^{-5}$ **13.** $6.696 \times 10^9$

**14.** $1.7\overline{2} \times 10^{-2}$

## Exercises 5.4, page 415

**1.** $\frac{7}{10}$ **3.** $\frac{5}{10}$ **5.** $\frac{16}{1000}$ **7.** $\frac{835}{100}$ **9.** $\frac{1}{8}$ **11.** $1\frac{4}{5}$ **13.** $-2\frac{3}{4}$ **15.** $\frac{33}{100}$ **17.** $0.\overline{6}$ **19.** $-0.\overline{142857}$

**21.** 0.6875 **23.** $0.\overline{63}$ **25.** 0.208 **27.** 0.917 **29.** 5.714 **31.** –2.273 **33.** 1.5 **35.** 1.888 **37.** 71.22 **39.** –2.875 **41.** – 0.625 **43.** 0.0567 **45.** –10.395 **47.** 120.31 **49.** – 17.55

**51.** – 1.7 **53.** $2\frac{3}{4}$ **55.** 0 **57.** $-\frac{5}{9}$ **59.** $15\frac{5}{8}$ **61.** $1.8 \times 10^5$ **63.** $1.24 \times 10^{-4}$ **65.** $8.9 \times 10^2$

**67.** $5.88 \times 10^{12}$ **69.** 40,000,000,000,000,000,000,000,000 **71.** 49,068,000,000; $4.9068 \times 10^{10}$ **73.** 9,915,000,000,000; 82,128,000,000,000 **75. a.** $7.65 \times 10^8$ **b.** 765,000,000 **77. a.** $9.0 \times 10^{-5}$ **b.** 0.00009 **79. a.** $3.15 \times 10^8$ **b.** 315,000,000 **81. a.** and **b.** Examples will vary. **c.** $|x + y| < |x| + |y|$ when $x$ and y are opposite in sign. **d.** $|x + y| = |x| + |y|$ when $x$ and $y$ have the same sign or one or both are 0. **83.** Explanations will vary. Digits are compared, place by place, until one digit is larger ( or smaller. )

## Margin Exercises 5.5

**1.** $x = -12.8$ **2.** $y = 0.3$ **3.** $x = 1820$

## Exercises 5.5, page 427

**1.** $x = 31$ **3.** $x = 74.1$ **5.** $x = -3.31$ **7.** $y = 81.2$ **9.** $y = 165.9$ **11.** $t = 10.3$ **13.** $t = 2503$

**15.** $x = 5.14$ **17.** $x = 30.25$ **19.** $y = -1.5$ **21.** $y = 11.44$ **23.** $z = 0$ **25.** $z = -4.\overline{66}$ **27.** $x = 0$ **29.** $x = 20.3$ **31.** \$17.50 ( manual ); \$12.95 ( disk ) **33.** –13.1 **35.** 240 miles **37. a.** 50.24 in. **b.** 200.96 in.$^2$ **39. a.** 150.72 mm **b.** 1808.64 mm$^2$ **41. a.** 10.71 cm **b.** 7.065 cm$^2$ **43. a.** 26.84 m **b.** 37.68 m$^2$ **45.** d = 16.8 in. r = 8.4 in. **47.** 2.95 in. **49. a.** Diameter is 2 inches. **b.** The perimeter of the square is 8 inches. Since the circle fits inside the square, its circumference must be less than 8 inches.

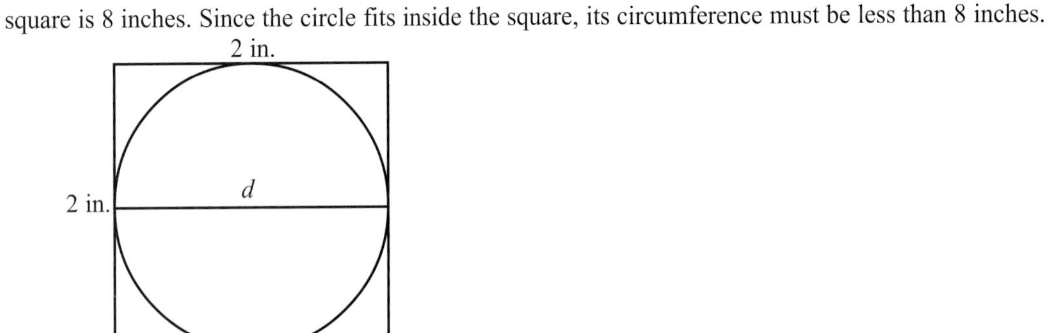

### Margin Exercises 5.6

**1. a.** 64 **b.** 10 **2.** $(2.82)^2 = 7.9524 < 8$ and $(2.83)^2 = 8.0089 > 8$ **3.** 8.660254038 **4.** 4.123105626
**5.** $\sqrt{18}\,ft. \approx 4.24\,ft.$ **6.** $\sqrt{52}\,in. \approx 7.211\,in.$ **7.** $\sqrt{200}\,in. \approx 14.14\,in.$

### Exercises 5.6, page 439

**1.** Yes, $121 = 11^2$ **3.** Not a perfect square **5.** Yes, $400 = 20^2$ **7.** Yes, $225 = 15^2$
**9.** Not a perfect square **11.** 7 **13.** 36 **15.** 16 **17.** 21 **19.** 206 **21. a.** Answers will vary. $\sqrt{39} > \sqrt{36} = 6$, so $\sqrt{39} > 6$ **b.** Answers will vary. $\sqrt{39} < \sqrt{49} = 7$, so $\sqrt{39} < 7$

**23.** $(1.4142)^2 = 1.99996164 < 2$ and $(1.4143)^2 = 2.0024449 > 2$. So, $1.4142 < \sqrt{2} < 1.4143$. **25. a.** 4 and 5 **b.** 4.7958 **27. a.** 3 and 4 **b.** 3.6056 **29. a.** 8 and 9 **b.** 8.4853 **31. a.** 7 and 8 **b.** 7.7460 **33. a.** 9 and 10 **b.** 9.7468 **35.** – 0.4142 **37.** 6.4641 **39.** 5.8990 **41.** –1 **43.** Yes, $5^2 + 12^2 = 13^2$ **45.** Yes, $6^2 + 8^2 = 10^2$ **47.** $c = 2.24$ **49.** $x = 14.14$ **51.** $c = 20.62$ **53.** $x = 6.00$ cm **55.** $x = 13.42$ ft **57. a.** $C = 94.2$ ft, $A = 706.5\text{ft}^2$ **b.** $P = 84.85$ ft, $A = 450\text{ ft}^2$ **59. a.** no **b.** home **c.** no **61.** 8.5 cm **63.** 58.31 ft **65. a.** 2 **b.** 3 **c.** 4 **d.** 7 **67. a.** 3.107232506 **b.** 4.562902635 **c.** 7.488872387 **d.** 20.000000000

### Margin Exercises 5.7

**1.** $-9\sqrt{10}$ **2.** $\frac{7}{5}$ **3.** $8\sqrt{3}$ **4.** $10\sqrt{2}$

### Exercises 5.7page 453

**1. a.** $2\sqrt{2}$ **b.** $2\sqrt{3}$ **c.** 4 **d.** $2\sqrt{5}$ **e.** $2\sqrt{6}$ **3. a.** $5\sqrt{2}$ **b.** $5\sqrt{3}$ **c.** 10 **d.** $5\sqrt{5}$ **e.** $5\sqrt{6}$
**5. a.** $7\sqrt{2}$ **b.** $7\sqrt{3}$ **c.** 14 **d.** $7\sqrt{5}$ **e.** $7\sqrt{6}$ **7.** $-5\sqrt{3}$ **9.** $-5\sqrt{10}$ **11.** $12\sqrt{3}$ **13.** $10\sqrt{10}$
**15.** 19 **17.** $\frac{11}{12}$ **19.** $\frac{3\sqrt{3}}{4}$ **21.** $-\frac{1}{3}$ **23.** 11 **25.** $6\sqrt{2}$ **27.** $7\sqrt{5}$ **29.** $-11\sqrt{6}$ **31.** $6\sqrt{11}$
**33.** $9\sqrt{15}$ **35.** 37 **37.** –1 **39.** $11\sqrt{5}$ **41.** $17\sqrt{5}$ **43.** $50\sqrt{2}$ **45.** $42\sqrt{6}$

### Chapter 5 Test, page 459

**1. a.** thirty-two and sixty-four thousandths **b.** $32\frac{8}{125}$ **2.** 2.032 **3. a.** 216.7 **b.** 216.70 **c.** 216.705 **d.** 220
**4. a.** 73.0 **b.** 73.01 **c.** 73.015 **d.** 70 **5.** 122.863 **6.** – 126.49 **7.** 205.27 **8.** 50.716 **9.** – 5.3075
**10.** 0.00006 **11.** 0.08217 **12.** 0.2182 **13.** 82,170 **14.** 10.80667 **15.** 25.81 **16.** 16.49
**17.a.** $6.7 \times 10^7$ **b.** 0.000083 **18. a.** 14.12 **b.** –1.9875 or $\left(-1\frac{79}{80}\right)$ **19.** $x = 2.5$ **20.** $x = -10$
**21.** $y = -1.1$ **22. a.** 121 **b.** 5 **23. a.** 17.3205 **b.** 4.4641 **24. a.** $2\sqrt{10}$ **b.** $13\sqrt{5}$ **c.** $13\sqrt{2}$
**25. a.** The Pythagorean Theorem **b.** Yes, $10^2 + 24^2 = 26^2$ **26.** 35.5 ft **27. a.** 157 yd **b.** 188.4 yd **c.** 863.5 yd$^2$
**28. a.** 359 mi **b.** 24 gal **29.** Pay cash, because he can save $5760. **30. a.** 6 and 7 **b.** 6.7823

### Cumulative Review Chapters 1 - 5, page 463

**1. a.** commutative property of addition ; $x = 15$ **b.** associative property of multiplication; $y = 3$ **c.** associative property of addition; $a = 6$ **2.** 141 **3.** 2,400,000 **4.** – 16 **5.** 58 **6.** 24 **7.** – 27 **8.** 57 **9.** $2 \cdot 2 \cdot 5 \cdot 19$
**10. a.** 560 **b.** $120a^2b^3$ **11. a.** $\frac{3}{200}$ **b.** $6\frac{8}{25}$ **c.** $-1\frac{1}{20}$ **12.** $12\frac{19}{24}$ **13.** 6 **14.** $157\frac{13}{20}$ **15.** 73.28 $\left(\text{or } 73\frac{7}{25}\right)$
**16.** 61.65 **17.** $\frac{13}{15}$ **18.** $6\frac{3}{32}$ **19.** $\frac{5x-3}{30}$ **20.** 280.7 **21.** 102.91 **22. a.** $8.65 \times 10^5$ **b.** 0.00341
**23.** $x = -1$ **24.** $y = -3.01$ **25.** $x = 0$ **26.** $n = 13$ **27. a.** $5\sqrt{6}$ **b.** $-20\sqrt{5}$ **c.** $\frac{2\sqrt{13}}{9}$ **d.** $20\sqrt{6}$

**28.** $2\sqrt{13}$ in. **29. a.** More, because ; $5^2 = 25 < 29$ **b.** 5.385 **30.** \$33,405 **31. a.** 7.85 ft. **b.** 4.90625 $ft^2$

**32.** 90 $cm^3$ **33. a.** More than \$1500 **b.** \$2250 **34. a.** $x = 1.8$ **b.** $y = \frac{5}{3}\left(\text{or } y = 1\frac{2}{3}\right)$ **35.** 2 in. by 2.8 in.

**36.** $x = 2.4$; $y = 1.8$ in. **37.** Responses will vary. The question is designed to help the students reflect on what they are learning.

## CHAPTER 6

### Margin Exercises 6.1

**1.** 11% **2.** 140% **3.** 8.25% **4.** 5% **5.** 35% **6.** 217% **7.** 0.213 **8.** 0.006 **9.** 26%

**10.** 175% **11.** 10% **12.** 3.75% **13.** $\frac{9}{10}$ **14.** $\frac{8}{25}$ **15.** $3\frac{7}{20}$

### Exercises 6.1page 481

**1.** 66% **3.** 20% **5.** 99% **7.** 30% **9.** 48% **11.** 16.3% **13.** $24\frac{1}{2}\%$ **15. a.** $\frac{19}{100} = 19\%$

**b.** $\frac{17}{100} = 17\%$ **c.** Investment a. is better. **17. a.** $\frac{5}{100} = 5\%$ **b.** $\frac{4}{100} = 4\%$ **c.** Investment a. is better. **19.** 3%

**21.** 300% **23.** 5.5% **25.** 175% **27.** 108% **29.** 36% **31.** 0.02 **33.** 0.22 **35.** 0.25 **37.** 0.101

**39.** 0.065 **41.** 0.8 **43.** 5% **45.** 96% **47.** $14\frac{2}{7}\%$ or 14.3% **49.** $137\frac{1}{2}\%$ or 137.5% **51.** $\frac{1}{20}$

**53.** $\frac{1}{8}$ **55.** $1\frac{1}{5}$ **57.** $\frac{1}{500}$ **59. b.** 0.875 **c.** 87.5% **61. a.** $\frac{3}{50}$ **c.** 6% **63. a.** $\frac{6}{25}$ **b.** 0.24

**65.** 0.15 **67.** 32% **69.** $\frac{7}{20}$ **71.** about 12% in California; about 0.18% in Wyoming**73. a.** 0.000575 **b.** 20.7%

**75.** Answers will depend on the student's calculator.

### The Recycle Bin, page 486

**1.** $x = 2.2$ **2.** $y = \frac{5}{2}$ (or $y = 2.5$) **3.** $w = 8$ **4.** $A = 0.345$ **5.** 10.6 gallons **6.** \$2.20

### Margin Exercises 6.2

**1.** 9 **2.** 144 **3.** 59.4

### Exercises 6.2, page 495

**1.** $A = 6$ **3.** $A = 10.5$ **5.** $A = 10$ **7.** $A = 90$ **9.** $B = 225$ **11.** $B = 2180$ **13.** $B = 620$

**15.** $B = 120$ **17.** $R = 1.5 = 150\%$ **19.** $R = 2 = 200\%$ **21.** $R = \frac{1}{3} = 33\frac{1}{3}\%$ **23.** $R = 0.252 = 25.2\%$

**25.** $A = 17.5$ **27.** $B = 75$ **29.** $R = 1.5 = 150\%$ **31.** $B = 36$ **33.** $A = 115$ **35.** $A = 58.56$
**37.** $R = 1.25 = 125\%$ **39.** $B = 80$ **41.** 42 **43.** 4 **45.** 25 **47.** 100% **49.** 61.5
**51.** 11,882,080 **53. a.** 14.6 million **b.** 72.3% **55. a.** Explanations will vary. The decimal equivalent of 10% is 0.10, and multiplication of a decimal number by 0.10 moves the decimal point one place to the left. **b.** Explanations will vary. Taking 10% of a number is the same as dividing by 10 or multiplying by $\frac{1}{10}$. **57.** Answers will vary. Students may suggest finding 1% of a number by moving the decimal point two places to the left and then finding the desired multiple.

**Exercises 6.3, page 501**

**1.** c **3.** a **5.** a **7.** b **9.** b **11.** b **13.** d **15.** a **17.** b **19.** d **21.** c **23.** b **25.** b **27.** d **29.** c **31.** a **33.** b **35.** b **37.** c **39.** c **41.** c **43.** c **45.** d **47.** c **49.** d **51.** 86.3 **53.** 258.9 **55.** 2.52 **57.** 810.0 **59.** 0.8456 **61.** 45.8 **63.** 14.0 **65.** 24.0 **67.** 550.0 **69.** 134.16

**Margin Exercises 6.4**

**1. a.** \$350 **b.** $28\frac{4}{7}\%$ (or 28.6%) **c.** 25% **d.** 20%

**Exercises 6.4, page 515**

**1.** \$7500 **3.** \$770 **5.** 1.5% **7.** 136 **9.** 40 **11. a.** \$9.38 **b.** \$3.38 **13.** \$37,500

**15. a.** \$180 **b.** 40% **c.** $28\frac{4}{7}\%$ (or 28.6%) **17.** 30%; 32%; 12% **19. a.** 86% **b.** 300 **c.** 42 **21. a.** \$2550 **b.** \$2703 **23. a.** \$875 **b.** \$700 **c.** \$750.75 **25. a.** \$156 **b.** \$5460 **27. a.** \$10.35 **b.** \$79.25 **29.** \$26.45 **31. a.** \$675 and \$775, respectively. **b.** The \$100 difference is 12.9% based on the second salesman's pay, or \$14.8% based on the first salesman's pay. **33. a.** \$8800 **b.** Jerry **c.** Wilma. Explanations will vary. For sales over \$8800, 8% of the sales is more than \$500 plus 3% of the sales over \$2000.

**Exercises 6.5, page 527**

**1.** \$1278 **3.** \$9362.50 **5.** \$6501.50 **7. a.** \$24,000 **b.** \$96,000 **c.** \$26,280 **9.** Yes. You will need \$27,300, and you have \$30,000. **11. a.** \$315,000 **b.** \$945,000 **c.** \$9450 **d.** \$329,500

**Margin Exercises 6.6**

**1.** \$480 **2.** \$150 **3.** 15% **4.** First quarter: $(\$4000)(0.08)\left(\frac{1}{4}\right) = \$80$,

Second quarter: $(\$4080)(0.08)\left(\frac{1}{4}\right) = \$81.60$, Third quarter: $(\$4161.60)(0.08)\left(\frac{1}{4}\right) = \$83.24$, Total interest = \$244.84

**5.** \$15,682.24; \$5682.24

**Exercises 6.6, page 543**

**1.** \$37.50 **3.** \$275 **5.** \$48 **7.** 72 days **9.** 10% **11. a.** \$470.25 **b.** \$29.75 **c.** To collect interest at 18% **13. a.** 16 **b.** 100 **c.** 30 days **d.** 8.5% **15. a.** \$7803 **b.** \$8118.25 **17. a.** \$1200 **b.** \$2346.65 **19. a.** \$450.77 **b.** \$459.23 **21. a.** \$26,620 **b.** \$26,801.92, No **c.** Explanations will vary. By compounding semiannually, interest is posted twice as frequently, so more interest is earned on interest accrued. **23. a.** \$13,498.03 **b.** about 7 years **25. a.** \$24,492.29 **b.** \$95,64 **27. a.** \$73,424.76 **b.** \$3313.07 **c.** \$170.14 **d.** \$2.24 **29.** \$1,078,673.79 **31. a.** \$1638.62 **b.** \$638.62 **33. a.** \$1648.61 **b.** \$648.61 **35. a.** \$123,803.81 **b.** \$98,803.81 **37.** Answers will vary. The time for doubling depends on the rate of interest and the period of compounding and is independent of the principal.

**Chapter 6 Test, page 551**

**1.** 102% **2.** 7.25% **3.** 0.5% **4.** 15% **5.** 37.5% **6.** 0.24 **7.** 0.068 **8.** 0.055 **9.** $1\frac{3}{5}$

**10.** $\frac{12}{125}$ **11.** $\frac{5}{8}$ **12.** 9.2 **13.** 50 **14.** 135.45 **15.** 20.3 **16.** c **17.** d **18.** b

**19.** \$5400 **20. a.** more **b.** 97% **c.** \$2500 **d.** \$75 **21.** \$85.30 **22. a.** 30% **b.** $23\frac{1}{13}\%$ (or 23.1%)

**23.** 9% **24. a.** \$13,828.18 **b.** \$3828.18 **25. a.** \$19,000 **b.** \$76,000 **c.** \$20,610

### Cumulative Review: Chapters 1- 6, page 555

**1.** identity; $a$ **2.** In general, $a - b \neq b - a$. Any counterexample will also serve as an explanation. **3.** 2.8; −3.6; −10.1
**4. a.** 90 (estimate) **b.** 89.22 **5.** 37.4 **6.** 4.248 **7. a.** $5 \cdot 13$ **b.** $2^4 \cdot 3^2$ **c.** $2 \cdot 3^3 \cdot 5$
**8.** 6, 12, 18, 24, 30, 36, 42, 48, 54, 60, 66, 72, 78, 84, 90, 96 **9. a.** Yes **b.** 63 **10.** $A = 0.25$ **11.** $\frac{1}{5}$
**12.** $\frac{2x+1}{4}$ **13.** $-6.55$ **14.** $-52 = x$ **15.** $a = \frac{11}{5}$ $(\text{or } a = 2.2)$ **16.** $x = -\frac{7}{5}$ **17.** 800
**18. a.** 330 yd **b.** 6500 yd$^2$ **19. a.** 30 in. **b.** 72 in. **c.** 216 in.$^2$ **20. a.** $\frac{11}{16}$ **b.** $8\frac{1}{4}$ ft **21. a.** 99 lb **b.** 99 lb
**c.** Explanations will vary. The result is the same, since $(100)(0.9)(1.1) = (100)(1.1)(0.9) = 99$. **22.** \$143.75
**23.** \$1734.91 **24. a.** \$30 **b.** $\frac{1}{2}$ yr **c.** 20% **25. a.** $28\frac{4}{7}\%$ (or 28.6%) **b.** 25% **c.** 20% **26. a.** 180 **b.** 252
**27.** 27 min **28.** \$38.92 **29. a.** right; $10^2 + 24^2 = 26^2$ **b.** 60 cm **c.** 120 cm$^2$ **30.** 15 ft and 18 ft
**31. a.** \$33,750 **b.** \$101,250 **c.** \$36,645 **32.** \$74.85

### Margin Exercises 7.1

**1.** The sum of a number and twice the number. **2.** Five times a number increased by 1. **3.** $3x - 6$ **4.** $7(x - 2)$

### Exercises 7.1, page 567 (Answers may vary. These are some possibilities.)

**1.** three times a number **3.** fifteen less than a number **5.** six times a number plus 4.9 **7.** twice a number increased by three times the number **9.** 2.5 times a number plus 3.5 times the same number **11.** the product of five and a number minus the number plus 2 **13.** four more than the product of 1.5 and a number **15.** seven decreased by the product of 3 and a number **17.** three times the difference of a number and 2 **19.** the product of −4 and the sum of a number and 3 **21.** the quotient of a number and 3 increased by 15 **23.** three-halves less than a quotient of a number and 5 **25. a.** $n - 3$ **b.** $3 - n$ **27.** $6x$ **29.** $\frac{7}{x}$ **31.** $4(y + 8)$ **33.** $6(x - 1)$
**35.** $\frac{8}{x+3}$ **37.** $5n - 2n$ **39. a.** four times a number decreased by five **b.** four times the difference between a number and 5 **41. a.** the product of a number and 3 increased by 1 **b.** three times the sum of a number and 1
**43. a.** twice a number minus 7 **b.** twice the difference between a number and 7 **45.** $60h$ **47.** $1.25x$ **49.** $\frac{m}{60}$
**51.** $365y$ **53. – 67.** Responses will be varied and imaginative.

### Exercises 7.2, page 575

**1.** $x = -6$ **3.** $x = -6$ **5.** $y = 22$ **7.** $y = -4$ **9.** $x = 0$ **11.** $n = 0$ **13.** $x = \frac{7}{3}$
**15.** $z = -7$ **17.** $y = 5$ **19.** $t = -\frac{3}{40}$ **21.** $x = -3$ **23.** $x = 4$ **25.** $y = 2$ **27.** $x = \frac{9}{5}$
**29.** $x = \frac{2}{3}$ **31.** $x = 1$ **33.** $x = 6$ **35.** $x = -3$ **37.** Answers will vary. Integers can be written as fractions with denominator 1.

### Exercises 7.3, page 579

**1.** $x = 75$ **3.** $x = -\frac{5}{2}$ **5.** $x = \frac{220}{3}$ **7.** $y = 0$ **9.** $x = -5$ **11.** $x = \frac{9}{14}$ **13.** $n = -7$
**15.** $y = \frac{6}{5}$ **17.** $x = \frac{19}{27}$ **19.** $y = \frac{10}{11}$ **21.** $x = \frac{27}{8}$ **23.** $x = -7$ **25.** $x = \frac{7}{60}$

#### Margin Exercises 7.4

**1.** $2n + 4 = n - 16; n = -20$ **2.** $n + 68 = 35n; n = 2$ **3.** $n = 6$ in.

#### Exercises 7.4, page 589

**1. a.** $x - 10$ **b.** $2x$; $x - 10 = 2x$ **3. a.** $3(y + 2)$ **b.** $2y$; $3(y + 2) = 2y$ **5. a.** Let $x$ = the number. **b.** $x + 32$; 90 **c.** $x + 32 = 90$ **d.** $x = 58$ **7. a.** Let x = the number. **b.** $-10 + x$; $-25$ **c.** $-10 + x = -25$ **d.** $x = -15$ **9. a.** Let n = the number. **b.** $-10 - n$; $-25$ **c.** $-10 - n = -25$ **d.** $15 = n$ **11. a.** Let $y$ = the number. **b.** $3(y + 4)$; $-60$ **c.** $3(y + 4) = -60$ **d.** $y = -24$ **13. a.** Let $a$ = the number. **b.** $a - 7$; $8a$ **c.** $a - 7 = 8a$ **d.** $-1 = a$ **15. a.** Let $y$ = the number. **b.** $2(y - 5)$; $6y + 14$ **c.** $2(y - 5) = 6y + 14$ **d.** $y = -6$ **17. a.** Let $x$ = length of third side. **b.** $8 + 8 + x$; 30 **c.** $8 + 8 + x = 30$ **d.** $x = 14$ in. **19. a.** Let $x$ = length of the pool. **b.** $2x + 2(x - 16)$; 168 **c.** $2x + 2(x - 16) = 168$ **d.** $x = 50$ m; $x - 16 = 34$ m **21. a.** Let $n$ = value of home when new. **b.** $2n + 80{,}000$; 260,000 **c.** $2n + 80{,}000 = 260{,}000$ **d.** $n = \$90{,}000$ **23. a.** Let $n$ = the first integer and $n + 1$ = the second integer. **b.** $n + n + 1$; 67 **c.** $n + n + 1 = 67$ **d.** $n = 33$; $n + 1 = 34$ **25. a.** Let $n$ = the first integer and $n + 1$ = the second integer **b.** $3(n + n + 1)$; $-15$ **c.** $3(n + n + 1) = -15$ **d.** $n = -3$; $n + 1 = -2$ **27. a.** Let $n$ = the first odd integer, $n + 2$ = the second odd integer and $n + 4$ = the third odd integer **b.** $n + n + 4$; $3(n + 2) - 27$ **c.** $n + n + 4 = 3(n + 2) - 27$ **d.** $n = 25$; $n + 2 = 27$, $n + 4 = 29$ **29. a.** Let $n$ = the first even integer, $n + 2$ = the second even integer and $n + 4$ = the third even integer **b.** $n + n + 2 + n + 4$; $n + 2 + 168$ **c.** $n + n + 2 + n + 4 = n + 2 + 168$ **d.** $n = 82$; $n + 2 = 84$; $n + 4 = 86$ **31. – 39.** Responses will be varied and imaginative. **41.** Discussions will vary. Students should follow the pattern of Polya's four-step process in problem solving.

#### Margin Exercises 7.5

**1.** \$225 **2.** 6 hr 40 mins **3.** $x = \dfrac{7 + y}{2}$ **4.** $d = \dfrac{C}{\pi}$

#### Exercises 7.5, page 599

**1.** $d = 90$ mi **3.** $C = 9.42$ ft **5.** $P = \$2500$ **7.** $A = \$1005$ **9.** $\alpha = 75°$

**11. a.** $y = -3x + 14$

| x | y |
|---|---|
| 0 | 14 |
| −1 | 17 |
| 5 | −1 |

**13. a.** $y = 4x + 6$

| x | y |
|---|---|
| 2 | 14 |
| 1.5 | 12 |
| −2 | −2 |

**15. a.** $y = \dfrac{5 - 2x}{2}$

| x | y |
|---|---|
| 0 | 2.5 |
| 2.5 | 0 |
| 3 | −0.5 |

**17. a.** $x = 3y + 15$

| x | y |
|---|---|
| 30 | 5 |
| 15 | 0 |
| 0 | −5 |

**19. a.** $x = y + 10$

| x | y |
|---|---|
| 10 | 0 |
| 0 | −10 |
| 13 | 3 |

**21. a.** $x = \dfrac{-2y + 5}{2}$

| x | y |
|---|---|
| 0 | 2.5 |
| 2.5 | 0 |
| 0.5 | 2 |

**23.** $m = \dfrac{f}{a}$ **25.** $r = \dfrac{C}{2\pi}$ **27.** $w = \dfrac{P - 2l}{2}$

**29.** $C = \dfrac{5}{9}(F - 32)$ **31. a.** $z = 0$ **b.** values less than 60 **c.** Answers will vary. Z-scores will be a measure of individual performance relative to that of the entire class.

## Margin Exercises 7.6

**1.** 

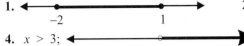

**2.** 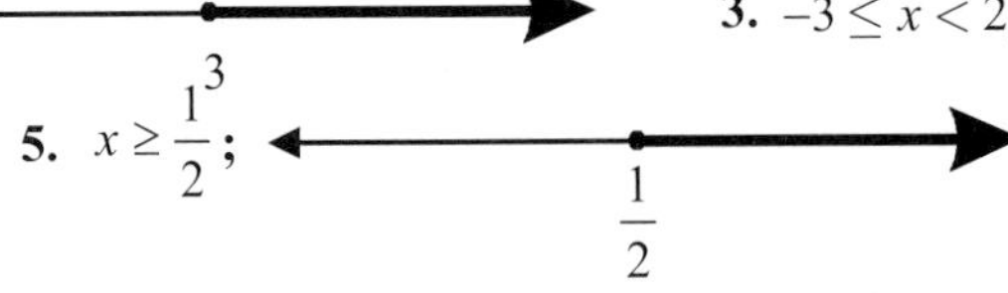

**3.** $-3 \le x < 2$; half-open interval

**4.** $x > 3$; 3

**5.** $x \ge \frac{1}{2}$; $\frac{1}{2}$

**6.** $-1 < y \le 2$; $-1$ $2$

## Exercises 7.6, page 611

**1.** Three does not equal negative three. True. **3.** Negative thirteen is greater than negative one. False. Correction: $-13 < -1$. **5.** The absolute value of negative seven is less than positive five. False. Correction: $|-7| > +5$ **7.** Negative one-half is less than negative three-fourths. False. Correction: $-\frac{1}{2} > -\frac{3}{4}$

**9.** Negative four is less than negative 6. False. Correction: $-4 > -6$ **11.** $-1 < x < 2$; open interval

**13.** $x < 0$; open interval **15.** $x \ge -1$; half-open interval **17.** open; 

**19.** closed; 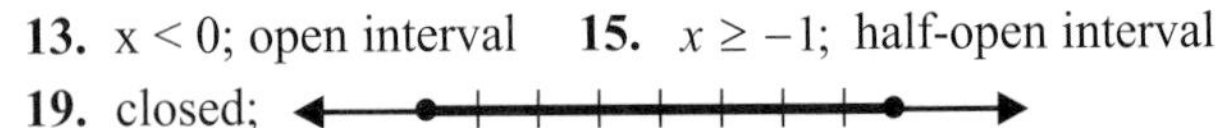

**21.** half-open; 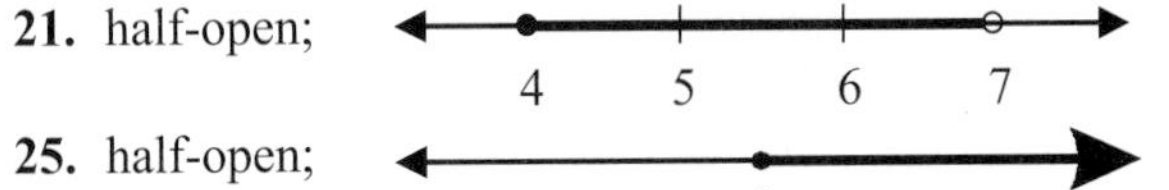

**23.** open; 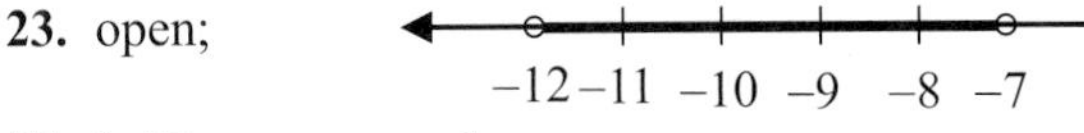

**25.** half-open; 0

**27.** half-open; $\frac{2}{3}$

**29.** $x > 4$; 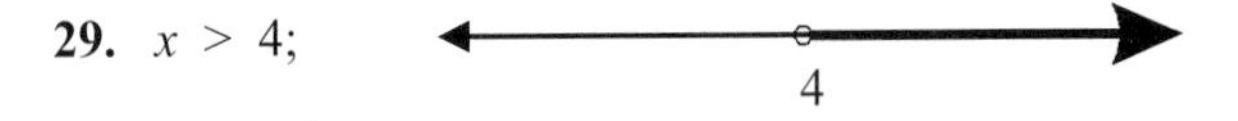

**31.** $y \le -3$; $-3$

**33.** $y > -\frac{6}{5}$; $-\frac{6}{5}$

**35.** $x < \frac{11}{7}$; 

**37.** $x < -5$; $-5$

**39.** $t \le 5$; 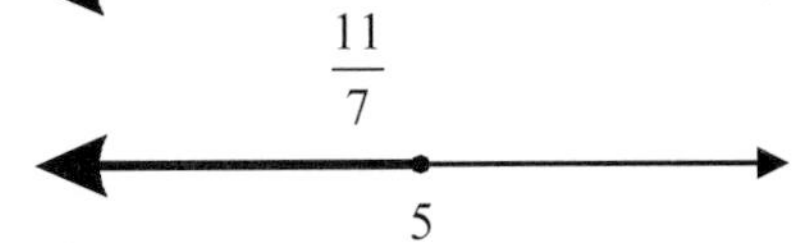

**41.** $x > -6$; $-6$

**43.** $t \le -1$; $-1$

**45.** $t > -1$; $-1$

**47.** $x < -5$; $-5$

**49.** $1 \le x \le 4$; 1 4

**51.** $\frac{1}{3} \le x \le 2$ 

**53.** $-3 \le y \le 1$; $-3$ 1

**55.** $-5 \le t < -3$ $-5$ $-3$

## Chapter 7 Test, page 619

**1.** five times a number plus 6 **2.** four decreased by twice a number **3.** three times the sum of a number and 1

**4.** $\frac{n}{7} + 6n$ **5.** $8n - 4$ **6.** $24x$ **7.** $\frac{x}{24}$ **8.** Answers will vary. **9.** $x = 12$ **10.** $y = 0$

**11.** $x = -5$ **12.** $x = -5$ **13.** $y = 20$ **14.** 83, 84, 85 **15.** 5 **16.** \$9000 **17.** 14, 16, 25

**18.** \$2460 **19.** $h = \frac{A - 2x^2}{4x}$ **20. a.** $y = \frac{18 - 4x}{3}$ **b.**

| $x$ | $y$ |
|---|---|
| 0 | 6 |
| 1.5 | 4 |
| 2.5 | $\frac{8}{3}$ |
| $-3$ | 10 |

**21.** $x > 4$; open

**22.** closed; 0.5 3

**23. a.** $-1 \le x$; $-1$

**b.** $x < 2$; 2

## Cumulative Review : Chapter1 - 7, page 623

**1.** 560,000,000 **2.** 32.79 **3.** 59.892 **4.** 9637.7 **5.** 0.01465 **6.** 24.4 **7.** 3 **8.** 9 **9.** – 48

**10. a.** $8.35\times10^4$ **b.** $6.71\times10^{-4}$ **11.** 35 **12.** 122.4 **13.** 84% **14. a.** \$15 **b.** $33.\overline{3}\%$ **c.** 25%

**15.** $\frac{37}{40}$ **16.** $\frac{-5}{6}$ **17.** $-\frac{7}{36}$ **18.** $-\frac{525}{512}$ **19.** $12\frac{1}{2}$ **20.** $\frac{34}{45}$ **21.** $37\frac{5}{8}$ **22.** $-5\frac{1}{4}$ **23.** $x=8.4$

**24.** $x=-15.75$ **25.** $x=25$ **26.** $x=-38$ **27. a.** $16\pi$ ft ( or 50.24 ft ) **b.** $64\pi$ ft² ( or 200.96 ft² )

**28.** $10\sqrt{5}$ m; 22.361 m **29.** –5, –3, –1 **30.** –7 **31.** \$750 **32.** \$11,618.23 **33. a.** \$144,000 **b.** \$1440

**34.** $x < -2$; 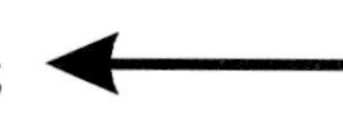 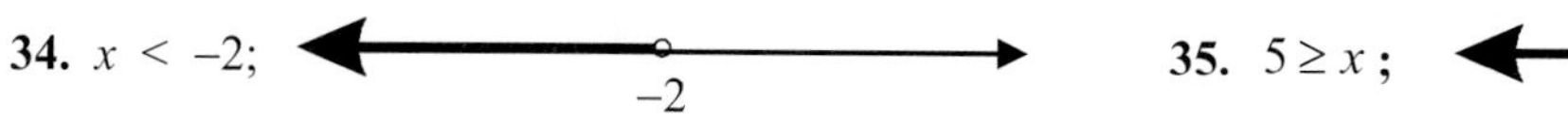


**35.** $5 \geq x$ ; 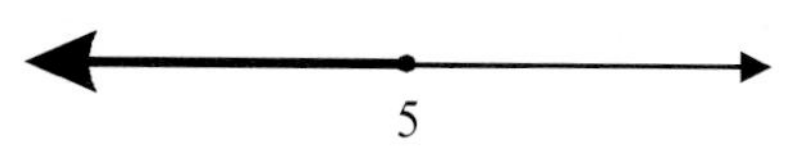


## CHAPTER 8

### Margin Exercises 8.1

**1.** $8^3$( or 512 ) **2.** $(-2)^7$ ( or –128 ) **3.** $-10x^6$ **4.** $3^6$ ( or 729 ) **5.** $2^{12}$ ( or 4096 ) **6.** $a^{14}$

**7.** $3^3$ ( or 27 ) **8.** 5 **9.** $2t^5$ **10.** $\frac{1}{4^2}=\frac{1}{16}$ **11.** $\frac{1}{(-3)^3}=-\frac{1}{27}$ **12.** $13x^2$ **13.** $\frac{1}{7}$

**14.** 6 **15.** $t^{15}$ **16.** 3

### Exercises 8.1, page 637

**1.** $4^4$ **3.** $x^{11}$ **5.** $15x^6$ **7.** $1.5^7$ **9.** $(-6)^5$ **11.** $x^{15}$ **13.** $7^4$ **15.** $5.2^{12}$ **17.** $(-3)^8$

**19.** $t^0=1$ **21.** $6^2$ **23.** $x^9$ **25.** $t^6$ **27.** $3.5^2$ **29.** $(-5)^3$ **31.** $\frac{1}{5^2}$ **33.** $\frac{1}{x^5}$ **35.** $\frac{1}{(-2)^3}$

**37.** $\frac{1}{(1.4)^2}$ **39.** $\frac{1}{y^6}$ **41.** $4^3$ **43.** $2^3$ **45.** $x^3$ **47.** $20x^5$ **49.** 1 **51.** $\frac{1}{y^8}$ **53.** $x^9$

**55.** $\frac{1}{t^9}$ **57.** $s^6$ **59.** $\frac{1}{x^8}$ **61.** $\frac{1}{4^2}$ **63.** 1 **65.** 12 **67.** $x^2$ **69.** $\frac{1}{p^2}$ **71. a.** $\frac{1}{x^6}$

**b.** $\frac{1}{x}$ **c.** $x^5$ **d.** $\frac{1}{x^5}$ **73. a.** $y^{10}$ **b.** $\frac{1}{y^7}$ **c.** $y^3$ **d.** $\frac{1}{y^3}$ **75. a.** $x^{18}$ **b.** $x^9$ **c.** $x^3$ **d.** $\frac{1}{x^3}$

### Margin Exercises 8.2

**1.** $3x^2 - 3x + 6$ **2.** $6x^4 - 11x^3 - 2x^2 - 4x - 2$ **3.** $-4x^3 + 6x^2 + 14x - 14$ **4.** $3x^3 - 20x - 9$

### Exercises 8.2, page 645

**1.** $11x - 7$; first-degree binomial **3.** $2x^2 + 6x + 2$; second-degree trinomial **5.** $5a^2$; second-degree monomial
**7.** $-3y^2 + 7y + 7$; second-degree trinomial **9.** $5x^3 - 3x + 17$; third-degree trinomial **11.** $5x - 10$
**13.** $2x^2 - 4$ **15.** $4y^3 + 3y^2 + 2y - 9$ **17.** $x - 2$ **19.** $a^2 + 5a$ **21.** $12x^3 + x^2 - 6x$
**23.** $-2x^2 + 7x - 8$ **25.** $-x^3 - x + 6$ **27.** $5x^2 - 2x - 5$ **29.** $3x^3 - 5x^2 + 13x - 6$ **31. a.** $5x^3 - 3x^2 - 2x + 30$

**b.**
$$2(1)^3-4(1)^2+3(1)+20=2-4+3+20=21$$
$$3(1)^3+(1)^2-5(1)+10=3+1-5+10=9$$
$$5(1)^3-3(1)^2-2(1)+30=5-3-2+30=30$$
$$21+9\overset{?}{=}30$$
$$30=30$$

**c.**
$$2(3)^3-4(3)^2+3(3)+20=54-36+9+20=47$$
$$3(3)^3+(3)^2-5(3)+10=81+9-15+10=85$$
$$5(3)^3-3(3)^2-2(3)+30=135-27-6+30=132$$
$$47+85\overset{?}{=}132$$
$$132=132$$

**d.** Answers may vary. This does provide a legitimate check, though it is not necessarily fool-proof. Using $x = 0$ is not a good idea, since it will merely cause all variable terms to have value 0.

### Exercises 8.3, page 657

**1.** $12x^4$ **3.** $18a^3$ **5.** $8y^3+4y^2+8y$ **7.** $-4x^3+2x^2-3x+5$ **9.** $-14x^4+21x^3-84x^2$ **11.** $x^2+7x+10$

**13.** $x^2-7x+12$ **15.** $a^2+4a-12$ **17.** $7x^2-20x-3$ **19.** $x^2-81$; difference of two squares **21.** $x^2-9$; difference of two squares **23.** $x^2-2.25$; difference of two squares **25.** $9x^2-49$; difference of two squares

**27.** $4y^2-81$; difference of two squares **29.** $y^2+2y+1$; perfect square trinomial **31.** $4x^2+12x+9$; perfect square trinomial **33.** $4y^2-4y+1$; perfect square trinomial **35.** $25x^2+60x+36$; perfect square trinomial **37.** $x^3+3x^2-9x+5$ **39.** $4y^4-25y^3-y^2+42y$ **41.** $2y^4-y^3-45y^2+22y+99$

**43.** $x^6-2x^5-2x^4+10x^3-11x^2+6x-1$ **45.** $2y^5-y^4+9y^3-11y^2+13y-12$ **47.** $a^8-3a^6+2a^4+11a^2-7$

**49.** $x^6-2x^5+3x^4-4x^3+3x^2-2x+1$ **51. a.** $10x^4-24x^3-17x^2+40x-12$

**b.**
$$2(2)^3-4(2)^2-5(2)+6=16-16-10+6=-4$$
$$5(2)-2=10-2=8$$
$$10(2)^4-24(2)^3-17(2)^2+40(2)-12=160-192-68+80-12=-32$$
$$-4\cdot 8\stackrel{?}{=}-32$$
$$-32=-32$$

**c.**
$$2(-2)^3-4(-2)^2-5(-2)+6=-16-16+10+6=-16$$
$$5(-2)-2=-10-2=-12$$
$$10(-2)^4-24(-2)^3-17(-2)^2+40(-2)-12=160+192-68-80-12=192$$
$$-12(-16)\stackrel{?}{=}192$$
$$192=192$$

**d.** Answers may vary. This does provide a legitimate check, though it is not necessarily fool-proof. Using x = 0 is not a good idea, since it will merely cause all variable terms to have value 0.

### Margin Exercises 8.4

**1.** $x^3(x+5)$ **2.** $-5(2x^2+4x+3)$ **3.** $6x^2(x+3)$ **4.** $4(x^2+25)$ **5.** $4(x+5)(x-5)$ **6.** $4(x-10)^2$
**7.** $5(x^2-3x-3)$ **8.** $8(2x^2+x+1)$

### Exercises 8.4, page 667

**1.** $2(x+7)$ **3.** $5(x-2)$ **5.** $x(3x-5)$ **7.** $5x(x+2)$ **9.** $2x(x-1)$ **11.** $5a(a^2+a+1)$
**13.** $6y(y^2-2y-1)$ **15.** $5x^3(3x+2)$ **17.** $9x^2(8x^2-3x+1)$ **19.** $6a^2(5a^4+3a^2+2a+2)$

**21.** $(x+1)(x-1)$ **23.** $(x+13)(x-13)$ **25.** $(y+15)(y-15)$ **27.** $(6y+1)(6y-1)$ **29.** $(4z+9)(4z-9)$

**31.** $(x+1)(x+1)$ **33.** $(y-5)(y-5)$ **35.** $(z+6)(z+6)$ **37.** $(x-9)(x-9)$ **39.** $(x+20)(x+20)$

**41.** $2(x+5)(x+5)$ **43.** $4(x-1)(x-1)$ **45.** $2(y+5)(y-5)$ **47.** $6x(x-1)(x-1)$ **49.** $9y(y-2)(y-2)$

**51.** not factorable **53.** $5(x^2-x+1)$ **55.** $4(x^2+9x-7)$ **57.** $7(x^2+4)$ **59.** $4y(y^2+25)$

### Exercises 8.5, page 677

**1.** {(–4, 0), (–2, 1), (–1, 0), (1, 1), (2, 0)} **3.** {(–3, –2), (–2, –1), (–2, 1), (0, 0), (2, 1), (3, 0)}
**5.** {(–3, –4), (–3, 3), (–1, –1), (–1, 1), (1, 0)} **7.** {(–1, –5), (0, –4), (1, –3), (2, –2), (3, –1), (4, 0)}
**9.** {(–1, 4), (0, 3), (1, 2), (2, 1), (3, 0), (4, –1), (5, –2)}

**11.**

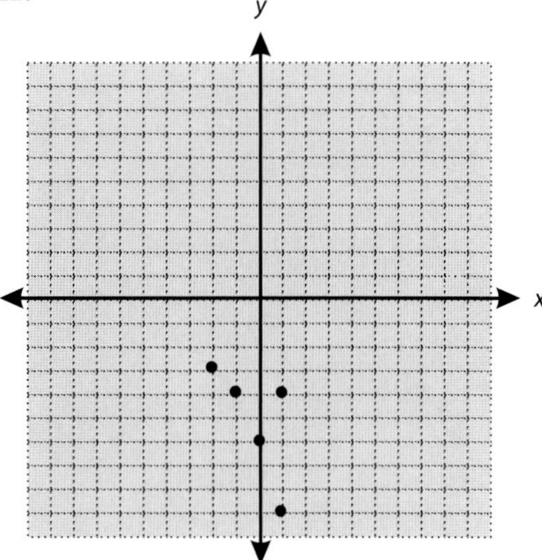


**13.**

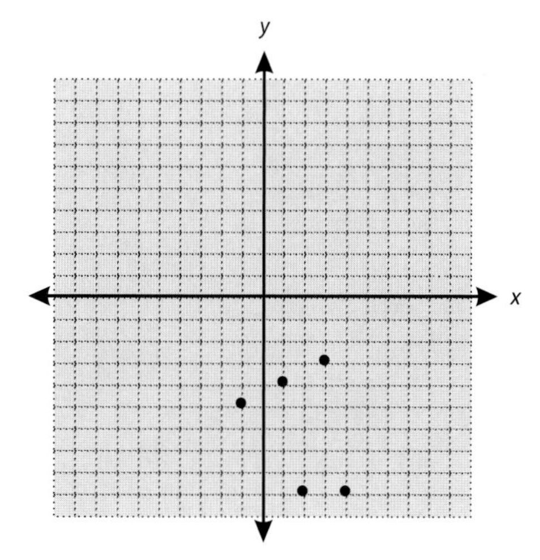


**15.**

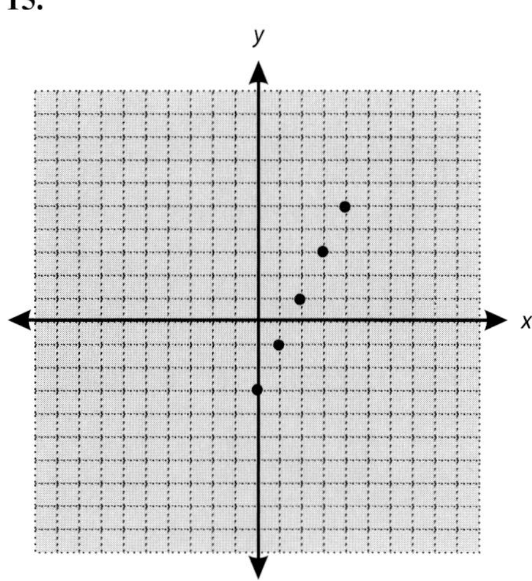


**17.**

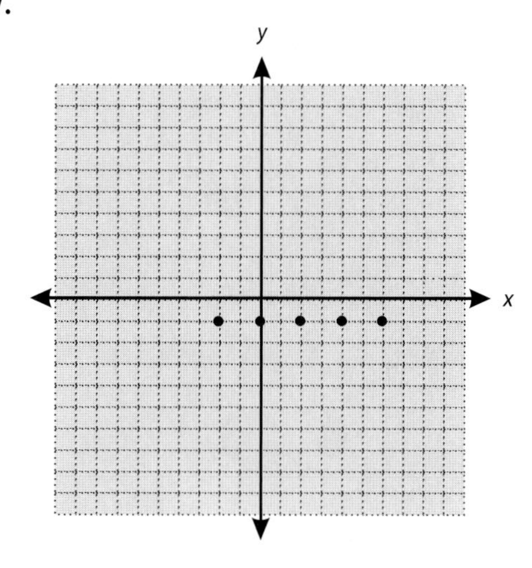


**19.**

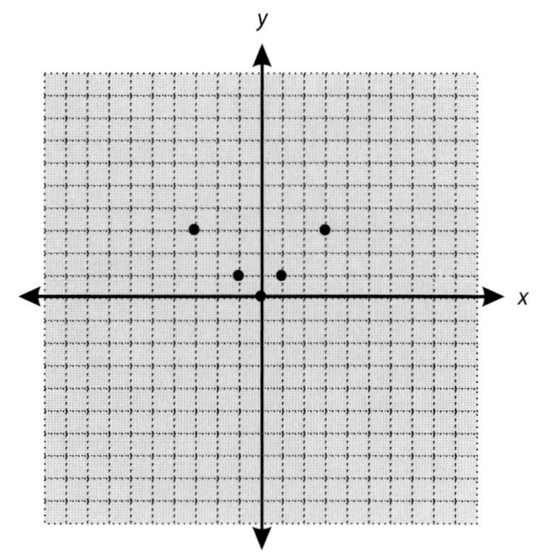


**21.**

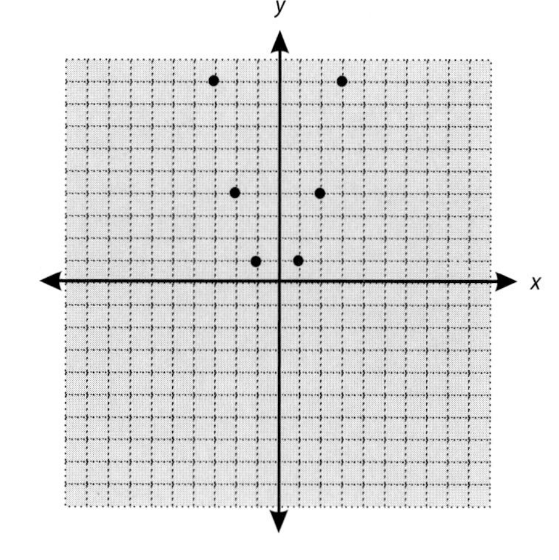

**23.**

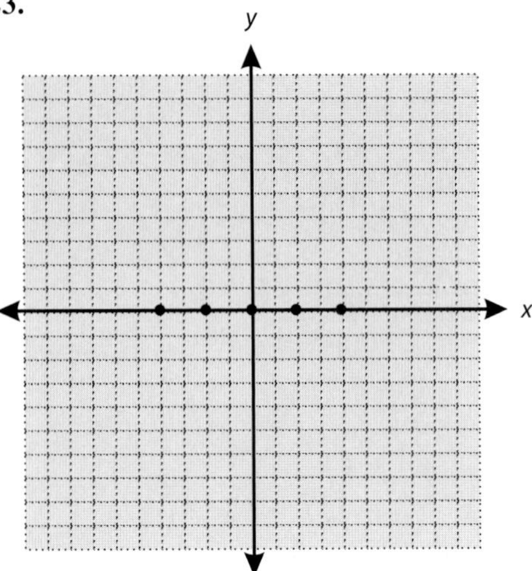


**25.**

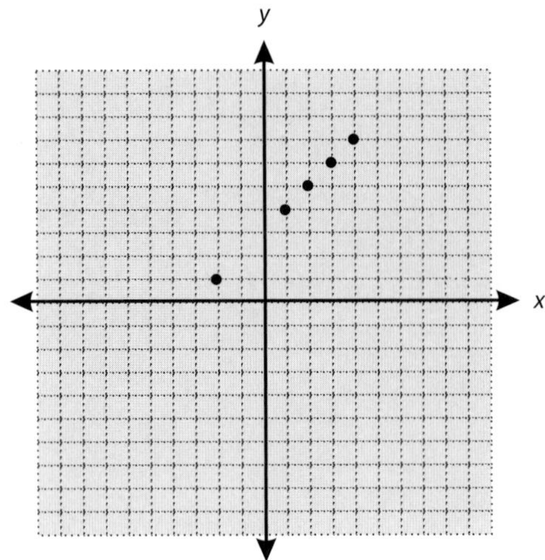


**27.**

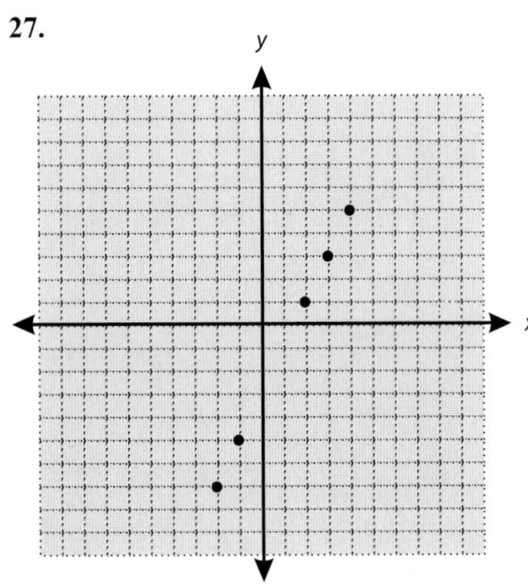


**29.**

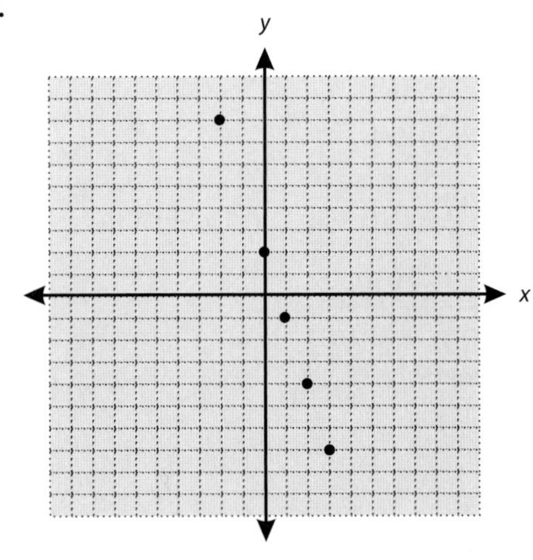


**31.**

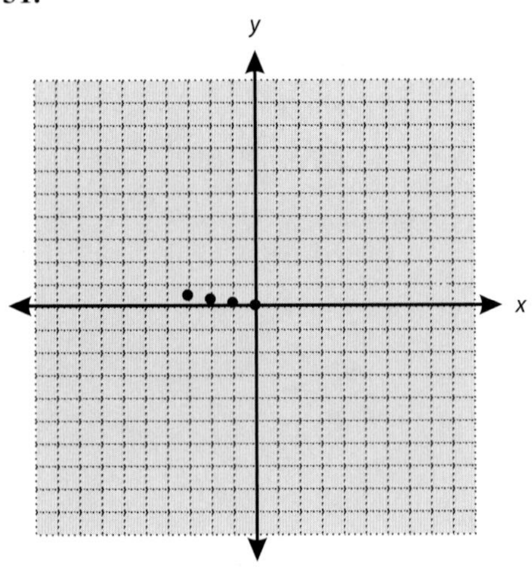


**33.**

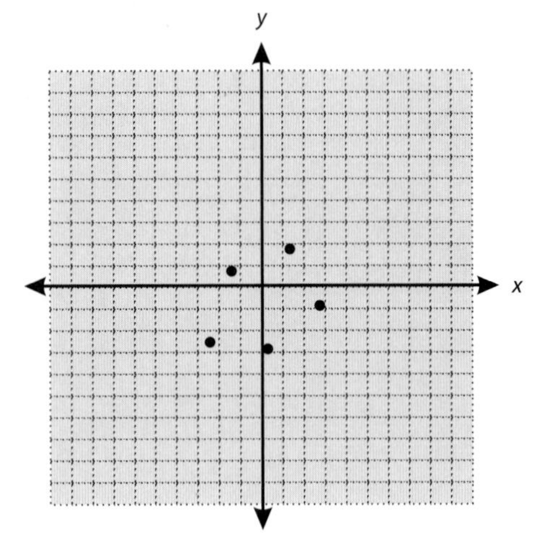

**35.**

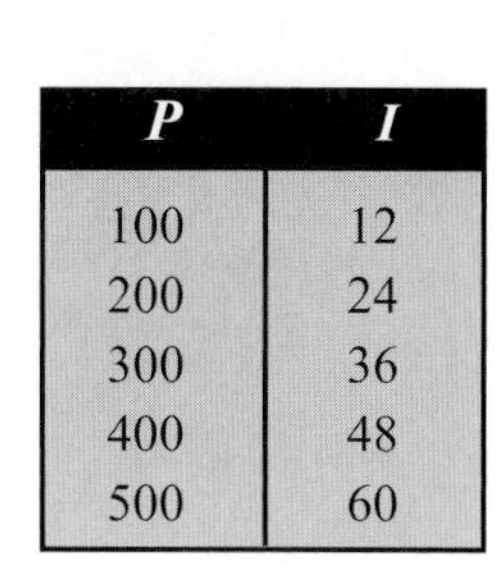

| $P$ | $I$ |
|---|---|
| 100 | 12 |
| 200 | 24 |
| 300 | 36 |
| 400 | 48 |
| 500 | 60 |

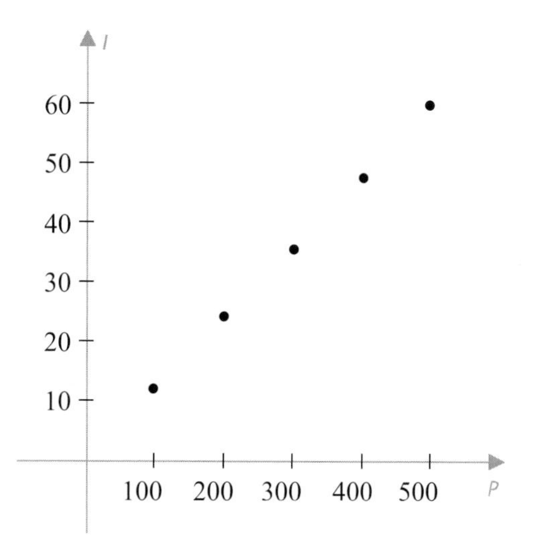


| $C$ | $F$ |
|---|---|
| −20 | −4 |
| −15 | 5 |
| −10 | 14 |
| −5 | 23 |
| 0 | 32 |
| 5 | 41 |
| 10 | 50 |

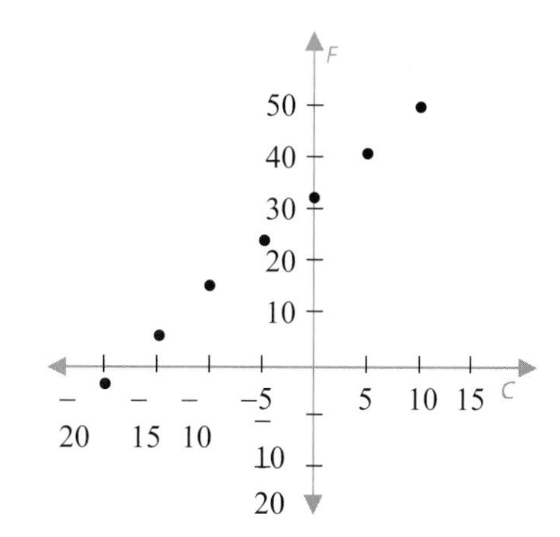


## Exercises 8.6, page 691

**1.**

| $x$ | $y$ |
|---|---|
| 2 | 7 |
| −1 | 1 |
| 0 | 3 |

**3.**

| $x$ | $y$ |
|---|---|
| −1 | −3 |
| 3 | 9 |
| 0 | 0 |

**5.**

| $x$ | $y$ |
|---|---|
| 0 | 2 |
| 3 | 0 |
| 6 | −2 |

**7.**

| $x$ | $y$ |
|---|---|
| −2 | 5 |
| −2 | −1 |
| −2 | 7 |

**9.**

| $x$ | $y$ |
|---|---|
| 0 | −4 |
| 6 | 0 |
| 12 | 4 |

**11.** (2, −12)

**13.** (2, 1), (3, −1) ; (5, −5)

**15.** (3, 1)

**17.** (0, 0), (−3, 6), (−1, 2)

**19.** (−3, 15), (−3, −1.2), (−3, 3)

**21.**

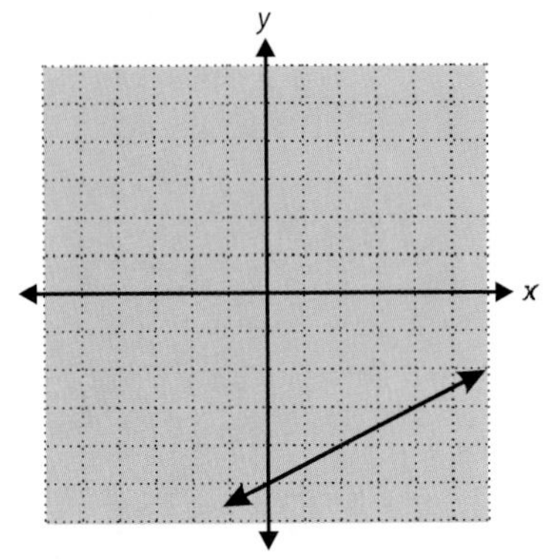


**b.** $x - 2y = 10$

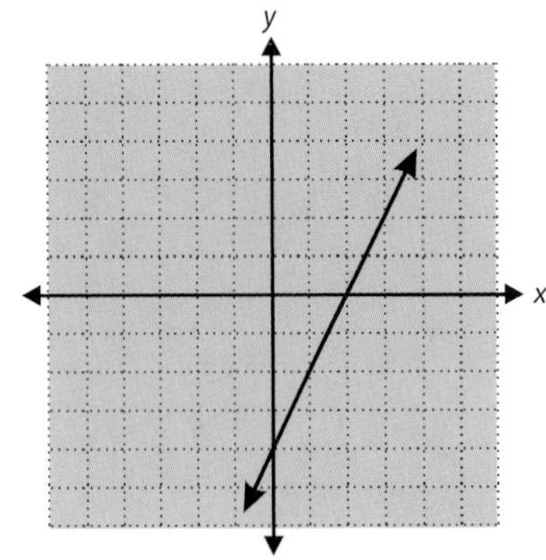


**d.** $2x - y = 4$

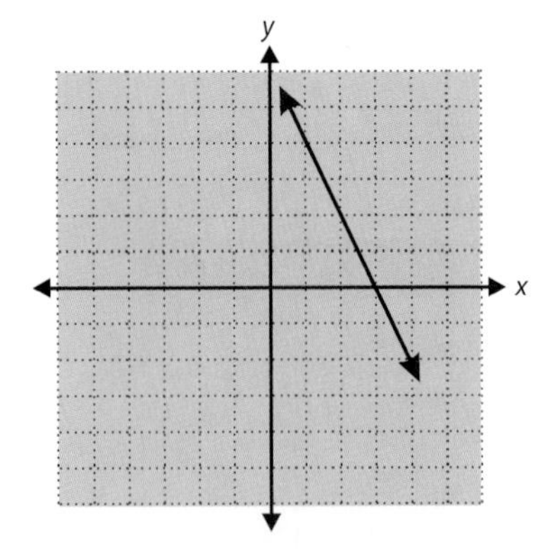


**a.** $2x + y = 6$

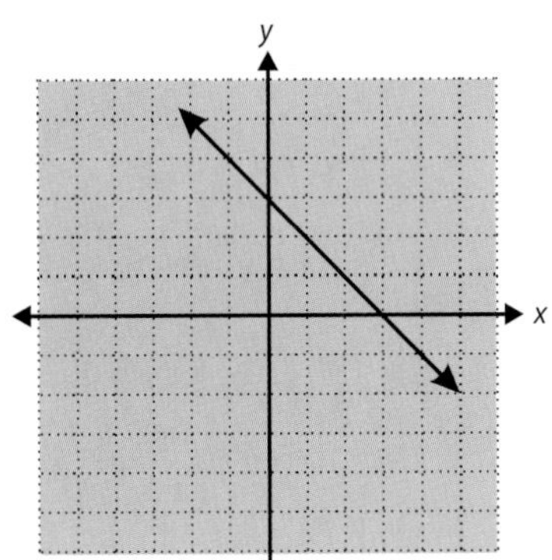


**c.** $2x + 2y = 6$

**23.**

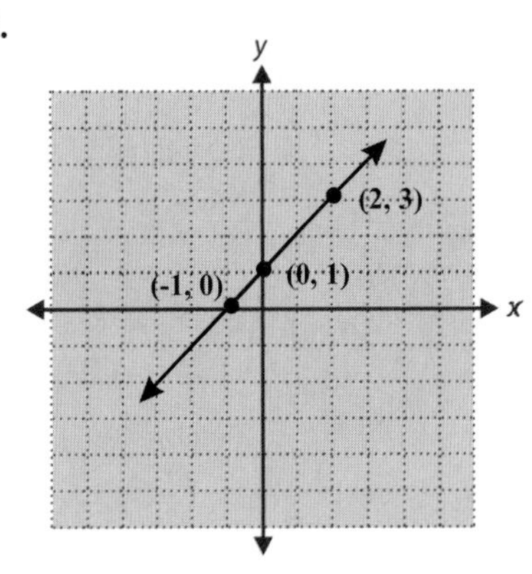


**25.**

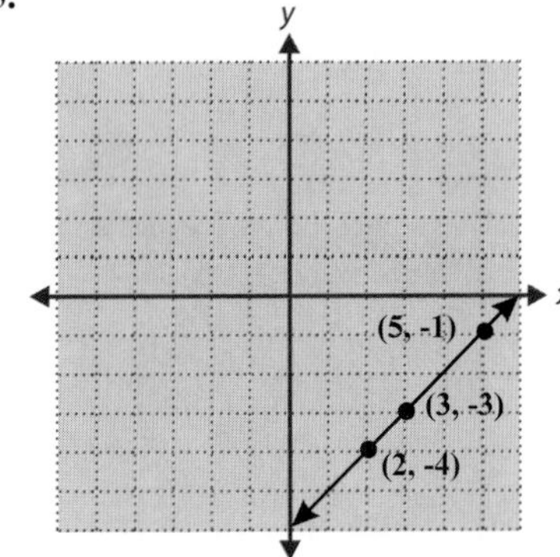

**27.**

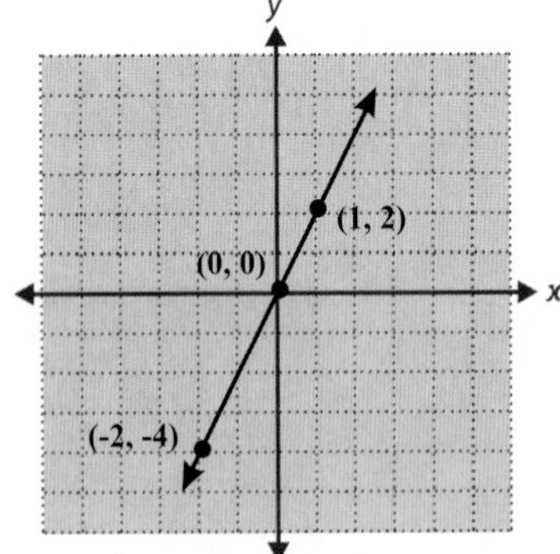


**29.**

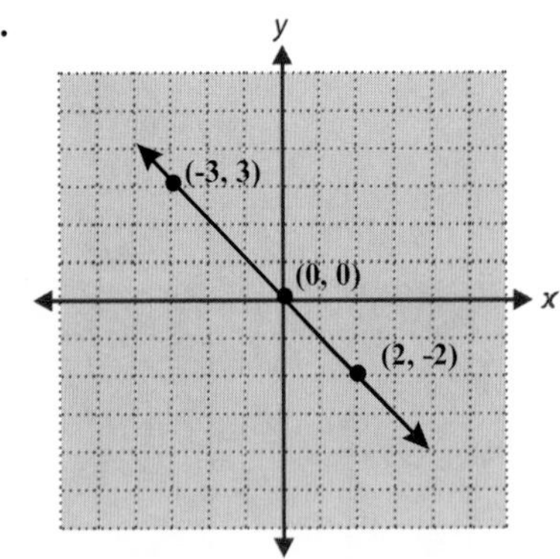


**31.**

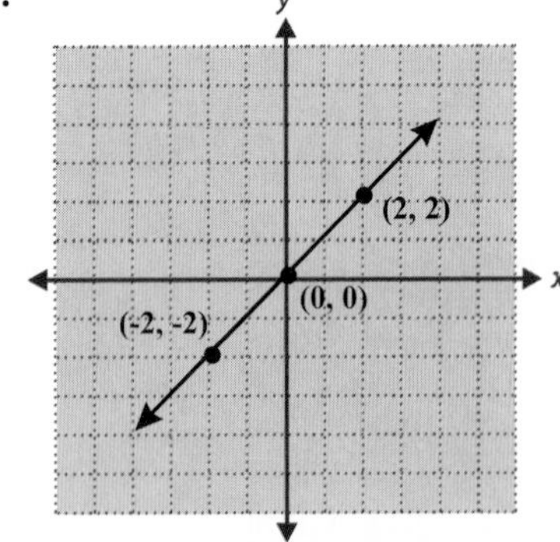


**33.**

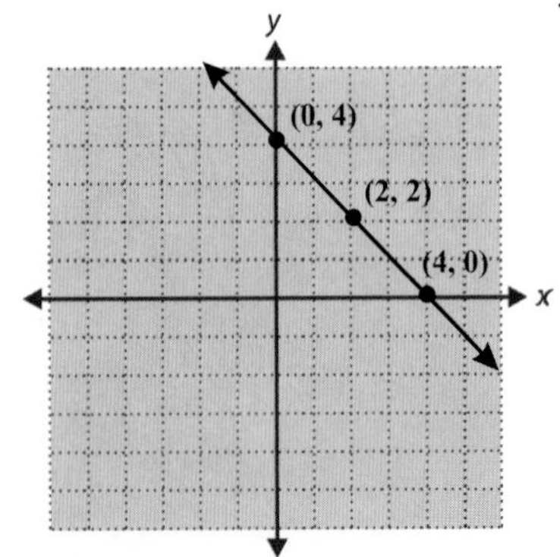


**35.**

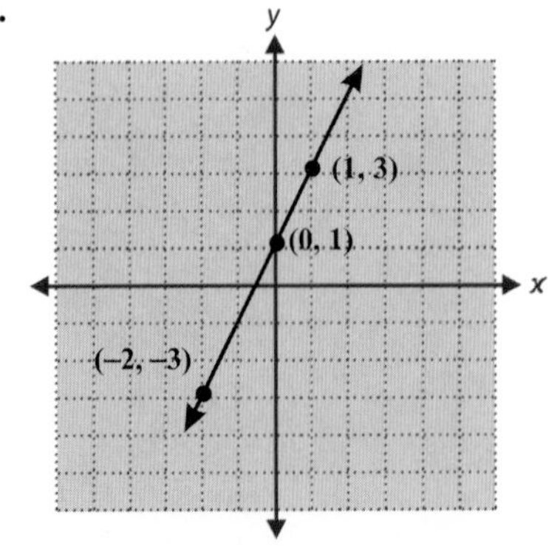


**37.**

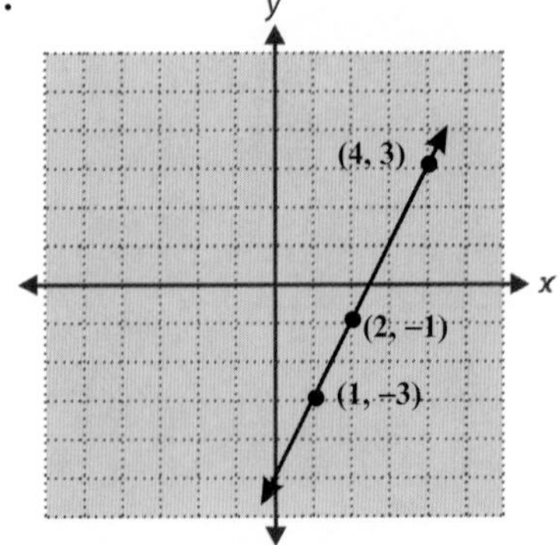

**39.**

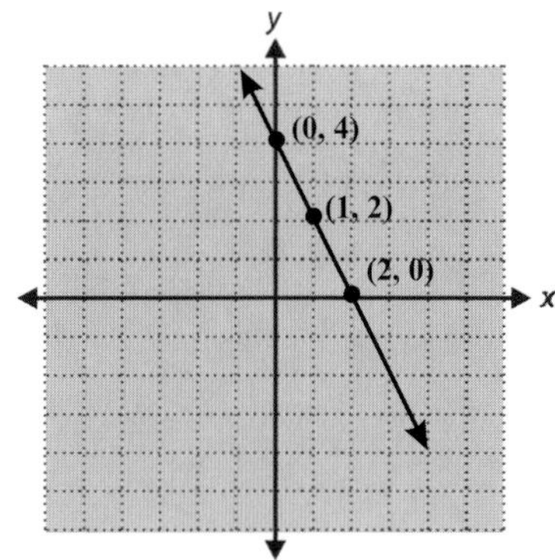


**41.**

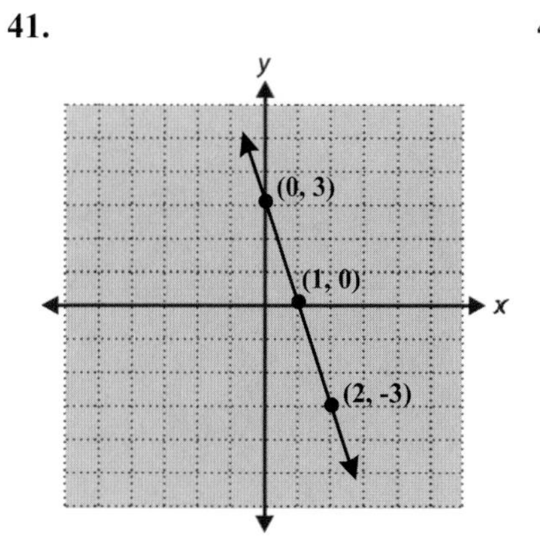


**43.**

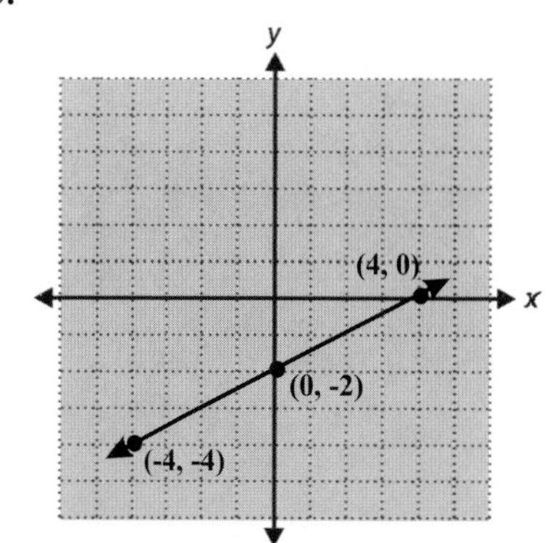


**45.**

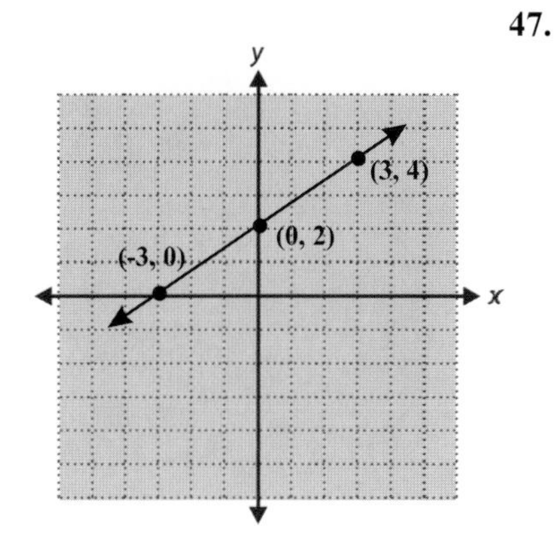


**47.**

y

(0, 4)

(-1, 2)

(-2, 0)

x

**49.**

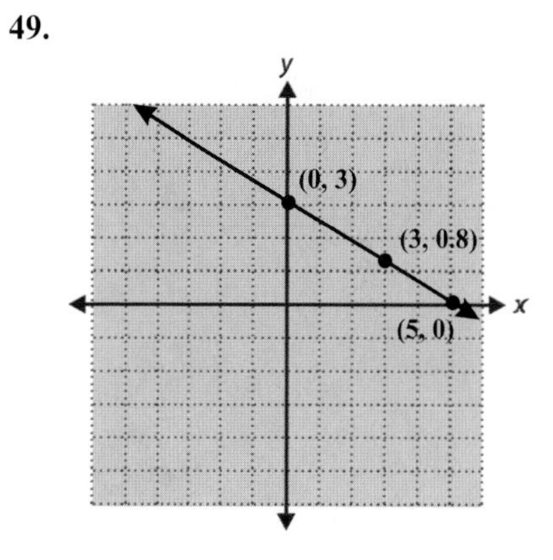


**51.**

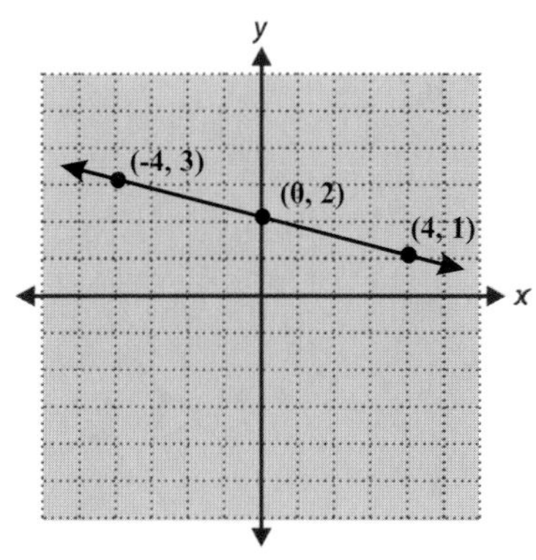


**53.**

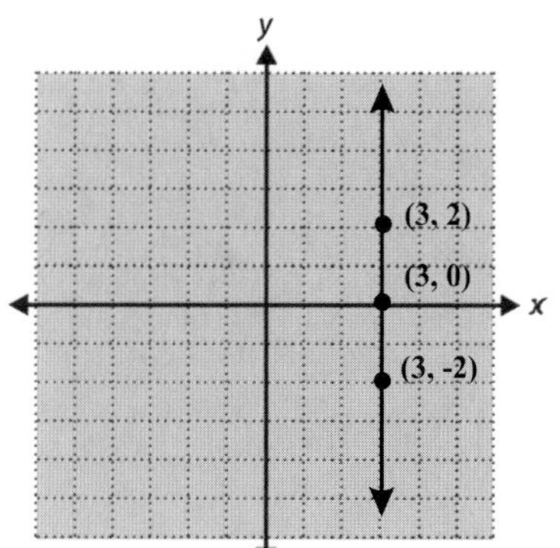


**55.**

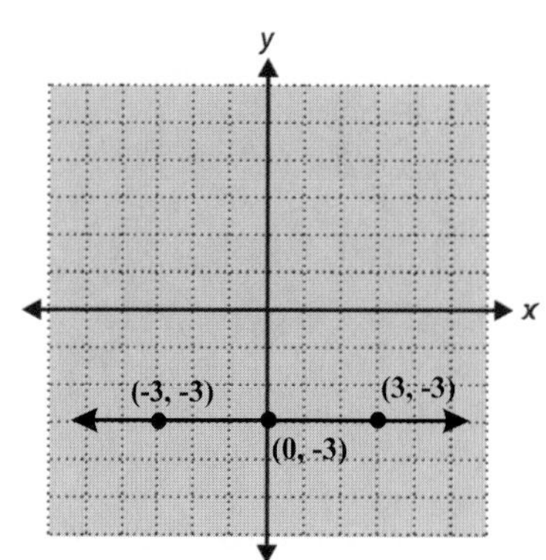


**57.**

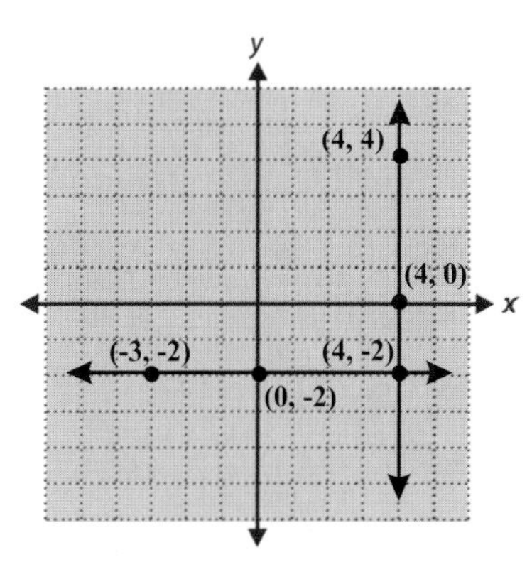


**59.**

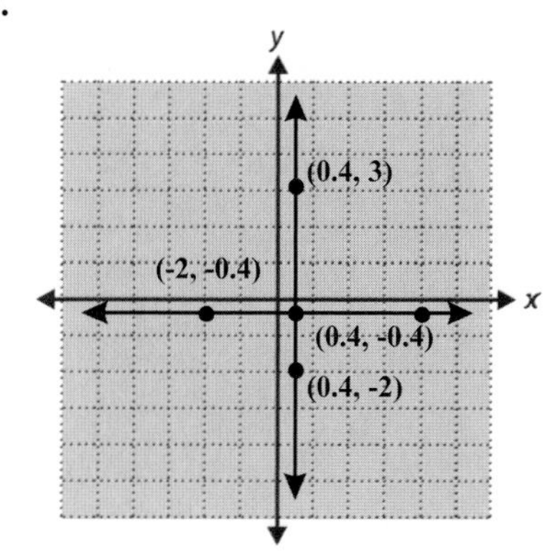




**61. a.** $(2, 7)$ **b.** $(1, 4)$ **c.** $(-1, 3)$

### Chapter 8 Test, page 703

**1.** $60x^5$ **2.** $\frac{1}{5^2}=\frac{1}{25}$ **3.** $\frac{1}{y}$ **4.** $\frac{1}{x^6}$ **5.** $y^{20}$ **6.** 1 **7.** $(-3)^6=729$ **8.** $x^2$ **9.** $\frac{1}{a^5}$

**10.** $4x^3+4x^2-7x$; third-degree trinomial **11.** $6x+4$; first-degree binomial **12.** $-52$ **13.** $-y^3+5y^2-2y-18$

**14.** $-2y^2+19y+16$ **15.** $4x+27$ **16.** $-2a^2-4a$ **17.** $15x^3+10x^2+20.5x$ **18.** $8a^2-10a-33$

**19.** $10x^4-23x^3+30x^2-20x+8$ **20.** $9x^4-9x^3+6x^2+15x+7$ **21. a.** $3(x-9)$ **b.** $2(x^2+25)$

**c.** $3x^2(x^2+9x-3)$ **22. a.** $(a-8)^2$ **b.** $(a+12)^2$ **c.** $(5a+1)(5a-1)$ **23. a.** $2(y+8)^2$ **b.** $3(y+5)(y-5)$

**c.** not factorable **24.** {(–4, 1), (–3, 0), (–2, –1), (0, 0), (1, 1), (2, 0)} **25.** {(–1, 1), (0, 1), (1, 1), (2, 1)}

**26.**

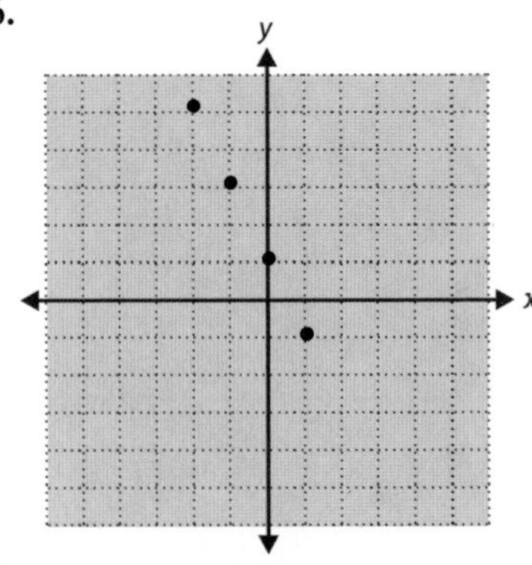


**27.**

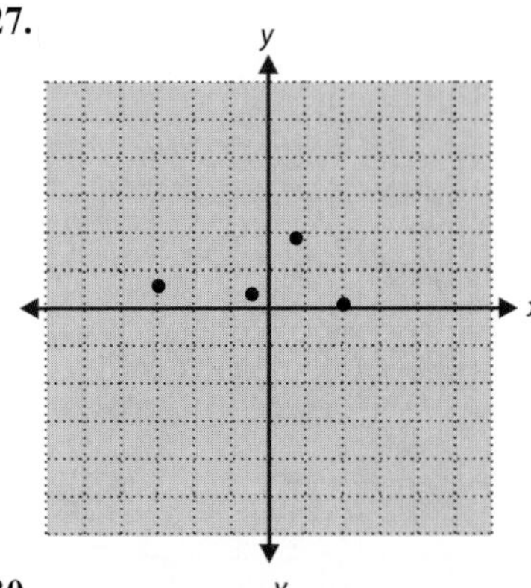


**28.**

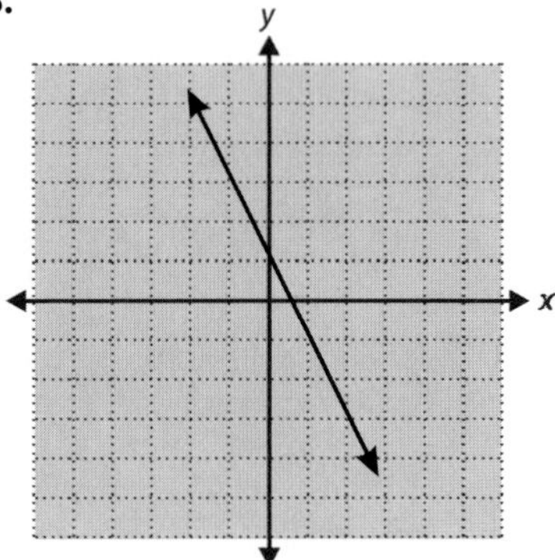


**29.**

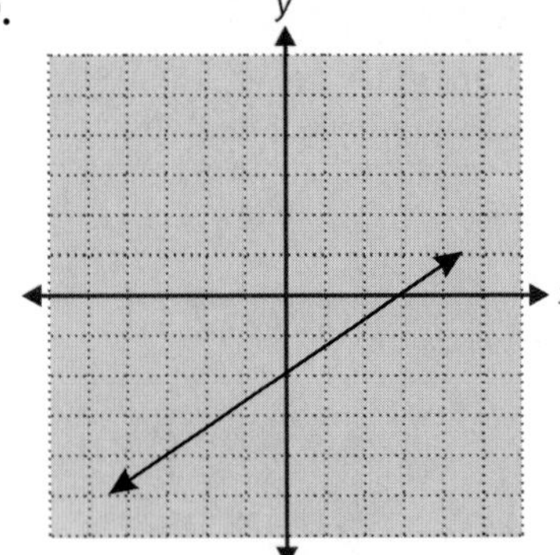


**30.**

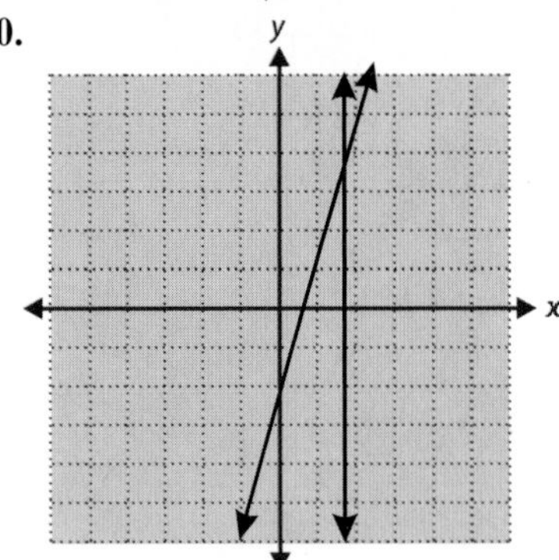


### Cumulative Review: Chapters 1 - 8, page 707

**1.** $1\frac{13}{18}$ **2.** $30\frac{13}{40}$ **3.** 23.69 **4.** 9.068 **5.** $\frac{7}{36}$ **6.** $1\frac{1}{2}$ **7.** 74.875 **8.** 4.47

**9.** 72,000,000 **10.** $\frac{1}{7}$ **11.** 16 **12.** $-1.9684$ **13.** $\frac{5}{6}$ **14.** $7\frac{1}{2}$ **15.** 0.608 **16.** undefined

**17.** $2^3\cdot 3^2\cdot 11$ **18.** 30 **19.** 5990 **20.** 723.1 **21.** 22 **22.** $\frac{5}{36}$ **23.** $3\sqrt{5}+20\sqrt{3}$ **24.** $<$

**25.** $2(n+6)-5$ **26.** $x=-12$ **27.** $x=18$ **28.** $x=\frac{y-b}{m}$ **29.** $-\frac{1}{2}\le x<\frac{1}{4}$

**30.** $5\sqrt{5}$ in. ( or 11.18 in. ) **31.** {(-1, 0), (0, 2), (1, 4), (2, 2), (3, 0)}

**32.**

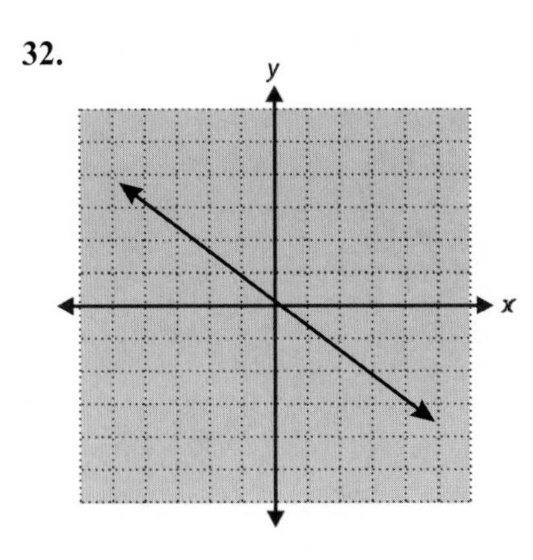


**33.**

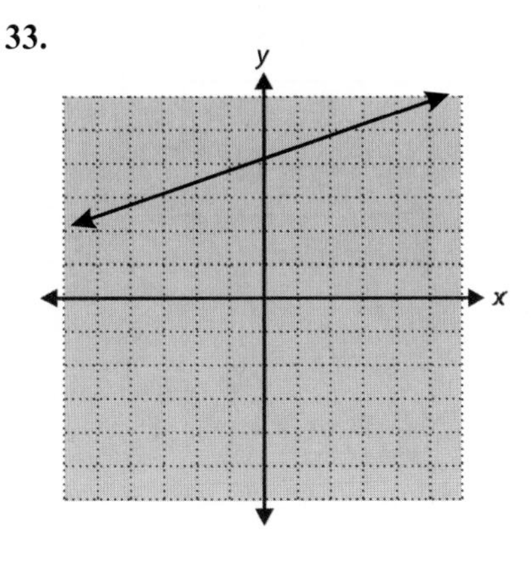


**34.** 36 **35.** 210% **36. a.** $P = 17.8$ ft **b.** $A = 15.6$ ft$^2$ **37. a.** $C = 125.6$ in. **b.** $A = 1256$ in.$^2$ **38.** 225 mi

**39. a.** \$1969 **b.** \$419 **40.** $-15, -14$ **41.** 21 **42.** $3.45 \times 10^6$ **43.** $7^2$ **44.** $\frac{1}{a^{10}}$ **45.** $\frac{1}{x^{10}}$

**46.** $\frac{18}{y^2}$ **47.** $-x^3 + 9x^2 - 11x + 23$ **48.** $5x^2 - 4x + 18$ **49.** $10x^2 - 9x - 9$ **50.** $x^4 - 2x^3 - 5x^2 - 10x - 8$

**51. a.** $3(x+6)(x-6)$ **b.** $5(x^2 + 2x + 7)$ $6x^2(x^2 + 6x - 2)$ **52. a.** $25(a-1)^2$ **b.** $2(a+6)^2$ **c.** $a(a+9)^2$

## CHAPTER 9

### Exercises 9.1, page 721

**1. a.** Mean = 80 **b.** Median = 82 **c.** Mode = 85 **d.** Range = 36 **3. a.** Mean = \$48,625 **b.** Median = \$46,500 **c.** Mode = \$63,000 **d.** Range = \$43,000 **5. a.** Mean = \$9701 **b.** Median = \$9280 **c.** Mode = none **d.** Range = \$3000 **7. a.** Mean = 1135 miles **b.** Median = 980 miles **c.** Mode = none **d.** Range = 2020 miles **9. a.** Mean = 301,124,000 m$^3$ **b.** Median = 238,200,000 m$^3$ **c.** Mode = none **d.** Range = 418,280,000 m$^3$ **11. a.** Mean = 16.0 **b.** Median = 15.5 **c.** Mode = 17 **d.** Range = 44 **13.** 79 **15.** 3.3; 2.4; answers vary

### Exercises 9.2, page 729

**1. a.** women over 55 **b.** men 18-24 **c.** over 55 **d.** 82 hours **3. a.** agriculture/biological sciences **b.** math **c.** about 1800 **d.** about 21% **5.** stocks: \$125,000; bonds: \$200,000; cash: \$100,000; real estate: \$75,000 **7.** pepperoni: 900; sausage: 450; vegetarian: 180; other: 270 **9. a.** 1992 **b.** about 2700 **c.** about 1050 **d.** about 2720 **11. a.** 4 **b.** 10 **c.** Monday and Tuesday **d.** 66 **13. a.** no **b.** 11% **c.** 28,600,000 ; 2,600,000 **d.** 6,650,000 ; 15,750,000

### Exercises 9.3, page 743

**1. a.** February and August **b.** \$190,000 **c.** October **d.** \$208,000 **e.** \$205,000 **f.** The general trend is a decline. **3. a.** central cities **b.** rural **c.** rural; suburbs **d.** 1970**5. a.** the third class **b.** 36 **c.** 1 **d.** 2%

**e.**

| | | | |
|---|---|---|---|
| 1 | \$18,001 - \$21,000 | 5 | \$19,500.50 |
| 2 | \$21,001 - \$24,000 | 13 | \$22,500.50 |
| 3 | \$24,001 - \$27,000 | 18 | \$25,500.50 |
| 4 | \$27,001 - \$30,000 | 10 | \$28,500.50 |
| 5 | \$30,001 - \$33,000 | 3 | \$31,500.50 |
| 6 | \$33,001 - \$36,000 | 0 | \$34,500.50 |
| 7 | \$36,001 - \$39,000 | 1 | \$37,500.50 |

**f.**

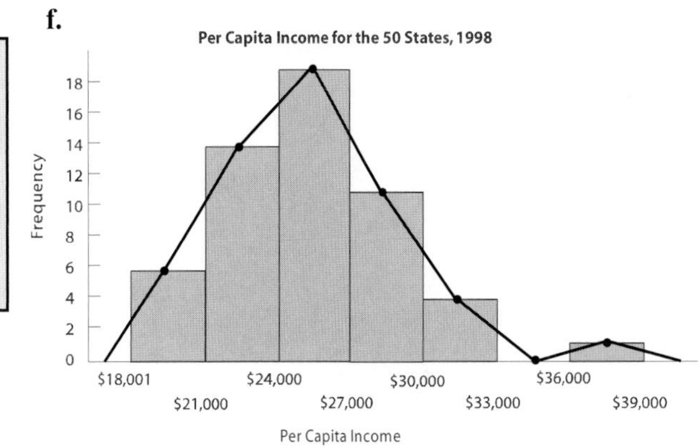

**7. a.** the first class **b.** 34.5 and 39.5 **c.** 23 **d.** 57.5%

**e.**

| | | | |
|---|---|---|---|
| 1 | 25 - 29 | 7 | 27 |
| 2 | 30 - 34 | 5 | 32 |
| 3 | 35 - 39 | 11 | 37 |
| 4 | 40 - 44 | 8 | 42 |
| 5 | 45 - 49 | 3 | 47 |
| 6 | 50 - 54 | 2 | 52 |
| 7 | 55 - 59 | 3 | 57 |
| 8 | 60 - 64 | 1 | 62 |

**f.**

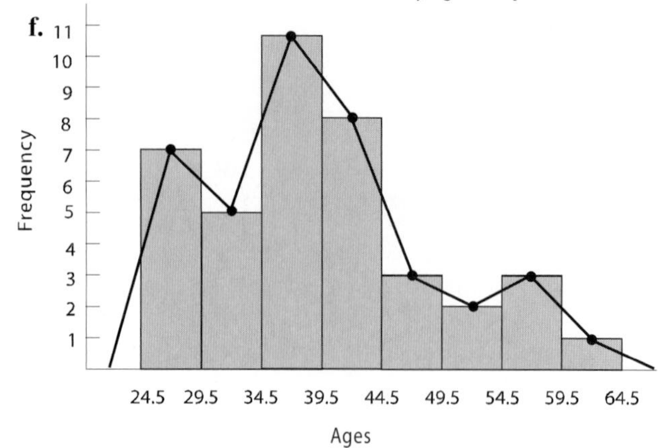


## Exercises 9.4, page 755

**1.**

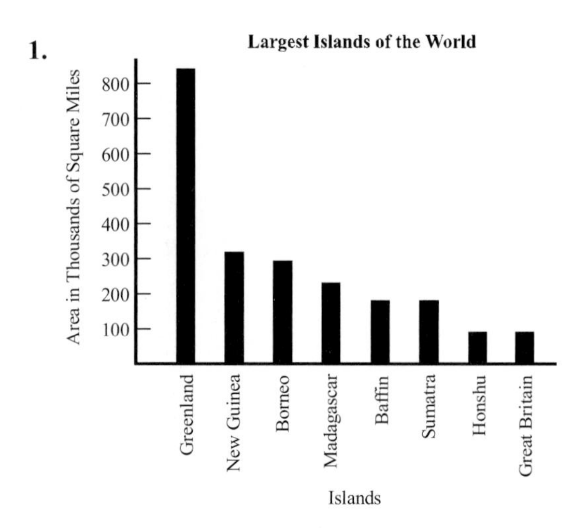


**3.**

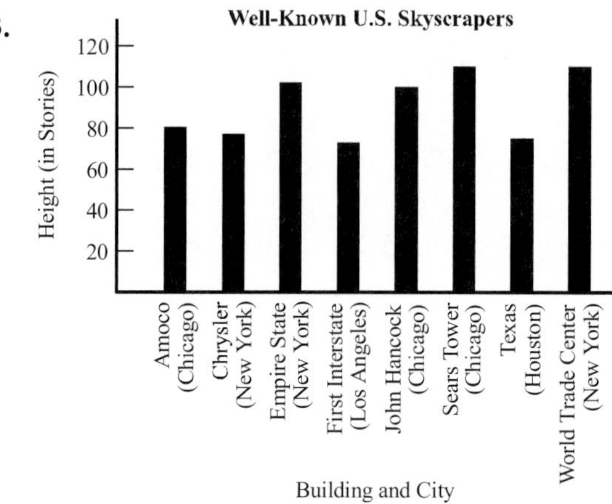


**5.**

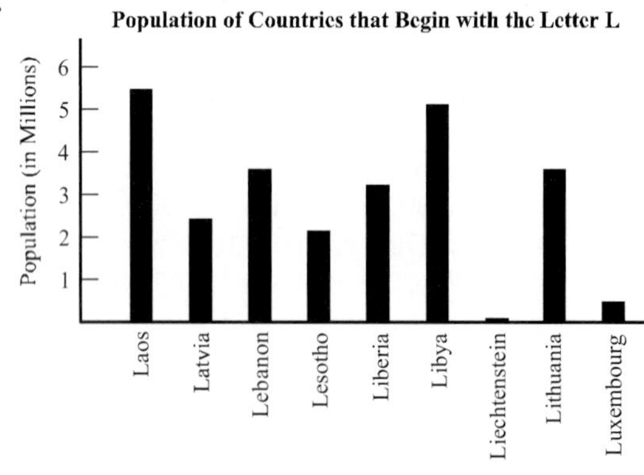


**7.**

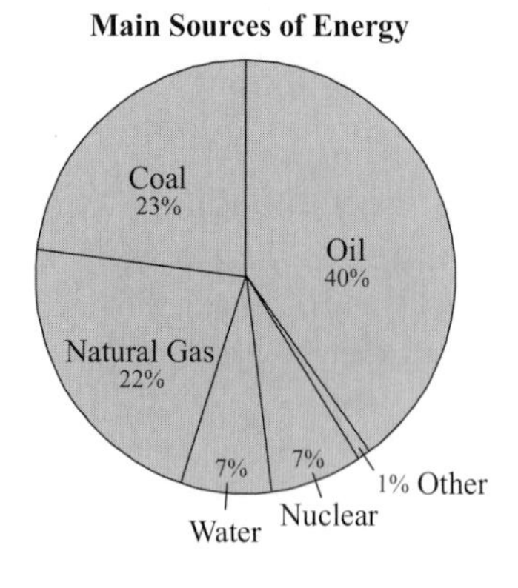


**9.**

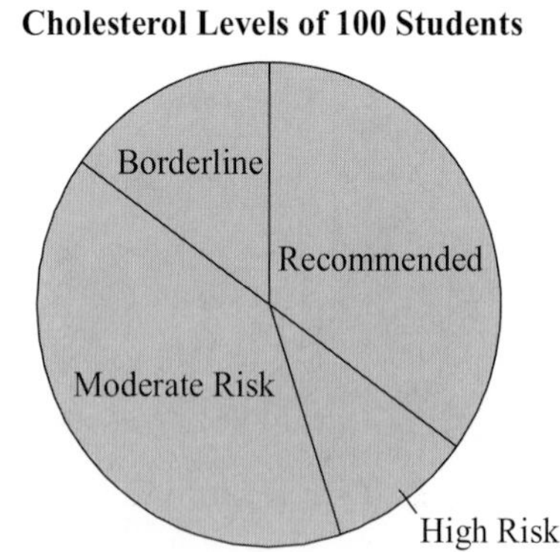


**11.**

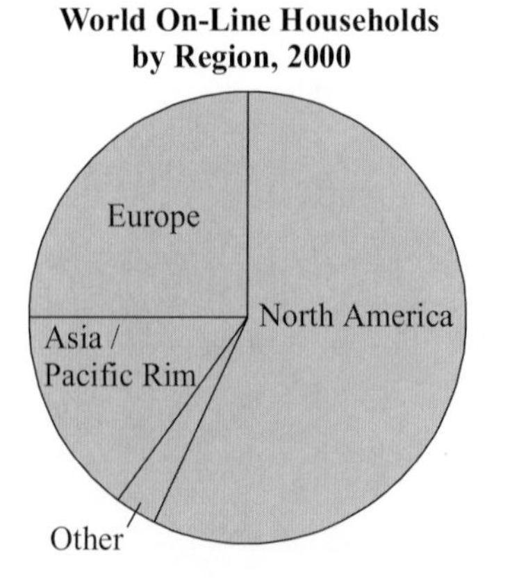

**13.**

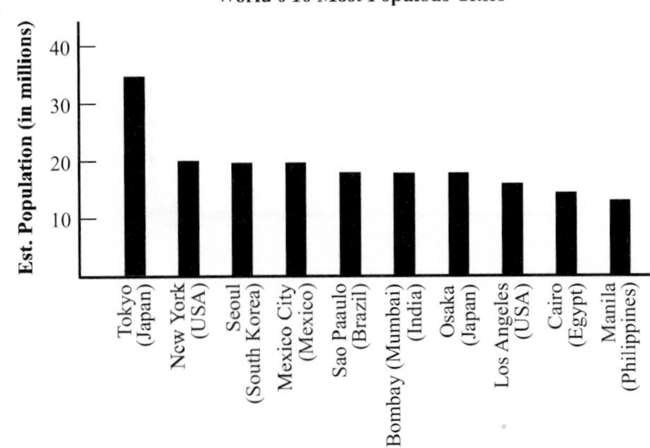


## Chapter 9 Test, page 765

**1. a.** Mean = 21.4 **b.** Median = 20 **c.** Mode = 20 **d.** Range = 14 **2. a.** Mean = 46.6 in. **b.** Median = 48 in. **c.** none **d.** Range = 37 in. **3.** 52% **4.** \$62,500 **5.** \$65,000 **6. a.** 58° **b.** Saturday **7. a.** 61° **b.** Friday **8.** 9.6° **9. a.** 5 **b.** 10 **c.** 6 **10. a.** 60% **b.** 85% **11. a.** 54.5; 64.5; 74.5; 84.5; 94.5

**b.**

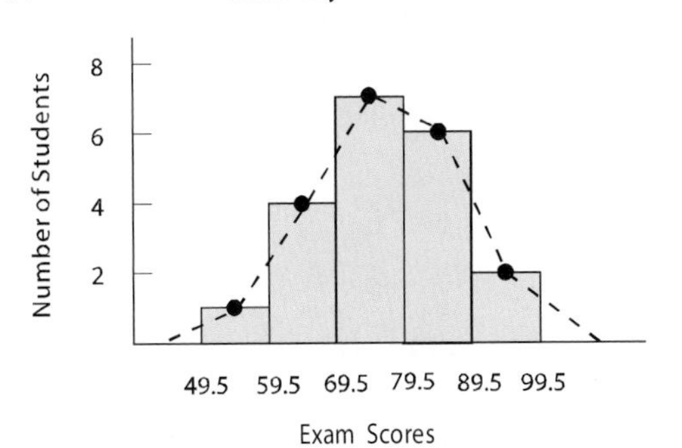


**12.**

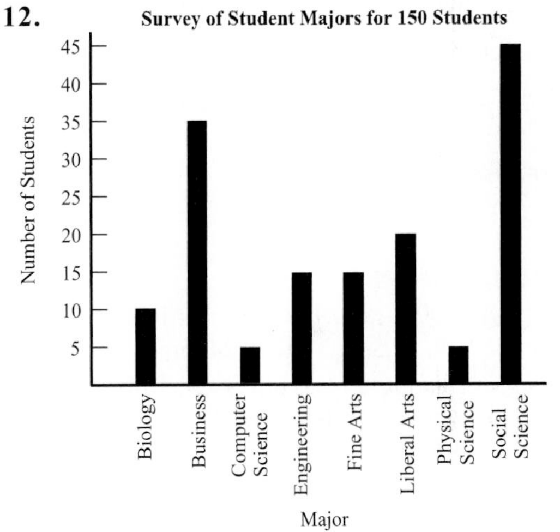


**13.**

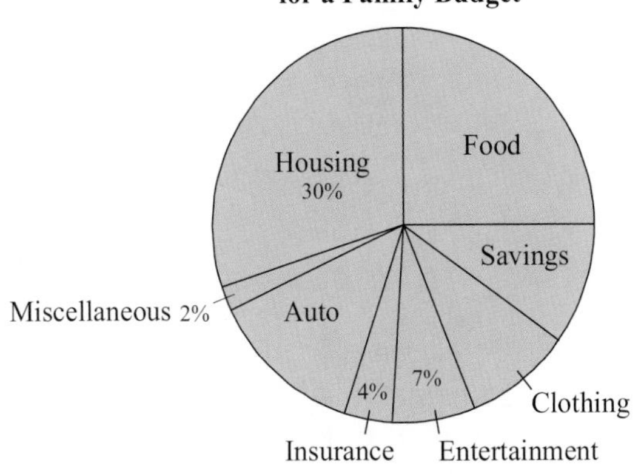




## Cumulative Review: Chapters 1 - 9, page 769

**1.** 200,016.04 **2.** 17.99 **3.** 0.4 **4.** 180% **5.** 0.015 **6.** $x=\frac{4}{3}$ **7.** $3\frac{1}{11}$ or $\left(\frac{34}{11}\right)$

**8.** $-\frac{4}{21}$ **9.** $10\frac{2}{15}$ **10.** $46\frac{5}{12}$ **11.** 12 **12.** $1\frac{4}{5}$ **13.** −560,000 **14.** −26.9 **15.** 75.74

**16.** 0.00049923 **17.** −80.6 **18.** 7 **19.** −11.11 **20.** −2.75 **21.** $2^2 \cdot 3^2 \cdot 11$ **22.** 210

**23.** $3.2 \times 10^7$ **24. a.** yes **b.** 336 **25.** 0 **26.** 50 **27.** 18.5 **28.** $x = 0.375$ **29.** $x = -9$

**30.** $l = \dfrac{P-2w}{2}$ **31. a.** \$13,488.51 **b.** \$3488.51 **32. a.** mean = 45 **b.** median = 36 **c.** range = 72

**33.** 196 **34.** 325 **35.** \$41 **36. a.** $C$ = 62.8 ft **b.** $A$ = 314 $ft^2$

**37. a.** $x < 6$; 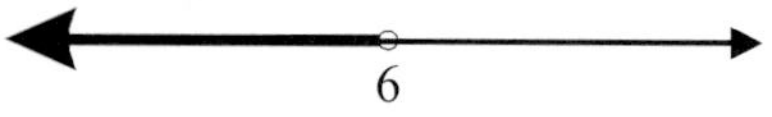 **b.** $x \geq \frac{1}{3}$; 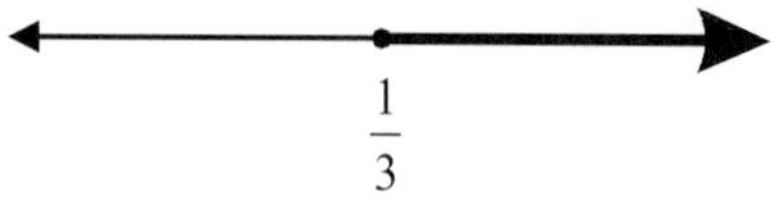

## CHAPTER 10

### Exercises 10.1, page 789

**1. a.** m∠2 = 75° **b.** m∠2 = 87° **c.** m∠2 = 45° **d.** m∠2 = 15° **3. a.** a right angle **b.** an acute angle **c.** an acute angle **5.** 25° **7.** 40° **9.** 55° **11.** 110° **13.** m∠$A$ = 55°; m∠$B$ = 51°; m∠$C$ = 74° **15.** m∠$L$ = 104°; m∠$M$ = 109°; m∠$N$ = 105°; m∠$O$ = 106°; m∠$P$ = 116° **17. a.** straight **b.** right **c.** acute **d.** obtuse **19. a.** m∠2 = 150° **b.** yes; ∠2 and ∠3 are supplementary **c.** ∠1 and ∠2; ∠1 and ∠4; ∠2 and ∠3 ; ∠3 and ∠4 **21.** m∠2 = 150°; m∠3 = 30°; m∠4 = 150° **23.** m∠1 = m∠4 = 140°; m∠4 = m∠6 = 140°; m∠1 = m∠6 = 140°; m∠3 = m∠5 = 40°; m∠2 = m∠5 = 40° **25.** equilateral, acute **27.** obtuse, scalene **29.** isosceles, obtuse **31.** acute, equilateral **33.** yes, since 25 < 12 + 15 **35. a.** m∠$Z$ = 80° **b.** acute, scalene **c.** $\overleftrightarrow{YZ}$ **d.** $\overleftrightarrow{XZ}$ and $\overleftrightarrow{XY}$ **e.** no; no ∠ = 90°

### Margin Exercises 10.2

**1.** 3500 **2.** 64 **3.** 0.0059 **4.** 0.32 **5.** 60.96 **6.** 4.57 **7.** 19.7 in. **8.** 49,104 ft

### Exercises 10.2, page 803

**1.** 300 **3.** 1500 **5.** 450 **7.** 13,600 **9.** 182.5 **11.** 0.48 **13.** 0.065 **15.** 1.3 **17.** 0.0525 **19.** 5.5 **21.** 0.245 m **23.** 10 km **25.** 200 m **27.** 0.32 cm **29.** 17,350 mm **31.** 7.62 **33.** 3.66 **35.** 39.4 **37.** 220.0 **39.** 12.4 **41.** 236.0 in. **43.** 9.84 ft **45.** 299.0 in. **47. a.** B **b.** C **c.** D **d.** A **e.** F **f.** E **49.** $P$ = 104 mm; $P$ = 4.10 in. **51.** $C$ = 18.8 yd; $C$ = 17.2 m **53.** $P$ = 107.0 mm; $P$ = 10.7 cm; $P$ = 4.22 in. **55.** $P$ = 27.4 mm; $P$ = 2.74 cm; $P$ = 1.08 in. **57.** $P$ = 254 mm; $P$ = 25.4 cm; $P$ = 10 in. **59.** $P$ = 600 mm; $P$ = 60 cm; $P$ = 23.6 in. **61. a.** 4 **b.** Answers will vary somewhat, but they should be about 3.3. **c.** Answers will vary. The numbers will become closer to pi as the number of sides increases.

### Margin Exercises 10.3

**1.** 2300 $mm^2$ **2.** 52 $cm^2$; 0.52 $dm^2$ **3.** 360 $m^2$ **4.** 495a; 4.95 ha **5.** 38.7 $cm^2$ **6.** 58.1 $m^2$ **7.** 538.2 $ft^2$ **8.** 7.41 acres

### Exercises 10.3, page 819

**1.** 1300 $mm^2$ **3.** 960 $mm^2$ **5.** 5 $cm^2$ **7.** 1400 $cm^2$ **9.** 5000 $mm^2$ **11.** 1300 $cm^2$ = 130 000$mm^2$ **13.** 1150 $dm^2$ = 115 000 $cm^2$ = 11 500 000 $mm^2$ **15.** 4 $dm^2$ = 400 $cm^2$ = 40 000 $mm^2$ **17.** 670 $m^2$ **19.** 20 000 $m^2$ **21.** 0.02 ha **23.** 575 ha **25.** 75 000 a **27.** 27 a = 2700 $m^2$ **29.** 62 500 a = 625 ha **31.** 25.8 $cm^2$ **33.** 46.5 $m^2$ **35.** 8.36 $m^2$ **37.** 161.5 $ft^2$ **39.** 2.17 $in.^2$ **41.** 810 ha **43.** 494 acres **45.** 269.1 $ft^2$ **47. a.** C **b.** B **c.** D **d.** A **e.** F **f.** E **49.** $A$ = 16 $cm^2$; $A$ = 2.48 $in.^2$ **51.** $A$ = 38.5 $mm^2$, $A$ = 0.385 $cm^2$ **53.** $A$ = 706.5 $mm^2$, $A$ = 7.065 $cm^2$, $A$ = 1.095 $in.^2$

**55.** $A$ = 36 840 $mm^2$, $A$ = 368.4 $cm^2$, $A$ = 57.12 $in.^2$ **57.** $A$ = 10 600 $mm^2$, $A$ = 106 $cm^2$, $A$ = 16.43 $in.^2$ **59.** $A$ = 50.13 $mm^2$, $A$ = 0.5013 $cm^2$, $A$ = 0.077 70 $in.^2$ **61.** $A$ = 304 400 $mm^2$, $A$ = 3044 $cm^2$, $A$ = 472 $in.^2$

### Margin Exercises 10.4

**1.** 9000 $mm^3$ **2.** 7 500 000 $cm^3$ **3.** 0.000 54 $cm^3$ **4.** 0.059 82 $m^3$ **5.** 0.34 L **6.** 700 $cm^3$ **7.** 4.752 gal **8.** 9.46 L

### Exercises 10.4, page 831

**1.** $1\text{cm}^3 = \underline{1000}\ \text{mm}^3$
$1\text{dm}^3 = \underline{1000}\ \text{cm}^3$
$1\text{m}^3 = \underline{1000}\ \text{dm}^3$
$1\text{km}^3 = \underline{1\ 000\ 000\ 000}\ \text{m}^3$

**3.** $1\text{ m} = \underline{10}\ \text{dm}$
$1\text{ m} = \underline{100}\ \text{cm}$
$1\text{ m}^2 = \underline{100}\ \text{dm}^2$
$1\text{ m}^2 = \underline{10\ 000}\ \text{cm}^2$
$1\text{ m}^3 = \underline{1000}\ \text{dm}^3$
$1\text{ m}^3 = \underline{1\ 000\ 000}\ \text{cm}^3$

**5.** 85 000 **7.** 325 000 **9.** 9 150 000

**11.** 0.075 **13.** 0.135 **15.** 0.035 $\text{cm}^3$ **17.** 0.000 029 $\text{dm}^3$ **19.** 12,000 $\text{cm}^3$ **21.** 4600
**23.** 0.38 kL **25.** 6700 mL **27.** 0.3 L **29.** 95 000 mL; 95 000 $\text{cm}^3$ **31.** 295.0 **33.** 6 **35.** 0.756
**37.** 35 320 **39.** 9.46 **41.** 21.2 **43.** 13.73 **45.** 75.7 **47. a.** C **b.** E **c.** D **d.** B **e.** A
**49.** $V = 1.178\text{ ft.}^3$, $V = 0.032\ 97\text{ m}^3$ **51.** $V = 12.56\text{ dm}^3$, $V = 0.012\ 56\text{ m}^3$ **53.** 96 mL
**55.** $V = 224{,}000\text{ mm}^3$, $V = 224\text{ cm}^3$, $V = 13.44\text{ in.}^3$ **57.** $V = 169\ 600\text{ mm}^3$, $V = 169.6\text{ cm}^3$, $V = 10.18\text{ in.}^3$
**59.** $V = 9106\text{ mm}^3$, $V = 9.106\text{ cm}^3$, $V = 0.546\ 4\text{ in.}^3$

### Chapter 10 Test, page 841

**1. a.** $m\angle 2 = 55°$ **b.** $m\angle 4 = 165°$ **2. a.** Obtuse **b.** Straight **c.** $\angle AOD$ and $\angle DOB$; $\angle AOC$ and $\angle COB$

**3. a.** isosceles, acute **b.** obtuse, scalene **c.** right, scalene **4. a.** $m\angle x = 30°$ **b.** obtuse, scalene **c.** $\overline{RT}$
**5.** 10 cm is longer; 90 mm longer **6.** The volume is the same. 10 mL = 10 $\text{cm}^3$ **7.** 0.56 m **8.** 3300 cm
**9.** 11 m **10.** 0.083 5 L **11.** 9.6 $\text{cm}^2$ **12.** 150 000 $\text{cm}^2$ **13.** 0.75 ha **14.** 7500 a **15.** 140 000 $\text{mm}^3$
**16.** 0.032 $\text{m}^3$ **17.** 19.7 in. **18.** 6.6 gal **19.** 124 mi **20.** 50.31 $\text{cm}^2$ **21.** 28.38 L **22.** 1.092 $\text{m}^3$
**23. a.** 94.2 cm **b.** 706.5 $\text{cm}^2$ **24. a.** 40.2 in. **b.** 98.6 $\text{in}^2$ **25.** 904.3 $\text{in.}^3$ = 14 820 $\text{cm}^3$ **26. a.** 11.5 mm
**b.** 6 $\text{mm}^2$ **27. a.** 13.71 m **b.** 11.23 $\text{m}^2$ **28. a.** 30 ft. **b.** 30 $\text{ft}^2$ **29.** 486 $\text{m}^2$ **30.** 90 $\text{cm}^3$ = 5.4 $\text{in.}^3$

### Cumulative Review: Chapters 1 - 10, page 845

**1.** 1 **2.** 520.065 **3.** 5.1 **4.** $16\sqrt{2}$ **5.** 7.937 **6.** 735 **7.** –25 **8.** 1.15 **9. a.** –10
**b.** –0.81 **c.** $-\frac{5}{7}$ **10.** \$8950.46 **11.** \$2000 **12.** Food: \$9000; Housing: \$11,250; Transportation: \$6750;
Miscellaneous: \$6750; Savings: \$2250; Education: \$4500; Taxes: \$4500 **13.** 20 **14.** 130 **15.** 38.775
**16.** 12,040 **17. a.** \$17.50 **b.** \$45 **18.** $x = -1.05$ **19.** $x = -32.5$ **20.** $y = \frac{10-5x}{3}$ **21.** 17
**22. a.** mean = 45 **b.** median = 46 **c.** mode = 41 **d.** range = 35 **23. a.** $1.93 \times 10^8$ **b.** $3.86 \times 10^{-4}$
**24.** $x = 3$ in.; $y = 25°$ **25. a.** $\sqrt{20}$ in. = $2\sqrt{5}$ in. **b.** 4.47 in. **26.** $x > 3$;

**27. a.**

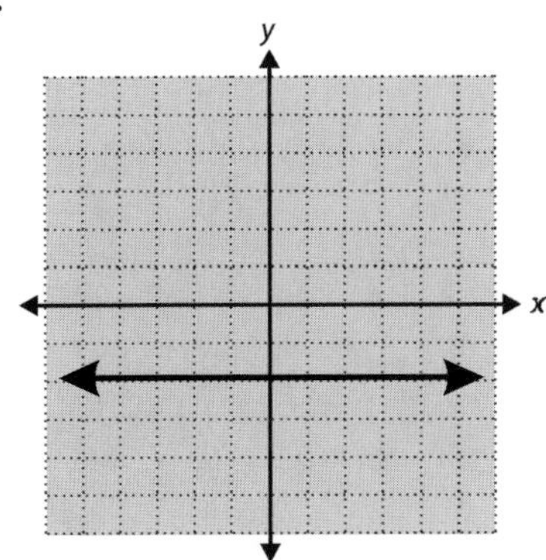


**b.**

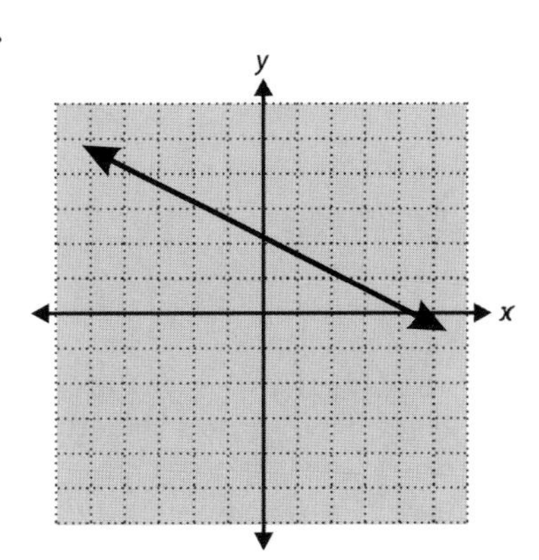


**28. a.** $\frac{1}{a^6}$ **b.** $\frac{1}{x^2}$ **c.** 1 **d.** $-10y^7$ **29. a.** $9x^2 - 7$; second-degree binomial **b.** $3x^3 + 2x^2 - 4x$;
third-degree trinomial **c.** $6x^5 + 12x^3 - 6x^2$; fifth-degree trinomial **d.** $6x^2 + x - 15$; second-degree trinomial

**30. a.** $3x(x+5)$ **b.** $5x(x-2)(x+2)$ **c.** $2(y+5)^2$ **31.** Answers will vary. For example: The sum of a number and ten is multiplied by five, and this product equals three times the number. What is the number? Since the solution is $x = -25$, any word problem must be such that a negative solution is sensible. **32.** Answers will vary. For example: A tile patio uses a simple pattern of large circles inscribed in large squares. The squares are 100 cm on a side and the circle inside each square is 1 m in diameter. How much area enclosed in each square is outside the inscribed circle? Answer: $100^2 - \pi(50)^2 = 10{,}000 - 2500(3.14) = 10{,}000 - 7850 = 2150 \text{ cm}^2$. **33.** Answers will vary. The problem should be to find the mean of these values. For example: Find Elliot's test average in European Literature if he has scored 72, 80, and 97 on three exams. **34.** Answers will vary.

## Appendix III

### Exercises III.1, page A9

**1.** 253 **3.** 21,327 **5.** 140 **7.** 53 **9.** 2362 **11.** 744 **13.** 978 **15. a.** 𓍢𓍢𓍢𓍢𓍢 ∩∩∩ ||

**b.** ∴ ·· ═ **c.** 𐅆ΔΔΔII **d.** DXXXII **17. a.** 𓍢𓍢𓍢𓍢𓍢𓍢𓍢𓍢 ∩∩∩∩ ||||||  **b.** ∴ ─ · ─ **c.** 𐅆HHHΔΔΔΔΓI

### Exercises III.2, page A15

**1.** 4983 **3.** 1802 **5.** 47,900 **7.** 46 **9.** 999 **11. a.** ∨∨∨∨∨ <<< ∨∨∨ / ∨∨∨∨ << ∨∨∨ **b.** ΦCF

**c.** 五 百 九 十 大 **13. a.** ∨∨∨ / ∨∨ ∨∨ **b.** /ΓΧΞCE **c.** 三 千 大 百 大 十 五 **15. a.** ∨∨∨ <<< ∨∨ / << **b.** A / ΜΩΝΒ **c.** 十 千 八 百 五 十 二

## Appendix IV

### Exercises IV, page A15

**1.** 4 **3.** 17 **5.** 10 **7.** 3 **9.** 6 **11.** 2 **13.** 8 **15.** 4 **17.** 12 **19.** 25
**21.** 1 **23.** 1 **25.** 35 **27.** relatively prime **29.** relatively prime **31.** relatively prime
**33.** relatively prime **35.** relatively prime

## Appendix V

### Exercises V, page A25

**1.** $x^2+3x+2$ **3.** $7x^2-4x+\frac{5}{2}$ **5.** $4x-5$ **7.** $2x+4-\frac{2}{3x}+\frac{2}{x^2}$ **9.** $2x-\frac{4}{x}-\frac{1}{7x^2}$

**11.** $5+\frac{11}{x}-\frac{2}{x^2}$ **13.** $-4x-3-\frac{1}{2x}+\frac{2}{x^2}$ **15.** $2x+1$; R 4 **17.** $y-19$; R 80 **19.** $4y-7$; R 12

**21.** $10x+3$; R 6 **23.** $x^2+3x-20$; R 42 **25.** $2x-3$ **27.** $4x-5$ **29.** $2y+3$ **31.** $x^2-8x+2$

**33.** $y^2+3y+4$

# INDEX

## Types of Angles

**1. Acute** $0° < m\angle A < 90°$

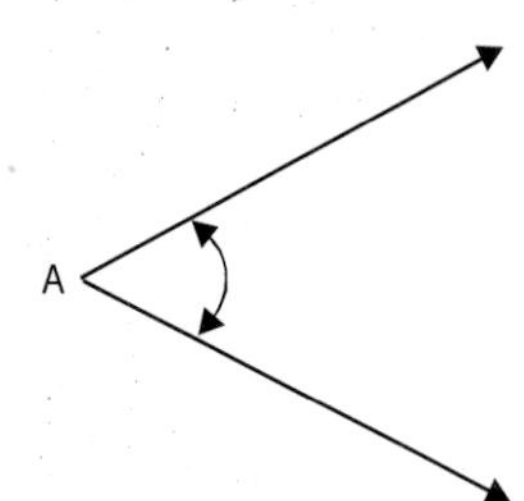


**2. Right** $m\angle A = 90°$

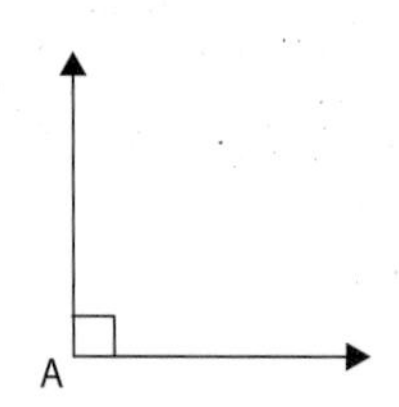


**3. Obtuse** $90° < m\angle A < 180°$

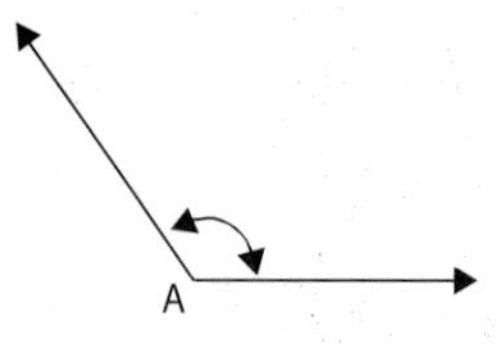


**4. Straight** $m\angle A = 180°$

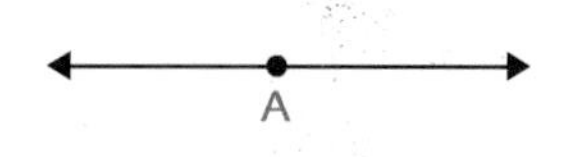


## Triangles

### Two Important Statements about Any Triangle:

1. The sum of the measures of the angles is 180
2. The sum of the lengths of any two sides must be greater than the third side.

### Triangles Classified by Sides

**1. Scalene** No two sides are equal.

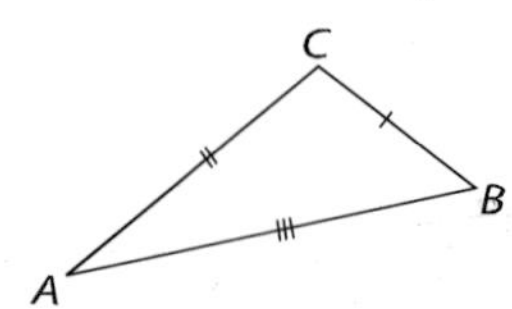


**2. Isosceles** Two sides are equal.

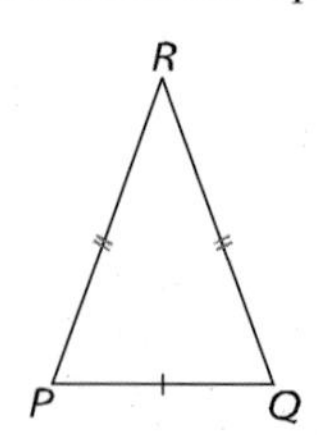


**3. Equilateral** All three sides are equal.

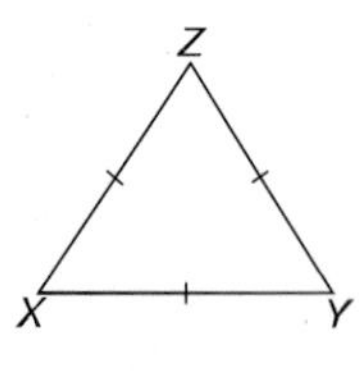


### Triangles Classified by Angles

**1. Acute** All three angles are acute.

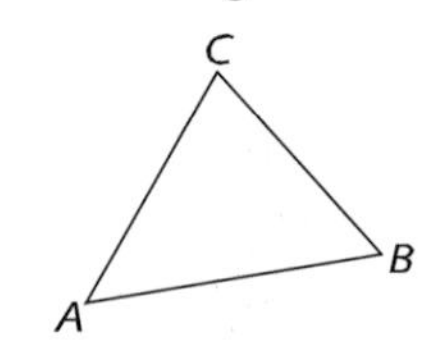


**2. Right** One angle is a right angle.

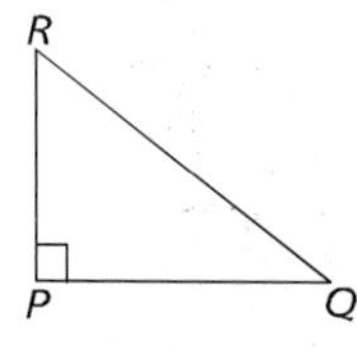


**3. Obtuse** One angle is an obtuse angle.

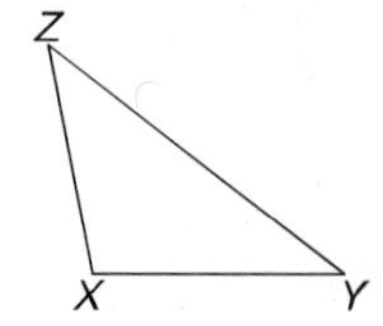